Lecture Notes in Computer Science 1112

Edited by G. Goos, J. Hartmanis and J. van Leeuwen

Springer

Berlin
Heidelberg
New York
Barcelona
Budapest
Hong Kong
London
Milan
Paris
Santa Clara
Singapore
Tokyo

C. von der Malsburg W. von Seelen
J.C. Vorbrüggen B. Sendhoff (Eds.)

Artificial Neural Networks – ICANN 96

1996 International Conference
Bochum, Germany, July 16-19, 1996
Proceedings

Springer

Series Editors

Gerhard Goos, Karlsruhe University, Germany

Juris Hartmanis, Cornell University, NY, USA

Jan van Leeuwen, Utrecht University, The Netherlands

Volume Editors

Christoph von der Malsburg
Jan C. Vorbrüggen
Ruhr-Universität Bochum, Institut für Neuroinformatik
Lehrstuhl Systembiophysik, ND 03/34
D-44780 Bochum, Germany

Werner von Seelen
Bernhard Sendhoff
Ruhr-Universität Bochum, Institut für Neuroinformatik
Lehrstuhl für Theoretische Biologie, ND 04/584
D-44780 Bochum, Germany

Cataloging-in-Publication data applied for

Die Deutsche Bibliothek - CIP-Einheitsaufnahme

Artificial neural networks : 6th international conference ;
proceedings / ICANN 96, Bochum, Germany, July 16 -19, 1996.
C. von der Malsburg ... (ed.). - Berlin ; Heidelberg ; New York
; Barcelona ; Budapest ; Hong Kong ; London ; Milan ; Paris ;
Santa Clara ; Singapore ; Tokyo : Springer, 1996
  (Lecture notes in computer science ; Vol. 1112)
  ISBN 3-540-61510-5
NE: Malsburg, Christoph von der [Hrsg.]; ICANN <6, 1996, Bochum>;
  GT

CR Subject Classification (1991): F.1.1, C.2.1, C.1.3, I.2, G.1.6, I.5.1, B.7.1,
J.1, J.2

ISSN 0302-9743
ISBN 3-540-61510-5 Springer-Verlag Berlin Heidelberg New York

© Springer-Verlag Berlin Heidelberg 1996
Printed in Germany

Typesetting: Camera-ready by author
SPIN 10513275   06/3142 – 5 4 3 2 1 0   Printed on acid-free paper

Preface

The *International Conference on Artificial Neural Networks,* ICANN, the European Neural Networks Society's official conference series, is a remarkable phenomenon. It was born together with ENNS in 1991 in Helsinki, Finland, and was continued in the following years in the UK (Brighton 1992), The Netherlands (Amsterdam 1993), Italy (Sorrento 1994), and France (Paris 1995), and will, after Bochum this year, take place in Switzerland (Lausanne) and Sweden.

The conference is remarkable for thriving without a trace of routine and continuous organization. This discontinuity has had its painful sides for the teams of organizers, who have had to re-invent ICANN from year to year, but it may actually be a virtue. As much as Europe is a concert of nations, the field of Artificial Neural Networks is a concert of disciplines — mathematics, statistics, physics, electrical engineering, computer science, psychology, neuroscience, cognitive science, ethology, medicine, and various branches of business. It would be a disaster to our field if it were seized by any one style or routine, as much as it would be a disaster to Europe if suspicion came up that it was going to be dominated by any one nation or group of nations. Our field would quickly lose its sources of creative energy if it defined itself as "just a set of methods of statistical estimation" or "just a fad of statistical physics" or "just a branch of brain theory", or any other narrow definition. Our European nations' diverse and rich roots and traditions form an ecosystem in which interdisciplinarity may thrive better than it would in a homogeneous, efficient market. It needs great efforts on our part, however, to bring the various traditions — of nations and of disciplines — into mutual contact so they cross-fertilize and prosper.

Bochum itself, and the Ruhr Valley megalopolis of which it is part, are in the midst of a difficult move out of their old routine based on heavy industries, coal and steel, towards a new self-definition as an economic and cultural region. The *Institut für Neuroinformatik* (Institute for Neural Computing) and the *Zentrum für Neuroinformatik,* which host this year's ICANN, have themselves been founded by the State of North Rhine-Westphalia at the Ruhr-Universität Bochum as part of efforts to foster high technology in the region.

As organizers of this year's conference, we have continued the attempt to establish ICANN as a high-quality event that can attract the best contributions Europe has to offer. We have therefore paid much attention to the reviewing process and have enforced high standards. Every paper was reviewed by three members of the Programme Committee. Final decisions regarding acceptance of papers, form of presentation, and session structure were then taken by the Executive Committee. We do know that even the best Programme Committee is fallible and apologize for any manuscript refusal due to misjudgement.

We would like to warmly thank the sponsors of our conference, the *Deutsche Forschungsgemeinschaft,* the *Ministerium für Wissenschaft und Forschung* of the State of North Rhine-Westphalia, *Daimler-Benz AG* (whose contribution supports attendance of two students from overseas), and the *Ruhr-Universität Bochum,* which generously let us use its premises and has assisted the conference organization in various and many ways.

Early in the game we made the choice of not employing an organizing company. This meant that the heavy burden and responsibility for organizing this conference was carried by a local team. The driving force and creative spirit behind this team was Frau Uta Schwalm, and we would like to express our heartfelt thanks to her, and to J.C. Vorbrüggen, B. Sendhoff, J. Gayko, and the many students of our Institute who have helped her, for making this conference happen.

Bochum, May 1996

Ch. von der Malsburg
W. von Seelen
ICANN96 Programme Co-Chairs

Organization

Advisory Board

Abeles, M.
Amari, S.
Amit, D.J.
Anninos, P.
Barlow, H.
Changeux, J.-P.

Eckmiller, R.
Goser, K.
Hepp, K.
Innocenti, G.M.
Koenderink, J.J.
Palm, G.

Rolls, E.T.
Schürmann, B.
Sompolinsky, H.
Taylor, J.G.
Toulouse, G.

Executive Committee

Fritzke, B.
Gayko, J.
Güntürkün, O.
Kappen, H.J.
Kunze, M.
Mallot, H.A.

Malsburg, C. von der
Martinetz, T.
Pawelzik, K.
Ritter, H.
Seelen, W. von
Sendhoff, B.

Schüffny, R.
Vorbrüggen, J.C.
Willshaw, D.
Würtz, R.P.

Programme Committee

Aertsen, A.
Andler, D.
Bichsel, M.
Blayo, F.
Bourlard, H.
Budinich, M.
Buhmann, J.
Cabestany, J.
Canu, S.
Chapeau-Blondeau, F.
Cirrincione, M.
Collobert, D.
Dinse, H.R.
Eckhorn, R.
Fogelman Soulié, F.
Fritzke, B.
Gallinari, P.
Garda, P.
Germond, A.
Gerstner, W.

Gielen, S.
Göppert, J.
Groß, H.-M.
Güntürkün, O.
Habel, C.
Hallam, J.
Hancock, P.
Herault, J.
Hertz, J.
Holmström, L.
Kaernbach, C.
Kappen, H.J.
Kawato, M.
King, I.
Kohonen, T.
König, P.
Kreiter, A.
Kunze, M.
Kurkova, V.
Lampinen, J.

Langner, G.
Lansky, P.
Li, Z.
Littmann, E.
Lowe, D.
Luttrell, S.
MacKay, D.J.C.
Mahowald, M.
Martinetz, T.
Mira, J.
Morasso, P.G.
Murray, A.
Neumann, H.
Neumerkel, D.
Niebur, E.
Oja, E.
Pawelzik, K.
Perrett, D.I.
Perretto, P.
Plumbley, M.

Plunkett, K.
Prieto, A.
Prinz, W.
Quinlan, P.T.
Ramacher, U.
Raudys, S.
Reyneri, L.
Ritter, H.
Schaal, S.

Schneider, W.X.
Schnelle, H.
Solla, S.A.
Sommer, G.
Sporns, O.
Tavan, P.
Thiria, S.
Thorpe, S.
Torras, C.

Verleysen, M.
Vittoz, E.
Weinfeld, M.
Willshaw, D.
Wiskott, L.
Wörgötter, F.
Würtz, R.
Zemel, R.

In addition to the members of the Programme Committee, the following individuals have also kindly reviewed contributions to the conference. We thank them for their hard work and would like to apologize in advance for any inadvertent omissions which might have occured.

Alché-Buc, F. d'
Alphey, M.
Anizan, P.
Arndt, M.
Artières, T.
Bauer, H.-U.
Beaudot, B.
Berardi, F.
Bogdan, M.
Böhme, H.-J.
Bonabeau, E.
Born, C.
Braithwaite, E.
Bruske, J.
Burwick, T.
Cechin, A.
Celaya, E.
Chalimourda, A.
Chiaberge, M.
Cibas, T.
Cichocki, A.
Cirrincione, G.
Collobert, M.
Cornu, T.
Dang, J.C.
Deco, G.
Denoena, T.

Gewaltig, M.-O.
Glover, M.
Goerick, C.
Govaert, G.
Guermeur, Y.
Holmström, P.K.
Jansen, M.
Janßen, H.
Joublin, F.
Jutten, C.
Kaiuen, P.
Kefalea, E.
Kennedy, M.P.
Koistinen, P.
Kopecz, K.
Kowalczyk, A.
Krüger, N.
Lehmann, T.
Leray, P.
Li, Y.
Lund, H.
Maël, E.
Marchal, P.
Masa, P.
Maurer, T.
Mayer, P.
Mei, Z.J.

Miranda, E.
Neruda, R.
Orr, M.
Palm, G.
Piras, A.
Pötzsch, M.
Rinne, M.
Rohwer, R.
Saad, D.
Schittenkopf, C.
Schüffny, R.
Schwenker, F.
Stagge, P.
Strube, G.
Strube, H.-W.
Thiran, P.
Triesch, J.
Viallet, J.-E.
Visetti, Y.-M.
Wacquant, S.
Werner, M.
Wienholt, W.
Yang, H.H.
Zaragoza, H.
Zhu, H.

Contents

Plenary Presentations

Oral Presentations: Theory

Theory I: Associative Memory   Session 1A

Theory VI: Time Series   Session 10A

Theory VII: Unsupervised Learning   Session 11A

Theory VIII: Self-Organizing Maps   Session 12A

Oral Presentations: Applications

Oral Presentations: Sensory Processing

Sensory Processing I: Classification   Session 9B

Sensory Processing II: Object Recognition   Session 10B

Oral Presentations: Cognitive Science and AI

Oral Presentations: Implementations

Oral Presentations: Neurobiology

Neurobiology III: Motor Control   Session 7A

Neurobiology IV: Temporal Processing   Session 8A

Poster Presentations 1

Theory I: Associative Memory   Section 1

Theory II: Learning Section 2

Theory III: Generalization Section 3

Scientific Applications   Section 4

Applications in Robotics and Industry

Section 5

Poster Presentations 2

Neurobiology I: Visual Cortex/Cerebellum

Section 1

Image Processing Applications Section 4

Implementations Section 5

Poster Presentations 3

Sensory Processing
Section 3

Cognitive Science and AI
Section 4

Indices

Application of Artificial Neural Networks in Particle Physics

Hermann Kolanoski

Humboldt-Universität zu Berlin and DESY-IfH Zeuthen, Germany
Platanenallee 6, D-15738 Zeuthen (kolanosk@ifh.de)

Abstract. The application of Artificial Neural Networks in Particle Physics is reviewed. Most common is the use of multi-layer perceptrons for event classification and function approximation. This network type is best suited for a hardware implementation and special VLSI chips are available which are used in fast trigger processors. Also discussed are self-organizing networks for the recognition of features in large data samples. Neural net algorithms like the Hopfield model have been applied for pattern recognition in tracking detectors.

1 Introduction

Modern experiments in Nuclear and Particle Physics are generating an information flow well above tens of TByte per second. In order to allow for a permanent storage of the data and further detailed analyses, this flow must be reduced by orders of magnitude. In this talk recent developments in using Artificial Neural Network (ANN) algorithms and hardware for processing, reduction and analysis of data from Particle Physics experiments will be reviewed [1]. I will start with a brief introduction into the research goals of Particle Physics which should explain the high demands on data processing capabilities.

Particle Physics and its Research Goals: Particle Physics deals with the structure of matter on the smallest spatial scale accessible today, which is about 10^{-18} m or 1/1000 of the radius of the proton. According to our present notion, matter is built up by elementary particles and by the forces which act between the particles. There are two groups of particles which could be elementary: leptons and quarks. To the group of leptons belong the electrons which are well known to constitute the shell of the atoms and which are, together with the electro-magnetic force, responsible for the chemical bindings of atoms. The nuclei of atoms are composed of protons and neutrons, the nucleons. These nucleons, however, were found to be not elementary. Each of them is composed of three quarks, the other particle species which appears to be elementary. The quarks are bound into the nucleons by the 'strong force'.

Each lepton and quark has as a partner its anti-particle. If a particle and its anti-particle meet, they annihilate into a lump of energy. According to Einstein's famous relation $E = mc^2$ new massive particles can be created out of the annihilation energy. Figure 1 shows the creation of a quark–anti-quark pair in a

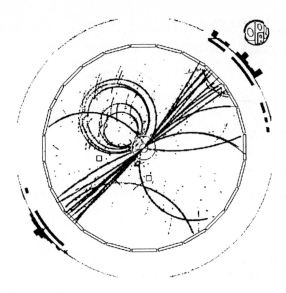

Fig. 1. Event with two back-to-back jets originating from a quark – anti-quark pair produced in electron-positron annihilation (OPAL experiment). The tracks of charged particles are curved by a magnetic field. The dark areas on the outer ring indicate energy deposited in the calorimeter.

high-energy collision of an electron and an anti-electron (= positron). Actually not the quark and anti-quark are visible but rather two jets of particles which particle physicists interpret as being produced by the energy in the strong field stretching between the quark and the anti-quark. With increasing energy more and heavier particles can be generated.

The heavy particles are unstable and decay into the stable particles which are usually around us, like the electrons, protons and photons. The stable and long-living particles can be detected, the properties of the unstable particles have to be inferred from their decay products.

The heavy, unstable particles seem not to influence at all our daily life. Why then do we want to investigate their nature? They probably played an important role at the very beginning of our universe when in a 'Big Bang' a fire ball of very energetic particles started to expand and cool down to the world of stars and galaxies which is around us today. To understand the development of the universe and the formation of matter particle physicists are trying to simulate the conditions of the 'Big Bang' in the laboratory.

Particle Accelerators: Particles like electrons and protons, after acceleration to high energies, are used to probe the microscopic structure of matter. The high energies are required for two reasons: firstly, the spatial resolution increases with the energy of the probe. Secondly, the mass spectrum of the particles can only be studied if sufficient energy is available. The heaviest particle known up to

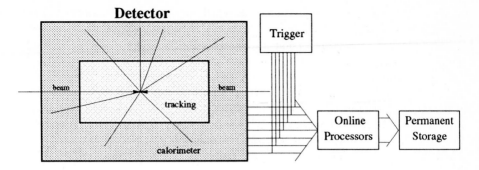

Fig. 2. Schematic view of the detector with the data acquisition and trigger systems.

now is the recently discovered top quark. Its mass is equivalent to an energy of about 180 GeV $= 180 \cdot 10^9$ eV (an electron gains the energy of 1 eV in a potential difference of 1 Volt).

Particle reactions can be observed by shooting an accelerated beam on a target ('fixed target', see e.g. Fig. 5). Accelerating two particle beams in opposite directions and colliding them head-on makes most efficient use of the energy. This is done in collider rings such as the electron-positron collider LEP at CERN in Geneva, the electron-proton collider HERA at DESY in Hamburg or the proton-anti-proton collider Tevatron at Fermi-Laboratory near Chicago.

Particle Detectors: Each collision point is surrounded by detectors which register the particles generated in a collision. In most cases one wants to reconstruct the whole kinematics of the event, which means that the momenta, energies, directions and identities of all particles must be determined.

A typical detector of a collider experiment (see Fig. 2) has an inner part in which the tracks of charged particles are measured. These detectors collect and amplify the ionisation charge the particles generate when passing through the detector. Bending the tracks by a magnetic field (see Fig. 1) allows to measure the momentum of the particles. The reconstruction of the tracks from the measured hit patterns is a problem of high combinatorial complexity and much work has gone into the development of reconstruction software. In recent years also ANN have been studied for this task.

Tracking detectors should have little material which could disturb the passage of the particles. Further outside, however, very dense material is placed in which the energy of most of the particles is totally absorbed. These devices allow to measure the energy of particles and are therefore called calorimeters. In calorimeters high-energy particles create a cascade of showers of particles with lower energies which can be finally absorbed. The analysis of the generated shower patterns, which are characteristic for the incident particle species, is a typical field of ANN applications.

Most of the detectors generate signals which can be read out and further processed electronically. The quality of the event reconstruction from the detector signals depends to a large extent on the granularity of the detector measurements. The largest existing detectors have about 1 million signal channels. The next generation of detectors, currently under development for the 'Large Hadron Collider' project at CERN, will have up to about 100 million channels. Each channel may have one or two bytes of information. With a collision frequency of several times 10 MHz this leads to a data flow of up to 100 Tbyte per second. However, the large majority of the data is background or 'uninteresting' physics which has to be recognized by very fast real-time trigger processors (Fig. 2).

2   Applications of Multi-Layer Perceptrons

Event Classification: A typical problem in experimental Particle Physics is to decide what class a measured event belongs to. The decision is derived from the usually high-dimensional pattern vector containing variables derived from processing the detector signals. The event variables include the momentum vectors, energies, identification tags and quality criteria of the measured particles. In addition to these basic quantities variables are derived which characterize the whole event, like the total energy in the event or the multiplicities of charged and neutral particles.

A very common classification procedure is to impose cuts on sensitive features, i. e. to require that the value of a variable is in a certain range. The cuts are tuned by studying the distributions of one- or two-dimensional projections of the pattern vectors. Correlations between variables can in principle be accounted for by defining a cut by a multi-dimensional surface. In practice, however, it is quite difficult to include correlations between more than two or three variables.

These limitations can be overcome by using multi-layer perceptrons (MLP) which can define nearly arbitrary volumes in the feature space, accounting for all correlations. The boundaries of the volumes containing the events of a certain class can be trained using either real or simulated data. Another advantage of MLP is that the output of a well-trained net has a precise statistical interpretation as a Bayesian discriminant (see below).

A MLP network has a layered structure, with an input layer, mostly one or two 'hidden' layers and an output layer. The signals are always sent to the next layer above (feed-forward). Mathematically the network maps the n-dimensional input vector **x** into the m-dimensional output vector **y** where the mapping is defined by the structure of the network and the weights connecting the nodes of one layer to those of the next layer.

The network is trained with N input patterns $\mathbf{x}^p$ ($p = 1, \ldots, N$) to approximate the expected output $\mathbf{y}^p$ ('supervised learning'). This is done by adjusting the weights such that the error function is minimized:

$$E = \frac{1}{2} \sum_{p=1}^{N} \sum_{i=1}^{m} (y_i{}^p - \hat{y}_i{}^p)^2 \tag{1}$$

For classification problems the net is trained with events of known classification such that the expected value $\hat{y}_j$ of the jth output is 1 if the event belongs to C_j, else 0. If the classes are well separated the trained net can assign the events to the proper classes with 100 % efficiency provided the net has enough degrees of freedom, i. e. enough nodes.

In most of our applications the classes have overlapping distributions. In this case it is in principle no longer possible to classify with 100 % efficiency. If the MLP net was trained with n_i events of class i it can be shown that the network outputs approximate the Bayesian discriminator:

$$y_i(\mathbf{x}) = \frac{n_i\, p_i(\mathbf{x})}{\sum_{j=1,m} n_j\, p_j(\mathbf{x})}. \tag{2}$$

The maximal y_i belongs to the class C_i for which the event has the highest probability. The advantage of ANN is that the probability densities $p_i(\mathbf{x})$ need not to be known explicitely because they are learnt by training using real or simulated data.

From (2) an interesting application follows: the net can learn multi-variate probability densities which are not known analytically. Training a MLP with a data sample of unknown distribution against an equally distributed data sample yields a net output which approximates the unknown probability density. This can be used for matching model simulations to measured data samples.

Example for Event Classification: Figure 3 shows an example for the output of a net which was trained with two classes [4]. Without going into the details just a short description of the experiment: The data are from the ARGUS experiment where in electron-positron collisions a heavy unstable object, here called 'resonance', was generated and the decay products were studied. The 'resonance' had to be separated from the usual 'continuum' of quark jets. This is a typical example where the distributions of single kinematical variables do not differ very much for the two classes. Differences become only significant if the multi-dimensional correlations between the variables are exploited. However, the network output in Fig. 3 is still not completely separated for the two classes. A compromise between efficiency and purity of the selected data sample has to be found by chosing a cut on the net output.

The Level-2 Trigger of the H1 Experiment: As an example of a real-time trigger application employing dedicated hardware we discuss the level-2 trigger based on MLP networks which has been developed for the H1 experiment at HERA [5, 6]. Using the data available at this trigger level (from calorimetry, tracking and others) several neural nets are trained to select the wanted physics events and reject the background, mainly from protons scattered outside the detector. The concept is to have one net for each physics process and in addition a positive background identification ('background encapsulator'). The used hardware (based on the CNAPS chip, see sect. 3) suggests networks of up to 64 inputs and 64 hidden nodes.

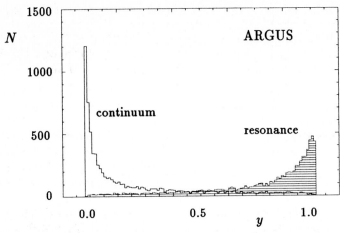

Fig. 3. Network output for simulated $\Upsilon(4S)$ resonance and continuum data [4].

Extensive studies have been performed to optimize the network architecture, the choice of variables and input data (real or simulated), learning strategies, efficiencies and many other topics [5]. Since the whole collaboration had to be convinced that a neural net is not a mysterious black box, particular emphasis was put on understanding how the network works. Questions which have been tackled are, for instance, the importance of each input variable, the visualisation of the network decision, or efficiencies in 'kinematical corners' (which may be less frequently trained).

Analysis of Shower Clusters: There are many examples which demonstrate that ANN can be very helpful for the analysis of calorimeter showers. The Crystal Barrel Collaboration, for instance, discriminates low-energy photons against 'split-offs', i. e. fluctuations from larger showers which look like single photons [7]. Another nice example is the use of ANN for the separation of photon and hadron induced cosmic air showers in the HEGRA experiment [8].

In the H1 collaboration work is in progress to discriminate electrons from pions using the pattern of the energy deposition in different sections of the liquid argon calorimeter [9]. At the same time (and with the same network) the deposited and incident energies are determined. The preliminary results suggest that calorimeters which compensate the fluctuating loss in the observed energy could be efficiently compensated by means of ANN processing.

Function Approximation: The determination of the incident energy from the energies deposited in calorimeter cells is an example for the use of MLP nets as function approximators. The network is trained to yield a continuous output. The goodness of the approximation is controlled by the number of contributing nodes. If a smoothing interpolation is desired the number of nodes must be appropriately restricted.

Another example for the use of MLP networks for function approximation is the determination of the W-boson mass from calorimetric information in a proton–anti-proton collider experiment [10]. The determination of a density function by network training is in fact also a function approximation. In these cases the functions are learnt from the sample events of a simulation.

3 Neural Net Hardware

The MLP algorithm is well suited for a hardware implementation. The processes at each node are local and can be executed consecutively and in parallel for all nodes of a layer. Dedicated VLSI chips have been developed which are optimized for fast matrix multiplications as needed for the computation of the activation of each node.

Table 1 lists some properties of a selection of analog and digital chips which have been used in Particle Physics applications. Compared to the digital approach, the analog processing is potentially faster and the I/O bandwidth of analog chips can be higher for a given number of connection pins. On the other hand analog circuits are more susceptible to instabilities. Also for general purpose neuro-computer systems only digital chips offer the required flexibility.

In Particle Physics the requirements on the speed of a chip may be more demanding than in other areas of real-time ANN applications. At HERA every 100 ns a beam-beam interaction occurs and a decision has to be taken whether the event has to be recorded or discarded (see Fig. 2). The trigger for keeping an event is usually generated in a hierarchy of different trigger levels. On each level the rate is reduced so that on the next level more time is available to apply more sophisticated algorithms. The condition for a deadtime-free system is that the input frequency into a trigger level times the decision time on this level must not exceed 1. On the first level this can often only be achieved by a trigger pipeline, i.e. the decision is subdivided in many small steps which are executed sequentially in a pipeline. The data generated by the experiment during the trigger execution must be stored in a data pipeline of the same depth as the trigger pipeline.

The interesting real-time applications for ANN are on the first and second trigger level, with decision times ranging from about 10 ns to several μs for the first level and from about 10 μs to several 100 μs for the second level. Except for the NeuroClassifier the chips in Table 1 are not fast enough for first level applications.

Level-2 Triggers with the ETANN Chip: ANN hardware has first been used for triggering in a high energy experiment at Fermi-Lab [11] employing the analog chip ETANN, e.g., for real-time tracking, calorimetry triggers and recognition of isolated electrons.

When the ANN activity started at Fermi-Lab the ETANN chip was the best choice. Today digital chips can perform similar tasks at the same speed while offering more stability and easier handling. However, it may well be that in the future analog circuits become interesting again.

Table 1. Characteristic properties of a selection of neuro-chips (PU = processor unit).

	ETANN	NeuroClassifier	CNAPS	MA16
arithmetic	6 bit analog	5 bit analog	16 bit fix	16 bit fix
PU per chip	4096	426	64	16
speed [MCPS]	1300	20000	1160	800
clock [MHz]	0.5	20 ns tot. delay	20	25

The Level-2 Trigger of the H1 Experiment: The experiments at the electron-proton storage ring HERA have to cope with a bunch crossing rate of 10 MHz. The H1 experiment has a four-level trigger architecture. At the 2nd level the allocated decision time is 20 μs, too short for a standard programmable processor but offering a niche for a dedicated neuro-processor.

In the previous section we have discussed the general concept of this trigger [5, 6]. The hardware which is designed to process feed-forward nets of a maximal size 64–64–1 is based on the CNAPS chip (table 1) available on a VME board environment together with specific software. The trigger system will finally consist of about 10 VME boards, each of these 'pattern recognition engines' processing the decision for one specific physics process. The system has been sucessfully tested during the 1995 running of the H1 experiment. In this year the system will be fully commissioned.

ANN Applications on the First Trigger Level: The first generation ANN applications for triggering experiments work on the time scale of some 10 μs and, frankly speaking, still on a level of quite moderate complexity. However, given the speed of electronic developments (at least if there is a commercial interest) we hope that a next generation of neuro-hardware would offer an order-of-magnitude improvement in speed and complexity for the same price.

Of great interest are ANN applications for high rate, high multiplicity experiments where complex pattern recognition tasks (tracking, calorimetry) have to be performed on a first-level trigger time scale. Replacing the highly specialized hard-wired processors by standardized 'neuro-hardware' could facilitate the design of triggers and potentially also lower the costs.

An interesting step in this direction is the development of the NeuroClassifier chip at the University of Twente [12] in close contact with Particle Physics groups at DESY and CERN. The chip accepts 70 inputs at a bandwidth of 4 GByte/s, has 6 hidden layers and 1 output. The decision time is only 20 ns. The high speed is achieved by analog computing while keeping moderate precision (5 bit). A group in the H1 Collaboration is currently implementing this chip into the first level trigger for deciding quickly if tracks are coming from the same point of the interaction region [13]. It thus discriminates against background coming mainly from protons of the beam halo which are scattered outside the detector and generate tracks which do not originate from the interaction region.

4 Self-Organizing Networks

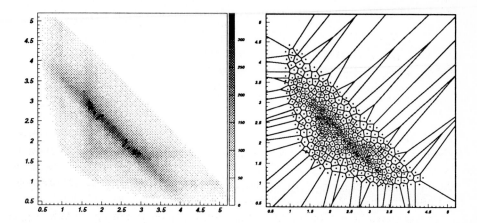

Fig. 4. Dalitz plot of a $\pi\pi\eta$ final state. Plotted are the squared masses of the two $\pi\eta$ combinations as explained in the text. The figure on the right shows the Voronoi regions as obtained with the GCS algorithm.

Feed-forward networks with supervised learning, while powerful for practical applications, are considered to have little 'neural appeal'. More guided by the biological examples are networks which are trained without a predefined output goal. In these self-organizing nets the training result is determined by the offered patterns and the learning rules.

Topographical mapping ('Kohonen map') of event vectors unto a one- or two-dimensional neuron space allows to discover features by visual inspection which may be otherwise difficult to detect in the high-dimensional space of the event vector. Different reactions with sufficiently different features cluster in different regions in the neuron space. The Voronoi regions belonging to a neuron can be selected for the classification of events.

Similar to the Kohonen mapping is the organisation of data by the 'Growing Cell Structure' (GCS) algorithm [14]. An application to data of the Crystal Barrel experiment at CERN has been reported in [15]. The Crystal Barrel experiment investigates the generation of particles in annihilations of protons and anti-protons. Figure 4a shows a so-called Dalitz plot of a specific three-particle final state, two pions (π) and one η-meson. For each two-particle combination one can calculate the mass an object would have if it decayed into these two particles. In the plot the squares of these masses for the two possible $\pi\eta$ combinations are plotted against each other. Within the kinematical boundaries the darker bands indicate that indeed the two particles can be the decay products of heavier objects. To reveal the nature of these objects, which are actually short-lived particles, one has to analyse the structures in the Dalitz plot. Since some million events contribute to the plot an event-by-event analysis becomes very time consuming. Therefore the data have been binned into cells using the GCS algorithm (Fig. 4b). The algorithm ensures that in each cell there is a similar number of events which is statistically optimal for fitting the plot.

5 Global Algorithms for Pattern Recognition in Tracking Detectors

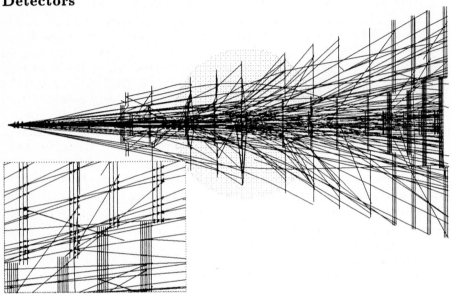

Fig. 5. Tracks in the HERA-B tracking detector generating hits in the planar detector layers (see also the enlarged detail in the inset). The tracks originate from a fixed targed on the left side which is hit by the HERA proton beam. The boost of all particles in the direction of the incident beam is typical for a 'fixed target' experiment. In the circled area the tracks are deflected by a magnetic field which allows to analyse the momentum of the tracked particles.

Tracking of Particles: The passage of charged particles through a detector can be made visible via the ionisation which the particles produce along their path. Usually tracking detectors have a layer structure. In each layer the ionisation charge of a passing particle is collected, amplified and converted to an electronic signal which tells where the particle passed the layer. In most cases one layer yields only the coordinate of one projection, the full three-dimensional track information then comes from the combination of several layers, sometimes combined into 'superlayers' (as in the inset of Fig 5).

Figure 5 shows tracks in the tracking detector of the HERA-B experiment. The HERA-B experiment is currently under construction and will start data taking in 1998 in the proton beam of HERA at DESY. The experiment aims to attack one of the most fundamental and yet unsolved questions: Why is matter so much more abundant in the universe than anti-matter. In order to reach this ambitious goal the detector must run under quite extreme conditions. As can be seen in Fig. 5 the track density in the detector is very high. Each electronic channel has a hit probability per event of up to about 30 % which seriously affects the reconstruction efficiency. With standard pattern recognition and track reconstruction methods, efficiencies larger than 90 % have been reached.

The classical algorithms for pattern recognition in tracking detectors, like the Kalman filter, work locally in the sense that they start from a seed in one layer and follow the track through the detector from one layer to the next. At each step all the combinatorial possibilities for the next hit on the track must be checked. The reconstruction probability depends in this case on the choice of the seeds and their order. A global optimisation of the assignment of hits to tracks can be done with ANN algorithms. Indeed, tracking in particle detectors is a problem of high combinatorial complexity like the 'travelling salesman' or the 'graph bisectioning' problems which have been successfully treated by ANN, e. g. by the Hopfield algorithms.

Denby-Peterson Nets: The application of Hopfield networks for track reconstruction has been suggested by Denby and Peterson [16] independently. In their model a neuron is a link between two measured hits i and j in a tracking detector. A neuron with a label ij should have the activation $S_{ij} = +1$ if it is on a track otherwise $S_{ij} = 0$. The neurons interact according to the dynamics of the Hopfield model leading to an energy function which has to be minimized:

$$E = -\frac{1}{2} \sum_{ij,kl} w_{ij,kl} S_{ij} S_{kl} + \sum_{m} \lambda_m f_m \quad (3)$$

The interaction strength is given by the weights $w_{ij,kl}$, the functions f_m are constraint terms and the factors λ_m Lagrange multipliers. In the case of the Denby-Peterson model the weights are some power of the cosine of the angle between two links to ensure smooth curvatures of the found tracks. Most of the weights can be set to zero because the corresponding neurons have no chance to be on the same track. The energy function contains two additional constraint terms, one to suppress bifurcations of tracks, the other to ensure that all or most hits are used for a track. The importance of the constraint terms can be regulated by the Lagrange multipliers. For instance, the fraction of hits belonging to a found track depends on the noise level in the detector.

The method has been applied to real data of the central tracking detector of the ALEPH experiment at CERN and found to be comparable in tracking efficiency and computing time to the standard tracking algorithm of ALEPH [17].

For this algorithm most of the computer time is actually used for the initialisation of the weights which are (in contrast to feed-forward nets) different for each event. Therefore the algorithm should become even more favourable for higher track multiplicities because the computing time increases not much more than linearly with the multiplicity in contrast to a power law for the conventional method.

The optimal selection and matching of track segments reconstructed in two tracking detectors of the DELPHI experiment is described in [18]. In this case the Hopfield algorithm replaced the standard reconstruction program because it turn out to be more efficient and faster.

In studies for tracking in HERA-B, good reconstruction efficiencies in dense track environments where found for a model detector with equal spacing of the detector layers. The unequal spacing in the real detector, however, caused major problems. Another possible disadvantage of the Hopfield model is the lack of an estimator for the goodness of the track reconstruction. In addition the algorithm seems not to be well suited for a hardware implementation since it is not inherently parallel and local. However, one should carefully investigate if this statement is biased by a too conservative notion of data processing.

Deformable Templates: As another global tracking algorithm the 'deformable template' method [19] has been applied to simulated HERA-B data. In this algorithm one starts with track specimens, the templates, which are defined in the parameter space of the tracks (for straight three-dimensional tracks these are 4 parameters, for curved tracks in a magnetic field 5 parameters). A reasonable initialisation of templates can be obtained by analysing the Hough-transforms of the hits in the parameter space.

For hit i, a distance d_{ia} to each template a has to be calculated. The template parameters are then adjusted such that for each template the distances for a group of hits is minimized. This is achieved by globally minimizing the function

$$E = \sum_{i,a} d_{ia}^2 \frac{e^{-\beta d_{ia}^2}}{\sum_b e^{-\beta d_{ib}^2}} . \qquad (4)$$

The distances are weighted by a Boltzmann-type exponential with a corresponding 'temperature' $T = 1/\beta$. A high temperature corresponds to a long-range interaction between hits and templates. Lowering the temperature during the optimization process ('simulated annealing') stabilizes the assignment of hits to the templates. The lowest reasonable temperature is obviously given by the square of the spatial resolution of the track hits.

The results for the 'deformable template' reconstruction are very encouraging: the algorithm is independent of the geometry of the detector and performs the track reconstruction in three dimensions, the minimal temperature supplies ω: minimise different t function + elastic term

$$E = \sum_i \ln \sum_j \frac{e^{-\beta d_{ij}}}{\sum e^{-\beta d_{jk}}} \quad {}^{13} + K \sum_i (y_i - y_{i+1})^?$$

a natural measure for the reconstruction quality and, last but not least, the achieved reconstruction efficiencies are very satisfactory. It appears also to be straightforward to extend the method to curved tracks in a magnetic field. The major drawback is a still too large computing time. However, there are some ideas how this can be improved in the future.

$$\Delta y_i = \alpha \sum_j (y_i - x_j) w_{ij} + \beta K (y_{j-1} - 2y_j + y_{j+1})$$

6 Conclusion

In recent years Neural Network algorithms have found a wide interest in the Particle Physics community. Various ANN models have been studied for their applicability for classification of events, particle identification, function approximation and pattern recognition.

Multi-Layer Perceptrons have proven to be valuable tools for data analysis. For the classification of events these networks have very advantageous properties, such as flexibility because of their trainability, the capability to solve high-dimensional problems and deterministic behaviour.

The intrinsic parallelism of these networks make them well suited for hardware implementations. Specialized analog and digital VLSI chips for neuroprocessing have found applications for triggering. One can expect a rapid development of the speed and complexity of the neuro-hardware over the next years opening the possibility to be used for first-level triggering in high-intensity experiments. The development of dedicated neuro-hardware is particularly essential for a full exploitation of neural algorithms.

Quite promising are the first studies of self-organising networks, like Kohonen maps and 'Growing Cell Structures', for feature extraction and the organisation of large data sets.

The ANN algorithms, in particular the Hopfield and 'deformable template' algorithms, appear very attractive for solving pattern recognition problems for tracking detectors. Indeed very promising studies using real tracking devices, e. g. of LEP experiments, have been carried out and partly included in the standard reconstruction software of these experiments. However, the state of the art in pattern recognition is highly developed and is extremely difficult to beat. In addition, the hardware implementation of Hopfield nets is not as straightforward as in the case of feed-forward nets.

Acknowledgements: I would like to thank all those who helped me in the preparation of this paper. In particular I want to mention: C. Borgmeier, M. Kunze, T. Lohse, R. Mankel, M. Medinnis and my collaborators on the level-2 trigger of the H1 experiment.

The work was supported by Bundesministerium für Bildung und Forschung under contract numbers 056DO57P, 056DO50I and 057BU35I.

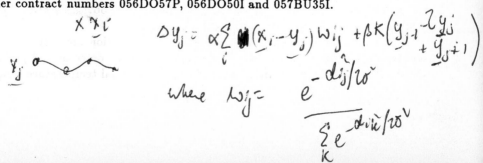

$$\Delta y_j = \alpha \sum_i (x_i - y_j) w_{ij} + \beta K \left(y_{j-1} - 2y_j + y_{j+1} \right)$$

$$\text{where } w_{ij} = \frac{e^{-d_{ij}/2v}}{\sum_k e^{-d_{ik}/2v}}$$

References

1. A good overview of ANN applications in Particle Physics can be obtained from the proceedings of the 'Workshops on Software Engineering and Artificial Intelligence for High Energy and Nuclear Physics (AIHENP)' which in this year takes place in Lausanne (Sept. 2-6). Previous workshops have been held at Lyon (1990), La Londe les Maures (1992), Oberammergau (1993) and Pisa (1995).
   Information also via WWW: http://www1.cern.ch/NeuralNets/nnwInHep.html.
2. Proceedings of AIHENP 1993, eds. K.-H. Becks and D. Perret-Gallix, World Scientific (1994).
3. Proceedings of AIHENP 1995, eds. B. Denby and D. Perret-Gallix, World Scientific (1995).
4. M. Joswig et al., in [2].
5. C. Kiesling et al.; P. Ribarics et al.; A. Gruber et al.; T. Krämerkämper et al., in [2].
6. D. Goldner et al.; J. Möck et al., in [3].
7. T. Degener, in [2].
8. S. Westerhoff and H. Meyer, in [3].
9. M. Höppner, contribution to the 1996 AIHEP Workshop.
10. L. Lönnblad et al., Phys. Lett. B278 (1992) 181.
11. B. Denby, Proceedings of AIHENP 1992, ed. D. Perret-Gallix, World Scientific (1992).
12. P. Masa et al., Proceedings 'Fourth Intern. Conf. on Microelectronics for Neural Networks and Fuzzy Systems', Turin 1994.
13. T. Tong Tran et al., in [3].
14. B. Fritzke, Intern. Computer Science Institute TR-93-026, Berkeley, 1993.
15. M. Kunze and J. Steffens in [3].
16. B. Denby, Comput. Phys. Commun. 49 (1988) 429; C. Peterson, Nucl. Instrum. Methods A279 (1989) 537.
17. G. Stimpfl-Abele and L. Garrido, Comput. Phys. Commun. 64 (1991) 46.
18. R. Frühwirth, Comput. Phys. Commun. 78 (1993) 23.
19. M. Ohlsson and C. Peterson, Comput. Phys. Commun. 71 (1992) 77.

Evolutionary Computation — History, Status, and Perspectives

H.P. Schwefel

Universität Dortmund
Germany

Abstract. Algorithmic models a la Darwin can be dated back to the beginning of the digital computer era. They mimic evolutionary principles like mutation, recombination, natural selection, and sometimes more, as operators working on a data structure, called individual, representing a possible solution to a given problem. According to the population principle, evolutionary algorithms are inherently parallel by handling several individuals at the same time. Besides of basic similarities, there are some strong differences between algorithms of this class, e.g., between genetic algorithms and evolution strategies. Essential for the latter is the collective auto- adaptation of internal strategy parameters. Though the underlying theories are still quite weak, the past decade has witnessed an exponential increase in diverse applications, from design synthesis, planning and control processes, to various other adaptation and optimization tasks. They cannot be reported adequately in an overview like this. But a summary of existing theoretical results will be given. A look back from existing practice of evolutionary computation to the natural prototype reveals a still substantial gap between current models of organic evolution and reality. This opens prospects for further advancements in the applicability and efficiency of the algorithms as well as a better understanding of some important real world phenomena, e.g. in ecosystems.

Temporal Structure of Cortical Activity

M. Abeles and Y. Prut

Dept. of Physiology and the Center for Neural Computation
The Hebrew University
POB 12272
Jerusalem 91120
Israel

Abstract. Usually, it is assumed that the spike trains represent stochastic point processes, for which the main parameter are the (possibly variable) firing rates. However, when evaluating the relations among the firing times of several neurons, it is common to find very precise timing. Activity of 8–15 single units was recorded in parallel from the frontal cortex of monkeys performing a variety of localization and problem solving tasks. Patterns of firing in which several (3 or more) spikes appear repeatedly with the same intervals are called here spatio- temporal firing patterns. The present study refers to spatio-temporal firing patterns lasting for up to 450 ms, and in which the firing time is accurate to within ≈1 ms.

In many cases the spatio-temporal firing patterns were associated with the monkey's behavior. In some cases, the entire response of a single unit was composed of such accurate firing patterns. The internal time-locking among spikes of any given spatio-temporal firing pattern were tighter then the time-locking of the pattern to the external stimulus (≈1 ms versus ≈50 ms). Often, the spikes which were involved in spatio-temporal firing patterns conveyed more specific information about the stimulus or the response as compared to all the spikes of a single unit.

A simple neural network that can generate such patterns is the syn-fire chain. Insight of the properties of syn-fire chains was gained by simulating networks of neurons in which chains of feed-forward, diverging converging connections were laid. The simulations revealed that such chains convert diffuse (asynchronous) excitation into synchronized volleys of spikes that can propagate over many links with very little jitter in time.

The experimental data shows that in approximately half of the patterns the same neuron fired twice. This suggest that the same neuron might participate twice (or more) in the same synfire chain. Simulations have shown that this may occur without disrupting the nature of the network. Under such conditions the synfire chain is not strictly a feed-forward network. In such synfire chains reverberations can occur. The experimental data and the simulations support the idea that cortical activity is generated and maintained by reverberations in syn-fire chains, and that binding between such syn-fire activities in different brain regions is the mechanism for generating a hierarchy of representations.

SEE-1 — A Vision System for Use in Real World Environments

U. Ramacher

Siemens AG
ZFE T ME4
Otto-Hahn-Ring 6
München-Perlach
Germany

Abstract. The Siemens Electronic Eye project is presented, in particular the HW-SW system architecture of the SEE-1. Key components are the Vision Instruction Processor VIP128, the C-compiler and the vision application library. Because of its general instruction set the SEE-1 computer seems to be of considerable advantage for neurocomputing as well as multi-media applications.

Towards Integration of Nerve Cells and Silicon Devices

P. Fromherz

Dept. Membrane and Neurophysics
Max-Planck-Institute for Biochemistry
D-82152 Martinsried/München

Abstract. We develop the interface between single neurons and silicon microstructures. The research has three prospects: (i) We may learn about the biophysics of neuron membranes. (ii) We may study neural networks by large scale stimulation and large scale recording of the response. (iii) We may see whether hybrid systems of neural tissue and silicon devices may be used for technological and medical purposes.

At present the investigations are concerned with the physical basis of interfacing. It is shown, how direct stimulation and recording of neurons from silicon is achieved by electrical induction. The AC-transmission of the junctions is studied by patch-clamp technique. The geometry of the interface is visualized by fluorescence interferometry. The spatial resolution of interfacing is tested with an array of closely packed transistors.

With respect to future developments we study the design of nets by guided growth of neurons from ganglia of leeches and snails and from the hippocampus of rats and the construction of a 2000-contact chip for multiple recording on the basis of modern CMOS-technology.

References

Science 252, 1290 (1991)

Phys. Rev. Lett. 71, 4079 (1993)

Phys. Rev. Lett. 75, 1670 (1995)

Phys. Rev. Lett. 76, 327 (1996)

Unifying Perspectives on Neuronal Codes and Processing

C.H. Anderson

Department of Anatomy and Neurobiology
Washington University School of Medicine
St. Louis
USA

Abstract. This talk will review a rapidly growing body of work on modeling neuronal ensembles that provides an understanding of a diverse body of experimental neurobiological observations and has generated new insights into how neurons can implement a rich repertoire of powerful information-processing functions. The starting point is the Georgopoulos 'Population Vector' for modeling motor cortex experimental results, which has been expanded upon by Abbott, Sanger, Salinas, and others. Miller's studies of the cricket cercal system provides the most detailed experimental example of a population code. This simple system exhibits many of the characteristics seen in primate retinal ganglion cells; visual, parietal and motor cortex pyramidal cells; as well as cells in the subcortical systems that generate eye movements. One form of population codes, which is closely related to Specht's probability neural networks, is based on the assumption that neuronal firing rates encode the amplitudes of representations of probability density functions (PDFs). The PDF framework can be utilized to generate models of very large neuronal cortical circuits that incorporate the statistical inference 'Pattern Theory' of Grenander as well as generalizations of the Anderson, Van Essen and Olshausen routing circuits.

Visual Recognition Based on Coding in Temporal Cortex: Analysis of Pattern Configuration and Generalisation across Viewing Conditions without Mental Rotation

D.I. Perrett

School of Psychology
University of St Andrews
Scotland

Abstract. A model of recognition is described that is based on cell properties in the ventral cortical stream of visual processing in the primate brain. At a critical intermediate stage in this system Elaborate feature sensitive cells respond selectively to visual features in a way that depends on size (± 1 octave), orientation (± 45) but does not depend on position within central vision (± 5). These features are simple conjunctions of 2-D elements (e.g. a horizontal dark area above a dark smoothly convex area). Such features can arise either as elements of an objects surface pattern or as 3-D component parts of the object. By requiring a combination of several such features without regard to their position within the central region of the visual image, Pattern sensitive cells at higher levels can become selective for complex configurations that typify objects experienced in particular viewing conditions. Given that input features are specified in approximate size and orientation, initial cellular 'representations' of the visual appearance of object type (or object example) are also selective orientation and size. Such representations are sensitive to object view (± 40-60) because visual features disappear as objects are rotated in perspective. Combined sensitivity to multiple 2-D features independent of their position establishes selectivity for configuration of object parts (from one view) because rearranged configurations yield images lacking some of features present in the normal configuration. Different neural populations appear to be tuned to particular components of the same biological object (e.g. face, eyes, hands, legs), perhaps because the independent articulation of these components gives rise to correlated activity in different sets of input visual features. Generalisation over viewing conditions for a given object can be established by hierarchically pooling outputs of view specific cells. Such pooling could depend on the continuity in experience across viewing conditions: different object parts are seen together and different views are seen in succession when the observer walks around the object. For any familiar object, more cells will be tuned to the configuration of the objects features present in the view(s) frequently experienced. Therefore, activity amongst the population of cells selective for the objects appearance will accumulate more slowly when the object is seen in an unusual orientation or view. This accounts for increased time to recognise rotated views without the need to postulate mental rotation or transformations of novel views to align with neural representations of familiar views. The model is in accordance with known physiological findings and matches the behavioural performance of the mammalian visual system which displays view, orientation and size selectivity when learning about new pattern configurations.

A Novel Encoding Strategy for Associative Memory

Han-Bing JI, Kwong-Sak LEUNG, and Yee LEUNG

Dept. of Computer Science and Engineering, The Chinese University of Hong Kong, Shatin, N.T., Hong Kong. E-mail: hbji@cs.cuhk.edu.hk.

Abstract: A novel encoding strategy for neural associative memory is presented in this paper. Unlike the conventional pointwise outer-product rule used in the Hopfield-type associative memories, the proposed encoding method computes the connection weight between two neurons by summing up not only the products of the corresponding two bits of all fundamental memories but also the products of their neighboring bits. Theoretical results concerning stability and attractivity are given. It is found both theoretically and experimentally that the proposed encoding scheme is an ideal approach for making the fundamental memories fixed points and maximizing the storage capacity which can be many times of the current limits.

1 Introduction

Hopfield proposed an outer-product based neural associative memory in his seminal paper [1]. Since its storage capacity was found to be severely constrained both empirically [1] and theoretically [2], much research has been done for the improvements of the storage capacity and associative recall by various approaches [3]-[8]. In this paper, we will propose a novel neural associative memory based on a new encoding strategy. Unlike the pointwise outer-product rule commonly used in the Hopfield-type models, the novel encoding method computes the connection weight between two neurons (including self-feedback connections) by summing up not only the products of the corresponding two bits of all the fundamental memories, but also the products of their neighboring bits as well. Theoretical analysis is conducted to investigate the performances in terms of stability and attractivity which are supported by simulation results.

2 A Novel Encoding Strategy

Assume there are m FMs $u^{(r)}$, $r = 1, 2, \cdots, m$, stored in the network, each with n-bit long. The connection weights computed by the novel encoding strategy are as follows:

$$w_{ij} = \sum_{t=1}^{L} \sum_{r=1}^{m} u_{i+t-1}^{(r)} \cdot u_{j+t-1}^{(r)}, \qquad i, j = 1, 2, \cdots, n, \tag{1}$$

where w_{ij} is the connection weight between neurons i and j; n is the dimension of the FMs; L is an integer in the range of $1 \le L \le n$; $u^{(r)}$, $r = 1, 2, \cdots, m$, are called fundamental memories (FMs); and $u_i^{(r)}$ is the ith element of FM $u^{(r)}$, $i = 1, 2, \cdots, n$. Observing eqn.(1), we can see that the connection weights are not just decided by the summed products of the corresponding two bits of all the FMs as in the Hopfield-type neural associative memories. Summed products of a number of neighboring bits of the corresponding two bits in all the FMs are also taken into account. This is the significant difference between the proposed neural associative memory and the Hopfield-type models. Nevertheless, the network structure of the proposed model is the same as the Hopfield model except that it is with self-feedback connections. The retrieval process is also identical to that of the Hopfield model which is as follows:

$$y_i = \sum_{j=1}^{n} w_{ij} \cdot x_j,$$

$$o_i = \text{sgn}(y_i),$$

(2)

where $x_j, j = 1, 2, \cdots, n$, is the network's input, y_i is the summed input of neuron i, o_i is the output of neuron i, $i = 1, 2, \cdots, n$, and $\text{sgn}(\cdot)$ is the sign function.

The integer parameter L in eqn.(1) is defined as the length of a neighboring range of each bit. The neighboring range of each bit can be chosen from one side (either right-hand or left-hand side) or both sides. It can also be specifically designed on the basis of concrete applications. In this paper, without the loss of generality, only the right-hand-side neighboring range is used for computation. Besides, in eqn.(1), when k > n, we set $u_k^{(r)} = u_{k-n}^{(r)}$. In eqn.(1), when L = 1, the proposed neural associative memory reduces to the first-order outer-product neural associative memory which is the same as the Hopfield model except for the self-feedback connections.

Suppose one of the FMs, $u^{(p)}$, is taken as a retrieval key input. By eqns.(1) and (2),

$$y_i = \sum_{t=1}^{L} \sum_{r=1}^{m} \sum_{j=1}^{n} u_{i+t-1}^{(r)} \cdot u_{j+t-1}^{(r)} \cdot u_j^{(p)}$$

$$= (n + L \cdot m - 1) \cdot u_i^{(p)} + \sum_{r=1, \neq p}^{m} \sum_{j=1, \neq i}^{n} u_i^{(r)} \cdot u_j^{(r)} \cdot u_j^{(p)} + \sum_{t=1}^{L} \sum_{j=1, \neq i}^{n} \sum_{r=1}^{m} u_{i+t}^{(r)} \cdot u_{j+t}^{(r)} \cdot u_j^{(p)}. \quad (3)$$

If $\text{sgn}(y_i) = u_i^{(p)}$, $i = 1, 2, \cdots, n$, then $u^{(p)}$ is called a fixed point (FP). The first term in the right-hand side of eqn.(3) can be regarded as a "signal" for the associative recall of $u_i^{(p)}$, while the last two terms can be regarded as a "noise". Assume all the FMs are randomly generated from symmetric Bernoulli trials, i.e., $\text{Prob}\{u_i^{(r)} = 1\} = \text{Prob}\{u_i^{(r)} = -1\} = 0.5$, i = 1, 2, $\cdots$, n, r = 1, 2, $\cdots$, m. Accordingly, we obtain an SNR (Signal to Noise Ratio) [4] for the associative recall of $u_i^{(p)}$ in the proposed model as follows:

$$\frac{E[\,|(n + L \cdot m - 1) \cdot u_i^{(p)}|\,]}{\sqrt{\left(\text{var}\left(\sum_{r=1, \neq p}^{m} \sum_{j=1, \neq i}^{n} u_i^{(r)} \cdot u_j^{(r)} \cdot u_j^{(p)} + \sum_{t=1}^{L-1} \sum_{j=1, \neq i}^{n} \sum_{r=1}^{m} u_{i+t}^{(r)} \cdot u_{j+t}^{(r)} \cdot u_j^{(p)}\right)\right)}}$$

$$= \frac{n + L \cdot m - 1}{\sqrt{((m-1)(n-1) + (L-1)(n-1)m)}} = \frac{n + L \cdot m - 1}{\sqrt{((n-1)(L \cdot m - 1))}}. \quad (4)$$

The corresponding SNR in the Hopfield model is [2]:

$$\frac{E[\,|(n-1) \cdot u_i^{(p)}|\,]}{\sqrt{\left(\text{var}\left(\sum_{r=1, \neq p}^{m} \sum_{j=1, \neq i}^{n} u_i^{(r)} \cdot u_j^{(r)} \cdot u_j^{(p)}\right)\right)}} = \frac{\sqrt{(n-1)}}{\sqrt{(m-1)}}. \quad (5)$$

Observing eqn.(3), we can see that on one hand, the "signal" is greatly enhanced when L and\or m are large; on the other hand, although the number of terms comprising the "noise" also increases when L and\or m are large, relatively, the "noise" is not enhanced as much as the "signal" so long as all the FMs are randomly generated from symmetric Bernoulli trials.

3 Analysis of the Stability and Attractivity

There are two fundamental requirements for associative memories: The first one is the stability of the FMs which should all be FPs; and the other is the attractivity of these FPs which should have a radius of attraction.

It is commonly known that when x is one of the FMs, $u^{(p)}$, we can say that $u^{(p)}$ is an FP if and only if $y_i \cdot u_i^{(p)} \geq 0$, $i = 1, 2, \cdots, n$; and when x is an "error" version of one of the FMs, $u^{(p)}$, we can say that $u^{(p)}$ can be correctly recalled if and only if $y_i \cdot u_i^{(p)} \geq 0$, $i = 1, 2, \cdots, n$. Based on such a concept, we will derive some theoretical results to compare the performances among the Hopfield model, the first-order outer-product model with self-feedback connections, and our proposed model in terms of the stability and attractivity.

3.1 Stability Analysis

Here, FM $u^{(p)}$ is taken as the network's input.

3.1.a The Hopfield Model

Connection weights of the Hopfield model are computed as follows:

$$w_{ij} = \begin{cases} \sum_{r=1}^{m} u_i^{(r)} \cdot u_j^{(r)}, & \text{if } i \neq j, \\ 0, & \text{if } i = j. \end{cases} \tag{6}$$

To the Hopfield model, a necessary and sufficient condition for the FMs to be FPs is

$$|C(u^{(r)}, u^{(p)})|_{\max}_{r \neq p} \leq \frac{n - m}{m - 1}, \tag{7}$$

or

$$d_{\min}(u^{(r)}, u^{(p)})_{r \neq p} \geq \frac{n(m - 2) + m}{2(m - 1)}, \tag{8}$$

And the maximum number of the FMs that can become FPs, given the minimum distance between any two FMs, $d_{\min}(u^{(r)}, u^{(p)})_{r \neq p}$, $r = 1, 2, \cdots, m, \neq p$, is

$$m \leq \frac{n - 1}{1 + n - 2d_{\min}(u^{(r)}, u^{(p)})_{r \neq p}} + 1, \tag{9}$$

where

$$|C(u^{(r)}, u^{(p)})|_{\max}_{r \neq p} = |\sum_{j=1}^{n} u_j^{(r)} \cdot u_j^{(p)}|_{\max}_{r \neq p} = n - 2d_{\min}(u^{(r)}, u^{(p)})_{r \neq p}, \tag{10}$$

$$d(u^{(r)}, u^{(p)}) = \min[H(u^{(p)}, u^{(r)}), H(u^{(p)}, -u^{(r)})], \tag{11}$$

and H stands for the Hamming distance.

3.1.b The First-Order Outer-Product Model with Self-Feedback Connections

The connection weights of this model are computed as follows:

$$w_{ij} = \sum_{r=1}^{m} u_i^{(r)} \cdot u_j^{(r)}, \qquad\qquad i, j = 1, 2, \cdots, n. \tag{12}$$

They are the same as those of the Hopfield model except the self-feedback connections being not set to zero. To this model, a necessary and sufficient condition for the FMs to be FPs is

$$|C(u^{(r)}, u^{(p)})|_{\substack{max \\ r \neq p}} \leq \frac{n}{m-1}, \tag{13}$$

or

$$d_{\substack{min \\ r \neq p}}(u^{(r)}, u^{(p)}) \geq \frac{n(m-2)}{2(m-1)}, \tag{14}$$

and the maximum number of the FMs that can become FPs, given the minimum distance between any two FMs, $d_{\substack{min \\ r \neq p}}(u^{(r)}, u^{(p)})$, $r = 1, 2, \cdots, m, \neq p$, is

$$m \leq \frac{n}{n - 2d_{\substack{min \\ r \neq p}}(u^{(r)}, u^{(p)})} + 1. \tag{15}$$

3.1.c The Novel Encoding Strategy Based Neural Associative Memory

The following theorem gives the theoretical relationship between the length (L) of the neighboring range of each bit and the requirement for the FMs to be FPs as well as the maximum number of the FMs that can be FPs.

Theorem 1: i, Assume $u^{(r)}$, $r = 1, 2, \cdots, m$, are m randomly generated FMs. An FM $u^{(p)}$ will be an FP if and only if

$$|C(u^{(r)}, u^{(p)})|_{\substack{max \\ r \neq p}} \leq \frac{n + m(L - 1)}{m - 1}, \tag{16}$$

or

$$d_{\substack{min \\ r \neq p}}(u^{(r)}, u^{(p)}) \geq \frac{n(m-2) - m(L-1)}{2(m-1)}, \tag{17}$$

where $|C(u^{(r)}, u^{(p)})|_{\substack{max \\ r \neq p}} = |\sum_{j=1}^{n} u_j^{(r)} \cdot u_j^{(p)}|_{\substack{max \\ r \neq p}} = n - 2d_{\substack{min \\ r \neq p}}(u^{(r)}, u^{(p)})$, $d(u^{(r)}, u^{(p)}) = \min[H(u^{(p)},$

$u^{(r)})$, $H(u^{(p)}, -u^{(r)})]$, and H stands for the Hamming distance.

ii, Given the minimum distance between any two FMs, $d_{\substack{min \\ r \neq p}}(u^{(r)}, u^{(p)})$, $r = 1, 2, \cdots, m,$

$\neq p$, the maximum number of the FMs that can become FPs is

$$m \leq \frac{2n - 2d_{\substack{min \\ r \neq p}}(u^{(r)}, u^{(p)})}{n - (L - 1) - 2d_{\substack{min \\ r \neq p}}(u^{(r)}, u^{(p)})}. \tag{18}$$

(Proof)

Comparing eqns.(7), (8), (9), (13), (14), (15), (16), (17) and (18), we can see that in the proposed model, the integer parameter L can be made to increase the stability of the FMs significantly. As a result, large storage capacity can be achieved in such a model. In principle, the larger the neighboring-range length L is, the looser the condition is required for the FMs to become FPs, as well as the larger the storage capacity becomes. By the loose condition required for the FMs to be FPs, we mean that the FMs can be allowed to be near in the distance defined by $d(u^{(r)}, u^{(p)})$, $r = 1, 2, \cdots, m, \neq p$. Finally, those equations also imply that the first-order outer-product model with the self-feedback connections performs better than the Hopfield model with respect to the stability.

3.2 Attractivity Analysis

Here, the network's input x is an "error" version of FM $u^{(p)}$. The Hamming distance between x and $u^{(p)}$ is k bits.

3.2.a The Hopfield Model

To the Hopfield model, a necessary and sufficient condition of the radius of attraction of each FM is

$$k \leq \frac{n - m - (m - 1) \cdot C_{max}}{2m} , \tag{19}$$

and given the maximum correlation among all the FMs, the maximum number of the FMs that have the radius of attraction defined by eqn.(19) is

$$m \leq \frac{n + C_{max}}{1 + C_{max} + 2k} , \tag{20}$$

where C_{max} is the maximum correlation among all the FMs.

3.2.b The First-Order Outer-Product Model with Self-Feedback Connections

To this model, a necessary and sufficient condition of the radius of attraction of each FM is

$$k \leq \frac{n - (m - 1) \cdot C_{max}}{2m} . \tag{21}$$

Given the maximum correlation among all the FMs, the maximum number of the FMs that have the radius of attraction defined by eqn.(21) is

$$m \leq \frac{n + C_{max}}{2k + C_{max}} . \tag{22}$$

3.2.c The Novel Encoding Strategy Based Neural Associative Memory

The following theorem gives the theoretical relationship between the length (L) of the neighboring range of each bit and the radius of attraction of each FM as well as the resulting maximum number of the FMs.

Theorem 2. i, Assume $u^{(r)}$, $r = 1, 2, \cdots, m$, are m randomly generated FMs. Suppose the network's input x is k bits away from the corresponding FM $u^{(p)}$. FM $u^{(p)}$ can be correctly recalled from x if and only if

$$k \leq \frac{n - m(L - 1) - (m - 1) \cdot C_{max}}{2m} ,$$ (23)

where C_{max} is the maximum correlation among all the FMs;

ii, Given the maximum correlation among all the FMs, the maximum number of the FMs that have the above radius of attraction is

$$m \leq \frac{n + C_{max}}{2k + L - 1 + C_{max}} .$$ (24)

(Proof)

Comparing eqns.(19), (20), (21), (22), (23) and (24), we can see that in the proposed model, the integer parameter L will decrease both the radius of attraction of each FM and the maximum number of the FMs that have the radius of attraction as in eqn.(23). Theoretically, the larger the L is, the smaller the radius of attraction of each FM will become as well as the smaller the maximum number of the FMs that have the radius of attraction as in eqn.(23). Moreover, those equations also imply that the first-order outer-product model with the self-feedback connections performs better than the Hopfield model with respect to the attractivity.

By the above theoretical results, we reach a conclusion that in the novel encoding strategy based neural associative memory, extreme stability of the FMs can be achieved at the cost of their attractivity. The more stable the FMs are, the less attractive they will be.

4 Computer Simulations

The synchronous mode is adopted for the proposed network's evolution. All the FMs are randomly generated from symmetric Bernoulli trials. The right-hand-side neighboring range of each bit is utilised. The computer simulations procede by gradually increasing m (the number of the FMs) with n (the dimension of the FMs) being fixed. Each point on all curves given is obtained by ten times average under the same n, m, and L but different sets of FMs. Here, only the stability of the FMs is examined.

Fig.1 shows the results concerning how many FMs can be FPs in the proposed model for the case of n=30. The results using different L (L = 1, 2, 3, 6, 9, 14 and 23) are given to demonstrate the effect of the neighboring ranges in the proposed model. From the figure, we can see that the larger the L is, the more stable the FMs will be. It is shown that when L is not smaller than 9, ideal results can be obtained (It should be pointed out that 9 may not be the smallest value of L which is necessary for the proposed model to achieve ideal results in this case). However, given an n, when L reaches a certain sufficient high value, ideal performance can be achieved without increasing L further. Results corresponding to the cases of other n demonstrate the same matter and are not listed here due to space limitation. In general, the neighboring range required for ideal performance is not very large comparing with the dimension of the FMs.

5 Conclusion

We have proposed a novel encoding strategy based neural associative memory. Unlike the conventional outer-product rule used in the Hopfield-type models, the proposed encoding method computes the connection weights by summing up not only the products of the corresponding two bits of all the FMs, but also the products of their neighboring bits as well. Theoretical analysis has been done to investigate the performances of the proposed model in terms of the stability and attractivity, and to compare them with those of the Hopfield-type models. Both the theoretical and experimental results show that the novel encoding strategy is an ideal approach for a neural associative memory to achieve maximum stability. The memory capacity of the proposed model is found to be many fold of the existing models.

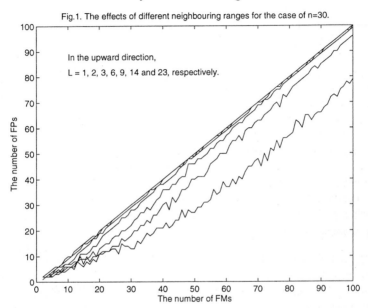

Fig.1. The effects of different neighbouring ranges for the case of n=30.

In the upward direction,
L = 1, 2, 3, 6, 9, 14 and 23, respectively.

References

1. J. J. Hopfield, "Neural networks and physical systems with emergent collective computational abilities," Proc. Natl. Acad. Sci., USA, vol.79, pp.2554-2558, April 1982.
2. R. J. McEliece, E. C. Posner, E. R. Rodemich and S. S. Venkatesh, "The capacity of the Hopfield associative memory," IEEE Trans. on Information Theory, vol.IT-33, no.4, pp.461-482, July 1987.
3. Y. -F. Wang, J. B. Cruz and J. H. Mulligan, "Guaranteed recall of all training pairs for bidirectional associative memory," IEEE Trans. on Neural Networks, vol.2, no.6, pp.559-567, Nov. 1991.
4. K. -S. Leung, H. -B. Ji, and Y. Leung, "Adaptive weighted outer-product learning associative memory," to be published in IEEE Trans. on Systems, Man, and Cybernetics.
5. T. -D. Chiueh and R. M. Goodman, "Recurrent correlation associative memories," IEEE Trans. on Neural Networks, vol.2, no.2, pp.275-284, March 1991.
6. M. Morita, "Associative memory with nonmonotone dynamics," Neural Networks, vol.6, pp.115-126, 1993.
7. H. Nishimori and I. Opris, "Retrieval process of an associative memory with a general input-output function," Neural Networks, vol.6, pp.1061-1067, 1993.
8. B. -L. Zhang, B. -Z. Xu and C. -P. Kwong, "Performances analysis of the bidirectional associative memory and an improved model from the matched-filtering viewpoint," IEEE Trans. on Neural Networks, vol.4, no.5, pp.864-872, Sept. 1993.

Autoassociative Memory
with high Storage Capacity

Paulo J. L. Adeodato and John G. Taylor

King's College London, Mathematics Department,
The Strand WC2R 2LS, London, UK.

Abstract. The general neural unit (GNU) [1] is known for its high storage capacity as an autoassociative memory. The exponential increase in its storage capacity with the number of inputs per neuron is far greater than the linear growth in the famous Hopfield network [2]. This paper shows that the GNU attains an even higher capacity with the use of pyramids of neurons instead of single neurons as its nodes. The paper also shows that the storage capacity/cost ratio increases, giving further support to this node upgrade. This analysis combines the modular approach for storage capacity assessment of pyramids [3] and of GNUs [4].
Keywords: Autoassociative memory, storage capacity, RAM-based neural networks, general neural unit (GNU), pyramidal architecture.

1 Introduction

Storage capacity is one of the most important aspects of associative memories — a fundamental topic for practical applications of neural networks. A great deal of research has been devoted to it, particularly to discover ways of increasing the storage in Hopfield networks. Using a different paradigm, Aleksander proposed a model of autoassociative memory (now referred to as the general neural unit — GNU) [5] with the generalising-RAM neuron (G-RAM) as node. In this architecture, training creates attractors (re-entrant states) with their basins of attraction forming an energy landscape for retrieval [5]. It was soon verified [2] that this is a high storage capacity network.

Storage capacity of this network has been thoroughly investigated. Wong and Sherrington [2, 6] used a statistical mechanical approach to prove that its storage capacity is much greater than that of its counterpart with McCulloch-Pitts neurons — the Hopfield network. Their analysis is only applicable to large dilute networks ($n \gg 1$ and $n \ll \log N$). Lucy [7] proved that the GNU achieves maximum storage capacity when fully connected. Braga [8] analysed the more realistic case of sparsely connected networks. All those, however, are architecture-dependent approaches. Recently, Adeodato & Taylor [4] developed a general modular approach to assess the storage capacity of RAM-based networks and applied it to GNUs [4] and pyramids [3]. The technique is broadly applicable and fits well with the previous theoretical models and experimental data [3, 4].

Later, Aleksander expanded and formalised this architecture naming it the general neural unit (GNU) [1]. He suggested that the typical RAM-based feedforward structures (discriminators [9] and pyramids [1]) should be used as GNU nodes but the topic has received no further attention. This paper analyses the storage capacity of the GNU composed of pyramids and configured as autoassociative memory, without external inputs. This problem is amenable to analysis by the modular approach developed by Adeodato & Taylor [4]. The analysis combines the storage capacity of pyramids [3] with that of GNUs [4] to obtain the storage capacity for this combined architecture. The exponential growth in cost as a function of the neuron fan-in has precluded the use of RAM-based neuron models in GNUs, except for G-RAM neurons, which use an emulator to achieve their *store + spread* algorithm. The practical relevance of this paper is that pyramids of neurons are a way of increasing the storage capacity of the GNU for all RAM-based neurons.

2 Storage Capacity of Single Neurons

RAM-based neural networks are characterised by their implementability in random access memory (RAM) chips. In RAM-based neurons, the binary address vector (*input pattern* $\mathbf{x}$) is presented to the n address-in terminals (*neuron inputs* $\mathbf{t}$) to access one of the 2^n *sites*. Each site stores the content (α_i) to produce the desired output for the input pattern that addresses it. Its cost grows exponentially with the fan-in (2^n). Being a look up table, this neuron is able to implement any function on its input space; an advantage over the McCulloch-Pitts neuron, which, on its own, can only implement the set of linearly separable functions of its inputs. However, this full-functionality feature is not transferred to the networks composed by these neurons due to their small fan-in, which allows each neuron to receive inputs from only a fraction of the input space. Therefore, patterns belonging to different classes might present the same inputs to some neurons, provoking conflicts of information during learning which affect the storage process.

In training a neuron, a *collision* of information is the storage of a target class different from the previous ones stored for the same input pattern. In the modular approach developed by Adeodato & Taylor [4], the term *probability of collision* in a neuron refers to the probability of at least 1 collision occurring in at least 1 site of the neuron. It is a measure of the risk of disruption of information in the storage process. *Probability of error-free storage* is the probability of not having any collision in a neuron (the complement of the probability of collision). *Storage capacity* is the maximum number of patterns randomly drawn from a uniform distribution that can be stored in a neuron or network at a given risk of disruption of information — at a given probability of collision.

For patterns (s) randomly drawn from uniform distributions, the probability of collision in a neuron [4] is expressed as:

$$P(nc) = 1 - (\frac{1}{2})^{(s-1)/2^n} \tag{1}$$

Storage capacity is the inverse of this function, having the number of storable patterns as a function of the probability of collision and the neuron fan-in:

$$s = 1 - 2^n \log_2(1 - P(nc)) \tag{2}$$

3 Storage Capacity of Pyramids

The pyramid is a tree of fixed connections composed of RAM-based neurons with fan-out of one arranged in layers [1]. The N input terminals of the pyramid are connected to the N input features in a 1-to-1 fashion so that the input space is divided in $\frac{N}{n}$ mutually exclusive sets of n features. The crucial point to apply the modular model of neuron collision to pyramids is how to overcome the lack of information about the contents of the sites of the hidden layers' neurons. Assuming storage optimality in the learning algorithm, a collision is only propagated to a neuron in a higher layer if *all* the neurons from the previous layer, which supply inputs to it, have suffered a collision. A single neuron without collision is enough to prevent the propagation of collisions to its successor in the next layer. This is because the non-colliding neuron's output divides the addressable sites of its successor in two disjoint sets; each set associated with a different class (target).

Since the patterns are randomly drawn from a uniform distribution (for storage capacity assessment), collisions in neurons of the same layer are independent events. Therefore the probability of collision — $P(C_l)$ — in a neuron of fan-in n_l of layer l is simply the product of the probabilities of collision in the neurons of the preceding layer, which supply input to it. $P(nc_l)$ below is just the neuron model for collision ($P(nc)$, equation 1). Mathematically:

$$P(C_1) = P(nc_1)$$
$$P(C_l) = P(nc_{(l-1)})^{n_l} \qquad \text{for } l = 2, 3, ...L \tag{3}$$

This is the building block for analysing pyramidal structures (tree structures, in general). The probability of collision in the whole pyramid — $P(C_{pyr})$ — is effect of neuron collisions propagated from the input to the output layer (L).

$$P(C_{pyr}) = P(nc_1) \cdot \prod_{l=2}^{L} P(C_l) \tag{4}$$

Substituting the values and taking the common case of equal neurons in all the pyramid, the probability of collision in the pyramid is expressed as a function of the probability of collision in a neuron, the neuron fan-in and the number of layers, as in the equation below

$$P(C_{pyr}) = P(nc)^{(n(L-1)+1)} \tag{5}$$

where, $P(nc)$ is a function of the number of patterns and the neuron fan-in. The storage capacity of the pyramid for any desired level of confidence can be expressed as a function of the probability of collision in a neuron by inverting the equation above. The inversion results in the equation below which is substituted in the storage capacity equation s for the neuron model:

$$P(nc) = 2^{\left(\frac{\log_2 P(C_{pyr})}{n(L-1)+1} \right)} \tag{6}$$

4 Storage Capacity of GNUs with Single Neurons

One important feature of the autoassociative GNU is that a neuron with input to itself (*self-feedback*) never produces collisions [8]. The neuron connections are randomly sampled from the feature space without replacement and this determines the probability of self-feedback ($P(sf)$). Therefore the probability of collision in a GNU ($P(C_{gnu})$) of dimension N is given by:

$$P(C_{gnu}) = 1 - P(\overline{C}_{gnu})$$
$$\text{where: } P(\overline{C}_{gnu}) = (P(sf) + P(\overline{sf})P(\overline{nc}))^N \qquad (7)$$
$$P(sf) = \frac{n}{N}$$

where $P(\overline{nc}) = 1 - P(nc)$, $P(nc)$ being the model for neuron collision ($P(nc)$, equation 1). The storage capacity of the GNU can be expressed as a function of the probability of collision in a neuron by just inverting the equation above. The inversion results in the equation below which is substituted in the storage capacity equation s for the neuron model:

$$P(nc) = 1 - \frac{(1 - P(C_{gnu}))^{1/N} - \frac{n}{N}}{1 - \frac{n}{N}} \qquad (8)$$

5 Storage Capacity of GNUs with Pyramids

Now that the foundation has been laid, this section answers the question that motivated this research [11]: *"Is there any gain in using pyramids instead of single neurons as GNU nodes?"* The modular approach developed makes the assessment of the storage capacity of GNUs with pyramidal nodes become a composition of the previous functions. As pyramids receive inputs from the whole N-dimensional space, there is always self-feedback. On the other hand, by not having full functionality, pyramids have a chance of provoking collisions in a GNU, even with self-connections. In this case, a collision in a GNU occurs when there is collision in at least 1 pyramid. This results in the equation below for the probability of collision in a GNU with pyramidal nodes ($P(C_{gnupyr})$) as a function of the probability of collision in its component pyramids

$$P(C_{gnupyr}) = 1 - (P(\overline{C}_{pyr}))^N \qquad (9)$$

where $P(\overline{C}_{pyr})$ is the error-free storage in a pyramid — the complement of $P(C_{pyr})$ defined in equation 6. Its storage capacity is obtained by inverting the equation above. The inversion results in the equation below, which is substituted in the storage capacity equation (6) for the pyramid, which is substituted in the storage capacity equation s for the neuron model:

$$P(C_{pyr}) = 1 - (1 - P(C_{gnupyr}))^{\frac{1}{N}} \qquad (10)$$

Figure 1 shows the probability of error-free storage for the single neuron-GNU (a) and for the pyramid-GNU (b) in a 256-dimensional space as a function

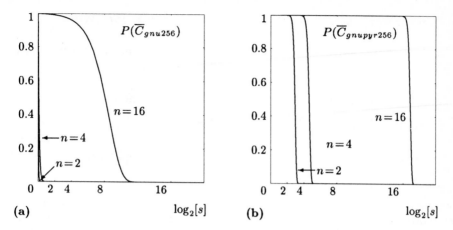

Fig. 1. Probability of error-free storage in a 256-GNU as a function of $\log_2$[number of patterns] for different fan-ins ($n = 2, 4, 16$). (a) single neurons, (b) pyramids.

of $\log_2$[number of patterns]. The gain with the pyramid nodes is impressive — a minimum magnitude of the order of 2^8 (the dimension of the space). It increases the storage capacity and creates an abrupt transition in the curve.

Naturally, this storage capacity increase comes accompanied by additional costs. This paper is concerned only with the hardware cost that is assessed by the total number of sites in the network. The costs are presented here without much detail for architectures with neurons of the same fan-in n:

$$
\begin{aligned}
cost(gnu) \quad &= N\,2^n \qquad\qquad\qquad \text{for single neuron nodes} \\
cost(gnupyr) &= N\,2^n \sum_{l=1}^{L} n^{(l-1)} \quad \text{for pyramidal nodes}
\end{aligned}
\tag{11}
$$

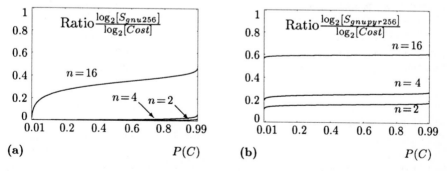

Fig. 2. Ratio of $\log_2$[storage capacity] and $\log_2$[cost of the network] of a 256-GNU with single neurons (a) and with pyramids (b) as a function of the probability of collision for neurons with different fan-ins ($n = 2, 4, 16$).

The storage capacity/cost ratio is an important feature in considering the viability of a network. Figure 2 shows that the use of pyramids represents a storage capacity increase much bigger than that of the cost, reaching a gain of the order of 10^3, for 16-input neurons at 99% probability of error-free storage.

6 Conclusion

This paper has presented the first theoretical results on the use of "large neurons" [1] as GNU nodes. It has proved that pyramid-GNUs are one of the highest storage capacity autoassociative memories. A similar approach with McCulloch-Pitts neurons would be to use N multilayer perceptrons as nodes in a Hopfield network. The paper has also proved that there is a big increase in the storage capacity/cost ratio highlighting the viability of the expansion.

From the practical point of view, the increase in storage capacity makes it viable to take advantage of the other RAM-based neuron models. pRAM neurons with noisy reinforcement learning [12] have the advantage of realising the *generalised majority rule* which attains the minimum output error [10]. GSN neurons [13] have the advantage of accepting *undefined* inputs (3-state neurons), an important feature for *pattern completion* tasks where, otherwise, several runs from different initial states would be necessary.

This architecture preserves the emergent properties of the GNU and the feature of fast training but more research has to be done to check the generalisation of the pyramid-GNU. This modular technique for storage capacity assessment is applicable to any network configuration and future work involves the development of algorithms to construct networks based on it.

References

1. Aleksander I.: *IEE Comp. & Cont. Eng. J.*, **1**, 1990, 259–265.
2. Wong K. Y. M., Sherrington D.: *Europhysics Letters*, **7** (3), 1988, 197–202.
3. Adeodato P. J. L., Taylor J. G.: *Neural Network World*, **6**, (3), 1996, 241–249.
4. Adeodato P. J. L., Taylor J. G.: In: *Proc. of Weightless Neural Networks Workshop*, D. L. Bisset (ed.), Canterbury, UK, Sep. 1995, 103–110.
5. Aleksander I.: In: *Neural Computing Architectures*, I. Aleksander (ed.), 1989, North Oxford Academic Publishers Ltd., London, UK.
6. Wong K. Y. M., Sherrington D.: *J. Phys.*, **22**, 1989, 2233–2263.
7. Lucy J.: *Electronics Letters*, **27** (10), 1991, 799–800.
8. Braga A. P.: *Electronics Letters*, **30** (1), 1994, 55–56.
9. Rower R., Morciniec M.: *Neural Computation*, **8**, (3), 1996, 629–642.
10. Wong K. Y. M., Sherrington D.: *Europhysics Letters*, **10** (5), 1989, 419–425.
11. Adeodato P. J. L., Taylor J. G.: In *Proc. of ICANN'95*, F. Fogelman-Soulié and P. Gallinari (eds.), Paris, France, Oct., 1995, 607–612.
12. Guan Y., Clarkson T., Taylor J. G., Gorse D.: *Neural Networks*, **7**, (3), 1994, 523–538.
13. Filho E. C. D. B. C., Bisset D. L., Fairhurst M. C.: *Patt. Rec. Lett.*, **12** (3), 1992, 131–138.

Information Efficiency of the Associative Net at Arbitrary Coding Rates

Bruce Graham and David Willshaw

Centre for Cognitive Science, University of Edinburgh
2 Buccleuch Place, Edinburgh EH8 9LW, Scotland, UK
E-mail: *B.Graham@cns.ed.ac.uk* & *D.Willshaw@cns.ed.ac.uk*
Phone: +44 131 650 4404; Fax: +44 131 650 6626

Abstract. The associative net is a neural network model of associative memory that is unusual in having binary-valued connections between units. This net can work with high information efficiency, but only if the patterns to be stored are extremely sparse. In this paper we report how the efficiency of the net can be improved for more dense coding rates by using a partially-connected net. The information efficiency can be maintained at a high level over a 2–3 order of magnitude variation in the degree of pattern sparseness.

1 Introduction

The associative net (Willshaw, Buneman, & Longuet-Higgins, 1969; Willshaw, 1971) is a distributed model of heteroassociative memory. It consists of a set of input units connected to a set of output units by modifiable synapses. Pairs of binary patterns are stored in the net using a "clipped Hebbian" learning rule that changes the weight of a synaptic connection from 0 to 1 if both input and output units are active for the same pattern pair. One performance measure of the associative net is how much information can be stored and successfully recalled from it. The information efficiency, η_o, of the net is the number of stored pattern pairs multiplied by the amount of information that can be recalled about each output pattern, divided by the amount of storage, which is the number of synaptic connections. The fully-connected net (every input unit is connected to every output unit) has an optimum information efficiency $\sim \ln 2$ ($= 0.693$) bits per synapse, when the patterns are suitably sparse. This is remarkably close to the limit of information efficiency, $1/(2 \ln 2)$ (~ 0.721), that can be achieved for sparse patterns using continuous-valued synapses (Gardner, 1987). So the lack of precision in the synapses of the associative net is not crucial for the storage of sparse patterns.

Unfortunately, optimum efficiency for reliable recall (at most one bit in error in the recalled output pattern) is only achieved when the number of active units, M_A, in an input pattern, is given by $M_A = \log_2(N_B)$, where N_B is the number of output units (Willshaw et al., 1969). For square nets, where the number of input units, N_A, is the same as the number of output units, the input patterns quickly become very sparse indeed as the net size increases. For example, if

$N_A = N_B = 2^{20}$, only 20 units out of 1,048,576 should be active in each input pattern. Though the exact coding rate is unspecified, the output patterns should also be very sparse if the limit of efficiency is to be approached.

It has been shown that the optimum efficiency of the associative net can be retained for even sparser coding ($M_A < \log_2(N_B)$), if errors are allowed during recall (error-full recall) (Nadal & Toulouse, 1990; Buckingham & Willshaw, 1992). However, as the density of input pattern coding is increased beyond the optimum, information efficiency drops rapidly and entering the error-full regime does not help. This contrasts with continuous-valued synapses, where an efficiency of 2 bits per synapse can be obtained theoretically at dense coding levels (Gardner, 1987). Nadal and Toulouse (1990) have shown that even simple Hebbian learning, in which a synaptic weight is incremented by 1 if the input and output units are active at the same time, saturates at an efficiency limit of $1/(\pi \ln 2)$ (~ 0.459) bits per synapse in the error-full regime, for dense coding.

In this paper, we present an approach that allows the associative net to retain near maximal information efficiency for input coding rates 2–3 orders of magnitude higher than the optimum for the fully-connected net. This is achieved by using a partially-connected net with a suitable thresholding strategy for recall. Optimum efficiency is obtained with reliable recall, with no advantage being available by entering the error-full regime. Though efficiency still drops to low levels as standard coding ($M_A = N_A/2$) is approached, it remains finite for the nets considered here (around 0.05 bits per synapse). For the comparable fully-connected net it is effectively zero.

2 Theoretical Information Efficiency

The information efficiency, η_o, of reliable recall from the fully-connected net is a function of the average probability, $\hat{\rho}$, that a synapse has been modified during pattern storage (Willshaw et al., 1969; Willshaw, 1971; Buckingham & Willshaw, 1992). Optimum efficiency is achieved when $\hat{\rho} = 0.5$ and is approximately $\eta_o = \ln 2$. Under the condition that at most one bit in the output pattern is allowed to be in error, the optimum is achieved when the number of active units in an input pattern is $M_A = \log_2(N_B)$.

The information efficiency of a partially-connected net can be derived in exactly the same manner, if the appropriate threshold strategy is used during recall. For the fully-connected net with a noise-free input cue, an output pattern is recalled by selecting those output units to be active that are connected to all M_A active input units by modified synapses. In a partially-connected net, an output unit that should be active is not necessarily connected to all M_A active input units, so a different threshold for output unit activity must be found. The appropriate threshold is the number of active input units an output unit is connected to, by modified or unmodified synapses. This threshold is different for each output unit. The use of this strategy for partially-connected nets has been investigated in detail elsewhere (Buckingham & Willshaw, 1993; Graham & Willshaw, 1994, 1995).

Given this method of pattern recall, the optimum efficiency is again $\eta_o = \ln 2$ when $\hat{\rho} = 0.5$. This is now achieved under the further condition that $ZM_A = \log_2(N_B)$, where Z is the fraction of input units connected to any given output unit (ZN_AN_B is the number of synapses in the net). So as the input coding rate is raised above $M_A = \log_2(N_B)$, it is simply a matter of lowering the connectivity level in proportion to retain optimum information efficiency.

The approximations used to derive these results are not entirely accurate, even at sparse coding rates (Buckingham & Willshaw, 1992). In order to investigate information efficiency over a wide range of coding rates, accurate numerical calculations of recall performance need to be used. The following information efficiency results for two large nets have been obtained using such calculations. The technical details of these calculations have been given previously (Buckingham & Willshaw, 1992, 1993; Graham & Willshaw, 1994). We only consider input coding rates $M_A \geq \log_2(N_B)$, as the efficiency at sparser rates has been dealt with elsewhere (Nadal & Toulouse, 1990; Buckingham & Willshaw, 1992).

3 Results

Calculations of information efficiency have been made for two different associative nets. The first net has $N_A = 2^{17} = 131072$ and $N_B = 2^{20} = 1048576$. This expansion in size from the input to the output layer represents that seen in the projection from the entorhinal cortex to the dentate gyrus in the mammalian hippocampus. The number of units corresponds roughly to that found in the rat brain (Amaral, Ishizuka, & Claiborne, 1990). For this net, the numbers of active input and output units are varied independently. The second net is square, with $N_A = N_B = 2^{18} = 262144$. This could represent the recurrent collaterals of CA3, or the projection from CA3 to CA1 in rat hippocampus. In this case, the numbers of active inputs and outputs are varied but kept equal.

Figure 1 shows the information efficiency of the first net for different coding rates and connectivities. The theory gives the optimality criterion, $ZM_A = 20$. For each value of the input and output coding levels, M_A and M_B, the information efficiency was calculated over a range of connectivities, Z. In all cases, optimum efficiency was obtained with a connectivity very close to that specifed by the above criterion. For $Z = 20/M_A$ (Fig. 1a), maximum efficiency is obtained at $M_A = 20$, and decreases as M_A is increased, regardless of the output coding rate. While η_o does drop rapidly as dense input coding levels are approached, it is maintained at a high level for a 2–3 order of magnitude increase in M_A beyond $M_A = 20$. It is still above 90% of maximum at 2 orders of magnitude, and above 60% at 3 orders of magnitude. This contrasts markedly with the efficiency of a fully-connected net (Fig. 1b). Now η_o decreases very rapidly with increasing M_A, and is only at about 10% of maximum after a 2 orders of magnitude increase in M_A. It is at this magnitude that the maximum improvement in efficiency is obtained by using partial connectivity. The difference between η_o at optimal partial connectivity and at full connectivity is shown in Fig. 1c. Note that more patterns can be stored in a fully-connected net than a partially-connected net.

For example, at $M_A = 1024$, the fully-connected net can store and reliably recall 7 times more patterns than the optimum partially-connected net, but requires roughly 50 times more connections to do so.

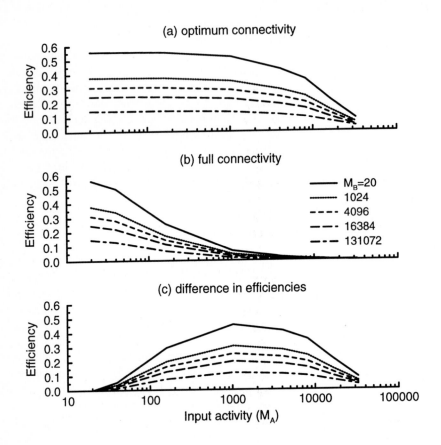

Fig. 1. Information efficiency of the first net ($N_A = 131072$, $N_B = 1048576$)

The sensitivity of η_o to variations in connectivity is demonstrated in Figure 2. It is only mildly sensitive to Z around the optimum connectivity, which is essentially $Z_o = \log_2(N_B)/M_A$. When M_A is held constant, at $Z = 2Z_o$ efficiency has dropped to around 90% of maximum, and at $Z = Z_o/2$ it is down to 80% of maximum.

The maximum efficiency is also dependent on the output coding rate, M_B, and is somewhat less than $\ln 2$. It decreases linearly with $\log(M_B)$ and, at least for $M_A = 20$, a reasonable approximation for η_o is

$$\eta_o = \left(1 - \frac{\log_2(M_B)}{\log_2(N_B)}\right) \ln 2 \qquad (1)$$

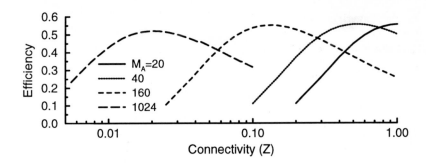

Fig. 2. Sensitivity of information efficiency to variations in connectivity around the optimum

The results for the second net are similar to those for the first net. The information efficiency obtained using either optimum or full connectivity, while coding levels M_A and M_B are varied equally, is plotted in Figure 3a. At optimum connectivity, η_o decreases approximately linearly as the log of the coding levels is increased. The maximum improvement by using partial connectivity is again obtained when M_A is increased by about 2 orders of magnitude from the initial value of $M_A = \log_2(N_B) = 18$ (Fig. 3b).

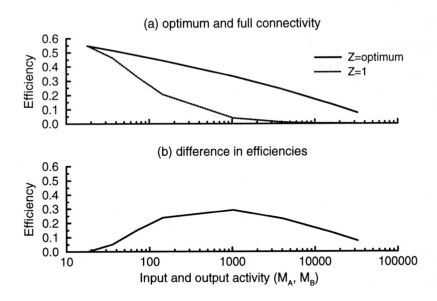

Fig. 3. Information efficiency of the second net ($N_A = N_B = 262144$; $M_A = M_B$)

4 Discussion

We have demonstrated, with numerical calculations, an associative net model with high information efficiency for input coding levels above the optimum for a fully-connected net of $M_A = \log_2(N_B)$. By lowering the connectivity level as M_A is increased, so that $ZM_A = \log_2(N_B)$, and using a suitable thresholding strategy for recall, information efficiency is held at high levels for a 2–3 order of magnitude increase in M_A. Optimum efficiency is achieved with reliable recall, and no advantage is gained by entering the error-full regime. This contrasts with the situation for input coding rates less than $\log_2(N_B)$, and other learning algorithms, such as Hebbian learning, where optimum efficiency may be found in the error-full regime (Nadal & Toulouse, 1990; Buckingham & Willshaw, 1992).

This work goes some way towards extending the associative net as a useful model for arbitrary coding rates. However, the performance of the associative net at standard coding ($M_A = N_A/2$, $M_B = N_B/2$) remains rather poor, with a maximum information efficiency of around 0.05 bits per synapse for the large nets studied here.

Acknowledgement: to the MRC for financial support under PG 9119632.

References

Amaral, D., Ishizuka, N., & Claiborne, B. (1990). Neurons, numbers and the hippocampal network. In Storm-Mathisen, J., Zimmer, J., & Ottersen, O. (Eds.), *Progress in Brain Research, Vol.83*, pp. 1–11. Elsevier Science Publishers B.V.

Buckingham, J., & Willshaw, D. (1992). Performance characteristics of the associative net. *Network, 3*, 407–414.

Buckingham, J., & Willshaw, D. (1993). On setting unit thresholds in an incompletely connected associative net. *Network, 4*, 441–459.

Gardner, E. (1987). Maximum storage capacity in neural networks. *Europhysics Letters, 4*, 481.

Graham, B., & Willshaw, D. (1995). Improving recall from an associative memory. *Biol. Cybern., 72*, 337–346.

Graham, B., & Willshaw, D. (1994). Capacity and information efficiency of a brain-like associative net. In Tesauro, G., Touretzky, D., & Leen, T. (Eds.), *Neural Information Processing Systems 7*, pp. 513–520. MIT Press.

Nadal, J.-P., & Toulouse, G. (1990). Information storage in sparsely coded memory nets. *Network, 1*, 61–74.

Willshaw, D. (1971). *Models of distributed associative memory*. Ph.D. thesis, University of Edinburgh.

Willshaw, D., Buneman, O., & Longuet-Higgins, H. (1969). Non-holographic associative memory. *Nature, 222*, 960–962.

Efficient Learning in Sparsely Connected Boltzmann Machines

Marcel J. Nijman and Hilbert J. Kappen

RWCP* Novel Functions, SNN** Laboratory
Dept. of Medical Physics and Biophysics, University of Nijmegen
Geert Grooteplein 21, NL 6525 EZ Nijmegen, The Netherlands

Abstract. We present a heuristical procedure for efficient estimation of the partition function in the Boltzmann distribution. The resulting speed-up is of immediate relevance for the speed-up of Boltzmann Machine learning rules, especially for networks with a sparse connectivity.

1 Introduction

Boltzmann Machines (BMs) [1] form an attractive group of Neural Networks for several reasons. The local learning rule, for instance, offers the possibility of parallel implementation. Their main disadvantage, however, is that computing the correlations $\langle S_i S_j \rangle$ exactly can only be done in a reasonable time for small networks. Although the correlations can be approximated with simulated annealing, this is very slow. Some good results have been reported about mean field learning, which speeds up the computation by approximating $\langle S_i S_j \rangle$ by $\langle S_i \rangle \langle S_j \rangle$. However, in most situations mean field learning leads to large errors and cannot be applied successfully. An approach which leads to tractable Boltzmann Machines is to restrict the state space by arranging the hidden units into layers that perform a winner-take-all competition [4] or many-take-all [9].

For some simple structures efficient learning rules exist. In [8] a decimation method is presented which leads to linear time learning rules for Boltzmann Trees. In Section 3 we explain the principle of decimation. Since tree-structured networks are of limited use, we would like to extend this method to apply to networks with a general architecture. This can be done as follows. For any network structure we can clamp some nodes, such that the remaining structure is a tree, which can be decimated. From this we can compute the partition function, as is shown in Section 4. The challenge is to identify a small set of nodes which to clamp. This problem is addressed in Section 5. The learning rules which result from this are very efficient when applied to networks with a sparse connectivity, as is shown in Section 6. Some extensions are discussed in Section 7. We finish with the conclusions in Section 8.

We are interested in Boltzmann Machines with sparse connectivity for several reasons. The use of sparse connectivity for Bayesian networks is well appreciated [7], and is argued to play an important role in probabilistic reasoning,

---

* Real World Computing Partnership
** Dutch Foundation for Neural Networks

as it makes many conditional independencies explicit, and facilitates interpretation, inference, and estimation and storage of the parameters. Also, prior knowledge concerning conditional independencies can be incorporated into the structure of the network. We believe that most of the motivations for using sparse connectivity in Bayesian networks also applies to Boltzmann Machines.

2 Boltzmann Machine Learning

Consider a Boltzmann Machine with binary units S_i. The probability to observe a state $S = \{S_i\}$ is

$$p(S) = \frac{1}{Z} \exp(-H(S)), \tag{1}$$

with

$$H(S) = -\sum_{(ij)} w_{ij} S_i S_j \tag{2}$$

and

$$Z = \sum_S \exp(-H(S)). \tag{3}$$

Learning is defined as gradient descent on the Kullback-Leibler divergence [6] and leads to the following learning rule [1]

$$\Delta w_{ij} \propto \overline{\langle S_i S_j \rangle}_{\text{clamped}} - \langle S_i S_j \rangle_{\text{free}}. \tag{4}$$

Since the two-point correlations are related to the partition function by $\langle S_i S_j \rangle = \frac{\partial}{\partial w_{ij}} \ln Z$, efficient computation of Z leads to efficient learning rules. Therefore, we will only concentrate on computing the partition function Z. Computing this partition functions by explicit summing over all the states of S involves an exponential number of terms, and is thus intractable in general. However, for some architectures more efficient learning rules exits, as we shall see.

3 Decimation

Recently an algorithm has been described [8] which computes the partition function of a BM with a tree-structure in linear time (for the free and the clamped phase). The algorithm is based on decimation [3], and can be described as follows.

Consider the two network fragments in Figure 1. In the left fragment S_2 and S_3 have an arbitrary number of connections with the rest of the network. S_1, however, is only connected to S_2 and S_3. By integrating out S_1 we obtain the network on the right. This introduces an additional weight w_{23} into the network. In order for the two networks to have the same probability distribution on the remaining units the following equation must hold:

$$\sum_{S_1} \exp(w_{12} S_1 S_2 + w_{13} S_1 S_3) = \sqrt{C} \cdot \exp(w_{23} S_2 S_3) \tag{5}$$

Fig. 1. Two network fragments. The right fragment is obtained from the left fragment by integrating the probability distribution over S_1.

for $S_2 = \pm 1$ and $S_3 = \pm 1$. The solution to this is [8]

$$\tanh w_{23} = \tanh w_{12} \tanh w_{13} \tag{6}$$

$$C = 2\cosh(w_{12} + w_{13}) \cdot 2\cosh(w_{12} - w_{13}). \tag{7}$$

For a BM with a tree-structure all units can be decimated out. This can be shown as follows. Introduce a bias unit S_0 permanently fixed to 1. Let w_{i0} represent the bias of S_i. Handle units which should be clamped by leaving them unclamped, but setting the bias w_{i0} to $\pm\infty$. The resulting structure will not be a tree anymore, because all units are connected to the same bias unit. However, a leaf in the original tree now has degree 2 (it is connected to its parent, and to the bias unit) and thus can be decimated out. This can be done iteratively until all the units have been decimated out. Thus, for a BM with a tree-structure the partition function (and thus the correlations $\langle S_i S_j \rangle$) can be computed exactly in linear time.

4 Divide and Conquer

The idea of decimation can be applied to a general BM in the following way. Partition the units S in two sets X and Y such that G_Y (the sub-graph defined on Y) contains no cycles. A trivial choice is $Y = \emptyset$ and $X = S$, but in general more interesting partitionings exist. If the units in X are clamped we can decimate the units in Y, since G_Y is a forest. We can rewrite (1) as follows

$$p(X, Y) = \frac{1}{Z} \exp(-H(X) - H(Y; X)), \tag{8}$$

with

$$H(X) = -\sum_{(ij)} w_{ij} X_i X_j \tag{9}$$

and

$$H(Y; X) = -\sum_{(ij)} w_{ij} Y_i Y_j - \sum_{(ij)} w_{ij} Y_i X_j, \tag{10}$$

where the partition function is

$$Z = \sum_X \exp(-H(X)) \sum_Y \exp(-H(Y; X))$$

$$= \sum_X Z_{Y;X}. \tag{11}$$

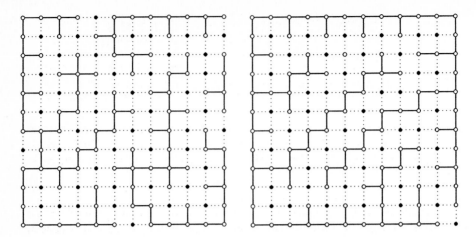

Fig. 2. Two partitionings of a 12 by 12 grid. The black units cut every cycle of the grid. The partitioning on the left was obtained with our heuristic. The partitioning on the right is optimal (to our knowledge).

Note that Eq. 11 holds independent of the partitioning (even if G_Y contains cycles), but that only something has been gained if $Z_{Y;X}$ can be computed efficiently.

So, the original partition function Z can be computed by summing over the states of the units in X, and computing $Z_{Y;X}$ (in linear time) for each state of X. If X has a limited size, this sum is tractable, and the partition function can be computed exactly. If X is a large set, we can estimate Z with Gibbs sampling by using

$$Z = 2^{|X|} \left\langle Z_{Y;X}^{-1} \right\rangle_X^{-1}. \tag{12}$$

5 Feedback Vertex Set

The efficiency of Eq. 11 or Eq. 12 depends on the size of X; the smaller the better. Therefore, we need an algorithm to identify a small set X. Such a set which cuts every cycle in a network is known as a feedback vertex set [2]. So, our optimization problem becomes a search for the minimal feedback vertex set, which, unfortunately, is NP-complete [5]. Therefore, we use the following heuristical algorithm. Start with an empty set X and with a graph representing the network structure. Prune the graph by iteratively deleting all nodes with degree less than 2 (leaves and unconnected roots). Then, identify the node with the highest degree, delete it from the graph, and insert it in the set X. Repeat this procedure until the graph contains no more nodes. This heuristic is fast (linear in the number of nodes), and generally gives a small set X. As an example, consider the two networks in Figure 2. The network on the left shows a typical partitioning obtained with our algorithm on a 12 by 12 grid. The set X consists of the black units, and the dotted lines symbolize the biases

edges	Equation 13	Equation 12	deleted nodes
100	218.7 ± 7.2	19.11 ± 0.51	11.678 ± 0.033
200	222.6 ± 7.9	90.9 ± 2.9	22.036 ± 0.034
400	232.8 ± 7.7	227.9 ± 6.5	31.287 ± 0.033
800	227.3 ± 6.7	231.1 ± 7.3	38.159 ± 0.025

Table 1. Average number of iteration needed for convergence for two methods of computing the partition function for a Boltzmann machine with 50 nodes, and the average number of nodes deleted, plus error bounds on the estimates.

which result from clamping the black units. This partitioning (with $|X| = 46$) is quite good, since we believe the optimum (shown on the right in Figure 2) to be 45. Typically, our heuristic finds solutions with 46 to 48 black units. Therefore, although finding the optimal set is NP-hard we can still find a good approximation with this heuristic.

6 Numerical Results

It is to be expected that Gibbs sampling with Eq. 12 converges faster than Gibbs sampling on the total space S using Eq. 13.

$$Z = 2^{|S|} \langle \exp(H(S)) \rangle_S^{-1} . \tag{13}$$

In order to show the significance of this gain we have generated random BMs with 50 nodes by randomly inserting a number of edges (100, 200, 400, and 800). All edges were given a weight uniformly drawn from $[-1, 1]$. Then, we computed a feedback vertex set X and estimated the partition function with both methods. We stopped when the standard deviation on the estimate was less than 1% of Z. Table 1 shows for both algorithms and for a varying number of edges the number of iterations needed until convergence. The results are averaged over 1000 networks. We conclude that the results of standard Gibbs sampling are hardly effected by the complexity of the network, whereas the new method leads to much faster convergence for sparse networks.

7 Discussion

In this paper we have derive learning rules for sparsely connected networks. We have not addressed the problem of identifying the structure of a sparce network which fits a given data set well.

Secondly, we note that G_Y doesn't have to be a forest. Decimation cannot only be used on trees, but also on a bit more complex structures like rings and ladders. This suggests a heuristical algorithm to determine X by decimating the original network as far as possible, deleting the node with the highest degree, and so on.

Another observation is that the efficiency of the resulting algorithm is not only determined by the number of units in X, but also by the number of connections in G_X. Less connections means that Gibbs sampling will sample the state space of X more evenly. How this can be incorporated in an algorithm to compute a set X is still an open question.

8   Conclusions

We have shown how the partition function of a Boltzmann Machine with a sparsely connected structure can be computed efficiently, which leads to efficient learning rules. This allows for a trade-off between the complexity of a network (and, consequently, the complexity of the problems that can be solved) and the computation time that one is willing to spend on training the network.

References

1. D.H. Ackley, G.E. Hinton, and T.J. Sejnowski. A learning algorithm for Boltzmann Machines. *Cognitive Science*, 9:147–169, 1985.
2. M.R. Garey and D.S. Johnson. *Computers and Intractability: A Guide to the Theory of NP-Completeness*. Freeman, San Francisco, 1979.
3. C. Itzykson and J. Drouffe. *Statistical Field Theory*. Cambrigde University Press, Cambridge, 1991.
4. B. Kappen and M.J. Nijman. Radial basis Boltzmann machines and learning with missing values. In *Proceedings of the World Congress on Neural Networks*, volume I, pages 72–75. Lawrence Erlbaum Associates, Mahwah, NJ, 1995.
5. R.M. Karp. Reducibility among combinatorial problems. In R.E. Miller and J.W. Thatcher, editors, *Complexity of Computer Computations*, pages 85–103. Plenum Press, 1972.
6. S. Kullback. *Information theory and statistics*. Wiley, N.Y., 1959.
7. J. Pearl. Probabilistic reasoning using graphs. In B. Bouchon and R.R. Yager, editors, *Uncertainty in Knowledge-Based Systems*, pages 200–202. Springer-Verlag, 1987.
8. L. Saul and M. Jordan. Learning in Boltzmann trees. *Neural Computation*, 6(6):1174–1184, 1994.
9. D.M.J. Tax and H.J. Kappen. Learning structure with many-take-all networks. In *proceedings of ICANN*, 1996.

Incorporating Invariances in Support Vector Learning Machines

Bernhard Schölkopf[1][2], Chris Burges[2], and Vladimir Vapnik[2]

[1] Max–Planck–Institut für biologische Kybernetik,
Tübingen, Germany, E–mail bs@mpik-tueb.mpg.de
[2] AT&T Bell Laboratories, Holmdel, NJ, USA

Abstract. Developed only recently, support vector learning machines achieve high generalization ability by minimizing a bound on the expected test error; however, so far there existed no way of adding knowledge about invariances of a classification problem at hand. We present a method of incorporating prior knowledge about transformation invariances by applying transformations to support vectors, the training examples most critical for determining the classification boundary.

1 Incorporating Invariances

In many applications of learning procedures, prior knowledge about properties of the function to be learned is available (for a review, see Abu–Mostafa, 1995). For instance, certain transformations of the input could be known to leave function values unchanged. Mostly, two different ways of exploiting this knowledge have been used: either the knowledge is directly incorporated in the algorithm, or it is used to generate artificial training examples ("virtual examples") by transforming the training examples accordingly.

In the first case, an additional term in an error function can force a learning machine to construct a function with the desired invariances (Simard et al., 1992); alternatively the invariance can be achieved by using an appropriate distance measure in the pattern space (Simard, Le Cun, and Denker, 1993). The latter is akin to changing the representation of the data by first mapping them into a more suitable space; an approach pursued for instance by Segman, Rubinstein, & Zeevi (1992), or Vetter & Poggio (1996).

In the second case, it is hoped that given sufficient time, the learning machine will extract the invariances from the artificially enlarged training data. Figure 1 contains illustrations of the different approaches.

Simard et al. (1992) compare the two techniques and find that for the considered problem — learning a function with three plateaus where the function values are locally invariant — training on the artificially enlarged data set is significantly slower, due to both correlations in the artificial data and the increase in training set size. Moving to real–world applications, the latter factor becomes even more important. If the size of a training set is multiplied by a number of desired invariances (by generating a corresponding number of artificial examples for each training pattern), the resulting training set can get rather large (as the ones used by Drucker, Schapire, and Simard, 1993). On the other hand, the method of generating virtual examples has the advantage of being readily implemented for all kinds of learning machines and symmetries. If instead of Lie groups of symmetry transformations one is dealing with discrete symmetries, such as the bilateral symmetries of Vetter, Poggio, & Bülthoff (1994), derivative-based methods such as the ones of Simard et al. (1992) are not applicable. In conclusion, it would be desirable to have an intermediate method which has the advantages of the virtual examples approach without its computational cost. In this paper, we will try to convince the reader that such a method can be realized if we build on a type of learning machine to be described in the following section.

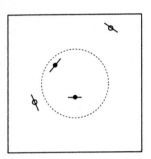 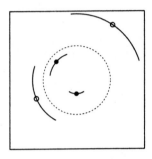 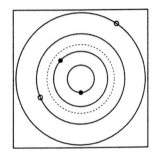

Fig. 1. Different ways of incorporating invariances in a decision function. The dotted line marks the "true" boundary, disks and circle are the training examples. We assume that prior information tells us that the classification function only depends on the norm of the input vector (the origin being in the center of each picture). Solid lines crossing an example indicate the type of information conveyed by the different methods of incorporating prior information. Left: incorporating a regularizer to learn tangent values (cf. Simard et al., 1992); middle: generating virtual examples in a localized region around each training example, right: changing the representation of the data by first mapping each example to its norm. If feasible, the latter method yields the most information. However, if the necessary nonlinear transformation cannot be found, or if the desired invariances are of localized nature, one has to resort to one of the former techniques. Finally, the reader may note from the right hand side picture that in particular, examples close to the boundary allow us to exploit prior knowledge very effectively: given a method to get a first approximation of the true boundary, the examples closest to it would allow good estimation of the true boundary. A similar two–step approach shall be pursued in this paper.

2 Support Vector Learning Machines

The support vector algorithm (Vapnik, 1995, Boser, Guyon & Vapnik, 1992, Cortes & Vapnik, 1995) uses the Structural Risk Minimization principle (Vapnik, 1979) to construct decision rules that generalize well. In doing so, they extract a small subset of the training data. Space does not permit to explain this algorithm in detail; we thus will merely outline its main ideas. The Method of Structural Risk Minimization is based on the fact that the test error rate is bounded by the sum of the training error rate and a term which depends on the so–called VC(Vapnik–Chervonenkis)–dimension of the learning machine. By minimizing the sum of both quantities, high generalization performance can be achieved. For linear hyperplane decision functions

$$f(\mathbf{x}) = \text{sgn}\left((\mathbf{w} \cdot \mathbf{x}) + b\right), \tag{1}$$

the VC–dimension can be controlled by controlling the norm of the weight vector $\mathbf{w}$ (Vapnik, 1995). Given training data $(\mathbf{x}_1, y_1), \ldots, (\mathbf{x}_\ell, y_\ell)$, $\mathbf{x}_i \in \mathbf{R}^N, y_i \in \{\pm 1\}$, a separating hyperplane which generalizes well can be found by minimizing (Cortes & Vapnik, 1995)

$$\|\mathbf{w}\|^2 + \gamma \cdot \sum_{i=1}^{\ell} \xi_i \tag{2}$$

subject to

$$\xi_i \geq 0, \quad y_i \cdot \left((\mathbf{x}_i \cdot \mathbf{w}) + b\right) \geq +(1 - \xi_i) \quad \text{for } i = 1, \ldots, \ell \tag{3}$$

(γ is a constant which determines the trade–off between training error and VC–dimension). The solution of this problem can be shown to have an expansion

$$\mathbf{w} = \sum_{i=1}^{\ell} \lambda_i \mathbf{x}_i, \tag{4}$$

where only those λ_i are nonzero which belong to an $\mathbf{x}_i$ precisely meeting the constraint (3) — these $\mathbf{x}_i$ lie closest to the decision boundary, they are called *Support Vectors*. The λ_i are found by solving the quadratic programming problem defined by (2) and (3).

Finally, this method can be generalized to nonlinear decision surfaces by first mapping the input nonlinearly into some high–dimensional space, and finding the separating hyperplane in that space (Boser, Guyon & Vapnik, 1992). This is achieved implicitly by using different types of symmetric functions $K(\mathbf{x}, \mathbf{y})$ instead of the ordinary scalar product $(\mathbf{x} \cdot \mathbf{y})$. This way one gets as a generalization of (1) and (4)

$$f(\mathbf{x}) = \text{sgn}\left(\sum_{i=1}^{\ell} \lambda_i \cdot K(\mathbf{x}, \mathbf{x}_i) + b\right). \tag{5}$$

3   Virtual Support Vectors

The choice of K determines the type of classifier that is constructed; possible choices include polynomial classifiers ($K(\mathbf{x}, \mathbf{y}) = (\mathbf{x} \cdot \mathbf{y})^n$), neural networks ($K(\mathbf{x}, \mathbf{y}) = \tanh(\kappa \cdot (\mathbf{x} \cdot \mathbf{x}_i) - \Theta)$) and radial basis function classifiers ($K(\mathbf{x}, \mathbf{y}) = \exp\left(-\|\mathbf{x} - \mathbf{y}\|^2 / \sigma^2\right)$). Already without the use of prior knowledge, these machines exhibit high generalization ability;[3] moreover, they extract almost identical sets of support vectors (cf. Eq. 4). It is possible to train any one of

---

[3] In a performance comparison, Bottou et al. (1994) note that the support vector machine "has excellent accuracy, which is most remarkable, because unlike the other high performance classifiers, it does not include knowledge about the geometry of the problem."

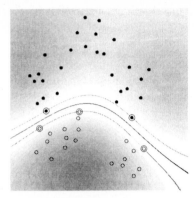

Fig. 2. Example of a support vector classifier found by using a radial basis function kernel $K(\mathbf{x}, \mathbf{y}) = \exp(-\|\mathbf{x} - \mathbf{y}\|^2)$. Both coordinate axes range from -1 to +1. Circles and disks are two classes of training examples; the middle line is the decision surface; the outer lines precisely meet the constraint (3). Note that the support vectors found by the algorithm (marked by extra circles) are not centers of clusters, but examples which are critical for the given classification task.

these machines solely on the support vector set extracted by another machine, with a test performance not worse than after training on the full data base (Schölkopf, Burges, and Vapnik, 1995). Using this finding as a starting point, we investigated the question whether it might be sufficient to generate virtual examples from the support vectors only. After all, one might hope that it does not add much information to generate virtual examples of patterns which are not close to the boundary. In our experiments, we proceeded as follows:

1. Train a support vector machine to extract the support vector set.
2. Generate artificial examples by applying the desired invariance transformations to the support vectors. In the following, we will refer to these examples as *virtual support vectors*.
3. Train another support vector machine on the generated examples.

The first set of experiments was conducted on a US postal service data base of handwritten digits, containing 7291 training examples and 2007 test examples. This data base has been used extensively in the literature, with a LeNet1 Convolutional Network achieving a test error rate of 5.0% (Le Cun et al., 1989). In the experiments, we used $\gamma = 10$ (cf. (2)), and a smoothing of the data with a Gaussian kernel of width 0.75.

To get a ten–class classifier for digit recognition, we combine ten binary classifiers of the described type (cf. Vapnik, 1995). In this case, virtual support vectors are generated for the set of all different support vectors of the ten classifiers. Alternatively, one can carry out the procedure separately for the ten binary classifiers, thus dealing with smaller training sets during the training of the second machine. Table 1 shows that incorporating only translational invariance already improves performance significantly, from 4.0% to 3.2% error rate.[4]

---

[4] It should be noted that the used test set is rather difficult — the human error rate is 2.5% (for a discussion, see Simard, Le Cun, and Denker, 1993).

Table 1. Comparison of support vector sets and performance for training on the original data base and training on the generated virtual support vectors. In both training runs, we used a polynomial classifier of degree 3. Virtual support vectors were generated by simply shifting the images by one pixel in the four principal directions. Adding the unchanged support vectors, this leads to a training set of the second classifier which has five times the size of the first classifier's overall support vector set (i.e. the union of the 10 support vector sets of the binary classifiers — note that due to some overlap, this is smaller than the sum of the ten support set sizes).

classifier trained on	size	average no. of SVs	no. of different SVs	test error
full training set	7291	274	1677	4.0%
overall SV set	1677	268	1582	4.1%
virtual SV set	8385	686	4535	3.2%

For other types of invariances, we also found improvements, albeit smaller ones: generating virtual support vectors by rotation or by the line thickness transformation of Drucker, Schapire, and Simard (1993), we constructed polynomial classifiers with 3.7% error rate (in both cases).

The larger a database, the more information about invariances of the decision function is already contained in the differences between patterns of the same class. To show that it is nevertheless possible to improve classification accuracies with our technique, we applied the method to the MNIST database of 60000+10000 handwritten digits. This database has become the standard for performance comparisons in our department; the error rate record of 0.7% is held by a boosted LeNet4 (Bottou et al. 1994). Using virtual support vectors generated by 1–pixel translations, we improved a degree 5 polynomial classifier from 1.4% to 1.0% error rate. In this case, we applied our technique separately for all ten support vector sets of the binary classifiers (rather than for their union) in order to avoid getting overly large support vector sets after retraining.

Further improvements can possibly be achieved by combining different types of invariances, and choosing the kind of transformations applied to individual support vectors according to whether they actually do provide new information about the decision boundary (c.f. (3)). Another intriguing extension of the scheme would be to use techniques based on image correspondence (e.g. Vetter & Poggio, 1996) to extract transformations from the training examples. Those transformations can then be used to generate virtual support vectors.

4 Discussion

We have shown that for support vector learning machines, invariances can readily be incorporated by generating virtual examples from the support vectors, rather than from the whole training set. The method yields a significant gain in classification accuracy at a moderate cost in time: it requires two training runs (rather than one), and it constructs classification rules utilizing more support vectors, thus slowing down classification speed (cf. Eq. 5) — in our case, both points amounted to a factor of about 2.

Given that support vector machines are known to allow for short training times (Bottou et al., 1994), the first point is usually not critical. Certainly, training on virtual examples generated from the whole data base would be significantly slower. To compensate for the second point, one could use the reduced set method of Burges (1996) to increase the speed.

Acknowledgements The ideas presented here were influenced by discussions with F. Girosi, P. Niyogi, T. Poggio, and K. Sung during a stay of B. S. at the Massachusetts Institute of Technology.

References

Abu-Mostafa, Y. S.: Hints. Neural Computation **7** (1995) 639–671

Boser, B. E., Guyon, I. M., Vapnik, V.: A training algorithm for optimal margin classifiers. Fifth Annual Workshop on Computational Learning Theory, Pittsburgh ACM (1992) 144–152.

Bottou, L., Cortes, C., Denker, J. S., Drucker, H., Guyon, I., Jackel, L. D., Le Cun, Y., Müller, U. A., Säckinger, E., Simard, P., Vapnik, V.: Comparison of classifier methods: a case study in handwritten digit recognition. Proceedings of the 12th International Conference on Pattern Recognition and Neural Networks, Jerusalem (1994)

Burges, C.: Simplified support vector decision rules. 13th International Conference on Machine Learning (1996)

Cortes, C., Vapnik, V.: Support Vector Networks. Machine Learning **20** (1995) 1–25

Drucker, H., Schapire, R., Simard, P.: Boosting performance in neural networks. International Journal of Pattern Recognition and Artificial Intelligence **7** (1993) 705–719

Le Cun, Y., Boser, B., Denker, J. S., Henderson, D., Howard, R. E., Hubbard, W., Jackel, L. J.: Backpropagation applied to handwritten zip code recognition. Neural Computation **1** (1989) 541–551

Schölkopf, B., Burges, C., Vapnik, V.: Extracting support data for a given task. In: Fayyad, U. M., Uthurusamy, R. (eds.): Proceedings, First International Conference on Knowledge Discovery & Data Mining, AAAI Press, Menlo Park, CA (1995)

Segman, J., Rubinstein, J., Zeevi, Y. Y.: The canonical coordinates method for pattern deformation: theoretical and computational considerations. IEEE Transactions on Pattern Analysis and Machine Intelligence **14** (1992) 1171–1183

Simard, P., Le Cun, Y., Denker, J.: Efficient pattern recognition using a new transformation distance. In: Hanson, S. J. , Cowan, J. D., Giles, C. L. (eds.): Advances in Neural Information Processing Systems 5, Morgan Kaufmann, San Mateo, CA (1993)

Simard, P., Victorri, B., Le Cun, Y., Denker, J.: Tangent Prop — a formalism for specifying selected invariances in an adaptive network. In: Moody, J. E., Hanson, S. J., Lippmann, R. P.: Advances in Neural Information Processing Systems 4, Morgan Kaufmann, San Mateo, CA (1992)

Vapnik, V.: Estimation of Dependences Based on Empirical Data. [in Russian] Nauka, Moscow (1979); English translation: Springer Verlag, New York (1982)

Vapnik, V.: The Nature of Statistical Learning Theory. Springer Verlag, New York (1995)

Vetter, T., Poggio, T.: Image Synthesis from a Single Example Image. Proceedings of the European Conference on Computer Vision, Cambridge UK, in press (1996)

Vetter, T., Poggio, T., Bülthoff, H.: The importance of symmetry and virtual views in three–dimensional object recognition. Current Biology **4** (1994) 18–23

Estimating the Reliability of Neural Network Classifications

Aarnoud Hoekstra, Servaes A. Tholen and Robert P.W. Duin

Pattern Recognition Group, Faculty of Applied Physics,
Delft University of Technology, Lorentzweg 1, 2628 CJ Delft, The Netherlands
e-mail: *aarnoud@ph.tn.tudelft.nl*

Abstract. A method for quantifying the reliability of object classifications using trained neural networks is given. Using this method one is able to give an estimation of a confidence value for a certain object. This reveals how trustworthy the classification of the particular object by the neural pattern classifier is. Even for badly trained networks it is possible to give reliable confidence estimations. Several estimators are considered. A *k*-NN technique has been developed to compare these using a learning set based artificially generated validation set. Experiments show that applying the developed estimators on a validation set gives the same results as applying the estimators on an independent test set. The method was tested on a real-life application, human chromosome classification, and gave good results which indicate the applicability of our method.

1 Introduction

In pattern recognition applications one is not only interested in the performance of a pattern classifier but also in the reliability of such a classifier. It has been shown in many applications that especially in nonlinear problems neural networks can be a good alternative for traditional pattern recognizers. We investigated the situations in which one has a given, trained, neural network. Given a set of unlabeled objects to be classified, one cannot determine from the neural classifier output whether it performs correctly or not. If the classifier is trained badly the network's performance cannot be judged from the net output values. Only perfectly trained networks will give reliable output values, due to the fact that these networks approximate the a posteriori probabilities sufficiently well. An interesting point therefore is to define

a reliability measure that still works in cases of badly trained networks. In this paper a method is proposed with which the reliability of a single classification of an object by a given neural classifier can be determined. This reliability measure, referred to as *confidence value*, enables further processing of an object or rejecting it.

The method presented here does not transform the network's output values in such a way that they fulfil the probability constraints, e.g. the network's ouput values constitute the aposteriori probabilities of the classes. The research started from the same point of view as the work by Denker and LeCun [2] but differs significantly in the developed technique. In their work it is assumed that a network is trained is such a way that some second order constraints are fullfilled. We do not make any assumptions on the way the network is trained. Furthermore also a method for generating a validation set, e.g. a calibration set as mentioned in [2] will be presented in this paper. Denker and LeCun use such a set as an independent test set to collect statistics about the classification results of the network. However, they do not state how a validation set was inferred, whereas we propose a k nearest neighbour data generation method for the case that an independent set is not available.

The neural networks under investigation are the normal feedforward networks trained with backpropagation and having sigmoidal transfer functions. Furthermore only networks consisting of one hidden layer are considered, although the results can be generalized to more than one hidden layer. The outline of the paper is as follows. First the problem of confidence in neural networks will be introduced. The next section is dedicated to the calculation of confidence values. The third section treats the generation of validation sets from a given learning set. In the following section the application of confidence values on a real-life problem, the classification of human chromosomes, is presented. The paper ends with conclusions and some remarks.

2 Confidence Values

When a trained neural network classifies an unlabeled object, the operator would like to know what confidence one should have in the classification. Much has been published about the overall error rate for a discriminant function [5, 3], but here the goal is to estimate the probability that a certain classification is correct, see also [2]. The confidence of a classification can be estimated

using different methods and in different layers of the network. The confidence, and its corresponding estimator, will be introduced more formally.

Assume a network finds a discriminant function $S(\mathbf{x}) : \mathbf{x} \rightarrow \hat{\omega}$. This discriminant function S assigns a class label $\hat{\omega}$ to a sample $\mathbf{x}$. For each sample $\mathbf{x}$ the classification based on S is either correct or incorrect, therefore a correctness function $C(\mathbf{x}, \omega)$ can be defined:

$$C(\mathbf{x}, \omega) = \begin{cases} 1 \text{ if } S(\mathbf{x}) = \omega \\ 0 \text{ if } S(\mathbf{x}) \neq \omega \end{cases}, \quad (1)$$

where ω is the true class label of sample $\mathbf{x}$. When an unlabeled sample is classified, the *confidence* in this classification is defined as the probability the classification is correct. Given a certain network, for an (in general unknown) object $\mathbf{x}$, the confidence is given by:

$$q(\mathbf{x}) = P(C(\mathbf{x}, \omega) = 1|\mathbf{x}). \quad (2)$$

The classification confidence is thereby the posteriori probability for the estimated class. For known distributions and given $S(\mathbf{x})$ the errors due to class overlap as well as imperfect training influence $q(\mathbf{x})$, therefore $q(\mathbf{x})$ can be expressed as:

$$q(\mathbf{x}) = \frac{P_{\hat{\omega}} f_{\hat{\omega}}(\mathbf{x})}{f(\mathbf{x})}, \quad (3)$$

where $f_{\hat{\omega}}$ is the class density function for class $\hat{\omega}$ as a function of the feature $\mathbf{x}$ and $P_{\hat{\omega}}$ the a priori class probability. The function $f(\mathbf{x})$ denotes the joint probability for $\mathbf{x}$. For unknown distributions $q(\mathbf{x})$ has to be estimated. As there are several possibilities for estimating $q(\mathbf{x})$ (see the next section), a criterion function J is defined to compare the different estimators for $\hat{q}_i(\mathbf{x})$:

$$J_i = \frac{1}{N_{\mathcal{T}}} \sum_{\mathcal{T}} [C(\mathbf{x}, \omega) - \hat{q}_i(\mathbf{x})]^2, \quad (4)$$

where $\mathcal{T}$ is an independent test set and $N_{\mathcal{T}}$ is the number of samples in the test set, the index i is used to indicate the J for the ith estimator $\hat{q}_i(\mathbf{x})$. The lower bound of J is usually not zero due to class overlap. However, when the underlying class densities are known J_{min} can be estimated by substituting (2) for each sample in (4). The aim is to minimize J over different estimators $\hat{q}_i(\mathbf{x})$ for a certain independent test set, yielding the best estimator for the confidence values.

3 Confidence Estimators

Since in real applications parameters such as distributions of the samples and a priori probabilities are unknown, there is a need for estimating confidence values. It was discussed in the previous section how the different estimators can be compared but nothing was said yet about which kind of estimators are used for $q(\mathbf{x})$. One can distinguish two kinds of estimations.

Let us first consider the layer in the network where the confidence is estimated. The confidence can be estimated in the input space, hidden unit space or output space. A feed forward network can be considered as series of data mappings, from the input feature space, being the input space of the network, followed by one or more hidden layers, towards the output space in which the final classification is done. These spaces differ in dimensionality and data configuration. Thereby, depending on the result of the training, it is open which space is the optimal one for confidence estimation. For sufficiently trained networks it can be advantageous to estimate in a different layer than the output space.

Second there are different methods for estimating the confidence value. We used three methods for estimating the value of $q(\mathbf{x})$, namely:

1. Nearest neighbour method
2. Parzen method
3. Logistic method

Each of the methods has its advantages and disadvantages. In the application to be presented next we restrict to the methods 1 and 3. Method 2 has been extensively tested, but appeared to be outperformed by the other methods. However, the Parzen method is suited for detecting outliers and therefore mentioned for the sake of completeness. Now each of these methods will be discussed.

3.1 Nearest Neighbour Method

Given a certain space (input, hidden or output) with a defined measure, the k-nearest neighbour (k-NN) method determines the k nearest learning samples to a test sample under consideration. For that sample the confidence is estimated as:

$$\hat{q}_{knn} = \frac{N_{\hat{\omega}}}{k}, \qquad (5)$$

where $N_{\hat{\omega}}$ is the number of correctly classified samples among the k nearest neighbours. For this number we just used $k = \sqrt{N_{\mathcal{L}}}$, where $N_{\mathcal{L}}$ is the number of samples in the learning set. Note that equation 5 needs to be altered when the a priori class probabilities differ.

3.2 Parzen Method

Here we use a Parzen estimator to estimate the confidence, by estimating the class densities and kernels. Experiments show that the Parzen estimator never outperforms the k-NN estimator ([7]). Denker and LeCun already indicated that the Parzen method would fail due to the high dimensionality of the space [2]. This was confirmed by our experiments. Therefore the Parzen estimator was not used in the application which is described in the next section.

3.3 Logistic Method

This method was chosen because of is simplicity, by which this method can be applied fast. The parameters can be calculated quite easily and need to be determined only once. It is assumed that the conditional class densities can be approximated by a sigmoidal function [1]. For a two-class problem, with classes A and B, this means that:

$$P(B|\mathbf{x}) = \frac{1}{1 + \exp(\boldsymbol{\beta}\mathbf{x} + \beta_0)}, \quad (6)$$

under the assumption that P_A equals P_B. The number of parameters that have to be estimated in equation 6 equals $d+1$, the dimension of the space to be investigated plus one bias term. These parameters can be estimated using a learning set and an optimization method. In the experiments the parameters for the logistic model were optimized using maximum likelihood estimation. The method is extended to more classes by taking one class as base class and estimate all parameters of the other classes with respect to this base class (g). By using our definition for confidence values, the logistic estimator $\hat{q}_{logistic}$ becomes:

$$\hat{q}_{logistic} = \{1 + \sum_{i=1}^{g-1} \exp(-\boldsymbol{\beta}_{gi}\mathbf{x}')\}^{-1}. \quad (7)$$

4 Validation sets

In the previous section three estimators for $q(\mathbf{x})$ have been introduced. Each of these estimators should be tested using an independent test set. Usually such independent test sets are not available, just one set is available from which both the learning set and test set are drawn. One could take the learning set for testing the estimators, however this is a very unhealthy situation. Both, the network and the estimator were created using the same set, which makes the estimator useless. To overcome this in practical situations a k-NN data generation method [4] was used to create a *validation set*. Such a validation set has roughly the characteristics of the learning set but is distinctive enough to be used a test set.

The k-NN data generation method is a fairly simple way to create a new dataset from a given one. In this method a learning set is used as starting point for a new set. Each point in the original set is used for generating an arbitrary number of new points by adding Gaussian noise in such a way that it follows the local probability density estimated from its k nearest neighbours. For k the dimensionality + 2 was used.

Note that a validation set will **not** be able to replace an independent test set entirely, but it is an acceptable alternative. Experiments show that the k-NN data generation method generates a validation set which is reliable enough to predict the same estimators and the network layer as an independent test set would do. In the application to be described next both a validation set and a test set were used. Many experiments not reported here can be found in [7]. These experiments show the validity of our data generation method for badly trained neural networks.

5 Application: Human Chromosome Classification

The problem of human chromosome classification has been under extensive investigation in our research group. Due to the increasing interest in neural networks in the recent years, also neural networks have been applied to chromosome classification. A normal human cell contains 46 chromosomes, which can be divided into 24 groups. There are 22 pairs of so called autosomes and two sex chromosomes. The chromosomes have different features that can be used for classification: *length, centromere position* and *banding pattern*. In this classification problem we only take the banding pattern into account. This banding pattern is sampled and then fed to a neural network for classification.

The sampling of a chromosomeband resulted in 30 features, ie. thirty samplings per band, which are used for classification of a chromosome. Earlier studies indicate that this is a reasonable value to take [6]. Consequently a neural network consisting of 30 input neurons, and 24 output neurons (one for each class) was used. The number of hidden units, however, is still arbitrarily chosen and was set to 100, a number derived from earlier experiments [6]. For training the network a total set of 32,000 training samples was used, the same number of samples was used for testing the network. Furthermore each sample was scaled between 0 and 1.

Network training was done for 200 epochs with momentum and adaptive learnrate, using the Matlab neural toolbox. The momentum term was set to 0.7, the initial learnrate was set to 0.1. Each time the mean squared error of the network increased by 10% the learnrate was halved. After training the network performance was 85% on the learn set and 83.5% on the test set (see also table 1). Two estimators plus the output of the network were used for the estimation of the confidence values. Table 1 states the different values of J which were found for the different estimators. Furthermore the k-NN and the logistic estimator were also applied in different parts of the network in order to estimate the confidence.

used set	learn set with l.o.o. est.	validation set	test set
estimator	J	J	J
k-NN			
input	0.167	0.165	0.161
hidden	0.119	0.119	0.119
output	0.086	0.088	0.095
logistic			
input	0.140	0.141	0.141
hidden	0.692	0.692	0.694
output	0.147	0.150	0.158
netout	0.482	0.479	0.474
avg. sd(J)	0.001	0.001	0.001
class perf	85.0%	84.5%	83.5%

Table 1. The criterion function values for all estimators, on the learning set, validation set and the test set.

The abbreviation l.o.o. means leave one out. Table 1 shows that the k-NN estimator in the output space has the best score over all the different sets. The logistic estimator performs worse in the output space than in the input space. Furthermore the network's output is not very useful as an estimator for the confidence. This is due to the fact that the network has only trained for 200 epochs. The data generation method was also tested, but it shows no advantage over using the learn set. However, other experiments showed that in some cases the validation set method is preferred over the learn set, especially in overtrained networks (see [7]).

6 Conclusions

In this paper an alternative method for calculating the confidence in classifications made by a neural network has been presented. This way uncertainties due to network training and class overlap are found. For the computation of the confidence three methods have been proposed: the nearest neighbour, the Parzen and the logistic method. From experiments (not reported here but to be found in [7]) we selected two methods for application on a real-life problem. These methods has been applied in the classification of human chromosomes. Moreover the estimates were made in different parts of the network. The results obtained from the chromosome experiments were very promising. From the experiments it was concluded that the k-NN method in the output space performs the best.

The method of generating a validation set has some shortcomings and needs therefore to be refined. A problem arises when the learning set contains some outliers. Since the data generation method is based on the learning set it will also generate new points around the outliers. However, in a test set these outliers will not be present at all. In this case the method fails. Another way to apply confidence values is to use them as a stopping criterion during training. At the moment the best confidence estimator is in the output layer the network can be expected to have an optimal generalization.

7 Acknowledgements

This research was partially sponsored by the SION (Dutch Foundation for Computer Science Research) and SNN (Foundation for Neural Networks, Nijmegen). Furthermore the authors would like to thank Martin Kraaijveld (Shell Research, Rijswijk) for his helpful discussions.

References

1. Anderson, J.A. *Logistic discrimination*, volume 2 of *Handbook of Statistics*, pages 169–191. North-Holland, 1982.
2. Denker, J.S. and Le Cun, Y. Transforming neural-net output levels to probability distribution. In R.P. Lippmann, J.E. Moody, and D.S. Touretzky, editors, *Advances in Neural information Processing Systems 3*, pages 853–859. Morgan Kaufmann, 1991.
3. Duda, R.O. and Hart, P.E. *Pattern classification and scene analysis*. John Wiley and sons, 1973.
4. Duin, R.P.W. Nearest neighbor interpolation for error estimation and classifier optimization. In *Proceedings of the 8th SCIA tromsø*, 1993.
5. Fukanaga, K. *Introduction to Statistical pattern recognition*. Academic Press, 1990.
6. Houtepen, J.D.J. Chromosome banding profiles: how many samples are necessary for a neural network classifier ? Master's thesis, Delft University of Technology, October 1994.
7. Tholen, S.A. Confidence values for neural network classifications. Master's thesis, Delft University of Technology, April 1995.

Bayesian Inference of Noise Levels in Regression

Christopher M. Bishop
Cazhaow S. Qazaz

Neural Computing Research Group
Aston University, Birmingham, B4 7ET, U.K.
ncrg@aston.ac.uk
http://www.ncrg.aston.ac.uk/

Abstract. In most treatments of the regression problem it is assumed that the distribution of target data can be described by a deterministic function of the inputs, together with additive Gaussian noise having constant variance. The use of maximum likelihood to train such models then corresponds to the minimization of a sum-of-squares error function. In many applications a more realistic model would allow the noise variance itself to depend on the input variables. However, the use of maximum likelihood for training such models would give highly biased results. In this paper we show how a Bayesian treatment can allow for an input-dependent variance while overcoming the bias of maximum likelihood.

1 Introduction

In regression problems it is important not only to predict the output variables but also to have some estimate of the error bars associated with those predictions. An important contribution to the error bars arises from the intrinsic noise on the data. In most conventional treatments of regression, it is assumed that the noise can be modelled by a Gaussian distribution with a constant variance. However, in many applications it will be more realistic to allow the noise variance itself to depend on the input variables. A standard maximum likelihood approach would, however, lead to a systematic underestimate of the noise variance. Here we adopt an approximate hierarchical Bayesian treatment (MacKay, 1995) to find the most probable interpolant and most probable input-dependent noise variance. We compare our results with maximum likelihood and show how this approximate Bayesian treatment leads to a significantly reduced bias.

In order to gain some insight into the limitations of the maximum likelihood approach, and to see how these limitations can be overcome in a Bayesian treatment, it is useful to consider first a much simpler problem involving a single random variable (Bishop, 1995). Suppose that a variable z is known to have a Gaussian distribution, but with unknown mean and variance. Given a sample $D \equiv \{z_n\}$ drawn from that distribution, where $n = 1, \ldots, N$, our goal is to infer values for the mean μ and variance σ^2. The likelihood function is given by

$$p(D|\mu, \sigma^2) = \frac{1}{(2\pi\sigma^2)^{N/2}} \exp\left\{ -\frac{1}{2\sigma^2} \sum_{n=1}^{N} (z_n - \mu)^2 \right\}. \tag{1}$$

A non-Bayesian approach to finding the mean and variance is to maximize the likelihood jointly over μ and σ^2, corresponding to the intuitive idea of finding the parameter values which are most likely to have given rise to the observed data set. This yields the standard result

$$\widehat{\mu} = \frac{1}{N}\sum_{n=1}^{N} z_n, \qquad \widehat{\sigma}^2 = \frac{1}{N}\sum_{n=1}^{N}(z_n - \widehat{\mu})^2. \tag{2}$$

It is well known that the estimate $\widehat{\sigma}^2$ for the variance given in (2) is *biased* since the expectation of this estimate is not equal to the true value

$$\mathcal{E}[\widehat{\sigma}^2] = \frac{N-1}{N}\sigma_0^2 \tag{3}$$

where σ_0^2 is the true variance of the distribution which generated the data, and $\mathcal{E}[\cdot]$ denotes an average over data sets of size N. For large N this effect is small. However, in the case of regression problems there are generally much larger number of degrees of freedom in relation to the number of available data points, in which case the effect of this bias can be very substantial.

By adopting a Bayesian viewpoint this bias can be removed. The marginal likelihood of σ^2 should be computed by *integrating* over the mean μ. Assuming a 'flat' prior $p(\mu)$ we obtain

$$p(D|\sigma^2) = \int p(D|\sigma^2, \mu)p(\mu)\,d\mu \tag{4}$$

$$\propto \frac{1}{\sigma^{N-1}}\exp\left\{-\frac{1}{2\sigma^2}\sum_{n=1}^{N}(z_n - \widehat{\mu})^2\right\}. \tag{5}$$

Maximizing (5) with respect to σ^2 then gives

$$\widetilde{\sigma}^2 = \frac{1}{N-1}\sum_{n=1}^{N}(z_n - \widehat{\mu})^2 \tag{6}$$

which is unbiased.

This result is illustrated in Figure 1 which shows contours of $p(D|\mu, \sigma^2)$ together with the marginal likelihood $p(D|\sigma^2)$ and the conditional likelihood $p(D|\widehat{\mu}, \sigma^2)$ evaluated at $\mu = \widehat{\mu}$.

2 Bayesian Regression

Consider a regression problem involving the prediction of a noisy variable t given the value of a vector $\mathbf{x}$ of input variables[1]. Our goal is to predict both a regression function and an input-dependent noise variance. We shall therefore

---

[1] For simplicity we consider a single output variable. The extension of this work to multiple outputs is straightforward.

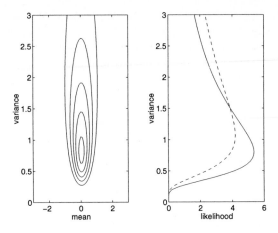

Fig. 1. The left hand plot shows contours of the likelihood function $p(D|\mu, \sigma^2)$ given by (1) for 4 data points drawn from a Gaussian distribution having zero mean and unit variance. The right hand plot shows the marginal likelihood function $p(D|\sigma^2)$ (dashed curve) and the conditional likelihood function $p(D|\hat{\mu}, \sigma^2)$ (solid curve). It can be seen that the skewed contours result in a value of $\hat{\sigma}^2$ which is smaller than $\tilde{\sigma}^2$.

consider two networks. The first network takes the input vector $\mathbf{x}$ and generates an output $y(\mathbf{x}; \mathbf{w})$ which represents the regression function, and is governed by a vector of weight parameters $\mathbf{w}$. The second network also takes the input vector $\mathbf{x}$, and generates an output function $\beta(\mathbf{x}; \mathbf{u})$ representing the inverse variance of the noise distribution, and is governed by a vector of weight parameters $\mathbf{u}$. The conditional distribution of target data, given the input vector, is then modelled by a normal distribution $p(t|\mathbf{x}, \mathbf{w}, \mathbf{u}) = \mathcal{N}(t|y, \beta^{-1})$. From this we obtain the likelihood function

$$p(D|\mathbf{w}, \mathbf{u}) = \frac{1}{Z_D} \exp\left\{ -\sum_{n=1}^{N} \beta_n E_n \right\} \tag{7}$$

where $\beta_n = \beta(\mathbf{x}_n; \mathbf{u})$,

$$Z_D = \prod_{n=1}^{N} \left(\frac{2\pi}{\beta_n} \right)^{1/2}, \qquad E_n = \frac{1}{2}(y(\mathbf{x}_n; \mathbf{w}) - t_n)^2 \tag{8}$$

and $D \equiv \{\mathbf{x}_n, t_n\}$ is the data set.

Some simplification of the subsequent analysis is obtained by taking the regression function, and $\ln \beta$, to be given by linear combinations of fixed basis functions, as in MacKay (1995), so that

$$y(\mathbf{x}; \mathbf{w}) = \mathbf{w}^{\mathrm{T}} \phi(\mathbf{x}), \qquad \beta(\mathbf{x}; \mathbf{u}) = \exp\left(\mathbf{u}^{\mathrm{T}} \psi(\mathbf{x}) \right) \tag{9}$$

and we assume that one basis function in each network is a constant $\phi_0 = \psi_0 = 1$ so that w_0 and u_0 correspond to bias parameters.

The maximum likelihood procedure chooses values $\hat{\mathbf{w}}$ and $\hat{\mathbf{u}}$ by finding a joint maximum over $\mathbf{w}$ and $\mathbf{u}$. As we have already indicated, this will give a biased result since the regression function inevitably fits part of the noise on the data, leading to an over-estimate of $\beta(\mathbf{x})$. In extreme cases, where the regression curve passes exactly through a data point, the corresponding estimate of β can go to infinity, corresponding to an estimated noise variance of zero.

The solution to this problem has already been indicated in Section 1 and was first suggested in this context by MacKay (1991, Chapter 6). In order to obtain an unbiased estimate of $\beta(\mathbf{x})$ we must find the marginal distribution of β, or equivalently of $\mathbf{u}$, in which we have integrated out the dependence on $\mathbf{w}$. This leads to a hierarchical Bayesian analysis.

We begin by defining priors over the parameters $\mathbf{w}$ and $\mathbf{u}$. Here we consider spherically-symmetric Gaussian priors of the form

$$p(\mathbf{w}|\alpha_w) = \left(\frac{\alpha_w}{2\pi}\right)^{1/2} \exp\left\{-\frac{\alpha_w}{2}\|\mathbf{w}\|^2\right\} \tag{10}$$

$$p(\mathbf{u}|\alpha_u) = \left(\frac{\alpha_u}{2\pi}\right)^{1/2} \exp\left\{-\frac{\alpha_u}{2}\|\mathbf{u}\|^2\right\} \tag{11}$$

where α_w and α_u are *hyperparameters*. At the first stage of the hierarchy, we assume that $\mathbf{u}$ is fixed to its most probable value $\mathbf{u}_{\mathrm{MP}}$, which will be determined shortly. The value of $\mathbf{w}_{\mathrm{MP}}$ is then found by maximizing the posterior distribution[2]

$$p(\mathbf{w}|D, \mathbf{u}_{\mathrm{MP}}, \alpha_w) = \frac{p(D|\mathbf{w}, \mathbf{u}_{\mathrm{MP}})p(\mathbf{w}|\alpha_w)}{p(D|\mathbf{u}_{\mathrm{MP}}, \alpha_w)} \tag{12}$$

where the denominator in (12) is given by

$$p(D|\mathbf{u}_{\mathrm{MP}}, \alpha_w) = \int p(D|\mathbf{w}, \mathbf{u}_{\mathrm{MP}})p(\mathbf{w}|\alpha_w) \, d\mathbf{w}. \tag{13}$$

Taking the negative log of (12), and dropping constant terms, we see that $\mathbf{w}_{\mathrm{MP}}$ is obtained by minimizing

$$S(\mathbf{w}) = \sum_{n=1}^{N} \beta_n E_n + \frac{\alpha_w}{2}\|\mathbf{w}\|^2 \tag{14}$$

where we have used (7) and (10). For the particular choice of model (9) this minimization represents a linear problem which is easily solved (for a given $\mathbf{u}$) by standard matrix techniques.

At the next level of the hierarchy, we find $\mathbf{u}_{\mathrm{MP}}$ by maximizing the marginal posterior distribution

$$p(\mathbf{u}|D, \alpha_u, \alpha_w) = \frac{p(D|\mathbf{u}, \alpha_w)p(\mathbf{u}|\alpha_u)}{p(D|\alpha_w, \alpha_u)}. \tag{15}$$

---

[2] Note that the result will be dependent on the choice of parametrization since the maximum of a distribution is not invariant under a change of variable.

The term $p(D|\mathbf{u}, \alpha_w)$ is just the denominator from (12) and is found by integrating over $\mathbf{w}$ as in (13). For the model (9) and prior (10) this integral is Gaussian and can be performed analytically without approximation. Again taking logarithms and discarding constants, we have to minimize

$$M(\mathbf{u}) = \sum_{n=1}^{N} \beta_n E_n + \frac{\alpha_u}{2} \|\mathbf{u}\|^2 - \frac{1}{2} \sum_{n=1}^{N} \ln \beta_n + \frac{1}{2} \ln |\mathbf{A}| \qquad (16)$$

where $|\mathbf{A}|$ denotes the determinant of the Hessian matrix $\mathbf{A}$ given by

$$\mathbf{A} = \sum_{n=1}^{N} \beta_n \phi(\mathbf{x}_n) \phi(\mathbf{x}_n)^{\mathrm{T}} + \alpha_w \mathbf{I} \qquad (17)$$

and $\mathbf{I}$ is the unit matrix. The function $M(\mathbf{u})$ in (16) can be minimized using standard non-linear optimization algorithms. We use scaled conjugate gradients, in which the necessary derivatives of $\ln |\mathbf{A}|$ are easily found in terms of the eigenvalues of $\mathbf{A}$.

In summary, the algorithm requires an outer loop in which the most probable value $\mathbf{u}_{\mathrm{MP}}$ is found by non-linear minimization of (16), using the scaled conjugate gradient algorithm. Each time the optimization code requires a value for $M(\mathbf{u})$ or its gradient, for a new value of $\mathbf{u}$, the optimum value for $\mathbf{w}_{\mathrm{MP}}$ must be found by minimizing (14). In effect, $\mathbf{w}$ is evolving on a fast time-scale, and $\mathbf{u}$ on a slow time-scale. The corresponding maximum (penalized) likelihood approach consists of a joint non-linear optimization over $\mathbf{u}$ and $\mathbf{w}$ of the posterior distribution $p(\mathbf{w}, \mathbf{u}|D)$ obtained from (7), (10) and (11). Finally, the hyperparameters are given fixed values $\alpha_w = \alpha_u = 0.1$ as this allows the maximum likelihood and Bayesian approaches to be treated on an equal footing.

3   Results and Discussion

As an illustration of this algorithm, we consider a toy problem involving one input and one output, with a noise variance which has an x^2 dependence on the input variable. Since the estimated quantities are noisy, due to the finite data set, we consider an averaging procedure as follows. We generate 100 independent data sets each consisting of 10 data points. The model is trained on each of the data sets in turn and then tested on the remaining 99 data sets. Both the $y(\mathbf{x}; \mathbf{w})$ and $\beta(\mathbf{x}; \mathbf{u})$ networks have 4 Gaussian basis functions (plus a bias) with width parameters chosen to equal the spacing of the centres. Results are shown in Figure 2. It is clear that the maximum likelihood results are biased and that the noise variance is systematically underestimated. By contrast, the Bayesian results show an improved estimate of the noise variance. This is borne out by evaluating the log likelihood for the test data under the corresponding predictive distributions. The Bayesian approach gives a log likelihood per data point, averaged over the 100 runs, of -1.38. Due to the over-fitting problem, maximum likelihood occasionally gives extremely large negative values for the

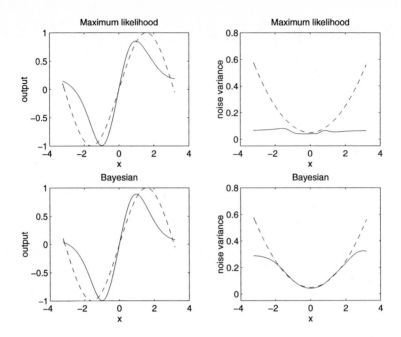

Fig. 2. The left hand plots show the sinusoidal function (dashed curve) from which the data were generated, together with the regression function averaged over 100 training sets. The right hand plots show the true noise variance (dashed curve) together with the estimated noise variance, again averaged over 100 data sets.

log likelihood (when β has been estimated to be very large). Even omitting these extreme values, the maximum likelihood still gives an average log likelihood per data point of -17.1 which is substantially smaller than the Bayesian result.

Recently, MacKay (1995) has proposed a different approach to treating this model, in which the numerical optimization to find the most-probable parameters is replaced with a fully Bayesian treatment involving Gibbs sampling from the posterior distribution. It will be interesting to compare these two approaches.

Acknowledgements: This work was supported by EPSRC grant GR/K51792, *Validation and Verification of Neural Network Systems*.

References

Bishop, C. M. (1995). *Neural Networks for Pattern Recognition*. Oxford University Press.

MacKay, D. J. C. (1991). *Bayesian Methods for Adaptive Models*. Ph.D. thesis, California Institute of Technology.

MacKay, D. J. C. (1995). Probabilistic networks: new models and new methods. In F. Fogelman-Soulié and P. Gallinari (Eds.), *Proceedings ICANN'95 International Conference on Artificial Neural Networks*, pp. 331–337. Paris: EC2 & Cie.

Complexity Reduction in Probabilistic Neural Networks

Ari Hämäläinen[1] and Lasse Holmström[2]

[1] NOKIA Research Center, P.O.Box 45, 00211 Helsinki, Finland
[2] Rolf Nevanlinna Institute, P.O.Box 4 , 00014 University of Helsinki, Finland

Abstract. Probability density estimation using the probabilistic neural network or the kernel method is considered. In its basic form this method can be computationally prohibitive, as all training data need to be stored and each individual training vector gives rise to a new term of the estimate. Given an original training sample of size N in a d-dimensional space, a simple binned kernel estimate with $O(N^{d/(d+4)})$ terms can be shown to attain an estimation accuracy only marginally inferior to the standard kernel method. This can be taken to indicate the order of complexity reduction generally achievable when a radial basis function style expansion is used in place of the probabilistic neural network.

1 Introduction

Probability density estimation involves approximating an unknown density function on the basis of a training sample drawn from it. The many applications of density estimation include exploratory data-analysis, visualization, simulation, and pattern recognition. A currently popular density estimation method is the kernel density estimate (e.g. [6]). Given a sample $\mathbf{x}_1, \ldots, \mathbf{x}_N$, of d-dimensional data vectors from an unknown probability density f, the kernel density estimate (KDE) is defined as

$$(1) \qquad \hat{f}(\mathbf{x}) = \frac{1}{N} \sum_{i=1}^{N} K_h \left(\mathbf{x} - \mathbf{x}_i \right),$$

where K is a fixed probability density function, called the kernel, $h > 0$ is a smoothing parameter, and K_h denotes the scaled kernel $K_h(\mathbf{x}) = h^{-d} K(\mathbf{x}/h)$, $\mathbf{x} \in \mathbb{R}^d$. The probabilistic neural network (PNN) [7] is the neural network counterpart of the KDE.

A drawback of kernel density estimation is that it suffers from the curse of dimensionality: the size of the data sample needed for given estimation accuracy grows exponentially with the dimension d. Besides, even if a sufficiently large data sample is available for accurate density estimation, the full kernel estimate with all N terms may be too expensive to use in practice. One solution is to use fewer kernels and to place them where they are needed most. The Reduced Kernel Density Estimate (RKDE) introduced in [3] is defined by choosing an $L \ll N$, and setting

$$(2) \qquad \hat{f}(\mathbf{x}) = \sum_{k=1}^{L} w_k K_{h_k}(\mathbf{x} - \mathbf{c}_k),$$

where $\mathbf{c}_1, \ldots, \mathbf{c}_L$ are the reference vectors of a Self-Organizing Map [4] trained with the full sample $\mathbf{x}_1, \ldots, \mathbf{x}_N$, $w_1, \ldots, w_L$ are nonnegative weights with $\sum_{k=1}^{L} w_k = 1$, and h_k is a smoothing parameter associated with the kth kernel. In the on-line version of the method, the weights w_k are be computed iteratively to reflect the fractions N_k/N of training data in the Voronoi regions of the reference vectors and the smoothing parameters h_k are optimized via stochastic gradient descent that attempts to minimize the integrated squared error $\int (\hat{f} - f)^2$. In a batch version of RKDE, the kernel locations $\mathbf{c}_k$ are first found using the SOM. The weight w_k corresponding to $\mathbf{c}_k$ is then set to N_k/N, and the smoothing parameters h_k (or just a single h) are found using standard kernel estimation methods, such as cross-validation. For references to other reduced kernel estimation methods, see [3]. We also note that (2) has the general appearance of a mixture model as well as a radial basis function expansion.

A particularly simple reduced kernel estimation method is the binned kernel density estimate (BKDE) that rounds the data to bin centers before the kernel estimate is computed. We extend (Section 2) the univariate analysis of [5, 1] to the multivariate case and show (Section 3) that for large N the BKDE needs only $O(N^{d/(d+4)})$ kernels to match essentially the accuracy of the standard KDE. This suggests that the more flexible RKDE and similar density estimate expansions might be able to achieve comparable or even more dramatic complexity reduction in kernel density estimation, a conjecture supported also by our simulations (Section 4).

2 Multivariate Binned Kernel Estimation

To define the binned estimator, let $\delta > 0$ and define for each d-tuple of integers $\mu = (\mu_1, ..., \mu_d) \in \mathbb{Z}^d$ the bin center $\mathbf{c}_\mu = \delta\mu$, and the associated bin as the d-dimensional cube centered at $\mathbf{c}_\mu$. Given a sample $\mathbf{x}_1, \ldots, \mathbf{x}_N$ from the unknown density f, let $w_\mu = N_\mu/N$, where N_μ is the number of data points in the μth bin. The binned kernel density estimate of f is then

$$(3) \qquad \hat{f}(\mathbf{x}) = \sum_{\mu \in \mathbb{Z}^d} w_\mu K_h(\mathbf{x} - \mathbf{c}_\mu).$$

Mathematical analysis of the estimation properties of the RKDE and other similar methods is complicated by the procedure used to compute the kernel locations from the data. The advantage of the BKDE from a theoretical point of view is that, apart from the random weights w_μ, it is defined by an explicit deterministic formula.

The usual measure of estimation error of a density estimator is the mean integrated squared error

$$\text{(4)} \qquad \text{MISE} = E \int (\hat{f}(\mathbf{x}) - f(\mathbf{x}))^2 d\mathbf{x},$$

where the integration is extended over the whole space $\mathbb{R}^d$. The mean or expectation is computed with respect to the random sample $\mathbf{x}_1, \ldots, \mathbf{x}_N$, on which the estimate $\hat{f}$ depends. Like for the standard KDE ([6]), it is also possible to derive an approximate expression for the MISE of the BKDE under certain regularity conditions on f and K. If we assume that $h = h_N$ satisfies $h_N \to 0$ and $N h_N^d \to \infty$, and that $\delta = \delta_N \to 0$, then the following holds.

Theorem 1. *Suppose that f and K are four times continuously differentiable and have compact supports. Assume also that K is a product of symmetric univariate kernels κ, $K(\mathbf{x}) = \prod_{i=1}^{d} \kappa(x_i)$, $\mathbf{x} = (x_1, \ldots, x_d)$. Then the asymptotic mean integrated squared error of the binned kernel density estimator as $N \to \infty$ is given by*

$$\text{(5)} \qquad \text{AMISE}_{\text{BKDE}} = \frac{R(\kappa)^d}{N h^d} + \frac{1}{4} R(\nabla^2 f) \left[\sigma_\kappa^2 h^2 + \frac{\delta^2}{12} \right]^2,$$

where the notations $R(g) = \int g(\mathbf{x})^2 d\mathbf{x}$ (g a function), and $\sigma_\kappa^2 = \int x^2 \kappa(x) dx$ are used.

The somewhat technical proof of the this result is given in [2]. We also note that (5) is closely related to the asymptotic mean integrated squared error of the KDE given by

$$\text{(6)} \qquad \text{AMISE}_{\text{KDE}} = \frac{R(\kappa)^d}{N h^d} + \frac{1}{4} R(\nabla^2 f) \sigma_\kappa^4 h^4.$$

The optimal smoothing parameter derived by minimizing (6) is

$$\text{(7)} \qquad h_{\text{KDE}}^* = \left[\frac{d R(\kappa)^d}{R(\nabla^2 f) \sigma_\kappa^4} \right]^{1/(d+4)} N^{-1/(d+4)}.$$

3   Complexity Reduction

We now consider the number of kernels needed in (3) in order to achieve an estimation accuracy comparable to that of the standard KDE. Let m be the number of bins that fit into a cube of side length h, that is, $m = h^d / \delta^d$. Given $\alpha > 0$, the square root of the asymptotic error $\sqrt{\text{AMISE}_{\text{BKDE}}}$ is at most $100 \cdot \alpha\%$ higher than $\sqrt{\text{AMISE}_{\text{KDE}}}$ if

$$\text{(8)} \qquad \frac{\text{AMISE}_{\text{BKDE}}}{\text{AMISE}_{\text{KDE}}} \leq (\alpha + 1)^2 =: \beta.$$

When the same K, N, and h are used for the KDE and the BKDE, then (8) is equivalent to

$$(9) \qquad \frac{\frac{R(\kappa)^d}{Nh^d} + \frac{1}{4}h^4 R(\nabla^2 f)\left(\sigma_\kappa^2 + \frac{1}{12m^{2/d}}\right)^2}{\frac{R(\kappa)^d}{Nh^d} + \frac{1}{4}h^4 \sigma_\kappa^4 R(\nabla^2 f)} \leq \beta.$$

The smallest m (corresponding to the largest bin size) that satisfies (9) is given by

$$(10) \quad m_N = \frac{\left(\frac{h^4}{4}R(\nabla^2 f)\right)^{d/4}}{2^d 3^{d/2}\left[\sqrt{(\beta-1)\frac{R(\kappa)^d}{Nh^d} + \beta\frac{\sigma_\kappa^4 h^4}{4}R(\nabla^2 f)} - \sqrt{\frac{\sigma_\kappa^4 h^4}{4}R(\nabla^2 f)}\right]^{d/2}},$$

which generally depends on the sample size N, both through the explicit N in the denominator and the smoothing parameter $h = h_N$. However, a natural choice for h is the error minimizing h^*_{KDE} of (7) for which m_N reduces to the constant value

$$(11) \qquad m_\ell = \left(\frac{\sqrt{1 + \sqrt{1 + 10\alpha + 5\alpha^2}}}{\sqrt{60}\sigma_\kappa\sqrt{2\alpha + \alpha^2}}\right)^d.$$

This suggests the use of the bin side length $\delta = m_\ell^{-1/d} h^*_{\text{KDE}}$. Another approach is to write (9) as

$$(12) \quad \frac{1}{4}h^4 R(\nabla^2 f)\left(\sigma_\kappa^2 + \frac{1}{12m^{2/d}}\right)^2 - \frac{1}{4}\beta\sigma_\kappa^4 h^4 R(\nabla^2 f) \leq (\beta-1)\frac{R(\kappa)^d}{Nh^d},$$

and to note that

$$(13) \qquad m = m_u = \left(12\alpha\sigma_\kappa^2\right)^{-d/2}$$

makes the left side of (12) zero. Thus, the choice $m = m_u$ satisfies (9), no matter how h is chosen. It is easy to see that $m_u > m_\ell$ (hence ℓ and u for "lower" and "upper", respectively).

The number of kernels needed in the BKDE to satisfy (8) can now be estimated. If the density function vanishes outside a cube with volume r^d, the maximum number of bins or kernels needed is given by $L = r^d/\delta^d = mr^d h^{-d}$. With $h = h^*_{\text{KDE}}$ and $m = m_u$, this gives

$$(14) \qquad L = \left[\frac{r}{\sqrt{12\alpha}}\right]^d \left[\frac{R(\nabla^2 f)}{d\sigma_\kappa^d R(\kappa)^d}\right]^{d/(d+4)} N^{d/(d+4)}.$$

For small d, this suggests a very dramatic reduction in kernel density estimate complexity. For $d = 1$, the number of kernels needed is only of the order $O(N^{1/5})$ where N is the number of kernels needed in the standard KDE. For high d, the curse of dimensionality sets in, as the number of bins needed has the asymptotic order $O(N)$ when $d \to \infty$.

	$N = 500$		$N = 5000$	
d	KDE	BKDE	KDE	BKDE
1	0.037	0.039 / 1.054 / 17	0.020	0.020 / 1.000 / 33
2	0.039	0.040 / 1.026 / 129	0.021	0.022 / 1.048 / 409
3	0.033	0.035 / 1.061 / 345	0.018	0.019 / 1.056 / 1983

Table 1. The result of the first test. For the KDE, the estimated $\sqrt{\mathrm{MISE}}$ is given in each case. For the BKDE, the estimated $\sqrt{\mathrm{MISE}}$, its ratio to the $\sqrt{\mathrm{MISE}}$ of the KDE, and the average number of nonempty bins are given.

4 Simulations

We report here simulations that support the theoretical results of the previous Section. In all cases the data were generated from the normal distribution and binning in the BKDE was chosen so that one bin was centered at the origin. The MISEs were estimated by testing each method in 10 independent trials and computing the mean value of the integrated squared errors. The error ratio $\sqrt{\widehat{\mathrm{MISE}}_{\{B/R\}KDE}}/\sqrt{\widehat{\mathrm{MISE}}_{KDE}}$ was also found in each case.

In the first test the KDE and the BKDE were compared. Dimensions $d = 1, 2$, and 3 with sample sizes $N = 500$, and $N = 5000$ were considered. We fixed $\alpha = 0.1$ in (13) and took $\delta = m_u^{-1/d} h_{KDE}^*$. The results are given in Table 1. As can be seen, the error ratio in each case is less than 1.1, as it should be. Note also that the number of nonempty bins is smaller than the original sample size, but increases rapidly with the dimension.

In the second test the BKDE was compared to the on-line and batch versions of the RKDE. The sample size N was 1000 and α was set to 3.0 in (13). In an attempt to make the comparison fair, the smoothing parameter and the bin size for the BKDE were estimated from the data as follows. An h was chosen for the KDE by least squares cross validation (LSCV,[6]), we set $\delta = m_u^{-1/d} h$, and picked the final smoothing parameter using unbiased LSCV ([2], Section 4.2). For the batch RKDE, h was also chosen using unbiased LSCV. The results are given in Table 2.

This simulation suggests that in order to reduce the number of kernels radically, it might be advisable to use an adaptive method like the RKDE instead of the fixed bin size BKDE. It is also seen that the simultaneous optimization of L smoothing parameters required in the on-line RKDE gets more difficult as L increases. This phenomenon is further depicted in Fig. 1, where RKDEs with a single smoothing parameter ($*$) and kernel dependent smoothing parameters ($\times$) are compared. In both cases the smoothing parameters were optimized iteratively. With only a few kernels, separate smoothing for each kernel works well, but as the number of kernels increases, the RKDE with a single smoothing parameter performs better.

d	NNB	BKDE	RKDE	Batch	KDE
1	4	0.078 / 2.229	0.066 / 1.886	0.068 / 1.943	0.035
2	11	0.068 / 2.000	0.064 / 1.882	0.060 / 1.765	0.034
3	26	0.053 / 1.893	0.048 / 1.714	0.044 / 1.571	0.028

Table 2. The result of the second test. The average number of nonempty bins (NNB) in the BKDE was used as the number of kernels, L, for the RKDE. The estimated $\sqrt{\text{MISE}}$ for the BKDE, the RKDE, the batch RKDE, as well as its ratio to the $\sqrt{\text{MISE}}$ of the KDE (in the last column) are given.

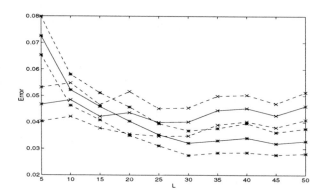

Fig. 1. Comparison of kernel dependent smoothing ($\times$) and the use of a single smoothing parameter ($*$) in the RKDE. Training sample size was $N = 10000$. The 99% confidence bounds are also shown.

References

1. P. Hall. The influence of rounding errors on some nonparametric estimators of a density and its derivatives. *SIAM J. Appl. Math.*, 42(2):390–399, April 1982.
2. A. Hämäläinen. Self-organizing map and reduced kernel density estimation. PhD Thesis, Research Reports A11, Rolf Nevanlinna Institute, University of Helsinki, 1995.
3. L. Holmström and A. Hämäläinen. The self-organizing reduced kernel density estimator. In *Proceedings of the 1993 IEEE International Conference on Neural Networks, San Francisco, California, March 28 - April 1*, volume 1, pages 417–421, 1993.
4. T. Kohonen. *Self-Organizing Maps*. Springer-Verlag, 1995.
5. D. W. Scott and S. J. Sheather. Kernel density estimation with binned data. *Comm. Statist. A-Theory Methods*, 14(6):1353–1359, 1985.
6. B. W. Silverman. *Density Estimation for Statistics and Data Analysis*. Chapman and Hall, 1986.
7. D. F. Specht. Probabilistic neural networks. *Neural Networks*, 3(1):109–118, 1990.

Asymptotic Complexity of an RBF NN for Correlated Data Representation

Lucia Sardo and Josef Kittler

Dept of Electronic & Electrical Engineering, University of Surrey,
Guildford, Surrey GU2 5XH, United Kingdom

Abstract. We address here the problem of architecture selection for an RBF network designed for classification purposes. Given a training set, the RBF network produces an estimate of the Probability Density Function (PDF) in terms of a mixture of l uncorrelated Gaussian functions, where l is the number of hidden neurons. Using uncorrelated Gaussians alleviates the heavy computational burden of estimating the full covariance matrix. However, the simplicity of such building blocks has to be paid for by the relatively large numbers of units needed to approximate the density of correlated data. We define two scalar parameters to describe the complexity of the data to be modelled and study the relationship between the complexity of the data and the complexity of the *best* approximating network.

1 Introduction

The problem of architecture selection for Neural Networks is a subject often undeservedly neglected. Choosing an appropriate architecture is of extreme importance if the network is to have good generalization properties. Provided a complex enough architecture and an appropriate learning algorithm are given, a network can reproduce the training set to a desired degree of accuracy, however this is not the aim of the training procedure. The main aim is to learn something about the real structure of the data in order to be able to predict future outcomes.

The usual procedure in this case is to have a test set to estimate the *generalization performance*. Unfortunately to obtain such an estimate a large test set is required. If the problem of architecture selection has to be addressed then a third set, for validation, is required to decide which architecture is to be preferred. The question addressed in this paper is how to select the network architecture in a non-subjective manner, when the design set (training and test set) is relatively small.

There are basically two types of approach to this problem:
- constructive algorithms
- information theoretic criteria of optimality

In the following we will deal with the latter.

There have been a few attempts to derive criteria of optimality based on already known statistical results ([1, 2, 4]) and even fewer examples of experimental application of such criteria([3, 5]). Nevertheless such criteria offer a methodology for automatic architecture selection at a much lower computational cost than the cross-validation technique. We will give a brief general description of information criteria with potential applicability to the problem of network architecture selection. Then we will describe our approach to this problem for the particular case of RBF networks trained to learn the PDF from a limited training sample. We show the results obtained in an unfavourable case when our network can only approximate the PDF, but not actually reproduce it. The results obtained have an implication on the definition of the information criterion we have been using.

2 Information criteria for network architecture selection

We define a network as a machine able to produce a function of its input

$$N(\theta) : \theta \mapsto f(x; \theta) \qquad \theta \in \Theta \tag{1}$$

where θ is the vector of all the weights of the network. Let us define the unknown mapping we are looking for

$$M : x \mapsto y \qquad x \in X, y \in Y \tag{2}$$

If a $\theta_* \in \Theta$ such that $N(\theta_*) = M$ exists, our aim is to find an estimate $\hat{\theta}_*$ of θ_*. Otherwise we look for an estimate $\hat{\theta}_0$ of θ_0 such that $\theta_0 = min_{\theta \in \Theta} D(N(\theta), M)$ where $D(\cdot)$ denotes a certain measure of discrepancy between the two mappings $N(\theta)$ and M ([9]). The training of the network is performed by minimizing a criterion function

$$J(\theta; X, Y) = F(\theta; X, Y) + G(\theta) \tag{3}$$

where $F(\theta; X, Y)$ is a function measuring the goodness of fit between the training data and the model, such as the sum of squared errors, and $G(\theta)$ is an optional regularization term. Equation (3) could be used in order to train the network and choose the architecture at the same time, but this leads to selecting the most complex architecture especially in the case when θ_* does not exist and only a limited training set size is available ([8, 5]). Akaike in [1] suggests to modify (3) by adding a term that is proportional to the number of parameters to be estimated. This results in a so-called AIC

$$J(\theta; X, Y) = F(\theta; X, Y) + G(\theta) + C(\theta) \tag{4}$$

where $C(\theta)$ is the term that takes into account the complexity of the network in relation to the training set size available ([4]).

were unexpected: the network complexity for correlated data is entirely determined by the intrinsic dimensionality and is independent of the degree of data correlation. Two important implications stem from these results. First, the results allowed us to define a functional form for the relaxation factor. Second, the findings suggest that in order to obtain a reliable PDF estimation it will be necessary, as a prerequisite, to establish the intrinsic dimensionality of the data. This may require a multipass procedure of PDF estimation and PDF estimation verification which will be the subject of our future research.

Acknowledgments

This study has been supported by the EPSRC Research Grant GR/J89255.

References

1. H. Akaike. A new look at the statistical model identification. *IEEE trans. on Automatic Control*, AC-19(6):716–723, 1974.
2. A. R. Barron and T. M. Cover. Minimum complexity density-estimation. *IEEE trans. on Information Theory*, 37(4):1034–1054, 1991.
3. D. B. Fogel. An information criterion for optimal neural network selection. *IEEE Trans. on Neural Networks*, 2(5):490–497, 1991.
4. N. Murata, S. Yoshizawa, and S. Amari. Network information criterion – determining the number of hidden units for an artificial neural network model. *IEEE Trans. on Neural Networks*, 5(6):865–872, 1994.
5. L. Sardo and J. Kittler. Maximum likelihood estimators for gaussian mixtures model selection. Submitted to *Neural Networks*.
6. L. Sardo and J. Kittler. Minimum complexity pdf estimation for correlated data. In *Proceedings of ICPR96 (International Conference on Pattern recognition), Vienna*, 25-30 August1996.
7. L. Sardo and J. Kittler. Minimum complexity estimator for rbf networks architecture selection. In *Proceedings of ICNN96 (International Conference on Neural Networks), Washington DC*, 3-6 June 1996.
8. G. Schwarz. Estimating the dimension of a model. *The Annals of Statistics*, 6(2):461–464, 1978.
9. V. Vysniauskas, F. C. A. Groen, and B. J. A. Kröse. The optimal number of learning samples and hidden units in function approximation with a feedforward network. Technical Report CS-93-15, Dept. of Comp. Sys, Univ. of Amsterdam, Nov. 1993.

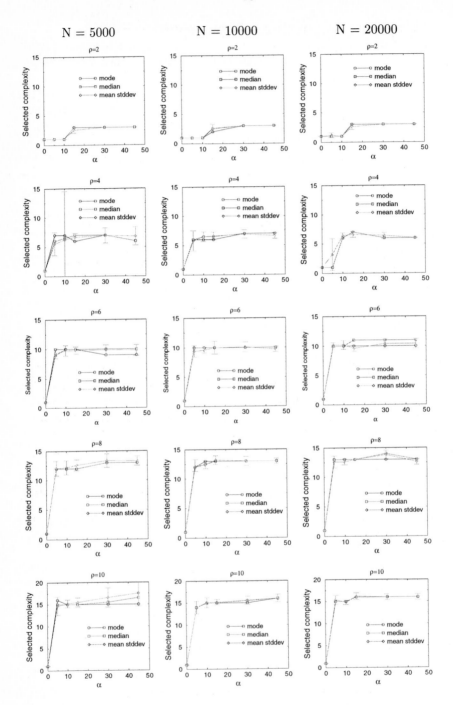

Fig. 1. The three column show the architecture selected by the log-likelihood in the large set size limit

Regularization by Early Stopping in Single Layer Perceptron Training

Sarunas Raudys[1] and Tautvydas Cibas[2]

[1]Department of Informatics, Vytautas Magnus University, Kaunas, Lithuania
e-mail:raudys@ktl.mii.lt
[2]Université de Paris-Sud, LRI, Bât.490, F-91405 Orsay Cedex, France
e-mail:cibas@lri.lri.fr

Abstract. Adaptative training of the non-linear single-layer perceptron can lead to the Euclidean distance classifier and later to the standard Fisher linear discriminant function. On the way between these two classifiers one has a regularized discriminant analysis. That is equivalent to the "weight decay" regularization term added to the cost function. Thus early stopping plays a role of regularization of the network.

1 Introduction and notations

A artificial feedforward neural networks (ANN) were found to be a solid tool in pattern recognition. Because of their nonlinearity and complex structures, their behavior is difficult to analyse analytically. To facilitate this task we analyse the single-layer perceptron (SLP) which is a key elementary grain in modern feedforward networks and has a strong tie with a classical discriminant analysis.

In classical statistical discriminant analysis, to discriminate between two pattern classes one uses a discriminant function

$$g(\mathbf{x}) = \mathbf{w}^T \mathbf{x} + w_0 \tag{1}$$

and performs a classification of the p-variate vector $\mathbf{x} = \left(x_1, x_2, ..., x_p\right)^T$ according to its sign. $\mathbf{w}^T$ denotes the vector transposition. The weights w_0, $\mathbf{w} = \left(w_1, w_2, ..., w_p\right)^T$ in (1) of the form

$$\mathbf{w}^F = S^{-1}\left(\overline{\mathbf{x}}^{(1)} - \overline{\mathbf{x}}^{(2)}\right), \qquad w_0 = -\tfrac{1}{2}\mathbf{w}^T\left(\overline{\mathbf{x}}^{(1)} + \overline{\mathbf{x}}^{(2)}\right) \tag{2}$$

are the weights of a *Fisher discriminant function* (Fisher, 1936),

where $S = \frac{1}{N_1+N_2-2}\sum_{i=1}^{2}\sum_{j=1}^{N_i}\left(\mathbf{x}_j^{(i)} - \overline{\mathbf{x}}^{(i)}\right)\left(\mathbf{x}_j^{(i)} - \overline{\mathbf{x}}^{(i)}\right)^T$ is a sample pooled covariance matrix, $\overline{\mathbf{x}}^{(i)} = \frac{1}{N_i}\sum_{j=1}^{N_i}\mathbf{x}_j^{(i)}$ is a sample mean vector, $\mathbf{x}_j^{(i)}$ is the j-th learning vector from the i-th class and N_i is the learning vector's number from class i.

When one omits the covariance matrix S in (2) then on has the *Euclidean distance classifier*.

The SLP itself can form linear discriminant hyperplane in a highdimensional feature space and discriminate complicated objects. In the notations used in the present paper, the SLP has p inputs and one output which is calculated by $output = o(\mathbf{w}^T\mathbf{x} + w_0)$, where $o(net)$ is a non-linear (activation) function. The weights $w_0, \mathbf{w}$ of the discriminant function will be learned during the training process. For two class classification problem we minimize the following mean squares cost function :

$$cost = \frac{1}{2(N_1+N_2)} \sum_{i=1}^{2} \sum_{j=1}^{N_i} \left(d_j^{(i)} - o\left(\mathbf{w}^T\mathbf{x}_j^{(i)} + w_0\right)\right)^2 \tag{3}$$

by adapting the weight vector according to a standard global gradient delta learning rule

$$\mathbf{w}_{(t+1)} = \mathbf{w}_{(t)} - \eta \frac{\partial cost_{(t)}}{\partial \mathbf{w}_{(t)}} \tag{4}$$

where $d_j^{(i)}$ is the desired output for input $\mathbf{x}_j^{(i)}$, $\mathbf{w}_{(t)}$ is a weight vector at instant t and η is a learning step.

It is known (Koford & Groner, 1966; Gallinari et al., 1991), that when the number of training vectors from two pattern classes is the same ($N_1=N_2$), training of cost function (3) with linear transfer function $o(net)=net$ (which becomes ADALINE, a prototype of modern SLP) leads to a weight vector $\mathbf{w}$ which is equivalent to (2), the weights of the standard Fisher linear discriminant function, who is asymptotically optimal when the classes are Gaussian with common covariance matrix.

The minimization of the cost function (3) with an additional "weight decay" regularization term $\lambda \mathbf{w}^T\mathbf{w}$ gives weights of the form (Raudys et al., 1994)

$$\mathbf{w}^{FR} = (S + \lambda I)^{-1}\left(\overline{\mathbf{x}}^{(1)} - \overline{\mathbf{x}}^{(2)}\right)k, \qquad w_0 = -\tfrac{1}{2}\mathbf{w}^T\left(\overline{\mathbf{x}}^{(1)} + \overline{\mathbf{x}}^{(2)}\right) \tag{5}$$

where λ is a positive regularization constant, I is a $p \times p$ identity matrix and k denotes a scalar which does not depend on $\mathbf{x}$, the vector to be classified. Vector $\mathbf{w}^{FR}$ is in fact a weight vector of the *regularized discriminant analysis* (see e.g. Friedman, 1989; McLachlan, 1992).

In (Raudys, 1995) it was indicated that in a case $N_1=N_2$ and with centred data in a way that $\overline{\mathbf{x}}^{(2)} = -\overline{\mathbf{x}}^{(1)}$, after the first back-propagation learning step in a batch mode of the SLP initialized with zero weights, one obtains the weights equivalent to the Euclidean distance classifier weight vector. On the way between the Euclidean distance classifier and the Fisher classifier we have regularized discriminant analysis.

An objective of the present paper is, in comparison of SLP training with the classical statistical discriminant analysis, to show that the *regularization of SLP by "weight decay" can be replaced by early stopping* in cost function (3) minimization case. We present here a more strict proof when in (Raudys, 1995) to demonstrate that in some conditions a SLP at the beginning of its training reacts like the Euclidean distance classifier and continuing its training until it recovers the functions of a Fisher classifier, it discriminates between two classes like in the regularized discriminant analysis case. We will show that *the criterion used to find weights of the SLP changes during the training process.*

2 Regularized discriminant analysis in SLP design

Consider the SLP's batch mode training rule where one uses a gradient of cost function (3) to make weight changes according to equation (4). Let's use an activation function of type $o(net) = \frac{e^{net} - e^{-net}}{e^{net} + e^{-net}}$. Suppose that $d_j^{(1)} = 1$, $d_j^{(2)} = -1$, data is centred in a way that $\bar{\mathbf{x}}^{(2)} = -\bar{\mathbf{x}}^{(1)}$ and $N_1 = N_2 = N$.

For very small initial weights $\mathbf{w}^T \mathbf{x}_j^{(i)} + w_0$ is close to zero, therefore $\frac{\partial o(\mathbf{x})}{\partial \mathbf{x}} = d\mathbf{x}$ and $o(net) = net$. With a condition $\bar{\mathbf{x}}^{(2)} = -\bar{\mathbf{x}}^{(1)}$ and $N_1 = N_2$ we have $\frac{\partial cost_{(t)}}{\partial w_{0(t)}} = w_{0(t)}$, so we will interesting only in the weights $\mathbf{w}$ evolution during the training.

From (3) we have

$$\frac{\partial cost_{(t)}}{\partial \mathbf{w}_{(t)}} = \frac{1}{2N} \sum_{i=1}^{2} \sum_{j=1}^{N} \left(d_j^{(i)} - \mathbf{w}_{(t)}^T \mathbf{x}_j^{(i)} \right) \left(-\mathbf{x}_j^{(i)} \right) = -\frac{1}{2} \Delta\bar{\mathbf{x}} + \frac{1}{2} K \mathbf{w}_{(t)} \tag{6}$$

where $\Delta\bar{\mathbf{x}} = \bar{\mathbf{x}}^{(1)} - \bar{\mathbf{x}}^{(2)}$ and $K = \frac{1}{N} \sum_{i=1}^{2} \sum_{j=1}^{N} \mathbf{x}_j^{(i)} \left(\mathbf{x}_j^{(i)} \right)^T$.

When the prior weights are very small one may assume $\mathbf{w}_{(0)} = 0$. Then after the first learning iteration for the weight vector $\mathbf{w}$ one obtains

$$\mathbf{w}_{(1)} = \mathbf{w}_{(0)} - \eta \frac{\partial cost_{(0)}}{\partial \mathbf{w}_{(0)}} = \frac{\eta}{2} \left(\bar{\mathbf{x}}^{(1)} - \bar{\mathbf{x}}^{(2)} \right) \tag{7}$$

It is the weight vector of the Euclidean distance classifier designed for centred data.

Now we will analyse the changes of the weight vector in the second and following iterations.

The use of total gradient adaptation rule (4) with (7) and gradient (6) results in:

$$\mathbf{w}_{(2)} = \mathbf{w}_{(1)} - \eta \frac{\partial cost_{(1)}}{\partial \mathbf{w}_{(1)}} = \frac{\eta}{2} \Delta\bar{\mathbf{x}} - \eta \left(-\frac{1}{2} \Delta\bar{\mathbf{x}} + \frac{1}{2} K \mathbf{w}_{(1)} \right)$$

$$= \left(I - \left(I - \frac{\eta}{2} K \right)^2 \right) K^{-1} \Delta\bar{\mathbf{x}}$$

and further,

$$\mathbf{w}_{(t)} = \left(I - \left(I - \frac{\eta}{2} K \right)^t \right) K^{-1} \Delta\bar{\mathbf{x}} \tag{8}$$

where K was defined above. By definition matrix K is not singular, so it has an inverse. The use of the first terms of the expansion $\left(I - \frac{\eta}{2} K \right)^t = I - t \frac{\eta}{2} K + \frac{1}{2} t(t-1) \left(\frac{\eta}{2} \right)^2 K^2 - \ldots$ in (8) for small η and t results in

$$\mathbf{w}_{(t)} \approx \left(t \frac{\eta}{2} K - \frac{1}{2} t(t-1) \left(\frac{\eta}{2} \right)^2 K^2 \right) K^{-1} \Delta\bar{\mathbf{x}}$$

$$= t \frac{\eta}{2} \left(I - \frac{1}{2} (t-1) \frac{\eta}{2} K \right) \Delta\bar{\mathbf{x}}.$$

Further, the use of the first terms of the expansion $(I - \beta K)^{-1} = I + \beta K + \ldots$, with supposition that η and t are small, gives

$$\mathbf{w}_{(t)} = t\frac{\eta}{2}\left(I + \frac{1}{2}(t-1)\frac{\eta}{2}K\right)^{-1}\Delta\overline{\mathbf{x}}$$

$$= t\frac{\eta}{2}\left(I + (t-1)\frac{\eta}{2}\left(\frac{N-1}{N}S + \frac{1}{4}\Delta\overline{\mathbf{x}}\Delta\overline{\mathbf{x}}^T\right)\right)^{-1}\Delta\overline{\mathbf{x}}. \tag{9}$$

Assuming matrix $I\lambda + S$ is not singular, after some matrix algebra (see e.g. the use of Bartlett formula in Raudys *et al.*, 1994) from (9) we obtain

$$\mathbf{w}_{(t)} = \left(I\frac{2}{(t-1)\eta}\frac{N}{(N-1)} + S\right)^{-1}\Delta\overline{\mathbf{x}}\frac{tN}{(t-1)(N-1)}k. \tag{10}$$

Weight vector $\mathbf{w}_{(t)}$ in (10) is equivalent to the weight vector $\mathbf{w}^{FR}$ of *regularized discriminant analysis* (5) with regularization parameter $\lambda = \frac{2N}{(t-1)\eta(N-1)}$; k is a same scalar as in (5). Equation (10) indicates explicitly that with an increase in t, the number of iterations, the weight vector $\mathbf{w}_{(t)}$ moves from an Euclidean distance classifier (when $t=1$, see eq.(7)) to a Fisher classifier, since $\lambda \to 0$ when $t \to \infty$.

3 Regularization and classifier complexity

In statistical pattern recognition it is well known that the sensitivity of a classifier to the learning sample size depends on the complexity of the classifier (Raudys, 1970). For example, in the case of classification of individuals into one of two multivariate Gaussian classes with different means μ_1, μ_2 and sharing common covariance matrix Σ it is known that for small learning sets it is preferable to use the simple structured Euclidean distance classifier while for larger sets it is better to use the complex structured Fisher classifier (Raudys, 1970; Jain & Chandrasekaran, 1982; Raudys & Jain, 1991).

The use of the regularized discriminant analysis (5) offers a great number of other algorithms. They differ in λ, the optimal value of which depends on N, the learning sample size: λ decreases as N increases (see e.g. Raudys *et al.*, 1994).

In our simulations we have used two class 20-variate Gaussian centred data with unit variance for all variables, correlation between all the variables was $\rho=0.1495$, and $(\mu_1-\mu_2)^T=(0.0125, 0.0778,...,1.1867, 1.2519)^T$.

In Fig.(a) we present the curves of the generalization errors versus sample size calculated theoretically for three classifiers (Euclidean distance, Fisher and regularized discriminant analysis), performed by 120 repetitions of the experiment with different randomly chosen learning sets of size $12 \le N \le 80$. Regularized discriminant analysis was performed with the optimal value of the regularization parameter for each individual learning set. As we see in Fig.(a) it is evident that in finite learning-set cases the regularized discriminant analysis with optimal λ (curve-3 in Fig.(a)) outperforms the both here used statistical classifiers.

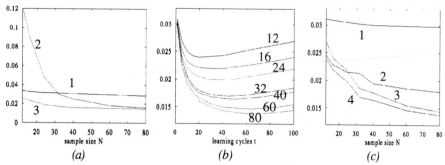

Figures: generalization error versus learning set size in (a),(c) and versus the number of learning iteration in (b). (a): (1) - Euclidean dist.classifier, (2) - Fisher classifier, (3) - regularized discr.analysis. (b): SLP training with N=12,16,24,32,40,60,80. (c): (1) - SLP after one training it. (⇔ Euclidean dist. classif.), (2) - SLP after 13 training it., (3) - SLP after 100 training it. and (4) - SLP after opt.number of training it. to obtain min.of EP_N

4 Simulation with the single layer perceptron

Numerous simulation experiments (see e.g. previous section, as well as Friedman, 1988; McLachlan, 1992) indicate that regularized discriminant analysis with optimal value of the smoothing parameters outperforms other parametric statistical classification rules. This is absolutely true when the pattern classes are Gaussian and one uses statistical classifiers designed for Gaussian models. For real non-Gaussian data, however, the classical parametric classifiers can fail to obtain good generalization. In this sense "nonparametric" classifiers based on single-layer and multi-layer perceptrons can appear to be better candidates. Experiments with feedforward ANN confirm this. Moreover, optimaly regularized networks usually help to reduce generalization error.

In section 2 it was shown that, in the sense of classical pattern recognition, while training the SLP, on the way between the Euclidean distance classifier and the Fisher classifier one discovers regularized discriminant analysis. Therefore one can expect that the generalization error of the optimally stopped SLP be lower than that at the very beginning and at the end of the iterative learning process. In Fig.(b) we present the evolution of the generalization error of SLP during learning for seven different learning-set sizes. Each of these curves is a mean value of 120 experiment with 120 different randomly chosen learning sets. We use the same data as in previous section. We initialized all weights to zero and used learning step $\eta=1$. For all graphs we found a minimum which depends on the sample size N: it increases with N.

In Fig.(c) we present the generalization error of the SLP versus the sample size obtained after a different number of learning iterations. Here we observe the same effect which is well known in statistical pattern recognition: for small sample sizes one needs to use simple structured classification rules (curve-2 in Fig.(c)) and only for large sample sizes one can use the complex ones (curve-3). The use an optimal number of learning iterations individually for each sample size (curve-4) results in the smallest generalization error which is very close to the generalization error of the regularized discriminant analysis with value λ_{opt} (curve 3 in Fig.(a)).

The above observations agree fairly well with the theory discussed in the section 2: *the number of learning iterations plays a role of the regularization parameter and thus helps reduce the generalization error.*

5 Conclusions

Regularization of statistical classifiers and neural networks is a powerful tool which helps to obtain the classifier of optimal complexity and to reduce the generalization error. The number of learning steps in the single-layer perceptron training as well as the step size play a role of the regularization parameter. The influence of these parameters however can be compensated by other regularization methods such as addition of a "weight decay" regularization term to the cost function, a change in the target value, noise injection, etc.

Thus one regularization method can be compensated by another one. The optimal values of the regularization parameters (including the number of learning iterations) depend on the data (the pattern classification problem to be solved), the complexity of the classifier (the number of inputs in the SLP design), and, it is very important to stress, the learning-set size. In the experimental results reported in this paper we have presented only mean values obtained from 120 repetitions of the experiments. It is worth to mention that in individual learning sets we observe significant deviations of the results and the optimal values of the regularization parameters (here, the number of learning iterations) varies with each individual learning set.

References

Fisher R.A.(1936). The use of multiple measurements in taxonomic problems. *Ann.of Eugenics*, Vol.7, No.2, London, 179-188.Record, Part 4, 96-104.

Friedman J.M.(1989). Regularized discriminant analysis. *J.American Statistical Association*, Vol.84, 165-175.

Gallinari P., Thyria S., Badran F., Fogelman-Soulie F.(1991). On relations between discriminant analysis and multilayer perceptrons. *Neural Networks*, Vol.4, 349-360.

Jain A., Chandrasekaran B.(1982). Dimensionality and size considerations in pattern recognition practice. *Handbook of Statistics*, Vol.2, 835-855, North Holland.

Koford J.S., Groner G.F.(1966). The use of an adaptive threshold element to design a linear optimal pattern classifier. *IEEE Trans.on Inf. Theory*, Vol.IT-12, 42-50.

McLachlan G.J.(1992). *Discriminant Analysis and Statistical Pattern Recognition*, Willey.

Raudys S.(1970). On the problems of sample size in pattern recognition. Proc. *2nd All-Union Conf.Statist.Methods in Control Theory*, Moscow, Nauka, 64-67 (in Russian).

Raudys S.(1995). A negative weight decay or antiregularization. Proc.of *ICANN'95*, Paris, Vol.2., 449-454.

Raudys S., Jain A.K.(1991). Small sample size effects in statistical pattern recognition: Recommendations for practitioners. *IEEE Trans.on Pattern Analysis and Machine Intelligence*, Vol.13, 252-264.

Raudys S., Skurichina M., Cibas T., Gallinari P.(1994). Optimal regularization of linear and nonlinear perceptrons. Proc.*Int.Conf.Neural Networks and Applications*, Marseilles.

Clustering in Weight Space of Feedforward Nets

Stefan M. Rüger and Arnfried Ossen

Informatik, Sekr. FR 5-9
Technische Universität Berlin
Franklinstr. 28/29, 10 587 Berlin, Germany

Abstract. We study symmetries of feedforward networks in terms of their corresponding groups and find that these groups naturally act on and partition weight space. We specify an algorithm to generate representative weight vectors in a specific fundamental domain. The analysis of the metric structure of the fundamental domain enables us to use the *location* information of weight vector estimates, e. g. for cluster analysis. This can be implemented efficiently even for large networks.

1 Introduction

A successful analysis of the weight space of feedforward neural networks can help to solve a number of problems in neural network research: assessment of uncertainty of network parameters and outputs, the interpretation of learning results, visualization of possible network parameter regions.

The analysis is difficult because of the non-trivial structure of the weight space. One problem is redundancy: the interchange of two hidden nodes within a hidden layer does not change the network function (other less obvious symmetries may still exist). Thus, two weight vectors with a large Euclidean distance can implement exactly the same network function.

Our goal was to develop a canonical metric — which reflects the similarity of network functions — in a non-redundant fundamental domain of the weight space. Although the computation of this metric turns out to be intractable in the number of hidden nodes, we were able to reduce the problem to the Traveling-Salesman Problem, for which efficient and sufficiently precise algorithms exist.

It is then feasible to apply, e. g., cluster algorithms to approach the problems mentioned above, see Section 3.

Throughout this article, we will consider neural feedforward networks. The nodes of the network are addressed by numbers 0, 1, 2, etc., 0 being the name of the bias node. All non-bias nodes are divided into $k \geq 1$ hidden layers $L_1, \ldots, L_k$, one input layer L_o and one output layer L_{k+1}. Every layer is fully connected to the next layer, which means that there is a weight w_{ab} assigned to every pair (a, b) of nodes from $L_i \times L_{i+1}$, $i = 0, 1, \ldots, k$. Hidden nodes and output nodes have a bias weight termed w_{oa}. All weights of the network form a weight vector $w \in \mathbb{R}^E$ that parameterizes a network function out_w.

Each activation function f of a hidden layer node is assumed to be sigmoidal; we require that every f exhibit the same type of symmetry $f(x) = e - f(-x)$, where, e. g., $e = 1$ for the logistic function or $e = 0$ for $f = \tanh$. This class of networks is quite universal: choosing the activation functions of the output layer as identity and using one hidden layer makes this type of network an universal approximator [6].

2 Metric Structure of Weight Space Owing to Symmetries

2.1 Symmetries

A symmetry of weight space $\mathbb{R}^E$ is a transformation $t\colon \mathbb{R}^E \to \mathbb{R}^E$ that leaves the network function invariant, i.e., $\mathrm{out}_w = \mathrm{out}_{t(w)}$. There have been several attempts to analyze the symmetries of the network function [5, 1]. We summarize an independent approach [4], in which the arising symmetries are described and analyzed in terms of their groups.

A permutation of the nodes in a hidden layer L_i induces a certain linear operation π_i on the weight space $\mathbb{R}^E$, which leaves the network function invariant. Thus π_i is an element of the group $\mathrm{GL}(E, \mathbb{R})$ of all bijective linear functions $\mathbb{R}^E \to \mathbb{R}^E$, the group operation being function composition. For a given layer L_i all possible π_i generate a group Π_i, i.e., the smallest subgroup of $\mathrm{GL}(E, \mathbb{R})$, which contains those π_i. Π_i turns out to be isomorphic to the permutation group of L_i.

Let $\tilde{w}(b, i)$ be the subvector of $w \in \mathbb{R}^E$ containing all weights that involve the hidden node b of layer L_i: $\tilde{w}(b, i) = (w_{ob}, w_{a_1 b}, w_{a_2 b}, \dots, w_{bc_1}, w_{bc_2}, \dots)$. a and c are enumerations of the nodes in the layer L_{i-1} and L_{i+1}, respectively. From the symmetry $f(x) = e - f(-x)$ of the activation functions it follows that the flipping of all signs of the components in the subvector $\tilde{w}(b, i)$ can be corrected by changing all bias weights of the nodes of the layer L_{i+1}, i.e., by $w_{oc_i} \mapsto w_{oc_i} + e \cdot w_{bc_i}$. Let $t_b\colon \mathbb{R}^E \to \mathbb{R}^E$ denote this linear operation, which leaves out_w invariant. Let T be the group generated by all t_b.

Let $S \subset \mathrm{GL}(E, \mathbb{R})$ denote the group generated by $\Pi_1, \dots, \Pi_k$ and T. One important result about the symmetry group S is that each element $s \in S$ has a *unique* representation $s = \pi_k \dots \pi_1 t$ with $\pi_i \in \Pi_i$ and $t \in T$. So far the symmetry group S of the weight space has been identified as a certain subgroup of $\mathrm{GL}(E, \mathbb{R})$.

As pointed out by [1] no analytic function other than an element of S can represent a symmetry in this context. However, there exist a lot of discontinuous functions that give rise to a symmetry: Fix two hidden nodes a, b from the same hidden layer. If the incoming weights of a coincide with the incoming weights of b, then all corresponding two outgoing weights might be replaced by their average value. These kinds of symmetries are probably not important in practice as they live in hyperplanes of $\mathbb{R}^E$ with zero Lebesgue measure. We therefore exclude them from our studies.

2.2 A Fundamental Domain

S acts on $\mathbb{R}^E$ in a natural way: distinct orbits $S(w) := \{x \in \mathbb{R}^E \mid x = s(w)$ for some $s \in S\}$ (with respect to the natural group action) partition $\mathbb{R}^E$. There is an interesting open (and even convex) set W of weight vectors which contains at most one representative of every orbit such that $S(W)$ is dense in $\mathbb{R}^E$. Such a remarkable set is called fundamental domain and may be constructed as follows:

Let $a_1^i, \dots, a_{|L_i|}^i$ denote the nodes of hidden layer L_i. Then,

$$W := \left\{ w \in \mathbb{R}^E \mid 0 < w_{oa_1^i} < \dots < w_{oa_{|L_i|}^i} \text{ for all } 1 \le i \le k \right\}$$

is a fundamental domain (note that W is a cone with apex 0, i.e., $cw \in W$ for all $w \in W$ and $c > 0$).

Sketch of a proof. Applying T on W allows each bias weight of a hidden node to change its sign, and applying $\Pi_1, \ldots, \Pi_k$ creates all possible orders of the bias weights within a hidden layer, thus removing any restriction given to specify W except the condition that no bias weight may vanish and except that no two bias weights of a hidden layer may coincide. Taking the closure removes these conditions leaving $\mathbb{R}^E$.

Hence, it suffices to optimize (estimate, learn) in W instead of the much larger space $\mathbb{R}^E$. Indeed, the idea of a fundamental domain is to define a convenient non-redundant subset of $\mathbb{R}^E$ with respect to S.

2.3 Algorithm for Representative Vectors

Let $\hat{w}$ denote a weight vector resulting from some learning algorithm. Beginning with layer L_1, apply the symmetry operation t_a of Section 2.1 whenever a bias weight of a hidden node a is negative until all hidden nodes have nonnegative bias weights. Then, for every hidden layer, apply permutation symmetry operations by interchanging subvectors of nodes in this layer such that the bias weights of each layer are in a definite order — say, ascending — thus arriving at a representative weight vector $r(\hat{w})$. Essentially, this is a simple sorting problem.

The function $r: \mathbb{R}^E \to \mathbb{R}^E$ as implemented by the above algorithm maps onto a region $W' \supset W$. The difference $W' \setminus W$ is caused by dealing with weight vectors that have vanishing or coinciding bias weights within a hidden layer. Fortunately, $W' \setminus W$ has zero Lebesgue measure.

2.4 W as a Metric Space

In W the Euclidean distance is no longer a canonical metric: two points $v \notin W$ and $w \in W$ might be very close in $\mathbb{R}^E$, their representatives $r(v)$ and $w = r(w)$ being separated. Let d_E be a metric in $\mathbb{R}^E$. One idea assigning a distance to two points in W is using the minimal distance d_E in $\mathbb{R}^E$ of two points of their corresponding orbits. It turns out that the compatibility of d_E with the symmetry group, i.e., for all $s \in S$ it holds that $d_E(sx, sy) = d_E(x, y)$, is a quite important prerequisite. If a metric d_E in $\mathbb{R}^E$ is compatible with S, then

$$d(x, y) := \min_{s \in S} d_E(sx, y)$$

is a canonical metric in W.

The Euclidean metric is compatible with S provided that the activation functions of the hidden layers exhibit the symmetry $f(x) = -f(-x)$, as is the case for $f = \tanh$: all points of the orbit of a weight vector w lie on a sphere with radius $|w|$. Using the logistic function as the activation function, a sign flip of a subvector causes the bias corrections to change the Euclidean norm of the equivalent vector. Fortunately, it is easy to construct a peculiar norm that distorts the unit sphere in a way that all points of the orbit of a certain vector lie on the same sphere. The metric induced by this norm is then compatible with S.

Having identified a metric that is compatible with S, the calculation of the corresponding metric d in W with the above formula is intractable for a large number of hidden units: Consider a network with one hidden layer using tanh as activation function; let $d_E(x, y) = |x - y|_1$ be the Manhattan distance. Then

$$d(v, w) - \sum_{c \in L_2} |w_{oc} - v_{oc}| = \min_{\pi} \sum_{a \in L_1} K(a, \pi(a))$$

has exactly the structure of the general Traveling-Salesman Problem with a non-symmetric cost function $K(a, b) = \min(|\tilde{w}(a, 1) - \tilde{v}(b, 1)|_1, |\tilde{w}(a, 1) + \tilde{v}(b, 1)|_1)$. Thus, it is possible to exploit the rich literature that deals with efficient approximations of a minimal tour in order to approximate d in large networks.

3 Clustering in Weight Space and an Application

Feedforward networks can be interpreted as a form of nonlinear regression. They offer great flexibility at the price of a complicated structure. However, the usual gradient-based parameter estimation, or, in the language of neural networks, learning procedures, may get stuck in local extrema. We propose applying a clustering algorithm for the weight vectors in the fundamental domain using its canonical metric in order to obtain several clusters of weight vectors.

3.1 Learning and Error Bars Using a Statistical Context

The data are generated using a "true" model, i.e., a neural network with a certain fixed weight vector w°. The actual data are modeled using

$$y_i = \text{out}_{w^\circ}(x_i) + \varepsilon_i,$$

where the "measurement" errors ε_i are independent identically distributed. The training multiset D is a set $\{(x_1, y_1), \ldots, (x_n, y_n)\}$ of n such data pairs.

Given the distribution of the random variable ε_i and a certain weight vector w, every data pair has a probability density and the joint density $D \mapsto L_D(w)$ factorizes by the stochastic independence of the errors. The function $w \mapsto L_D(w)$, a.k.a. likelihood, describes how likely the data have been generated by w.

Under quite general assumptions, the weight vector $\hat{w}_n$ that maximizes the likelihood is asymptotically (w.r.t. n) unbiased, consistent, asymptotically efficient and asymptotically normal-distributed [6].

The bootstrap method [3] is based on re-estimations of the parameter vector on B bootstrap samples of the training set. The bth bootstrap sample is a random multiset $D^{*b} = \{(x_1^{*b}, y_1^{*b}), \ldots, (x_n^{*b}, y_n^{*b})\}$ drawn from the training data *with replacement*. The standard error of a predicted value $\text{out}_{\hat{w}}(x)$ is approximately given by

$$\sqrt{\frac{1}{B-1} \sum_{b=1}^{B} \left(\text{out}_{\hat{w}^{*b}}(x) - \frac{1}{B} \sum_{b=1}^{B} \text{out}_{\hat{w}^{*b}}(x) \right)^2}, \tag{1}$$

where $\hat{w}^{*b} = \underset{w}{\text{argsup}}(L_{D^{*b}}(w))$ is the estimated weight vector.

It is also possible to estimate confidence intervals using the bootstrap approach. Let $\text{out}^{\alpha}_{\hat{w}^*}(x)$ be the α quantile of the empirical distribution of $\text{out}_{\hat{w}^*}(x)$. The approximate $1 - 2\alpha$ confidence interval of a predicted value $\text{out}_{\hat{w}}(x)$ is

$$[\text{out}^{\alpha}_{\hat{w}^*}(x), \text{out}^{1-\alpha}_{\hat{w}^*}(x)].$$

3.2 An Application

For the sake of clarity, we demonstrate our techniques using a 1-3-1 network; a more advanced application may be found in [4]. A data set with 32 points was generated by a true model (dotted line in Fig. 3). Several weight vectors were obtained by different runs of a standard algorithm to maximize the likelihood of the data. We calculated a respective canonical distance matrix for a hierarchical clustering algorithm [2]. In contrast to the raw projection of weight vectors (Fig. 1a), we obtained several clusters (Fig. 1b) indicating the respective local minima of the error surface.

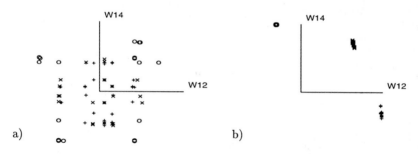

Fig. 1. Projection of several weight vectors after learning (a) and corresponding representatives of their orbits under S (b).

We then generated bootstrap weight vectors and respective clusters (see Figure 2). The *location* information of these cluster makes it possible to restrict the averaging in Eq. (1) to the cluster containing the best maximum likelihood weight vector (see Fig. 2). Figures 3a and 3b show the improvement of the error bars owing to the exclusion of local minima that were caused by deficiencies of our gradient-based learning algorithm.

We have to emphasize that our experiments have also shown that the likelihood values $L_{D^{*b}}(\hat{w}^{*b})$ cannot sufficiently detect those deficiencies.

4 Discussion

We have shown how the location information of weight vectors in a fundamental domain can be successfully exploited.

We observed that if we deliberately allowed for too much network flexibility, i.e., too many hidden nodes, the cluster structure collapses into less well detectable substructures. In our opinion that may potentially be used to assess, whether the chosen network structure is adequate. This issue and others are left for further studies.

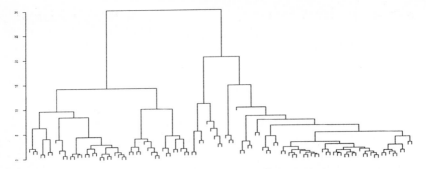

Fig. 2. Plot of hierarchical clustering (complete linkage method) of bootstrap estimates of network weight vectors: the height of a junction represents the distance of the subclusters. The entire left subtree corresponds to a global optimum.

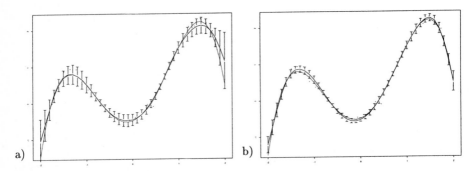

Fig. 3. Simple bootstrap error bar estimation (a) of the estimated network function and clustered bootstrap: bootstrap estimates confined to global optimum cluster (b).

References

1. An Mei Chen, Haw-minn Lu, and Robert Hecht-Nielsen. On the geometry of feedforward neural network error surfaces. *Neural Computation*, 5(6):910–927, 1993.
2. Richard O. Duda and Peter E. Hart. *Pattern Classification and Scene Analysis*. Wiley, 1973.
3. Bradley Efron and Robert J. Tibshirani. *An Introduction to the Bootstrap*. Chapman & Hall, 1993.
4. Arnfried Ossen and Stefan M. Rüger. Weight space analysis and forecast uncertainty. Submitted to *Journal of Forecasting*, 1996.
5. Héctor J. Sussmann. Uniqueness of the weights for minimal feedforward nets with a given input-output map. *Neural Networks*, 5:589–593, 1992.
6. Halbert White. *Artificial Neural Networks — Approximation & Learning Theory*. Blackwell, Oxford, Cambridge, 1992.

Learning Curves of *On–line* and *Off–line* Training

Siegfried Bös

Lab for Information Representation, RIKEN
Hirosawa 2–1, Wako–shi, Saitama, 351–01, Japan
Tel: +81–(0)48-, phone: -467–9625, fax: -462–9881
email: boes@zoo.riken.go.jp

Abstract. The performance of on–line training is compared with off–line or batch training using an unrealizable learning task. In naive off–line training this task shows a tendency to strong overfitting on the other hand its optimal training scheme is known. In the regime, where overfitting occurs, on–line training outperforms batch training quite easily. Asymptotically, off–line training is better but if the learning rate is chosen carefully on–line training remains competitive.

1 Introduction

On–line training attracted much attention in recent time, for up–to–date references see [4]. It is the main advantage of on–line training that it uses only the last example and does not retrain with the previous ones. Furthermore, theoretical examinations of on–line learning are easier. Very recent works like [5] claim that optimized on–line learning schemes yield the same asymptotical convergence rate as *off–line* or batch training.

On–line learning is very attractive, since it has several advantages. For each new example the weights are updated exactly once using only the newest example. There is no retraining with the older examples, therefore its computational expense is much lower compared to off–line learning. Also the storage of the older examples is not necessary, so a smaller storage capacity can be sufficient. Furthermore, it is less susceptible to overfitting, which is a serious problem in off–line training. But the performance of on–line learning depends crucially on the way how the learning rate is chosen. A good determination scheme for the learning rate can cost some of these advantages. It can increase the computation dramatically or it makes the storage of all examples necessary in order to determine the training error or similar measures.

In this paper we want to compare on–line and off–line training not only in the asymptotical regime but for any number of examples. It can happen that the asymptotical considerations are valid only for very high absolute numbers of examples.

2 The model system

The paper is based on a specific model which was introduced in previous works [2] and [3]. It is easy enough to allow a full analytical description, but it is still a characteristical example for learning in feedforward networks. The system consists of a single–layer perceptron, i.e. it has only one layer of adjustable weights W_i (with $i = 1, \ldots, N$) between input and output layer. To compute the output z a continuous function g is applied on the weighted sum of the inputs x_i, i.e.

$$z = g(h), \qquad \text{with} \qquad h := \frac{1}{\sqrt{N}} \sum_{i=1}^{N} W_i x_i. \tag{1}$$

Here we examine *supervised* learning, where a number of examples $\mathbf{x}^\mu$ is learnt, for which the correct output z_* is known. It makes theoretical examinations easier, if we assume furthermore that the examples are provided by another network, the so–called *teacher network*.

We use the mean squared error to measure the performance of the *student network*. Training attempts to minimize the training error, the error averaged over all examples, i.e. $E_T := < (z_*^\mu - z^\mu)^2 >_{\{\mathbf{x}^\mu\}}$. But actually we are interested in the typical performance over all possible inputs $\mathbf{x}$, which is given by the *generalization error*, i.e. $E_G := < (z_* - z)^2 >_{\{\mathbf{x} \in I\}}$.

The concept of the teacher network allows a convenient monitoring of the training process. Training is fully described by the *order parameters*,

$$Q := \frac{1}{N} \sum_{i=1}^{N} (W_i)^2 = q^2, \qquad R := \frac{1}{N} \sum_{i=1}^{N} W_i^* W_i = \frac{r}{q}. \tag{2}$$

With a '*' we indicate the variables belonging to the teacher network. In on–line learning the non–normalized order parameters, i.e. Q and R, are more convenient. But the normalized ones, i.e. q and r, have a more obvious meaning: q is the norm of the student's weights, and r is the cosine of the angle between the two weight vectors of teacher and student.

In this paper we compare the performance of on–line and off–line training for an unrealizable learning task. Both networks, teacher and student, are one–layer perceptrons and the unrealizability results from a different choice for the output functions, i.e. $g_*(h_*) = \tanh(\gamma h_*)$ and $g(h) = h$. The gain γ allows to tune $g_*(\gamma h_*)$ between a linear and a highly nonlinear function. It also takes the norm of the teacher weights into account, so that we can assume $||\mathbf{W}^*|| = 1$.

Then the generalization error can be expressed by the order parameters (2). We should make some remarks about this averaging process, since we will need it again in the calculation of the on–line order parameters. We assume that the inputs $\mathbf{x}$ are uniformly distributed random variables, with $< x_i > = 0$ and $< (x_i)^2 > = 1$, and random weights $\mathbf{W}$. Then the weighted sums, h and h_* from Eq.(1), are correlated Gaussians with the correlations, i.e. $< (h_*)^2 > = 1$, $< (h)^2 > = Q$ and $< h_* h > = R$. The averaging can be transformed in an integral over the uncorrelated Gaussians $\tilde{h}_*$ and $\tilde{h}$,

$$E_G(R, Q) = \frac{1}{2} \left\langle \left[g_*(\tilde{h}_*) - g(R\tilde{h}_* + \sqrt{Q - R^2}\, \tilde{h}) \right]^2 \right\rangle_{\tilde{h}_*, \tilde{h}}, \tag{3}$$

which reduces in the case of the linear student network, i.e. $g(h) = h$, to

$$E_{\mathrm{G}} = \frac{1}{2}(G - 2HR + Q), \quad \text{with} \quad < f(\tilde{h}) >_{\tilde{h}} := \int_{-\infty}^{\infty} \frac{d\tilde{h}}{\sqrt{2\pi}} \exp(-\frac{\tilde{h}^2}{2})f(\tilde{h}). \quad (4)$$

The two constants, $G(\gamma) :=< g_*^2(\gamma\tilde{h}_*) >_{\tilde{h}_*}$ and $H(\gamma) :=< g_*(\gamma\tilde{h}_*)\tilde{h}_* >_{\tilde{h}_*}$, express the dependence on the teacher. These constants can be defined for many learning tasks with a linear student. Therefore our theory is not restricted to this special unrealizable task. It can easily be extended to noisy teachers, different output functions and other tasks.

From the results on off–line training we just want to recall that a strong overfitting appears if training is continued until the absolute minimum of E_{T} is reached. If the training is stopped earlier, the overfitting can be avoided. Suitable schemes to avoid overfitting were discussed in [2] and [3].

3   On–line training

In on–line training each example is used exactly once to update all the weights. There is no retraining with the previous examples. Since we have continuous outputs we use gradient descent update. Following [1], we find

$$W_i^{\mu+1} = W_i^{\mu} + \frac{\eta}{\sqrt{N}}(z_*^{\mu} - z^{\mu})x_i^{\mu}, \quad (5)$$

in which the learning rate η determines the relative strength of the update. Inserted in the order parameter equations (2) this defines directly two difference equations for R and Q. These equations can be transformed in continuous equations, if we introduce the 'time' $\alpha = \mu/N$. Then we have to average over the realizations μ,

$$\frac{dR}{d\alpha} = \eta < (z_* - z)h_* >, \quad \frac{dQ}{d\alpha} = 2\eta < (z_* - z)h > +\eta^2 < (z_* - z)^2 > . \quad (6)$$

This involves the above mentioned averaging. As result we find the following differential equations,

$$\dot{R} + \eta R = \eta H, \quad \dot{Q} + \eta(2 - \eta)Q = 2\eta(1 - \eta)HR + \eta^2 G. \quad (7)$$

Their solution, with the initial conditions $R(0) = 0$ and $Q(0) = 0$, is,

$$R(\alpha; \eta) = (1 - e^{-\eta\alpha})H, \quad (8)$$

$$Q(\alpha; \eta) = \frac{2H^2 - \eta G}{2 - \eta}e^{-\eta(2-\eta)\alpha} - 2H^2 e^{-\eta\alpha} + \frac{2(1 - \eta)H^2 + \eta G}{2 - \eta},$$

with $\eta \in [0, 2]$. Inserted in the generalization error (4) we receive

$$E_{\mathrm{G}}(\alpha; \eta) = \frac{G - H^2}{2 - \eta} + \frac{2H^2 - \eta G}{2(2 - \eta)}e^{-\eta(2-\eta)\alpha}. \quad (9)$$

In the limit $\alpha \to \infty$ only the first term does not vanish. In order to reach the absolute minimum in this learning task, which is $E_{\mathrm{inf}} := (G - H^2)/2$, the learning rate η has to vanish with $\alpha \to \infty$.

What can we learn from this result? First, we can plot the generalization error as a function of the number of learnt examples, i.e. $E_G(\alpha)$, for fixed learning rates η as shown in Fig. 1. If the learning rate is higher than $2H^2/G$ the generalization error will increase. Obviously, there is no choice for η that is optimal over the whole range of α. The envelope, which resembles the optimal choice for each α, is also shown in Fig. 1 (solid line). It is already quite near to the result of the optimal off–line training, which is given by the dotted lines.

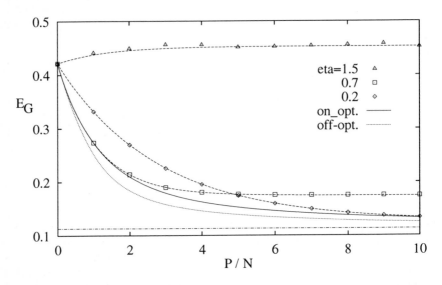

Fig. 1. The generalization error as a function of the number of examples for optimal off–line training (dotted line) and on–line training (dashed lines). Different fixed learning rates η were used in on–line training, i.e. $\eta = 1.5, 0.7, 0.2$ (on the right side from the top). Also the envelope, which is defined as the minimum over all η at each $\alpha = P/N$, is shown as a solid line. The points represent simulation results for a system with $N = 200$ weights averaged over 100 different runs. (gain $\gamma = 5$).

Now the question arises, how the optimal learning rate can be determined. Therefore, we plot the generalization error as a function of the learning rate, i.e. $E_G(\eta)$, for different fixed values of α, see Fig. 2. Here we can see clearly that there is a value of η, which minimizes E_G. Also the training error E_T should have a minimum. From the simulations — which correspond to the theory very well — the training error $E_T(\eta)$ can be determined. The difference between the two optimal learning rates, which minimize E_G respectively E_T, becomes small with increasing α. This is not surprising in the asymptotical limit ($\alpha \to \infty$), but it is fulfilled already for quite small α. Since lack of space we can not present

$\eta_{opt}(\alpha)$ in an extra figure, but illustrate the parameter α by stars on the solid lines in Fig. 2. The solid lines connect the positions of the minima of $E_G(\eta)$ respectively $E_T(\eta)$. The stars are plotted at certain α values. A comparison of the corresponding stars on the E_G–line and the E_T–line shows that the optimal η–values become identical with increasing α. So the optimal learning rate can be determined from the minimum of the training error.

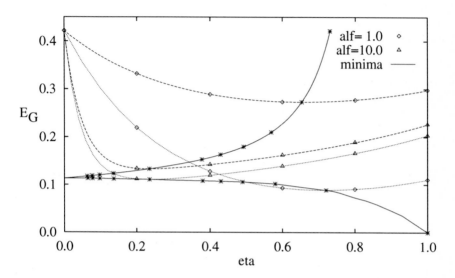

Fig. 2. The generalization error E_G (dashed lines) and the training error E_T (dotted lines) as a function of the learning rate η. For two examples, $\alpha = 1.0$ and 10.0, theory (lines) and simulations (squares and triangles) compared. Both errors show a minimum. The location of these minima are connected by the solid lines corresponding to E_G respectively to E_T. The stars indicate the location of the minimum for $\alpha = 0, 1, 2, 3, 4, 5, 10, 20, 30, 40$ and 50. Already for $\alpha \geq 3$ the optimal learning rate is more or less the same for both errors. (Parameters: $N = 200$, 100 runs, gain $\gamma = 5.0$)

Finally, we want to compare the asymptotical behavior of the optimal on–line training and the off–line training. The optimal learning rate has to be determined numerically. Asymptotically we can find an analytical expression,

$$\eta^{opt}(\alpha \gg 1) = \frac{\log \alpha + c}{2\alpha}, \qquad E_G^{opt}(\alpha \gg 1) = E_{inf}\left(\frac{\log \alpha + c + 1}{4\alpha - \log \alpha - c}\right), \quad (10)$$

with $c := \log(4H^2) - \log(G - H^2)$. In Fig. 3 these two approximations are compared with the numerical solution.

So far the asymptotical behavior of the on–line training is not as good as the α^{-1}–behavior of the off–line training. A recent publication [5] claims that on–line training can asymptotically be as efficient as off–line training if the learning rate is adapted during the training process. These question will be addressed in a future work.

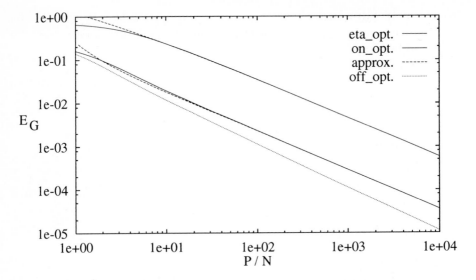

Fig. 3. Asymptotical behavior of on–line training with an optimized fixed learning rate. The two solid lines are the optimal learning rate η^{opt} and the corresponding generalization error $E_{\mathrm{G}}^{\mathrm{opt}}$. The approximations, equation (10), are given by the dashed lines. Also the optimized off–line training curve is plotted as a dotted line.

4 Summary

In this short paper an on–line learning scheme with fixed learning rates η was studied. The learning rate can be optimized if most of the examples are available in order to determine the training error E_{T}. Then the performance of the on–line training is competitive compared to off–line training, but the necessary computation is much lower.

Acknowledgment: We thank Shun-ichi Amari for useful discussions, and Herbert Wiklicky for comments on the manuscript.

References

1. M. Biehl, and H. Schwarze (1995), "Learning by on–line gradient descent", *J. Phys. A* **28** p.643–656.
2. S. Bös (1995), "Avoiding overfitting by finite temperature learning and cross–validation", in *ICANN'95*, edited by EC2 & Cie, Vol.2, p.111–116.
3. S. Bös (1996), "Optimal weight decay in a perceptron", this volume.
4. D. Saad, and S. Solla (1995), *Workshop at NIPS'95*, see World–Wide–Web page: http://neural-server.aston.ac.uk/nips95/workshop.html and references therein.
5. J.W. Kim, and H. Sompolinsky (1995) "On–line Gibbs–learning", preprint.

Learning Structure with Many-Take-All Networks

D.Tax and H.J.Kappen

RWCP*** Novel Functions SNN[†] Laboratory
Dept. of Medical Physics and Biophysics, University of Nijmegen
Geert Grooteplein 21, NL 6525 EZ Nijmegen, The Netherlands

Abstract. It is shown that by restricting the number of active neurons in a layer of a Boltzmann machine, a sparse distributed coding of the input data can be learned. Unlike Winner-Take-All, this coding reveals the distance structure in the training data and thus introduces proximity in the learned code. Analogous to the normal Radial Basis Boltzmann Machine, the network uses an annealing schedule to avoid local minima. The annealing is terminated when generalization performance deteriorates. It shows symmetry breaking and a critical temperature, depending on the data distribution and the number of winners. The learned structure is independent of the details of the architecture if the number of neurons and the number of active neurons are chosen sufficiently large.

1 Introduction

In biological systems it is assumed that objects in the outside world are encoded by a sparse distributed code where the input data is encoded by several feature vectors, one for each active neuron in the code. The important advantage of this sparse coding is the ability of obtaining knowledge about the underlying structure of received data by the coding of the objects. When two objects share an active neuron in their coding, they share a property encoded by the feature vector belonging to the neuron. By the number of active neurons that overlap, the distances in a high dimensional pattern space of the these objects can be found. Thus the complete topology of places of objects and distances between objects in pattern space can be deduced. Koenderink [4] showed that this can be done in biological systems by fully using the modalities (and the cohesion within the modalities) of the perceptual data the biological brain perceives.

In 1994 Kappen [2] introduced lateral inhibition in a Boltzmann machine to reduce the computational costs of executing this Boltzmann machine. By allowing a restricted number of neurons to become active, the Boltzmann Machine encodes all objects by a sparse distributed code. In this paper we will show that with the use of this architecture, called a Many-Take-All network, the extra distance information from the input data can be encoded by a set of neurons.

*** Real World Computing Partnership
† Foundation for Neural Networks

In Section 2 we will start with a general introduction of a Boltzmann machine with a restricted number of winners. In Section 3 the annealing schedule and symmetry breaking will be explained. We will show in Section 4 that in contrast with the Winner-Take-All network a Many-Take-All network can learn all topological information of data which can easily be inspected by looking at the coding In addition we show that in high dimensional data spaces the Many-Take-All network can offer a more compact representation.

2 The restricted Boltzmann machine

The Boltzmann Machine we will consider consists of a set of neurons $\mathbf{x} = (x_1, ...x_n)$, $x_i \in \mathbf{R}$, and a set of hidden neurons $\mathbf{s} = (s_1, ..., s_h)$, $s_j \in [0, 1]$. We call the connections between $\mathbf{x}$ and $\mathbf{s}$ w_{ij}. We will also use thresholds in the hidden layer, which will be called θ_j. By presenting training patterns $\mathbf{x}$ this Boltzmann machine can learn the probabilities associated with these training patterns (the probability of pattern $\mathbf{x}$ will be called $q(\mathbf{x})$). During the training of this Boltzmann machine a partition function has to be calculated. The number of terms in this partition function depends in an exponential manner on the size of the network so it is very time consuming to train this network.

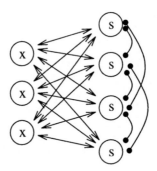

Fig. 1. The architecture of a Radial Basis Boltzmann machine. $\mathbf{x}$ are input neurons and $\mathbf{s}$ are hidden neurons. The connections with circles are inhibitory weights, the connections with arrows are learnable.

We can, as shown by Kappen [2], introduce lateral inhibition in the hidden layer of a Boltzmann machine. Then the number of permissible states in the network and so the number of terms in the partition function will be reduced. This results in a serious reduction of the required training time. We add an threshold $J(2h_0 - 1)$ to each neuron, where $J > 0$ the strength of the lateral inhibition (see Figure 1). If h_0 neurons are permitted to be 'on', then the local field of hidden unit j becomes

$$-\beta \sum_{j=1}^{h} \|\mathbf{w}_j - \mathbf{x}\|^2 s_j + \sum_{j=1}^{h_0} \theta_j s_j - J \left(\sum_{j' \neq j}^{h} s_j - 2h_0 + 1 \right). \tag{1}$$

In the limit of an infinite strong lateral inhibition the $p(\mathbf{x})$ reduces to:

$$p(\mathbf{x}) = \frac{1}{Z} \sum_{(j_1 \cdots j_{h_0})} \exp \left\{ \sum_{\alpha=1}^{h_0} \left[-\beta \|\mathbf{w}_{j_\alpha} - \mathbf{x}\|^2 + \theta_{j_\alpha} \right] \right\} \tag{2}$$

Here, $\sum_{(j_1 \cdots j_{h_0})}$ means we have a summation over all combinations of h_0 hidden units out of h hidden units (so this summation will consist of $\binom{h}{h_0}$ terms). The factor $\exp(Jh_0^2)$ has disappeared because of the normalization.

Now the learning rules for the Many-Take-All network can be derived by performing a gradient descent on the Kullback divergence K between the target probability density $q(\mathbf{x})$ and the network probability density $p(\mathbf{x})$ by variating the weights w_{ij}. The complexity of the learning rule for this Boltzmann Machine is by the introduction of lateral inhibition diminished from a 2^h dependency to a $\binom{h}{h_0}$.

3   Annealing schedule and symmetry breaking

By applying gradient descent on the Kullback divergence one hopes to reach the global minimum of K. Unfortunately, this learning rule often causes the network to converge not to the global, but to a local minimum. To prevent this and as discussed in Kappen, Nijman [3], an annealing schedule can be used, where β plays the role of inverse temperature. Here one starts with random weights at a high temperature (small β). Then repeatedly the network is trained until the weights are converged and the temperature is lowered (or β is increased). The procedure is repeated until the Kullback divergence on an independent test set of patterns is at a minimum and starts to increase. Then the best modeling of the data is achieved. A check on the test set in contrast to a check on the train set is used to avoid overfitting on the training patterns. When the network is trained in this way, local minima can be avoided effectively, as was shown for $h_0 = 1$ in [2].

When β is increased, the weight vectors will shift, and thus specialize, from the average of the big cluster to the average of a certain sub cluster. These shifts do not occur smoothly, but there will be several symmetry breakings at critical temperatures. This mechanism of symmetry breaking resembles the clustering in statistical mechanics (see for instance Rose et al. in [7]) and is seen before in the RBBM (Kappen, [2]). The critical β for the first symmetry breaking can be calculated.

To do this calculation we will take the average of all patterns in the origin. Thus $w_{ij} = 0$ for small β. At some $\beta > 0$ this solution gets unstable and we will get the first phase transition. We expand the learning rule for small β. Assuming that the weights will stay small and that $h \gg h_0$, we can derive to lowest order in w_{ij}:

$$(I - 2\beta h_0 C_{xx})\mathbf{w}_j = 0 \tag{3}$$

where C_{xx} is the covariance matrix of the training patterns. Thus the critical temperature will be:

$$\beta_c = \frac{1}{2\lambda_m h_0} \tag{4}$$

where λ_m is the largest eigenvalue of C_{xx}. This is in agreement with Rose et al.[7] for $h_0 = 1$.

We see that the number of winners and the variance along the largest principal axis of the training data, determine the value of β_c. The more winners the network has, the faster the first phase transition will occur.

4 Results

4.1 16 clusters in 4 dimensions

The properties of this Many-Take-All network can best be showed by demonstrating an example: a data distribution in four dimensions. The data clusters are placed on the angles of a four dimensional hypercube (so it consists of sixteen clusters). A winner-Take-All network requires $h = 16$ and $h_0 = 1$. In this dataset distances $0\ldots4$ between clusters can be distinguished. A minimal code that contains these distances is $h = 8$ and $h_0 = 4$, which is smaller than the Winner-Take-All. After learning, we see that the sparse representation correctly gives the distance distribution between the corresponding data clusters. Furthermore, we observe that pairs of neurons encode cartesian axes in the data space. For instance, neuron 1 and 5 always have opposite value and encode whether the first coordinate of the particular cluster is 0 or 1 (see Table 1).

cluster	s	cluster	s	cluster	s	cluster	s
1	 1111	5	..1. 11.1	9	...1 111.	13	..11 11..
2	1.... .111	6	1.1. .1.1	10	1..1 .11.	14	1.11 .1..
3	.1.. 1.11	7	.11. 1..1	11	.1.1 1.1.	15	.111 1...
4	11.. ..11	8	111. ...1	12	11.1 ..1.	16	1111 ....

Table 1. The encoding of a four dimensional cubic distribution by eight hidden units with four winners.

4.2 Scaling of learning properties

The number of distances the network can distinguish with this coding, depends on the number of neurons overlap and thus on the number of winners in the network. In general, with h_0 winners and enough hidden units available, the distances $0, \ldots, h_0$ can be encoded. Unfortunately, the complexity of the learning rule is roughly proportional to $\binom{h}{h_0}\binom{h_0}{2}nh$ (see eq (2)), so the more winners are taken, the more time the learning will require (see Table 2).

dim	Winner-Take-All		Many-Take-All		
---	hidden	time (s)	hidden	winners	time (s)
1	2	1	2	1	1
2	4	2	4	2	5
3	8	6	6	3	72
4	16	37	8	4	947
5	32	192	10	5	11511
6	64	1292	12	6	88855

Table 2. Comparison of learning times of Winner-Take-All and Many-Take-All network for cubic distributions.

When more than one winner is introduced a new phenomenon occurs. Although in principle the network can encode $\binom{h}{h_0}$ different clusters, in practice some combinations are not used. This is because the neurons not only encode the clusters but also encode information over the distances between the clusters. This distance constraint makes it for instance impossible that in the first example neurons will fire simultaneously. Although in small and low dimensional problems the number of neurons for Many-Take-All are sometimes higher than Winner-Take-All, in higher dimensions and larger numbers of hidden units, the Many-Take-All becomes more efficient. To encode for example the cubic distribution of clusters in a n-dimensional space, a Winner-Take-All network requires 2^n units, while a Many-Take-All needs $2n$ (see Table 2).

4.3 Hierarchical data structures

We show here how Many-Take-All network can be used to learn structure in hierarchical data sets. Ultrametric sets can be represented as the leafs of a tree, where distance is defined as the number of nodes to a common parent [6]. Clearly, ultrametric structure is not represented in the output coding of Winner-Take-All networks, and Many-Take-All offers a clear advantage. As an example we take an ultrametric data set, consisting of 8 clusters in 4 dimensions.

The results are shown in Table 3 (left) for a network with $h = 6$ and $h_0 = 2$. After learning, the distances in the code display the same hierarchical structure

cluster	coding $h = 6, h_0 = 2$	coding $h = 14, h_0 = 3$
1	110 000	1110 000 0000 000
2	110 000	1101 000 0000 000
3	101 000	1000 110 0000 000
4	101 000	1000 101 0000 000
5	000 110	0000 000 1110 000
6	000 110	0000 000 1101 000
7	000 101	0000 000 1000 110
8	000 101	0000 000 1000 101

Table 3. Coding of an ultrametric data set consisting of eight clusters in four dimensions.

as the original data, although not in great detail, using $h = 6, h_0 = 2$ can distinguish only four different clusters. This result is robust for changes in h and h_0. In Table 3 (right) shows the results on the same data for a network with $h = 14$ and $h_0 = 3$. As can be seen, the fact that we have used a larger network does not affect the basic distance structure of the coding that we obtained, but adds more details. The network with $h = 14, h_0 = 3$ distinguishes all eight clusters and their distances.

References

1. Földiák, P.: Forming sparse representations by local anti-Hebbian learning. Biological Cybernetics **64** (1990) 165-170
2. Kappen, H.: Deterministic Learning Rules for Boltzmann Machines. Neural Networks **8** (1994) 537-548
3. Kappen, H., Nijman, M.: Radial Basis Boltzmann Machines and Learning of Missing Values. Proceedings of the World Congres on Neural Networks **1** (1995) 72-75
4. Koenderink, J.J.: Simultaneous Order in Nervous Nets form a Furnctional Standpoint. Biological Cybernetics **50** (1984) 35-41
5. Koenderink, J.J.: (1984). Geomtrical Structures Determined by the Functional Order in Nervous Nets Biological Cybernetics **50** (1984) 41-50
6. Rammal, R., Toulouse, G., Virasoro, M.A.: Ultrametricity for physicists. Rev.Mod.Phys. **3** (1986) 765-787
7. Rose, K., Gurewitz, E. and Fox, G.: Statistical Mechanics and Phase Transitions in Clustering. Physical Review Letters **65** (1990) 945-948
8. Saund, E.: A Multiple Cause Mixture Model for Unsupervised Learning. Neural Computation **7** (1995) 51-57

Dynamic Feature Linking in Stochastic Networks with Short Range Interactions

Bert Kappen[1] and Pablo Varona[2]

[1] RWCP Novel Function SNN Laboratory, Department of Biophysics, University of Nijmegen, Geert Grooteplein 21, NL 6525 EZ Nijmegen, The Netherlands
[2] Departamento de Ingenieria Informática, Universidad Autónoma de Madrid, Canto Blanco, 28049 Madrid, Spain

Abstract. It is well established that cortical neurons display synchronous firing for some stimuli and not for others. The resulting synchronous subpopulation of neurons is thought to form the basis of object perception. In this paper this 'dynamic linking' phenomenon is demonstrated in networks of binary neurons with stochastic dynamics. Feed-forward connections implement feature detectors and lateral connections implement memory traces or cell assemblies.

1 Introduction

It is well established that cortical neurons display synchronous firing for some stimuli and not for others [1, 2]. In particular, it has been shown experimentally, that correlations depend on the amount of conflict in the stimulus presented [3].

The resulting synchronous subpopulations of neurons (cell assemblies) are thought to form the basis of segmentation and object perception [4, 5]. The role of individual cells is to represent important 'atomic' visual features, such as edges, corners, velocities, colors, etc. Objects can be defined as a collection of these features: The cell assembly is a neural representation of the entire object. Thus if the local features are part of a *coherent* global stimulus, the corresponding neurons synchronize. If the same local features are not part of a global stimulus, no such synchronization occurs.

The observed synchronous firing in animal experiments has in fact two components: one is the presence or absense of an oscillatory component in the auto- and cross-correllograms. The second phenomenon is the presence or absence of a central peak in the cross-correllograms. We will refer to these phenomena as oscillations and correlations, respectively. Models for feature linking have been proposed for feature linking by various authors, either based on oscillating neurons [6, 7] or integrate-and-fire or bursting neurons [8, 9, 10].

In this paper we want to study stimulus dependent assembly formation in networks of stochastic binary neurons. Feed-forward connections implement feature detectors and symmetric lateral connections implement memory traces. We define correlation as the presence of a central peak in the cross-correllograms. Neurons that are synaptically connected will display correlated fire under almost all stimulus conditions. Therefore, we use a network with short-range connections and make use of the effect of long range correlations that are present in spin systems near the critical temperature.

These networks provide an attractive model to study feature linking for two reasons. The stochastic dynamics of these networks lead asymptotically to the Boltzmann-Gibbs distributions, which gives insight in the conditions under which long-range correlations occur. In this way the correlated firing can be related to well know equilibrium properties of spin systems. Secondly, these networks offer an immediate solution to learning based in correlated activity using the Boltzmann Machine learning paradigm [11].

2   The model

The basic architecture that we consider is given in Fig. 1. A sheet of sensory

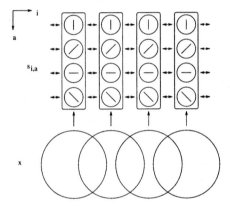

Fig. 1. Boltzmann Machine architecture with feed-forward and lateral connections. Input is provided by $x_1, \ldots, x_n$, with x_i real-valued or binary valued. Hidden units are $s_1, \ldots, s_h$, with $s_j = \pm 1$.

neurons in the visual cortex is modeled as a two-dimensional grid. At each grid location, a column of neurons with identical receptive fields is present. Neurons in one column respond optimally to different features in the receptive field.

The equilibrium distribution of the sensory neurons s *given* a stimulus x is given by

$$p(s|x) = \frac{1}{Z(x)} \exp\{w \sum_{(i,j),\alpha} s_{i,\alpha} s_{j,\alpha} + v \sum_{i,\alpha,\alpha'} s_{i,\alpha} s_{i,\alpha'} + \sum_{i,\alpha} h_{i,\alpha}(x) s_{i,\alpha}\} \quad (1)$$

$s_{i,\alpha} = \pm 1, i = 1, \ldots, n, \alpha = 1, \ldots, m$ denote the firing of the neuron with feature preference α at grid location i. $s_{i,\alpha}$ is a stochastic variable subject to Glauber dynamics, x denotes the external stimulus. It consists of a two-dimensional array of feature values $x_i = 1, \ldots, m, i = 1, \ldots, n$. w and v denote the strength of the lateral nearest neighbor interaction and the intra-column interaction, respectively.

$h_{i,\alpha}(x)$ is the external field (stimulus) component for neuron $s_{i,\alpha}$ in the presence of stimulus x. We assume that $h_{i,\alpha}(x)$ only depends on the local stimulus value x_i, i.e. $h_{i,\alpha}(x) = h_\alpha(x_i)$. By definition, neuron $s_{i,\alpha}$ has a prefered stimulus value α but it is usually also activated with nearby feature values. Here we will assume $h_\alpha(x_i) = h\delta_{\alpha,x_i} + h_0$. h is the overal strength of the stimulus and h_0 is a neuron threshold, which are free parameters of our model.

In the case that the intra-column interaction $v = 0$, Eq. 1 becomes a product of independent models, one for each feature value α:

$$p(s|x) = \Pi_\alpha p_\alpha(s_\alpha|x)$$

$$p_\alpha(s|x) = \frac{1}{Z(x)} \exp\{w \sum_{(i,j)} s_i s_j + \sum_i (h\delta_{\alpha,x_i} + h_0)s_i\}$$

Thus we can study the behaviour for one value of α.

3   Correlation lengths

Consider a visual stimulus x. The external input to neuron i in layer α is either h or 0 depending on whether $x_i = \alpha$. The task of the network is to represent these inputs in the various feature layers, such that 1) neurons locally represent the presence or absence of a feature value and 2) the activity between the neurons that encode one stimulus are correlated in regions where the stimulus is coherent.

A convenient quantity that expresses correlation between neurons j and k is the correlation function:

$$\Gamma_{jk}(t) = \langle s_j(0)s_k(t) \rangle - \langle s_j(0) \rangle \langle s_j(0) \rangle \tag{2}$$

The correlation function depends on the temperature β and on the connectivity of the network. For instance, in a d-dimensional Ising spin system, the connections are only between nearest neurons in a d-dimensional grid. Γ_r can be calculated in the Landau approximation and takes the form

$$\Gamma_r \propto r^{2-d} \exp(-r/\xi)$$

with r the distance in the grid. ξ depends on the temperature of the system. Around the critical temperature T_c, $\xi \propto |T - T_c|^{-\frac{1}{2}}$.

In the absence of a stimulus to layer α, we want to have low firing rates of the neurons. Therefore, we require $h_0 < 0$. In the presence of a coherent stimulus, we would like long range correlations. Correlations are maximal when $h_0 + h = 0$, i.e. when the total external field is absent. This is because for $h_0 + h \neq 0$, the second term in Eq. 2 strongly increases, making the correlation function effectively zero.

In Fig. 2 we show the auto-correlations $\langle s_i(0)s_i(t) \rangle$ as a function of time t for 9 neurons in a 3×3 sub-grid of a 10×10 grid of neurons for various coupling strengths w and coherent stimulus. Note that $\langle s_i s_i \rangle$ approaches the squared mean firing rate $\langle s_i \rangle^2$ for large times. $\langle s_i \rangle = 0$ for small coupling w. There exists a critical coupling above which there is a coexistence of two phases, each with non-zero $\langle s_i \rangle$. The critical coupling is approximately $w = 0.4$.

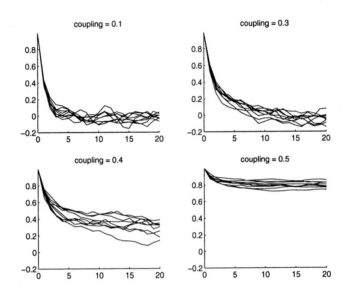

Fig. 2. Auto-correlations as a function of time for different coupling strength for coherent stimulation. The network consists of a 10×10 grid of neurons with periodic boundary conditions. External field is zero.

In Fig. 3 we show the equal time cross correlation function $\Gamma_r(0)$ for various coupling strengths w and coherent stimulus as a function of distance in the network.

In [12], we applied this mechanism to a stimulus that consists of 2 objects. It was shown, that all cells belonging to the same object are highly correlated, whereas cells belonging to different objects are not correlated.

4   Discussion

We have shown how long range correlations in spin models can be used to signal coherence in stimuli. In particular, we have proposed a two dimensional Ising model for feature linking in a visual task. This model allows to study the response of a network to stimuli with varying degree of coherence.

Whereas the mean firing rate indicates the local evidence for a stimulus feature, the correlations signal whether these local features are part of a coherent

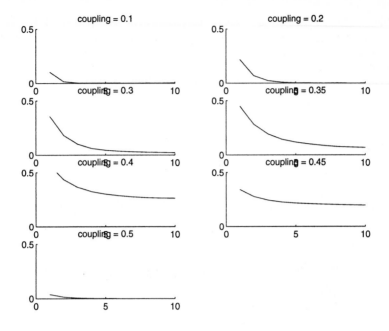

Fig. 3. Equal time cross-correlation function as a function of distance in the grid for different coupling streng th. The network consists of a 10×10 grid of neurons with periodic boundary conditions. External field is zero. The network was iterated 10000 times to remove transient effects. Correlation function was calculated over 3000 τ times. Results are averaged over all neuron pairs with equal distance.

stimulus or not. In this sense, stochastic networks can be used to solve the dynamic linking problem as stated in the introduction.

This model contains a sensory layer and a hidden layer. The hidden layer represents an interpretation of the sensory input. Sensory input provides local evidence. The lateral connections in the hidden layer provide global correlations between features that belong to the same stimulus and no correlations between features from different stimuli.

In this paper, we have only studied correlations at 0 time delay. In [13], delayed correlations were studied in networks composed of fully connected subpopulations, but the issue of dynamic linking was not addressed there. Nevertheless, one expects similar time-delayed correlation as were reported there.

Clearly, we are not proposing the Ising model as a serious computational model for the cortex and it should be investigated whether and how this mechanism can be extended to other network architectures. In the present model, neighboring cells with identical receptive fields are connected. In a more realistic network, the lateral connectivity would arise from learning and more complex connectivity patterns arrise. When learning results in a combination of excitatory and inhibitory connections, the resulting network will contain frustration, and may behave more like a spin glass than like ferromagnetic model. It is known, that long range correlations also exist in such frustrated systems.

A straightforward way to learn the lateral connections strengths from an environment is given by the Boltzmann Machine learning framework [11]. It is interesting to note that this rule is based on correlated activity $\langle s_i s_j \rangle$ instead of mean field activity $\langle s_i \rangle \langle s_j \rangle$.

References

1. C.M. Gray, P. König, A.K. Engel, and W. Singer. Oscillatory responses in cat visual cortex exhibit inter-columnar synchronization which reflects global stimulus properties. *Nature*, 338:334, 1989.
2. R. Eckhorn, R. Bauer, W. Jordan, M. Brosch, W. Kruse, M. Munk, and H.J. Reitboeck. Coherent oscillations: A mechanism of feature linking in the visual cortex? *Biological Cybernetics*, 60:121–130, 1988.
3. A.K. Engel, P. König, and W. Singer. Direct physiological evidence for scene segmentation by temporal coding. *Proceedings of the National Academy of Science of the USA*, 88:9136–9140, 1991.
4. B Julesz. *Foundations of cyclopean perception*. University of Chicago Press, Illinois, 1971.
5. D Marr. *Vision: A computational investigation into the human representation and procesing of visual information*. Freeman, San Francisco, 1982.
6. P. König and T.B. Schillen. Stimulus-dependent assembly formation of oscillatory responses: I synchronization. *Neural Computation*, 3:155–166, 1991.
7. H. Sompolinsky, D. Golomb, and D. Kleinfeld. Cooperative dynamics in visual processing. *Physical Review A*, 43:6990–7011, 1991.
8. Ch. Malsburg and W. Schneider. A neural cocktail-party processor. *Biological Cybernetics*, 54:29–40, 1986.
9. T. Chawanya, T. Aoyagi, I. Nishikawa, K. Okuda, and Y. Kuramoto. A model for feature linking via collective oscillations in the primary visual cortex. *Biological Cybernetics*, 68:483–490, 1993.
10. M. Arndt, P. Dicke, M. Erb, R. Eckhorn, and H.J. Reitboeck. Two-layered physiology-orineted neuronal network models the combine dynamic feature linking via synchronization with a classical associative memory. In J.G. Taylor, editor, *Neural Network Dynamics*, pages 140–155. Springer Verlag, 1992.
11. D. Ackley, G. Hinton, and T. Sejnowski. A learning algorithm for Boltzmann machines. *Cognitive Science*, 9:147–169, 1985.
12. H.J. Kappen and M.J. Nijman. Dynamic linking in stochastic networks. In R. Moreno-Diaz, editor, *Proceedings W.S. McCullock: 25 years in memoriam*, pages 294–299, Las Palmas de Gran Canaria, Spain, 1995. MIT Press. F-95-043.
13. I Ginzburg and H Sompolinsky. Theory of correlations in stochastic neural networks. *Physical Review E*, 50:3171–3191, 1994.

Collective Dynamics of a System of Adaptive Velocity Channels

U. Hillenbrand and J. L. van Hemmen

Physik-Department der TU München, D-85747 Garching bei München, Germany

1 Motivation

Our key assumption is that local, visual *motion* is represented by velocity channels (VCs), i.e. populations of neurons that respond optimally to edges or dots that move through their receptive field at some speed lying in a small range. This very general idea is suggested by the simple fact that the visual system is able to track moving objects within certain limits of velocity without motion blur, i.e. with a good signal to noise ratio [7]. Furthermore, it is supported by physiological studies [4, 5] as well as theoretical considerations [6].

The problem that is posed by this general setting is an economical one. In order to represent a wide range of velocities, a large number of channels would be required for every retinal locus and every orientation of edges [7]. The hypothesis elaborated here circumvents this by proposing a small number of channels that can *adapt* their sensitivity to the most important velocity component within an area spanning several receptive fields. The question then is: What is the "most important velocity component"?

From a different point of view, having a limited number of VCs could be an advantage for treating the general task of the visual system of focusing attention to object-related information and suppressing noise and less significant input. It is certainly desirable to accomplish this at a processing stage as early as possible. That is to say, in early vision.

The model system of VCs proposed below adapts its sensitivity to the mean speed of an object moving on a noisy background into predominantly one direction. The object pops out from the background; see Fig. 1. Noise is assumed to be made up of some isotropic distribution of speeds. This kind of noise can be e.g. induced by micro-saccades or could be part of the external world. In addition, our model naturally accounts for a number of psychophysical observations, which can now be re-interpreted in a new light.

2 The model

Instead of the speed we consider the time s it takes a local stimulus (edge of the appropriate orientation or dot) to pass through the receptive field of a VC.

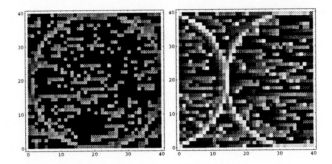

Fig. 1. Pop-out of an object (periodic circles), moving across an array of velocity channels from left to right at a constant speed. The object is superimposed with random dots moving left to right and right to left at various speeds. Shown are the activities as greylevels of our model velocity channels coding the motion from left to right (dark: low activity, light: high activity). Left: earlier stage of adaptation. Right: later stage of adaptation.

Each VC has a preferred passage time l for a stimulus – equivalently, a preferred velocity –, so that it will respond most vigorously, if the stimulus passage time s equals l. We model this by a response tuning function $R(s - l)$ representing the peak firing rate during the time course of a neuronal response. The function R has a maximum at 0 and decreases to 0 as $|s - l|$ becomes large. The preferred passage time l is meant to allude to the lag time of some thalamic neurons projecting – together with non-lagged thalamic neurons – on cortical simple cells in V1 and giving rise to their speed preference [6]. Alternatively, one might think of a motion detector à la Hassenstein and Reichardt.

Adaptation is mediated by the dynamics of the lag times $l(t)$ at time t of all VCs belonging to a spatially extended population; the extension is measured in retinal coordinates. On exposition to a stimulus comprising isotropically moving noise plus some non-isotropically moving object the lag values of the VCs approach the mean passage time of the object; see Fig. 2.

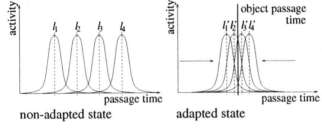

Fig. 2. Adaptation of the tuning curves of velocity channels by shifting of the lag values l_i to l'_i on exposition to a moving object superimposed with noise.

We consider only VCs that all have the same preferred orientation. For such a population any movement is effectively one-dimensional. The passage time s of a stimulus and the lag l of a VC can assume positive and negative values, representing the two directions of movement. In the following, quantities belonging to VCs coding the positive direction are superscripted by "+", quantities belonging to those coding the negative direction by "−".

Adaptation occurs in a population of neurons that receive input from a pool of receptive fields spanning an extended retinal region. For simplicity, we assume the same set of VCs for each receptive field in the pool – i.e., the same set of initial lag values $l_{vc}^{0,\pm} \equiv \pm l_{vc}^0$ ($l_{vc}^0 > 0$) in every hypercolumn – distributed coarsely over the range of interest. The subscript vc hence refers to the VC subpopulation comprising all VCs with the initial lag time $+l_{vc}^0$ or $-l_{vc}^0$ in all hypercolumns.

Of course, ideally an adaptation of a VC to a *local* stimulus with passage time s would consist in the correction of its lag $l_{vc}^{\text{sgn}(s)}$ by $s - l_{vc}^{\text{sgn}(s)}$. The following local computations attempt to approximate this "obvious solution" and can be implemented by a biological or artificial neural network (such as analog VLSI). These *local* correction signals are then averaged over the pool of receptive fields to result in a *global* correction signal applied to all the lags in the pool. Thus the pool is adapted as a whole to some (biologically sensible) velocity.

We now turn to the mathematical details of our model. Each VC computes a local correction signal for its lag value that can bring it closer to the actual passage time of the local stimulus. Its *amplitude* is extracted from the response R of the channel and is formally written as A [R] with some adequate function A. Ideally, A $[R(s - l)] = |s - l|$. Since the response tuning function R is not invertible, the *sign* of the correction signal cannot be computed from the response of one VC alone. It must be computed from the response of the other channels receiving input from the same spatial receptive field and coding for the same direction of motion, i.e., responding to the same local stimulus at the same time. To this end, appropriate connections $J_{vc,vc'}^{\pm}$ between the channels within a hypercolumn and belonging to one direction of motion ("+" or "−") are needed; see Fig. 3. An approximation to the sign of correction is then given by a sigmoidal function σ with $\sigma(\mathbb{R}) = (-1, +1)$ and

$$\text{sgn}\left\{\sigma\left[\sum_{vc'} J_{vc,vc'}^{\text{sgn}(s)} \, R\big(s - l_{vc'}^{\text{sgn}(s)}\big)\right]\right\} \overset{!}{=} \text{sgn}(s - l_{vc}^{\text{sgn}(s)}) . \tag{1}$$

That is, the sign of σ is the sign of the correction of the lag $l_{vc}^{\text{sgn}(s)}$ needed for approaching the local stimulus passage time s. In order to fulfill (1) and with $|l_{vc}^{\pm}|$ increasing with vc the weights can e.g. be chosen to be

$$
\begin{aligned}
J_{vc,vc'}^{\pm} &= \pm\delta_{vc+1,vc'} \mp \delta_{vc-1,vc'} , \\
J_{vc,vc'}^{\pm} &= \pm\delta_{vc+1,vc'} \mp \delta_{vc,vc'} \quad \text{for } vc \text{ minimal,} \qquad \text{with} \quad \delta_{i,j} := \begin{cases} 1 & \text{if } i = j, \\ 0 & \text{if } i \neq j. \end{cases} \\
J_{vc,vc'}^{\pm} &= \pm\delta_{vc,vc'} \mp \delta_{vc-1,vc'} \quad \text{for } vc \text{ maximal,}
\end{aligned}
$$
$$\tag{2}$$

As will be shown, under certain circumstances the order of the $l_{vc}^{\pm}$ never changes in the course of the adaptation process. Thus, condition (1) is always fulfilled and each local correction signal brings the lag values $l_{vc}^{\pm}$ closer to the passage time of the local stimulus.

Altogether, if a stimulus enters a receptive field of a VC from the subpopulation $vc^{\mathrm{sgn}(s)}$ at time t_0 and passes through in time s, the initiated *local adaptation signal* (LAS) is for $t > t_0$

$$\mathrm{LAS}_{vc}^{\mathrm{sgn}(s)}(t-t_0) := \mathrm{A}\Big[\mathrm{R}\big[s - l_{vc}^{\mathrm{sgn}(s)}(t)\big]\Big] \, \sigma\Bigg[\sum_{vc'} J_{vc,vc'}^{\mathrm{sgn}(s)} \, \mathrm{R}\big[s - l_{vc'}^{\mathrm{sgn}(s)}(t)\big]\Bigg] \, \lambda(t-t_0) \, .$$

(3)

The function λ describes the time course of the adaptation response to s. It should reach its maximum 1 after a short rise time and decay to 0 not too fast after that. The computation performed for each receptive field in the pool is depicted functionally in Fig. 3.

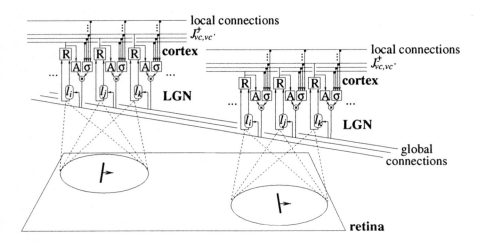

Fig. 3. Wiring diagram of the velocity channels belonging to one of the direction populations. Two receptive fields and three velocity channels each are shown. Visual information from the local receptive fields converges via non-lagged and lagged LGN neurons (lags $l_i, l_j, l_k, \ldots$) onto cortical neurons giving rise to a response R as a function of the stimulus passage time through the receptive field. The amplitude A of the local lag correction signal is computed from the response R of a single velocity channel, its sign σ from the activities of all local velocity channels via appropriate local connections $J_{vc,vc'}^{\pm}$. Amplitude and sign are multiplied and the results – the total correction signals – are passed to the LGN lagged neurons of *all* the velocity channels with the *same* actual lag $l_i, l_j, l_k, \ldots$ via global connections. Not shown: opposite direction population.

The *total adaptation signal* (TAS) for all the VCs in the population at any instant of time is given by the sum of all local adaptation signals (3) initiated in the past by VCs belonging to the same subpopulation vc (*both signs*); see Fig.

3. The dynamic lag of VCs from the subpopulation vc (for all retinal loci) is therefore

$$l_{vc}^{\pm}(t) = \pm l_{vc}^0 + \gamma \underbrace{\sum_{f=1}^{F(t)} \text{LAS}_{vc}^{\text{sgn}(s^f)}(t - t^f)}_{=: \text{TAS}_{vc}(t)} \tag{4}$$

with $F(t)$ as the number of stimulus events in the entire pool until time t. All the stimulus events at all places and for both directions are ordered in time: $t^1 \leq t^2 \leq \ldots \leq t^F$. Finally, $\gamma \in (0, 1]$ is a parameter for the strength of the total adaptation signal.

For obtaining analytical results we introduce the following idealizations:

- The amplitude function A is optimally adjusted to the passage time tuning R, i.e. $A[R(s - l)] = |s - l|$.
- The sigmoidal function σ changes sign infinitely fast, i.e. $\sigma = \text{sgn}$.

In what follows we drop the index vc for brevity. The lag values immediately before the response to the local stimulus event F, that is the lag values significant for the response R to the local stimulus, can be written as $l^{\pm}(t^F) = \pm l^0 + x^F$ with the total adaptation signal $x^F := \text{TAS}(t^F)$.

3 The case of a short rise time or a low stimulation rate

Let r^f be the time elapsed between the initiations of a local adaptation signal by the stimuli with passage times s^f and s^{f+1}. In case r^f is always larger than the rise time of λ, we can proceed as if λ jumps to its peak value instantaneously before decaying to 0 with a time constant τ. We therefore set $\lambda(t) := \exp(-t/\tau)$ and understand r^f in this section as the inter-stimulus time minus the rise time of λ.

We assume that r^f and $s^{f'}$ for all f, f' are independently distributed according to stationary densities $u(r)$ on $\mathbb{R}_+$ and $v(s)$ on $\mathbb{R}$, respectively. Stationarity is ensured by considering a pool of receptive fields receiving some stationary noise input and/or covering the same visual objects at all times. Independence is a good approximation assuming some random distribution of receptive field places and sizes.

The averages we are interested in are $\langle x^n \rangle_f := \int_{-\infty}^{\infty} dx\, p(x, t^f)\, x^n$, with $p(x, t)$ as the density of the total adaptation signal at time t. One can show that all the moments $\langle x^n \rangle_f$ obey recursive equations and can be determined successively for increasing n. We will present the results for the mean. For the variance it is sufficient to say here that it always converges to some finite value.

The kind of scenario we want to analyze is one of a stimulus consisting out of two components: one that is symmetric in its distribution of speeds with respect to the two directions and another that has a density in one of the two directions only, which we choose to be the "+"-direction. Indicating the significance of these two components we call the former the "noise component", the latter the "object component". Let $\mu_{\pm} := \int_{-\infty}^{\infty} ds\, v(s)\, \Theta(\pm s)$ and $\mu := \mu_{+} - \mu_{-}$, which is in $(0,1]$. We can write the density v as the sum

$$v(s) = (1 - \mu)v_N(s) + \mu v_O(s) \tag{5}$$

where $v_{N/O}$ are the densities of the noise and object component, respectively. Let $\langle \cdot \rangle_{N/O}$ denote the averages taken with respect to $v_{N/O}$. Since $\langle s \rangle_N = 0$ we have $\langle s \rangle_O = \langle s \rangle /\mu$. We parameterize the lag time as

$$\pm \langle l^{\pm} \rangle_f = l^0 \pm \langle x \rangle_f =: (1 - \alpha_f^{\pm})l^0 + \alpha_f^{\pm} \langle s \rangle_O . \tag{6}$$

The quantities $\alpha_f^{\pm}$ are the *degrees of adaptation* of the lags of the two direction populations to the mean passage time of the object. It can be shown that the $\alpha_f^{\pm}$ are *the same for all* l^0 and converge monotonically to their asymptotic values

$$\alpha_f^{\pm} \xrightarrow{f \to \infty} \alpha^{\pm} = \pm \frac{\mu\gamma}{\langle e^{-r/\tau} \rangle^{-1} - 1 + \gamma}, \quad \alpha^+ \in (0,1), \quad \alpha^- \in (-1,0). \tag{7}$$

This means that the l_{vc}^+ approach the mean passage time of the object monotonically as a function of the stimulus event f while the l_{vc}^- monotonically diverge from it, each keeping their relative distances; cf. Fig. 5. Therefore, the lag values of different VCs never cross and the signs of the local adaptation signals are computed correctly at all times. The averages in equation (7) are taken with respect to the density u.

The degrees of adaptation $\alpha_f^{\pm}$, and hence the lag times, approach their asymptotic values exponentially with the time constant

$$\tau' = - \langle r \rangle \left[\ln(1 - \gamma) + \ln \langle e^{-r/\tau} \rangle \right] . \tag{8}$$

(Remember: τ is the relaxation time constant.) Both τ and τ' can be chosen to lie in the millisecond range.

For a fixed distribution of velocities, the quantity $\langle e^{-r/\tau} \rangle \in (0,1)$ is a measure of the retinal stimulus density and approaches 1 for a high density and 0 for a low one. So it makes sense that as $\langle e^{-r/\tau} \rangle \to 1$, according to (8), τ' approaches its smallest value 0 ($\langle r \rangle \to 0$) whereas, according to (7), $|\alpha^{\pm}|$ tends to its largest value μ.

Figure 4 is a plot of α_f^+ as a function of the stimulus event f for different retinal densities and unidirectional motion of the whole stimulus ($\mu = 1$). Smaller values of μ just result in linearly squeezed curves.

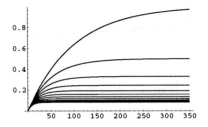

Fig. 4. Degree of adaptation α_f^+ as a function of the stimulus event f. The curves have been plotted for $\mu = 1$, $\gamma = 0.01$ and, from bottom to top, $\langle e^{-r/\tau} \rangle = 0.9, 0.91, 0.92, \ldots, 1$.

4   The motion-after-effect

It is well-known that after having observed a moving object for a sufficiently long time the observer, when suddenly viewing a stationary target, has the impression of motion in the reverse direction. This effect is naturally incorporated in our model; see Fig. 5.

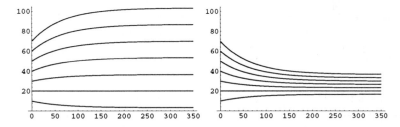

Fig. 5. Demonstration of the motion-after-effect. The plots show the (modulus of the) lag values according to (6) of seven velocity channels as a function of the stimulus event f. Left: for the channels representing motion in the direction opposite to the object motion. Right: for the channels representing motion parallel to the object motion. The latter population is dragged towards the mean object passage time, the former is repelled from it. Consequently the population encoding the moving object has no channels representing low velocities (high passage times), the other population is enriched with low velocity channels. For instance, asymptotically we have four channels with a lag time above 40 time units on the left, none on the right. Hence stationary objects will be mainly represented in the no-object population, i.e. in the direction opposite to the object motion, until the state of the system has relaxed. The parameters are $\langle s \rangle_O = 20$, $\mu = 1$, $\langle e^{-r/\tau} \rangle = 0.995$ and $\gamma = 0.01$.

5   Existence of an optimal range of temporal frequencies

From (7) it follows that the adaptation to $\langle s \rangle_O$ becomes better with increasing rate of stimulation. This is only true up to a certain limit; cf. beginning of §3. For any rise time of λ it can be shown rigorously for a grid-like stimulus that there is always a temporal frequency above which the degree of adaptation α_f^+ exceeds one. Since this implies a crossing of the lag values of different velocity channels, the signs of the local adaptation signals are no longer computed correctly. Consequently, the adaptation becomes worse. This is consistent with psychophysical results on the existence of a unique temporal frequency window for the sensitivity to contrast [3] and direction of motion [2] and for the strength of the motion-after-effect [1]. Simulation results are presented in Fig. 6.

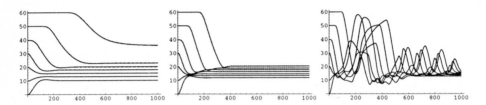

Fig. 6. Demonstration of the temporal frequency window. The plots are simulated lag values of seven velocity channels as a function of time. A grid is moved across a random array of 100 receptive fields. Its mean passage time is 16 time steps. Its wavelength is halved successively from left to right. The middle picture shows the best adaptation to the stimulus. In all three cases the inter-stimulus intervals are much shorter than the rise time of $\lambda(t) = (1/2)^{-1/2} \exp(1/2)(t/\tau)^{1/2} \exp(-t/\tau)$ with $\tau = 100$. Furthermore, $R(s) = \exp(-s^2/36)$, $A(a) = 5(1-a)$, $\sigma(a) = \tanh(15a)$, the connections are as in (2), and $\gamma = 0.01$.

6 Conclusion

We have presented a simple model which solves the problem of limited resources in the brain, shows the capability of object segmentation mediated by motion and at the same time gives a unified explanation of some seemingly rather different psychophysical observations. Full details will be published elsewhere.

References

1. A. PANTLE, Vision Research **14** (1974) 1229 - 1236
2. D. J. TOLHURST, J. Physiology **231** (1973) 385 - 402
3. D. C. BURR, J. ROSS, Vision Research **22** (1982) 479 - 484
4. G. A. Orban, in A. G. Leventhal (ed.), *Quantitative electrophysiology of visual cortical neurons*, Vol. 4 (1991) 173 - 222
5. G. C. DEANGELIS, J. OHZAWA, R. D. FREEMAN, TINS **18** (1995) 451 - 458
6. S. WIMBAUER, K. D. MILLER, J. L. VAN HEMMEN, to be published
7. D. C. VAN ESSEN, C. H. ANDERSON, in E. L. Schwartz (ed.) *Computational Neuroscience*, A Bradford Book, MIT Press (1990) 278 - 294

Local Linear Model Trees for On-Line Identification of Time-Variant Nonlinear Dynamic Systems

Oliver Nelles

Technical University of Darmstadt, Institute of Automatic Control
Laboratory of Control Engineering and Process Automation
Landgraf-Georg-Str. 4, D-64283 Darmstadt, Germany
Phone: +49/6151/16-4524, Fax: +49/6151/293445
E-mail: `nelles@irt1.rt.e-technik.th-darmstadt.de`

Abstract - This paper discusses on-line identification of time-variant nonlinear dynamic systems. A neural network (LOLIMOT, [1]) based on local linear models weighted by basis functions and constructed by a tree algorithm is introduced. Training of this network can be divided into a structure and a parameter optimization part. Since the network is linear in its parameters a recursive least-squares algorithm can be applied for on-line identification. Other advantages of the proposed local approach are robustness and high training and generalisation speed. The simplest recursive version of the algorithm requires only slightly more computations than a recursive linear model identification. The locality of LOLIMOT enables on-line learning in one operating region without forgetting in the others. A drawback of this approach is that systems with large structural changes over time cannot be properly identified, since the model structure is fixed.

1 Introduction

Nonlinear system identification is an important task in many disciplines. Nonlinear dynamic models are the foundation for e.g. prediction, simulation, model based control and fault diagnosis. In most cases a derivation of such models by first principles (physical, chemical, biological, etc. laws) is expensive, time-consuming and involves many unknown parameters and heuristics. Hence, methods for data-driven modelling (identification) are of big interest. Let's denote the physical input and output of a nonlinear time-discrete dynamic system at time instant k as $u(k)$ and $y(k)$, respectively. Then the identification of such a system with dynamic order m is equivalent to the approximation of the following function $f(\cdot)$

$$y(k) = f(\underline{x}(k)), \quad \underline{x}(k) = [u(k-1) \ ... \ u(k-m) \ y(k-1) \ ... \ y(k-m)]^T \quad (1)$$

Therefore, to determine the model's inputs besides the physical input variables also the dynamic order m has to be known or estimated. In principle, nonlinear dynamic systems with an auto-regressive part (i.e. $\underline{x}(k)$ contains $y(k-i)$-regressors) can be modelled with two different approaches. The first one is called series-parallel or NARX model, the second one is called parallel or NOE model. For the series-parallel model previous *process* outputs are used as regressors in $\underline{x}(k)$, while for the parallel model previous *model* outputs are fed back [2]. The LOLIMOT algorithm proposed in this paper applies the series-parallel model for parameter optimization to exploit the linear relationships and it utilizes the parallel model for structure optimization to avoid most problems discussed above. Thus the advantages of both approaches can be combined.

2 Local Linear Model Trees

The local linear model tree (LOLIMOT) is based on the idea to approximate a nonlinear function with piece-wise linear models. The algorithm has an outer loop (upper level) that determines the input partitions (structure) where the local linear models are valid and an inner loop (lower level) that estimates the parameters of those local linear models. However, the partitions where the linear models are valid are not crisp but fuzzy, i.e. the local linear models are interpolated by weighting functions. In this paper normalized Gaussian weighting functions are applied. The LOLIMOT output y is calculated by summing up the contributions of all M linear models (hyper-planes)

$$y = \sum_{i=1}^{M} (w_{oi} + w_{1i}x_1 + ... + w_{ni}x_n) \cdot \Phi_i(\underline{x}, \underline{c}_i, \underline{\sigma}_i) \tag{2}$$

where w_{ji} are the parameters of the ith linear regression model, $\underline{x}$ is the input vector and Φ_i is the normalized Gaussian weighting function for the ith model with centre coordinates $\underline{c}_i$ and standard deviations $\underline{\sigma}_i$.

$$\Phi_i(\underline{x}, \underline{c}_i, \underline{\sigma}_i) = \frac{z_i}{\sum_{j=1}^{M} z_j}, \qquad z_j = \exp\left(-\frac{1}{2} \left(\frac{(x_1 - c_{1j})^2}{\sigma_{1j}^2} + ... + \frac{(x_n - c_{nj})^2}{\sigma_{nj}^2} \right) \right) \tag{3}$$

These equations can be interpreted alternatively as tree structure, as Sugeno-Takagi fuzzy system [3] or as local model network. Assume the weighting functions would have been already determined. Then the parameters of each linear model are estimated separately (local estimation) by a weighted least-squares technique. With the data matrix X (each row represents one measurement of $\underline{x}^T$ at time instant k, i.e. $\underline{x}_k^T$), the diagonal weighting matrix Q_i (each entry q_k is the weighting function value of the corresponding data $\underline{x}_k$) and desired outputs $\underline{y}$ the optimal parameters $\underline{w}_i$ of the ith model are

$$\underline{w}_i = \left(X^T Q_i X \right)^{-1} X^T Q_i \underline{y} \tag{4}$$

The overlapping of the weighting functions is ignored in this local estimation approach. This may lead to interpolation errors [4] that grow with increasing standard deviations of the weighting functions [5]. On the other hand local estimation is very fast and robust. Instead of optimizing all $M(n+1)$ parameters in (2) globally, only a $(n+1)$ parameter estimation is performed M times. Since a least-squares estimation has cubic complexity the global estimation approach is of $O(M^3)$ while the local estimation is of $O(M)$.

The tree construction algorithm exploits ideas from CART [6] and MARS [7] to determine the centres and standard deviations of the weighting functions. The LOLIMOT algorithm partitions the input space into hyper-rectangles. Each local linear model belongs to one hyper-rectangle in whose centre the weighting function is placed. The standard deviations are set proportional to the size of the hyper-rectangle. This makes the size of the validity region of a local linear model proportional to its hyper-rectangle extension. A model may be valid over a wide operating range of one input variable but only in a small area of another one. At each iteration, the worst performing local linear model is subdivided into two new ones. Cuts in all dimensions are tested and the one with the highest performance improvement is chosen. Note that the evaluation of the model's performance involves generalisation since it is run with a parallel model, while parameter tuning applies a series-parallel model to utilize linear least-squares estimation techniques. Fig. 1a shows six iterations of the LOLIMOT algorithm for a two dimensional input space ($n=2$). It can be represented in a tree structure, see Fig. 1b. For more details on the off-line identification of nonlinear dynamic systems with LOLIMOT see [1] and for a real-world application with a mulitple input process see [8].

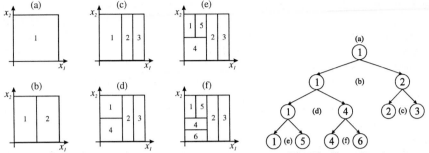

Fig. 1a: 6 iterations of the LOLIMOT algorithm   **Fig. 1b: Tree structure**

3 Recursive Least-Squares Updating of LOLIMOT

There are at least two important reasons for applying on-line identification. First, a simple (e.g. linear) model may be used that is only capable of describing the process behaviour within a small operating range. This is one classical approach of linear adaptive control. The need for on-line adaptation then emerges from the process nonlinearities that are not represented by the model. A good nonlinear model should make such an approach superfluous. However, the second reason for the requirement of on-line adaptation is time-variant behaviour of the process. This problem holds for linear and nonlinear models and is addressed in this paper.

In linear system identification a powerful on-line version of the least-squares parameter optimization exists. This is the recursive least-squares (RLS) algorithm with weighting and forgetting. It computes a new parameter estimate $\underline{w}_k$ at time instant k as follows

$$\underline{w}_k = \underline{w}_{k-1} + \underline{\gamma}_k\left(y_k - \underline{x}_k^T \underline{w}_{k-1}\right) \tag{5}$$

$$\underline{\gamma}_k = \frac{1}{\underline{x}_k^T P_{k-1} \underline{x}_k + \dfrac{\lambda}{q_k}} P_{k-1} \underline{x}_k \tag{6}$$

$$P_k = \frac{1}{\lambda}\left[I - \underline{\gamma}_k \underline{x}_k^T\right] P_{k-1} \tag{7}$$

In (5) the parameter vector $\underline{w}$ is dated up by adding a correction vector to the old estimate $\underline{w}_{k-1}$. In (6) and (7) λ is a forgetting factor that allows exponential forgetting of the old measurements and q_k is the weighting of the actual data. The forgetting factor allows a trade-off between tracking performance and noise suppression. Since LOLIMOT consists of M local linear models each of these models can be separately adapted with an RLS by following the local optimization approach.

Usually for the off-line identification a quite representative training data set can be collected that covers a wide frequency and amplitude range since measurements from many operating points and conditions can be included and badly exciting data can be excluded. Hence, a model identified off-line can be expected to be of good quality. However, when performing on-line identification the distribution of the data usually cannot be influenced. Therefore two difficulties may arise: insufficient excitation in the frequency and in the amplitude range.

The first problem of insufficient dynamic excitation is well known from linear model identification. Often processes are operated at one set point for a long time and it is obvious that without excitation of frequencies no dynamic model can be estimated. Thus to avoid deterioration of the already estimated dynamic model during static operating phases a supervisory level freezes the identification algorithm. This technique can also be applied for nonlinear dynamic system identification.

However, the second problem of insufficient static excitation arises exclusively for nonlinear models. In many cases the process will remain within one operation region for a long time. The challenge for nonlinear on-line identification is to prevent that all parameters of the model adapt to this single operating point, consequently unlearning all previously trained nonlinear process behaviour. In principle, this problem can be solved by a local approach since the model's parameters represent different operating regions of the process. Then only the parameters responsible for the operating region presently driven by the process can be adapted, leaving all other parameters unchanged and therefore conserving the behaviour in all other operating regions. LOLIMOT meets all these requirements. Hence, it is recommended to make the recursive LOLIMOT algorithm robust against insufficient static excitation by adapting only the most active local linear model by an RLS. Besides the increasing robustness another advantage of this strategy is the small computational demand. It allows the application of the recursive LOLIMOT algorithm in all applications where a linear model RLS is implemented today.

4 Examples

In the following examples the LOLIMOT structure is determined by a preceding off-line identification phase. The RLS is optimally initialized with the off-line estimated parameters. However, in the examples presented below the parameters are set to zero in order to show the rate of convergence of the algorithm. The initial P-matrix is chosen as $P_0 = \alpha I$ with $\alpha = 100$ and a forgetting factor of 0.98 is applied. The test process is an arctan($\cdot$)-function in series with a linear second order dynamic system (Hammerstein model) with a time constant between $T = 2s$ and 6s, a damping factor of $D = 0.5$ and a gain between $K = 0.5$ and 1. Thus with input $u(k)$ and output $y(k)$ the following black-box identification approach is taken

$$y(k) = f\big(u(k-1), u(k-2), y(k-1), y(k-2)\big) \tag{8}$$

Processes of other types have been tested and led to similar results. This process is sampled with $T_0 = 1s$ and excited with an amplitude modulated pseudo random signal (APRBS) [9] with 1260 samples and an amplitude between -4 and 4. For all Figures the training was split into three parts. During the first part of the data (samples 1-420) the process parameters are constant, during the second part (samples 421-840) the process is time-variant and during the third part (samples 841-1260) the process parameters are kept fixed again. Note that all off-line models perform best around sample 630 since this region represents the average parameter values (the parameters are change linearly from 420 to 840).

Fig. 2 compares the performance of the off-line LOLIMOT model (a) and the recursive on-line version (b) (8 local linear models were estimated). The time constant of the process increases linearly from 2s to 6s. It can be clearly seen that the recursive LOLIMOT keeps track with the parameter change. Fig. 3 shows three of five para-meters (a) of one local linear model and the corresponding model activation, i.e. weighting function value. The parameters obviously are only adapted if the model activation is significant. After 300 samples the parameters converged to their optimal values for the process with $T = 2s$. Then the parameters track the time-variant process behaviour. However, some parameter adaptation takes place after sample 840 were the process is held constant ($T = 6s$). This effect is due to the fact that the local model was not active from sample 770-880 and therefore could not track the time constant changes within that interval.

In Fig. 4 and 5 the same plot as in Fig. 2 and 3 are shown but the process gain was decreased from $K = 1$ to 0.5 with a fixed time constant of $T = 4s$. The interpretation of this Figure is basically the same as discussed above. In Fig. 6 and 7 the gain and the time constant of the process are fixed. Instead the nonlinearity is changed. The positive part ($x > 0$) of the arctan(αx)-function is changed by increasing α from 0.2 to 2.0 while the negative part is kept constant. Therefore, the nonlinear structure of the process is changed ($\alpha = 0.2$ leads to weakly, $\alpha = 2.0$ to strongly nonlinear behaviour) and the off-line LOLIMOT structure determination is becoming sub-optimal. It is interesting to note that only the parameters of those local linear models are changed that lie in the space with $u(k-1) > 0$ or $u(k-2) > 0$. This represents the fact that only the positive part of the arctan(αx)-function is changed. (Note that LOLIMOT can recognize and exploit the Hammerstein structure of the process by dividing the space exclusively in the $u(k-1)$- and $u(k-2)$-directions [1].)

5 Conclusions

The proposed recursive local linear model tree (LOLIMOT) algorithm can be applied for on-line identification of nonlinear dynamic systems. After a preceding off-line identification that determines the model structure and initial parameter values the algorithm can be run on-line. Due to the local estimation approach the recursive LOLIMOT algorithm has many appealing properties. The parameter adaptation is almost as fast as for linear models. It is robust against insufficient excitation in the amplitude range since no unlearning takes place. Because the parameters are linear a mature theory and sophisticated RLS algorithms can be utilized. Finally, at least for simple LOLIMOT models an interpretation as Sugeno-Takagi fuzzy system can reveal insights about the process, since the dynamic (poles) and static (gain) behaviour can be extracted for different operation regions. Drawbacks of the algorithm are the restriction to axis-orthogonal cuts and the predetermination of the model structure in the off-line identification phase. However, it has been shown that even structural changes in the process characteristics can be identified properly.

References

[1]   Nelles O., Isermann R.: "Radial Basis Function Networks for Interpolation of Local Linear Models", submitted to IEEE Conference on Decision and Control, Kobe, Japan, 1996
[2]   Narendra K.S., Parthasarathy K.: "Identification and Control of Dynamical Systems Using Neural Networks", IEEE Transactions on Neural Networks, Vol. 1, No. 1, March 1990
[3]   Takagi, T., Sugeno M.: "Fuzzy Identification of Systems and its Application to Modelling and Control", IEEE Transactions on Systems, Man and Cybernetics Vol. 15, No. 1, 1985
[4]   Johansen T.A., Foss B.A.: "Constructing NARMAX Models Using ARMAX Models", International Journal of Control, Vol. 58, No. 5, 1993
[5]   Murray-Smith R.: "A Local Model Network Approach to Nonlinear Modelling", Ph.D. Thesis, University of Strathclyde, UK, 1994
[6]   Breiman L., Friedman J., Olshen R., Stone C.J.: "Classification and Regression Trees", Wadsworth Belmont, CA, 1984
[7]   Friedman J.H.: "Multivariate Adaptive Regression Splines (with discussion)", Annals of Statistics, March, 1991

[8] Nelles O., Isermann R., Sinsel S.: "Local Basis Function Networks for Identification of a Turbocharger", IEE UKACC, Exeter, 1996

[9] Nelles O.: "On the Identification with Neural Networks as Series-Parallel and Parallel Models", International Conference on Artificial Neural Networks, Paris, Oct. 1995

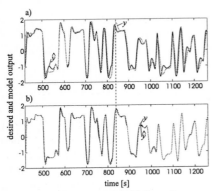

Fig. 2: (a) Off-line LOLIMOT performance
(b) On-line LOLIMOT performance

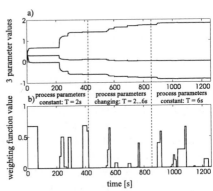

Fig. 3: (a) Tracking of three parameters
(b) Activation of this local model

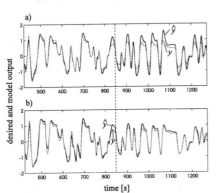

Fig. 4: (a) Off-line LOLIMOT performance
(b) On-line LOLIMOT performance

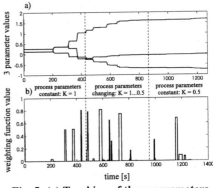

Fig. 5: (a) Tracking of three parameters
(b) Activation of this local model

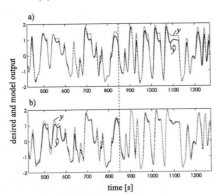

Fig. 6: (a) Off-line LOLIMOT performance
(b) On-line LOLIMOT performance

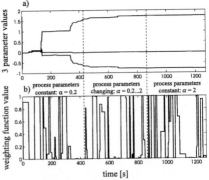

Fig. 7: (a) Tracking of three parameters
(b) Activation of this local model

Nonparametric Data Selection for Improvement of Parametric Neural Learning: A Cumulant-Surrogate Method

Gustavo Deco and Bernd Schürmann

Siemens AG, R&D, Otto-Hahn-Ring 6, 81739 Munich, Germany

Abstract

We introduce a nonparametric cumulant based statistical approach for detecting linear and nonlinear statistical dependences in non-stationary time series. The statistical dependence is detected by measuring the predictability which tests the null hypothesis of statistical independence, expressed in Fourier-space, by the surrogate method. Therefore, the predictability is defined as a higher-order cumulant based significance discriminating between the original data and a set of scrambled surrogate data which correspond to the null hypothesis of a non-causal relationship between past and present. In this formulation nonlinear and non-Gaussian temporal dependences can be detected in time series. Information about the predictability can be used for example to select regions where a temporal structure is visible in order to select data for training a neural network for prediction. The regions where only a noisy behavior is observed are therefore ignored avoiding in this fashion the learning of irrelevant noise which normally spoils the generalization characteristics of the neural network.

I. Introduction

During this decade the interest in nonlinear time series has increased remarkably [1-3]. The main problem consists in determining the underlying dynamics of the process when only the measured data are given. Particularly, the knowledge of the kind of dynamics which generates the data opens the possibility to answer essential questions like theoretical predictability and forecasting horizon in time series, i.e. possibility and reliability of modeling. Information about the predictability can be used for example to select regions where a temporal structure is visible in order to select data for training a neural network for prediction. The regions where only a noisy behavior is observed are therefore ignored avoiding in this fashion the learning of irrelevant noise which normally spoils the generalization characteristics of the neural network. A task which has to be solved first before the final goal of extracting the dynamics of the system can be achieved, is the reliable detection of statistical dependence in the data. In the context of time series, the statistical dependence between the past points and the considered present point therefore yields a measure of the predictability of the corresponding system. An information-theoretic based redundancy in data has also been formulated in the context of neural networks (see [4-6] for a rigorous mathematical formulation). Deco and Schürmann [4] have applied this theory of redundancy minimization to the unsupervised parametric extraction of statistical dependences in time series. In the same paper [4], Deco and Schürmann formulated also a cumulant based method for the parametric extraction of statistical dependence in time series which does not involve the calculation of entropy but the estimation of higher order moments.

The aim of the present paper is to formulate a nonparametric cumulant based statistical approach for detecting linear and nonlinear statistical dependences in time series. The statistical dependence is detected by measuring the discriminating significance which tests the null hypothesis of statistical independence, expressed in Fourier-space, by the surrogate method. The surrogate data herein used correspond to the null hypothesis of a non-causal relationship between past and present, i.e. of a random process with statistical properties equivalent to the original time series. The formulation of statistical independence in Fourier-space leads automatically and consistently to a cumulant based measure. In this formulation nonlinear and non-Gaussian temporal dependences can be detected in time series. This cumulant based predictability offers a method for the estimation of temporal redundancy which is reliable even in the case of a small number of data. This fact permits us to avoid the assumption of stationarity and to use the slicing window based method for measuring the non-stationarity that can only be reliably employed when the estimation of predictability can be performed with a small number of data points.

II. Theory

Let us consider a time series $x(t)$ and let us define an embedding vector constructed from the observable x by $\dot{x}(t) = (x(t), x(t-\Delta), ..., x(t-(m-1)\Delta)) = (x_1, ..., x_m)$ of dimensionality m and time-lag Δ . The vector $\dot{x}$ is a random variable distributed according to the probability density $P(\dot{x})$. A measure of the predictability of the present from the past is given by how strong the probability of the presence (say the point $x(t)$) is conditioned by the past (i.e. by $(x(t-\Delta), ..., x(t-(m-1)\Delta))$). In other words, the predictability is a measure of how different from zero is the substraction

$$P(x_1) - P(x_1|x_2, ..., x_m) \tag{1}$$

Multiplying this quantity by $P(x_2, ..., x_m)$ yields

$$P(x_1) P(x_2, ..., x_m) - P(x_1, x_2, ..., x_m) \tag{2}$$

We see that the predictability is nothing more than a measure of the statistical independence between the presence and the past. If this quantity is zero, then the presence is independent of the past and therefore unpredictable from the past, i.e. there is no statistical structure underlying the time series data meaning that the data are just uncorrelated and therefore defined by a Bernoulli process. Due to this fact we will measure in this work the predictability by testing the null hypothesis $H = \{P(x_1, x_2, ..., x_m) = P(x_1) P(x_2, ..., x_m)\}$ of statistical independence. In order to be able to estimate from the data the null hypothesis H we would like to express H in terms of empirically measurable entities like higher order cumulants, which take into account the effect of non-Gaussianity and nonlinearity underlying the data. Therefore, let us consider the Cumulant expansion of the Fourier transforms of the joint and marginal probability distributions:

$$\phi(K_1, K_2, ..., K_m) = \int dx_1 dx_2 ... dx_m \ e^{i\left(\sum_{j=1}^{m} x_j K_j\right)} \ P(x_1, x_2, ..., x_m) = e^{\sum_{n=1}^{\infty} \frac{i^n}{n!} \sum_{i_1 \cdots i_n = 1} \aleph_{i_1 \cdots i_n} K_{i_1} ... K_{i_n}} \qquad ,(3)$$

$$\varphi(K_1) = \int dx_1 \ e^{i(K_1 \cdot x_1)} \ P(x_1) = e^{\sum_{n=1}^{\infty} \frac{i^n}{n!} \aleph_1^{(n)} K_1^n} \qquad , \qquad (4)$$

$$\lambda(K_2, ..., K_m) = \int dx_2 ... dx_m \ e^{i\left(\sum_{j=2}^{m} x_j K_j\right)} \ P(x_2, ..., x_m) = e^{\sum_{n=1}^{\infty} \frac{i^n}{n!} \sum_{i_1 \cdots i_n = 2} \aleph_{i_1 \cdots i_n} K_{i_1} ... K_{i_n}} \qquad (5)$$

where $i = \sqrt{-1}$. In Fourier space the independence condition is given by

$$\ln(\phi(K_1, K_2, ..., K_m)) = \ln(\varphi(K_1)) + \ln(\lambda(K_2, ..., K_m)) \qquad . \ (6)$$

Substituting eqs. (3), (4) and (5) into eq. (6), we find that in the case of independence the following equality is should be satisfied

$$\sum_{n=1}^{\infty} \frac{i^n}{n!} \sum_{i_1, ..., i_n = 1}^{m} \aleph_{i_1 ... i_n} K_{i_1} ... K_{i_n} = \sum_{n=1}^{\infty} \frac{i^n}{n!} \aleph_1^{(n)} K_1^n + \sum_{n=1}^{\infty} \frac{i^n}{n!} \sum_{i_1, ..., i_n = 2}^{m} \aleph_{i_1 ... i_n} K_{i_1} ... K_{i_n} \qquad (7)$$

and therefore

$$\sum_{n=1}^{\infty} \frac{i^n}{n!} \sum_{i_2, ..., i_n = 1}^{m} (1 - \delta_{1 i_2 ... i_n}) \aleph_{1 i_2 ... i_n} K_1 K_{i_2} ... K_{i_n} = 0 \qquad (8)$$

$\delta_{1 i_2 ... i_n}$ is the Kroenecker's delta i.e $\delta_{1 i_2 ... i_n} = 1$ if $i_2 = i_3 = ... = i_n = 1$ and 0 otherwise. Due to the fact that eq. (8) should be satisfied for all $\aleph$, all coefficients in each summation of eq. (8) must be zero. Therefore, a measure that can be used for testing statistical independence in time series can be defined by the cost function

$$D = \sum_{n=1}^{\infty} \sum_{i_2, ..., i_n = 1}^{m} (1 - \delta_{i_2 ... i_n}) \ \{\aleph_{1 i_2 ... i_n}\}^2 \qquad (9)$$

The quantity D is such that $D \geq 0$.

In the framework of statistic testing, an hypothesis about the data can be rejected or accepted based on a measure which discriminates the statistics underlying a distribution consistent with the hypothesis from the original distribution of the data. The hypothesis to be tested is called *null hypothesis*. The measure which quantifies the above mentioned distinction between the statistics of the distributions is called *discriminating statistics*. In general it is very difficult to derive analytically the distribution of a given statistics under a given null hypothesis. The surrogate method proposes to estimate such distributions empirically by generating different versions of the data, called *surrogates*, such that they share some given properties of the original data and are at the same time consistent with the null hypothesis. The rejection of a null hypothesis is therefore based on the computation of the discriminating statistics for the original time series D_O and the discriminating statistics D_{S_i} for the i-th surrogated time series generated under the null hypothesis. The null hypothesis is rejected if the *significance* s given by

$$S = \frac{|D_O - \mu_S|}{\sigma_S} \qquad (10)$$

is greater than $\hat{s}$ corresponding to a p-value $p = erfc(\hat{s}/(\sqrt{2}))$ (p being usually 0.05 which corresponds to $\hat{s} = 1.645$). This means that the probability of observing a significance $\hat{s}$ or larger is p if the null hypothesis is true under the assumption of gaussianity in the surrogates distribution (which is a realistic assumption if the number of surrogates are equal or larger than 100). In eq. (10)

$$\mu_S = \frac{1}{N}\sum_{i=1}^{N} D_{S_i} \qquad \text{and} \qquad \sigma_S = \sqrt{\frac{1}{N}\sum_{i=1}^{N}(D_{S_i} - \mu_S)^2} \qquad (11)$$

are the estimated mean value and standard deviation of the discriminating statistics under the surrogates distribution. In our case the concrete null hypothesis that we would like to test is the one corresponding to the assumption of non-causal relationship between past and presence, i.e. of statistical independency in time. In other words we are trying to answer the most modest question whether there is any dynamics at all underlying the data. The surrogates data used are generated from the original time series by mixing the temporal order, so that if originally any temporal dependence was present it is destroyed by this mixing process. The generated surrogates are scrambled surrogates and have the same statistical characteristics as the original data. The histogram of the original data is preserved but with the particularity that these data do not possess any statistical temporal correlation. The discriminating statistics used here measures the statistical correlations by means of the cumulant expansion, i.e. by calculating the non-diagonal elements of the cumulants according to equation (9). In this paper we include cumulant terms up to fourth order.

If the assumption of stationarity of the observed time series cannot be accepted, then the test of the existence of a dynamics should be performed by the method of overlapping windows or slicing windows. In this method the statistical test of temporal independency at time t is performed in a window of the data of length N_W, i.e by using the data from time $t - N_W \Delta$ to t. Subsequently we slice the window into N_S time steps and perform again the statistical test by using the data from $t + N_S - N_W \Delta$ to $t + N_S$ and so on. In this case it is useful to plot the significance s as a function of the time t where the slicing window finished. It is clear that for stationary time series the value of s will be approximately the same for all times. In other cases this method can be used for the detection of regions in which a stationary behavior is observed or regions in which a dynamics is really underlying the process.

III. Data Selection: Experiments

In this section we would like to demonstrate the selection of data in a time series so that they can be used for learning the underlying dynamic. We use the method of slicing windows in combination with our cumulant-surrogates based testing of statistical correlations. We use the chaotic time series of Hénon [7] distorted by non-stationary additive Gaussian noise defined by the following equation

$$x_{t+1} = 1 - Ax_t^2 + Bx_{t-1} + n_v v \qquad (12)$$

with $A = 1.4$ and $B = 0.3$ and

$$n_v = 0 \qquad \text{if} \quad t < 1000 \lor t > 2000,$$

$$n_v = \left(\frac{t - 1500}{500}\right) \qquad \text{if} \quad t > 1000 \land t < 1500, \qquad (13)$$

$$n_v = \left(\frac{2000 - t}{500}\right) \qquad \text{if} \quad t > 1500 \land t < 2000 \cdot$$

v denotes standard Gaussian noise. In order to select the data in the region with moderate noise where the underlying dynamic is learnable, we calculate the significance, including cumulant terms up to fourth order. The number of surrogates used in this experiments is $N = 100$. In Figure 1 the significance for the noisy Hénon time series is plotted as a function of time for a slicing window with $N_W = 256$ and $N_S = 4$. The dimension of the embedding vector is $m = 3$ and the time lag is $\Delta = 1$. A non-stationary evolution of the significance is observed indicating that between t=1000 and t=2000 the evidence of nonlinear correlations disappears: due to a small signal to noise ratio only uncorrelated Gaussian noise is observed. Therefore the significance is approximately zero meaning that the null hypothesis of independence cannot be rejected.

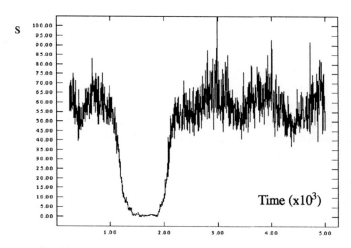

Figure 1: Significance as a function of time for the chaotic noisy Hénon series.

We suppose now that the available training data are the first 2000 and the generalization set is given by the last 3000 data. We train a two-layer perceptron with 2 input units for x_t and x_{t-1}, 10 hidden units and 1 output unit for x_{t+1}, first without using any data selection criterion and then selecting the training data according to the nonparametric results given in Figure 1; i.e. the data selected are the first 1117 which correspond to the region where the null hypotheses of statistical independence can be rejected. The results are presented in Table 1. Clearly, data selection improves the results obtained on the same generalization set.

TABLE 1. : MSE Error

	Training Error	Generalization Error
Without Data Selection	1.6	0.15
With Data Selection	0.00027	0.0012

IV. Conclusions

In this paper we introduce a nonparametric cumulant based statistical approach for detecting statistical dependences in non-stationary time series. The statistical dependence is detected by measuring the predictability which tests the null hypothesis of statistical independence, expressed in Fourier-space, by the surrogate method. Information about the predictability can be used for example to select regions where a temporal structure is visible in order to select data for training a neural network for prediction. The regions where only a noisy behavior is observed are therefore ignored avoiding in this fashion the learning of irrelevant noise which normally spoils the generalization characteristics of the neural network.

References:

[1]   M. Palus, *Physica* D **80**, 186 (1995).
[2]   M. Palus, V. Albrecht and I. Dvorák, *Physics Letters* A **175**, 203 (1993).
[3]   J. Theiler, S. Eubank, A. Longtin, B. Galdrikian and J. Farmer, *Physica* D **58**, 77 (1992).
[4]   G. Deco and B. Schürmann, *Physical Review* E **51**, 1780 (1995).
[5]   G. Deco and W. Brauer, *Neural Networks* **8**, 525 (1995).
[6]   G. Deco and D. Obradovic, "An Information-Theoretic Approach to Neural Computing" (Springer Verlag, New York, 1996).
[7]   M. Hénon, *Comm. Math. Phys.* **50**, 69 (1976).

Prediction of Mixtures

K. Pawelzik[1], K.-R. Müller[2], J. Kohlmorgen[2]

[1] Inst. f. theo. Physik, Universität Frankfurt, 60054 Frankfurt/M., Germany
[2] GMD FIRST, Rudower Chaussee 5, 12489 Berlin, Germany

Abstract. The problem of predicting time series originating from mixtures of signals from independent dynamical systems is considered. We show that the problem of finding representations for the dynamics of such systems is hard if the mixing structure of the system is not taken into account. If, on the contrary, the sources can be unmixed in a preprocessing step the complexity of system identification may be drastically reduced. This is demonstrated using chaotic maps. It is shown that applications of methods for blind separation of sources can substantially improve both: prediction performance and prediction horizon.

1 Introduction

Time series from real systems originate from unique autonomous dynamical systems only under ideal circumstances. More common is the presence of additional noise, but also nonstationarities and the fact that data often are superpositions of different sources may challenge attempts to model the systems by compact representations e.g. using large neural nets (see e.g. [14, 16]).

For systems which switch their dynamics, methods for unsupervised segmentation have already been presented and it has been demonstrated that they can improve predictions [6, 7, 12, 13, 3, 17].

Here we are concerned with the other paradigmatic situation of compositional systems: mixtures of independent sources. This situation is present in many problems of time series analysis; most prominent maybe is the example of speech recognition where the relevant signal often is superimposed by other voices or non-speech signals.

When analysing the problem of system identification and prediction of mixtures, we found that its complexity may grow dramatically, if the mixed nature of the signal is not taken into account. If, in contrast, the sources can be separated, system identification can become feasible in cases which are otherwise intractable for the given data.

2 Curse of Dimensionality

Here we demonstrate this general effect with the simple example of mixtures of two chaotic time series. Consider the time series $x^1_{t+1} = f_1(x^1_t)$ and $x^2_{t+1} = f_2(x^2_t)$, where $f_1(x) = 4x(1 - x)$, and $f_2(x) = 2x$, if $0 \leq x \leq 1/2$, and $f_2(x) = 2 - 2x$, if $1/2 < x \leq 1$, are the logistic map and the tent map respectively.

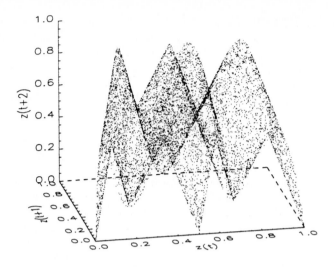

Fig. 1.

Next, consider the simple mixture signal $z_{t+1} = f_1(x_t^1) + f_2(x_t^2)$, $t = 1, \ldots, T$. Obviously there is no unique one dimensional map representing this situation, i.e. there is no function $h : z_t \to z_{t+1}$. On the other hand it has been proven that a system can in principle be reconstructed from an observable using an embedding of dimension $m^* = 2d+1$ where d is the dimension of the underlying dynamics [15]. In our case the dimensionality of the system is the sum of the dimensions of the individual subsystems, i.e. $d = \sum d_i$ and we have $m^* = 5$. This means that there is a representation $z_{t+1} = h(z_t, z_{t-1}, \ldots, z_{t-m+1})$ as soon as m is large enough.

Considering $m = 2$ we see that the problem is far from solved for our example (Fig. 1). We observe not only a 'cloud' instead of a function in this case, but also that the representation is strongly modulated. This comes as no surprise considering that already the complexity of representing the dynamics of one source (expressed e.g. by the order of the polynomial) grows exponentially with iteration time, the exponent being the positive Lyapunov exponent.

These two effects, the necessity of high dimensional input spaces together with the increasing complexity of the representation make system identification and prediction highly tedious even for this simple example. To show this, we use a simple radial basis function network $\hat{h}$ with 100 centers [11] trained on the mixture signal z_t using T=2000 points. The root mean squared one-step prediction errors (RMSE) are $e = 0.314, 0.198, 0.231, 0.289, 0.304$ for the embeddings $m = 1, 2, 3, 4, 5$ respectively. Note that the nominally sufficient embedding of $m = 5$ did not entail an improvement of prediction. This is not surprising, considering the fact that the complexity of representing the dynamics in time delay coordinates grows *exponentially* with the embedding dimension, the exponent being 2 in this case.

This result is particularly striking because predicting the individual maps in $m = 1$ using two radial basis function networks of 100 nodes each leads to errors which are $2 - 3$ orders of magnitude smaller, namely $e = 0.000222$ and $e = 0.000327$ for the logistic map and the tent map, respectively. Obviously, mixing induces a severe dimensionality problem, which can be avoided, if the underlying sources are separated prior to modeling.

3   Blind separation

Our previous example demonstrates the importance of separating the sources underlying a signal, if the dynamics of the time series has to be modeled. Fortunately, several powerful methods for blind separation [1, 2, 4, 5, 10] have been recently developed, which are very useful when several observables are available which represent different mixtures of the underlying systems. More explicitly, let the time series be represented by a vector $\mathbf{z}_t = (z_t^1, \ldots, z_t^n)^T$ with $\mathbf{z}_t = M\mathbf{x}_t$, i.e. a mixture of the source vector $\mathbf{x}_t = (x_t^1, \ldots, x_t^n)^T$. Then the above mentioned methods may reconstruct the original source signals without further assumptions except that M is invertible and that the sources are mutually independent. The method from [10] applies to linearly independent sources and exploits linear autocorrelations. The other approaches cited above rely on higher moments of mutual correlations and ignore the temporal coherence of the sources. Recent results indicate that modifications of these methods may be useful also in cases where the number of available signals is smaller than the number of underlying sources (T. Gramß, personal communication).

4   Predicting Mixtures

Instead of discussing the advantages and drawbacks of these methods we will here demonstrate that their application can substantially improve predictions. As a first example we used a mixture of the maps discussed in section 2. We generated series of $T = 2000$ points each and mixed them using

$$M = \begin{pmatrix} 1 & -0.53 \\ -0.87 & 1 \end{pmatrix}$$

Applying a blind separation method [2, 10], we found a separation matrix B yielding good estimates for the sources $\tilde{x}_t^1$ and $\tilde{x}_t^2$. We then trained two radial basis function networks $\tilde{f}_1$ and $\tilde{f}_2$ with 20 nodes each on $\tilde{x}_t^1$ and $\tilde{x}_t^2$. Finally, we computed the prediction errors e_1 and e_2 for z^1 and z^2, respectively using

$$\tilde{\mathbf{z}}_{t+1} = B^{-1}\tilde{F}[B\,\mathbf{z}_t], \quad \text{where} \quad \tilde{F}(\tilde{\mathbf{x}}_t) = (\tilde{f}_1(\tilde{x}_t^1), \tilde{f}_2(\tilde{x}_t^2)). \tag{1}$$

We found that this approach improved predictions by a factor of 4 compared to the best results for the direct prediction method. As a more realistic example

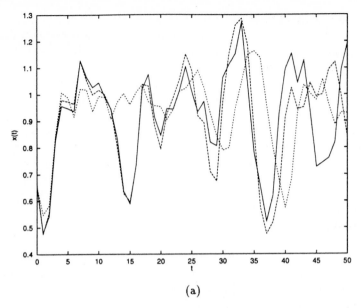

(a)

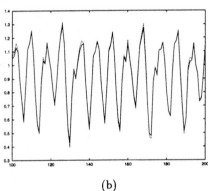

(b)

Fig. 2. *Prediction of mixed Mackey-Glass sources. (a) Mixed signal z^1 (solid line), iterative prediction of a global radial basis function network (dashed), and iterative prediction by the divisive approach which involved blind separation (long dashes). Clearly the divisive strategy yields a better result than the global prediction. (b) Original source x^1 (solid line) in comparison to the estimate $\tilde{x}^1$ from blind separation (dashes). Similar results were obtained for z^2 and x^2 (not shown here).*

we used the well-known Mackey-Glass equation [9]

$$\frac{dx(t)}{dt} = -0.1x(t) + \frac{0.2x(t-\tau)}{1 + x(t-\tau)^{10}}, \qquad (2)$$

which was originally introduced as a model for the irregular dynamics of blood cell production. We generated two time series of $T = 1000$ points each, sampled at time steps of length $\Delta t = 6$ using two parameter values $\tau = 17$ and $\tau = 23$ (Fig. 2b). The mixed sources of these two dynamical systems using

$$M = \begin{pmatrix} 0.3 \ 0.7 \\ 0.8 \ 0.2 \end{pmatrix}$$

then provided the observables z_t (Fig. 2a). First, these signals were modeled globally using a single radial basis function network (m= 6, 120 nodes each). The RMSE were computed from a test sample as $e = 0.068$ and $e = 0.055$ for z^1 and z^2, respectively.

A subsequent blind separation of the mixtures [2] proved to be very effective here (see dashed line in Fig. 2b). As before, we trained two radial basis function networks (m = 6, 120 nodes) on the reconstructed source signals and then computed the RMSE for the remixed signals $\tilde{z}_t^1, \tilde{z}_t^2$ using the inverse of the estimated unmixing matrix as in Eq. 1, which gave $e = 0.016$ and $e = 0.014$ respectively, i.e. the prediction improved by a factor of roughly 4 again. This is highly relevant, because it also substantially increases the prediction horizon as seen from Fig. 2a.

5 Summary and Discussion

It was demonstrated that mixing of dynamical sources can impose severe problems for system identification and prediction tasks. We showed that applications of blind separation methods can alleviate these problems, in particular the curse of dimensionality. Clearly, prior information, e.g. knowing that there is a mixing structure inherent in the signal, leads to a *divide-and-conquer* strategy, which in this case is in fact an *unmix-predict-mix* algorithm. The performance gain of this strategy for a prediction task is large and depends crucially on (a) the severity of the curse of dimensionality and (b) the accuracy of the estimation of the mixture matrix M. It should be noted that also the prediction horizon is increased, i.e. we obtain a more stable long term prediction (cf. Fig. 2a).

Future research will focus on the application of our approach to real data.[3]

Acknowledgment: K.P. is supported by DFG (grant Pa 569/1-1). We thank T.Gramß and A.Ziehe for fruitful discussions. We acknowledge T.Bell, H.Yang and J.F.Cardoso for providing source code and valuable help.

References

1. Amari, S., Cichocki, A., Yang, H., A new learning algorithm for blind signal separation, *Advances in Neural Information Processing Systems 8* (NIPS 95), D.S. Touretzky, M.C. Mozer and M.E. Hasselmo (eds.), MIT Press: Cambridge, MA (1996)
2. Bell, A.J., Sejnowski, T., An information-maximization approach to blind separation and blind deconvolution, *Neural Computation* 7, 1129-1159 (1995)
3. Bengio, Y., Frasconi, P. (1994). Credit assignment through time: Alternatives to backpropagation. NIPS 93, Morgan Kaufmann.
4. Cardoso, J.F., Laheld, B., Equivariant adaptive source separation, to appear in *IEEE Trans. on Signal Processing* (1996)

[3] Further information on related research can be found at:
http://www.first.gmd.de/persons/Mueller.Klaus-Robert.html.

5. Jutten, C., Herault, J., Blind separation of sources, Part I: An adaptive algorithm based on neuromimetic architecture, *Signal Processing* 24, 1-10 (1991)

6. Kohlmorgen, J., Müller, K.-R., Pawelzik, K. (1994). Competing Predictors Segment and Identify Switching Dynamics. ICANN'94, Springer London, 1045–1048.

7. Kohlmorgen, J., Müller, K.-R., Pawelzik, K. (1995). Improving short-term prediction with competing experts. ICANN'95, EC2 & Cie, Paris, 2:215-220.

8. Liebert, W., Pawelzik, K., Schuster, H.G., Optimal embeddings of chaotic attractors from topological considerations, Europhys. Lett. **14**, 521 (1991).

9. Mackey, M., Glass, L., Oscillation and chaos in a physiological control system, Science **197**, 287 (1977).

10. Molgedey, L., Schuster, H.G., Separation of a mixture of independent signals using time delayed correlations, *Phys. Rev. Lett.* 72, 23, 3634-3637 (1994)

11. Moody, J., C. Darken (1989). Fast Learning in Networks of Locally-Tuned Processing Units. Neural Computation **1**, 281-294, 1989.

12. Müller, K.-R., Kohlmorgen, J., Pawelzik, K. (1995). Analysis of Switching Dynamics with Competing Neural Networks, *IEICE Transactions on Fundamentals of Electronics, Communications and Computer Sciences*, E78-A, No.10, 1306-1315.

13. Pawelzik, K., Kohlmorgen, J., Müller, K.-R. (1996). Annealed Competition of Experts for a Segmentation and Classification of Switching Dynamics, *Neural Computation*, 8:2, 342-358.

14. Rumelhart, D.E., McClelland, J.L., *Parallel distributed processing*, MIT Press, Cambridge Massachusetts (1984).

15. Takens, F., Detecting strange attractors in turbulence, in: Rand, D., Young, L.-S., (Eds.), *Dynamical Systems and Turbulence*, Springer Lecture Notes in Mathematics, **898**, 366 (1981).

16. Weigend, A.S., Gershenfeld, N.A. (Eds.), *Time Series Prediction: Forecasting the Future and Understanding the Past*, Addison-Wesley (1994)

17. Weigend, A.S., Mangeas, M., Nonlinear gated experts for time series: discovering regimes and avoiding overfitting, University of Colorado tech. Report. CU-CS-764-95 and submitted to *Machine Learning* (1995)

Learning Dynamical Systems Produced by Recurrent Neural Networks

Masahiro KIMURA and Ryohei NAKANO

NTT Communication Science Laboratories
2-2 Hikaridai, Seika-cho, Kyoto 619-02, Japan

Abstract. In this paper, the concept of a neural dynamical system is proposed to investigate mathematically the learning of a dynamical system by a recurrent neural network with hidden units. We present the conditions for our recurrent neural network to produce a neural dynamical system, and also discuss the possibility of identifying a neural dynamical system via the learning of given its temporal trajectories.

1 Introduction

A recurrent neural network (RNN) usually consists of visible units (i.e., output units) and hidden units. The visible space, the space of the visible units, corresponds to the configuration space of an object to be controlled, and the temporal trajectories that the RNN generates on the visible space model the dynamical behavior of the object. In this paper, only continuous-time RNNs and continuous dynamical systems (DS) [7] are considered.

It is known [3] that an RNN can approximate an arbitrary DS in a finite time and a compact space. For an RNN with no hidden units, we have proved that such a simple RNN always produces a DS on the phase space, and also have obtained a geometric criterion for the possibility of identifying the DS via the learning of its temporal trajectories [6]. However, an RNN with hidden units may not produce a DS on the visible space in general, because on the visible space, such an RNN can generate infinitely many trajectories through the same point and these trajectories can have self-intersections. Thus, the extension of our previous result [6] to the case with hidden units is never straightforward.

We consider making an RNN learn temporal trajectories generated from a DS. The learning of temporal trajectories is basically performed under the gradient descent method for the error function which measures the discrepancy between the trajectories generated from the RNN and the desired trajectories [1]. However, for complicated tasks, the straightforward application of exisiting general-purpose algorithms to train the RNN has been less successful than their feedforward counterparts [9]. This problem is mainly caused by the fact that the role of hidden units is crucial for RNN training [9]. Therefore, for developing effective learning algorithms, it is significant to analyze mathematically what an RNN with hidden units actually produces on the visible space.

In order to investigate mathematically the learning of a DS by an RNN with hidden units, this paper proposes a *neural dynamical system* (NDS), that is, such a DS that an RNN produces on the visible space, and also extends our previous result [6] to the case with hidden units. In section 2, the class of RNNs to be considered is specified, and NDS is formally defined. In section 3, we present the conditions for our RNN to produce a NDS, and also refer to a phase space learning scheme [9] of a DS by our RNN. In section 4, given temporal trajectories generated from an NDS, we present a geometric interpretation for RNNs that generate the temporal trajectories. This interpretation leads to a geometric criterion for whether or not the NDS can be identified when the given temporal trajectories are perfectly learned.

2 Preliminaries

2.1 Recurrent Neural Networks

In this paper, the following additive type network is considered as an RNN because it is widely used [4]; that is, our RNN consists of $n + k$ units (the units 1 to n are visible and the units $n + 1$ to $n + k$ are hidden), and the state $u_i(t)$ of unit i at time t is determined by the system of the ODEs

$$\frac{du_i}{dt}(t) = -\frac{1}{\tau}u_i(t) + \sum_{j=1}^{n+k} W_{ij}g(u_j(t)) + I_i, \qquad (i = 1, \cdots, n+k),$$

where τ is a positive constant, g is a bounded monotone increasing C^∞-function on $\mathbb{R}$, $W = (W_{ij})$ is an element of the set $M_{n+k}(\mathbb{R})$ of all $(n+k) \times (n+k)$ real matrices, and $I = (I_1, \cdots, I_{n+k}) \in \mathbb{R}^{n+k}$. Note that the weight matrix W is the only parameter adjustable for learning while the time constant τ, the activation function g, and the external input I are given. Since our RNN is parametrized by an element of $M_{n+k}(\mathbb{R})$, we denote by $\mathcal{RN}_W$ the RNN corresponding to $W \in M_{n+k}(\mathbb{R})$. The activation function g is extended to the C^∞-isomorphism g of $\mathbb{R}^d$ by $g(y_1, \cdots, y_d) = (g(y_1), \cdots, g(y_d))$. Let the external input I be $I = (I^{(1)}, I^{(2)})$, $(I^{(1)} \in \mathbb{R}^n, I^{(2)} \in \mathbb{R}^k)$.

The phase space of an RNN is $\mathbb{R}^{n+k}$, the visible space is $\mathbb{R}^n$, and the hidden space is $\mathbb{R}^k$; they have the relation $\mathbb{R}^{n+k} = \mathbb{R}^n \times \mathbb{R}^k$. Let $W \in M_{n+k}(\mathbb{R})$. The following notation is used:

$$W = \begin{pmatrix} W^{(1,1)} & W^{(1,2)} \\ W^{(2,1)} & W^{(2,2)} \end{pmatrix},$$

where $W^{(1,1)}$, $W^{(1,2)}$, $W^{(2,1)}$, and $W^{(2,2)}$ are the $n \times n$, $n \times k$, $k \times n$, and $k \times k$ real matrices, respectively. We define the vector field F_W on $\mathbb{R}^n \times \mathbb{R}^k$ as follows: For $x \in \mathbb{R}^n$ and $v \in \mathbb{R}^k$, $F_W(x, v) = \left(F_W^{(1)}(x, v), F_W^{(2)}(x, v)\right) \in \mathbb{R}^n \times \mathbb{R}^k$,

$$F_W^{(1)}(x, v) = -\frac{1}{\tau}x + W^{(1,1)}g(x) + W^{(1,2)}g(v) + I^{(1)},$$

$$F_W^{(2)}(x, v) = -\frac{1}{\tau}v + W^{(2,1)}g(x) + W^{(2,2)}g(v) + I^{(2)},$$

where matrix operations are used. Throughout this paper, an element of $\mathbb{R}^d$ is regarded as a $d \times 1$ real matrix whenever matrix operations are used.

Remark A: Suppose that $(x(t), v(t))$ is an integral curve (a solution curve) [8] of the vector field F_W, where $x(t) \in \mathbb{R}^n$, $v(t) \in \mathbb{R}^k$ for any t. Then, $(x(t), v(t)) \in \mathbb{R}^{n+k}$ is the state of the RNN $\mathcal{RN}_W$ at time $t \in \mathbb{R}$ under the initial condition $(x(0), v(0)) \in \mathbb{R}^{n+k}$, and the C^∞-curve $x(t)$ is a temporal trajectory generated from the $\mathcal{RN}_W$ on the visible space $\mathbb{R}^n$.

2.2 Neural Dynamical Systems

First, we recall the definition of a DS on $\mathbb{R}^d$ [7]. Let $\psi : \mathbb{R} \times \mathbb{R}^d \to \mathbb{R}^d$ be a C^∞-map. For each $t \in \mathbb{R}$, we define a C^∞-map $\psi^t : \mathbb{R}^d \to \mathbb{R}^d$ by $\psi^t(y) = \psi(t, y)$, $(y \in \mathbb{R}^d)$. The pair $(\mathbb{R}^d, \psi)$ is called a *dynamical system* (DS) on the phase space $\mathbb{R}^d$ if the following conditions are satisfied: $\psi^0 = \mathrm{id}_{\mathbb{R}^d}$, and $\psi^s \circ \psi^t = \psi^{s+t}$, $(s, t \in \mathbb{R})$. As is well-known [7, 8], a DS $(\mathbb{R}^d, \psi)$ *induces* a complete vector field Y on $\mathbb{R}^d$ by $Y(y) = (d/dt)|_{t=0}\psi^t(y)$, $(y \in \mathbb{R}^d)$, and this correspondence is one-to-one. For the DS $(\mathbb{R}^d, \psi)$ corresponding to a complete vector field Y on $\mathbb{R}^d$, the C^∞-map ψ is called the *flow* of Y.

We propose the concept of a *neural dynamical system* (NDS) which means such a DS that an RNN produces on the visible space. It is known [6] that an RNN $\mathcal{RN}_W$, $(W \in M_{n+k}(\mathbb{R}))$ produces a DS on the phase space $\mathbb{R}^{n+k}$ using the states of all $n + k$ units, and this correspondence gives a one-to-one map from the set of our RNNs to the set of DSs on $\mathbb{R}^{n+k}$. The DS produced by the $\mathcal{RN}_W$ is denoted by $(\mathbb{R}^{n+k}, \Phi_W)$. Note that the temporal trajectories that the $\mathcal{RN}_W$ generates through $x \in \mathbb{R}^n$ on the visible space $\mathbb{R}^n$ are written in the form of projection $\pi(\Phi_W^t(x, v))$, $(v \in \mathbb{R}^k)$, where $\pi : \mathbb{R}^n \times \mathbb{R}^k \to \mathbb{R}^n$ is the natural projection. Hence, a DS produced by $\mathcal{RN}_W$ on $\mathbb{R}^n$ can be written in the form $\pi(\Phi_W^t(x, h(x)))$, where $h : \mathbb{R}^n \to \mathbb{R}^k$ being not necessarily continuous.

Throughout this paper, we fix a C^∞-map $h : \mathbb{R}^n \to \mathbb{R}^k$. For $W \in M_{n+k}(\mathbb{R})$, a C^∞-map $\varphi_W : \mathbb{R} \times \mathbb{R}^n \to \mathbb{R}^n$ is defined by

$$\varphi_W^t(x) = \pi(\Phi_W^t(x, h(x))), \qquad (t \in \mathbb{R}, \ x \in \mathbb{R}^n). \tag{1}$$

When $(\mathbb{R}^n, \varphi_W)$ forms a DS, we call the pair $(\mathbb{R}^n, \varphi_W)$ the *neural dynamical system* (NDS) produced by the RNN $\mathcal{RN}_W$ under the C^∞-map h.

3 Existence of Neural Dynamical Systems

This section investigates the conditions for an RNN with hidden units to produce an NDS on the visible space, and also present examples of RNNs that produce DSs on the visible spaces. Moreover, we specify the form of the vector field induced by an NDS. Note that a DS is determined by a complete vector field.

Theorem 1. *Let $W \in M_{n+k}(\mathbb{R})$. An RNN $\mathcal{RN}_W$ produces the NDS $(\mathbb{R}^n, \varphi_W)$ under the C^∞-map h if and only if the state $(x(t), v(t)) \in \mathbb{R}^n \times \mathbb{R}^k = \mathbb{R}^{n+k}$ of the RNN $\mathcal{RN}_W$ at time $t \in \mathbb{R}$ under the initial condition $(p, h(p)) \in \mathbb{R}^n \times \mathbb{R}^k = \mathbb{R}^{n+k}$ satisfies*

$$g(h(x(t))) - g(v(t)) \in \mathrm{Ker}\, W^{(1,2)}, \qquad (t \in \mathbb{R}),$$

in the real vector space $\mathbb{R}^k$ *for any* $p \in \mathbb{R}^n$, *where the weight matrix* $W^{(1,2)}$ *from the hidden units to the visible units is regarded as a linear map from* $\mathbb{R}^k$ *to* $\mathbb{R}^n$.

Moreover, for the NDS $(\mathbb{R}^n, \varphi_W)$, *the* C^∞*-map* φ_W *is the flow of the vector field* X_W *on* $\mathbb{R}^n$ *defined by* $X_W(x) = F_W^{(1)}(x, h(x))$, $(x \in \mathbb{R}^n)$.

Proof. From definition (1) of φ_W, it is easily seen that $\varphi_W^0 = \mathrm{id}_{\mathbb{R}^n}$ and $(d/dt)|_{t=0}$ $\varphi_W^t(x) = F_W^{(1)}(x, h(x)) = X_W(x)$, $(x \in \mathbb{R}^n)$ since $\Phi_W^0 = \mathrm{id}_{\mathbb{R}^{n+k}}$ and $(d/dt)|_{t=0}$ $\Phi_W^t(x, h(x)) = F_W(x, h(x))$, $(x \in \mathbb{R})$. Then, as is well-known [8], $\varphi_W^s \circ \varphi_W^t = \varphi_W^{s+t}$, $(s, t \in \mathbb{R})$ if and only if

$$\frac{d}{dt}\varphi_W^t(x) = X_W(\varphi_W^t(x)) \left(= F_W^{(1)}(\varphi_W^t(x), h(\varphi_W^t(x)))\right), (t \in \mathbb{R}, \ x \in \mathbb{R}^n). \quad (2)$$

Hence, the pair $(\mathbb{R}^n, \varphi_W)$ is a DS if and only if condition (2) is satisfied. For $p \in \mathbb{R}^n$ and the integral curve $(x(t), v(t))$ of the vector field F_W on $\mathbb{R}^n \times \mathbb{R}^k$ with $x(0) = p$ and $v(0) = h(p)$, it follows that $\varphi_W^t(p) = x(t)$ and $(d/dt)\varphi_W^t(p) = F_W^{(1)}(x(t), v(t))$, $(t \in \mathbb{R})$. From these results and Remark A, we can easily prove the theorem. □

Let $\mathcal{H}$ be the subset of $M_{n+k}(\mathbb{R})$ such that an RNN $\mathcal{RN}_W$, $(W \in \mathcal{H})$ produces the NDS $(\mathbb{R}^n, \varphi_W)$ on the visible space $\mathbb{R}^n$ under the C^∞-map h. Theorem 1 shows a necessary and sufficient condition for $W \in M_{n+k}(\mathbb{R})$ to be an element of $\mathcal{H}$, and also gives the general form of NDSs.

Next, we consider an important special case; $h(x) = Hx + \theta$, $(x \in \mathbb{R}^n)$, where H is a $k \times n$ real matrix and $\theta \in \mathbb{R}^k$. The subset $\mathcal{H}_1$ of $M_{n+k}(\mathbb{R})$ is defined by

$$\mathcal{H}_1 = \left\{ W \in M_{n+k}(\mathbb{R}) \ \middle| \ HW^{(1,1)} = W^{(2,1)}, \ HW^{(1,2)} = W^{(2,2)} \right\}.$$

The following corollary shows $\mathcal{H}_1 \subset \mathcal{H}$, and supplies examples of RNNs that produce NDSs.

Corollary 2. *Suppose that* $HI^{(1)} + (1/\tau)\theta = I^{(2)}$. *For* $W \in \mathcal{H}_1$, *the RNN* $\mathcal{RN}_W$ *produces the NDS* $(\mathbb{R}^n, \varphi_W)$ *on the visible space* $\mathbb{R}^n$ *under the* C^∞*-map* $Hx + \theta$, *and the induced vector field* X_W *on* $\mathbb{R}^n$ *is as follows:*

$$X_W(x) = -\frac{1}{\tau}x + W^{(1,1)}g(x) + W^{(1,2)}g(Hx + \theta) + I^{(1)}, \quad (x \in \mathbb{R}^n). \quad (3)$$

Proof. Suppose that $(x(t), v(t))$ is the integral curve of the vector field F_W with $x(0) = p$ and $v(0) = h(p)$ for $p \in \mathbb{R}^n$. Then, we can easily show that the C^∞-curve $(x(t), h(x(t)))$ in $\mathbb{R}^n \times \mathbb{R}^k$ is also the integral curve of the vector field F_W with the same initial condition as above. Hence, $h(x(t)) = v(t)$, $(t \in \mathbb{R})$. From Theorem 1 and Remark A, we get the corollary. □

As for learning a DS $(\mathbb{R}^n, \psi)$, the phase space learning scheme [9] is proposed to overcome the problem of existing algorithms to train an RNN. The phase space learning is accomplished through both the reconstruction of the induced vector field Y and the learning by a feedforward network with the universal approximation capability [2, 5]. Note that Corollary 2 interprets the phase space learning scheme by our RNN (see equation (3) and the following remark).

Remark B: Suppose that $Y(x) + (1/\tau)x = Bg(Hx + \theta)$, ($x \in \mathbb{R}^n$), where B is an $n \times k$ real matrix, H is a $k \times n$ real matrix and $\theta \in \mathbb{R}^k$. Note that the C^∞-map $\mathbb{R}^n \ni x \mapsto Bg(Hx + \theta) \in \mathbb{R}^n$ represents a three-layer feedforward neural network. We consider the RNN $\mathcal{RN}_W$ such that

$$W = \begin{pmatrix} 0 & B \\ 0 & HB \end{pmatrix}, \quad I = \left(0, \frac{1}{\tau}\theta\right) \in \mathbb{R}^n \times \mathbb{R}^k.$$

From Corollary 2, it is easily seen that the DS $(\mathbb{R}^n, \psi)$ is the NDS that this RNN produces on the visible space $\mathbb{R}^n$ under the C^∞-map $Hx + \theta$.

4 Temporal Trajectories of Neural Dynamical Systems

This section discusses the possibility of identifying an NDS by learning given its temporal trajectories, and presents an extension of our previous result [6] to the case with hidden units. We fix $A = (A_{ij}) \in \mathcal{H}$. Let $\xi^{(1)}, \cdots, \xi^{(m)}$ be temporal trajectories generated from the NDS $(\mathbb{R}^n, \varphi_A)$. The following theorem presents a geometric interpretation for the set $\Omega(\xi^{(1)}, \cdots, \xi^{(m)})$ of RNNs $\mathcal{RN}_W$, ($W \in \mathcal{H}$) such that the produced NDS $(\mathbb{R}^n, \varphi_W)$ generates the temporal trajectories $\xi^{(1)}, \cdots, \xi^{(m)}$.

Theorem 3. *For $W = (W_{ij}) \in \mathcal{H}$, $\mathcal{RN}_W \in \Omega(\xi^{(1)}, \cdots, \xi^{(m)})$ if and only if*

$$(W_{i1}, \cdots, W_{in+k}) \in (A_{i1}, \cdots, A_{in+k}) + \mathcal{V}(\xi^{(1)}, \cdots, \xi^{(m)})^\perp, \quad (i = 1, \cdots, n),$$

in the real vector space $\mathbb{R}^{n+k}$, where $\mathcal{V}(\xi^{(1)}, \cdots, \xi^{(m)})$ is the minimum vector subspace of $\mathbb{R}^{n+k}$ including the set

$$\bigcup_{\alpha=1}^{m} \left\{ \left(g(\xi^{(\alpha)}(t)), g(h(\xi^{(\alpha)}(t)))\right) \in \mathbb{R}^n \times \mathbb{R}^k \mid t \in \mathbb{R} \right\},$$

and the symbol "$\perp$" means the orthogonal complement with respect to the canonical inner-product $\langle , \rangle$ on $\mathbb{R}^{n+k}$.

Proof. Let $W \in \mathcal{H}$. By Theorem 1, a temporal trajectory $x(t)$ of an NDS $(\mathbb{R}^n, \varphi_W)$ is a solution of the ODE $(d/dt)x(t) = F_W^{(1)}(x(t), h(x(t)))$. Hence, a necessary and sufficient condition for $\mathcal{RN}_W \in \Omega(\xi^{(1)}, \cdots, \xi^{(m)})$ is that for $t \in \mathbb{R}$ and $\alpha = 1, \cdots, m$,

$$F_W^{(1)}(\xi^{(\alpha)}(t), h(\xi^{(\alpha)}(t))) = F_A^{(1)}(\xi^{(\alpha)}(t), h(\xi^{(\alpha)}(t))). \tag{4}$$

A straightforward calculation yields that condition (4) is equivalent to the following condition: For $t \in \mathbb{R}$ and $\alpha = 1, \cdots, m$,

$$(W^{(1,1)} - A^{(1,1)})g(\xi^{(\alpha)}(t)) + (W^{(1,2)} - A^{(1,2)})g(h(\xi^{(\alpha)}(t))) = 0. \tag{5}$$

Condition (5) is also equivalent to the condition; for $t \in \mathbb{R}$ and $\alpha = 1, \cdots, m$,

$$\left\langle \left(W_{i1} - A_{i1}, \cdots, W_{in+k} - A_{in+k}\right), \left(g(\xi^{(\alpha)}(t)), g(h(\xi^{(\alpha)}(t)))\right) \right\rangle = 0,$$

$(i = 1, \cdots, n)$. From these results, we can easily prove the theorem. $\qquad\square$

Suppose again that $h(x) = Hx + \theta$, where H is a $k \times n$ real matrix and $\theta \in \mathbb{R}^k$. We also assume that the external input I and the time constant τ satisfy $HI^{(1)} + (1/\tau)\theta = I^{(2)}$, and consider only the RNNs constructed from the elements of $\mathcal{H}_1$. The following corollary is easily derived from Theorem 3.

Corollary 4. *Let $A \in \mathcal{H}_1$. Then, the NDS $(\mathbb{R}^n, \varphi_A)$ is identified via the perfect learning of the temporal trajectories $\xi^{(1)}, \cdots, \xi^{(m)}$ by the RNN $\mathcal{RN}_W$, $(W \in \mathcal{H}_1)$, that is, $\{\mathcal{RN}_W \in \Omega(\xi^{(1)}, \cdots, \xi^{(m)}) \mid W \in \mathcal{H}_1\} = \{\mathcal{RN}_A\}$, if and only if the images of the C^∞-curves $(g(\xi^{(\alpha)}(t)), g(H\xi^{(\alpha)}(t) + \theta))$, $(\alpha = 1, \cdots, m)$ in $\mathbb{R}^{n+k}$ span the real vector space $\mathbb{R}^{n+k}$, that is, $\mathcal{V}(\xi^{(1)}, \cdots, \xi^{(m)}) = \mathbb{R}^{n+k}$.*

Corollary 4 shows a geometric criterion for the possibility of identifying an NDS produced by an RNN with the special architecture via learning given its temporal trajectories.

5 Conclusion

We have proposed and mathematically investigated a neural dynamical system (NDS), that is, such a dynamical system (DS) that a recurrent neural network (RNN) with hidden units produces on the visible space, and also investigated the possibility of identifying an NDS via the learning of given its temporal trajectories. We have obtained the conditions for our RNN to produce an NDS, and also presented a geometric interpretation for RNNs that generate given temporal trajectories of an NDS.

References

1. Baldi, P., Gradient descent learning algorithm overview: A general dynamical systems perspective, *IEEE Transactions on Neural Networks* **6** (1995), 182–195.
2. Funahashi, K., On the approximate realization of continuous mappings by neural networks, *Neural Networks* **2** (1989), 183–191.
3. Funahashi, K., and Nakamura, Y., Approximation of dynamical systems by continuous time recurrent neural networks, *Neural Networks* **6** (1993), 801–806.
4. Hertz, J., Krogh, A., and Palmer, R. G., *Introduction to the Theory of Neural Computation*. Addison Wesley, 1991. Lecture Notes, Santa Fe Institute Studies in the Sciences of Complexity
5. Hornik, K., Stinchcombe, M., and White, H., Multilayer feedforward networks are universal approximators, *Neural Networks* **2** (1989), 359–366.
6. Kimura, M., and Nakano, R., Learning dynamical systems from trajectories by continuous time recurrent neural networks, *Proceedings of the 1995 IEEE International Conference on Neural Networks* **6** (1995), 2992–2997.
7. Matsumoto, T., Komuro, M., Kokubu, H., and Tokunaga, R., *Bifurcations*. Springer-Verlag, 1993.
8. Matsushima, Y., *Differentiable Manifolds*, Marcel-Dekker, 1972.
9. Tsung, F-S., and Cottrell, G. W., Phase-space learning, *Advances in Neural Information Processing Systems* **7** (1995), 481–488.

Purely Local Neural Principal Component and Independent Component Learning

Aapo Hyvärinen

Helsinki University of Technology
Laboratory of Computer and Information Science
Rakentajanaukio 2 C, FIN-02150 Espoo, Finland
Email: aapo.hyvarinen@hut.fi

Abstract. New algorithms for neural Principal Component and Independent Component Analysis (PCA and ICA) are introduced. Special emphasis is laid on the locality of learning. This enables a simpler hardware implementation, and may provide a more plausible model of biological neurons. To achieve this, the algorithms feature a new kind of feedback which is multiplicative and anti-Hebbian. The convergence of the ICA algorithm is proven analytically in the general case; the convergence of the PCA algorithm is proven for Gaussian data.

1 Introduction

Several "neural" algorithms for Principal Component Analysis (PCA) and Independent Component Analysis (ICA) have been proposed by different authors [1, 2, 5, 6, 7]. PCA is a widely used signal processing paradigm (see [6]). ICA is a very recent extension of PCA which is explained in Section 4.

In most PCA and ICA algorithms, explicit knowledge of the weight vectors of the other units is required to determine the change in the weight vector $\mathbf{w}^i$ of the i-th unit. This is due to the form of feedback used. In a completely neural algorithm such feedback is to be avoided for two reasons.

First, in a VLSI implementation, such feedbacks lead to a rather cumbersome wiring. If the network contains m neurons and the weight vectors are n-dimensional, the number of scalar feedback connections becomes of the order nm^2, which may be prohibitive as n can be very large. Second, locality seems to be an intrinsic property of biological neural networks. If neural networks are used for modelling biological neurons, non-local architectures may not be considered very plausible.

Therefore, we present in this paper neural PCA and ICA algorithms that are truly local in the following sense: Every neuron receives, in addition to the original input, at most one scalar feedback from each neuron. The number of feedback connections is thus reduced to the order of m^2, and can further be reduced to the order of m, if simple multiplicative units are added to aggregate feedbacks. The algorithms use a normalization term adapted from [7], and a new kind of feedback. This multiplicative anti-Hebbian feedback makes the neuron learn slower when other neurons fire strongly. It is purely local in the sense defined above. Moreover, it requires no additional adaptive feedback weights.

2 Network Configuration

Assume we have a network of m PCA/ICA neurons. Every PCA/ICA neuron has an n-dimensional $(n \geq m)$ weight vector $\mathbf{w}^i$, where i is the index of the neuron. An n-dimensional input $\mathbf{x} = (x_1, x_2, ..., x_n)^T$ is fed to every neuron. The input stream is assumed to be stationary and zero-mean.

The output of a neuron is linear, and does not depend on the feedback, which is used only in learning. Thus, the output y^i of the i-th neuron is simply the scalar product $y^i = \mathbf{x}^T \mathbf{w}^i$. Therefore, if the $\mathbf{w}^i, i = 1...m$ converge to an orthogonal basis of the input space, the outputs of the neurons will simply be the coordinates of the input signal in that basis.

Moreover, the neurons are arranged in a hierarchical manner, so that the i-th neuron receives as feedback the outputs of the $i - 1$ preceding neurons, as well as its own output. No other feedback information is transmitted.

In fact, in our algorithms, the i-th neuron only needs to know the product of the outputs of the preceding $i - 1$ neurons. Therefore, we can reduce the complexity of the network even further by adding $m - 1$ simple multiplicative units, one between every two neurons, to aggregate the feedbacks. Such a multiplicative unit takes as input the output of the i-th unit and the product of the outputs of the $i - 1$ preceding units, multiplies them and feeds the product back to the following $(i + 1)$-th neuron, as well as to the following multiplicative unit. Then no direct feedback connections between the neurons are needed, and the number of feedback connections becomes a linear function of the number of neurons.

3 PCA Learning Rule

3.1 Learning rule

In the network described in Section 2, consider, for the weight vector $\mathbf{w}^i$ of the i-th unit, the following averaged differential equation:

$$d\mathbf{w}^i/dt = E\{\mathbf{x}y^i\} + M(t)(1 - E\{(y^i)^2\})\mathbf{w}^i - bE\{[\prod_{s=1}^{i-1}(y^s)^2]\mathbf{x}y^i\} \qquad (1)$$

where $M(t)$ is a penalty function fulfilling $\lim_{t\to\infty} M(t) = \infty$, $1/3 < b < 1$ is a constant, and for $i = 1$, the last term is defined to be 0.

The first term of the learning rule (1) is the classical linear Hebbian term, known to perform PCA [6]. The second term is a bigradient normalization term, as proposed in [7]. The new feature of this algorithm is in the third term, which performs what we call multiplicative, anti-Hebbian feedback. We use this expression because in the feedback term, the anti-Hebbian term $-\mathbf{x}y^i$ is multiplied by the energies of the outputs of other neurons. As can easily be seen, this feedback is purely local in the sense defined in the Introduction. It is this *local feedback* that clearly distinguishes our learning rule from those presented for example in the survey [6].

Below, we prove analytically that a network whose architecture is as in Section 2, and whose learning is described by (1), performs PCA. This means that y^i converges (up to the norm) to the projection of the input data $\mathbf{x}$ on the i-th principal component of $\mathbf{x}$. The weight vectors will be normalized in the sense that $\operatorname{Var} y^i = 1$ for all i. In particular, if $m = n$, the network spheres the data.

In the proof it will be assumed that the data is Gaussian. If the algorithm is used for non-Gaussian data, the parameter b must be chosen according to the shape of the input distributions. The relevant measure here is the kurtosis of the inputs. The kurtosis of a scalar random variable x is defined as $\operatorname{kurt} x = E\{x^4\} - 3(E\{x^2\})^2$. As can be seen in the proof below (see (2)), if the input data has a strong negative kurtosis (in any direction), b must be chosen so that it is near to 1, which also slows down the convergence considerably. However, most real-world signals seem to have a non-negative kurtosis [2], and b can be chosen as in the Gaussian case. Simulations performed by us confirm these theoretical results.

3.2 Convergence Proof

In this subsection, we prove the convergence of the PCA learning rule described in continuous-time form in (1). In this proof, we suppose that $\mathbf{x}$ has a multivariate *Gaussian* distribution, with $E\mathbf{x} = 0$.

To begin, we diagonalize $E\{\mathbf{x}\mathbf{x}^T\} = \mathbf{B}\mathbf{D}\mathbf{B}^T$, where $\mathbf{B}$ is orthogonal and the columns of $\mathbf{B}$ are the principal components, of unit norm. Furthermore, $\mathbf{D} = \operatorname{diag}(d_1, d_2, ..., d_n)$, where $d_1 > d_2 > ... > d_n > 0$ (for simplicity, we suppose the eigenvalues are distinct). Making the changes of variables $\mathbf{z} = \mathbf{B}^T\mathbf{w}^i$ and $\mathbf{u} = \mathbf{B}^T\mathbf{x}$ (we drop the index i for simplicity), and using the fact that non-correlated Gaussian variables, e.g. the components of $\mathbf{u}$, are independent, and the symmetry of the Gaussian distribution, we can calculate explicitly the expectations, and get

$$d\mathbf{z}/dt = \mathbf{A}\mathbf{z} + M(t)(1 - \mathbf{z}^T\mathbf{D}\mathbf{z})\mathbf{z}$$

where $\mathbf{A} = \operatorname{diag}(a_1, a_2, ..., a_n)$ with $a_j = d_j(1 - bE\{u_j^4\}/d_j^2)$ for $j < i$ and $a_j = d_j(1 - b)$ for $j \geq i$ (for $i = 1$, by definition $\mathbf{A} = \mathbf{D}$).

The proof is inductive. First, for $i = 1$ we adapt the proof in [7]. Defining $\theta_{jk} = z_j/z_k$ we get $d\theta_{jk}/dt = (d_j - d_k)\theta_{jk}$. Taking $k = 1$, we see that all other components z_j become infinitely small compared to z_1. On the other hand, the penalty term ensures that asymptotically $\mathbf{z}^T\mathbf{D}\mathbf{z} \to 1$. This implies that $z_1^2 \to 1/d_1$ and $z_j \to 0$ for $j > 1$. But this proves that $\mathbf{w}^1$ converges, up to the norm, to the first principal component. In addition, $\operatorname{Var} y^1 = \mathbf{z}^T\mathbf{D}\mathbf{z} \to 1$. This completes the first part of the inductive proof.

Next, assume that the first $i - 1$ units have converged, finding the $i - 1$ first principal components, which means that $y^j = \pm u_j/\sqrt{d_j}$ for $j < i$. Then we calculate $d\theta_{jk}/dt = (a_j - a_k)\theta_{jk}$ and choose b so that

$$\frac{d_j^2}{E\{u_j^4\}} < b < 1 \tag{2}$$

for all j. (Note that $d_j^2 \leq E\{u_j^4\}$ always). For Gaussian data, this means $1/3 < b < 1$, as was assumed above. For non-Gaussian data, (2) shows the connection between the kurtoses of the distributions of u_j and the choice of b. If these kurtoses are strongly negative, i.e. the $E\{u_j^4\}$ are small, then b must be chosen near to 1.

Now, take $j < i$ and $k \geq i$. Then we have $a_j < 0$ and $a_k > 0$, which means $\theta_{jk} \to 0$. On the other hand, for $k > i$ we get, as in the first part of the proof, $\theta_{ki} \to 0$. Taking into account the penalty term, we see that $z_i^2 \to 1/\sqrt{d_i}$ and $z_l \to 0$ for $l \neq i$. But this means that $\mathbf{w}^i$ converges to the i-th principal component, normalized as desired. This completes the second part of the inductive proof.

4 ICA Learning Rule

4.1 Definition of ICA

Independent Component Analysis is an important extension of PCA that has received increasing attention recently [1, 2, 3, 5]. In the simplest form of ICA [3, 5], it is assumed that we observe an n-dimensional signal $\mathbf{v}_t = (v_t^1, v_t^2, ..., v_t^n)^T$ which is a linear, invertible combination of n independent components u_t^j

$$\mathbf{v}_t = \mathbf{H}\mathbf{u}_t$$

Here, $\mathbf{H}$ is an unknown $n \times n$ invertible matrix, called the mixing matrix. The stochastic process $\mathbf{u}_t = (u_t^1, u_t^2, ..., u_t^n)^T$ is stationary and zero-mean, and its components are mutually independent, at any point of time t.

The problem is then to estimate the independent components u_t^j, or, equivalently, to estimate the mixing matrix $\mathbf{H}$. The solution of the problem is highly non-unique; we must impose additional constraints. Here, we impose the signals u_t^j to have unit variance. This makes the mixing matrix unique, up to a permutation and a change of signs of the columns [3].

The problem is simplified if we sphere (whiten) the data, i.e. transform it linearly to $\mathbf{x} = \mathbf{M}\mathbf{v}$ so that $E\{\mathbf{x}\mathbf{x}^T\} = \mathbf{I}$ (henceforth, we drop the index t for simplicity). Incidentally, this is exactly what algorithm (1) of the preceding section does, if $n = m$. Then we have $\mathbf{x} = \mathbf{B}\mathbf{u}$ where the transformed mixing matrix $\mathbf{B} = \mathbf{M}\mathbf{H}$ is *orthogonal*.

4.2 Learning Rule

Assume that the sphered input $\mathbf{x}$, as defined above, is fed as input to the network described in Section 2. Then, to find the independent components, we can use the following learning rule for the i-th neuron:

$$d\mathbf{w}^i/dt = sE\{\mathbf{x}(y^i)^3\} + M(t)(1 - \|\mathbf{w}^i\|^2)\mathbf{w}^i - bE\{[\prod_{s=1}^{i-1}(y^s)^2]\mathbf{x}y^i\} \quad (3)$$

where $M(t)$ is a penalty function as in the preceding section, b is a positive constant, $s = \pm 1$ is a sign that depends on the kurtoses of the independent components u_j, and for $i = 1$, the third term is defined to be 0.

The first term is a non-linear (anti-)Hebbian learning term, widely used in neural ICA [5, 7]. The second term is a bigradient normalization term as proposed in [7]. The third term, which distinguishes this algorithm from other ICA algorithms, is a multiplicative anti-Hebbian feedback term, which is *purely local*.

The constant b must be chosen so that

$$b > \frac{|\operatorname{kurt} u_j|}{\operatorname{kurt} u_j + 2}, \quad \text{for all } j \tag{4}$$

Note that for sphered data, $\operatorname{kurt} u_j + 2 > 0$ always. For simplicity (this is not strictly necessary), we presume that all components of $\mathbf{u}$ have kurtoses of the same sign. Then we must choose s to have that same sign.

Below, we prove that for every i, $\mathbf{w}^i$ converges to one of the columns of the transformed mixing matrix $\mathbf{B}$ (up to the sign). Note that in contrast to PCA, no order is defined between the independent components. Moreover, we prove that two weight vectors do not converge to the same column of $\mathbf{B}$. This means that the outputs of the neurons y^i become, up to the sign, the independent components u_j.

It must be noted that, as can be seen in (4), the algorithm does not work well if the independent components have a strongly negative kurtosis, because then b should be very large. In our simulations, separation of source signals of positive kurtosis was quick enough, whereas in the case of negative kurtosis, it was fairly slow.

4.3 Convergence Proof

Now, we prove the convergence of the learning rule (3). For $i = 1$, the convergence is directly implied by the well-known results of the connection between ICA and the extrema of the kurtosis of $\mathbf{w}^T\mathbf{x}$, see, e.g., [4]. The second term performs the normalization $\|\mathbf{z}\| = 1$ and the first term finds a minimum or a maximum of the kurtosis of the output.

The proof proceeds inductively. Assume that the $i - 1$ first weight vectors have converged to $i-1$ columns of the transformed mixing matrix $\mathbf{B} = \mathbf{MH}$. Let S be the set of indices j such that $\mathbf{w}^k = \mathbf{b}_j$ for some $k < i$, i.e. S contains the indices of the independent components found by neurons higher in the hierachy. Next, make the change of variable $\mathbf{z} = \mathbf{B}^T\mathbf{w}^i$ (where, again, we drop the index i for simplicity). Computing explicitly the expectations in (3), and taking into account the independence and normalization of the independent components u_j, we get for the j-th component of $\mathbf{z}$ the equation

$$dz_j/dt = s((\operatorname{kurt} u_j)z_j^3 + 3\|\mathbf{z}\|^2 z_j) + M(t)(1 - \|\mathbf{z}\|^2)z_j - bc_j z_j \tag{5}$$

where $c_j = E\{u_j^4\}$ for $j \in S$, and $c_j = 1$ for $j \notin S$.

As in the preceding section, we now calculate, for $j \in S$ and $k \notin S$

$$d\theta_{jk}/dt = [s((\text{ kurt } u_j)z_j^2 - (\text{ kurt } u_k)z_k^2) - b(E\{u_j^4\} - 1)]\theta_{jk}$$

The penalty term implies that $\|\mathbf{z}\| \to 1$. Then we see by some elementary manipulations that the time derivative of θ_{jk} is (uniformly) negative , if we have (4). Thus, all z_j with $j \in S$ tend to zero.

Therefore, we only need to consider the system of z_k for $k \notin S$, supposing $z_j = 0$ for all $j \in S$. But this system is equivalent to the system with no feedback. The only difference is the additional term $-b\mathbf{z}$, but this does not change the convergence as this term can be absorbed in the penalty term. Therefore, using the same proof as in the first step of the induction, we complete the proof of convergence of (3).

5 Conclusions

New algorithms for neural PCA and ICA were introduced. It was claimed that these algorithms can be considered "more neural" than most other neural PCA or ICA algorithms, because every neuron only uses information available locally. The novel technique introduced is the use of multiplicative anti-Hebbian feedback. Such a feedback is purely local, in contrast to the well-known projective feedback [5, 6, 7], or other kinds of feedbacks based on the inversion of the weight matrix [2]. Furthermore, no adaptation of feedback weights is needed.

The convergence of the algorithms has been rigorously proven. The convergence of the ICA algorithm requires some prior information on the shapes of the distribution of input data. To be able to prove analytically the convergence of the PCA algorithm, we must suppose that the data is Gaussian.

References

1. S. Amari, A. Cichocki, and H.H. Yang. A new learning algorithm for blind source separation. In *Advances in Neural Information Processing 8 (Proc. NIPS'95)*, Cambridge, MA, 1996. MIT Press.
2. A.J. Bell and T.J. Sejnowski. An information-maximization approach to blind separation and blind deconvolution. *Neural Computation*, 7:1129–1159, 1995.
3. P. Comon. Independent component analysis – a new concept? *Signal Processing*, 36:287–314, 1994.
4. N. Delfosse and P. Loubaton. Adaptive blind separation of independent sources: A deflation approach. *Signal Processing*, 45:59–83, 1995.
5. J. Karhunen, E. Oja, L. Wang, R. Vigario, and J. Joutsensalo. A class of neural networks for independent component analysis. Technical Report A 28, Helsinki University of Technology, Laboratory of Computer and Information Science. Submitted to a journal, 1995.
6. E. Oja. Principal components, minor components, and linear neural networks. *Neural Networks*, 5:927–935, 1992.
7. L. Wang and J. Karhunen. A unified neural bigradient algorithm for robust PCA and MCA. *Int. J. of Neural Systems*, to appear.

How Fast Can Neuronal Algorithms Match Patterns?

Rolf P. Würtz [*][1], Wolfgang Konen[2], and Kay-Ole Behrmann

[1] Computing Science, University of Groningen, The Netherlands
[2] Zentrum für Neuroinformatik GmbH, Bochum, Germany

Abstract. We investigate the convergence speed of the Self Organizing Map (SOM) and Dynamic Link Matching (DLM) on a benchmark problem for the solution of which both algorithms are good candidates. We show that the SOM needs a large number of simple update steps and DLM a small number of complicated ones. A comparison of the actual number of floating point operations hints at an exponential vs. polynomial scaling behavior with increased pattern size. DLM turned out to be much less sensitive to parameter changes than the SOM.

1 Introduction

For visual perception in a biological or artificial system the *visual correspondence problem* is of central importance: "Given two images of the same physical object decide which point pairs belong to the same point on the object." A generic solution to this problem will at least greatly alleviate many of the difficulties encountered by computer vision research. Invariant object recognition, e.g., becomes easy if a correspondence map of sufficient density and reliability can be constructed between objects and stored prototypes [1]. The study [2] is another illustration of the power of such an algorithm. It describes a network based on Dynamic Link Matching (DLM) that learns to evaluate input patterns for the presence of one out of three mirror symmetries. That approach has been compared with a standard backpropagation network that needed some 10^4 examples [3]. The basic idea of DLM dates back to [4].

For a solution of the correspondence problem an ordered mapping from one plane to another has to be established with a combination of two constraints: Matching points must carry *similar features*, and *neighborhood relations* must be preserved. A good candidate to solve that problem is a self-organizing algorithm that develops from an unordered initial state to a clean one-to-one mapping. As self-organization is a notoriously slow process the *convergence speed* is an important detail. Short of sound analytical results we have used the problem from [2] as a benchmark to evaluate the relative performance of DLM [5] and the Self Organizing Map (SOM) algorithm [6, 7].

[*] Please address correspondence to rolf@cs.rug.nl. Funding from the HCM network "Parallel modeling of neural operators for pattern recognition" by the European Community is gratefully acknowledged.

2 Definition of a benchmark problem

A fair comparison of algorithms that were developed to solve different problems and whose full range of applicability is still subject of intensive research is not easy. The least one can do is to define problem and simulations very explicitly and leave it to the reader to judge if justice has been done to both algorithms, which are specified in sections 3.1 and 4.1, respectively.

If a square lattice is mapped onto a continuous input square (a typical problem for the SOM), the correct solution is not obvious. For a fair comparison, however, the quality must be assessed by objective means. The decision will be influenced considerably by boundary effects. Furthermore, there may be multiple solutions of identical intuitive quality (e.g., mirror reflections, rotations by 90°).

To avoid these problems we have chosen the above-mentioned mirror-problem as a benchmark. The setup consists of two square layers X and Y of $N \times N$ neurons that in addition carry features $f \in \{1, \ldots, F\}$. The feature distribution in X is chosen at random, but patterns with multiple symmetries are discarded. The distribution in Y is identical to the one in X except for either a mirror reflection or a rotation by 90°. Now the feature distributions induce a unique neighborhood-preserving mapping from X to Y. The benchmark task for the self-organizing algorithms is to find these mappings given only the feature distributions. *Similarity* of features of neurons $x \in X$ and $y \in Y$ is defined as all-or-nothing for this benchmark (for practical applications smooth similarity functions are usually more suitable):

$$T(x, y) = \delta\left(f_x, f_y\right) = \begin{cases} 1 & \text{if } f_y = f_x \\ 0 & \text{otherwise} . \end{cases} \tag{1}$$

A nearest-neighbor topology with wrap-around borders is imposed on both X and Y (this makes them 2D-tori rather than squares). The mirror axis may thus be any line and the center of rotation any point on the layers.

Due to the discrete lattices in both layers neighborhood preservation is clearly defined. Thus, the optimal solution is known beforehand, which gives a straightforward error measure that can be monitored through the whole process. Let $\mathbf{w}_y(t)$ be the position where neuron y points in layer X at time t, and $\mathbf{w}_y^{opt}$ the optimal mapping. Then the error will be

$$E(t) = \sum_y \left(\mathbf{w}_y^{opt} - \mathbf{w}_y(t)\right)^2 . \tag{2}$$

For the $N \times N$-size benchmark problem a particular solution is said to have *converged* if the average position error $E(t)/N^2$ is below the threshold $\varepsilon = 1/640$. The number of features F is a useful parameter to control the difficulty of the problem. Here we have used $F = 10$ equally distributed features.

3 Solution with the SOM

3.1 Method description

In order to apply the SOM to our benchmark we have identified the discrete neuron layer with our layer Y and the input space with X. Furthermore, we have included the feature similarity $T(x, y)$ into the learning rule of an otherwise unmodified SOM-algorithm [6, 7]:

$$\Delta \mathbf{w}_i = \lambda \exp\left(-|i - i_\mu|^2/2\sigma^2\right) (\mathbf{v}^\mu - \mathbf{w}_i) T(\mu, i_\mu). \tag{3}$$

Here, $T(\mu, i_\mu) \in [0, 1]$ is the similarity between the features of neurons $\mu \in X$ and its best matching counterpart $i_\mu \in Y$, respectively, i.e. the adaptation rate is weighted by the feature similarity. In our benchmark, T can only take the binary values 0 and 1, so that iteration steps for neurons with unequal features are without effect. Therefore, we have optimized the algorithm by skipping all such iteration steps and choosing i_μ directly among the neurons with the same features as μ. Only those "effective" iteration steps are counted in our results.

3.2 Parameter tuning

Many applications of the SOM require considerable care in adjusting the parameters. Generally, a decrease of the learning rate λ and the width σ is necessary to assure convergence. Unfortunately, there is no problem-independent rule on how to find these parameters. We have chosen a linear decreasing scheme. Different schemes (e.g. exponential ones) are in use, but are known [7] to yield the same general behavior of the algorithm. Extensive experiments have shown that the performance could not be improved by decreasing λ. We have thus kept it constant at $\lambda = 1$, but the width parameter σ and its decrease schedule (start/stop-value, decrease rate) had to be chosen carefully. Both were optimized individually for each problem size ($N = 4, \ldots, 20$) by scanning the reasonable parameter range (40 values) with 10 executions of SOM each. Only the 8 best results are shown in figure 1. The results are consistent with an exponential scaling of the number of update steps with the problem size (see figure 1).

4 Solution with DLM

4.1 Method description

In the DLM scheme layer X is fully connected with Y by a matrix $J(x, y)$ of dynamical links. Their development is governed by a Hebbian rule with competition and influence of feature similarity. In other words, links between pairs of neurons which have *similar features* and are *active at the same time* will be strengthened, others decay. Neighborhood preservation in DLM is achieved by ensuring that in each layer only *one connected subregion* of a given form and size, which we will call a *blob*, can be active at one time. This is a way to code

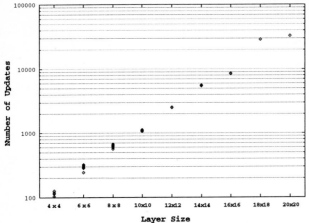

Fig. 1. Scaling behavior of the SOM algorithm on the mirror benchmark. Only runs with the optimal parameter set for each problem size are shown. The vertical spread of data points is due to different random seeds. A straight line fits the data well, hinting at exponential scaling with problem size.

neighborhood in the layer as common activity in the same time slot. A blob in layer X excites layer Y by means of the dynamic links $J(x, y)$. Layer Y supports only blobs of the same form and size as X. The links only influence the *position* of the blob, which can be calculated. Now only the neurons in one blob in X and one blob in Y are active and can strengthen their links in the following update step according to their feature similarities. An activity blob is a unimodal non-negative function b of the layer neurons. In our simulations we have chosen it to be 1 inside a square of size $B \times B$ and 0 outside. Then, the concrete algorithm runs as follows:

1. All links are initialized to $1/N^2$.
2. A position $x_0 \in X$ is chosen at random, a blob is placed there, and the resulting blob position y_0 in Y is calculated such as to minimize the *potential*

$$V(y_0) = - \sum_y \sum_x J(x, y)\, T(x, y)\, b(x - x_0)\, b(y - y_0)\,. \qquad (4)$$

3. The activities in X and Y are now blobs positioned at x_0 and y_0, respectively, and the links $J(x, y)$ are updated by the learning rule:

$$\Delta(x, y) := \lambda\, (J(x, y) + J_0)\, T(x, y)\, b(x - x_0)\, b(y - y_0)\,. \qquad (5)$$

4. The updated links are first normalized by $\sum_x J(x, y)$ and then by $\sum_y J(x, y)$.

Steps 2 through 4 are iterated until convergence. For the pointer vector $\mathbf{w}_y$ that enters the error function (2) the *center of mass* for all links $J(x, y)$ projecting onto neuron y has been used:

$$\mathbf{w}_y = \frac{\sum_x \mathbf{p}(x)\, J(x, y)}{\sum_x J(x, y)}\,, \qquad (6)$$

where $\mathbf{p}(x)$ is the vector in the unit square specifying the position of neuron x in layer X.

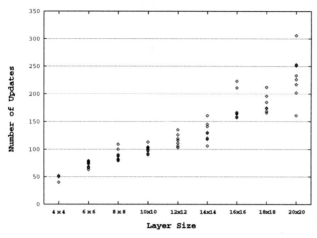

Fig. 2. Scaling behavior of the DLM algorithm on the mirror benchmark. The vertical spread of data points is due to different random seeds. A straight line is compatible with the data well, this time suggesting linear scaling with problem size.

4.2 Parameter tuning

The simulations of DLM on the mirror benchmark have shown that its parameters are fairly simple to adjust. There is no necessity to decrease the learning rate λ or change the blob size B during iterations in order to assure convergence. The form of the blob (circle, square, ...) does not influence convergence. J_0 serves only to prevent very small links from being suppressed for good, its value makes no difference. The only relevant parameter is the *blob size B*. Simulations in [8] have shown that convergence is fastest if a blob covers half the layer area. This blob size, which we used for all DLM simulations, can be shown to maximize the average information gain per iteration step (see [9] for details).

The experimental conditions have been identical to those for the SOM. The scaling behavior, however, was different. On variation of the problem sizes N the iteration steps needed for convergence increase only linearly with N (figure 2), in sharp contrast with the exponential behavior of the SOM (figure 1).

5 Results and conclusions

During our experiments we have encountered fewer difficulties in adjusting the parameters for DLM than for the SOM. Nevertheless, we have invested considerable effort to tailor both algorithms for the benchmark problem. Figures 1 and 2 show the number of iterations required to solve the mirror problem. The comparison of these two figures is not completely fair, because the single update steps for DLM are much more complicated ($O(N^4)$) than the ones for the SOM ($O(N^2)$). In order to show the actual execution time for concrete layer sizes we have plotted the number of floating point operations required to reach convergence in figure 3. This figure indicates that DLM converges faster once the layer size exceeds 16×16. We conclude that the convergence time for DLM will scale as N^4 and the one for the SOM as $\exp(N)$. Experiments that check this trend for higher values of N are currently performed and will be reported in [9].

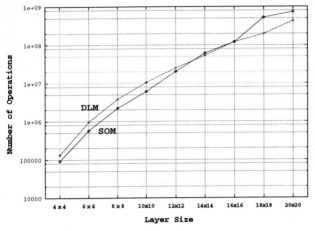

Fig. 3. Floating point operations required for the SOM and DLM on the mirror benchmark. The data indicate that DLM is faster for layers larger than 16×16.

A question that can not be ignored in the comparison of neuronal algorithms as models for perception is the time required on parallel machines. Given arbitrarily many parallel processors the single update steps can be executed in constant time for both algorithms. The update steps, however, can not be parallelized completely. To what extent partial parallelization (epoch learning) can be applied is unclear at this point. We therefore expect that on massively parallel machines DLM will scale linearly with the layer size, whereas the SOM will probably retain its exponential behavior.

References

1. R.P. Würtz. *Multilayer Dynamic Link Networks for Establishing Image Point Correspondences and Visual Object Recognition*, volume 41 of *Reihe Physik*. Verlag Harri Deutsch, Thun, Frankfurt am Main, 1995.
2. W. Konen and C.v.d. Malsburg. Learning to generalize from single examples in the dynamic link architecture. *Neural Computation*, 5:719–735, 1993.
3. T.J. Sejnowski, P.K. Kienker, and G.E. Hinton. Learning symmetry groups with hidden units: Beyond the perceptron. *Physica D*, 22:260–275, 1986.
4. D.J. Willshaw and C. v.d. Malsburg. How patterned neural connections can be set up by self-organization. *Proceedings of the Royal Society, London B*, 194:431–445, 1976.
5. W. Konen, T. Maurer, and C. v.d. Malsburg. A fast dynamic link matching algorithm for invariant pattern recognition. *Neural Networks*, 7(6/7):1019–1030, 1994.
6. T. Kohonen. Self-organized formation of topologically correct feature maps. *Biological Cybernetics*, 43:59–69, 1982.
7. T. Kohonen. The self-organizing map. *Proc. IEEE*, 78:1464–1480, 1990.
8. K.-O. Behrmann. Leistungsuntersuchungen des Dynamischen-Link-Matchings und Vergleich mit dem Kohonen-Algorithmus. Technical Report IR-INI 93-05, Ruhr-Universität Bochum, 1993.
9. R.P. Würtz, W. Konen, and K.-O. Behrmann. On the performance of neuronal matching algorithms. Manuscript in preparation.

An Annealed 'Neural Gas' Network for Robust Vector Quantization

Thomas Hofmann & Joachim M. Buhmann *

Rheinische Friedrich–Wilhelms–Universität
Institut für Informatik III, Römerstraße 164, D-53117 Bonn, Germany
email:{th,jb}@cs.uni-bonn.de, http://www-dbv.cs.uni-bonn.de

Abstract. Vector quantization, a central topic in data compression, deals with the problem of encoding an information source or a sample of data vectors by means of a finite codebook, such that the average distortion is minimized. We introduce a common framework, based on maximum entropy inference to derive a deterministic annealing algorithm for robust vector quantization. The objective function for codebook design is extended to take channel noise and bandwidth limitations into account. Formulated as an on–line problem it is possible to derive learning rules for competitive neural networks. The resulting update rule is a generalization of the 'neural gas' model. The foundation in coding theory allows us to specify an optimality criterion for the 'neural gas' update rule.

1 Introduction

Vector quantization is a lossy data compression method which aims at encoding a given set of data vectors, $\mathcal{X} = \{\mathbf{x}_i \in \mathbb{R}^d | 1 \leq i \leq N\}$, with a set of prototypes, also called a codebook $\mathcal{Y} = \{\mathbf{y}_\alpha \in \mathbb{R}^d | 1 \leq i \leq K \ll N\}$. Information reduction is achieved by transmitting only codebook indices α instead of real–valued data vectors to the receiver. The receiver decodes the transmitted index by look up of the corresponding codebook vector in the codebook. The interested reader may be refered to the extensive presentation in Gersho and Gray [3]. The main problem in vector quantization is to select an appropriate codebook $\mathcal{Y}$, such that the information loss or mean *distortion* induced by the data reduction is minimal. The most prominent conventional methods for optimizing the mean distortion are the K–means clustering algorithm [8] and the similar generalized Lloyd or LBG algorithm [6]. Applications of algorithms for vector quantization are of course not limited to communication theory. Data reduction or *data clustering* plays an important role in many other areas, like pattern recognition, classification, unsupervised learning, data analysis and data mining. In the neural network community online clustering algorithms are often discussed as competitive learning methods with additional features, e.g. the topology preserving mapping of Kohonen [4] or the neural gas model of Martinetz et al. [10]. The data compression context has the advantage to provide a precise formalization of the problem and to rely on an explicit objective function, while many 'neural' models for vector quantization are only heuristically motivated without

---

* Supported by the Federal Ministry for Education, Science and Technology (BMBF) under grant # 01 M 3021 A/4

an optimization principle or are founded on assumed principles of biological neural systems. The aim of this paper is to show in what sense the neural gas model can be derived as a *robust* vector quantizer. This allows us to specify an optimality criterion for this update rule. Moreover, we will show how to address the problem of robust vector quantization in the maximum entropy framework and how to derive a deterministic annealing algorithm. To evaluate and compare different quantization algorithms we measure the performance on multiresolution data of a real–world teleconferencing application.

2 Robust Vector Quantization

We start by formalizing the classical vector quantization of finding a codebook $\mathcal{Y}$ for a given set of data vectors $\mathcal{X}$ under the assumption of an unperturbed encoding of data samples and a noise-free transmission of prototype indices. For denotational convenience we introduce a Boolean assignment matrix $M \in \{0, 1\}^{N \times K}$ with coefficients M_α^i, where $M_\alpha^i = 1$ indicates an assignment of $\mathbf{x}_i$ to the prototype $\mathbf{y}_\alpha$. The assignment matrix has to fulfill the constraints $\sum_{\alpha=1}^{K} M_\alpha^i = 1$ for all $1 \leq i \leq N$ to assure a unique mapping of data vectors to prototypes. Data reduction has furthermore to rely on a distortion measure $\mathcal{D}(\mathbf{x}_i, \mathbf{y}_\alpha)$, which quantifies the 'loss' of an encoding. A widely applied distortion measure is the squared Euclidean distance $\mathcal{D}(\mathbf{x}_i, \mathbf{y}_\alpha) = \|\mathbf{x}_i - \mathbf{y}_\alpha\|^2$. This motivates the following objective function for vector quantization,

$$\mathcal{H}(M, \mathcal{Y}) = \sum_{i=1}^{N} \sum_{\alpha=1}^{K} M_\alpha^i \, \mathcal{D}(\mathbf{x}_i, \mathbf{y}_\alpha) \, , \tag{1}$$

which has to be minimized with respect to M and $\mathcal{Y}$. Luttrell [7] and Buhmann et al. [1] discuss an extension of vector quantization to noisy transmission channels. Additional knowledge about transition probabilities for prototype indices which are sent through a noisy channel was incorporated in the codebook optimization procedure and led to a codebook design of topologically organized vector quantizers, which are closely related to self-organizing feature maps [4]. We consider a different extension of vector quantization, which can also be combined with a noisy channel model. More precisely we define a communication model, such that for encoding a vector $\mathbf{x}_i$, parts of the codebook are inaccessible with a certain probability. In the simplest case, we assume that every prototype $\mathbf{y}_\alpha$ is eliminated with a probability ϵ_α, independent of the elimination of other prototypes. To simplify the notation we restrict ourselves to the special case $\epsilon_\alpha = \epsilon, \forall \alpha$. Similar situations occur in practical coding situations, where the codebook size has to match and scale with an instantaneously available channel capacity. Moreover in adaptive codebook design with codebook replenishments, prototypes might be eliminated and reinserted to capture the non–stationary nature of an information source. In these situations it is clearly advantageous to use a codebook, which possesses a high *robustness* with respect to these code vector erasures. Another motivation arises in the biological context, where the robustness of a neuronal subsystem with respect to single neuron defects is an essential or even vital feature of an organism. As in the case of noisy channels, knowledge about the stochastic properties of involved noise processes can already be taken into consideration in the optimization phase. This yields codebooks which minimize the expected distortion at the receiver. Formally the objective function for robust vector quantization is given by

$$\mathcal{H}(M,\mathcal{Y}) = \sum_{i=1}^{N}\sum_{\alpha=1}^{K} \left(\sum_{r=1}^{K} M_{\alpha r}^{i}\,(1-\epsilon)\epsilon^{r-1}\right) \mathcal{D}_{i\alpha}(\mathbf{x}_i,\mathbf{y}_\alpha) \ . \tag{2}$$

Here we have introduced Boolean variables $M_{\alpha r}^{i}$, which encode the rank r of a prototype α for encoding data vector $\mathbf{x}_i$, e.g. $M_{\alpha 1}^{i} = 1$ specifies the first choice for encoding $\mathbf{x}_i$ to be $\mathbf{y}_\alpha$. For every data index i, $M^i \in \{0,1\}^{K \times K}$ is a permutation matrix defining a total order on the set of prototype indices. Obviously all rows and columns of M^i have to sum to 1. For $\epsilon = 0$ the distortion depends only on the first choices $M_{\alpha 1}^{i}$ which can be identified with assignments M_α^i in Eq. (1).

3 Maximum Entropy Vector Quantization

Minimizing objective functions like Eq. (1) or (2) is a hard combinatorial optimization problem caused by the potentially large number of local minima. Following Rose et al. [11] and Buhmann et al. [2] we regard the assignments M_α^i and $M_{\alpha r}^i$ as random variables and apply the maximum entropy principle by introducing a temperature parameter T. Justifications and extensive motivations for the maximum entropy approach is given in Buhmann et al. [2]. The quantity which has to be minimized at temperatures $T > 0$ with respect to the codebook $\mathcal{Y}$ is known from statistical physics as the free energy $\mathcal{F}(\mathcal{Y}) = -T \log \sum_{i=1}^{N} \sum_{M^i} \exp\left[-\mathcal{H}(M^i,\mathcal{Y})/T\right]$. The sum runs over all permutation matrices for every data point $\mathbf{x}_i$. Since the cost function is linear in the Boolean rank variables, the free energy can be written as

$$\mathcal{F}(\mathcal{Y}) = -T \sum_{i=1}^{N} \log \sum_{M^i} \exp\left[-\frac{1}{T}\sum_{\alpha=1}^{K}\left(\sum_{r=1}^{K} M_{\alpha r}^{i}(1-\epsilon)\epsilon^{r-1}\right)\mathcal{D}(\mathbf{x}_i,\mathbf{y}_\alpha)\right] \ . \tag{3}$$

For every data vector $\mathbf{x}_i$ we have to independently solve the subproblem of performing the sum over permutation matrices M^i. Rewriting the sums in the exponent as $\mathcal{H}^i(M^i,\mathcal{Y}) = \sum_{\alpha=1}^{K}\sum_{r=1}^{K} M_{\alpha r}^{i} h_{\alpha r}^{i}$, $h_{\alpha r}^{i} = (1-\epsilon)\epsilon^{r-1}\mathcal{D}(\mathbf{x}_i,\mathbf{y}_\alpha)$, this can be q seen to be equivalent to a bipartite matching problem between prototype indices and ranks. Of course, we know the solution in the zero temperature case, where the rank order is simply given by the increasing order of distortions. The prototype with the lowest distortion costs is the first choice, the one with the second lowest distortions the second choice and so on. Differentiating $\mathcal{F}$ with respect to prototype coordinates $\mathbf{y}_\alpha$ for squared Euclidean distortions yields the well–known centroid equations, where the weights depend on the probability that $\mathbf{y}_\alpha$ is actually used for encoding a data vector $\mathbf{x}_i$,

$$\mathbf{y}_\alpha = \frac{\sum_{i=1}^{N}\langle M_\alpha^i\rangle \mathbf{x}_i}{\sum_{i=1}^{N}\langle M_\alpha^i\rangle}, \quad \langle M_\alpha^i\rangle = \sum_{r=1}^{K}\langle M_{\alpha r}^i\rangle(1-\epsilon)\epsilon^{r-1} \ . \tag{4}$$

The relevant quantities $\langle M_{\alpha r}^i\rangle$ represent the probability of the rank of $\mathbf{y}_\alpha$ for $\mathbf{x}_i$ to be r and are Gibbs averages with respect to the matching cost functions $\mathcal{H}^i(M^i,\mathcal{Y})$. To obtain estimates for $\langle M_{\alpha r}^i \rangle$ we apply a method known as Sinkhorns algorithm [12], which has been successfully applied to matching problems by Kosowsky and Yuille [5]. We initialize a matrix A^i with elements $A_{\alpha r}^i = \exp\left[-\frac{1}{T}h_{\alpha r}^i\right]$ and perform an alternating normalization of rows and columns of A^i, a process which is known to converge towards a doubly stochastic matrix. A justification for this procedure using an argument known as the saddle–point approximation can be found in [5]. Since the influence of higher ranks r on the expected distortions vanishes exponentially with ϵ, we may approximate the cost function by neglecting contributions of order $r > r^{max}$. For small ϵ it might even be sufficient to restrict to the case of $r^{max} = 2$. For this special case the cost function can be equivalently rewritten in the form

$$\mathcal{H}(M,\mathcal{Y}) = \sum_{i=1}^{N}\sum_{\alpha=1}^{K}\sum_{\nu=1,\nu\neq\alpha}^{K} M_{\alpha 1}^i M_{\nu 2}^i \left[(1-\epsilon)\,\mathcal{D}_{i\alpha} + (1-\epsilon)\epsilon\,\mathcal{D}_{i\nu}\right] , \qquad (5)$$

where $\mathcal{D}_{i\mu} = \mathcal{D}(\mathbf{x}_i, \mathbf{y}_\mu)$. Only the first and second rank are considered, and there are exactly $K(K-1)$ 'pair assignment' possibilities for every data vector. It is straightforward to derive the following equations for the expected rank variables by marginalization,

$$\langle M_{\alpha 1}^i \rangle = \sum_{\nu=1,\nu\neq\alpha}^{K} \frac{\exp\left[-\frac{1-\epsilon}{T}\mathcal{D}_{i\alpha}\right]\exp\left[-\frac{(1-\epsilon)\epsilon}{T}\mathcal{D}_{i\nu}\right]}{\sum_{\mu=1}^{K}\sum_{\gamma=1,\gamma\neq\mu}^{K}\exp\left[-\frac{1-\epsilon}{T}\mathcal{D}_{i\mu}\right]\exp\left[-\frac{(1-\epsilon)\epsilon}{T}\mathcal{D}_{i\gamma}\right]} \qquad (6)$$

and a similar equation for $\langle M_{\alpha 2}^i \rangle$. In this approximation $\langle M_{\alpha r}^i \rangle = 0$, for $r > 2$. Inserting Eq. (6) into Eq. (4) yields the corresponding prototype update rule.

4   Online Equations for Vector Quantization

To obtain the online equations for the proposed robust vector quantization model we perform a Taylor expansion of Eq. (4) in $1/N$ [2]. With an arriving new data vector $\mathbf{x}$, prototypes are slightly moved towards $\mathbf{x}$ by an amount w_α,

$$\mathbf{y}_\alpha^{t+1} = \mathbf{y}_\alpha^t + w_\alpha(\mathbf{x},\mathcal{Y})\,(\mathbf{x} - \mathbf{y}_\alpha), \quad w_\alpha \sim \sum_{r=1}^{K}\langle M_{\alpha r}^x \rangle(1-\epsilon)\epsilon^{r-1}\|\mathbf{x} - \mathbf{y}_\alpha\|^2 . \qquad (7)$$

Usually w_α will also scale inversely with the number of data already encoded by $\mathbf{y}_\alpha$ so far, to achieve convergence in the infinite data limit. For non–stationary sources it might be appropriate, to use a constant learning rate η. Two limit cases of Eq. (7) are interesting to consider. For $\epsilon = 0$ the ordinary maximum entropy vector quantizer is retrieved. In the limit of $T \to 0$ the averages $\langle M_{\alpha r}^x \rangle$ converge towards Boolean variables representing the linear distortion order induced on $\mathcal{Y}$ by $\mathbf{x}$. This is exactly the update rule for the neural gas, as proposed in [10] with an exponentially decaying neighborhood function. Performing both limits simultaneously yields a winner–take–all rule, which only adapts the prototype closest to the new data $\mathbf{x}_i$. We draw the most important conclusion, that maximum entropy vector quantization and vector quantization using the neural gas model complement each other and have to be combined in the described communication scenario. The neural gas model minimizes the cost function in Eq. (2) which was interpreted as a robust variant of vector quantization. The maximum entropy method is (in principle) applicable to any optimization problem and can be extended to cover the neural gas algorithm, as demonstrated in this paper.

5   Results

We have tested the proposed algorithm on a data set obtained from a wavelet–transformed video sequence of 'Miss America' (174×144 pixels). The wavelet decomposition was performed according to Mallat's algorithm [9] at three resolution levels. Neighboring wavelet coefficient were grouped in blocks (4×4 for the higher levels, 2×2 at the lowest level) and for every subband a particular codebook was created. For the performance study the codebook was trained using a subsequence of 4 frames. In addition we have eliminated codebook vectors according to the assumed masking model. The experiments demonstrate the superior performance of the robust vector quantization. Table 1 summarizes the results of this series of experiments for the detail images, high–pass filtered in $x-$direction. The displayed mean squared error is the median out of 21 performed runs with randomized initializations.

Method	Subband	N	d	K	MSE, $\epsilon = 0.0$	MSE, $\epsilon = 0.02$	MSE, $\epsilon = 0.1$
K–means	Level 1, x	1584	16	16	76.8	93 ± 15	190 ± 50
	Level 2, x	396	16	32	469.2	650 ± 110	1315 ± 320
	Level 3, x	396	4	32	313.7	680 ± 100	2090 ± 800
annealed	Level 1, x	1584	16	16	75.4	90 ± 12	140 ± 22
K–means	Level 2, x	396	16	32	448.2	630 ± 90	1280 ± 280
	Level 3, x	396	4	32	290.8	430 ± 50	790 ± 100
neural gas	Level 1, x	1584	16	16	76.2	93 ± 10	145 ± 30
	Level 2, x	396	16	32	466.0	625 ± 20	1120 ± 310
	Level 3, x	396	4	32	305.4	450 ± 40	1085 ± 150
annealed	Level 1, x	1584	16	16	74.9	79 ± 4	89 ± 5
neural gas	Level 2, x	396	16	32	431.7	508 ± 15	780 ± 180
	Level 3, x	396	4	32	286.3	420 ± 35	642 ± 50

Table 1. Performance of vector quantization algorithms

As expected the annealed vector quantizers have a better performance than the zero temperature algorithms. The overall best result was obtained with the annealed neural gas algorithm. It was significantly better than all other algorithms in the presence of masking noise and even possesses the best performance for $\epsilon = 0$, where the annealing was performed by decreasing T and ϵ in a joint schedule. To visualize the advantages of robust vector quantization Fig. (1) displays a reconstructed image (masking probability $\epsilon = 0.1$) after quantizing with a typical codebook obtained by the K–means and the annealed neural gas network. The codebooks were optimized on a sequence of 10 video frames.

Fig. 1. 'Miss America' reconstructed with a 10% masking of codebook vectors. The left image was quantized with a codebook obtained by K–means, the right image with a codebook optimized with the annealed neural gas.

References

1. J. M. Buhmann and H. Kühnel. Complexity optimized data clustering by competitive neural networks. *Neural Computation*, 5:75–88, 1993.
2. J. M. Buhmann and H. Kühnel. Vector quantization with complexity costs. *IEEE Transactions on Information Theory*, 39(4):1133–1145, July 1993.
3. A. Gersho and R. M. Gray. *Vector Quantization and Signal Processing*. Kluwer Academic Publisher, Boston, 1992.
4. T. Kohonen. *Self–organization and Associative Memory*. Springer, Berlin, 1984.
5. J.J. Kosowsky and A.L. Yuille. The invisible hand algorithm: solving the assignment problem with statistical mechanics. *Neural Computation*, 7(3):477–490, 1994.
6. Y. Linde, A. Buzo, and R. M. Gray. An algorithm for vector quantizer design. *IEEE Transactions on Communications*, 28:84–95, 1980.
7. S.P. Luttrell. Hierarchical vector quantizations. *IEE Proceedings*, 136:405–413, 1989.
8. J. MacQueen. Some methods for classification and analysis of multivariate observations. In *Proceedings of the 5th Berkeley Symposium on Mathematical Statistics and Probability*, pages 281–297, 1967.
9. S. Mallat. A theory for multidimensional signal decomposition: the wavelet representation. *IEEE Transactions on Pattern Analysis and Machine Intelligence*, 11(7):674–693, 1989.
10. T.M. Martinetz, S. G. Berkovich, and K. J. Schulten. 'neural–gas' network for vector quantization and its application to time–series prediction. *IEEE Transactions on Neural Networks*, 4(4):558–569, 1993.
11. K. Rose, E. Gurewitz, and G. Fox. Statistical mechanics and phase transitions in clustering. *Physical Review Letters*, 65(8):945–948, 1990.
12. R. Sinkhorn. A relationship between arbitrary positive matrices and doubly stochastic matrices. *Ann. Math. Statist.*, 35:876–879, 1964.

Associative Completion and Investment Learning Using PSOMs

Jörg Walter and Helge Ritter

Department of Information Science
University of Bielefeld, D-33615 Bielefeld, FRG
Email: walter@techfak.uni-bielefeld.de
http://www.techfak.uni-bielefeld.de/~walter/

Abstract

We describe a hierarchical scheme for rapid adaptation of context dependent "skills". The underlying idea is to first invest some learning effort to specialize the learning system to become a rapid learner for a restricted range of contexts. This is achieved by constructing a "Meta-mapping" that replaces an slow and iterative context adaptation by a "one-shot adaptation", which is a context-dependent skill-reparameterization.

The notion of "skill" is very general and includes a task specific, hand-crafted function mapping with context dependent parameterization, a complex control system, as well as a general learning system.

A representation of a skill that is particularly convenient for the investment learning approach is by a *Parameterized Self-Organizing Map* (PSOM). Its direct constructability from even small data sets significantly simplifies the investment learning stage; its ability to operate as a continuous associative memory allows to represent skills in the form of "multi-way" mappings (relations) and provides an automatic mechanism for sensor data fusion.

We demonstrate the concept in the context of a (synthetic) vision task that involves the associative completion of a set of feature locations and the task of one-shot adaptation of the transformation between world and object coordinates to a changed camera view of the object.

1 Introduction

The data-driven construction of non-linear mappings for interpolation and classification has been among the most widely used features of artificial neural networks (ANN). This has led to a strong emphasis of feed-forward networks and to much fruitful work at elucidating their capabilities from the perspective of interpolation, statistics, and approximation theory [3, 11].

However, there is clearly more to ANNs than the approximation of a nonlinear function[1]. As applications of neural networks have become more demanding, the issue of how to build neural *systems* that are composed of multiple mapping modules has received increasing attention [4], raising the question of how such systems can be trained without requiring inordinately large training sets. This is particularly urgent in areas where training samples are not abundant or are costly, such as, e.g., in robotics. Related with this is the important question of how to achieve *"one-shot"-learning*, i.e., the ability of a system to adapt to a particular context within a single or at least a very small number of trials. Finally, we should weaken the emphasis of "one-way" mappings.

---

[1]the reader will be aware of the fact that *any* computation can be viewed as a sufficiently complex vector valued function; however, this theoretical possibility does not mean that this is always the most productive viewpoint.

A much more natural and flexible formulation of many tasks, such as, e.g., flexible sensor fusion, is in terms of *continuous relations* and the corresponding required ability is that of a *continuous associative memory* instead of the much more rigid one-way mappings of feedforward networks.

In the present paper we attempt to address some of these issues. First, we show how the recently introduced Parameterized Self-Organizing Maps ("PSOMs",[6, 8]) can be used as a flexible continuous associative memory that offers a very natural approach to the problem of *sensor fusion* [1].

As an illustration we consider the task to associate for a continuous set of camera views of a rigid 3D-object the spatial positions of occluded point features from varying subsets of visible point features. Sensor fusion occurs automatically when the PSOM is offered more than a minimally necessary set to determine the orientation of the object. In this case, the redundant information is exploited for an increased accuracy of the inferred point locations.

We then describe a learning architecture in which a PSOM can be combined with other mapping modules (some or all of which may be PSOMs again) in such a way that the resulting system can be prestructured for the ability of "one-shot-learning" or "one-shot-adaptation" by means of a prior "investment learning" phase. Here, the underlying idea is that one-shot adaptation can be achieved with a "Meta-mapping" for the dynamic reparameterization of one or several mapping module(s) that represent(s) the "skill" that is to be adapted; this "Meta-mapping" is constructed during the investment learning phase and its input is a sufficient (small) number of sensor measurements that characterise the current context.

To make the paper self-contained, we give a brief summary of the main characteristics of the PSOM approach in the next section, and append a condensed description in the appendix. A more detailed account, together with result of applications in the domain of vision and robotics, see [7, 9, 10].

2 Properties of PSOMs

The PSOM is characterized by the following important features:

- The PSOM is the continuous analog of the standard discrete self-organizing map ([5]). It shows *excellent generalization* capabilities based on *attractor manifolds* instead of just attractor points.

- For training data sets with the *known topology* of a multi-dimensional cartesian grid the map manifold can be constructed *directly*. The resulting PSOM is immediately usable – without any need for time consuming adaptation sequences. However, further on-line *fine-tuning* is possible, e.g., in the case of coarsely sampled data or when the original training data were corrupted by noise.

- The price for rapid learning is the cost when using. The PSOM needs an iterative search for the best-matching parameter location s^* (Eq. 1 in the appendix).

- The *non-linear associative completion* ability of the PSOM can be used to represent continuous relations instead of just functions ("dynamic multi-way mapping"). This is illustrated in Fig. 1. The choice of an input variable set is made by specifying a suitable diagonal projection matrix P whose non-zero elements select the input subspace(s) together with a possible relative weighting of the input components for the non-linear least-square best-match (see Eq. 1 in the appendix for the best-match vector $w(s^*)$.)

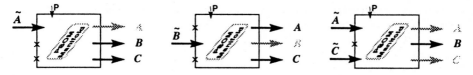

Figure 1: "Continuous associative memory" supports multiple mapping directions. The specified **P** matrices select different subspaces (here symbolized by $\tilde{A}$, $\tilde{B}$ and $\tilde{C}$) of the embedding space as inputs. Values of variables in the selected input subspaces are considered as "clamped" (indicated by a tilde) and determine the values found by the iterative least square minimization (Eq. 1). for the "best-match" vector $\mathbf{w}(\mathbf{s}^*)$. This provides an associative memory for the flexible representation of continuous relations.

3 Image Completion and Object Orientation Identification

To illustrate the flexibility of the PSOM-approach, we consider a task from the vision domain. Here, we are often confronted with the following *sensor-fusion* problem: given the identity of a geometric object together with a number of measurements of continuous valued features (such as the 2D-image locations of a subset of salient point features), use this information to infer a set of "missing" variable values, such as the location of occluded object parts or the 3D-orientation and/or location of the object.

Although the approach is not restricted to the use of camera sensors, we consider the case when the sensor measurements is a set of 2D locations of salient object features (of known identity). Such information would have to be obtained by some suitable preprocessing stage; in the present context we consider a synthetic vision task in which the set of 2D-locations has been computed from a perspective projection of a geometric object model. We also assume that the object has already been centered (by some suitable tracking process), leaving us with four remaining degrees of freedom: orientation and depth, which we describe by roll ϕ, pitch θ, yaw ψ and z [2].

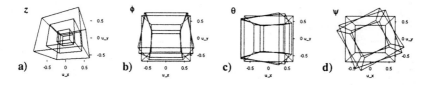

Figure 2: *(a)* The cubical test object seen by the camera when rotated and shifted in several depths z ($\phi=10°$, $\theta=20°$, $\psi=30°$, $z=2\dots6L$, cube size L.) *(b–d)* 0°, 20°, and 30° rotations in the roll ϕ, pitch θ, and yaw ψ system.

Fig. 2 shows the camera view of the test object (a unit cube, but note that the method works in the same way with any (fixed) shape) for different orientations and depths. This provides a different characterization of object pose by the coordinate pairs $\vec{u}_{P_i}$ of a set of salient feature points P_i, $i = 1, 2, \dots n$ on the object (for the cube, we choose its eight corners, i.e., $n = 8$). Therefore, a convenient embedding space X is spanned by the $4 + 2n$ variables $\mathbf{x} = (\phi, \theta, \psi, z, \vec{u}_{P1}, \vec{u}_{P2}, \dots, \vec{u}_{P_n})$. For the construction of the PSOM, X was sampled at the 81 points of a regular $3\times3\times3\times3$ grid (in ϕ, θ, ψ, z-space) covering a range of $150°$ for each orientation and twice the edge length L of the cube for the depth dimension.

Fig. 3 depicts some examples, in which orientation and depth of the cube were inferred using the constructed PSOM with the image positions of four of its corners (indicated by asterisks) chosen as input (dotted lines indicate true, solid lines indicate reconstructed object pose).

The achieved root mean square (RMS) errors for recovering the object pose were $2.7°$, $3.2°$, $2.8°$, and $0.12L$ (for ϕ, θ, ψ, z). The remaining four corner point locations could be predicted within an accuracy of 1.3% of the mean edge image size. A more detailed analysis of the impact of parameter variations, such as the range of the involved variables, the number of training vectors, the sensor noise and the number of available input points will be presented elsewhere [10].

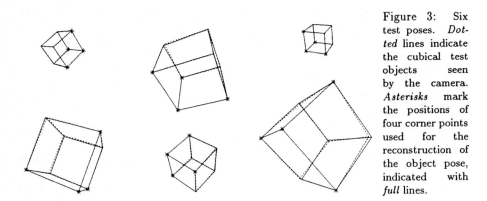

Figure 3: Six test poses. *Dotted* lines indicate the cubical test objects seen by the camera. *Asterisks* mark the positions of four corner points used for the reconstruction of the object pose, indicated with *full* lines.

4 Investment Learning in Prototypical Contexts

In this section we consider the problem of efficiently learning "skills" and there adaptation to changing context. To be concrete, we consider a "skill", in form of a multivariate function mapping $T : \vec{x}_1 \rightarrow \vec{x}_2$, transforming between two task variable sets $\vec{x}_1$ and $\vec{x}_2$. We assume: *(i)* that the "skill" can be acquainted by a "transformation box" ("T-Box"), which is a suitable building block with learning capabilities; *(ii)* the mapping "skill" T is internally modeled and determined by a set of parameters ω (which can be accessed from outside the "black box", which makes the T-Box rather an open, "white box"); *(iii)* the correct parameterization ω changes smoothly with the *context* of the system; *(v)* the situational context can be observed and is associated with a set of suitable sensors values $\vec{c}$ (some of them are possibly expensive and temporarily unavailable); *(vi)* the context changes only from time to time, or on a much larger time scale, than the task mapping T is employed.

The conventional, integrated approach is to consider the problem of learning the mapping from all relevant input values, $\vec{x}_1, \vec{c}$ to the desired output $\vec{x}_2$. This leads to a larger, specialized network. The disadvantages are *(i)* the possible catastrophic inter-

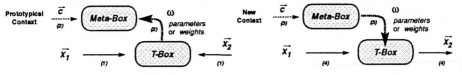

Figure 4: The transformation T-BOX maps between task input $\vec{x}_1$ and output $\vec{x}_2$ spaces and gets encoded in the parameter set ω.

ference (after-learning in a situated context may effect other contexts in an uncontrolled way); *(ii)* low modularity and re-usability.

Here, we approach a solution in a modular way and suggest to *split learning*: *(i)* (structurally) among two modules, the META-BOX and the T-BOX, and *(ii)* (temporally) in two phases. The first, the *investment learning* stage may be slow and has the task of learning prototypical context situations. It does not yet produce the final solution, but instead pre-structures the system such that in the subsequent, the *one-shot adaptation phase*, the specialization to a particular solution (within the chosen domain) can be achieved extremely rapidly.

As illustrated in Fig. 4, the META-BOX is responsible for providing the mapping from sensory context observations $\vec{c}$ to the parameter set ω. Therefore, each T-BOX parameter/weight set ω together with its associated context information $\vec{c}$ can serve as a high-dimensional training data vector for constructing the META-BOX mapping during the investment learning stage.

To obtain these training data vectors, we must choose a suitable set of *prototypical contexts* $j = 1, 2, \ldots$ and determine for each of them the parameter set ω_j of an adapted T-BOX *(1)* and the context sensor observation $\vec{c}_{,j}$ *(2)*. After the META-BOX has been trained, the task of adapting the "skill" to a new system context is tremendously accelerated. Instead of any time-consuming re-learning of the (T) mapping this adjustment now takes the form of an *immediate* META-BOX $\rightarrow$ T-BOX mapping (*one-shot adaptation*): The META-BOX maps a new (unknown) context observation $\vec{c}_{,new}$ *(3)* into the parameter/weight set ω_{new} for the T-BOX. Equipped with ω_{new}, the T-BOX provides the desired mapping T_{new} *(4)*.

In contrast to a *mixture-of-experts* architecture [4], suggesting a linear combination of multiple, parallel working "expert" networks, the described scheme could be viewed as *non-linear "interpolation-of-expertise"*. It is efficient w.r.t. network requirements in memory and in computation: the "expertise" (ω) is gained and implemented in a single "expert" network (T-BOX). Furthermore, the META-BOX needs to be re-engaged only when the context is changed, which is indicated by a deviating sensor value $\vec{c}$.

However, this scheme requires from the learning implementation of the T-BOX, that the parameter/weight set ω is represented in a -with $\vec{c}$ smoothly varying- "non-degenerated" manner. E.g. a regular multilayer perceptron allows many weight permutations ω. Employing a MLP in the T-BOX would additionally require a suitable stabilizer to avoid grossly inadequate interpolation between prototypical "expertises" ω_j, denoted in different kinds of permutations.

5 Object Coordinate Transformation by Investment Learning

To illustrate this approach let us revisit the vision example of Sec.3. For a robot, an interesting skill in that setting could be to transform coordinates from a camera centered (world or tool) frame (yielding coordinate values $\vec{x}_1$) to the object centered frame (yielding coordinate values $\vec{x}_2$). This mapping would have to be represented by the T-Box and its "environmental context" would be the current orientation of the object relative to the camera. Fig. 5 shows three ways how the investment learning scheme can be implemented in that situation. All three share the same PSOM network type as the META-Box building block. The "Meta-PSOM" bears the advantage, that the architecture can cope easily with situations, where various (redundant) sensory values are un-available.

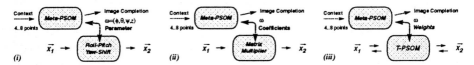

Figure 5: Three different ways to solve the context dependent, or investment learning task.

The first solution (i) uses the Meta-PSOM for the reconstruction of object pose in roll-pitch-yaw-depth values from Sec. 3. The T-Box is given by the four succesive homogeneous transformation on the basis of the ϕ, θ, ψ, z values obtained from the Meta-PSOM.

The middle solution (ii) represents the coordinate transformation as the product of the four successive transformations. Thus, in this case the Meta-PSOM controls the coefficients of a matrix multiplication. As in (i), the required ω are gained by a suitable calibration, or system identification procedure.

When no explicit ansatz for the T-Box is already available, we can use method (iii). Here, for each prototypical context, the required T-mapping is learned by a network and becomes encoded in its weight set ω. For this, one can use any trainable network that meets the requirement stated at the end of the previous section; however, PSOMs are a particularly convenient choice, since they can be directly constructed from a small data set and since they offer the advantage of multi-way mappings.

To illustrate this for the present example, we chose for the T-Box a $2 \times 2 \times 2$ "T-PSOM" that implements the coordinate transform for *both directions simultaneously*. Its training required eight training vectors arranged at the corners of a cubical grid.

In order to compare approaches (i) − (iii), we computed the transformation T-Box accuracy averaged over a set of 50 contexts (given by 50 randomly chosen object poses), each with 100 object volume points $\vec{x}_2$ to be transformed into camera coordinates $\vec{x}_1$. For the computed RMS value for the camera x, y, and z direction method (i) gave the result $[0.025, 0.023, 0.14]L$. Method (ii) led to the slightly better values of $[0.016, 0.015, 0.14]L$. Approach (iii), although not requiring any geometric analysis, led to practically the same accuracy, namely $[0.015, 0.014, 0.12]L$.

6 Discussion and Conclusion

To be practical, learning algorithms must provide solutions that can compete with solutions hand-crafted by a human who has analyzed the system. The criteria for success can vary, but usually the costs of gathering data and of teaching the system are a major factor on the side of the learning system, while the effort to analyze the problem and to design an algorithmn is on the side of the hand crafted solution.

Here we suggest the PSOM as a versatile module for the rapid learning of high-dimensional, non-linear, smooth relations. Based on a strong bias, introduced by structuring the training data in a topological order, the PSOM can generalize from very few examples - if this model is a good approximation to the system. The multi-way mapping ability PSOM is generated by the *associative completion* of partial inputs. We showed an example in which this ability was used to infer the positions of occluded feature points. In addition, this property leads to an interesting approach for the task of *sensor fusion*, since redundant data are exploited to achieve a higher accuracy for the inferred variable values.

As a general framework to achieve one-shot or very rapid learning we introduced the idea of *investment learning*. While PSOMs are extremely well suited for this approach, the underlying idea to "compile" the effect of a longer learning phase into a one-step META-BOX is more general and is independent from the PSOMs. The META-BOX controls the parameterization of a set of context specific "skills" which are implemented in the form of one or several transforms or T-BOX. Thus, learning at the level of the skills is replaced *by their context-sensitive dynamic re-parameterization* through the META-BOX-mapping. This emphasises an important point for the construction of more flexible and more powerful learning systems: in addition to focusing on output *values* we should increasingly embrace mappings that *produce as their result other mappings* – a point of view that has in recent years also become increasingly emphasized in the realm of functional programming languages.

We illustrated three versions of this approach when the output mapping was a coordinate transform between a camera centered and an object centered frame of reference. They differed in their choice of the T-BOX that was used. Comparing the three choices we found that the neural network based T-PSOM can fully compete with dedicated one-way function mapping boxes while at the same time offering the additional advantage of providing forward and backward transform simultaneously.

Appendix: The PSOM Algorithm

A PSOM is a parameterized, m-dimensional hyper-surface $M = \{\mathbf{w}(s) \in X \subseteq \mathbb{R}^d | s \in S \subseteq \mathbb{R}^m\}$ that is embedded in some higher-dimensional vector space X. M is used in a very similar way as the standard discrete self-organizing map: given a distance measure $dist(\mathbf{x}, \mathbf{x}')$ and an input vector $\mathbf{x}$, a best-match location $\mathbf{s}^*(\mathbf{x})$ is determined by minimizing

$$\mathbf{s}^* := \operatorname*{argmin}_{s \in S} dist(\mathbf{x}, \mathbf{w}(s)) \qquad (1)$$

The associated "best-match vector" $\mathbf{w}(\mathbf{s}^*)$ provides the best approximation of input $\mathbf{x}$ in the manifold M. If we require $dist(\cdot)$ to vary only in a input sub-space X^{in} of X (i.e., $dist(\mathbf{x}, \mathbf{x}') = (\mathbf{x} - \mathbf{x}')^T \mathbf{P}(\mathbf{x} - \mathbf{x}') = \sum_{k=1}^{d} p_k (x_k - x_k')^2$, where the diagonal matrix $\mathbf{P} = \operatorname{diag}(p_1, \cdots, p_d)$ projects into X^{in}), $\mathbf{s}^*(\mathbf{x})$ actually will only depend on $\mathbf{Px}$. The components of $\mathbf{w}(\mathbf{s}^*(x))$ with zero diagonal elements p_k can be viewed as the output. The best-match vector $\mathbf{w}(\mathbf{s}^*)$ is the (non-linear) *associative completion of a fragmentary input* $\mathbf{x}$ of which only the *part* $\mathbf{Px}$ *is reliable*. It is this associative mapping that we will exploit in applications of the PSOM, see Fig. 1.

M is constructed as a manifold that passes through a given set D of data examples. To this end, we assign to each data sample a point $\mathbf{a} \in S$ and denote the associated data sample by $\mathbf{w_a}$. The set $\mathbf{A}$ of the assigned parameter values $\mathbf{a}$ should provide a good discrete "model" of the topology of our data set. The assignment between data vectors and points $\mathbf{a}$ must be made in a topology preserving fashion to ensure good interpolation by the manifold M that is obtained by the following steps.

For each point $\mathbf{a} \in \mathbf{A}$, we construct a "basis function" $H(\cdot, \mathbf{a}; \mathbf{A})$ or simplified[2] $H(\cdot, \mathbf{a}) : S \mapsto \mathbb{R}$ that obeys (i) $H(\mathbf{a}_i, \mathbf{a}_j) = 1$ for $i = j$ and vanishes at all other points of $\mathbf{A}$ $i \neq j$ (orthonormality condition,) and (ii) $\sum_{\mathbf{a} \in \mathbf{A}} H(\mathbf{s}, \mathbf{a}) = 1$ for $\forall \mathbf{s}$ ("partition of unity" condition.) We will mainly be concerned with the case of $\mathbf{A}$ being a m-dimensional rectangular hyper-lattice; in this case, the functions $H(\cdot, \mathbf{a})$ can be constructed as products of Lagrange interpolation polynomials, see [9]. Then,

$$\mathbf{w}(\mathbf{s}) = \sum_{\mathbf{a} \in \mathbf{A}} H(\mathbf{s}, \mathbf{a}) \, \mathbf{w_a}. \tag{2}$$

defines a manifold M that passes through all data examples. Minimizing $dist(\cdot)$ in Eq. 1 can be done by some iterative procedure, such as gradient descent or – preferably – the Levenberg-Marquardt algorithm [9]. This makes M into the attractor manifold of a (discrete time) dynamical system. Since M contains the data set D, any at least m-dimensional "fragment" of a data example $\mathbf{x} = \mathbf{w} \in D$ will be attracted to the correct completion $\mathbf{w}$. Inputs $\mathbf{x} \notin D$ will be attracted to the interpolating manifold point.

References

[1] B. Brunner, K. Arbter, and G. Hirzinger. Task directed programming of sensor based robots. In *Intelligent Robots and Systems (IROS-94)*, pages 1081–1087, September 1994.

[2] K. Fu, R. Gonzalez, and C. Lee. *Robotics : Control, Sensing, Vision, and Intelligence.* McGraw-Hill, 1987.

[3] K. Hornik, M. Stinchcombe, and H. White. Multilayer feedforward networks are universal approximators. *Neural Networks*, 2:359–366, 1989.

[4] M. I. Jordan and R. A. Jacobs. Hierarchical mixtures of experts and the EM algorithm. *Neural Computation*, 6(2):181–214, 1994.

[5] T. Kohonen. *Self-Organizing Maps*, volume 30 of *Springer Series in Information Sciences*. Springer, Berlin, Heidelberg, 1995.

[6] H. Ritter. Parametrized self-organizing maps. In S. Gielen and B. Kappen, editors, *Proc. Int. Conf. on Artificial Neural Networks (ICANN-93), Amsterdam*, pages 568–575. Springer Verlag, Berlin, 1993.

[7] J. Walter and H. Ritter. Investment learning with hierarchical PSOM. In *NIPS*95*, page (in press). MIT Press, 1995.

[8] J. Walter and H. Ritter. Local PSOMs and Chebyshev PSOMs – improving the parametrised self-organizing maps. In *Proc. Int. Conf. on Artificial Neural Networks (ICANN-95), Paris*, volume 1, pages 95–102, 1995.

[9] J. Walter and H. Ritter. Rapid learning with parametrized self-organizing maps. *Neurocomputing, Special Issue*, (in press), 1996.

[10] J. A. Walter. *Rapid Learning in Robotics*. Infix Verlag, Sankt Augustin, Germany, 1996. (in preparation).

[11] H. White. Learning in artificial neural networks: a statistical perspective. *Neural Computation*, 1:425–464, 1989.

---

[2] In contrast to kernel methods, the basis functions may depend on the relative position to all other knots. However, we drop in our notation the dependency $H(\mathbf{s}, \mathbf{a}) = H(\mathbf{s}, \mathbf{a}; \mathbf{A})$ on the latter.

GTM: A Principled Alternative to the Self-Organizing Map

Christopher M. Bishop, Markus Svensén and Christopher K. I. Williams

Neural Computing Research Group
Aston University, Birmingham, B4 7ET, U.K.
ncrg@aston.ac.uk
http://www.ncrg.aston.ac.uk/

Abstract. The Self-Organizing Map (SOM) algorithm has been extensively studied and has been applied with considerable success to a wide variety of problems. However, the algorithm is derived from heuristic ideas and this leads to a number of significant limitations. In this paper, we consider the problem of modelling the probability density of data in a space of several dimensions in terms of a smaller number of latent, or hidden, variables. We introduce a novel form of latent variable model, which we call the GTM algorithm (for *Generative Topographic Map*), which allows general non-linear transformations from latent space to data space, and which is trained using the EM (expectation-maximization) algorithm. Our approach overcomes the limitations of the SOM, while introducing no significant disadvantages. We demonstrate the performance of the GTM algorithm on simulated data from flow diagnostics for a multi-phase oil pipeline.

1   Introduction

The Self-Organizing Map (SOM) algorithm of Kohonen (1982) represents a form of unsupervised learning in which a set of unlabelled data vectors $\mathbf{t}^n$ ($n = 1, \ldots, N$) in a D-dimensional data space is summarized in terms of a set of reference vectors having a spatial organization corresponding (generally) to a two-dimensional sheet. While this algorithm has achieved many successes in practical applications, it also suffers from some major deficiencies, many of which are highlighted in Kohonen (1995) and reviewed in this paper[1].

From the perspective of statistical pattern recognition the fundamental goal of unsupervised learning is to develop a representation of the distribution $p(\mathbf{t})$ from which the data were generated. In this paper we consider the problem of modelling $p(\mathbf{t})$ in terms of a number (usually two) of *latent* or *hidden* variables. By considering a particular class of such models we arrive at a formulation in terms of a constrained Gaussian mixture which can be trained using the EM

---

[1] Biological metaphor is sometimes invoked when motivating the SOM procedure. It should be stressed that our goal here is not neuro-biological modelling, but rather the development of effective algorithms for data analysis, for which biological realism is irrelevant.

(expectation-maximization) algorithm. The topographic nature of the representation is an intrinsic feature of the model and is not dependent on the details of the learning process. Our model defines a *generative* distribution $p(\mathbf{t})$ and will be referred to as the GTM (*Generative Topographic Map*) algorithm.

2   Latent Variables

The goal of a latent variable model is to find a representation for the distribution $p(\mathbf{t})$ of data in a D-dimensional space $\mathbf{t} = (t_1, \ldots, t_D)$ in terms of a number L of latent variables $\mathbf{x} = (x_1, \ldots, x_L)$. This is achieved by first considering a non-linear function $\mathbf{y}(\mathbf{x}; \mathbf{W})$, governed by a set of parameters $\mathbf{W}$, which maps points $\mathbf{x}$ in the latent space into corresponding points $\mathbf{y}(\mathbf{x}; \mathbf{W})$ in the data space. Typically we are interested in the situation in which the dimensionality L of the latent variable space is less than the dimensionality D of the data space, since our premise is that the data itself has an intrinsic dimensionality which is less than D. The transformation $\mathbf{y}(\mathbf{x}; \mathbf{W})$ then maps the hidden-variable space into an L-dimensional non-Euclidean manifold embedded within the data space.

If we define a probability distribution $p(\mathbf{x})$ on the latent variable space, this will induce a corresponding distribution $p(\mathbf{y}|\mathbf{W})$ in the data space. We shall refer to $p(\mathbf{x})$ as the prior distribution of $\mathbf{x}$ for reasons which will become clear shortly. Since $L < D$, the distribution in $\mathbf{t}$-space would be confined to a manifold of dimension L and hence would be singular. Since in reality the data will only approximately live on a lower-dimensional manifold, it is appropriate to include a noise model for the $\mathbf{t}$ vector. We therefore define the distribution of $\mathbf{t}$, for given $\mathbf{x}$ and $\mathbf{W}$, to be a spherical Gaussian centred on $\mathbf{y}(\mathbf{x}; \mathbf{W})$ having variance β^{-1} so that $p(\mathbf{t}|\mathbf{x}, \mathbf{W}, \beta) \sim \mathcal{N}(\mathbf{t}|\mathbf{y}(\mathbf{x}; \mathbf{W}), \beta^{-1}\mathbf{I})$. The distribution in $\mathbf{t}$-space, for a given value of $\mathbf{W}$, is then obtained by integration over the $\mathbf{x}$-distribution

$$p(\mathbf{t}|\mathbf{W}, \beta) = \int p(\mathbf{t}|\mathbf{x}, \mathbf{W}, \beta) p(\mathbf{x}) \, d\mathbf{x}. \tag{1}$$

For a given a data set $\mathcal{D} = (\mathbf{t}^1, \ldots, \mathbf{t}^N)$ of N data points, we can determine the parameter matrix $\mathbf{W}$, and the inverse variance β, using maximum likelihood, where the log likelihood function is given by

$$L(\mathbf{W}, \beta) = \sum_{n=1}^{N} \ln p(\mathbf{t}^n|\mathbf{W}, \beta). \tag{2}$$

In principle we can now seek the maximum likelihood solution for the weight matrix, once we have specified the prior distribution $p(\mathbf{x})$ and the functional form of the mapping $\mathbf{y}(\mathbf{x}; \mathbf{W})$, by maximizing $L(\mathbf{W}, \beta)$. The latent variable model can be related to the Kohonen SOM algorithm by choosing $p(\mathbf{x})$ to be a sum of delta functions centred on the nodes of a regular grid in latent space $p(\mathbf{x}) = 1/K \sum_{i=1}^{K} \delta(\mathbf{x} - \mathbf{x}^i)$. This form of $p(\mathbf{x})$ allows the integral in (1) to be performed analytically. Each point $\mathbf{x}^i$ is then mapped to a corresponding point $\mathbf{y}(\mathbf{x}^i; \mathbf{W})$ in data space, which forms the centre of a Gaussian density function,

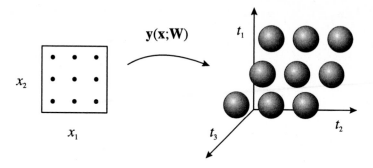

Fig. 1. We consider a prior distribution $p(\mathbf{x})$ consisting of a superposition of delta functions, located at the nodes of a regular grid in latent space. Each node $\mathbf{x}^i$ is mapped to a point $\mathbf{y}(\mathbf{x}^i; \mathbf{W})$ in data space, which forms the centre of the corresponding Gaussian distribution.

as illustrated in Figure 1. Thus the distribution function in data space takes the form of a Gaussian mixture model $p(\mathbf{t}|\mathbf{W}, \beta) = 1/K \sum_{i=1}^{K} p(\mathbf{t}|\mathbf{x}^i, \mathbf{W}, \beta)$ and the log likelihood function (2) becomes

$$L(\mathbf{W}, \beta) = \sum_{n=1}^{N} \ln \left\{ \frac{1}{K} \sum_{i=1}^{K} p(\mathbf{t}^n|\mathbf{x}^i, \mathbf{W}, \beta) \right\}. \tag{3}$$

This distribution is a *constrained* Gaussian mixture since the centres of the Gaussians cannot move independently but are related through the function $\mathbf{y}(\mathbf{x}; \mathbf{W})$. Note that, provided the mapping function $\mathbf{y}(\mathbf{x}; \mathbf{W})$ is smooth and continuous, the projected points $\mathbf{y}(\mathbf{x}^i; \mathbf{W})$ will necessarily have a topographic ordering.

2.1 The EM Algorithm

If we choose a particular parametrized form for $\mathbf{y}(\mathbf{x}; \mathbf{W})$ which is a differentiable function of $\mathbf{W}$ we can use standard techniques for non-linear optimization, such as conjugate gradients or quasi-Newton methods, to find a weight matrix $\mathbf{W}^\star$, and inverse variance $\beta^\star$, which maximize $L(\mathbf{W}, \beta)$. However, our model consists of a mixture distribution which suggests that we might seek an EM algorithm (Dempster *et al.*, 1977). By making a careful choice of model $\mathbf{y}(\mathbf{x}; \mathbf{W})$ we will see that the M-step can be solved exactly. In particular we shall choose $\mathbf{y}(\mathbf{x}; \mathbf{W})$ to be given by a generalized linear network model of the form

$$\mathbf{y}(\mathbf{x}; \mathbf{W}) = \mathbf{W}\phi(\mathbf{x}) \tag{4}$$

where the elements of $\phi(\mathbf{x})$ consist of M fixed basis functions $\phi_j(\mathbf{x})$, and $\mathbf{W}$ is a $D \times M$ matrix with elements w_{kj}. Generalized linear networks possess the same universal approximation capabilities as multi-layer adaptive networks, provided the basis functions $\phi_j(\mathbf{x})$ are chosen appropriately.

By setting the derivatives of (3) with respect to w_{kj} to zero, we obtain

$$\boldsymbol{\Phi}^{\mathrm{T}}\mathbf{G}\boldsymbol{\Phi}\mathbf{W}^{\mathrm{T}} = \boldsymbol{\Phi}^{\mathrm{T}}\mathbf{R}\mathbf{T} \qquad (5)$$

where $\boldsymbol{\Phi}$ is a $K \times M$ matrix with elements $\Phi_{ij} = \phi_j(\mathbf{x}^i)$, $\mathbf{T}$ is a $N \times D$ matrix with elements t_k^n, and $\mathbf{R}$ is a $K \times N$ matrix with elements R_{in} given by

$$R_{in}(\mathbf{W}, \beta) = \frac{p(\mathbf{t}^n | \mathbf{x}^i, \mathbf{W}, \beta)}{\sum_{i'=1}^{K} p(\mathbf{t}^n | \mathbf{x}^{i'}, \mathbf{W}, \beta)} \qquad (6)$$

which represents the posterior probability, or *responsibility*, of the mixture components i for the data point n. Finally, $\mathbf{G}$ is a $K \times K$ diagonal matrix, with elements $G_{ii} = \sum_{n=1}^{N} R_{in}(\mathbf{W}, \beta)$. Equation (5) can be solved for $\mathbf{W}$ using standard matrix inversion techniques. Similarly, optimizing with respect to β we obtain

$$\frac{1}{\beta} = \frac{1}{ND} \sum_{i=1}^{K} \sum_{n=1}^{N} R_{ni}(\mathbf{W}, \beta) \|\mathbf{y}(\mathbf{x}^n; \mathbf{W}) - \mathbf{t}^n\|^2 . \qquad (7)$$

Here (6) corresponds to the E-step, while (5) and (7) correspond to the M-step. Typically the EM algorithm gives satisfactory convergence after a few tens of cycles. An on-line version of this algorithm can be obtained by using the Robbins-Monro procedure to find a zero of the objective function.

3   Relation to the Self-Organizing Map

The list below describes some of the problems with the SOM procedure and how the GTM algorithm solves them.

1. The SOM algorithm is not derived by optimizing an objective function, unlike GTM . Indeed it has been proven (Erwin *et al.*, 1992) that such an objective function cannot exist for the SOM algorithm.
2. In GTM the neighbourhood-preserving nature of the mapping is an automatic consequence of the choice of a smooth, continuous function $\mathbf{y}(\mathbf{x}; \mathbf{W})$. Neighbourhood-preservation is not guaranteed by the SOM procedure.
3. There is no assurance that the code-book vectors will converge using SOM. Convergence of the batch GTM algorithm is guaranteed by the EM algorithm, and the Robbins-Monro theorem provides a convergence proof for the on-line version.
4. GTM defines an explicit probability density function in data space. In contrast, SOM does not define a density model. Attempts have been made to interpret the density of codebook vectors as a model of the data distribution but with limited success.
5. For SOM the choice of how the neighbourhood function should shrink over time during training is arbitrary, and so this must be optimized empirically. There is no neighbourhood function to select for GTM .

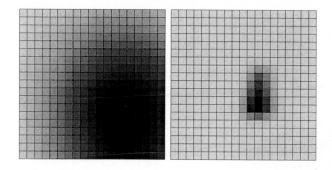

Fig. 2. Examples of the posterior probabilities (responsibilities) of the latent space points at an early stage (left) and late stage (right) during the convergence of the GTM algorithm, evaluated for a single data point from the training set in the oil-flow problem discussed in Section 4.

6. It is difficult to know by what criteria to compare different runs of the SOM procedure. For GTM one simply compares the likelihood of the data under the model, and standard statistical tests can be used for model comparison.

Notwithstanding these key differences, there are very close similarities between the SOM and GTM techniques. Figure 2 shows the posterior probabilities (responsibilities) corresponding to the oil flow problem considered in Section 4. At an early stage of training the responsibility for representing a particular data point is spread over a relatively large region of the map. As the EM algorithm proceeds so this responsibility 'bubble' shrinks automatically. The responsibilities (computed in the E-step) govern the updating of $\mathbf{W}$ and β in the M-step and, together with the smoothing effect of the basis functions $\phi_j(\mathbf{x})$, play an analogous role to the neighbourhood function in the SOM procedure.

4   Experimental Results

We present results from the application of this algorithm to a problem involving 12-dimensional data arising from diagnostic measurements of oil flows along multi-phase pipelines (Bishop and James, 1993). The three phases in the pipe (oil, water and gas) can belong to one of three different geometrical configurations, corresponding to stratified, homogeneous, and annular flows, and the data set consists of 1000 points drawn with equal probability from the 3 classes. We take the latent variable space to be two-dimensional, since our goal in this application is data visualization. Each data point $\mathbf{t}^n$ induces a posterior distribution $p(\mathbf{x}|\mathbf{t}^n, \mathbf{W}, \beta)$ in $\mathbf{x}$-space. However, it is often convenient to project each data point down to a unique point in $\mathbf{x}$-space, which can be done by finding the mean of the posterior distribution.

Figure 3 shows the oil data visualized with GTM and SOM. The CPU times taken for the GTM, SOM with a Gaussian neighbourhood, and SOM with a

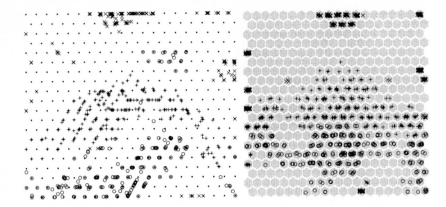

Fig. 3. The left plot shows the posterior-mean projection of the oil flow data in the latent space of the non-linear model. The plot on the right shows the same data set visualized using the batch SOM procedure, in which each data point is assigned to the point on the feature map corresponding to the codebook vector to which it is nearest. In both plots, crosses, circles and plus-signs represent the three different oil-flow configurations.

bubble neighbourhood were 644, 1116 and 355 seconds respectively. In each case the algorithms were run for 25 complete passes through the data set.

In conclusion, we have provided an alternative algorithm to the SOM which overcomes its principal deficiencies while retaining its general characteristics. We know of no significant disadvantage in using the GTM algorithm in place of the SOM. While we believe the SOM procedure is superseded by the GTM algorithm, is should be noted that the SOM has provided much of the inspiration for developing GTM. The relationships between GTM and a number of other algorithms are discussed in a longer paper available from:
http://www.ncrg.aston.ac.uk/Papers/postscript/NCRG_96_015.ps.Z

Acknowledgements This work was supported by EPSRC grant GR/K51808: *Neural Networks for Visualisation of High-Dimensional Data.*

References

Bishop, C. M. and G. D. James (1993). Analysis of multiphase flows using dual-energy gamma densitometry and neural networks. *Nuclear Instruments and Methods in Physics Research* **A327**, 580–593.

Dempster, A. P., N. M. Laird, and D. B. Rubin (1977). Maximum likelihood from incomplete data via the EM algorithm. *J. Roy. Stat. Soc B* **39** (1), 1–38.

Erwin, E., K. Obermayer, and K. Schulten (1992). Self-organizing maps: ordering, convergence properties and energy functions. *Biological Cybernetics* **67**, 47–55.

Kohonen, T. (1982). Self-organized formation of topologically correct feature maps. *Biological Cybernetics* **43**, 59–69.

Kohonen, T. (1995). *Self-Organizing Maps.* Berlin: Springer-Verlag.

Creating Term Associations Using a Hierarchical ART Architecture

Alberto Muñoz*

Department of Statistics and Econometrics
University Carlos III, 28903 Getafe, Madrid, Spain

Abstract. In this work we address the problem of creating semantic term associations (key words) from a text database. The proposed method uses a hierarchical neural architecture based on the Fuzzy Adaptive Resonance Theory (ART) model. It exploits the specific statistical structure of index terms to extract semantically meaningful term associations; these are asymmetric and one-to-many due to the polysemy phenomenon. The underlying algorithm is computationally appropriate for deployment on large databases. The operation of the system is illustrated with a real database.

Key Words: Knowledge extraction, information retrieval, neural ART models, text databases

1 Introduction

A well known problem in Information Retrieval (IR) research is the difficulty faced by users who try to accurately formulate queries to retrieve the documents they want. This problem has to do with the user's lack of familiarity with the base lexicon, and often results in a not so satisfactory retrieved document set [4, 8]. In order to overcome this problem, many commercial databases provide a thesaurus, which helps users find the precise terms that match the documents in the base. Since many databases (as in the case of networked information in Internet) change its contents rapidly, it is unfeasible to update thesauri manually; often, heterogeneity in database contents makes this task even harder.

A typical thesaurus provides, for every term in it, a list of broader, narrower and related terms. Our objective here is to automatically produce such a list of associated terms for every index term in the database under consideration. To this aim, we should first note that associations between terms are asymmetric in nature; this asymmetry means that, if s_{ij} denotes the strength of the association (t_i, t_j), it is not necessarily true that $s_{ij} = s_{ji}$. Consider, for instance, the pair

---

* e-mail: albmun@est-econ.uc3m.es. Financial support from DGICYT grant PB94-0374 (Spain) is gratefully appreciated. Thanks are due to J. Muruzábal and L. Tenorio for some useful comments.

'fuzzy set'; most people will relate 'fuzzy' to 'set' more strongly than conversely. This assymetry has been pointed out before in [2], where numerical measures for s_{ij} and s_{ji} are proposed; given a term t_i in a base, s_{ij} must be computed for each of the remaining terms in the base. However, no list of closely related terms is given.

Our specific target here is to detect only the most relevant related terms for each term, and measure the strength of such relationships. The paper is organized as follows. Section 2 introduces the neural ART architecture used in the sequence. Section 3 describes the performance of the system on a real base and finally, section 4 summarizes and points out some directions for future work.

2 Hierarchical Fuzzy ART model for semantic classes extraction

In order to model the hierarchical term relationships present in thesauri, we need the concept of subsethood. We use here the vector space model [7] to represent terms: $t_i \in \mathbb{R}^n \ \forall i$, where the base contains n documents, and $t_{ik} = 1$ if t_i occurs in document k, 0 otherwise. We say that $t_i \subset t_j$ when the topic represented by t_i is a subtopic of the topic represented by t_j. This is a fuzzy relation (and notion, too), and we will measure its strength by:

$$s_{ij} = \text{degree}\,(t_i \subset t_j) = \frac{|t_i \wedge t_j|}{|t_i|} \tag{1}$$

where, in coordinates, $(\mathbf{t}_i \wedge \mathbf{t}_j)_k \equiv \min(t_{ik}, t_{jk})$ and $|\mathbf{t}_i| = \sum_{k=1}^{n} |t_{ik}|$. Both, the interpretation of vectors as fuzzy sets and the justification for the election of s_{ij} are based on Kosko's subsethood theorem [5]. Note that it is implicitly assumed that if t_i appears in every document whenever t_j does, then $t_i \subset t_j$.

Fuzzy ART is a neural network architecture that makes explicit use of measure (1) (see [1] for a more detailed exposition). We can consider Fuzzy ART as a clustering algorithm (in the sense of [6]), where clusters are represented by prototype vectors. It uses two distance measures, essentially s_{ij} and s_{ji} (see measure (1) above), and a control for each measure to decide when a pattern X is too far from the cluster prototype T. These controls are $\frac{|T \wedge X|}{\beta + |T|} < \frac{|X|}{\beta + n}$, $\beta \in \mathbb{R}$, for measure s_{ij}, and $s_{ji} < \rho$, where $0 \leq \rho < 1$, for s_{ji} (ρ is called the vigilance parameter). The operation is as follows [6, 1]: (a) Start with an empty list of prototype vectors. (b) Let X be the next input vector. (c) Find the closest cluster prototype vector using s_{ij}; let T be this vector. (d) If T is too far from X using measure s_{ij} (or if there are no cluster prototype vectors), then create a new cluster with prototype vector X. (e) If T is too far from X by measure s_{ji}, deactivate T and go to step (c) to try another prototype. (f) If T is close enough to X according to both measures, then modify T by moving it closer to X. Go to step (b). Prototype vectors are adapted using the equation:

$$T^{(new)} = \lambda(X \wedge T^{(old)}) + (1 - \lambda)T^{(old)} \quad \text{where } 0 < \lambda \leq 1 \tag{2}$$

Thus, after training, clusters (nodes of F_2 in Fuzzy ART terminology) contain patterns that are close enough according to the two distances. This means that, given two patterns x and y in the same class, both $x \subset y$ and $y \subset x$ to the extent determined by Fuzzy ART parameters.

Because of the measures used and our definition of subtopic, Fuzzy ART seems to be a suitable tool for constructing semantic classes. There is a problem in applying Fuzzy ART to index terms, though: it is known that, after training, every prototype vector becomes the intersection of all learned patterns [3]. By Zipf law [9], if terms are arranged in descending occurrence order, the word frequency of term in position r ($r = 1, 2, 3, \ldots$), $f(r)$ say, verifies $f(r) \cdot r \simeq k$, where k is a constant. In particular, it follows that there are many rare terms but just a few very frequent terms (for an example, see fig. 1(a)). Therefore, if Fuzzy ART is trained on the full index term set, two things may happen: (1) The most frequent words will form isolated classes (provided that the ART vigilance parameter is high enough); (2) if a class contains both very frequent and rare terms, the intersection will usually boil down to a non zero component (that is, a single document in common). Moreover, the grouping of terms in the class will be likely to be haphazard. We have verified experimentally the occurrence of both situations.

In order to solve this problem, the whole term set T is divided into three groups: high frequency terms (T_1), moderate frequency terms (T_2), and low frequency terms (T_3). Now, we use a hierarchy of four Fuzzy ART models: Fuzzy ART A_1, for term set T_1; Fuzzy ART A_2, for set T_2; Fuzzy ART A_{12} for term set $T_{12} = T_1 \cup T_2$, and Fuzzy ART A_T for the complete term set $T = T_1 \cup T_2 \cup T_3$. No single Fuzzy ART model is used for term set T_3, as these are terms that appear in very few documents and, hence, their vector representation is very sparse. Use of Fuzzy ART in this situation will lead to the category proliferation phenomenon [1]. In this case, fuzzy complement coding [1] is not a solution due to the strong asymmetry between 1's and 0's: the fact that two terms have a zero in common (they are both absent from a given document) contains no useful information in itself. Therefore, fuzzy complement coding will not be used. We minimize this problem by first ranking the terms by descending norm. Due the way ART operates, unnecessary proliferation of ART classes is avoided. The architecture is shown in fig. 1(b).

Thus, terms organize into semantic classes at different levels of frequency. A particular term t can belong to up to four different semantic classes: one for $t \in T_3$, three for $t \in T_2$ and four for $t \in T_1$. Note that the number of related terms obtained in this way varies depending on the term frequency. In this way, rare words attain fewer related terms than common ones. The hierarchy structure can be straightforward generalized to work with finer term frequency partitions.

To create term associations, we can not relate each term to every other in the same class; that would produce, in general, too many associations. We must choose, for every term, only a few significant related terms. Given a term $t \in T_i$ (where T_i can be either T_1, T_2, T_{12}, or T), we will accept four associations (t, t_j), whose indexes j are given by:

$$j_{1,T_i}^* = \arg\max_j \frac{|t \wedge t_j|}{|t|}, \quad t_j \in T_i; \quad j_{1,c_i(t)}^* = \arg\max_j \frac{|t \wedge t_j|}{|t|}, \quad t_j \in c_i(t) \quad (3)$$

$$j_{2,T_i}^* = \arg\max_j \frac{|t \wedge t_j|}{|t_j|}, \quad t_j \in T_i; \quad j_{2,c_i(t)}^* = \arg\max_j \frac{|t \wedge t_j|}{|t_j|}, \quad t_j \in c_i(t) \quad (4)$$

where $c_i(t) = $ class of t in Fuzzy ART A_i. In words, t_{j_{1,T_i}^*} is the most closely related broader term for t in T_i; $t_{j_{1,c_i(t)}^*}$ is the most closely related broader term for t in the semantic class of t (regarding ART A_i). Similarly, t_{j_{2,T_i}^*} represents the closest narrower term for t in T_i, and $t_{j_{2,c_i(t)}^*}$ the closest narrower term for t within the class of t. Moreover, we can numerically measure the strength of each association.

A final computational consideration is in order. The adaptation law (2) implies that a zero vector prototype component will remain zero during training. Hence, only non zero prototype components need to be considered during Fuzzy ART operation. Besides, by Zipf law, most of the terms in any database have a sparse representation (they are rare terms). Given the initialization rule for prototype vectors in Fuzzy ART, the prototype sparsity is guaranteed too. Thus, for most of the terms, only a few non-zero components need to be considered. Therefore increasing the number of documents (and therefore, the dimension of the vector space) has little practical influence in the processing speed. The number of terms (number of patterns) is determined by the number of documents and the ART training time is not considerably affected by this parameter. For instance, the training time for the bigger ART A_T in the base used in the experimental section is only 5". In a larger base, with 5150 documents and 5803 terms, the time is 45". Thus, obtaining semantic classes with ART is always a fast process. The computational requirements grow when similarities are computed, because s_{ij} and s_{ji} have to be computed for all pairs (t_i, t_j), $i \neq j$. Anyway, the time required for about 6000 terms in $\mathbb{R}^{5150}$ does not take more than 2 hours in a 125 Mhz workstation. In summary, the proposed ART architecture is suitable for processing relatively large databases.

3 Experimental work

In this section we describe work on a real text base. This database contains 2939 documents on the topic 'information retrieval', retrieved from the commercial CD-ROM database "Information Science Abstracts Plus" (from Silver Platter Information Inc). Articles date from 1966 to June 1993. For each document, only the title and the abstract are used (descriptors are eliminated so the results are not distorted). We want to test the ability of the proposed neural system for

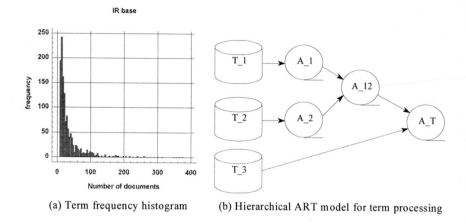

(a) Term frequency histogram   (b) Hierarchical ART model for term processing

Fig. 1. Term frequency distribution and related ART model

detecting the foremost relationships in the IR field. Examples of these relevant associations are 'information retrieval', 'information science', 'search strategies', and others. These relationships involve relatively common words within the IR field. Hence, we are only taking into account words appearing at least in 10 documents (there are 1361 of these in total, excluded empty words, like 'this', 'they', etc). These words are represented as binary vectors in $\mathbb{R}^{2939}$.

Let $ndoc$ be the number of documents in which a word occurs. As explained in section 2, we divide the term set into three frequency regions: T_1 (words with $ndoc \geq 100$), T_2 (words with $30 \leq ndoc < 100$), T_3 (words with $10 \leq ndoc < 30$). There is no method to precisely determine frequency regions. The above cut-points are chosen so that, approximately, very frequent terms (T_1) correspond to the 10-upper percentil term frequency distribution, moderately frequent terms (T_2) correspond to the 50-upper percentil (excluded T_1), and not so frequent terms correspond to the 50-lower percentil. In this work, ART parameters are set to $\lambda = 1$ (fast learning), $\beta = 0$ in control devices for similarity measures, and $\rho = 0$, so that the system can stabilize at the minimal attainable number of classes. After Fuzzy ART training, associations are made as explained in section 2, and joined together in a common data base.

Term set	Number of terms	ART module	Number of nodes
T_1	111	A_1	16
T_2	393	A_2	68
$T_{12}=T_1 \cup T_2$	504	A_{12}	70
$T=T_1 \cup T_2 \cup T_3$	1163	A_T	202

Table 1. Term subsets by frequency, and the corresponding ART modules

The rise in the number of ART nodes (see Table 1) is not only due to a higher number of terms (compare A_2 to A_{12}), but also to the increasing sparsity of terms. Therefore, the hierarchical division of the term set is reflected by the number of clases in the corresponding Fuzzy ART modules.

Next, we calculate the percentage of expert-assigned compound keywords that are present in the automatically obtained relationships set. In general, there are no fixed rules to assign keywords to documents. Hence, different authors may assign different keywords for the same article. That is why in this study only those terms used by several authors are taken into account. The expert agreement on these terms guarantees their validity and generality.

To select these generally accepted terms, we arrange the keywords by descending occurrence frequency. We then select those that are used in 10 or more documents. Besides, we remove keywords that do not appear in our term set (1361 words). After this selection, 109 compound high-frequency keywords remain. Then we find out how many of these are present in our automatically constructed set: the percentage is 59%. This is a remarkable satisfactory rate in any case, and it is worth noting that the method uses only document coocurrence information (no contiguity information, for example). To complete this report, we study the hit rate obtained with the set of (less important) remaining descriptors divided into two subsets: those used in more than 4 documents (and less than 10), and those used in more than 2 documents (and less than 5). The hit rates are 35% and 31% respectively. This drop is not surprising, for our method is based on frequency statistics and looks only for the strongest (more frequent) associations – words occurring only in three documents can not have strong associations. There are some other considerations that add worth to the proposed neural based method:

(a) The method is capable of detecting relationships between non adjacent words. That is the case of the pair 'recall - precision' (not present in the manual thesaurus). (b) For the terms whose manually-assigned relationships are not detected, other similar ones are. For example: 'elementary education', 'decision making' or 'state agencies' are not detected, but 'elementary school', 'decision theory' and 'government agencies' are given instead. (c) The method detects significant relationships for relevant terms that the manual thesaurus fails to include. A few examples are: OPAC, keyword, MARC, query, term, fuzzy, boolean or probabilistic. (d) For some terms, the thesaurus gives only non-specific relationships: 'logic philosophy', 'logic thinking'. The method detects context-dependent relationships: 'boolean logic', 'logic skills'. (e) We have a numerical measure for the strength of each association. A more complete study of this numerical strength and the underlying hierarchies of terms will be made in a forthcoming paper.

We conclude by showing some strong relationships for the term 'retrieval':

(information $\rightarrow$ retrieval, .57), (retrieval $\rightarrow$ information, .69), (text $\rightarrow$ retrieval, .62), (document $\rightarrow$ retrieval, .67), (indexing $\rightarrow$ retrieval, .50), (evaluation $\rightarrow$

retrieval, .57), (language → retrieval, .55), (query → retrieval, .68), (system → retrieval, .37), (probabilistic → retrieval, .79).

The symbol → indicates that the first term is a subset, or narrower term, of the second (all associations are obtained using eq. (3) for the first term; many of them are also obtained using eq. (4) for the second term). It is worth noting that 'retrieval' corresponds to a very general topic: its only broader term is 'information'. We can see that 'information' is a subset of 'retrieval' too, but to a lower degree.

4 Summary and directions for future work

In this paper we have proposed a hierarchical neural architecture for detecting term associations in a document database. For every term in the base, the method obtains narrower and broader terms, and gives the corresponding degrees of association. A relatively high number of expert-assigned keywords are discovered, and many meaningful new relationships (not present in human-made thesaurus) are given. Furthermore, the architecture is well suited for large databases, as the processing speed is not affected much by the base size. Future work will include the design of an algorithm for constructing a hierarchy in terms.

References

1. G.A. Carpenter, S. Grossberg, and D.B. Rosen. Fuzzy ART: Fast stable learning and categorization of analog patterns by an adaptive resonance system. *Neural Networks*, 4:759–771, 1991.
2. H. Chen and K.J. Lynch. Automatic construction of networks of concepts characterizing document databases. *IEEE Transactions on System, Man and Cybernetics*, 22(5):885–902, Sept.-Oct. 1992.
3. M. Georgiopoulos, G.L. Heilemann, and J. Huang. Properties of learning related to pattern diversity in ART 1. *Neural Networks*, 4:751–757, 1991.
4. L.M. Gomez, C.C. Lochbaum, and T.K. Landauer. All the right words:finding what you want as a function of richness of indexing vocabulary. *Journal of the American Society for Information Science*, 41(8):547–559, 1990.
5. B. Kosko. *Neural networks and fuzzy systems: A dynamical approach to machine intelligence*. Prentice Hall, Englewood Cliffs, New Jersey, 1991.
6. B. Moore. ART 1 and pattern clustering. In G. Hinton D. Touretzky and T. Sejnowski, editors, *Proceedings of the 1988 Connectionist Model Summer School*, pages 174–185, San Mateo, C.A., 1989. Morgan Kaufmann.
7. V.V. Raghavan and S.K.M. Wong. A critical analysis of vector space model for information retrieval. *Journal of the American Society for Information Science*, 37(5):100–124, 1986.
8. T. Saracevic and P. Kantor. A study of information seeking and retrieving. II. Users, questions, and effectiveness. *Journal of the American Society for Information Science*, 39(3):177–196, 1988.
9. G.K. Zipf. *The Psycho-biology of language: An introduction to dynamic philology*. M.I.T. Press, Cambridge, Mass., 1965.

Architecture Selection Through Statistical Sensitivity Analysis

Thomas Czernichow[1][*], [**]

[*]Electricité de France (EDF-DER),
1 Av. du Général De Gaulle,
Clamart, France.

[**]Institut National des Télécommunications,
Département EPH,
9 Av Charles Fourier, 91011 Evry Cedex,
France.

Abstract

In this paper, a method for pruning hidden neurones is presented, and illustrated on two different problems. It is based on the statistical study of the derivatives of the outputs of the model with regards to each hidden neurone. We claim that if the model is not using a particular neurone to estimate its outputs, then the corresponding sensitivities will have a low degree of significance. This article is an extension of a previous work dedicated to the selection of input variables. We consider each hidden layer as the input layer of a smaller network made of all the remaining layers between this one and the output. The aim of this analysis is the selection of an appropriate subset of neurones for each layer, to finally obtain a more parsimonious model.

1. Introduction

This article is the extension of a previous work dedicated to variable selection ([1]). The technique has proved to be efficient on toy series and hard financial data ([2]). We will now show how it can be extended to internal neurones, by considering them as input variables of a smaller model.

It is based on the analysis of the partial derivatives of the outputs of the model with regards to its inputs. Under certain assumptions and for a particular class of models (See [3-5]), it is possible to prove that when the model converges towards the underlying process, then all the model derivatives also converge towards the derivatives of the underlying process. We use this property to show that the proposed sensitivities are a very useful insight in the dynamics of the underlying process, since they are a measure of the influence input variables have on the output of the process. On the other hand, we argue with the fact that even though the model derivatives are not close to the derivatives of the underlying process, these sensitivities are still a useful information on the use the model is making of the input variables. For example, if you do know that a certain variable has an influence on the output, and the study of the derivatives tells you that the model is not using it, then your model is not a good estimation of the problem.

---

[1]The work of T. Czernichow has been supported by a CIFRE grant N°91/93 with Electricité de France, Direction des Etudes et Recherches (EDF-DER).

2. Statistical Sensitivity Analysis

The model we will use is a n hidden layer Multi-Layer Perceptron modelized in fig. 1. Each hidden layer is numbered from the input (I or H_0) to the Output (O or H_{n+1}), and has N_k neurones. To train the network, we minimize the usual mean square error.

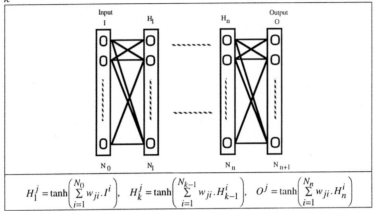

$$H_1^j = \tanh\left(\sum_{i=1}^{N_0} w_{ji}.I^i\right), \quad H_k^j = \tanh\left(\sum_{i=1}^{N_{k-1}} w_{ji}.H_{k-1}^i\right), \quad O^j = \tanh\left(\sum_{i=1}^{N_n} w_{ji}.H_n^i\right)$$

Fig. 1 : *An MLP (up), with its corresponding equations (below).*

The weight estimation process is a low memory quasi-Newton algorithm (see [6] for a complete review on classical optimization algorithms, and [7] for the description of an efficient low memory quasi-Newton algorithm), where the gradient is computed with the back-propagation algorithm. We use three databases during the estimation process. The first one is the training set, the second the validation set, and the third is the test set. The latter is not used in the estimation process.

Each input and output variable is normalized by linear transformations in order to have zero mean and unit variance, because the bounds of a variable should not be taken into account in its relevance rating. If the inputs are normalized, and if all the weights are initialised randomly with the same uniform distribution, then the outputs of each layer are in the same range. We will assume it stays like that all along the training phase.

In function analysis, the usual way to study the influence of a variable x on a function $f(x,y,z)$ is to compute the derivative: $\partial f(x,y,z)/\partial x$. It represents the sensitivity of the output $f(x,y,z)$ with respect to the variable x, at the point (x,y,z). If this sensitivity is insignificant all over the input space, then x is an irrelevant input variable which does not influence the output. In a real world estimation problem we will only have a finite number of samples with known outputs where the validity of the model may be proven.

The work previously done ([1, 2]) aimed at selecting the relevant input variables of a model to build a more robut model, and to give hints on the underlying process. We now want to prune hidden neurones. To do so, we will first select the relevant variables of the input layer, and then consider layer H_1 as the input of a (n-1) hidden layers MLP. We will then iteratively select the relevant inputs of a n-2, ..., 0 hidden layers MLP. This heuristic is driven by the idea that at each step, it is crucial to be sure that the inputs of the sub-model are all relevant. What we expect of a neurone that has no importance in the underlying process is that its derivative has zero mean and small variance. To each output neuron corresponds a matrix with as many lines as the number of patterns in the learning database and as many columns as the number of neurones (N_k) of the layer H_k we are dealing with. We will only restrict ourselves to one output problems for clarities' sake, but the extention to multiple outputs is straightforward. We thus want to compute the derivatives of the outputs O^j with regards to an input variable I^i or to $H^i_k, i = 1,...,N_k$ as we extend the concept to the hidden neurones. They can be computed recursively as follows:

$$\frac{\partial O^i}{\partial H^j_k} = \tanh'\left(\sum_{p=1}^{N_n} w_{ip}.H^p_n\right).\sum_{p=1}^{N_n} w_{ip}.\frac{\partial H^p_n}{\partial H^j_k},$$

$$\frac{\partial H^i_\alpha}{\partial H^j_k} = \tanh'\left(\sum_{p=1}^{N_{\alpha-1}} w_{ip}.H^p_{\alpha-1}\right).\sum_{p=1}^{N_{\alpha-1}} w_{ip}.\frac{\partial H^p_{\alpha-1}}{\partial H^j_k}, \quad while\,(\alpha-1)\neq k$$ (1).

$$\frac{\partial H^i_{k+1}}{\partial H^j_k} = \tanh'\left(\sum_{p=1}^{N_k} w_{ip}.H^p_k\right).w_{ij}$$

All the derivatives are computed on the learning set, and are pre-filtered to take off the matrix, all the derivatives (that is of an input pattern) which correspond to an error of more than twice the standard deviation of the model estimation error (also computed on the learning database). This will minimize the effect of bad estimation, and of outliers, on the measure of the influence of the variables. We then compute, for each column i of each derivatives matrix (only one matrix in our case because we deal with one output networks), the mean (E_i) and the standard deviation (S_i). If the derivatives corresponding to a neurone (each neurone correspond a column in the matrix) is close to zero, then its mean and variance will be small, and the neurone is deleted. To measure this, we compute, for the layer H_k we are treating, the vector C. It measures the distance to the origine (0,0) for each neurone:

$$C_i = \left(E^2_i + S^2_i\right) \bigg/ \max_{j=1,...,N_k}\left(E^2_j + S^2_j\right)$$ (2).

The algorithm will then be : to choose a layer H_k, compute the criteria C_i for all its neurones, discard the least relevant ones, and re-estimate the model from the current weights. This process goes on while the error on the validation base decreases.

3. Simulations

Example 1 : The chaotic Henon map: $y_t = 1 + 0.3 y_{t-2} - 1.4 y^2_{t-1}$

We have generated 1500 points with the previous formula, and splitted them in three equal parts (learning, validation, and test). We have tested three models (one, two, and three hidden layers), each of them having a 10 neurones input which correspond to the delais from y_{t-1} until y_{t-10}. All hidden layers have 20 neurones, which makes the models have respectively 241, 661, and 981 weights, which deliberately build over-parametrized models. The decision boundary used to cut the neurones is found by trying to locate a drop in the level of influence measured by C. We sort decreasingly the C_i, and build the vector composed of the higher value divided by the second higher one, then the second divided by the third and so on. The peaks in this vector identify the drops of influence. Keeping a variable in between two peaks implies to keep all the variable until the right peak, because they all have the same relevancy level.

We present the results by drawing the Average Relative Variance ($ARV = \dfrac{1}{N\hat{\sigma}^2} \displaystyle\sum_{i=1}^{N}(y_i - \hat{y}_i)^2$) against the total number of weights cut by the deletion of the irrelevant neurones at the current step since the beginning. The graphs are then to be read from left to right, one point for a derivative computation and a neurone deletion step.

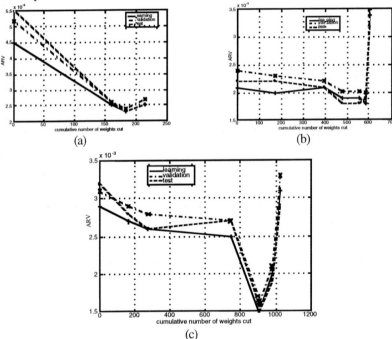

Fig. 2 : *Cumulative number of weights cut since the beginning at each iteration, against the Average Relative Variance computed on the three database. (a) on the 10-20-1 MLP (b) on the 10-20-20-1 MLP, and (c) on the 10-20-20-20-1 MLP.*

The two first input neurones are very easily selected at the first step. For the one hidden layer network we stopped at 184 weights cut, after only two iterations. It corresponds to a percentage of deletion of 76% for a more precise estimation. On figure 2.b, the last network kept, after 3 iterations of the algorithm has only 75 neurones left, which is equivalent to a cut percentage of 88%. On figure 2.c, after a quit drop to 0.0015, the error increases again. 909 weights have been cut (84%), through 3 passes of the algorithm.

One hidden layer	10-20-1	2-14-1
Two hidden layers	10-20-20-1	2-3-13-1
Three hidden layers	10-20-20-20-1	2-4-7-14-1

Table 1. Architecture at the beginning (left), and at the end, after deletion (right)

Example 2 : Load forecasting.

Load forecasting has been a key issue since the last decades, moreover after the economic crisis has struck western countries. Adjusting the production plan to the precisely estimated demand is very important for the energy reduction price, and for the mastering of the production, the transmission and the distribution of electrical energy. We present here a small model in order to illustrate our approach, but this model would not be robust enough for a real load forecasting application. For a review of neural network applications to load forecasting, we recommend [8].

We basically deal with three kinds of series. The first one is the load of the French national grid (EDF), recorded each half hour since some years. The second one is composed of the weather variables, the temperature and the cloud cover each three hours. These variables are supposed to partially explain the behaviour of the first series in some ways. The last kind of variables are the calendar ones.

To estimate the models, we have used the years 1989 to 1990 for learning, year 1991 for validation, and year 1992 for testing. The exogeneous variables are coded binary to know whether it is a day off or not, and the day after and the day before as well. We also code the hour of the day and special tariffs. We have also used the load at time t, t-48 (a day before) and t-336 (a week before), the temperature, and the cloud cover with the same lags. The model is a 25-20-20-1 MLP.

Amazingly, some of the inputs which we had taken for granted to be important were discarded : the cloud cover for delays longer than a few hours. The temperature is selected with a high influence. The minimum and maximum temperatures are barely used, conversely to the coding of the hour, and the day in the year. Finally, it appeared that the network was utterly able to use the binary inputs. At the end of this first step, only 19 input neurones were kept.

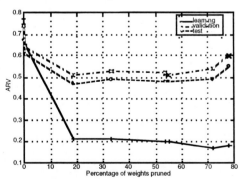

Fig. 3 : *Percentage of weights pruned since the beginning at each iteration, against the Average Relative Variance computed on the three databases.*

On figure 3, the ARV on the validation set clearly increases on the last point. It corresponds to the cutting of about 72% of the weights. It is interesting to notice that the first cut had a very large impact on the network. We end the simulation with a 19-10-6-1 Multi Layer Perceptron.

4. Conclusion

We have presented a method to prune hidden neurones in a MLP, and have tested it one toy problem and a real world problem. We have shown that it was possible to preserve the performances on the validation and test set while decreasing the complexity of the model by deleting the non-relevant neurones. These neurones are the units the model is not using in the modelization of the underlying process. We haven't made any comparison with other pruning technics like the Optimal Brain Damage. Indeed, most of these technics aim at destroying connections while our approach is devoted to neurones pruning. These two methods could be used sequentially : One could first use our approach to select a small architecture, then use an OBD like technic to prune some more connections.

One future work will be to build a more precise decision test for the selection of the useless neurones, to compare different cutting strategies, and to extend our approach to recurrent networks. For now on, it clearly improves the generalisation capabilities by the selection of a more parsimonious model.

Acknowledgement

I would like to thank A. Muñoz and A. Piras for sharing ideas, and for improving critics. I would also like to thank B. Dorizzi for her support.

References

[1] T. Czernichow, A. Muñoz, "Variable Selection through Statistical Sensitivity Analysis: Application to feedforward and recurrent networks.," INT-SIM, http://www-sim.int-evry.fr/People/Czernichow.html, Technical Report 95-07-01, 1995.
[2] B. Dorizzi, G. Pellieux, F. Jacquet, T. Czernichow, A. Muñoz, "Selecting the relevant variables to forecast the French T-Bond," presented at Third Chemical Bank/Imperial College Conference on Forecasting Financial Markets, London , 1996.
[3] P. Cardaliaguet, G. Euvrard, "Approximation of a Function and its Derivatives with a Neural Network," *Neural Networks*, vol. 5, pp. 207-220, 1992.
[4] K. Hornik, M. Stinchcombe, H. White, "Universal Approximation of an unknown Mapping and its Derivatives Using Multilayer Feedforward Networks," *Neural Networks*, vol. 3, pp. 551-560, 1990.
[5] A. R. Gallant, H. White, "On learning the derivatives of an unknown Mapping with Multilayer Feedforward Networks," *Neural Networks*, vol. 5, pp. 129-138, 1992.
[6] D. G. Luenberger, *Linear and non-linear programming*, 2nd ed: Adddison-Wesley, 1984.
[7] C. D. Liu, J. Nocedal, "On the limited memory BFGS method for large scale optimization," *Mathematical programming*, vol. 45, pp. 503-528, 1989.
[8] T. Czernichow, A. Piras, K. Imhof, P. Caire, Y. Jaccard, B. Dorizzi, A. Germond, "Short Term Electrical Load Forecasting with Artificial Neural Networks," *Int. Journal of Eng. Int. Syst.*, To appear, 1996.

Regression by Topological Map: Application on Real Data

DAIGREMONT Ph.[1], de LASSUS H.[3], BADRAN F.[1], THIRIA S.[1,2]

[1] CEDRIC, Conservatoire National des Arts et Métiers,
292 rue Saint Martin - 75003 PARIS, France
[2] Laboratoire d'Océanographie et de Climatologie (LODYC), Université de Paris VI,
4 place Jussieu, T14 - 75005 PARIS, France
[3] Laboratoire ARPEGES, CNRS URA 1757,
Observatoire de Paris-Meudon - 92195 MEUDON cedex, France

Abstract. We address the problem of oceanographic data regression with constrainted Kohonen self-organizing maps. Using constrainted topological mapping algorithm on real data, we show that it is well suited to geographic needs. It appears as an elegant way to overcome uneven spatial sampling problems.

Keywords: Constrainted topological mapping, Nonparametric regression, Oceanographic data analysis.

1 Introduction

In order to draw an accurate map of sea surface temperature, oceanographers have to derive spatial regression from data collected since the beginning of the century. Unfortunately oceanographic data are often very badly sampled in space and time. We propose a method to lessen the influence of spatial distribution of sampled data on regression analysis. We suggest the use of Constrainted topological mapping algorithm (CTM) and we study its performance on real data. To begin with, we describe the CTM algorithm and its links with classic statistical methods. Then, we present the characteristics of sea temperature data, emphasizing the difficulty of regression analysis in such a context. Finally, we perform CTM learning on these data and discuss results.

2 Constrainted Topological Mapping

Often regression problems use the Nadaraya-Watson estimator [2] [3] computed from kernel functions. But this technique is CPU consuming, and has to be simplified. An approximated solution of the Nadaraya-Watson estimator can be made using a partition of the regression domain. A simple way to do this is to use a regular grid as in WARPing methods [3]. When data are unevenly sampled, this partition is not optimal, and results are not accurate in undersampled regions.

Intuitively, one would like to use a method designed specificaly for irregularly sampled data. Oceanographers commonly begin with implementing a kriging technique and later smoothing their data. Doing so they take into account the uneven distribution of the physical measurements.

We propose here to use the Constrainted Topological Mapping (CTM) algorithm which combines, in a single approach, irregular partitioning (like kriging) and smoothing.

The CTM algorithm is basically a Self-Organizing Map (SOM) [4] with the modifications suggested by Ritter [5] and Cherkasski & al [1].

We present now the CTM algorithm in the particular case of a regression from R^2 to be valued in R. Let $s_i = (x_i, y_i, z_i)$ is an observation of the learning set S and $w_c = (w_c^1, w_c^2, w_c^3)$ the weight vector related to unit c of the M topological map.

1. Given a sample $s_i = (x_i, y_i, z_i)$, consider the projection of s_i in the subspace (x, y). In this projection space find the best matching unit q_i.
2. For all units of the map, do an unsupervised learning iteration:

$$w_c^1(t+1) = w_c^1(t) + \epsilon(t) K^{h_1(t)}(\delta(c, q_i))(x_i - w_c^1(t))$$

$$w_c^2(t+1) = w_c^2(t) + \epsilon(t) K^{h_2(t)}(\delta(c, q_i))(y_i - w_c^2(t))$$

3. For all units of the map, do a constrainted learning iteration:

$$w_c^3(t+1) = w_c^3(t) + \eta(t) K^{h(t)}(\delta(c, q_i))(z_i - w_c^3(t))$$

4. $t = t + 1$, stop the algorithm when $t = t_{max}$.

$\epsilon(t)$ and $\eta(t)$ are learning rates. K is a kernel function, defined by $K(d) = exp(-d^2)$. For a smoothing parameter h, $K^h(d) = \frac{1}{h}K(\frac{d}{h})$. In the algorithm, h_1, h_1 and h decrease as a function of t. $\delta(c, q)$ is the distance between unit c and the best matching unit q on the topological map. K^h controls the neighborhood of unit c.

In this algorithm, step 2 makes quantization in space (x, y), step 3 computes the regression values.

A similar version of CTM can be proposed if we decompose it into two phases. The fisrt one performs quantization by iterating step 2. The second one performs regression by iterating step 3 on this quantization. For a constant smoothing parameter $h = h(t)$, the second phase can be seen as a stochastic gradient descent for the cost function:

$$E(W) = \frac{1}{2} \sum_{s_i \in S} \sum_{c \in M} K^h(c, q_i)(w_c^3 - z_i)^2$$

The global minimum of this function is thus defined by the relationship:

$$w_c^3 = \frac{\sum_{r \in M} K^h(c, r) T_r}{\sum_{r \in M} K^h(c, r) n_r}$$

where F_c is the voronoi corresponding to (w_c^1, w_c^2),
n_c is the cardinal of voronoi F_c
and $T_c = \sum\limits_{(x_i, y_i) \in Fc} z_i$.

w_c^3 is closely related to the Nadaraya estimator and can be interpreted as a WARPing using an irregular partition defined by voronoi F_c [2].

In order to prevent topological order violation of the map and to give the regression by piecewize linear function, we implement the modification proposed by Cherkasski: the best matching unit q_i is found among all units that form the triangle enclosing the sample s_i in the projection space (x, y).

3   Experiment on Oceanographic Data

Data Analysis.
The database of sea surface temperatures collected for a century in July on the Northern Atlantic is made of 54,853 samples. Each observation s_i consists of a temperature measurement of the sea surface z_i, its location (longitude, latitude) (x_i, y_i) and its year. As can be seen on figure 1(a) the spatial distribution on (x, y) of the database is heterogeneous. For instance, along European coasts high concentration of grey dots gives evidence of repeated intensive measuring campaigns, while huge areas facing West Africa are nearly empty.

What's more, the histogram in figure 2(a) shows that time distribution of data is also uneven. Data belonging to the first half of the century is especially rare, whereas years 1962 to 1964 are overrepresented (8000 samples for 1963, versus an average of 600 samples).

In addition, figure 2(b) enhances the fact that mean temperature increases starting from 1950. This phenomenon seems to be related to more interest given to southern measurements in recent years.

Preprocessing.
We have shown that observations are badly sampled in time. This is a problem in places where punctual intensive campaigns have been made in the same year. In these locations, the same year is overrepresented. The systematic noise due to climatic conditions of this year induces a bad smoothing. Therefore, an ideal database should contain the same sample quantity per year in each region. The learning set (S_{learn}) and test set (S_{test}) we used for the regression are made as follow:

1. make a regular grid of rectangles in the (x, y) subspace (in our case a 40x40 grid of the sea surface)
2. for each rectangle, randomly select two samples per year to build S_{learn}
3. for each rectangle, randomly select two samples per year not previouly selected to build S_{test}

S_{learn} is made of 16,464 samples and S_{test} of 6,308. Their spatial densities are in figures 1(b) and 1(c).

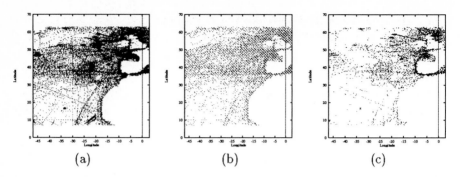

(a)   (b)   (c)

Fig. 1. Spatial distribution in (x, y) of the full data set (a), learning set S_{learn} (b), test set S_{test} (c)

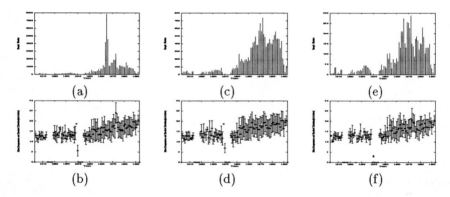

(a)   (c)   (e)

(b)   (d)   (f)

Fig. 2. The first row (a) (c) (e) show the distribution of the data with respect to year and the second (b) (d) (f) the statistics: the dots represents the observation mean of the year and the line the standard deviation

The mean and standard deviation with respect of the time are given in figures 2(d) and 2(f). These figures demonstrate that S_{learn} and S_{test} does not changed significantly statistics compared with the global database.

Performance criteria.

In order to evaluate the regression performances of the CTM, we use as a criterium the root mean square error.

$$RMS(\hat{f}((x, y), S) = \sqrt{\frac{\sum\limits_{k=1}^{p} \left(z_k - \hat{f}(x_k, y_k)\right)^2}{p}}$$

where $\hat{f}(x, y)$ is the regression function to qualify,
S is a set of p observations,
z_k is the third componant of the sample s_k,
and $\hat{f}(x_k, y_k)$ is the value of the regression function for s_k.

A perfect quantization assign the same number of observations in the learning set to each unit. In order to test the quality of a given quantizer, we compute its contrast given by

$$C(Q, S) = \sqrt{\frac{\sum_{i=1}^{N} \left(\frac{n_i}{\bar{n}} - 1\right)^2}{N}}$$

where Q is a quantizer of N vectors q_i,
S a set of p observations,
n_i is the observations number coded by vector q_i,
and $\bar{n}$ is the mean observations number per vector $(\frac{p}{N})$. The less the contrast, the best the quantizer Q is for a given data set S. In the CTM algorithm, the quantizer Q is the set of references vectors $q_c = (w_c^1, w_c^2)$; in the WARPing, they are the centers of the rectangles. In our experiment, the contrast is computed on space (x, y).

Results and Discussion.

We compare the performances provided by CTM and WARPing on the sea surface temperature regression. We run both methods using the same learning and test set and the same number of vectors q_i (40x40). The smooothing used in the WARPing has been optimized using a cross validation technique.

Results of both criteria are:

CTM RMS	Warping RMS	CTM Contrast on Z_{test}	Warping Contrast on Z_{test}
$1.19°C$	$1.25°C$	1.22	9.00

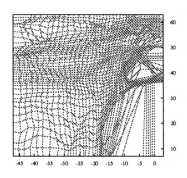
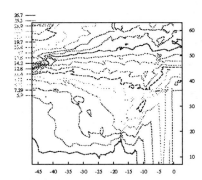

Fig. 3. Topological map in the (x, y) space. Each dot represent a unit c and a straight line connects neighbor units of the map

Fig. 4. Isolines of the sea surface temperature given by the topological map

For both criteria CTM is better than WARPing. Figure 3 gives the topological map in space (x, y) and figure 4 shows the results of the regression given by the map on space (x, y). This figure shows that the CTM regression provides realistic isolines which are compatible with kriging techniques used by oceanographers.

4   Conclusion

We analysed the Constrainted Topological Map when dealing with regression problems and showed the relationship between CTM and WARPing with kernel smoothing and WARPing. The good performances reached on a real world application chosen from the oceanographic domain prove the efficiency of the approach. Clearly CTM can be seen as an alternative approach for regression.

References

1. Najafi Cherkasski. Constrainted topological mapping for nonparametric regression analysis. *Neural Network*, 4:27–40, 1991.
2. Hardle. *Applied Nonparametric Regression*. Cambridge University Press, Cambridge, 1990.
3. Hardle. *Smoothing techniques with implementation in S*. Springer-Verlag, New York, 1991.
4. Kohonen. *Self-organisation and associative memory*. Springer-Verlag, 3rd edition, 1995.
5. Schulten Ritter, Martinetz. *Neural Computation and Self-Organizing Maps, An Introduction*. Addison-Wesley Publishing Company, 1992.

Signal Processing by Neural Networks to Create "Virtual" Sensors and Model-Based Diagnostics

Kenneth A. Marko, John V. Jamès, Timothy M. Feldkamp,
Gintaras V. Puskorius and Lee A. Feldkamp

Ford Research Laboratory, Ford Motor Company
Dearborn, Michigan   USA

Abstract. This paper discusses the application of advanced neural network methods to the development of diagnostics for complex, nonlinear dynamical systems, for which accurate, first-principles models either do no exist or are difficult to derive. We consider two approaches to detect and identify failures in these systems. First, neural networks are trained to act as *virtual* sensors that emulate the performance of laboratory-quality sensors; this approach provides higher quality diagnostic information than is available directly from production sensors. Second, neural networks are trained to emulate nominal (fault-free) system behavior; model-based fault diagnosis is subsequently achieved by detecting significant deviations between actual and predicted system performance. We present experimental evidence of the viability of both approaches for a difficult automotive diagnostic task.

1   Introduction

The problem under investigation is the real time and continuous detection of engine combustion failures, commonly known as engine misfire, under virtually all operating conditions, for a wide variety of powertrain systems. Frequent misfires will lead to deterioration of a vehicle's emissions control system, ultimately resulting in unacceptable levels of emitted pollutants. This is a significant problem, since U.S. government regulations require that misfire detection capability be provided on every gasoline powered car and light truck sold in the U.S. after 1998. Naturally, automotive manufacturers desire to provide misfire detection capability that is robust, yet inexpensive.

Since nearly all existing vehicles incorporate microprocessor engine control systems which utilize crankshaft position sensing to effect event based engine control, a preferred method of misfire monitoring involves observation of crankshaft dynamics to detect velocity fluctuations or acceleration deficits on a firing-by-firing basis. Although this approach has been demonstrated to be effective, the range of excellent capability has been limited mostly to steady state conditions, to engines with fewer than 8 cylinders, and to limited patterns of misfire [1, 2]. The task is complicated by both the dynamics of engine operation (one must distinguish unexpected deceleration due to misfire from an intended engine deceleration) and the dynamics of the crankshaft (it is not infinitely stiff and undergoes constant and frequently severe torsional oscillations). The diagnostic task is to decouple these unavoidable dynamics from true misfire dynamics, utilizing information derived from the production-level position sensors, and then to develop a means of deciding when a fault has occurred.

2 Virtual Sensors

Conceptually, the stated problem of misfire detection is somewhat analogous to the problem of inferring the forcing function delivered at some location to a set of coupled oscillators when the observation point is physically removed from the driving force. In the case of a rotating crankshaft, there are additional complications: the driving force is delivered along different points in the system (according to which cylinder is firing), and the driving force function is not constant, but rather a calculable function of several observable variables. The misfire detection problem is most easily addressed if the effects of the forcing function can be observed at a location along the crankshaft where the torsional oscillations are minimized, removable by some artifices, or absent. In this ideal case, engine dynamics can be easily handled, leading to a straightforward and conventional means of detecting engine misfire as described in [3].

Design and manufacturing considerations often preclude the possibility of ideal placement of crankshaft position sensing systems. On the other hand, for development purposes, we can construct a situation in which a second, laboratory-quality sensor is mounted along the crankshaft at a point where the torsional oscillations are minimized. Data is then carefully gathered for a wide range of engine operating conditions, as well as misfire patterns. The measured crankshaft accelerations from the laboratory sensor are post-processed to remove any additional artifacts, resulting in a nearly idealized data stream that should only contain the effects of engine operation dynamics. Thus, the problem is reduced to creating a transformation of the torsionally contaminated measurement to an estimate of an ideal signal.

Early attempts to create *virtual* crankshaft acceleration sensors using feedforward networks with relatively steady-state data were generally unsuccessful, due to architectural and training algorithm limitations. On the other hand, application of a higher-order training procedure with time-lagged recurrent networks [4] to this data provided initial, promising evidence that the torsional oscillations had a deterministic relationship with observable engine variables [5]. In this work, we investigate the capability of these training techniques on a larger and more comprehensive data set. The training data consist of over 500,000 examples of firing and misfiring events from a single vehicle. The driving conditions were varied considerably, representing both steady-state and transient conditions. In addition, the data sets contain a large variety of regular and irregular misfire patterns, as well as periods of no misfire. A second set of comprehensive data was gathered from a second, similar vehicle for purposes of testing.

Network inputs consist of engine speed, engine load, cylinder-specific crankshaft acceleration, and a binary flag indicating the beginning of a new engine cycle. The network output is an estimate of the idealized crankshaft acceleration, delayed by one engine cycle, since the effects of distinct engine events may persist for a number of events into the future in the measured acceleration. We use a recurrent multilayered perceptron (RMLP) consisting of 4 inputs, two recurrent hidden layers with 15 and 7 nodes, respectively, and a summing output node, where the nodes of each hidden layer are fully connected with one another via unit time delay operators. This time-lagged network was trained by the multi-stream DEKF procedure [4]. In this procedure, the tendency of the network to learn the most recent input sequence at the expense of previously learned patterns is reduced by performing weight updates that are simultaneously consistent for a number (e.g., 20) of data streams.

Figure 1 shows representative behavior of the *virtual* neural network sensor. One should note that the problem solved here is more complicated than it appears, because the torsional oscillations are affected by the presence of misfire events (indicated by the open symbols), and that the proper transformation is realized only through correct compensation for prior behavior of the system, i.e., the current transformation values depend on when and at what location the last misfire occurred.

3 Model-Based Diagnostics

Although improvement of the sensor information is crucial to diagnostic tasks, it is important to note that a comprehensive diagnostic system must be able to detect deviations from nominal behavior. In the misfire task, these deviations are restricted to be the total absence of power delivered by a firing stroke. However, smaller deficits might arise from a number of other causes (valve leakage, piston ring damage, etc.), and it is possible that excesses could also be a problem (e.g., detonation or engine "knock").

In this application, we would like to predict individual cylinder accelerations expected as a function of engine speed and load and to compare those to the observed accelerations. An extraordinary deficit between predicted and observed accelerations would imply that a misfire had occurred. A smaller deficit hints at a lesser, but real problem (e.g., leaky valve) which might have to be observed over a longer period of time to substantiate the existence of the problem. Alternatively, an acceleration excess might imply the wrong fueling or a "knock" problem causing the engine to produce more power than is required or acceptable.

In this modeling problem, the training data consists exclusively of normal engine firings, gathered over a wide range of operating conditions. Since engine speed and load change slowly, we model the accelerations for all cylinders simultaneously (i.e., 8 separate predictions for an 8-cylinder engine). Network inputs are taken to be speed, load and acceleration, where all quantities are averaged over an engine cycle consisting of 8 events. The network outputs are cylinder specific accelerations for each firing under normal conditions. A feed-forward network is found to perform this task adequately; the results are shown on a representative sequence of test data in Figure 2 for data gathered from the production crankshaft position sensing system. It is particularly noteworthy that the target values as shown in panel (c) appear to be quite noisy for certain input conditions, due to factors such as crankshaft dynamics and clock resolution, while the corresponding network outputs shown in panel (d) demonstrate that the average, context dependent torsional behavior of the crankshaft has been modeled.

Given these neural network modeling capabilities, we see that there are several means to implement model-based diagnostics, using either production or *virtual* sensor information from the vehicle to compare with predicted normal behavior. At this time, we find it most convenient to perform the comparison at the idealized level, which requires that we combine the two neural network techniques to arrive at a solution to the diagnostic problem. The decision function for the diagnostic is generated by subtracting the crankshaft acceleration as predicted by the *virtual* sensor from the crankshaft acceleration as predicted by the static model of nominal behavior, and setting a decision threshold to signal the occurrence of a misfire. An alternative approach would be to use a single recurrent network to accomplish both the transformation and diagnostic classification task simultaneously; this comprehensive and compact solution is addressed in a subsequent paper [6].

4 Conclusions

We have argued that it is possible to improve diagnostics for complex systems through a combination of methods incorporating neural networks to create virtual sensors and implicit models of system operation with which to compare observed performance. We have applied these approaches to a difficult diagnostic problem which has resisted solution by conventional means and found that a combination of recurrent and feedforward networks may be used to construct a solution utilizing these methods. These two approaches individually open new possibilities for providing real-time continuous diagnostics for a wide range of complex systems.

Acknowledgements: We would like to thank Danil Prokhorov for carrying out some of the neural network training exercises during his summer internship at Ford Research Laboratory.

References

1. G. Rizzoni, Diagnosis of Individual Cylinder Misfire by Signature Analysis of Crankshaft Speed Fluctuations, **SAE** No. 890884.
2. W. B. Ribbens and G. Rizzoni, Onboard Diagnosis of Engine Misfires, **SAE** No. 901768.
3. J. V. James, James M. Dosdall and Kenneth A. Marko, Misfire Detection in an Internal Combustion Engine, U.S. Patent 5,044,194. September 3, 1994.
4. L. A. Feldkamp and G. V. Puskorius, Training Controllers for Robustness: Multi-stream DEKF, *Proceedings of the IEEE Conference on Neural Networks*, vol. IV, pp. 2377-2382.
5. G. V. Puskorius and L. A. Feldkamp, Signal Processing by Dynamic Neural Networks with Application to Automotive Misfire Detection. To appear in *Proceedings of the 1996 World Congress on Neural Networks*, San Diego.
6. K. A. Marko, J. V. James, T. M. Feldkamp, G. V. Puskorius and L. A. Feldkamp, Training Recurrent Neural Networks for Classification: Realization of Engine Diagnostics. To appear in *Proceedings of the 1996 World Congress on Neural Networks*, San Diego.

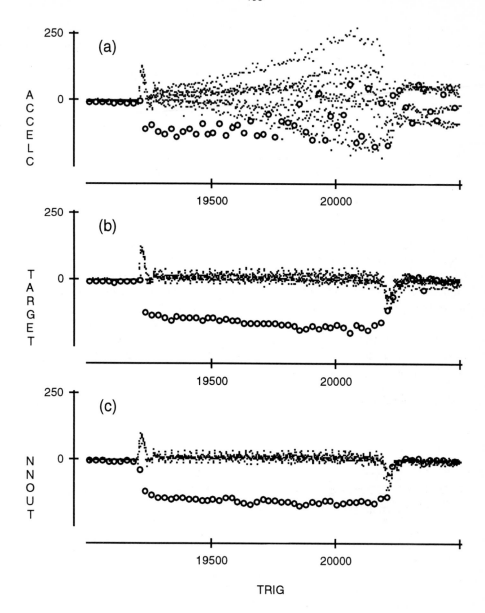

Fig. 1. Virtual sensor results: measured crankshaft acceleration (top), acceleration obtained from corrected values measured with a precision sensor ideally located on the crankshaft (middle), and network output that models the input-output relationship between the torsionally corrupted and ideal crankshaft accelerations (bottom), plotted as a function of time. Misfires and normal firings are indicated by open circles and dots, respectively, in these panels.

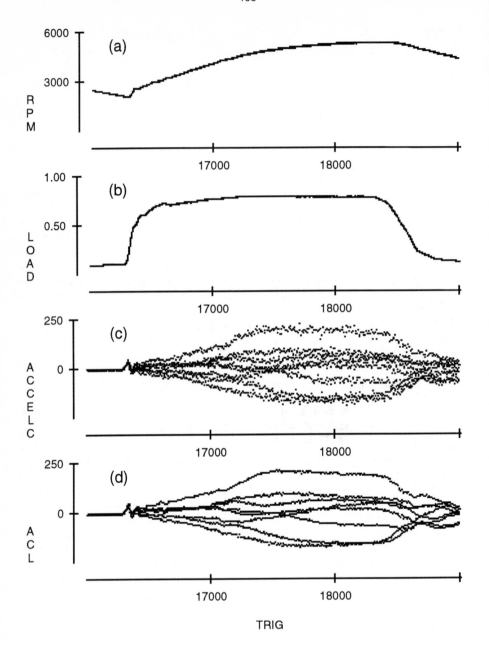

Fig. 2. Nominal crankshaft dynamics modeling results: panels (a) and (b) show the engine speed and load, which are used as network inputs; panel (c) shows the noisy target values for training derived from the production sensor, and panel (d) shows the output of the network for each of the eight cylinders.

Development of an Advisory System Based on a Neural Network for the Operation of a Coal Fired Power Plant

R.M.L. Frenken, C.M. Rozendaal, H.E. Dijk (KEMA) and P.C. Knoester (EZH) ,
KEMA, P.O. box 9035, 6800 ET Arnhem
email: R.M.L.Frenken@MTA4.KEMA.NL

Abstract

This paper describes the application of neural networks to the operaton of a coal fired power plant. The performance of a pulverized coal fired power plant depends on the design and operation of the plant and the quality of the coals. Incorrect operation of a power plant may lead to lower efficiencies and higher emissions of toxic gaseous elements. During the operation of the boiler the performance has to be controlled and optimized by adjusting the various boiler parameters. This is a difficult task for the operators because of the large number of parameters and the mutual dependence of the performance parameters. The complexity is even more enhanced ub case the quality of the coals changes frequently. In order to support the operator in performing the above task, KEMA, in cooperation with EZH and Schelde is developing an advisory system. Core of the advisory system will be a neural network model of the boiler. The model predicts accurately the various performance variables for a wide range of operating conditions. The applicability of neural networks to power plants is herewith demonstrated.

1   Introduction

At the beginning of 1995, KEMA discussed with the staff of Maasvlakte Power Plant (a plant owned by EZH, one of the 4 large Dutch electricity generating companies) the applicability of advanced computer techniques, especially neural networks. On the one hand, plant staff was eager to use new applications to improve plant performance; on the other hand, they had serious doubts whether a neural network could be of use for the optimization of plant operation, since the implementation of neural networks in the electricity generating industry is still very limited. At the same time, several Dutch power plants intended to build an advisory system for the operators of their plant. Basis of most such systems is a model of the boiler. However, due to the strong non-linear relations and the mutual dependence of various variables, building such a model is not an easy task. Therefore, Maasvlakte asked KEMA to demonstrate the applicability of neural networks by developing a model of the boiler. When successful, an advisory system based on this Neural Network (NN) model should be developed.

This paper deals with the development of the advisory system and focuses on the NN model. Chapter 2 gives an overview of the plant. In order to demonstrate the need for an advisory system, some difficulties during operation of the plant are discussed in detail. Chapter 3 discusses the lay out of the advisory system. This system is based on a NN model of the plant. The development of this model is described in chapter 4. Chapter 5 is devoted to the performance of the model for various operating conditions.

2 Operating a coal fired power plant

In the Netherlands, some 35% of the electricity is generated with 6 coal fired power plants. One of those plants is Maasvlakte Power Plant, located close to Rotterdam, which has a total capacity of 1080 MW (enough to supply several large cities with electricity). When operating on full load, ±180.000 kg coal per hour is burned. These numbers suggest that operating such a plant is not a trivial task. To illustrate this, a brief description of Maasvlakte power plant will be given. Maasvlakte power plant consists of 2 identical units with a capacity of 540 MW each. Each unit consists of a boiler, several turbines and a generator. In the boiler, water is heated to steam by the combustion of pulverized coal. The steam drives several turbines. These turbines drive the generator, which in turn generates the electricity. As the boiler is the subject of this paper, some further explanation is given.

Figure 1 gives a simplified overview of the boiler. The pulverized coal is injected in to the furnace through the burner (position A). Most of the air necessary for the combustion is injected together with the fuel, at point A and B. However, for environmental reasons, some air is injected above the burners (the so called over fire air, point C). The tilt (angles) at which the fuel and overfire air are injected can be varied. Also the amount of air and fuel can be varied. In figure 1, only one corner with one burner and one overfire air layer are sketched. In practice, the Maasvlakte boiler has 5 layers of burners at each of the 4 corners of the boiler. Above the burners are 4 layers of overfire air. Not all the above parameters can be varied independently,

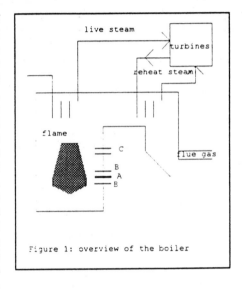

Figure 1: overview of the boiler

but in total the fuel and air injection give the operator some 24 parameters to influence the combustion process. Except from these parameters, the operator also has to deal with the fineness of the pulverized coal and the quality of the coal itself. These coal parameters have a large impact on the combustion process. However, the coal quality cannot be influenced by the operator. Every week a new coal blend is fired, which means that once a week the operators have to adapt the boiler settings to a new coal blend. The operator can influence the fouling (pollution) in the boiler by sootblowing actions. The load (amount of electricity, varying with the load demand in the Netherlands) is also a factor the operator has to take into account when optimzing the performance of the boiler. The performance is judged by the thermal efficiency (steam temperatures, attemperation water), the fly-ash quality (i.e. unburned carbon in ash) and the environmentally hazardous emissions (i.e. NO_x-emission).

When operating the plant, the operator tries to optimize the thermal behaviour ($\approx$ efficiency) and minimize the NO_x emission and the unburned carbon. Due to various reasons, a change in plant operation (for example increasing the overfire air) normally improves some of the performance parameters, but has a negative impact on at least one of the others parameters. This is the so called diabolic triangle (NO_x, fly ash quality and thermal behaviour [1]). For example, an increase of overfire air will cause a decrease of unburned carbon, but an increase of NO$_x$ emission. This demonstrates that only small margins are available to operators within which they can optimize boiler performance. Due to the large scale of the plant, small improvements in boiler performance already mean a significant cost reduction. This implies that there is a need for an advisory system to help the operators optimizing the boiler performance.

3   Advisory system

In recent years, several advisory systems for power plants have been developed. Normally, these systems are based on a model of the boiler. This model can be derived from the physical processes in the boiler. This can become quite difficult, due to the large number of parameters and the strong non-linear relationships. Therefore, some recently developed advisory systems are based on a black box model of the boiler, for example quadratic regression models [2] or neural networks [3]. Most of the existing advisory systems take only one "standard" coal quality into account. For strategic and financial reasons (prices of coal depend on the quality), at Maasvlakte a wide variety of coal blends (each blend has a different quality) is fired. The impact of coal quality on plant performance is large. This is the reason why a new advisory system is being developed by KEMA which takes coal quality into account.

The advisory system is called *TUNON*, which stand for *T*hermal behaviour, *U*nburned carbon, *N*O$_x$ emission *O*ptimized using *N*eural Networks. The advisory system basically consists of three parts, namely the interface, the neural network model and the optimizing algorithm. Development of the interface between the plant and the neural network model should not be underestimated. One needs to deal with missing values, averaging, gauging and drifting meters. Because these are plant specific problems, we will not discuss them in detail. The optimization algorithm will be based on the neural network model. The quality (reliability for a wide range of operating conditions) of this model is very important, for this reason next chapter is devoted to the development of the model.

4   Development of the NN model

To model the boiler, 41 parameters were identified which influence the boiler performance. The performance is measured by 7 parameters. Table 1 summarizes the input and output parameters. Please note that these parameters have been explained in chapter 2. Of course, among the parameters are various constraints, for example the relation between the load and injected coal (more electricity means that more coal has to be injected).
The 41 parameters are assumed to determine the 7 performance parameters. One set with the above 48 parameters (41 in, 7 out) is called a pattern. In total, some 2200 patterns were supplied by Maasvlakte. Each pattern covers 10 minutes averaged values. These patterns cover a wide range (13) of coal blends. After data validation and removing transient situations some 1800 patterns were available. At first, this set was random divided into a learning set of 1500 and a test set of 300 patterns (experiment 1). This learning set was used to train a standard NN [4]. The NN consisted of 41 input neurons, 10 hidden neurons and 7 output neurons. Experiments with a random test and learning set allowed us to tune the network and to get some feeling of the results we could achieve. However, practical relevance was limited, because in practice every week a new (and not seen before) coal

Table 1: parameters

In (external)	In (controlled)	Out (performance)
coal quality (5)	sootblowing (6)	Thermal behaviour (steam and attemperation water, 5)
load (1)	amount and finess of coal per mill (10)	
	total air (1)	NOx emission (1)
	aux. air (4)	Unburned Carbon (1)
	overfire air (4)	
	burner air (5)	
	tilt (5)	

blend is fired. For operating purposes it is necessary to know with which accuracy the model can predict the output parameters for a *new* coal blend. For this reason, the 1800 patterns, covering 13 coal blends, were divided into a learning set of 12 coal blends (± 1600 patterns) and a testset of 1 coal blend (± 200 patterns). These experiments are referred to as experiment 2.

Some experiments were done with 7 networks (one for every ouput parameter), consisting of 41 input, 6-10 neurons in the hidden layer and 1 output neuron. These limited experiments showed that results obtained with 7 "specialized" networks were worse than with 1 "general" network. Although interesting, this subject was not studied further.

5   Results

Together with the Maasvlakte staff, criteria were developed for quantifying the quality of the model. For example, NOx had to be determined within 5% margin to be called good. The measurement error was the main guideline for determining the criteria. A simple performance index was defined in order

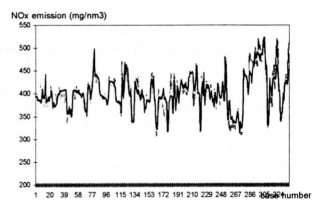

Figure 2 NOx emission (random testset)

to describe the quality of the prediction in term of "very good/good/acceptable/bad".

In case of experiment 1, some 85% of the test set was predicted very good or good. In 3% of the cases the prediction was bad. These experiments showed that the selected parameters (table 1) had been a good choice. Figure 2 gives an overview of the NO_x prediction (thick line) against actual values.

In case of experiment 2, about 63% of the test set was predicted very good or good. Some 7% was bad. These results are worse than in case of experiment 1, but still quite good. Figure 3 illustrates this by giving an overview of the actual versus the predicted NO_x emission.

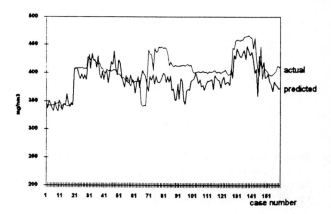

Figure 3 NOx emission (new coal blend)

Although no extensive comparison with classical solutions was made, it can be stated that for normal operating conditions the quality of the prediction of the NN model is at least the same as "conventional" models. An advantage of the NN model over models derived from the physical processes is the fast calculation time (<0.5 second versus several seconds).

After having validated the reliability of the model for normal operating conditions, a prediction of some "exceptional" operating regimes was asked from the NN model. This resembled situations which differed considerably from the learning set. Unfortunately, the quality of the prediction was very low in this case. This illustrated the importance of the selection of the test- and training set. In practice it is impossible to carry out a large number of experiments in order to cover a wide range of operating conditions. Therefore, the NN should be combined with some algorithm, which decides whether the current situation resembled the previously learned situations. When the current situation differs too much from the learned situations, some warning should be given. By such a "self test", it is avoided that confidence in the model is weakened by a wrong predictions in extraordinary operating conditions. Such a test is now being developed.

6 Conclusions

Neural networks are suitable to model the combustion process in a coal fired boiler. In general, the model performed very well for operating regimes closely related to situations presented in the learning set. Also, the model could cope quite well with different coal qualities. On the other hand, the quality of the prediction sharply decreased with complete new operating regimes. In order to develop a good model, one has to make sure that the learning data covers a wide range of operational conditions. Apart from that, in the interface in which the neural network is embedded some check has to be done whether the model is in its working range. According to our experiences, such a "self test" algorithm for feedforward neural networks, operating on line, still has to be developed.

Acknowledgement

The authors thank J. Arends, A. Meerkerk, M. Meier, C. van Uffelen, H. van Vliet and W. Winter for their discussions and supplying of the data. This project was undertaken by order of the Dutch Electricity Generating Companies.

REFERENCES

[1]   The diabolic triangle in modern low-NO_x coal firing, C.M. Rozendaal e.a., 11th International Conference on Power stations, September 1993, Liege, Belgium.

[2]   A test for an optimization method applied to controlling NO_x emissions in a PC fired boiler, J.J. Catasus-Servia e.a., 3rd International Joint POWID/EPRI Controls and Instrumentation Conference, June 1993, Phoenix, Arizona, USA.

[3]   GNOCIS: an update of the Generic NO_x Control Intelligent System. R. Holmes e.a., EPRI/EPA 1995 Joint Symposium on Stationary Combustions NO_x Control, May 1995, Kansas city, USA

[4]   Artificial Neural Networks, E. Sanchez-Sinencio and C. Lau, IEEE Press, New York, USA, 1992

Blast Furnace Analysis with Neural Networks

Joachim Angstenberger
MIT - Management Intelligenter Technologien GmbH
Promenade 9
52076 Aachen, Germany
Phone: +49 - 2408 - 9 45 80
Fax: +49 - 2408 - 9 45 82
email: jang@mitgmbh.de

Abstract: Nowadays blast furnace operation is supervised by extensive measurements and controlled accordingly. Characteristic indications concerning process quality are given by the analysis of the radial temperature profile in the upper part of the furnace. Optimising this temperature distribution would lead to considerable savings of input material. To achieve an optimisation, quantitative relations between furnace parameters are needed. As those relationships are unknown, a process model can be provided using neural networks and fuzzy methods. In this paper we show the application of fuzzy clustering and neural networks to classify temperature profiles and to build a model of the interdependence betwenn process operation parameters and the resulting temperature profiles. These investigations have been carried out in a plant of a German steel producer.

Keywords: blast furnace, process analysis, fuzzy clustering, multilayer perceptron, identification of temperature profiles

1 Introduction

During the blast furnace process, ferric oxides contained in the input materials ore and sinter are reduced and melted. The output material is liquid pig-iron.

Figure 1 shows a schematic overview of this process. The furnace is charged from its top. The input materials are sinking downwards while gas is flowing upwards through the furnace. The melting heat is supplied by this gas stream and by coke (which is also charged from above).

Although studies have been made about the internal operating conditions of the furnace [Omori, 1987], the blast furnace operating personnel relies mainly upon measurements at the inputs and outputs of the process.

A wide range of measuring data is acquired at the blast furnace. Blast furnace engineers check

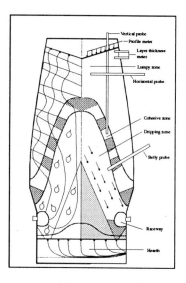

Fig. 1: Schematic view of the blast furnace

these data in order to find out if the change of a controllable furnace parameter is necessary. This inspection is done by expert knowledge, i.e. control of the blast furnace is based on the experience of the operating personnel. Quantitative knowledge of functional dependencies between furnace parameters is not absolutely required for that purpose.However, experience has shown that missing knowledge of quantitative functional interdependencies makes an optimised furnace operation almost impossible.The operation is considered to be optimal, if the fuel consumption (coke and coal) is minimised (cost savings) and the furnace shows a steady cycle without greater fluctuations of some parameters.

One of these essential parameters is the radial temperature distribution in the upper part of the blast furnace. According to the experts' knowledge, this magnitude is the most important parameter to judge the condition of the furnace. The temperature profile should show a determined characteristic shape: Maximal value in the centre of the furnace, fall off reaching the wall with a minimum briefly after the half radius and again rising reaching the wall. Past has shown that at the considered blast furnace among all of these profiles one specific shape is always connected with the best furnace operation (in the above-mentioned sense). The question indeed is how to receive such a shape and how to maintain it in the long run? For that purpose the shape influencing quantities have to be known as well as the order of magnitude of their quantitative influence on the furnace process. Due to the large number of furnace parameters and their highly non-linear coupling the identification of these influencing factors is difficult, which thus also holds for the modelling of the process itself.

As there is no possibility to generate an adequate physical model of the blast furnace, the relationship between these influencing parameters on one hand and the resulting temperature profile on the other hand were modelled using a neural network

By interviewing blast furnace experts the characteristics of the input materials sinter (iron supply) and coke (energy supply) as well as the charging program were identified as the main influence factors on the temperature profile. The charging program determines the order and the amount of the input materials brought into the furnace, thereby producing a series of sinter and coke layers. The thickness and the structure of these layers (shown at the upper left side of figure 1 as different hatchings) have great influence on the rising gas streams in the furnace and therefore determine indirectly the measured temperature profile.

The two objectives of this analysis were:

1. A classification of the temperature profiles with the objective to assess how much the current state of the furnace corresponds to determined quality criteria.

2. The determination of the quantitative dependency of the cross-sectional temperature profile on the decisive magnitudes that constitute this profile (i.e. modelling the blast furnace).

These two parts of the analysis can later on be combined into a complete blast furnace control system. In a first step the classification of the temperature profile gives information about the current state of the blast furnace. A suitable influencing of the profile can then be carried out using the furnace model.

All examinations have been carried out with the help of the data analysis software DataEngine®, which includes methods for fuzzy clustering as well as neural networks. Together with its data pre-processing, statistics and visualisation facilities this tool enables a complete solution of the whole task.

2　Classification of Temperature Profiles in the Blast Furnace by Fuzzy Clustering

For splitting the temperature profiles into classes 143 data records, each containing a profile of 8 temperature values recorded previously have been used. At the beginning, the number of different classes of profiles included in the data material was unknown, so the optimum number of classes had to be determined first.

For the purpose of classification the fuzzy c-means algorithm was used, a fuzzy clustering procedure that allows gradual association of the considered objects (here: temperature profiles) to the respective classes [Bezdek 1981]. The essential information this algorithm produces are the centres of the clusters found during clustering. Each cluster centre represents a (fictitious) sample that characterises the respective class best. Here, fuzzy clustering results in typical temperature profiles that can be interpreted as prototypes of the respective classes.

To determine the optimal number of classes, cluster validity measures can be used [Windham 1981]. The analysis of the given data records leads to an optimal partition of five clusters, whose centres are displayed in the following diagrams (Figure 2). In each diagram the temperature is plotted against the measuring position (seen from the centre of the furnace).

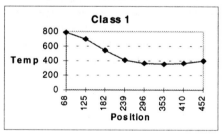
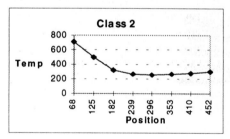
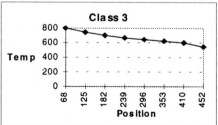
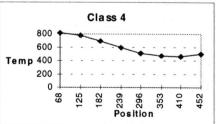

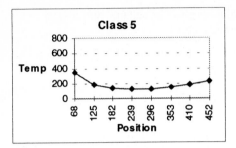

Fig. 2: Types of temperature profiles (classification into 5 classes)

Using the criteria mentioned in the introduction above, it can be recognised that the profiles of classes 1, 2 and, with certain reservations, class 4 represent good furnace states, whereas classes 3 and 5 characterise undesired states of the furnace.

The received class partition was validated by blast furnace experts. It turned out to represent a good characterisation of typical states of the furnace (cf. also [Bulsari, Saxen 1995]).

3 Modelling of the Dependency Between Temperature Profiles and Input Parameters by Neural Networks

The available data records for modelling the blast furnace comprised the charging program, the distributions of the grain sizes of coke and sinter as well as several parameters of the material strength. As the grain spectrum of coke covers a relatively narrow area, the evaluation considered only the medium grain diameter. The distribution of the grain size of sinter plays an essential part for the furnace process, as it bears strongly upon the gas flow in the furnace. For that reason the distribution of the sinter grain size was used completely as an input parameter. Altogether 10 input parameters were used.

Because of material transportation time and other effects the material parameters affect the temperature profile with a delay of approximately 24 hours. On the other hand, the charging program bears on the temperature profile with a delay of 8 hours due to process characteristics. An appropriate synchronisation of the data had to be carried out.

After further pre-processing (scaling, etc.) these data were used as inputs of a multi-layer perceptron. Some data records were extracted from the total amount of data in order to be used for validation of the network performance. The output of the network consists of 8 values modelling the temperature at the respective measurement positions of the measuring system. Figure 3 represents a schematic view of the neural network's structure.

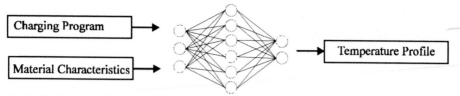

Fig. 3: Schematic structure of the neural network for blast furnace modelling

Several network configurations (number of hidden layers, number of hidden neurons, learning parameters) have been trained and tested with regard to their efficiency. The neural network is able to approximate the temperature profiles with good precision. This is proven by the comparisons shown in figure 4. The diagrams show the temperature profiles calculated by the neural network (indicated as "Network" in figure 4) as opposed to the actually measured profiles ("Measurement" in figure 4) for four data records. The cases shown in figure 4 have not been used for training the neural network. The neural network proved to be able to recognise relatively unusual temperature profiles as well (like the one displayed in figure 4 on the lower right).

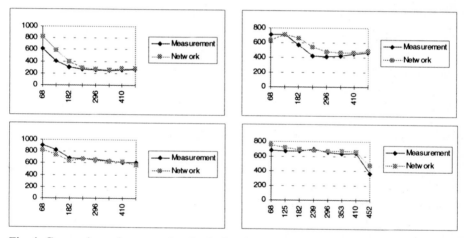

Fig. 4: Comparison of measured and calculated temperature profiles

However, the test results made clear as well that there are certain situations where the neural network is not able to achieve a reasonable result. Therefore, additional examinations have to be performed to quantify the influence of several additional magnitudes that have not been considered yet. In particular, these are magnitudes influencing the gas flow in the furnace and more detailed information about the input material. With inclusion of this information modelling and optimisation of the blast furnace process can be largely improved.

4   Conclusions

In this contribution modelling of a blast furnace by neural networks and fuzzy logic has been investigated. The neural network model achieved a high correlation between actual and estimated temperature profiles. Based on these results an improved process control with respect to input material savings and quality of the produced steel can be implemented.

5   References

[Bezdek 1981] Bezdek, J. C.: Pattern Recognition with Fuzzy Objective Function Algorithms. Plenum Press, New York London, 1981.

[Bulsari, Saxen 1995] Bulsari, A., Saxen, B.: Classification of blast furnace probe temperatures using neural networks. Steel Research 66, 1995, No. 6, pp. 231-236.

[Omori 1987] Omori, Y. (ed.): Blast Furnace Phenomena and Modelling. The Iron and Steel Institute of Japan. Elsevier, London, 1987.

[Windham 1981] Windham, M.P.: Cluster Validity for Fuzzy Clustering Algorithms. Fuzzy Sets and Systems 5, 1981, pp. 177-185.

Diagnosis Tools for Telecommunication Network Traffic Management

Philippe LERAY[1], Patrick GALLINARI[1] and Elisabeth DIDELET[2]

[1] LAFORIA - IBP
Université Paris 6 - boite 169
4, Place Jussieu
75252 Paris cedex 05
France
{leray , gallinari} @laforia.ibp.fr

[2] France Télécom - CNET
CNET PAA/ATR
38-40 rue du général Leclerc
92131 Issy les Moulineaux cedex
France
didelet@issy.cnet.fr

With the rapid evolution of telecommunication networks, real-time network traffic management is becoming more and more crucial. We propose here a modular system for performing diagnosis at different levels of the network. It is designed as an aid to the operator. We present results of different experiments with a first version of this system which operates at a local level.

1 Introduction

With the increasing complexity of telecommunication networks and the demand for new and sophisticated services, the role of real-time network traffic management is becoming more and more crucial. Both statistical and artificial intelligence tools have been proposed for network traffic management [1,2,3]. They are usually aimed at problems like traffic prediction, diagnosis, control or routing. Although this field is expanding rapidly, only a few systems have been implemented to deal with operational conditions, most of them having limited abilities. Reasons for this are the complexity of such tasks, the difficulty to analyse the large amount of data collected on a telecommunication network and to automate parts of the process in order to help the operator. We will focus here on the implementation of a diagnosis system for the detection of abnormal situations via the observation of the telecommunication network.

Treating this problem requires the analysis of a large amount of data collected at the different nodes of the network. This analysis involves low level processing techniques, for information selection or detection of local events, and high level methods for correlating different events, appearing at different times or locations, or for decision making. We have for now focused only on the low level stages of such a system, using neural network techniques. These methods have been used recently in different areas of telecommunications [2,4].

To build a diagnosis chain, neural networks are combined into modular systems where different specialized modules cooperate together to solve a global task. Such modular systems can be easily adapted to new configurations of the network structure by replacing, updating or adding specialized modules. In the following, we first describe the diagnosis problem and the simulation tool which has been used to emulate the network (Section 2). We then describe our modular architecture (Section 3) and present different experiments performed using a preliminary version of the system (Section 4).

2 Case study

Real world data are not yet available for training real time diagnosis systems. In principle they could be collected at the management centres where all the events appearing on the network are sent to the operator screen. However the corresponding amount of data is tremendous since information about network elements arrive every 1 to 5 minutes. For now, only data for off-line diagnosis, whose periodicity is larger, are available at telecommunication centres. For real time diagnosis, people rely on the use of network simulators and try to emulate real traffic conditions. We briefly describe the one which has been used in this study and the diagnosis problems we have been dealing with.

2.1 The network
The network model we consider here is based on the French long-distance network. It consists of 73 centres: 5 main transit centres (MTCs) and 68 secondary transit centres (STCs). In network management centres, diagnosis and control rely heavily on operators. Measurements from MTCs and STCs are aggregated in order to analyse their status, to detect any abnormal conditions such as traffic overloads and/or network failures and to activate traffic controls so as to minimize the effects of the disruption.

The various conditions we will study for each centre correspond to the following situations :

O_1 - Nominal situation: no abnormal condition.
O_2 - Outgoing overload: concentrated calls from a centre.
O_3 - Incoming overload: concentrated calls towards a centre.
O_4 - Overall overload: increase of traffic over the whole network.
O_5 - Regional overload: traffic increase in an MTC zone.

In the STC case, O_4 and O_5 situations are quite similar: in the space of data, these two classes are strongly overlapping.

2.2 The data
We have used data generated via the SuperMac simulator developed at CNET (Centre national d'études des télécommunications - France Télécom's Research Centre). This software permits to emulate the main characteristics of a telecommunication network. In particular, it enables us to:

- set the nominal traffic at each centre. Figure 1 shows a nominal traffic profile which is typical of a weekday. For now we have used a similar profile for all days of the week. This is a simplification of the real traffic conditions. However, observation of real data clearly shows the existence of characteristic profiles for weekdays, saturdays and sundays. Nevertheless, the fluctuations from one day of the week to another day are small compared to those corresponding to overload (abnormal) situations.

- generate data corresponding to the different overload situations described in 2.1 with different overload levels. The latter are expressed as a percentage of the nominal traffic conditions as they are defined at a given time of the day. For example we have generated data corresponding to O_2 or O_3 overloads ranging from 150 % to 1000 % perturbation of the nominal traffic. Similarly we made O_4 percentage vary between 125 % and 300 % and O_5 between 125 % and 225 %. These values correspond to observed situations.

For each type of overload, 7 days have been simulated with measurements every 4 minutes. Disturbances of 16 minutes are generated randomly during this period with a uniformly distributed overload percentage. All other measurements do correspond to a nominal traffic situation. Since these measurements are much more abundant than all others, some of them have been discarded in order to equilibrate the data for the different situations. We have distributed the 7 days measurements in two databases which have been used respectively for training and testing. The first one contains 4 days, i.e. 4480 examples (1280 from O_1, and 800 from each of the other classes), the second one corresponds to 3 days and is made of 3680 examples (1280 from O_1, and 600 from each of the other classes).

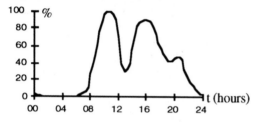

Fig.1. Nominal traffic profile for 24 hours. The y axis corresponds to a deviation (in %) of a 'standard' traffic.

3 A modular architecture

Our modular system, inspired from the telephone network structure, is composed of two levels. At the local level, data from STCs and MTCs are processed in order to detect and classify perturbations which may be identified at this level. The global level will be used as a network management centre in order to make a final diagnosis as shown in figure 2.

For each STC and MTC, the diagnosis system can be divided in two modules (figure 3):
- a classification module (CLASSIF) which determines the centre status (O_1 to O_5).
- one module dedicated to each overload situation (EXPERT-i, where i=2, 3, 4, 5) which indicates the corresponding overload percentage.

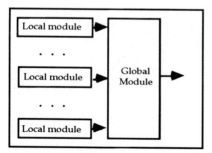 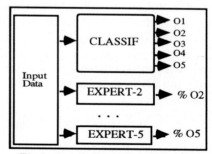

Fig.2. General diagnosis architecture. Each local module corresponds to a local diagnosis system (figure 3).

Fig.3. Local diagnosis architecture.

4 The STC diagnosis module

We now present the performances of our system for local diagnosis operations (CLASSIF and EXPERT modules) for STCs. The neural network used is a multilayer perceptron with one hidden layer. The learning algorithm is a batch conjugate gradient. Learning is stopped either when a plateau is reached for the error on test data, or when overtraining is detected.

4.1 Classification of perturbations (CLASSIF module)

Our telephone network simulator produces a set of 18 indicators describing the STC status. Let $X(t)$ denote the vector of these indicators measured at time t and $X_0(t)$ the vector corresponding to a nominal situation at t, as shown in figure 1.

Since we are interested in the detection of deviations from normal behavior, our overloads are relative to the nominal situation in a network location at a given time of the day, i.e. $X_0(t)$. We performed experiments with different sets of input variables with or without $X_0(t)$ so as to test the importance of different input information for diagnosis. A first statistical variable selection was performed via the procedure DISCRIM of the statistical package SAS [5].

We give below (table 1) performances for the following choices of inputs:
- I1: 18 indicators of $X(t)$,
- I2: 34 indicators of $X(t)$ and $X_0(t)$ (2 indicators of $X_0(t)$ are equal to zero),
- I3: 12 indicators: 7 selected by SAS in I_1 and the 5 corresponding ones in $X_0(t)$,
- I4: 8 indicators selected by SAS in set I_2.

Network structure	after 200 iterations	after 1000 iterations
18-30-5 (I1)	72.4 %	–
34-50-5 (I2)	81.6 %	**83.8 %**
12-25-5 (I3)	77.2 %	82.6 %
8-20-5 (I4)	77.3 %	82.2 %

Table 1. Best classification percentage on test data for different sets of input variables.

The number of hidden units in the experiments has been roughly set by cross validation. The best classifier in table 1 (34-50-5 (I2)) uses a large number (2005) of parameters. The Optimal Brain Damage (OBD) pruning method [6] has allowed us to decrease down to 385 this number without any loss of performance.

Several variable selection methods have been proposed in the NN literature. In [7] a measure inspired from OBD is used for pruning input units. In this technique, the inputs are ordered according to a saliency measure S_i (E1), and then pruned by comparing S_i to a threshold set by cross validation. C denotes the cost function and W_{ij} the weight from unit j to unit i. Other methods based on probabilistic dependence measures have been proposed. For example in [8] it is proposed to measure the mutual information between a set of inputs and the output. The expression $MI(X_i, Y)$ is given in (E2) in the case of one input variable X_i, $P(z)$ denotes a probability density function over z.

$$S_i = \sum_{j \in \text{fan\_out}(i)} \frac{1}{2} \frac{\partial^2 C}{\partial^2 W_{ji}} W_{ji}^2 \quad \text{(E1)} \qquad MI(X_i, Y) = \sum_{y} \sum_{x_i} P(x_i, y) \log\left(\frac{P(x_i, y)}{P(x_i)P(y)}\right) \quad \text{(E2)}$$

All these techniques are heuristics since the problem of variable selection is combinatorial in the dimension of the input space. Their respective merits are discussed in [7]. We show in table 2 the importance of our 34 variables according to three different techniques: the statistical method (SAS results on I_4), the saliency (S_i) and the mutual information (MI). Performances with input sets selected by SAS (I_3 and I_4) are not so good compared to performances with the whole input set (I_2). MI and S_i selections perform quite similarly for $X(t)$ components but there are significative differences on $X_0(t)$: MI does not select any variable in $X_0(t)$ whereas S_i does.

i	1	2	3	4	5	6	7	8	9	10	11	12	13	14	15	16	17
SAS																	
Si																	
MI																	

i	18	19	20	21	22	23	24	25	26	27	28	29	30	31	32	33	34
SAS																	
Si																	
MI																	

Table 2. Importance of each variable X_i, computed by 3 different methods. White, grey and black denote respectively little, fair and high signification. Variables 1 to 18 correspond to $X(t)$ and 19 to 34 to $X_0(t)$.

We can also use S_i in an iterative way (OCD [7]): we pruned the input with the weakest saliency then we train a new network, etc... With this method, we stopped after having pruned 14 inputs. We used the remaining 20 indicators, and obtained a **84.3 %** classification rate on test data and the confusion matrix of table 3.

	O_1	O_2	O_3	O_4	O_5
O_1	**97.4**	0.5	0.9	0.6	0.6
O_2	2.5	**96.5**	0.0	0.8	0.2
O_3	2.8	0.0	**96.4**	0.5	0.3
O_4	9.5	1.2	0.0	**67.7**	21.6
O_5	18.0	0.0	0.3	33.0	**48.7**

Table 3. Confusion matrix on test data for the best classifier (inputs selected with OCD).

We have analysed the errors of this classifier . The confusion matrix gives us some clues for improving the system. Good performances may be observed for the classification of the nominal situation and incoming or outgoing overloads. Two causes of error can be observed:
- regional and overall overloads (O_4 and O_5) with low or average percentage are confused (bottom-right of the confusion matrix). This can be easily explained by the definition of these overloads: some situations cannot be distinguished at the STC level. However, this could be possibly done at the MTC level.
- in low traffic periods (22h-8h) the error number increases: the system does not easily distinguish low percentage overloads. However, those errors are not really important since small perturbations during the night are usually not relevant.

4.2 EXPERT modules
We describe below the experiments carried out with the EXPERT modules. We will only give as an example the performances for the O_2 expert whose role is to determine the overload percentage corresponding to the situation "outgoing overload". Figure 4 illustrates the results of a MLP in predicting 24h on the test data (MSE=0.044). During the day, as for the CLASSIF module, errors are more

important during low traffic periods since data corresponding to different percentages may overlap. As it is the case for the classification of overloads, predicting the degree of overload is not crucial during these periods.

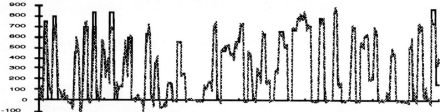

Fig.4. O_2 expert predicted value vs. real value for our best actual architecture. The real value is in solid line, and the predicted value in dashed line.

5 Conclusion

We have presented results on the development of the first stages of a telecommunication diagnosis system and on the importance of our different parameters. Both data generation and system implementation are still to be improved. We are currently working in producing more realistic scenarios to simulate real world conditions. We will also take into account previous measurements of our parameters so as to improve our current results and to correct some errors which are inherent to the first system. For the EXPERT modules, it will be useful to predict the conditional probability distribution of overload percentage instead of predicting simply the mean of this percentage as it is the case for now. This will allow a better use of these outputs in subsequent modules of the system. The next step will be the development of diagnosis tools at the global level.

Acknowledgement: this work has been performed with France Télécom - CNET under Grant 94 1B 003. We would like to thank the PAA/ATR team and particularly D. Stern and L. De Bois.

References

[1] E. Didelet, B. Dubuisson, D. Stern, P. Chemouil, *AIP approaches to diagnosis in network traffic management,* Qualitative Reasoning and Decision Technologies, Carreté/Singh ed., 1993.

[2] P. Chemouil, J. Filipiak, *Supporting Network Management with Real-Time Traffic Models,* IEEE journal on selected areas in Communications, vol. 9, n°2, pp. 151-156, 1991.

[3] D. Stern, *A Statistical Study of Real-Time Telephone Traffic Variations for Network Management,* ITC Krakow, 1991.

[4] Proceedings of the International Workshop on Applications of Neural Networks to Telecommunications 2, Alspector/Goodman/Brown ed., 1995.

[5] *SAS User's Guide: Statistics,* 1982 edition.

[6] Y. LeCun, J. Denker, S. Sola, *Optimal Brain Damage,* in NIPS, vol.2, pp. 598-605, 1990.

[7] T. Cibas, F. Fogelman Soulie, P. Gallinari, S. Raudys, *Variable Selection with Neural Networks,* to appear in Neurocomputing.

[8] R. Battiti, *Using Mutual Information for Selecting Features in Supervised Neural Net Learning,* IEE Transactions on Neural Networks, vol. 5, n°4, 1994.

Adaptive Saccade Control of a Binocular Head with Dynamic Cell Structures

Jörg Bruske, Michael Hansen, Lars Riehn, Gerald Sommer

Lehrstuhl für Kognitive Systeme, Christian Albrechts Universität zu Kiel,
Preußerstr. 1-9, 24105 Kiel

Abstract. In this article we report how Dynamic Cell Structures (DCS) [1] can be utilized to learn fast and accurate saccade control of a four-degrees-of-freedom Binocular Head. We solve the order selection problem by incremental growing of a DCS network until the controller meets a pre-specified precision. Calculation of the controller output is very fast and suitable for realtime control since the resulting network is as small as possible and only the best matching unit and its topological neighbors are activated on presentation of an input stimulus. Training of the DCS is based on error feedback learning and proceeds in two phases. In the first phase we use a crude model of the cameras and the kinematics of the head to learn the topology of the input submanifold and a rough approximation off-line. In a second phase, the operating phase, we employ error feedback learning for on-line adaptation of the linear output units. Besides our TRC binocular head we use a Datacube image processing system and a Stäubli R90 robot arm for automated training in the second phase. The controller is demonstrated to successfully correct errors in the model and to rapidly adapt to changing parameters.

1   Introduction

Saccades are fast eye movements used by animates or robots to change fixation from one point in the visual scene to another. Their speed must be very high (actually up to 1000 deg/s for human eye movements) and it has long been recognized that this speed is much to high to be influenced by visual feedback alone. Hence the saccade control systems must solve the inverse kinematics problem. According to an idea put forward by Kawato et al. [4] the inverse kinematics can be learned by error feedback learning, i.e. by using an error signal proportional to the retinal error as training feed back for a neural network. In his plenary lecture on ICANN'95 Kawato not only gave further neurobiological evidence that this is indeed the way the monkey visual system learns saccade control but that the principle of error feedback learning can be successfully applied to a variety of robotic tasks as well.

In a number of simulations Dean and Mayhew [2] have systematically studied different conventional neural net architectures (Linear nets, CMACs, Multilayer Backpropagation nets) for learning saccade control from error feedback. In [5] Mayhew et al. report a successful implementation of a layered control system for a four-degree-of-freedom stereo camera head utilizing their CMAC based PILUT architecture for learning the inverse kinematics from error feedback.

We have selected DCS [1] for saccade control because it ideally meets the demands of this control task: First, the calculation of the controller has to be as fast as possible to allow control at video rate. The incremental growing DCS network meets this requirement by growing the network only as large as to meet a pre-specified precision and, furthermore, by utilizing only a small subset of its neural units for output calculation. Second, it is well known that the angular rotations required to fixate are a linear function of the retinal coordinates of the target only in case of zero tilt and small angles. With increasing tilt the non linearity of the control law increases. Hence if we use an approximation scheme based on locally linear approximation, the density of neural units should be high in regions of the input space with a high tilt component. Contrary to Kohonen type networks or PILUT, the growing DCS is able to achieve this by allocating new neural units in regions of the input space where the approximation error is high. This is exactly the case for regions of the input space deviating from the linear control law.

Finally, the phase space trajectories of multidimensional systems like the head-eye system usually lie on a submanifold which locally may be of very much lower dimensionality than the system. The DCS just attempts a similar reduction in dimensionality in that it places its units in the input submanifold and adapts its lateral connection structure towards a perfectly topology preserving map which is utilized for improved adaptation and approximation.

2 Dynamic Cell Structures

Dynamic Cell Structures (DCS) as introduced in [1] denote a class of RBF-based approximation schemes attempting to concurrently learn and utilize perfectly topology preserving feature maps (PTFMs). DCS are a subclass of Martinetz's Topology Representing Networks (TRN) [6] defined to contain any network using competitive Hebbian learning for building PTFMs.

The architectural characteristics of a DCS network are a) one hidden layer of radial basis functions (possibly growing/shrinking) b) a dynamic lateral connection structure between these units and c) a layer of (usually linear) output units. Training algorithms for DCS adapt the lateral connection structure towards a PTFM by employing a competitive Hebbian learning rule and activate and adapt rbf units in the neighborhood of the current stimulus, where "neighborhood" relates to the simultaneously learned topology.

The particular DCS network used in this article is similar to the one introduced in [1] in that incremental growing of the network is performed similar to B. Fritzke's Growing Cell Structures [3] using a local error variable attached to each neural unit called the resource of this unit.

3 Training scheme

Training proceeds in two stages. First, we use a crude model of the cameras and the kinematics of the head to learn saccade control up to a pre-defined precision off-line. In the second phase, the operating phase, we continue to adapt the output layer of the DCS network on-line to cope with deviations of the real head-eye-system from the model and possibly changing parameters. The first phase could be on-line as well but using a model has the advantage that training takes less time (no real head movement involved) and that, in the second phase, the real system is under reasonable control right from the start.

3.1 Off-line training phase

In this article we restrict ourselves to the so called head movement problem, i.e. for fixating an object the cameras must verge symmetrically and the head must directly point at the target. We use a 5 dimensional input vector $\bar{u} = (x_l, y_l, x_r, \theta_l, \phi)$ which we associate with a 3×3 (Jacobian) matrix A, the output of the DCS network. Here, x_l, y_l, x_r denote retinal coordinates of the target on the left and right camera, θ_l is the vergence angle of the left camera and ϕ the tilt angle of the head, see Fig. 1 (d). The output of the controller $\bar{v}$ is calculated as

$$\bar{v} = (\Delta\chi, \Delta\phi, \Delta\theta_l) = A\bar{p}, \tag{1}$$

where $\Delta\chi, \Delta\phi, \Delta\theta_l$ denote the changes (rotations) in pan, tilt and vergence necessary to fixate the target, and $\bar{p} = (x_l, y_l, x_r)$ is the retinal coordinate vector.[1]

The matrix A is computed as a normalized weighted sum of the matrices A_i attached to the rbf units of the DCS network

$$A = \sum_{i \in Nh(bmu)} A_i h_i \text{ with } h_i = \frac{rbf(\|\bar{u} - \mu_i\|)}{\sum_{j \in Nh(bmu)} rbf(\|\bar{u} - \mu_j\|)}, \tag{2}$$

where Nh(bmu) denotes the best matching unit and its direct topological neighbors.

After fixation we use the output of a simple proportional feedback controller $\bar{v}^p = \left(\Delta\chi^p, \Delta\phi^p, \Delta\theta_l^p\right)$ to adapt the matrices A_i. The components $\Delta\chi^p, \Delta\phi^p, \Delta\theta_l^p$ are proportional to the mean target's retinal x-coordinates, the left cameras y-coordinate and the difference in the x-coordinates after fixation. If fixation was perfect, $\bar{v}^p$ would be the null vector.

Applying an α-LMS rule we obtain

$$\Delta A_i = \alpha(\bar{v}^p - A\bar{u})\frac{\bar{u}^T}{\|\bar{u}\|^2}h_i. \tag{3}$$

The centers μ_i of the rbf units are adapted according to a Kohonen type rule,

$$\Delta\mu_i = \varepsilon\lambda_{(bmu,i)}(\bar{u} - \mu_i), i \in Nh(bmu), \tag{4}$$

and the fixation error in retinal coordinates is used as a resource value. A new unit is inserted whenever the resource value of the bmu *and* the current fixation error exceed the required precision. Between successive insertions we require at least n ln (n) training steps without insertion, n the current number of neural units, to allow the lateral connection structure of the DCS to build a PTFM. Due to this strategy we need a relatively large number of trials for meeting high precision demands. However, in the off-line phase emphasis is on as small a number of neural units and as good a PTFM as possible since these will no longer be adapted in the on-line phase. Training in the off-line phase stops when the averaged fixation error falls below the pre-specified precision.

In off-line training we use a model of the head-eye system to calculate the new retinal coordinates of the target after applying the controller output $\bar{v}$. Input vectors are generated randomly. The only constraints are that the vergence and tilt angle, θ_l and ϕ, are

---

1. We assume $y_l = y_r$, and, because of symmetrical vergence, $\theta_l = \theta_r$. Note that $\bar{v}$ does not depend on χ.

restricted to an interval of interest and that the retinal coordinates x_l, y_l, x_r may not exceed the field of view of the cameras.

3.2   On-line training in the operating phase

In the operating phase we continue to adapt the matrices A_i in the output layer. This is done by error feedback, eq. (3), just as in the off-line phase. We do no longer grow the network nor do we further adapt the centers or the lateral connection structure.

Targets for fixation are generated by randomly moving our Stäubli R90 robot arm, Fig. 1 (b), in its work space with a light source attached to its gripper, Fig. 1 (c). Relying on our Datacube image processing system we can calculate the retinal coordinates of the target w.r.t the two cameras of our TRC binocular head, Fig. 1 (a), at video rate. The controller output as calculated according to (1) and (2) is then applied to fixate the target, and the output of the proportional controller is used for error feedback learning. The latter is also used for a correctional saccade.

(a) The TRC binocular head

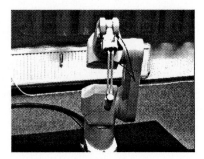

(b) The Stäubli R90 robot arm

(c) Setup for on-line training

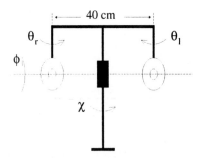

(d) Diagram of the TRC head

Fig. 1: Binocular Head (a), (c), (d) and Robot Arm (b), (c)

4 Experimental Results

Fig. 2 shows the average pixel error and the number of neural units in the DCS network versus the number of trials in the off-line training phase for a pre-defined precision of 0.5% (2.5 pixel). On reaching this precision after 40000 simulated saccades only 120 neural units have been inserted into the network. The tilt angle ϕ was restricted to the interval $[-30°, 30°]$ and the vergence angle θ_1 to $[0, 10°]$.

We could have reached this high precision with fewer training steps but since we want as few neural units and as good a PTFM as possible, we take our time. After all, the whole off-line training phase takes only 3 minutes for 40000 saccades on a Sparc 4 workstation. The very reasonable accuracy of 1% (5 pixel) is reached after 3500 steps with only 40 neural units.

Now we use the pre-trained controller for saccade control of the real TRC binocular head. As demonstrated in Fig. 3 the error at the start of the operating phase is about 12 pixel (2.5%) which is due to deviations of the model from the real system. However, due to on-line learning by error feedback the error drops down to 2.6 pixel within 1000 saccades. After 1000 saccades we changed the zoom of one of the cameras as reflected by a peak in the error plot. The controller is able to adapt to the new parameter setting of the cameras within the next 1000 saccades.

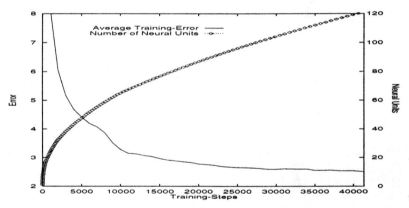

Fig. 2: Average fixation error in pixel and number of neural units in the DCS network versus number of training steps in off-line training phase

5 Discussion

Utilizing a DCS network for learning saccade control from error feedback we were able to substantially extend the work of Dean and Mayhew. By incrementally growing the network up to the size where it meets the pre-specified accuracy it utilizes as few neural units as possible. Since only a minor fraction of these units is involved in calculating the output of the DCS network, the generation of a controller output is very fast (even on conventional hardware) and suitable for real-time saccade control. By virtue of building perfectly topology preserving maps, the centers of the neural (rbf) units and

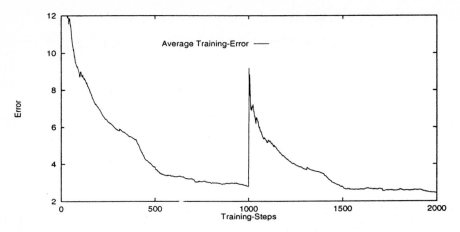

Fig. 3: Average fixation error in pixel versus number of training steps in operating phase. After 1000 saccades the zoom of the left camera changed.

the lateral connection structure are restricted to the relevant regions of the input space. Furthermore, the density of neural units is high in regions of the input space where the required sensory-motor mapping is difficult to approximate, i.e. is non-linear[2].

The controller has been successfully demonstrated to learn saccade control with high precision and to adapt to changing parameters on a real four-degree-of-freedom binocular head.

References

1. J. Bruske and G. Sommer: Dynamic Cell Structure learns Perfectly Topology Preserving Map, Neural Computation, Vol. 7, No. 4 (1995) 845-865
2. P. Dean, J.E. Mayhew, N. Thacker and P.M. Langdon: Saccade control in a simulated robot camera-head system: neural net architectures for efficient learning of inverse kinematics, Biol. Cybern., Vol. 66 (1991) 27-36
3. B. Fritzke: Growing cell structures - a self-organizing network for unsupervised and supervised learning, Neural Networks, Vol. 7, No. 9 (1995) 1441-1460
4. M. Kawato: Feedback-error-learning neural network for supervised motor learning, In: Advanced Neural Computers, Elsevier, Amsterdam (1990) 365-372
5. J.E. Mayhew, Y. Zheng and S. Cornell: The adaptive control of a four-degrees-of-freedom stereo camera head, Phil. Trans. R. Soc. Lond., Vol. 337 (1992) 315-326
6. Thomas Martinetz and Klaus Schulten: Topology Representing Networks, Neural Networks, Vol 7 (1994) 505-522

---

2. Due to lack of space this has not been demonstrated in this article. Together with additional experiments and further experimental details it will be the subject of a forthcoming publication of the authors.

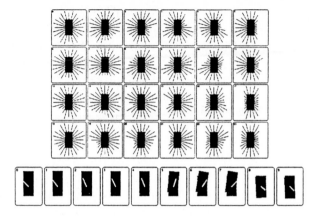

Fig. 1. The obstacle perceptions learnt by the super-net (upper drawing). Goal directions and actions learnt by sub-net 17 (lower drawing).

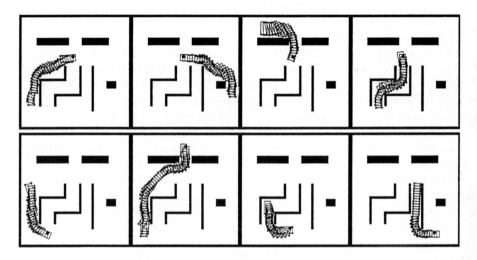

Fig. 2. The robot solving the path finding problem with the fixed goal used during the training phase (first row) and with new goal positions (second row).

the perception of free-space, unit #5 represents the perception of a wall on the right-hand side, unit #7 represents the perception of a wall behind the robot's back. It is possible to observe the *data topology-preserving* character of the KM: perception similarity varies in a continuous way on the map. The lower drawing in Figure 1 shows the weights of sub-net #17, which is associated to the perception of a narrow corridor. For each neuron, we represent the learnt goal direction (as a white vector) and the learnt action (the gray rectangle is the robot's initial configuration, the black rectangle is the robot's configuration after having performed the action). Again, the data topology-preserving character of the KM can be appreciated in this sub-net.

Figure 2 show some instances of path finding solved by the HEKM in co-operation with the planner. In these trajectories the planner takes control only when the action proposed by the HEKM would lead to a collision. In the first row of the Figure, the goal position (black circle) is fixed and it is the same used to generate the training examples for the HEKM. The second row depicts other trajectories with new goal positions. These runs prove that the motion skill acquired by the HEKM is independent from the chosen goal.

3 Why to Use a SOM-like Network?

We would like now to discuss the following claim: the data topology-preserving character of the HEKM could favor the learning of fine motion.

This statement can be proved experimentally by performing two separate training sessions. In the first session, the neighborhood parameters (one for the super-net, one for the sub-nets) are set to 0, while in second session they are set to values other than 0 (4 and 5, respectively). In this way, we can study the effect of cooperation during learning.

To evaluate the two methods, an error criterion and a performance criterion are used. The error measure is the mean squared error between the network output action and the target action proposed by the planner, while the performance criterion is the percentage of optimal actions learnt by the network. By definition, the optimal actions are those proposed by the planner.

Let us comment on the plots of error and performance as a function of the number of training cycles (Figure 3). As far as the error is concerned (left plot), one can see that without cooperation (curve with black dots) a certain error level is reached quite rapidly, but afterwards, no significant improvement is observed. On the contrary, with cooperation (curve with white dots) it takes more time to reach the very same error level, but the final error is lower. This type of behavior seems to be typical for cooperating agents, as it reported in [1]. In our experiment, a possible explanation for this could be that, when the cooperation between the neurons is active, it takes more time to find a good "compromise" to

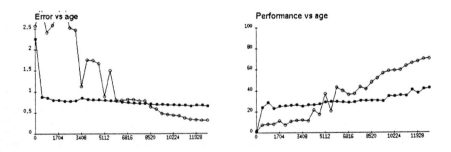

Fig. 3. Error (left) and performance (right) without cooperation (black dots) and with cooperation (white dots).

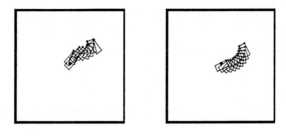

Fig. 4. The planner (left) and the HEKM (right) working as stand-alone systems.

satisfy competing learning needs. However, once the compromise is met, the final result gets improved. A corresponding behavior is observed in the performance curves (right plot). With no cooperation a certain performance level is achieved quite rapidly (42%), but after that point no further improvement occurs. With cooperation, the same performance level is obtained later, but the final result is more satisfactory (65%).

4 Planner versus HEKM

We conclude by highlighting an interesting side-effect which can be obtained by transferring motion knowledge from the planner to the HEKM.

Our planner is a *discrete* system. By the term "discrete" we refer to the fact that, at each step of the robot trajectory, the planner generates a finite number of neighboring configurations, and chooses, among them, the one which approaches the goal closest while avoiding collisions. The HEKM, on the contrary, tends to produce actions which look like being *continuous*. That is because the action learnt by the network for a given perception is a kind of average action performed by the planner in similar perceptual states. To illustrate this point, we let the planner and the HEKM solve the same path finding problem as *stand-alone* systems (Figure 4). One can immediately appreciate qualitative differences in the two paths. The discrete nature of the planner is evident in the left plot: the robot motion is optimal in terms of path length, but quite abrupt. On the contrary, in the HEKM path (right plot) smoothness has been traded against optimality. This observation can also account for the sub-optimal performance level reached by the HEKM (Figure 3) at the end of training.

5 Conclusions

We have presented a HEKM which learns fine motion under the control of a planner. *First*, we have discussed the utility of using a hierarchical KM instead of the usual "flat" version. The HEKM is more economic in terms of the way memory cells are used. It avoids unnecessary weight repetitions and allows for compact input representations. Clearly, one limitation of the current architecture is the fixed number of neurons. A growing network could be used instead

[5, 2]. *Second,* we have measured the effect of cooperative learning due to the interaction between adjacent neurons. We found that with cooperation learning is slowed down on the short run. But the benefits appear later on, resulting in a more satisfactory final performance. Our interpretation is that, at the beginning of learning, neighboring neurons work to meet a compromise to competing needs: this effort becomes rewarding on the long run. *Third,* we have pointed out the complementary nature of the paths generated by the planner and by the HEKM as stand-alone systems. The HEKM produces sub-optimal but smooth solutions, whereas the planner seeks for optimality while sacrificing the continuity of motion. The integration of these two philosophies leads to fruitful results. Our future work will include the implementation of these ideas on a physical robot.

Acknowledgements

Thanks to José del R. Millán, Vicente Ruiz de Angulo and Jürgen Schmidhuber for their comments. Thanks to *Neuristique* (France) for providing the SN neural network simulator.

References

1. Clearwater, S.H., Hogg, T., Huberman, B.A. (1992) Cooperative Problem Solving. In Huberman, B.A., Editor, *Computation: The Micro and the Macro View.*, World Scientific.
2. Fritzke, B. (1995) A Growing Neural Gas Network Learns Topologies. In Tesauro, G., Touretzky, D.S., Leen, T.K., Editors, *Advances in Neural Information Processing Systems 7*, MIT Press, Cambridge MA, pp. 625–632.
3. Gambardella, L.M., Versino, C. (1994) Learning High-Level Navigation Strategies from Sensor Information and Planner Experience. *Proc. PerAc94, From Perception to Action Conference*, Lausanne, Switzerland, September 7–9, pp. 428–431.
4. Heikkonen, J., Koikkalainen, P., Oja, E. (1993) Motion Behavior Learning by Self-Organization. *Proc. ICANN93, International Conference on Artificial Neural Networks*, Amsterdam, The Netherlands, September 13–16, pp.262–267.
5. Heikkonen, J., Millán, J. del R., Cuesta, E. (1995) Incremental Learning from Basic Reflexes in an Autonomous Mobile Robot. *Proc. EANN95, International Conference on Engineering Applications of Neural Networks*, Otaniemi, Espoo, Finland, August 21–23, pp. 119–126.
6. Knobbe, A.J., Kok, J.N., Overmars, M.H. (1995) Robot Motion Planning in Unknown Environments Using Neural Networks. *Proc. ICANN95, International Conference on Artificial Neural Networks*, Paris, France, October 9–13, pp. 375–380.
7. Millán, J. del R. (1995) Reinforcement Learning of Goal-Directed Obstacle-Avoiding Reaction Strategies in an Autonomous Mobile Robot. *Robotics and Autonomous Systems*, 15(3), pp. 275–299.
8. Ritter, H., Martinetz, T., Schulten, K. (1992) *Neural Computation and Self-Organizing Maps. An Introduction.* Addison-Wesley Publishing Comp.
9. Versino, C., Gambardella, L.M. (1995) Learning Fine Motion in Robotics by Using Hierarchical Neural Networks. *IDSIA-5-1995 Technical Report.*

Subspace Dimension Selection and Averaged Learning Subspace Method in Handwritten Digit Classification

Jorma Laaksonen and Erkki Oja

Helsinki University of Technology
Laboratory of Computer and Information Science
Rakentajanaukio 2C, FIN-02150 ESPOO, Finland

Abstract. We present recent improvements in using subspace classifiers in recognition of handwritten digits. Both non-trainable CLAFIC and trainable ALSM methods are used with four models for initial selection of subspace dimensions and their further error-driven refinement. The results indicate that these additions to the subspace classification scheme noticeably reduce the classification error.

1 Introduction

In a recent study [1], we demonstrated that subspace classifiers, especially the error correcting *Averaged Learning Subspace Method* (ALSM) perform excellently in classification of handwritten digits. In the comparison, some of the most popular neural and non-neural classifiers were included, like MLP, LVQ, k-NN, LLR, and various discriminant analysis and tree classifier methods. In the current paper, we present some modifications which further increase the achieved classification accuracy of the subspace classifier.

Subspace classifiers have earlier been intensively used in optical character recognition [2], but in recent years they seem to have been largely forgotten. In the mean time, the interest in the field of character recognition has shifted its focus to the recognition of handwritten text. We present results of experiments on classifying handwritten digits with subspace classifiers. The classical non-trainable CLAFIC (*CLAss-Featuring Information Compression*) classifier is a statistical classifier [3], the error correcting ALSM is a neural classifier [4],[5].

The experiments were done using data collected from 894 Finnish writers each giving two samples of each of the ten digits. Each writer was advised to type the digits in separate boxes on a printed form. The forms were scanned in as binary images with resolution 300 dpi. The digit images were size and slant normalized to size 32×32 using routines similar to those described in [6]. The first half of the total of 17880 digit images were used as the training material and the second half for testing. A feature extraction was performed first by subtracting the estimated mean of the training set from each image and Karhunen-Loève transforming to 64 dimensional vectors using transform coefficients obtained from the covariance matrix of the training set. These 64 dimensional vectors, representing 10 classes, are now the material for our classification experiments.

2 Subspace Classifiers

Subspace classifiers form a group of semi-parametric classifiers. They are based on the assumption that each of the classes in the feature vector mixture forms a lower-dimensional linear subspace distinct from the subspaces spanned by the other classes. A subspace $\mathcal{L}_j$ of class j is represented by a matrix $\mathbf{U}_j$ formed of vectors $\mathbf{u}_{1j}, \cdots, \mathbf{u}_{\ell_j j}$. These basis vectors are mutually orthonormal within each class, i.e., $\mathbf{U}_j^T \mathbf{U}_j = \mathbf{I}$. Any vector $\mathbf{x}$ of the pattern space $\mathbb{R}^d$ can then be represented as a sum of two vectors, one belonging to the subspace $\mathcal{L}_j$, the other orthogonal to it,

$$\mathbf{x} = \widehat{\mathbf{x}}_j + \widetilde{\mathbf{x}}_j = \mathbf{U}_j \mathbf{U}_j^T \mathbf{x} + (\mathbf{I} - \mathbf{U}_j \mathbf{U}_j^T) \mathbf{x}. \tag{1}$$

The length of $\widehat{\mathbf{x}}_j$ can be used as the discriminant function, giving the classification rule for c classes:

$$g(\mathbf{x}) = \operatorname*{argmax}_{j=1,\cdots,c} \|\mathbf{U}_j^T \mathbf{x}\|^2 = \operatorname*{argmax}_{j=1,\cdots,c} \sum_{i=1}^{\ell_j} (\mathbf{u}_{ij}^T \mathbf{x})^2. \tag{2}$$

As the basis vectors $\mathbf{u}_{ij}$, the ℓ_j first eigenvectors of the class j correlation matrix $\widehat{\mathbf{R}}_j$ are normally used, corresponding to the largest eigenvalues $\lambda_{1j}, \cdots, \lambda_{\ell_j j}$ in decreasing magnitude. The correlation matrix can be estimated as $\widehat{\mathbf{R}}_j = 1/n_j \sum_{i=1}^{n_j} \mathbf{x}_{ij} \mathbf{x}_{ij}^T$ where n_j is the number of class j vectors $\mathbf{x}_{ij}$ in the training set. As such, this type of classifier is called the CLAFIC [3] method.

Several learning variants of the CLAFIC algorithm have been suggested; for a review, see [4], [7]. One of these, the Averaged Learning Subspace Method will be reviewed below in Section 2.3, after first discussing two other modifications to the classical subspace method: the iterative selection of subspace dimensions, and the weighting of the basis vectors in computing the projections.

2.1 Iterative Selection of Subspace Dimensions

The subspace dimensions $\ell_1, \cdots, \ell_c$ need to be somehow fixed. Generally, there exists a unique combination of the dimensions which produces optimal classification accuracy with the testing data. In the current setting there are 10 classes and 64 possible dimensions for each of them. This totals to 64^{10} combinations which is too much to be evaluated in any practical experiment. Therefore some statistical or heuristic criteria need to be used in determining the dimensions.

It is easy to device a set of methods testing a portion of the possible combinations of subspace dimensions: by varying some relevant parameter, an U-shaped error curve is usually obtained and the index producing the minimum error level can be utilized. The simplest approach is to select a global dimension $D = \ell_1 = \cdots = \ell_c$ to be used for all the classes. In this case, all the values $D = 1, \cdots, d$, $d = 64$ can be evaluated, producing an error curve like those plotted in Figure 1.

Another set of selection schemes can be based on the decreasing eigenvalues λ_{ij} of each individual class correlation matrix $\widehat{\mathbf{R}}_j$, as already proposed by [3]: the selection of the subspace dimensions $\ell_1, \cdots, \ell_c$ can be based on thresholding either $\lambda_{\ell_j+1,j}$ or $\sum_{i=\ell_j+1}^{d} \lambda_{ij}$ below an interclass limit Λ after normalizing $\sum_{i=1}^{d} \lambda_{ij} = 1$.

All the three criteria described above can be thought of as initial guesses and the subspace dimensions can be further modified by observing how classification errors are distributed among the classes. If we denote with A_j the set of indices of vectors which are classified to class j but belong to some other class, and with B_j correspondingly the set of indices of vectors belonging to class j but erroneously classified to some other class, we may express a rule for iterative tuning of ℓ_j's:

$$j = \operatorname*{argmax}_{i=1,\cdots,c} |\#B_i - \#A_i| \qquad (3)$$

$$\Delta\ell_j = \frac{\#B_j - \#A_j}{|\#B_j - \#A_j|} \qquad (4)$$

Thus, in each iteration step one ℓ_j is either increased or decreased by one. The iteration may be stopped after a fixed number of trials or when the process has ended in a loop. The combination which gave the best accuracy is restored.

2.2 Weighting of Basis Vectors

The classes may be treated individually also by weighting the components in the projection (2) with a factor w_{ij}, i.e.,

$$g(\mathbf{x}) = \operatorname*{argmax}_{j=1,\cdots,c} \sum_{i=1}^{\ell_j} w_{ij} (\mathbf{u}_{ij}^T \mathbf{x})^2. \qquad (5)$$

If the w_{ij} are computed from the correlation matrix eigenvalues as $w_{ij} = \lambda_{ij}/\lambda_{1j}$, we are led to the so-called *Multiple Similarity Method* (MSM) [2]. A more general form of the same idea is to select a parameter ρ and to set

$$w_{ij} = (\lambda_{ij}/\lambda_{1j})^\rho. \qquad (6)$$

Thus, setting $\rho = 0$ equals to the conventional CLAFIC, $\rho = 1$ to MSM, whereas $\rho = -1$ leads to a sort of variance normalization within each subspace. In our experiments, the best results were obtained with ρ just above zero. This may be explained with the smooth fading of w_{ij}'s in each class. Figure 1 displays classification errors for varying the common basis dimension D when $\rho = 0$, i.e., $w_{ij} = 1$, and for $\rho = 0.05$ which produced the best results.

It is also possible to formulate the selection of the subspace dimensions $\ell_1, \cdots, \ell_c$ in the previous section in the terms of w_{ij}'s. Thus, for each class j, w_{ij} was 1 for $i = 1, \cdots, \ell_j$, and 0 for $i = \ell_j + 1, \cdots, d$. These two methods may also be combined, i.e., the optimal ρ can first be selected while all the subspaces have a common dimension, thereafter letting the ℓ_j's change individually as described above.

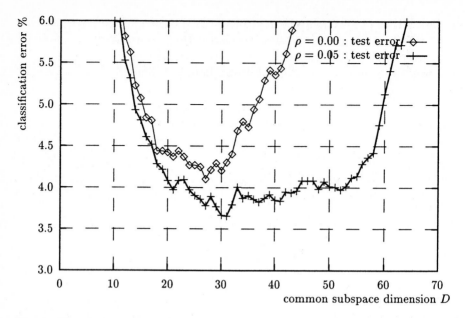

Fig. 1. Classification errors for unweighted projection length and projection length weighted with $(\lambda_{ij}/\lambda_{1j})^{0.05}$. The percentages 4.1 and 3.7 in the first column of Table 1 can be seen as the minimum values of the two curves.

2.3 The Averaged Learning Subspace Method

The methods described this far have not rotated the subspaces used in classification. A better performance can be obtained if the subspaces are modified in an error-driven method. This method – know as the *Averaged Learning Subspace Method* (ALSM) [7] – aims to change the subspace during the training period. When an error occurs, the correct subspace is rotated towards the misclassified vector and the wrong subspace is rotated away from it. This is achieved by modifying the correlation-type matrices and then updating the basis vectors:

$$\widehat{\mathbf{S}}_j(k+1) = \widehat{\mathbf{S}}_j(k) + \alpha \sum_{i \in A_j} \mathbf{x}_i \mathbf{x}_i^T - \beta \sum_{i \in B_j} \mathbf{x}_i \mathbf{x}_i^T, \quad \text{where} \tag{7}$$

A_j and B_j were the sets of incorrectly classified vectors, defined in Section 2.1, and the initial matrix $\widehat{\mathbf{S}}_j(0)$ is the scatter or unnormalized correlation matrix, i.e., $\widehat{\mathbf{S}}_j(0) = \sum_{i=1}^{n_j} \mathbf{x}_{ij} \mathbf{x}_{ij}^T$. The basis vectors $\mathbf{u}_{ij}$ are re-solved after each epoch. In our experiments, the dimensions $\ell_1, \cdots, \ell_c$ were kept constant during the ALSM iteration. Likewise, the ρ parameter was fixed to value 0.05 in the last case. The α and β values in (7) control the steepness of the error correction. In our experiments they both were fixed to value 3.0 which was found to be good in the previous experiments [1].

3   Results and Discussion

The classification accuracies of the test set in the experiments described in previous sections are summarized here. Table 1 contains four ways of selecting initial subspace dimensions. The columns display error percentages initially, after iteration described in Section 2.1, and finally after the ALSM training. In the last line, a weighting of $\rho = 0.05$ according to (6) was used throughout.

Table 1. Classification error levels with different subspace dimension selections.

Initial ℓ_j Selection	Error %'s		
	Initial	Iterated	ALSM
$\ell_j = 27$	4.1	—	3.0
$\ell_j = 27$	4.1	3.5	2.9
$\lambda_{ij} < \Lambda$	4.4	3.8	2.9
$\sum \lambda_{ij} < \Lambda$	5.1	3.5	2.8
$\ell_j = 31, \rho = 0.05$	3.7	3.5	2.7

As can be seen, the initial classification accuracies vary a lot, whereas the results after the iteration – and even more clearly after ALSM – are quite near each other. The first two rows of Table 1 indicate that dimension optimization prior to ALSM training is still somewhat beneficial. The advantage achieved by using weighted projection length calculation and the iterative tuning of subspace dimensions together with ALSM training as described in this paper is of the order of 0.3 percentage units.

Though the current results were not produced with cross-validated selection of parameters like those reported in [1], and are thus not as such directly comparable, the advantages reported here make ALSM compare favorable to the accuracy attained with the computationally highly complex Local Linear Regression (LLR) algorithm, which was the only classifier superior to ALSM in the earlier comparison.

References

1. Holmström, L., Koistinen, P., Laaksonen, J., Oja, E. Neural network and statistical perspectives of classification. In *Proceedings of 13th International Conference on Pattern Recognition* (1996), 1996 to be published
2. Iijima, T., Genchi, H., Mori, K. A theory of character recognition by pattern matching method. In *Proceedings of the 1st International Joint Conference on Pattern Recognition*, Washington, DC (October 1973). 1973 50–56
3. Watanabe, S., Lambert, P. F., Kulikowski, C. A., Buxton, J. L., Walker, R. Evaluation and selection of variables in pattern recognition. In J. Tou, editor, *Computer and Information Sciences II.* Academic Press, New York, 1967

232

4. Oja, E., Kohonen, T. The Subspace Learning Algorithm as a formalism for pattern recognition and neural networks. In *Proceedings of the International Conference on Neural Networks*, SanDiego, California (July 1988). IEEE, 1988 I–277–284
5. Oja, E. Neural networks, principal components, and subspaces. *International Journal of Neural Systems* (1989) 1:61–68
6. Garris, M. D., Blue, J. L., Candela, G. T., Dimmick, D. L., Geist, J., Grother, P. J., Janet, S. A., Wilson, C. L. NIST form-based handprint recognition system. Technical Report NISTIR 5469, National Institute of Standards and Technology (1994)
7. Oja, E. *Subspace Methods of Pattern Recognition.* Research Studies Press Ltd., Letchworth, England (1983)

A Dual Route Neural Net Approach to Grapheme-to-Phoneme Conversion[*]

Maria Wolters

Department of Linguistics, University of Edinburgh
Adam Ferguson Building, GB–EH8 9LL Edinburgh

Abstract. For multilingual text-to-speech synthesis, it is desirable to have reliable grapheme-to-phoneme conversion algorithms which can be easily adapted to different languages. I propose a flexible dual-route neural network algorithm which consists of two components: a constructor net for exploiting regularities of the mapping from graphemes to phonemes and a self-organizing map (SOM) for storing exceptions which are not captured by the constructor net. The SOM transcribes one word at a time, the constructor net one phoneme at a time. The constructor net output is then classified by mapping it onto a set of codebook vectors generated by Learning Vector Quantisation which capture the net's concept of each phoneme.

1 Introduction

Text-to-speech synthesis (TTS) converts a string of words into speech output. Before generating that output, however, the orthographic form has to be converted into a broad phonetic transcription.

In most current systems, each word is first looked up in a lexicon; if it cannot be found there, it is transcribed by a set of grapheme-to-phoneme conversion (G2P) rules. When extending a TTS system to another language, such a transcription module is cumbersome to adapt. This paper presents a Neural Network (NN) algorithm for deriving a G2P conversion module for any given language from a corpus of orthographic words and their phonetic transcriptions. Previous NN solutions (see [11] for a comprehensive summary) are not as reliable as one would wish, with the best performances at around 95% accuracy on the training data (NETtalk, [13]). This score could be improved by taking into account the *net's concept* of each phoneme, how humans *read aloud*, and the *phonology* of the language to be transcribed, which defines the rules that the output has to comply with. In this paper, only the first two potential sources of performance improvement are discussed, and the first one is evaluated.

---

[*] Thanks to T. Mark Ellison, Joachim Buhmann, Paul Taylor, and David Willshaw for valuable comments. The financial support of the Studienstiftung des deutschen Volkes and of ERASMUS programme ICP 95 NL 1186 is gratefully acknowledged.

2 Background

2.1 Models of Reading Aloud in Humans

Reading aloud involves mapping visual orthographic information onto an abstract phonological representation. There are two main models of reading aloud: The *dual-route* approach [1] assumes that each word is processed by two parallel pathways: look-up in a phonological input lexicon and application of a set of conversion rules. The first pathway is taken to be the norm, the second pathway is used for reading unknown words. In the connectionist model [12, 9], a feed forward neural net both generalises and stores exceptions implicitly in the weights of the connections. Both models emphasise the importance of a semantic component, which stores word meanings and is concurrently activated when reading a word. The dual route model provides an elegant interface to that semantic component and deals well with exception words. The connectionist approach, on the other hand, is very suitable for automatically extracting regularities, but less so for storing exceptions, because once a net starts to encode specific input patterns in its weights, it is very likely to overfit the training data, losing its ability to generalise.

The model presented here attempts to combine the advantages of both approaches. It consists of two components, a neural net that extracts regularities from its input and a lexicon-like component for storing those words which cannot be captured by these regularities.[2]

As children become more proficient at reading, most of them overgeneralise less until they have mastered the fine details of the orthography ([12]). This development can be mirrored by constructor nets, which, as they add more neurons, become able to capture finer regularities and can adapt their net structure to the problem.[3] Therefore, they are suitable for modelling processes which involve learning generalisations ([10, 2]), like learning to read aloud.

2.2 Phoneme Description

Phonemes are modelled as sets of 21 binary features. These features are taken from the Halle and Sagey model of phoneme features (cited after [5]). These features are not only inherently structured, but have also been used for expressing phonotactic rules. Choosing features and not phonemes allows us to tailor the net more closely to TTS needs. For example, errors on features like [± consonantal] are more grave than errors on features like [± nasal]. For each feature, the error propagated back through the net could therefore be multiplied by a weight encoding its relative importance. It is supposed that a feature is not present (value: − or 0) unless it is specified (value: + or 1). For example, the phoneme

---

[2] If semantic and syntactic information is available to the TTS system, the lexicon component should be able to use this information e.g. for determining the pronunciation of words that are written the same, but pronounced differently.

[3] Potentially, the less complex the orthography, the smaller the net.

/a/ would be specified as [+syllabic], [+dorsal], [+back], [+low], [+constricted glottis] and [+voiced].

It remains to be seen whether this admittedly very redundant encoding yields any advantages over a more efficient encoding using only 6-7 bits.

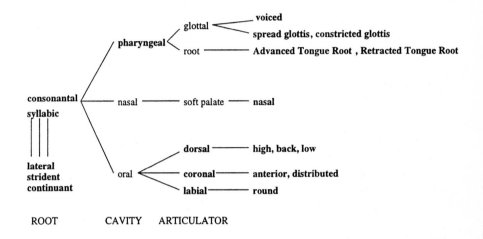

Fig. 1. The Halle and Sagey model of phoneme features. Root features specify the sound class (vowel, consonant or glide), the features on the side of the root describe the manner of articulation. The root node daughters correspond to cavities of the vocal tract, the daughters of these nodes represent further partitions and specifications of the states of these cavities. Features used in the net representation of phonemes are in bold type.

3 The Algorithm

3.1 Generalising: The Constructor Net

The input layer of the constructor net consists of 7 × 5 units, the output layer of 44 units. The 7 × 5 input unit represent a window of 7 graphemes, coded in an arbitrary 5 bit code, the middle grapheme of which is the one to be transcribed. The output is a feature-based representation of two phonemes with 21 units each plus a 2 unit encoding of word stress. The second set of 21 units is used only for diphthongs and affricates.

The phoneme which corresponds to a given output feature vector is determined by mapping this vector onto a set of codebook vectors. These codebook vectors are extracted from a classified sample of net output by Learning Vector

Quantisation (LVQ) ([7]) These codebook vectors mirror how a net represents a given phoneme, so that for recall, we rely on the categories that have been developed by the net itself, not on those with which it was taught.

3.2 Storing Exceptions: The Self-Organizing Map

Each word that is not transcribed correctly is to be stored in a Self-Organising Map (SOM, [6]). For storage, a word is converted into a unique vector representation by concatenating its grapheme codes. The node where a certain word has been stored is then associated with the correct transcription. A crucial advantage of the SOM is its suitability for interfacing purposes [8].

4 Evaluation

4.1 Method

For a first evaluation of the contribution of the codebook vector filter, two setups were compared: a three-layered feed-forward backpropagation network (B) and B with its output mapped onto a set of codebook vectors (BLVQ).[4]

The data used for evaluating the model was taken from the dictionary corpus used for training NETtalk ([13]), which contains 20011 words. Since this data is already aligned, the alignment problem is avoided here. The training set consisted of every $4n + 2$th word from that corpus (total: 5003), the test set of every $2n + 1$th word (total: 10005). [5] These sets were used both for training the network and for codebook generation. The networks were trained until the change of the error on the training set reaches an asymptote or the error on the test set began to rise. The error on the test set was computed every 10 epochs. The initial codebook was derived from net output for the training set with $40*N_{codebook}$ epochs of OLVQ1 and 1 epoch of LVQ3[6], where $N_{codebook}$ is the number of codebook vectors. Recognition accuracy was evaluated with the SNNS utility **analyse** for B and with the LVQ_PAK utility **accuracy** for BLVQ.

4.2 Results

Net B was trained for 70 epochs. Allowing for an output error of 0.2, it achieved about 61–65% correct classifications on both training and test set; the remaining 35–39% of output vectors could not be classified at all.

With BLVQ, all output vectors were classified. In general, accuracy was by about 30% lower for vowels than for consonants and depended on the sample size for that class. Furthermore, for 15% of all classes, classification accuracy

---

[4] The networks were simulated using SNNS version 4.1, the LVQ codebook vectors were calculated using the LVQ_PAK vector quantisation software.

[5] 5000 words seem to be a reasonable training corpus size, since for many languages, this corpus will have to be entered, at least partially, by hand.

[6] LVQ3 was run only once to avoid overfitting.

increased from about 0% for codebook size 100 to above 50% for codebook size 2000. However, for 10% of all classes, classification accuracy actually decreased with increasing codebook size, in one case by as much as 70%. For about 4%, accuracy remained constant when increasing the codebook size from 1000 to 2000.

Table 1. Performance of the algorithms: B: Backpropagation with learning rate $\alpha = 0.1$ and momentum $\nu = 0.5$, **BLVQ:** Backpropagation with LVQ, **Train:** performance on the training set, **Test:** performance on the test

Algorithm	No. of Codebook Vectors	Train % phonemes correct	Test % phonemes correct
B	0	65.61	61.63
BLVQ	100	64.36	65.76
	1000	83.71	78.59
	2000	84.35	78.77

4.3 Discussion

Although Tab. 1 might suggest that 1000 is a reasonable codebook size, 2000 is nevertheless preferable because there are significant improvements for some vowels at that size. The problems in classifying the vowels reflect that English vowel phonemes are notoriously hard to predict from the orthographic representation (cf. [12]). If there is no further increase in accuracy for a given class after increasing the codebook size, the net's concept of that class may be assumed to be optimally captured by the codebook vectors; if this accuracy is well below 90%, then this suggests that the class boundaries and therefore the net's concept of that class is still too fuzzy. A drop in accuracy after an increase of codebook size can occur for two reasons: overfitting and very close or overlapping class boundaries.[7] Because of the increase in the number of vectors, more vectors are placed in the regions of potential overlap, which distorts classification.

Nevertheless, filtering the net output through a codebook yields significant improvements in classification accuracy. But what would be the effect of replacing a Backpropagation net with a Cascade Correlation (CC, [3]) net? Preliminary tests indicate that CC results are significantly worse. There are several possible reasons for this: First, as previous work has shown ([12, 13]), broad and shallow nets[8] are well suited to the problem of G2P, but CC nets are typically deep and narrow[9]. Therefore, with constructor algorithms resulting in broad nets like Upstart ([4]), the results might be better. Secondly, since the mapping is quite

---

[7] It is possible for classes to overlap because one phoneme may correspond to different graphemes and vice versa.

[8] nets with few, but large hidden layers

[9] i.e. they have many very small hidden layers

complex, it takes some time for CC to build and add the required number of hidden units. Thirdly, CC has proven very sensitive to the number of epochs E_{update} allocated for updating the links to the output units after a new hidden unit has been added. If the error has not yet approached its asymptote, the output error starts to increase instead of decreasing. Either E_{update} has to be increased with number of hidden units, or it has to be fixed in advance by experimenting with large CC nets. It was found that a large E_{update} can outweigh performance advantages gained e.g. through the lack of a backpropagation cycle.

So although a constructor net can adapt its structure to match the complexity of the orthography and, furthermore, might be psychologically more realistic, this advantage can be outweighed by the disadvantages of the particular algorithm chosen. It remains to be seen if there is a constructor algorithm which is a reasonable substitute for Backpropagation.

References

1. Coltheart, M.; Curtis, B.; Atkins, P and Haller, M. (1993): Models of Reading Aloud. *Psychological Review* 100, pp. 589–608
2. Elman, J. (1993): Learning and development in neural nets: the importance of starting small. *Cognition* 48, pp. 71–99
3. Fahlman, S. and LeBiere, Ch. (1990): The Cascade-Correlation learning architecture. In: *Advances in Neural Information Processing Systems 2*, D.S. Touretzky (ed.) Morgan Kaufman, pp. 524–532
4. Frean, Marcus (1990): *Small Nets and Short Paths: Optimising Neural Nets.* Ph.D. thesis, University of Edinburgh
5. Kenstowicz, M. (1994): *Phonology in Generative Grammar.* Oxford: Basil Blackwell
6. Kohonen, T. (1989^3): *Self-Organisation and Associative Memory.* Berlin, Heidelberg, New York: Springer
7. Kohonen, T.; Kangas, J.; Laaksonen, J. and Torkkola, K. (1992): LVQ_PAK: A program package for the correct application of Learning Vector Quantisation algorithms. In: Proc. Int. Joint Conf. on Neural Networks, pp. I 725–730
8. Miikkulainen, R. (1993): *Subsymbolic Natural Language Processing* Cambridge, Mass.: MIT Press
9. Plaut, D.; Seidenberg, M.; McClelland, J. and Patterson, K.: (1994) Understanding Normal and Impaired Word Reading: Computational Principles in Quasi-Regular Domains. Technical Report PDP.CNS.94.5
10. Quartz, Stephen (1993): Neural networks, nativism, and the place of constructivism. *Cognition* 48, pp. 223–242
11. Rosenke, Katrin (1995): Verschiedene neuronale Strukturen für die Transkription von deutschen Wörtern. In: Proc. 6. Konferenz Elektronische Sprachsignalverarbeitung, R. Hoffmann and R. Ose (eds), TU Dresden, Institut f. Technische Akustik, pp. 159–166,
12. Seidenberg, M. and McClelland, J. (1989): A distributed, developmental model of word recognition and naming. *Psychological Review* 96, pp. 523–568
13. Sejnowski, T. and Rosenberg, C. (1987): Parallel networks that learn to pronounce English text. *Complex Systems* 1, pp. 145–168

Separating EEG Spike-Clusters in Epilepsy by a Growing and Splitting Net

Matthias Dümpelmann and Christian Erich Elger

University Clinic of Epileptology, Sigmund Freud Str. 25, 53105 Bonn, Germany

Abstract. The presurgical evaluation of epilepsy patients relies on an exact localization and delineation of the generators of epileptic seizures. During the registration of the electroencephalogram (EEG) sharp transient signals called spikes can be observed. These spikes give hints for the so called epileptogenic zone in the brain. In order to decide whether these spikes derive from single or multiple generators an incrementing topology preserving map with insertion and deletion of units was trained for the EEG data of individual patients. By deleting of units the net was separated into subnets. Thus it could be further used for vector quantization. The spatial distributions of the peak amplitude of the spikes in all channels as well as the time differences of their peaks were used as input signals. The separation of spatio-temporal clusters of the spikes was compared with those clusters identified by a human reviewer.

1   Introduction

A great number of epilepsy patients suffer from seizures that cannot be sufficiently controlled by antiepileptic drugs. In a part of these patients seizure freedom can be obtained by resecting the area of the brain responsible for seizure generation. For the determination of the area called epileptogenic zone [14], the registration of the EEG with surface electrodes and intracranial electrodes lying directly on the surface of the brain or inside suspicious structures is of major importance. Besides of the registration of epileptic seizures, epileptiform spikes that occur in the interval between seizures give hints for possible generators of the seizures. Averaging of the amplitude of nearly simultaneously occurring spikes in different EEG channels and of the time-shifts between their peaks can help to describe pathways of spike propagation [3,8]. However, to decide whether this averaging has to be done for several spatio-temporal spike-clusters indicating different generators is a crucial task for the EEG reviewer, since no exact rules exist for defining independent clusters. We tried to mimic the clustering done by the EEG reviewer with a topology preserving artificial neural net. The choice of a topology preserving net was motivated by the topological and anatomically arranged placement of the electrodes. This topological information is difficult to implement in ordinary statistical clustering algorithms. Self-organizing topology preserving maps proposed by Kohonen [10,11,12] found a great number of applications and modifications. Examples for applications in EEG analysis are given in [2,9]. Fritzke proposed a modified self-organizing net with an incrementing structure [4,5,6,7]. In addition to the insertion of units, rare selected units are deleted. As a consequence of the deletion of units, independent subnets are generated which can be used for vector quantization. In this study we tried to evaluate whether the vector quantization obtained by a growing and splitting net can be used to define independent spatio-temporal spike-clusters already defined by a human reviewer.

2 Methods

2.1 Network Architecture

The architecture of the network corresponds to the growing and splitting net proposed by Fritzke [4]. At each presentation of an input vector ξ one unit is selected. Selecting a unit relies on the minimum distance of its weight vector w_c to the input vector ξ (1). The weights w_s of the selected unit will be adapted with the learning factor ϵ_b that they will get closer to the input vector ξ (3). In order to attain a topological ordering, the neighbouring units are updated into the same direction, however with a smaller learning factor ϵ_n. In addition, the value τ_s is assigned to each unit, which sums up the quantization error (2).

$$\|\xi - w_s\| = \min_{c \in A} \|\xi - w_c\| \tag{1}$$

$$\Delta \tau_s = \|\xi - w_s\|^2 \tag{2}$$

$$\Delta w_s = \epsilon_b (\xi - w_s) \tag{3}$$

$$\Delta w_n = \epsilon_n (\xi - w_n) \qquad (\forall n, (n \neq s) \in N)$$

$$0.0 \leq \epsilon_n \leq \epsilon_b \leq 1.0$$

The network starts with 3 units. After λ_1 steps a new element is inserted. The position of the new unit is between the unit with the maximum quantization error τ_{max} and its neighbour exhibiting the most different weight vector. After insertion the mean of the weight vector of its two neighbours is assigned to the weight vector of the new element. The cumulated errors of the new unit and the unit with the maximum quantization error before inserting will be $\tau_{max}/2$.

After λ_2 ($\lambda_2 > \lambda_1$) steps the unit with the smallest number of selections is deleted. In special cases it is sometimes necessary to delete further units and edges between units to preserve a consistent net with complete triangles. As a result separated clusters of units occur. After training the net all input vectors ξ are again presented to the net and assigned to one of the clusters.

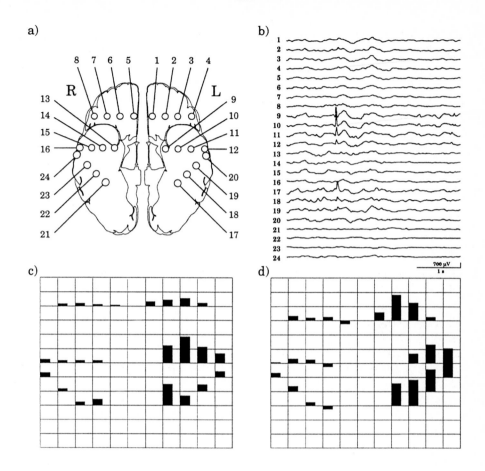

Fig. 1. (a) Schematic view of the bottom of the brain with electrode contacts, (b) EEG containing a spike in the left temporo-basal electrodes, (c) array with amplitude values presented to the net, (d) as (c) but for time-shift values.

2.2 Spatio-temporal EEG Spike Clustering with the Net

Spikes are detected by use of an automatic spike detection algorithm, whose two subsequent stages are derived from [13] and [1]. In a small time frame around a detected spike event peak amplitudes are determined for each channel. In addition the time-shifts between the peaks of all channels are computed. These values are presented to the net in a schematically anatomical ordered way (Fig.1). Thus, the input patterns consist of two quadratic arrays. In order to achieve a great number of training patterns, the spike events are presented several times to the net in their natural order of occurrence. Training is stopped when the number of clusters is equal to the number of clusters defined by the human reviewer or if they exceed the latter by one. After training, the spike events are presented to the net again. The events are assigned to the different clusters and for each cluster the peak amplitudes and time-shifts are averaged.

2.3 Interactive Spatio-Temporal EEG Spike-Clustering

The result of the spatio-temporal clustering of the neural net was compared to the visual clustering done by a human reviewer. Using an interactive software package the number of spike clusters has to be determined iteratively. One starts with a channel exhibiting a great number of spikes (called leading channel in the following), and computes the amplitude and time-shift distribution for this class of spikes. This is done by averaging amplitudes and time-shifts of peaks for all channels in a short time frame around the peak of the spikes in the leading channel. If not all spikes can be explained by this class of spikes the procedure is repeated until all spike events are assigned to one spike-cluster.

3 Results

The results of the analyses of 8 EEG segments (10 to 30 min. duration, one per patient) are summarized in table 1, which compares net and human reviewer clustering. In 4 cases all clusters were identical for both methods. However, two categories of differences in the clustering were obvious. First, a cluster defined by the human reviewer is represented by two or more clusters generated by the net. This was found in 4 cases in ($n = 6$) clusters. Second, the net generated a cluster that should be separated into two or more clusters. This was obtained in the same 4 cases in ($n = 4$) clusters. Fig 2. shows, as an example, amplitude and time-shift distributions of the spike peaks of case no. 7 as they were determined by the net. The net has separated two clusters, each of them lying in one hemisphere of the brain.

Table 1. Comparison of spike clustering by a neural net and a human reviewer for 8 cases

Case No.	Training patterns	Automatic clusters	Visual clusters	Identical clusters	Too detailed separations	Missing separations
1	25000	5	4	3	1	1
2	130000	7	7	3	3	1
3	70000	3	3	1	1	1
4	15000	3	3	3	0	0
5	50000	4	4	2	1	1
6	50000	1	1	1	0	0
7	10000	2	2	2	0	0
8	10000	3	3	3	0	0

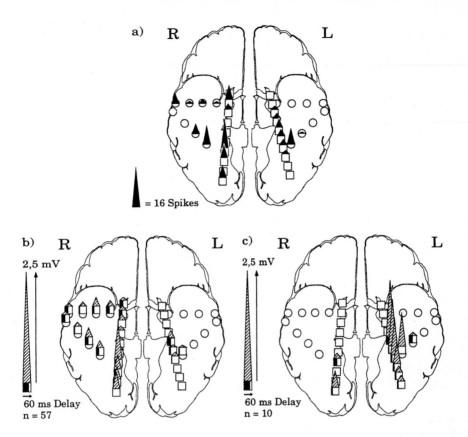

Fig. 2. Results of spike analyses of case no. 7, (a) number of spikes detected in each electrode (represented by the height of the triangles), (b) spike cluster in the right hemisphere (amplitudes are represented by the height of the triangles, time-shifts by the width of the black rectangle below the triangle). Amplitudes and time-shifts indicate a spike propagation from the inner to the outer electrode contacts, (c) spike-cluster in the left hemisphere with a smaller number of spikes. Amplitudes and time-shifts indicate a spike-cluster in a more restricted area.

4 Discussion

A self-organizing net with a growing and splitting topology was used for clustering of amplitude and time shift distributions of epileptiform spikes. To date no exact rules exist for the definition of independent clusters from the registered EEG. The results show that clusters generated by the net correspond to clusters evaluated by the human reviewer. However, the two categories of differences between both clustering procedures have to be judged differently. In the case of too detailed separations summing up clusters or assigning a higher accuracy to the net seems to be appropriate. In contrast, missing separations of clusters are more severe, as probably independent spike generators are not recognized. Therefore, new rules for the splitting of the net should be established, and a criterion when to stop the training should be developed. After solving these problems and examination of the results obtained from a greater number of representative selected cases, clustering of spikes using an artificial neural net can be discussed as an additional tool for the presurgical evaluation of epilepsy.

References

1. Blume, W., Lemieux, J., Morphology of spikes in spike-and-wave complexes, Electroencephalograhy and clinical Neurophysiology **69** (1988), pp 508-515
2. Elo, P., Saarinen, J., Värri, A., et. al., Classification of Epileptic EEG by Using Self-Organizing Maps, in: Artificial Neural Networks 2, Aleksander, I., Taylor, J., Elsevier, Amsterdam, (1992)
3. Dümpelmann, M., Hufnagel, A., Burr, W., Elger, C., Determination of the Epileptogenic zone by use of Spike Clustering and Spike-Amplitude-Latency-Topograms, Epilepsia, **35(Sup. 8)** (1994), p 28
4. Fritzke, B., Vector Quantization in a Growing and Splitting Elastic Net, in: Proc. ICANN'93, Int. Conf. on Artificial Neural Networks, Gielen, S., Kappen, B., Springer, London, (1993)
5. Fritzke, B., Growing Feature Maps and Growing Cell Structures - a Performance Comparison, in: Advances in Neural Information Processing Systems 5, Giles, L., Hanson, S., Cowan, J., Morgan Kaufmann Publishers, San Mateo, CA, (1993)
6. Fritzke, B., Growing cell structures – a self-organizing network for unsupervised and supervised learning, Neural Networks **7(9)**(1994), pp 1441-1460
7. Fritzke, B., Fast learning with incremental RBF networks, Neural Processing Letters, **1(1)** (1994), pp 2-5
8. Hufnagel, A., Burr, W., Elger, C., et al., Localization of the epileptic focus during methohexital induced anesthesia, Epilepsia, **33(2)** (1992), pp 271-284
9. Kaski, S., Joutsiniemi, S., Monitoring EEG Signal with the Self-Organizing Map, in: Proc. ICANN'93, Int. Conf. on Artificial Neural Networks, Gielen, S., Kappen, B., Springer, London, (1993)
10. Kohonen, T., Self-Organization and Associative Memory, Springer Series in Information Sciences 8, Springer, Heidelberg, (1984)
11. Kohonen, T., The Self-Organizing Map, Proc. IEEE **78** (1990), pp 1464-1480
12. Kohonen, T., Self-Organizing Maps, Springer Series in Information Sciences 30, Springer, Heidelberg, (1995)
13. Lopes da Silva, F., Dijk, A., Smits, H., Detection of Nonstationarities in EEGs using the Autoregressive Model. An application to the EEGs of Epileptics, in: CEAN: Computerized EEG-Analysis, Dolce, G., Künkel, H., Fischer, Stuttgart, (1975)
14. Lüders, H., Awad, I., Conceptual Considerations, in: Epilepsy surgery, Lüders, H., Raven Press, New York, (1992)

Optimal Texture Feature Selection for the Co-occurrence Map

Kimmo Valkealahti and Erkki Oja

Helsinki University of Technology, Laboratory of Computer and Information Science,
Rakentajanaukio 2 C, FIN-02150 Espoo, Finland

Abstract. Textures can be described by multidimensional co-occurrence histograms of several pixel gray levels and then classified, e.g., with nearest-neighbors rules. In this work, multidimensional histograms were reduced to two dimensions using the Tree-Structured Self-Organizing Map, here called the Co-occurrence Map. The best components of the co-occurrence vectors, i.e., the spatial displacements minimizing the classification error were selected by exhaustive search. The fast search in the tree-structured maps made it possible to train about 14 000 maps during the feature selection. The highest classification accuracies were obtained using variance-equalized principal components of the co-occurrence vectors. Texture classification with our reduced multidimensional histograms was compared with classification using either channel histograms or standard co-occurrence matrices, which were also selected to minimize the classification error. In all comparisons, the multidimensional histograms performed better than the two other methods.

1 Introduction

We have previously selected features for image classification from a large set by minimizing the classification error of a nearest-neighbor classifier trained with the Learning Vector Quantization algorithm (Valkealahti and Visa 1995). In another approach (Oja and Valkealahti 1995), textures were described by reduced multidimensional co-occurrence histograms of n-tuples of pixels. The histogram reduction was carried out by the Self-Organizing Map (Kohonen 1995). For a given set of spatial displacements, a collection of n-tuples of pixels, called co-occurrence vectors, were sampled from the textures, and a self-organizing map was trained for these vectors. The ensuing map was called the Co-occurrence Map. A texture was then described by the histogram of the best-matching map reference vectors determined for all the sample co-occurrence vectors of that texture.

A problem in this approach is how to choose the displacements. Ideally, the best features selected from a variety of features should minimize the probability of error classification. This problem was addressed in the present work. Starting from all 9 pixels in 3 by 3 windows, all combinations of pruned pixel sets were tried as displacement sets for the Co-occurrence Map. The best displacement set was the one that minimized the classification error. Almost 14 000 maps were trained to obtain the results in this study. This exhaustive search among the displacement sets was made possible by using the tree-structured variant (Koikkalainen 1995) of the Self-Organizing Map.

The map reference vectors tend to concentrate along those directions in the co-occurrence space for which the variances are largest. This may have adverse effects on the classification. Therefore, feature selection was also done with decorrelated co-occurrences whose variances were equalized. The performance of features selected for reduced co-occurrence histograms was compared with the performance of the best features selected for channel histograms (Unser 1986) and for standard two-dimensional co-occurrence matrices.

2 Texture Images and Co-occurrence Vectors

The experiments were carried out with the twelve Brodatz textures shown in Fig. 1. The texture images of 256 by 256 pixels were divided into 64 nonoverlapping 32-by-32-pixel blocks, one half of which were randomly selected into a design set and the other half into a test set. Ten different randomly selected design and test sets were used in the statistical evaluation of each feature set. To remove characteristics due to variation in illumination, the first-order pixel distribution of each texture block was equalized to 256 gray levels. The gray-level range was set to $[-127.5, 127.5]$.

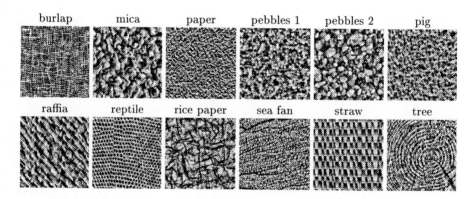

Fig. 1. Twelve Brodatz textures used in the study.

As the starting point for forming the co-occurrence vectors, consider a 3 by 3 square of pixels in a texture block, centered at location $\mathbf{x}$. Let us denote the gray levels of the 9 pixels, scanned row by row, by $s_i(\mathbf{x})$, $i = 1, 2, ..., 9$. They define a 9-dimensional co-occurrence vector $\mathbf{s}$. In some experiments, the vector was averaged, i.e., the local pixel average was subtracted from each vector component:

$$s_i := s_i - \sum_{j=1}^{9} s_j/9 \ . \tag{1}$$

Altogether 120 000 such co-occurrence vectors were randomly sampled from a design set of all twelve textures and used for the unsupervised training of a tree-structured self-organizing map. The sample covariance matrix was also determined from the same sample vectors.

3 Texture Classification

3.1 Histogram Collection with the Co-occurrence Map

The tree-structured self-organizing map was realized with four square layers $l = 1, 2, 3, 4$, each containing $U_l = 4^l$ nodes. Each node in a layer was connected with four neighboring child nodes in the next layer. The top layer was trained first. At the end of the training, the reference vectors in the layer were fixed and the values of the vectors were inherited by the child nodes in the next layer which was then trained. The next layer was initialized by giving each reference vector the average value of its four-connected neighbor nodes. During the training, the search for the best-matching reference vector always started from the top layer and proceeded downwards by limiting the search to the child nodes of the current best-matching node and its four neighbor nodes. During the training of the layers, the neighbors of the best-matching node were not updated. However, ordering of the reference vectors emerged due to the initialization and limited search procedures in the tree structure (Koikkalainen 1995).

The textures were described as histograms using the trained map nodes in the lowest layer as bins. Each new co-occurrence vector was mapped to the best-matching reference vector in the lowest layer found by the tree-search. A histogram was collected as

$$P(n) = \#\{\mathbf{x} \mid \|s(\mathbf{x}) - \mathbf{m}_n\| = \min_i \|s(\mathbf{x}) - \mathbf{m}_i\|\}, n = 1, \ldots, U_4 , \qquad (2)$$

with $\mathbf{x}$ moving over all possible locations in a texture block.

3.2 Feature Selection

In this study, all combinations of pruned pixel sets were used to train a number of Co-occurrence Maps. Also, three possibilities were tried for the matching between a sample co-occurrence vector and the map nodes: the Euclidean distance, the Mahalanobis distance, and the Euclidean distance in a rotated coordinate system determined by the principal component transformation. Both pixel set pruning and the modified distance calculations can be conveniently expressed under a unified mathematical notation.

Pruning of the displacement set can be done by introducing a binary weight matrix

$$\mathbf{W} = \begin{bmatrix} w_1 & & 0 \\ & \ddots & \\ 0 & & w_9 \end{bmatrix} , w_i = 0, 1 , \qquad (3)$$

which determined the components of vector s included in the Euclidean distance computation for the best-matching unit:

$$D_E^2 = (s - m_i)^T W^2 (s - m_i) \ . \tag{4}$$

Setting one or more of the weights w_i to zero is equivalent to removing the corresponding pixels from the co-occurrence vector (and the corresponding elements from the map weight vectors) and thus produces pruned displacement sets from the original set. Altogether, there were $2^9 = 512$ different weight matrices (displacement sets), but only those sets in which at least three pixels remained were considered.

For the Euclidean distance, the average-subtracted co-occurrence vectors defined in (1) were used, while the Mahalanobis and Euclidean distance in a rotated coordinate system used the original pixel values. The Mahalanobis distance equalizes variances in the directions of the principal axes of the co-occurrence distribution. Therefore, it may emphasize discriminating features with small variance. The binary diagonal matrix W then chooses the variance-equalized principal components included in the Mahalanobis distance computation:

$$D_M^2 = (s - m_i)^T U W \Lambda^{-1} W U^T (s - m_i) \ . \tag{5}$$

Matrices U and Λ are obtained by diagonalizing the sample covariance matrix of the distribution, $U^T \Sigma U = \Lambda$. The variance equalization by the Mahalanobis distance was further compared with the distance function which only rotated the coordinate axes to match the principal axes:

$$D_P^2 = (s - m_i)^T U W^2 U^T (s - m_i) \ . \tag{6}$$

If $W = I$ then (4) and (6) are equal.

Binary values w_i in W were selected with exhaustive search by evaluating all weight combinations with three or more values set to one (if only two values are nonzero, this corresponds to commonly used co-occurrence matrices, see Sect. 3.3). The minimized cost function was the classification error computed for texture blocks in a design set. For each W, a co-occurrence map was trained with the vectors sampled from the design set. Texture-descriptive histograms were collected according to (2) for all texture blocks. The classification error was measured with the leave-one-out method: one sample histogram was selected at a time and it was tested using the other histograms as models. Sample histogram P was identified with model P_m using the log-likelihood function

$$L(P, P_m) = \sum_{n=1}^{U_4} P(n) \log P_m(n) \tag{7}$$

and the 3-nearest-neighbors classification rule.

3.3 Channel Histograms and Co-occurrence Matrices

Channel histograms estimate one-dimensional marginal densities of a decorrelated feature distribution. They reduce a k-dimensional co-occurrence distribution to k one-dimensional histograms. The reduction would be almost lossless if the decorrelated co-occurrences were independent as in the case of Gaussian distribution. The 3-by-3-pixel co-occurrence vectors were decorrelated to nine channels with variances λ_i (which also form the diagonal of Λ in (5)). The channel deviations were roughly $[\sqrt{\lambda_1} \cdots \sqrt{\lambda_9}] \approx [190\ 76\ 69\ 31\ 25\ 20\ 14\ 12\ 9]$. Channel histograms with thirty-two bins were collected from each texture block. The bins for channel i were uniformly located within the range $[-3\sqrt{\lambda_i}, 3\sqrt{\lambda_i}]$. A subset of the nine histograms was concatenated to form texture descriptors. The subset selection was carried out to minimize classification error in the same way as was done with the map histograms.

In a 3-by-3-pixel neighborhood, a pixel pair for co-occurrence matrix computation can be chosen from twelve different displacements whose best subset was searched for. Texture histograms were formed by concatenating 8-by-8-bin co-occurrence matrices collected with the displacements. The subset selection was carried out to minimize classification error in the same way as for the other texture descriptors.

4 Results

Ten exhaustive searches for the best components of the co-occurrence vectors were carried out, using ten different divisions of the texture blocks into design and test sets. The ensuing selection of individual vector components with the different methods is shown in Table 1. The co-occurrence map method with the Mahalanobis distance and the channel histogram method, both of which gave an equal weight for each principal direction, preferred the direction with the smallest variance (w_9) to most other directions. The co-occurrence map method with distance D_P always discarded the direction of the highest variance (w_1), i.e., the direction corresponding roughly to the local pixel average.

Table 1. Component occurrences in ten selections and average accuracies with selected components.

Method	Components									Accuracy (std)	
	w_1	w_2	w_3	w_4	w_5	w_6	w_7	w_8	w_9	Design set	Test set
co-occurrence map with D_E	1	9	2	5	9	4	6	3	4	89.9 (0.9)	87.4 (1.0)
co-occurrence map with D_M	9	9	9	10	5	8	3	2	10	92.1 (0.9)	90.9 (1.0)
co-occurrence map with D_P	0	10	10	9	9	9	3	6	5	88.3 (1.2)	86.2 (1.6)
channel histograms	8	10	10	8	2	9	4	5	9	86.7 (0.9)	83.8 (1.7)
co-occurrence matrices										84.9 (1.6)	82.3 (1.3)

With the selected features, the test blocks were classified with a 3-nearest-neighbors classifier formed from the design blocks. Average classification accuracies obtained with the ten optimally selected sets for each method are shown in Table 1. The results obtained with the methods differed significantly from each other ($p < .01$; two-sided Wilcoxon signed-rank test for paired observations) with one exception: the test set accuracies of co-occurrence map methods with distances D_E and D_P were similar. All three map methods performed clearly better than channel histograms or co-occurrence matrices. The highest accuracies were obtained with the co-occurrence map method and the Mahalanobis distance; not only the average but also each of the ten accuracies determined for the design and test sets exceeded the corresponding accuracies obtained with the other methods.

5 Conclusions

The Tree-Structured Self-Organizing Map was successfully applied to the collection of reduced multidimensional histograms during feature selection. The extensive feature selection was carried out with rather complicated texture descriptors obtained with the map to maximize the performance of the final classifier. The selection method required no assumptions about feature distributions. In the present study, features were selected using binary weighting. We have also made feature selections with several weight levels (e.g., $w_i = 1.0, 0.7, 0.5, 0.35, 0.25, 0$). Evaluations of tens of thousands of weight combinations, searched with a genetic algorithm, showed no improvement of classification accuracy, however. The results show that multidimensional co-occurrence histograms reduced by the Self-Organizing Map gave higher classification accuracies than channel histograms or co-occurrence matrices. It was also observed that decorrelation and equalization of variances of gray-level vector components benefited the texture classification.

References

Kohonen, T.: Self-organizing maps. Springer-Verlag (1995).

Koikkalainen, P.: Fast deterministic self-organizing maps. Proceedings of the International Conference on Artificial Neural Networks, Paris, France, October 9–13, 1995, pp. 63–68.

Oja, E., Valkealahti, K.: Compressing higher-order co-occurrences for texture analysis using the self-organizing map. Proceedings of the IEEE International Conference on Neural Networks, Perth, Western Australia, November 27 – December 1, 1995, pp. 1160–1164.

Unser, M.: Local linear transforms for texture measurements. Signal Processing **11** (1986) 61–79.

Valkealahti, K., Visa, A.: Simulated annealing in feature weighting for classification with learning vector quantization. Proceedings of the 9th Scandinavian Conference on Image Analysis, Uppsala, Sweden, June 6–9, 1995, pp. 965–971.

Comparison of View-Based Object Recognition Algorithms Using Realistic 3D Models

V. Blanz[1,2], B. Schölkopf[1,2], H. Bülthoff[1], C. Burges[2], V. Vapnik[2], T. Vetter[1]

[1] Max–Planck–Institut für biologische Kybernetik,
Tübingen, Germany, E–mail volker@mpik-tueb.mpg.de
[2] AT&T Bell Laboratories, Holmdel, NJ, USA

Abstract. Two view-based object recognition algorithms are compared: (1) a heuristic algorithm based on oriented filters, and (2) a support vector learning machine trained on low-resolution images of the objects. Classification performance is assessed using a high number of images generated by a computer graphics system under precisely controlled conditions. Training- and test-images show a set of 25 realistic three-dimensional models of chairs from viewing directions spread over the upper half of the viewing sphere. The percentage of correct identification of all 25 objects is measured.

In computer vision, view–based models of object recognition have become more and more influential in recent years. Moreover, psychophysical evidence has been found for a view–based representation of objects in humans (Bülthoff and Edelman, 1992). Unlike viewpoint-invariant representations using structural descriptions (e.g. Marr and Nishihara, 1978), viewpoint-dependent models do not require a three-dimensional representation (Poggio, Edelman 1990, Lades et.al., 1993). The present study compares two recognition algorithms that are explained in the following sections.

1 Recognition by Oriented Filters

If a three-dimensional object is rotated about a frontoparallel axis, orthographic projections of surface points will move in the image plane in a direction perpendicular to the axis. To a great extent this also applies to perspective projection under realistic viewing conditions. Thus, images of an object can be made insensitive to rotations about a particular frontoparallel axis by lowpass filtering in one direction.

In order to compensate for relatively large displacements, the lowpass filter operation extinguishes much of the high spatial frequency structure in one direction. Due to a centering process described below, the lowpass filtering has to account also for displacement components along the axis of rotation. As a consequence, performance cannot be improved significantly by choosing image resolutions higher than 16x16 pixels. In order to retain some of the high spatial frequency information from the initial image, the representation also contains images with an edge detection performed before downsampling.

The algorithm uses a set of stored views of each object. They are preprocessed and stored in a representation of low resolution. To classify a test image, it is preprocessed in the same way and compared one by one with all of the stored views.

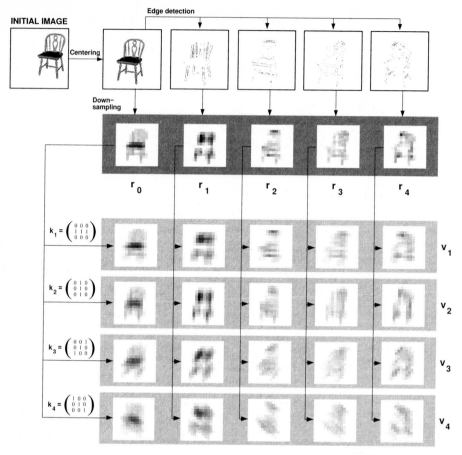

Fig. 1. Recognition by Oriented Filters: In preprocessing, a representation consisting of five low-resolution images $r_0...r_4$ is generated. For a comparison, these are convoluted with matrices $k_1...k_4$. The results are combined to $5*16*16$-dimensional vectors $v_1...v_4$. Euclidean distance between vectors v_i is used to measure similarity between images.

The algorithm performs the following steps:

Preprocessing:

1. **Centering:** The picture is centered with respect to the center of mass of the binarized image. All objects are shown on a white background, so the binarized image segregates figure from ground.
2. **Edge detection:** Four one-dimensional differential operators (vertical, horizontal, diagonal) are applied to the image and the modulus is taken.
3. **Downsampling:** Reducing resolution of all five images from 256x256 to 16x16 pixels, we obtain images $r_0...r_4$. In this representation, each view requires $5*16*16=1280$ bytes. In our simulation, we stored 25 views per object, summing up to a total of 32kB per object.

Image comparison: To compare a given image that has been preprocessed to vectors $r_0...r_4$ with a stored view $r'_0...r'_4$, we perform the following steps:

1. **Oriented filters:** Images are lowpass filtered in four directions, using the filter matrices $k_1...k_4$ shown in figure 1. Each of them is applied to all five low resolution images of a view. The resulting images are combined to a vector

$$v_i = (k_i \otimes r_0, \ k_i \otimes r_1, \ k_i \otimes r_2, \ k_i \otimes r_3, \ k_i \otimes r_4), \ i = 1...4. \qquad (1)$$

2. **Euclidean distance:** As a measure of similarity of two views, we compute sums of squared differences of corresponding pixel values. This yields four distance values $$d_i = \|v_i - v'_i\|, \ i = 1..4. \qquad (2)$$

Training: During training, a set of views of each object is preprocessed and stored. For each object, the same viewing directions are used. The selection of these views is done externally, but different ways for an automatic selection process are conceivable. For each of the four distances d_i of each view of each object, a threshold is calculated and stored. It is found by comparing the given view with all views of all other objects and choosing a value that leads to a false alarm rate below a fixed value on the training set.

Decision rule: If two images show the same object, at least one distance value of one object should be below threshold. It has proved to be most reliable to compute distance values with all stored views and then choose the object with the highest number of below-threshold distances.

2 Support Vector Learning Machines

To construct decision rules that generalize well, the support vector algorithm uses the Structural Risk Minimization (SRM) principle (Vapnik, 1995). SRM is based on the result that the error rate on an independent test set is bounded by the sum of the training error rate and a term which depends on the so–called VC(Vapnik–Chervonenkis)–dimension of the learning machine. By minimizing the sum of both quantities, high generalization performance can be achieved. For linear hyperplane decision functions $f(\mathbf{x}) = \text{sgn}((\mathbf{w} \cdot \mathbf{x}) + b)$, the VC–dimension can be controlled by controlling the norm of the weight vector $\mathbf{w}$ (Vapnik, 1995). Given training data $(\mathbf{x}_1, y_1), \ldots, (\mathbf{x}_\ell, y_\ell)$, $\mathbf{x}_i \in \mathbf{R}^N, y_i \in \{\pm 1\}$, a separating hyperplane which generalizes well can be found by minimizing (Cortes & Vapnik, 1995)

$$\|\mathbf{w}\|^2 + \gamma \cdot \sum_{i=1}^{\ell} \xi_i \qquad (3)$$

subject to
$$\xi_i \geq 0, \quad y_i \cdot ((\mathbf{x}_i \cdot \mathbf{w}) + b) \geq +(1 - \xi_i) \quad \text{for } i = 1, \ldots, \ell \qquad (4)$$

(γ is a constant which determines the trade–off between training error and VC–dimension).

The solution of this problem can be shown to have an expansion $\mathbf{w} = \sum_{i=1}^{\ell} \lambda_i \mathbf{x}_i$, where only those λ_i are nonzero which belong to an $\mathbf{x}_i$ precisely meeting the constraint (4) — these so–called *Support Vectors* lie closest to the

Fig. 2. The dataset of 25 3D-models of chairs.

decision boundary. The λ_i are found by solving the quadratic programming problem defined by (3) and (4). Finally, this method can be generalized to non-linear decision surfaces by first mapping the input nonlinearly into some high–dimensional space, and finding the separating hyperplane in that space (Boser, Guyon & Vapnik, 1992). This is achieved implicitly by using different types of symmetric functions $K(\mathbf{x}, \mathbf{y})$ instead of the ordinary scalar product $(\mathbf{x} \cdot \mathbf{y})$. This way one gets decision functions

$$f(\mathbf{x}) = \mathrm{sgn}\left(\sum_{i=1}^{\ell} \lambda_i \cdot K(\mathbf{x}, \mathbf{x}_i) + b\right).\tag{5}$$

In the present study, we used polynomial classifiers $K(\mathbf{x}, \mathbf{y}) = (\mathbf{x} \cdot \mathbf{y})^n$ of degree $n = 5$, and a value $\gamma = 10$. Other choices of K allow the implemetation of neural networks and radial basis function classifiers. In handwritten digit recognition, the support vector set has empirically been shown to be largely independent of the type of support vector machine constructed, and it contains all the information necessary to solve the classification task (Schölkopf, Burges, and Vapnik, 1995).

To construct the multi–class classifier needed for our purposes, we simply combined binary classifiers which were trained to recognize individual objects. This is done by choosing as the output of the multi–class classifier the class where the argument of the decision function (5) is maximal.

3 Experimental Results

The object database consisted of 25 different 3D-models of chairs (figure 2). All of them had a uniform grey surface. They were rendered in perspective projection in front of a white background on a Silicon Graphics workstation using Inventor software. The initial images had a resolution of 256x256 pixels.[1] In all viewing directions, image plane orientation was such that the vertical axis of the object was projected in an upright orientation. Both in training and test set, only views on the upper half of the viewing sphere were used. The training set consisted of 89 equally spaced views of each object. The test set contained 100 random views of each object. In both algorithms, the images ($r_0...r_4$) were rescaled based on their variances on the training set. Performance was measured in terms of correct identification of all 25 objects from all viewing directions.

Results for oriented filters: We stored only 25 equally spaced views per object, but used the full training set for calculation of the thresholds. Under these conditions, a classification error of 4.7% was achieved. Ignoring all data ($r_1...r_4$) from edge detection and relying only on r_0 increased the error to 21%.

Results for a Support Vector Learning Machine: Trained on the rescaled images, the support vector machine had an error rate of 1.0%. Without rescaling, error rate increased to 1.6%. Using only the images, i.e. r_0, the error was 8.4%.

Discussion: Images generated by means of computer graphics provide a useful basis for studying and comparing object recognition algorithms. However, generalizations of absolute performance values from simulations to real-world problems may be problematic. For the algorithms used in this work, noise should have only small effect because most of the processing is performed on a low spatial frequency domain. Much more impact has to be expected from a realistic, non uniform background. On the other hand, objects with different albedo and color can faciliate recognition significantly.

For both algorithms, performance data for this relatively difficult classification task were below 5% – with a fully connected feed-forward neural network with one hidden layer, we were not able to get error rates below 10%. Given the simple design of the oriented filter algorithm, its recognition rate was surprisingly high. A closer investigation of some of the image vectors ($r_0...r_4$) shows

---

[1] For benchmarking with other recognition algorithms, we will make the set of images available on our ftp-server.

that vectors of a single object change drastically with viewpoint. As it seems, the support vector machine is very much capable of dealing with such a complicated decision surface.

References

Boser, B. E., Guyon, I. M., Vapnik, V.: A training algorithm for optimal margin classifiers. Fifth Annual Workshop on Computational Learning Theory, Pittsburgh ACM (1992) 144–152.

Bülthoff, H. H., Edelman, S.: Psychological support for a two–dimensional view interpolation theory of object recognition. Proc. Natl. Acad. Sci. USA **89** (1992) 60–64

Cortes, C., Vapnik, V.: Support Vector Networks. Machine Learning **20** (1995) 1–25

Lades, M, Vorbrüggen, J. C., Buhmann, J., Lange, J., von der Malsburg, C, Würtz, R. P., Konen, W. Distortion Invariant Object Recognition in the Dynamic Link Architecture. IEEE Trans. Comp. **42** 3 (1993) 300–311

Marr, D., Nishihara, H. K. Representation and recognition of the spatial organization of three dimensional structure. Proceedings of the Royal Society of London B, **200** (1978) 269–294

Poggio, T., Edelman, S. A network that learns to recognize three-dimensional objects. Nature **343** (1990) 263–266

Schölkopf, B., Burges, C., Vapnik, V.: Extracting support data for a given task. In: Fayyad, U. M., Uthurusamy, R. (eds.): Proceedings, First International Conference on Knowledge Discovery & Data Mining, AAAI Press, Menlo Park, CA (1995)

Vapnik, V.: The Nature of Statistical Learning Theory. Springer Verlag, New York (1995)

Color-Calibration of a Robot Vision System Using Self-Organizing Feature Maps

Hubert Austermeier, Georg Hartmann, Ralf Hilker

University-GH Paderborn, FB-14 GET, 33098 Pohlweg 47-49, Germany
email: {austermeier,hartmann,gethilk}@get.uni-paderborn.de

Abstract. This paper presents an application of Kohonen's self-organizing feature maps (SOM) for solving the problem of color constancy. The main problem is to evaluate the transformation between collections of color-points forming differently shaped clouds in color space under changing illumination. The main idea is to embed appropriate 3D coordinate systems into these clouds by self-organization, and so to be able to find corresponding color points within different clouds. The difference of the locations of corresponding neurons in two SOMs is then an approximation for the particular color shift belonging to the difference between a reference illumination and a given illumination. The observed shifts provide a table of vectors in the color space, which in correction steps can be applied to color images taken under a given illuminant.

1 Introduction

Color image processing is a strong growing field in the environment of computer vision, not at least due to the rapidly increasing power of computer systems. Working in this field, one soon encounters the problem of color constancy. Color constancy refers to the ability to correctly classify color patches independent of the given illumination. This ability is limited of course, e.g. in the case of monochromatic illumination.

E. Land (the inventor of the Polaroid film) has made lots of experiments with human observers to study color constancy. He developed his well-known retinex theory which suggests a detailed procedure to produce color constancy in flat mondrian-like color pictures [Land 83]. [Zeki 83] has made investigations in different layers of the visual cortex of primates and found color-specific cells in V4 which show a somehow illuminant independent behaviour. [Courtney 94] developed an ANN with 3 different layers resembling the most important stages of the primate color perception system. [Marszalec 94] worked on a more technical approach to color calibration of camera systems. He used suitable basis functions to express a given illumination as linear combination and derives a illuminant independent color description. Gross analyses shape, position and orientation of clouds of colorpoints in a 3D color space and derives global linear mappings onto a normalised representation [Gross 95]. Our contribution shows a procedure to approximate a local nonlinear mapping between corresponding points of color clouds under different illumination conditions. Embedding of appropriate coordinate systems by self-organization provides these corresponding points within differently shaped clouds.

2 Topology preservation under changing illumination

Under different illumination conditions we take images of a calibration color table with all available colors. If we omit the spatial information of a so builded color image and only look at the distribution of the colors, this distribution in the color space (e.g. RGB) has a cloud-like shape. The detailed shape and location of such a cloud of colorpoints in the color space is representative for a special illumination.

We made the following observation: Although the location, size and shape of the clouds may change with different illuminations, the inner structure and topology is stable. So there exists a topology preserving mapping between two clouds of colorpoints, that means neighbouring color points in the first cloud are neighbouring in the second cloud too. In other words, if know a point in a cloud under given illumination, we can find the corresponding point in the cloud with reference illumination. By this procedure we can calibrate a given RGB value to the RGB value under reference illumination.

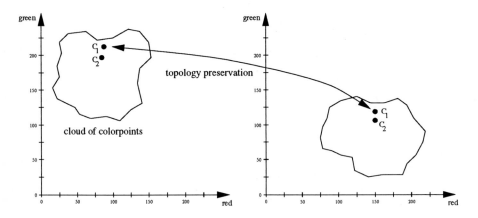

Fig. 1. Clouds of colorpoints derived from two color images taken under different illumination. Only the shape of the clouds in a 2D-projection of the RGB color space is shown. The topology preservation between the two clouds is 0 by the arrow.

3 Applying SOMs to the problem

The main problem is to find corresponding points within different clouds. Though there are different possibilities to find these correspondences (e.g. by referencing the position in the calibration table), it proved to be most reliable to embed appropriate coordinate systems into these clouds. We start with orthogonal $25x25x25$ grids and adapt these grids to the shapes of the individual clouds. We solve this adaption problem by using Kohonen's self-organizing feature maps [Kohonen 89], assigning volume elements of the cloud to neurons of the deformed grid (Fig. 2). It should be noted that we do not use SOMs to achieve dimensionality reduction, we only use the property of topology preservation. An 3D-SOM adapted to a cloud of colorpoints provide an accurate as possible representation of the inner parts of the cloud. In the actual implementation there are about 15.000 neurons representing 200.000 and more color points.

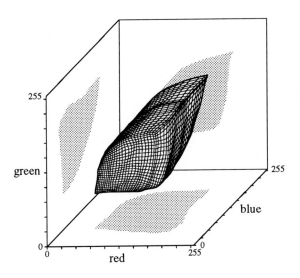

Fig. 2. A 3D feature map adapted to a cloud of colorpoints. Only the shape of the cloud is visible, not the color points themselves.

Let us now have a look at one particular adaption step in the self-organization phase. There is a fixed set of color points (that of the clouds) and iteratively we take a random one $f_{in} = (f_1, f_2, f_3)$ to present it to the feature map. The neuron s with highest activation to that input signal is called the winner neuron. The weight vector of s and those of its local neighborhood will then be modified to adapt themselves to the given signal f_{in}. The equation for weight vector modification of $\mathbf{w} = (w_1, w_2, w_3)^T$ is given as follows:

$$w_i(t+1) = w_i(t) + \epsilon(t) \cdot h(r, s, t) \cdot (f_i - w_i(t)) \quad \text{where} \quad i \in \{0, 1, 2\}. \quad (1)$$

The learning rate $\epsilon(t)$ and the size of local influence $h(r, s, t)$ are dependent on the actual number of learning steps t:

$$\epsilon(t) = \epsilon_{max} \left(\epsilon_{base}\right)^{t/T_{max}} \tag{2}$$

$$h(r, s, t) = e^{-\dfrac{(\|r - s\|)^2}{2 \cdot \sigma(t)^2}} \quad \text{where} \quad \sigma(t) = \sigma_{max} \left(\sigma_{base}\right)^{t/T_{max}}. \tag{3}$$

As mentioned earlier and in opposition to many applications using SOMs, we start with orthogonal grids (Fig. 3) whose nodes $\mathbf{w}$ are initially placed in the center region of the RGB color space:

$$\mathbf{w} = \begin{pmatrix} x \left(50/(d-1)\right) + 100 \\ y \left(50/(d-1)\right) + 100 \\ z \left(50/(d-1)\right) + 100 \end{pmatrix} \quad \text{where} \quad 0 <= x, y, z <= d - 1, \tag{4}$$

and where d is one grid dimension (25 here). As we use this form of initialisation the adaption process takes less computation time than using a random initialised map because of the already presorted embedding.

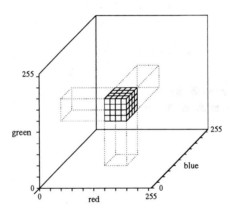

Fig. 3. Initial orthogonal grid.

We now build two SOMs both of which adapt themselves to two specific clouds of colorpoints; one of the above mentioned calibration table under a reference illumination and the other of the same table under an actual illumination for which the correction has to take place. The distance of corresponding grid locations in the two maps represents a correction vector for a set of color points belonging to that volume element. The main steps of the method are:

1. From a suitable calibration table under reference illumination (e.g. daylight) an image is taken and the cloud of colorpoints in the color space is build.

2. A SOM with dimension $25x25x25$ adapts it selves to this specific cloud of colorpoints. We call this map the reference map M_{ref} and use it in a later correction step.

3. For an actual illumination where a color correction has to be performed, we take an image of the same calibration table as in step 1 and build the cloud of colorpoints in the color space.

4. The most important step now is to adapt M_{ref} to this new distribution. This resulting map contains the information about the color shifts under the changing illumination. These shifts correspond to translation vectors of the lattice neurons. A translation vector connects two lattice neurons, one in each map with same grid coordinates.

5. In order to perform corrections on color images taken under the actual illumination, i.e. to achieve color constancy, we next put all translation vectors in a correction table. For each particular color in a given image the nearest translation vector in the table is determined and the color is transformed according to this translation vector.

4 Results

In our experiments we took color images under various daylight environments, under various artificial light sources and even with corruptions made by the camera system.

An example of a successful correction is shown in Fig. 4. The mean error between the correct and the corrupted image was about 80 units in the color space (the average euclidian distance between corresponding colors). After the correction, this error was reduced to under 5 units. For the SOMS we use the following parameters:

SOM :	25 x 25 x 25
T_{max} :	40.000
ϵ_{max} :	0.3
ϵ_{base} :	0.1
σ_{max} :	6.0
σ_{base} :	0.1

We also adapted grids with more and less than 25^3 neurons. The accuracy of color correction decreased rapidly with lower resolution, and on the other hand there was only a marginal increasing accuracy with higher grid resolution.

The calculations were performed on a Sparc Station 20 (2 Hyper-Sparc CPUs, 125 MHZ). The correction of one color image takes 35 seconds, but this speed will be increased by appropriate selection of pre-shaped SOMs.

As a conclusion we can say that this calibration scheme provides almost illumination independent operation of our robot vision system.

Fig. 4. The correction of a corrupted advertising poster. The figure only shows the blue channel of the poster under normal daylight (left), a reddish corruption performed by the camera (middle) and the corrected image after applying the SOM correction (right). As we can only see one color channel as a grey image, we can make a qualitative judgment only. Nevertheless the efficiency of the mechanism is evident.

References

[Austermeier 95] Austermeier H., Hartmann G. (1995). *Farbkonstanz mit künstlichen neuronalen Netzen.* 40. Int. Wiss. Kolloq. Illmenau 1995, Bd. 1, 755-760.

[Bollmann 95] Bollmann M., Mertsching B. (1995). *Entwicklung eines Gegenfarbemodells für das Neuronale-Active-Vision-System NAVIS.* DAGM 1995, 456-463

[Courtney 94] Courtney S., Finkel L.H. & Buchsbaum G. (1994). *Network Simulations of Retinal and Cortical Contributions to Color Constancy.* Vision Research, 35, 413-434.

[Kohonen 89] Kohonen T. (1989). *Self-Organization and Associative Memory.* Springer-Verlag.

[Land 83] Land H.E (1983). *Recent advances in retinex theory and some implications for cortical computations: Color vision and the natural image.* Proc. Nat. Acad. Sci. U.S. 80, S. 5163ff

[Marszalec 94] Marszalec E. & Pietikäinen M.(1994). *On-Line Color Camear Calibration.* IAPR 1994, 232-237.

[Gross 95] Pomierski T., Gross H.M. (1995). *Verfahren zur empfindungsgemäßen Farbumstimmung.* DAGM 1995, 473-480

[Ritter 92] Ritter H., Martinez T., Schulten K. (1995). *Neural computing and self-organizing maps.* Addison Wesley, Reading, Mass.

[Young 1802] Young T. (1802). *On the theory of light and colours.* Philos. Trans. Royal Society London, 1802, 12-48

[Zeki 83] Zeki S.M. (1983). *Colour coding in the cerebral cortex: The reaction of cells in the monkey visual cortex to wavelength and colours.* The Journ. of Neurosc. 9.

Neural Network Model for Maximum Ozone Concentration Prediction

Gonzalo Acuña[1] , Héctor Jorquera[2] and Ricardo Pérez[2]

[1]Universidad de Santiago de Chile, CECTA; [2]Pontificia Universidad Católica de Chile, Departamento de Ingeniería Química y Bioprocesos.

Abstract- A neural network dynamic model was used for predicting maximum ozone (O_3) concentration at Santiago de Chile. Learning and test data were collected during summer and springtime periods of 1990, 1992 and 1993. A neural network having $O_{3\,t}$, T_{t+1} (maximum air temperature) and T_t as inputs for predicting $O_{3\,t+1}$ was chosen because of its low test error. This neural network model greatly reduces the error coming from a pure persistence model when applied to the generalization set of data (1994). Long-term predictions results confirm the good concordance obtained between the observed and forecasted values thus showing the adequacy of neural networks to model the dynamics of this complex environmental phenomena.

Keywords- Neural networks, ozone, forecasting, dynamic modeling, predictive model.

1 Introduction

Ozone is a pollutant that abounds in large cities and is associated with respiratory problems, specially in children (Lippmann, 1989; Tilton, 1989). To minimize health harmful effects of high ozone levels, it is convenient to count on a system that permits to estimate the maximum impact values with sufficient precedence. In this way, persons particularly sensitive could take adequate measures to protect themselves.

It is well known that the chemistry of ozone generation is very complex, which includes more than 160 reactions and 70 species (Atkinson et al., 1982). Since these reactions are mainly photochemical and given that their kinetic constants are temperature sensitive, maximum ozone levels are reached during the summer, within and downwind the cities. That is to say, ozone level depends on air temperature, solar radiation, wind speed, other chemical species concentration, and on the mixing length.

There are in the literature some articles that describe and evaluate predictive systems for daily ozone concentrations forecasting. In many of these articles, the predictors are based only on the previous day maximum ozone concentration (McCollister and Wilson, 1975). No additional information is normally included in those predictors, such as atmospheric conditions or concentration of other pollutants. Furthermore, simple linear statistical models are used, despite that the phenomenology of the process is complex and highly non-linear. Robeson and Steyn (1990) compare three predictive models for maximum daily ozone concentration. They found that a first order linear dynamic model, with maximum air temperature as exogenous variable, presented the best results. In particular, this predictor reduces the forecasting error in about 10% compared with pure persistence. The model proposed by Robeson and Steyn requires the value of the maximum air temperature of the next day in order to predict the maximum ozone level.

In this work a simple model based on neural nets is developed to predict the maximum ozone daily concentration in the city of Santiago de Chile. As in the work of Robeson and Steyn cited above, only values of maximum O_3 concentration and maximum air temperature will be used as inputs in the model structure. It is assumed that a good temperature predictor is available. The objective of this work is to analyze the potential advantage of using neural nets to model environmental phenomena, which are characterized by highly non-linear and complex dynamics.

2 Methods

2.1 Data

Data used in this study were collected from one particular station (C) of a network of air quality and meteorological monitoring stations in Santiago's valley during springtime and summer periods (those of higher O_3 concentrations) of 1990, 1992, 1993 and 1994. Daily maximum ozone concentration (O_3) and maximum air temperature (T) were chosen as the relevant variables to be included in the model.

2.2 Neural Networks

A classical three layers (input-hidden-output) feedforward neural network (NN) with a sigmoidal non-linear activation function (for hidden and output neurons) (Bishop, 1994, Su *et al.*, 1992) was used as a dynamic model for ozone concentration prediction. Input and hidden layers included a bias node.

Dynamic Modeling

The canonical form for feedback networks, as presented by Nerrand *et al.* (1993), was used to model process dynamics. Following their nomenclature, maximum ozone concentration was chosen as the state variable, while maximum air temperature was the input variable.

Learning procedure

The learning algorithm included back-propagation of the output error associated with a quasi-Newton second order method to perform minimisation (Latrille *et al.*, 1994). Available data were splitted into three sets:

-learning set (01/01/90-03/31/90; 11/01/90-12/31/90; 01/01/92-03/31/92; 11/01/92-01/06/93) with 271 patterns,

-test set (01/07/93-03/31/93; 11/05/93-12/31/93) with 138 patterns,

-generalization set (G1: 01/01/94-03/04/94; G2: 11/01/94-12/19/94) with 112 patterns.

The number of neurons in the hidden layer (NH) was selected in order to have no more than 27 weights in the neural network (less than 1/10 of the number of learning patterns). Five different random initial values of weights were used in order to avoid bad local minima convergences. The learning procedure was stopped when the test error began to increase, thus preventing overtraining. A directed algorithm or series-parallel model was used for identification purposes (Nerrand et al., 1993). That is to say, the neural network was trained as a one-step ahead predictor associated with a NARX model of the process as defined by Sjoberg et al, 1995.

Statistical methods

Two indices were chosen for evaluating model performance. The Root Mean Square (RMS) and the Index of Agreement (IA), as proposed by Robeson and Steyn (1990).

$$RMS = \sqrt{\frac{\sum_{i=1}^{n}(o_i - p_i)^2}{N}} \qquad\qquad IA = 1 - \frac{\sum_{i=1}^{n}(o_i - p_i)^2}{\sum_{i=1}^{n}(|o'_i| - |p'_i|)^2}$$

where o_i and p_i are the observed and predicted values respectively at time i, while N is the total number of data. $p_i'=p_i-o_m$ and $o_i'=o_i-o_m$, with o_m being the observed mean value.

3 Results and Discussion

3.1 Model Selection

Like Robeson and Steyn (1990), the next day maximum temperature value was assumed to be well known. So, four input configurations were tested to predict next day maximum ozone concentration. They are presented in Table 1. The one including $O_{3\,t}$, T_{t+1} and T_t as inputs for predicting $O_{3\,t+1}$ was chosen because it had the least residual quadratic error over the normalized test set (TE).

Table 1: different NN models for predicting $O_{3\,t+1}$. The underline shows the selected model.

Inputs	$O3_t$, T_{t+1}	$O_{3\,t}$, $O_{3\,t-1}$, T_{t+1}	$\underline{O_{3\,t}, T_{t+1}, T_t}$	$O_{3\,t}$, $O_{3\,t-1}$, T_{t+1}, T_t
TE	0.7713	0.6840	0.4655	0.4938
Configuration	3-7-1	4-6-1	4-6-1	5-5-1

3.2 Model Performance

Being selected the neural network model over the data included in the learning and test sets, only the generalizing set will be used in order to evaluate model performance. As Robeson and Styne did, a comparison with a persistence forecast $(p_i=o_{i-1})$ will be also included.

Examining Table 2 it is clear that the neural network predictor (NN) greatly reduces the forecasting error coming from the pure persistence model. This one was absolutely unable to achieve acceptable results, which was not the case for Robeson and Styne. This fact reveals that the data were perhaps originated by very different underlying phenomena, thus preventing to do more accurate comparisons with them. We can only state that the RMS and IA values are slightly better than those obtained by Robeson and Styne (1990) although more input variables were necessary to achieve this performance. A better IA was obtained for the G2 set, thus making the regression estimates be near to the perfect prediction, as it can be seen in Figure 1.

Table 2: model performance for short-term and long-term (between brackets) predictions.

Generalization set	Index	Model	
		NN	Persistence
G1	RMS (ppb)	13.7 (13.8)	30.0
(01/01/94-03/04/94)	IA	0.83 (0.84)	0.49
G2	RMS (ppb)	15.4 (16.4)	32.2
(11/01/94-12/19/94)	IA	0.87 (0.87)	0.44

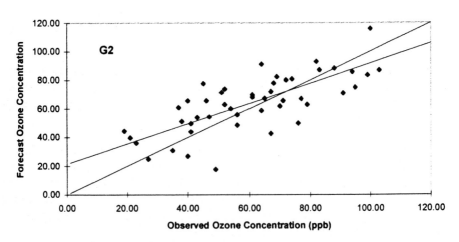

Figure 1: Short-term predictions of maximum ozone concentration vs. observed maximum ozone concentration during 1994, using a dynamic neural network model. Solid line indicates perfect prediction while the dashed line is an ordinary least square regression estimate.

These results show a good concordance between the observed and the short-term predicted values. Indeed, if we explore the capacity of the model to do long-term predictions only by providing the first observed ozone concentration and the observed daily temperatures, the RMS and IA obtained (Table 2) are as good as those coming from short-term predictions, thus confirming the adequacy of the neural network dynamic model.

Figure 2 shows the good results obtained over the G2 set when performing a short-term prediction. The neural network model was able to forecast the extreme values of the ozone variable.

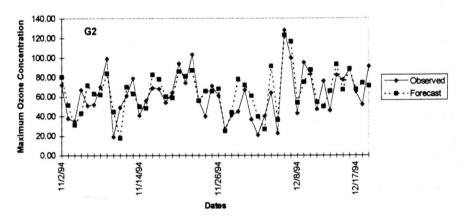

Figure 2: Observed (solid line) and short-term predictions (dashed lines) of maximum ozone concentration for the generalization set G2 (1994).

4 Conclusions

A dynamic neural network model was used to forecast maximum ozone concentration in Santiago de Chile. After testing different configurations the one which have $O_{3\,t}$, T_{t+1}, T_t as inputs, to predict $O_{3\,t+1}$ was chosen because of its low residual quadratic test error.

This neural network model has proven to greatly reduce the error coming from a pure persistence model. The RMS and IA obtained for short-term predictions over the generalization set of data were between 13.7 (ppb) and 15.4 (ppb) and between 0.83 and 0.87, respectively. These good results were confirmed when performing long-term predictions. In this case the RMS and IA obtained ranged respectively from 13.8 (ppb) to 16.4 (ppb) and from 0.84 to 0.87.

Taking into account the complexities of the underlying environmental phenomena the good results obtained show that the considered variables (O_3 and temperature) are the best adapted to perform O_3 forecasting, as it was already claimed by Robeson and Steyn (1990). Neural networks have proven to be a valuable tool for dynamic modeling of complex and bad known phenomena provided that a good set of data (rich in information about the underlying phenomena) is available and taking care of the number of parameters considered and the iterations performed in order to prevent overtraining during the identification phase.

Acknowledgements- this work was partially supported by a DICYT grant 95-95 AL.

References

R. Atkinson, A.C. Lloyd and L. Winges (1982), An updated chemical mechanism for hydrocarbon/NOx/SO2 photooxidants suitable for inclusion in atmospheric simulation models, *Atmos. Environ.*, 16, 1341-1355.

C. M. Bishop (1994), Neural networks and their applications, *Rev. Sci. Instrum*, 54(6), June 1994.

E. Latrille, G. Corrieu and J. Thibault (1994), Neural network models for final process time determination in fermented milk production, *Computers chem. Engng.* 18, 1171-1181.

M. Lippmann (1989), Health effects of ozone: To critical review, *J. Air and Waste Manage. Assoc.*, 39, 672.

G.M. McCollister and K.R. Wilson (1975), Linear Stochastic Models for Forecasting Daily Maxima and Hourly Concentrations of Air Pollutants, *Atmos. Environ.*, 9, 417-423.

O. Nerrand, P. Roussel-Ragot, L. Personnaz and G. Dreyfus (1993), Neural networks and non-linear adaptive filtering: unifying concepts and new algorithms. *Neural Comput.* 5, 165-199.

S.M. Robeson and D.G. Steyn (1990), Evaluation and Comparison of Statistical Forecast Models or Daily Maximum Ozone Concentrations, *Atmos. Environ.* 24B, 303-312.

J. Sjöberg, Q. Zhang, L. Ljung, A. Benveniste, B. Delyon, P. Glorennec, H. Hjalmarsson and A. Juditsky (1995), Nonlinear Black-box Modeling in System Identification: a Unified Overview, *Automatica*, 31,1691-1724.

H. Su, T. McAvoy and P. Werbos (1992), Long-Term Predictions of Chemical Processes Using Recurrent Neural Networks: A Parallel Training Approach, *Ind. Eng. Chem. Res.*, 31, 1338-1352.

B. Tilton (1989), Health effects of tropospheric ozone, *Environ. Sci. Technol.*, 23, 257.

Very Large Two-Level SOM for the Browsing of Newsgroups

Teuvo Kohonen, Samuel Kaski, Krista Lagus, and Timo Honkela

Helsinki University of Technology
Neural Networks Research Centre
Rakentajanaukio 2 C, FIN-02150 Espoo, Finland

Abstract. On January 19, 1996 we published in the Internet a demo of how to use Self-Organizing Maps (SOMs) for the organization of large collections of full-text files. Later we added other newsgroups to the demo. It can be found at the address http://websom.hut.fi/websom/. In the present paper we describe the main features of this system, called the WEBSOM, as well as some newer developments of it.

1 Introduction

When organizing large collections of free-form full-text document files that contain no keywords, e.g. the newsgroups in the Internet, it is difficult to base their analysis on traditional search expressions. The main information one can resort to in the classification of such documents is statistical.

SOMs of document collections have previously been constructed on the basis of their word histograms (published works are [5], [6], [7], [10], [11], [12]). Thereby, however, the size of the selected vocabulary cannot be large.

In other studies (cf., e.g., [1], [2], [8], [9], [11], and several others) it has also been found that short segments of text, such as triples of successive words, and in particular their statistical frequencies can effectively distinguish words according to their semantic roles. Self-organized maps of meaningful clusters of words are then formed.

In this work, in order to encode a document, we first formed a histogram of the *clusters* of its words on a SOM of the above type. Such histograms of different documents were then organized by a second SOM, which created another clustered display, namely, the document map. The various nodes in this second SOM can be seen to contain closely related documents, such as discussions on the same topics, answers to the same questions, calls for papers, publications of software, related problems (such as financial applications, ANNs and the brain), etc.

The first map contained 315 neurons with 270 inputs each. The second map had 49 152 neurons with 315 inputs each. The number of documents used for training and being mapped in this experiment was 131 500.

When provided with suitable means for communication, our system (dubbed the *WEBSOM*), can also be used as a kind of "agent" for the automatic searching of documents.

2   Detailed Description

2.1   Preprocessing of Text

First we eliminated some non-textual information (e.g., ASCII drawings and automatically included signatures) from the newsgroup articles. Numerical expressions and special codes were replaced by special symbols using heuristic rules.

To reduce the computational load, the words that occurred less than 50 times in the whole data base were neglected and treated as empty slots.

In order to emphasize the subject of an article and to reduce erratic variations due to different discussion styles, a number of common words was discarded from the vocabulary. There were 31 000 000 words processed. The size of the vocabulary, after discarding the rare words, was 22 000, from which 3 500 common words were still removed manually.

2.2   Formation of the Word Category Map

The first SOM, with 270 inputs and 315 map units, was formed and labeled using the whole text material as training data. Each word of the vocabulary was represented by a random code. Each "code," relating to word position i, was a random vector $x_i \in \Re^{90}$, every component of which was drawn from a uniform scalar distribution. The encoded words were concatenated into a single string of word symbols.

For each different (remaining) word in the corpora we then computed its *averaged statistical feature vector*

$$\begin{bmatrix} \mathrm{E}\{x_{i-1}|x_i\} \\ \varepsilon x_i \\ \mathrm{E}\{x_{i+1}|x_i\} \end{bmatrix} , \tag{1}$$

where i can now be any position in the string where the same code x_i of this word is found, and ε is a small numerical constant, e.g. equal to 0.2. These feature vectors were applied as inputs to the first SOM.

The nodes of the SOM were labeled by inputting the feature vectors once again and finding the winner node for each. A node was thus labeled by all the words the corresponding feature vector of which selected this node for a winner.

2.3   Formation of the Histograms

In the encoding of documents, the text of each document separately was preprocessed as described in Sec. 2.1. When the encoded string of its words was scanned, the occurrence of each word was counted and recorded at that node of the first SOM which was labeled according to this word.

If the documents belong to different groups, such as the newsgroups in the Internet, the counts can be further *weighted* by the information-theoretic *entropies* (Shannon entropies) of the words, defined in the following way. Denote by $n_g(w)$ the frequency of occurrence of word w in group g $(g = 1, \ldots, 20)$, and by $P_g(w)$ the probability that the word w belongs to group g. The entropy H of this word is defined as:

$$H(w) = -\sum_g P_g(w) \log P_g(w) \approx -\sum_g \frac{n_g(w)}{\sum_{g'} n_{g'}(w)} \log \frac{n_g(w)}{\sum_{g'} n_{g'}(w)} , \quad (2)$$

and the weight $W(w)$ of word w is defined as

$$W(w) = H_{\max} - H(w) , \quad H_{\max} = \log 20 . \quad (3)$$

2.4 Formation of the Document Map

Before using the *histograms* obtained in Sec. 2.3 as inputs to the second SOM, the *document map*, they were further *blurred* using a convolution with a symmetric Gaussian kernel, the full width at half maximum of which was two lattice units. The blurring increases invariance in classification. This map, with 315 inputs and 49 152 map units, was then computed as explained in Sec. 4.

2.5 Practical Computation of Large Maps

With large maps, both winner search and updating (especially of large neighborhoods in the beginning) are time-consuming tasks. With a parallel SIMD computer, such as the 512-processor neurocomputer CNAPS at our disposal, this can be made fairly rapidly, in a few dozens of minutes.

The local-memory capacity of the CNAPS, however, has so far restricted our computations to 315-input, 768-neuron SOMs. Recently [4] we have been able to multiply the sizes of the SOMs by two solutions: 1. Good initial values for a much larger map can be *estimated* on the basis of the asymptotic values of a smaller map, like the one computed with the CNAPS, by a local interpolation procedure. There is room for a much larger map in a general-purpose computer, and the number of steps needed for its fine tuning is quite tolerable. 2. In order to accelerate computations, the winner search can be speeded up by storing with each training sample an address pointer to the old winner location. During the next updating cycle, the approximate location of the winner can be found directly with the pointer, and only a *local search* around it needs to be performed. The pointer is then updated. In order to guarantee that the asymptotic state is not affected by this approximation, updating with a full winner search was performed intermittently, after every 30 training cycles.

3 Browsing Interface

The document space is presented at three basic levels of the system hierarchy: the map, the nodes, and the individual documents (Fig. 1). Any subarea of the map can be selected and zoomed by "clicking." One may explore the collection by following the links from hierarchy level to another. It is also possible to move to neighboring areas of the map, or to neighbors at the node level directly. This hierarchical system has been implemented as a set of WWW pages. They can be explored using any standard graphical browsing tool. A complete demo is accessible in the Internet at the address http://websom.hut.fi/websom/.

(a)   (b)   (c)

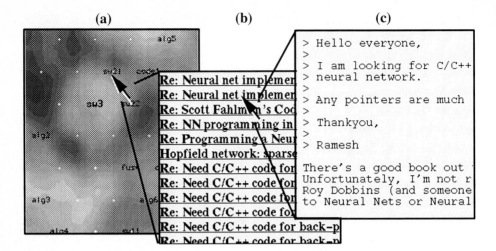

Fig. 1. Sample scenes of the WEBSOM interface. (a) Part of a zoomed document map display. The clustering tendency is visualised using the gray scale. (b) The map node contents. (c) An example of a newsgroup article picked up from the node.

4 Experiments

The word category map was computed with the CNAPS and fine-tuned on a general-purpose computer. The document map contained initially 768 units and it was computed with the CNAPS as discussed in Sec. 2.5 using 131 500 documents for training. It was then enlarged into 49 152 units by interpolation. Finally the whole document material was used to fine-tune the map on a general-purpose computer. The resulting document map is presented in Fig. 2, with separate images displaying the distribution of each group on the same map. The separation of the groups is presented by the confusion matrix in Table 1.

References

1. S. Finch and N. Chater, "Unsupervised methods for finding linguistic categories," in *Proc. of ICANN-92*, vol. 2, pp. 1365-1368, North-Holland, 1992.
2. T. Honkela, V. Pulkki, and T. Kohonen, "Contextual relations of words in Grimm tales analyzed by self-organizing map," in *Proc. of ICANN-95*, vol. 2, pp. 3-7, EC2 et Cie, 1995.
3. T. Kohonen, *Self-Organizing Maps*. Berlin, Heidelberg: Springer, 1995.
4. T. Kohonen, *Speedy SOM*, Report A33, Laboratory of Computer and Information Science, Helsinki University of Technology, 1996.
5. X. Lin, D. Soergel, and G. Marchionini, "A self-organizing semantic map for information retrieval," in *Proc. 14th. Ann. Int. ACM/SIGIR Conf. on R & D In Information Retrieval*, pp. 262–269, 1991.
6. D. Merkl, A. M. Tjoa, and G. Kappel, "A Self-Organizing Map that Learns the Semantic Similarity of Reusable Software Components," in *Proc. ACNN'94, 5th Australian Conference on Neural Networks*, Brisbane, Australia, pp. 13–16, 1994.

	1)	2)	3)	4)	5)	6)	7)	8)	9)	10)	11)	12)	13)	14)	15)	16)	17)	18)	19)	20)
1)	4157	48	143	135	72	288	151	23	52	148	167	165	114	29	225	178	194	126	99	189
2)	141	547	30	89	26	276	120	9	8	41	34	93	37	6	91	63	94	48	33	82
3)	263	23	1137	53	14	115	42	9	8	28	17	51	24	7	81	44	43	43	27	62
4)	345	100	62	770	33	115	103	10	25	102	125	71	97	42	119	60	84	79	40	103
5)	128	20	41	26	790	961	248	40	14	29	50	99	50	7	126	97	129	63	33	83
6)	234	84	79	56	277	5935	1095	186	40	110	137	323	162	21	274	209	291	158	94	224
7)	123	63	56	58	122	1389	3988	420	44	123	169	282	147	23	290	170	319	134	70	243
8)	49	23	37	18	53	597	846	660	20	43	64	95	57	13	110	68	134	66	31	95
9)	85	20	7	23	9	58	65	20	1751	570	547	196	238	99	184	182	137	33	31	200
10)	164	38	16	52	16	136	116	31	368	5688	931	350	464	173	291	321	289	72	66	407
11)	114	20	20	42	29	143	136	33	301	910	5632	390	606	188	353	316	304	81	52	327
12)	203	77	41	44	69	501	351	59	164	399	456	4967	394	91	484	383	590	108	106	499
13)	120	24	26	74	28	160	145	30	127	434	630	354	5535	293	593	705	254	63	38	352
14)	48	9	5	17	11	34	28	10	81	242	306	137	474	1608	234	205	95	23	23	143
15)	170	33	41	75	57	318	241	37	99	267	361	387	572	150	5736	597	286	68	75	416
16)	145	43	15	38	39	174	147	35	111	290	352	385	668	139	759	5833	250	55	57	448
17)	172	49	29	55	51	397	260	45	89	239	248	398	167	41	300	252	6578	142	126	355
18)	155	25	48	62	30	285	172	26	17	66	78	152	72	16	122	82	181	1414	72	115
19)	211	24	44	44	30	159	69	15	26	68	73	159	62	7	119	121	198	60	1145	129
20)	152	30	35	49	31	180	173	40	118	320	383	371	292	76	429	436	427	91	70	6263

Table 1. This confusion matrix indicates how the articles from a newsgroup are distributed to the map units dominated by the various groups (numbered as in Fig. 2). Each map unit was labeled according to the most frequent group in that unit. Each row describes into which groups the articles were distributed. Some similar groups like the philosophy groups (6, 7, and 8) and the movie-related groups (9, 10, and 11) contain similar discussions and are easily mixed.

7. D. Merkl, "Content-Based Document Classification with Highly Compressed Input Data," in *Proc. of ICANN-95*, vol. 2, pp. 239-244, EC2 et Cie, 1995.
8. R. Miikkulainen, *Subsymbolic Natural Language Processing: An Integrated Model of Scripts, Lexicon, and Memory*. Cambridge, MA: MIT Press, 1993.
9. H. Ritter and T. Kohonen, "Self-organizing semantic maps," *Biol. Cyb.*, vol. 61, no. 4, pp. 241-254, 1989.
10. J. C. Scholtes, "Unsupervised learning and the information retrieval problem," in *Proc. IJCNN'91*, pp. 18-21, IEEE Service Center, 1991.
11. J. C. Scholtes, *Neural Networks in Natural Language Processing and Information Retrieval*. PhD thesis, University of Amsterdam, Amsterdam, Netherlands, 1993.
12. J. Zavrel, *Neural Information Retrieval*, MA Thesis, University of Amsterdam, Amsterdam, Netherlands, 1995.

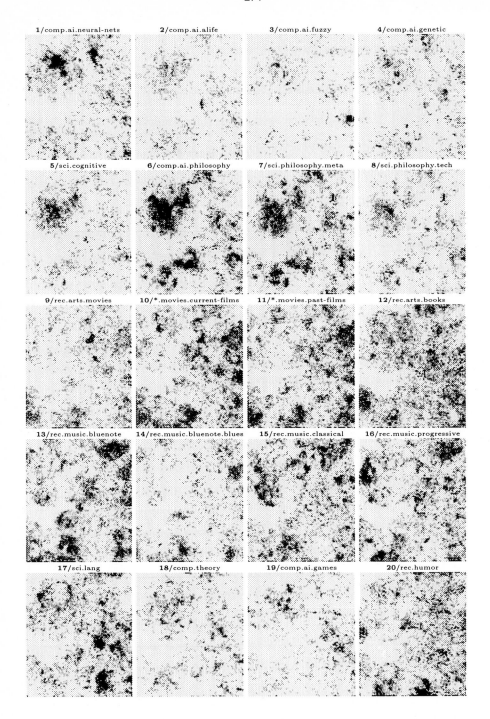

Fig. 2. Distributions of documents in different newsgroups on the WEBSOM of size $192 \times 256 = 49\,152$ units. Each small display only contains articles from a single newsgroup. The shade of the dots indicates the number of articles the unit contains: the darker the dot, the larger the number.

Automatic Part-Of-Speech Tagging of Thai Corpus Using Neural Networks

Qing Ma, Hitoshi Isahara, Hiromi Ozaku

Communications Research Laboratory
Ministry of Posts and Telecommunications
588-2, Iwaoka, Nishi-ku, Kobe, 651-24, Japan

Abstract. A cost-effective method for Part-Of-Speech (POS) tagging of a Thai corpus using neural networks is proposed. Computer experiments show that this method has a success rate of over 80% for tagging text of untrained data, and an error rate below 8%. These results are much better than those obtained by conventional table lookup methods. Some experiments comparing original and various modified back-propagation algorithms for training the neural network tagger are also conducted. Results of these experiments show that the learning algorithm with DBDB adaptation rule at a semi-batch update mode is the best one for tagging text in terms of convergence rate and computaional complexity.

1 Introduction

Since natural language texts vary considerably, it is next to impossible to precode all knowledge necessary for systems that will handle these texts. One solution to this problem is to compile knowledge needed by the system directly from the corpora. For this, it is essential to have a corpus which is properly tagged instead of one that consists only of plain texts. In English, such tagged corpora already exist, to which several kinds of tags have been added (Macleod et al., 1994). A project to develop a large tagged corpus was also started in Japan (Isahara et al., 1995). Tagged corpora are needed for each language used in natural language processing systems, e.g., machine translation systems. Tagging a corpus, however, is normally a very complex and costly endeavor, so it is very important to develop an automatic tagging method for natural language processing.

This paper presents a cost-effective method for tagging text of a Thai corpus with Part-Of-Speech (POS) tags using neural networks. Input texts have been segmented and tagged with POS using an electronic Thai dictionary. The POS of each word, however, is often ambiguous. Our neural network tagger analyzes the POS of each word based on contexts. Our computer experiment showed that the neural network tagger was highly successful in identifying untrained Thai texts, and has much better results than the conventional table lookup method. To find an optimal learning method for tagging that has a fast convergence rate and low computational complexity, we also conduct some experiments comparing the original and various modified back-propagation learning algorithms at different update modes.

2 Identifications of POS using Table Lookup Methods and Neural Network Methods

Table 1 shows examples of texts in a Thai corpus to which POS tags have been added. These tags are denoted by strings such as FIXP or VACT, and are separated by the symbol '@' or '/'. Some Thai words serve as more than one POS. This is true not only of Thai, but of other natural languages as well. The POS of a word in a particular sentence, however, can be determined by context, i.e., by the POSs of words near the target word. In Table 1, for example, the POS immediately following the symbol '@' is a correct POS which can only be determined by context. Thus, the identification of a POS by context should be a task of classification: a word in context C_1 should be classified as POS-1, whereas the same word in context C_2 should be classified as POS-2. Words can be classified using either conventional table lookup methods or neural networks.

Table 1. Examples of texts in the Thai corpus with POS tags

---

1 การ@FIXP ค้าขาย@VACT ของ@RPRE เขา/NCMN@PPRS สำเร็จ@VSTA/VATT
   แล้ว/CSBR/ADVN@XVAE   //

2 อาจารย์/NTTL@NCMN กำลัง/NCMN@XVBM ศึกษา@VACT เครื่องปั้นดินเผา@NCMN
   นี้@DDAC/PDMN ในแง่ที่เป็น@RPRE โบราณวัตถุ@NCMN   //

   .
   .
   .

where FIXP denotes prefix, VACT denotes active verb,
RPRE denotes other preposition except RVBP,....

2.1 Table Lookup Methods

Table lookup methods only take into consideration the POSs of the two words immediately left (l-POSs) and right (r-POSs) of the target word (tar-POSs) as contexts. Thus, a *triple*, which consists of non-ambiguous POSs of each word can be extracted from sample data [e.g., (RPRE, PPRS, VSTA) from Table 1]. Using sample triples, a lookup table (or matrix), W, can be constructed as follows.

(i) Initially, $w_{ij} = 0$ for all i and j,
(ii) When a sample triple, (POS-i, POS-j, POS-k), is presented,

$$w_{ij} = w_{ij} + \Delta w \text{ and } w_{jk} = w_{jk} + \Delta w \quad (\Delta w > 0) \qquad (L-1)$$

where w_{ij} is the possibility that POS i is followed by POS j. After constructing the matrix, for an ambiguous triple input $(a_1/a_2/\cdots, b_1/b_2/\cdots, c_1/c_2/\cdots)$ [e.g., (RPRE, NCMN/PPRS, VSTA/VATT) from Table 1], the triple (a_l, b_m, c_n) [e.g., (RPRE, PPRS, VSTA)] can be selected by calculating $w_{ab} + w_{bc}$ for all possible triples (a, b, c) so that $w_{a_l b_m} + w_{b_m c_n}$ has the maximum value. Note that this method can only be used in linearly separable problems and that it is difficult to construct a valid table when the context needed for identification is very complex.

2.2 Multilayer Feedforward Neural Networks

Multilayer feedforward neural networks with back-propagation learning algorithm can be used for identifying POS as shown in Figure 1. The input triples (l-POSs, tar-POSs, r-POSs) may be ambiguous both in training and identifying phases according to the actual data [e.g., (PRRE, NCMN +PPRS, VSTA+VATT) from Table 1] and the desired outputs in training phase are non-ambiguous ones played by the target words (e.g., PPRS in this example). The merits of using neural networks for tagging POS are (1) they can handle non-linear separable data, and (2) the complexity of the context used in the identification of POS does not require any changes in the learning algorithm.

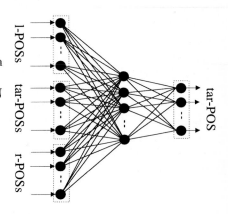

Fig. 1. The neural network for POS identification

3 The Modified Back-Propagation Algorithm

The original back-propagation algorithm for training multilayer neural networks is often found to be too slow in convergence for practical applications (Haykin and Deng, 1991). To accelerate the convergence, learning rate adaptation method, DBD (Jacobs, 1988) and a modified version DBDB (Ochiai et al. [1], 1994), as well as other methods such as Kick-Out correction (Ochiai et al. [2], 1994) have been incorporated in the back-propagation algorithm. Since these methods increase the complexity of the computations, Haykin and Deng (1991) proposed a modified version of the algorithm in which a batch update method is combined with DBD to reduce the complexity of the computations. They were successful in applying this modified algorithm to the classification of radar clutter. In other applications, such as the tagging of POS, however, we find that the batch update method slows down the convergence considerably.

To address this problem, we have introduced a semi-batch update method that reduces the computational complexity and yet prevents a slowing down of the convergence. We adopted DBDB instead of DBD to update the learning rate so that we could avoid having to carefully tune the increment and decrement factors in the update rule of learning rate. The modified back-propagation algorithm is as follows.

$$w_{ij}(k+1) = w_{ij}(k) - \eta_{ij}(k+1)G(k) + \alpha \Delta w_{ij}(k-1) \qquad (1)$$

$$\eta_{ij}(k+1) = \eta_{ij}(k) + \Delta \eta_{ij}(k) \qquad (2)$$

$$\Delta \eta_{ij}(k) = \begin{cases} \beta & \text{if } S(k-1)S(k) > 0 \\ -\phi \eta_{ij}(k) & \text{if } S(k-1)S(k) < 0 \\ 0 & \text{otherwise} \end{cases} \qquad (3)$$

$$S(k) = (1 - \theta)G(k) + \theta S(k - 1) \tag{4}$$

$$G(k) = \sum_{b=1}^{P} \frac{\partial E_b(k)}{\partial w_{ij}(k)} \tag{5}$$

$$E_b(k) = \frac{1}{2} \sum_i (t_{bi} - o_{bi})^2 \tag{6}$$

The subscript b refers to the presentation of the bth input pattern to the network in an epoch contaning P patterns. Note that $P = 1$ means that the training is in a *pattern mode* and $P = P_{all}$, which is the number of the whole training data, means that the training is in a *batch mode* (Haykin, 1994). Here, we call the case of $1 < P < P_{all}$ a *semi-batch mode*.

4 Experimental Results

The data used in the following experiments are obtained directly from the Thai corpus. The corpus contains 549 sentences which are originally in two files: the first 396 sentences are in one file and the remaining 153 sentences are in the other file. To avoid a bias on the selection of training and testing data, we simply use the sentences in the first file as training data and those in the other file as testing data. In the training data, since there are 1032 words that serve as more than one POS, we have 1032 patterns that can be used for training. In the testing data, since there are 377 words that serve as more than one POS, we have 377 patterns that can be used for testing. Considering that there are about 49 types of POS in Thai (Tantisawetrat and Sirinaovakul, 1993), we have encoded each POS into a 49-dimension pattern for the neural network tagger, which consists of a three-layer network with a size of 147-74-49.

4.1 Experiment on Identification of POS

Table 2 shows experimental results of POS identification using the table lookup and neural network methods. For the table lookup method, the sucess rates of POS identification both for the training and testing data are much lower than 80%. With the neural network method, however, the success rates are higher than 94% with an error rate of 1.6% for the training data, and higher than 80% with an error rate of 7.7% for the testing data.

Table 2. Results of POS Identification for Training and Testing Data

	Neural Network		Table Lookup	
	Training data	Testing data	Training data	Testing data
Success rate	0.948	0.801	0.707	0.756
Undetermined rate	0.036	0.122		
Error rate	0.016	0.077	0.293	0.244

It should be noted that the error rate of POS identification for the neural network method can be controlled by adjusting the threshold of the units of the output layer. The results of Table 2 are obtained for a threshold of 0.5. If we increase it to 0.85, for example, the error rate for the testing data can be reduced to 2.65%, whereas the undetermined rate will be increased to 28%.

When we compare each incorrect POS obtained by the neural network to the correct POS for the testing data, we find that about 90% of the incorrect tags are possible POSs of the target words. This indicates that the POS identification of the neural network matches the linguistic roles actually played by each word. For example, when we target the 4th word in the 1st sentence shown in Table 1, if an incorrect POS is obtained by the neural network, there is a 90% possibility that the POS will be NCMN, which denotes *common noun*, instead of one of the others. As shown in Table 1, NCMN is also a candidate for the POS of the 4th word. It is also clear that NCMN is much closer to the correct answer, PPRS, denoting *personal pronoun*, than other POSs such as FIXP, denoting prefix, or VACT, denoting *active verb*.

4.2 Comparison of Learning Algorithms

To find a good learning algorithm for tagging that has a fast convergence rate and low computational complexity, we conducted some experiments comparing the original and various modified back-propagation learning algorithms at different modes. In these experiments, the initial learning rates, $\eta_{ij}(k)$, are respectively set to 0.25, 0.125, and 0.001 for pattern, semi-batch, and batch update methods. The increment factor, β, is respectively set to 0.25, 0.0125, and 0.0001 for pattern, semi-batch, and batch update methods. The decrement factor, ϕ, is set to 0.1 for all the cases. The control parameter, θ, is respectively set to 0.7, 0.95, and 0.95 for pattern, semi-batch, and batch update methods. The moment factor, α, is set to 0.9 for the original back-propagation algorithm at pattern update mode and 0.5 for all the other cases. The epoch size P is set to 18 in semi-batch update mode.

Figure 2 shows the experimental results. The DBDB+KO in the figure denotes a variety of the modified learning algorithm described in Section 3, in which a Kick-Out correction for updating weights is added. The error shown in the figure is measured by $\frac{1}{N_P N_O} \sum_{p=1}^{N_P} \sum_{i=1}^{N_O} |t_{pi} - o_{pi}|$, where N_P and N_O are respectively the total number of training patterns and the number of units in output layer. The figure indicates that the learning algorithms using the DBDB rule and the DBDB+KO rule in pattern update mode are the fastest in convergence. For a large network and a sizable training database, however, the pattern updating results in a large increase in memory storage and computational complexity when the DBDB(+KO) rule is incorporated into the back-propagation algorithm. As shown by figure 2(c), the learning algorithm using the DBDB rule at semi-batch update mode also has relatively fast convergence. Since we adopted the semi-batch update form, memory storage and computational complexity may be reduced considerably compared to the former learning method.

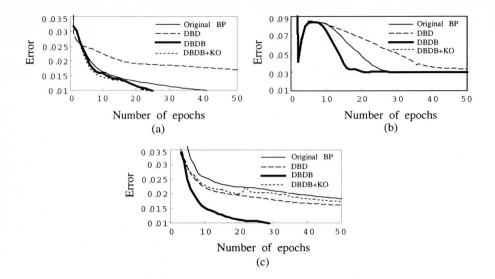

Fig. 2. Learning curves at pattern update mode (a),
batch update mode (b), and semi-batch update mode (c)

5 Conclusion

In this paper we proposed a cost-effective neural network method of tagging words in a Thai corpus with correct POS tags. Although a much larger Thai corpus, which we are now preparing, is necessary to confirm our results, it appears that this neural network tagger may be useful in pratical applications.

References

1. Macleod, C., et al.: Developing multiply tagged corpora for lexical research, *International Workshop on Directions of Lexical Research*, Beijing, China, 1994.
2. Isahara, H., et al.: Development of tagged text database by real world computing partnership, *First Annual Meeting of the Association for Natural Language Processing*, Japan, 1995 (in Japanese).
3. Haykin, S. and Deng, C.: Classification of radar clutter using neural networks, *IEEE Tran. on Neural Networks*, Vol. 2, pp. 589-600, 1991.
4. Jacobs, R. A.: Increased rates of convergence through learning rate adaptation, *Neural Networks*, Vol. 1, pp. 295-307, 1988.
5. Ochiai, K., et al.: A new renewal rule of learning rate in neural networks, *Transactions of Information Processing Society of Japan*, pp. 1081-1090, 1984 (in Japanese).
6. Ochiai, K., et al.: Kick-out learning algorithm to reduce the oscillation of weights, *Neural Networks*, Vol. 7, pp. 797-807, 1994.
7. Haykin, S.: *Neural Networks*, Macmillan College Publishing Company, Inc., 1994.
8. Trantisawetrat, N. and Sirinaovakul, B.: An electronic dictionary for multilingual machine translation, *Proceedings of the Symposim on Natural Language Processing in Thailand*, pp. 377-411, 1993.

Reproducing a Subjective Classification Scheme for Atmospheric Circulation Patterns over the United Kingdom using a Neural Network

G. C. Cawley[1] and S. R. Dorling[2]

[1] School of Information Systems,
[2] School of Environmental Sciences,
University of East Anglia, Norwich, U.K.

Abstract. Atmospheric circulation patterns are currently classified manually according to subjective schemes, such as the Lamb catalogue of circulation patterns centred on the United Kingdom. However, the sheer volume of data produced by General Circulation Models, used to investigate the effects of climatic change, makes this approach impractical for classifying predictions of the future climate. Furthermore, classification extending over long periods of time may require numerous authors, possibly introducing unwelcome discontinuities in the classification. This paper describes a neural classifier designed to reproduce the Lamb catalogue. Initial results indicate the neural classifier is able to out-perform the currently used rule-based system by a modest, but significant amount.

1   Introduction

Atmospheric circulation patterns can exhibit a great deal of day to day variability both in the lower layers of the troposphere and at so-called upper levels ($\approx$ 5km above ground level). Analysis shows however, that there are a finite number of such patterns which neatly summarise the range of this variability; each circulation pattern for a given day can be allocated to one such class. Subjective schemes have been developed for the United Kingdom and Europe, extending back to the nineteenth century on a daily basis [7, 2]. These schemes are known as the Lamb catalogue and the Grosswetterlagen respectively. While criticism can be made of any subjective index, Principle Component Analysis [6] confirms that, for example, the Lamb catalogue does provide a very useful indicator of changes in circulation both on the day to day scale and on the climatic time-scale. Furthermore, the Lamb index has been utilised in numerous studies relating atmospheric circulation to a suite of environmental problems (e.g. [1]).

Atmospheric scientists are increasingly interested in predictions of atmospheric circulation made by General Circulation Models (GCMs). To subjectively classify the profusion of such output now available would be impossible; an automated approach is required. GCMs are observed to reproduce today's climate more satisfactorily in terms of circulation patterns than temperature and precipitation patterns. Since a strong relationship exists between circulation on the regional scale and local temperature and precipitation conditions, some studies are using predicted changes in circulation patterns to *down-scale* what those circulation changes mean for local temperature and rainfall changes [3]. Subjective classifications may suffer from discontinuities in the record. Such discontinuities may have a number of sources, such as a change in the person carrying out the classification or a change in the raw data used for the purposes of the classification. The availability of upper air data, starting around the time of the Second World War, may represent such a discontinuity in both the Lamb Catalogue and Grosswetterlagen.

In this first study, we aim to implement an objective neural classifier designed to replicate, as closely as possible, the subjective Lamb catalogue. Being an automatic method, it is well suited to classifying the output of general circulation models in the study of climate change. The added benefit of a system designed to closely reproduce a subjective classification scheme, in this case the Lamb catalogue, would be to allow the existing catalogue to be continued seamlessly for an indefinite period. Here we report on the development of the neural classifier and compare the ability of the net to reproduce the Lamb classification on a day to day basis against an objective algorithm known as the Jenkinson scheme [4] as utilised by Jones *et. al.* [5].

2   The Lamb Catalogue

The Lamb catalogue describes twenty-seven distinct atmospheric circulation patterns centred on the United Kingdom, which can be:

- Purely directional from a given point of the compass (N, NE, E, SE, S, SW, W, NW)
- Non-directional but described by an Anticyclonic or Cyclonic circulation pattern (A, C)
- A hybrid of the above two types (AN, ANE, AE, ASE, AS, ASW, AW, ANW, CN, CNE, CE, CSE, CS, CSW, CW, CNW)
- Unclassified (U)

The classification is performed on the basis of sea-level pressure and 500mBar height charts for Northern Europe, taken at midnight and midday, with reference also to the charts for the preceding days. As a result, a given day might be left as unclassified, either because the circulation pattern for that day was weak or poorly defined, or because a transition between two distinct circulation patterns occurred during the day. The record shows that some types are considerably more common than others and that all exhibit inter-annual variations in their frequency.

3   Methodology

The neural weather pattern classifier consists of a single multi-layer perceptron network. Each input neuron monitors the sea-level atmospheric pressure measured twice daily at sites arranged over a grid with a resolution of 5 degrees latitude and 10 degrees longitude, as shown in figure 1. Before training, the input data is linearly scaled so that it has a zero mean and unit variance to simplify weight initialisation. The hidden and output units incorporate the standard symmetric logistic activation function. Each output unit corresponds to one of the seven major Lamb types: anticyclonic (A), cyclonic (C), northerly (N), easterly (E), southerly (S), westerly (W) and north-westerly (NW). Hybrid types, such as anticyclonic northerly, are represented by a combination of output units, in this case A and N. The north-westerly patterns though considered sufficiently distinct to merit a major type, are represented by the activation of three directional units N, W and NW as it was felt that this may simplify the learning task of the N and W output units. The target values for each output unit takes on values of −0.45 and 0.45. This prevents the pointless growth of output layer weights in a futile attempt to generate an asymptotically high output value.

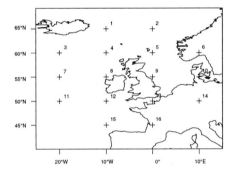

Fig. 1. Location of grid points over the British Isles. From [5]

The Rprop training algorithm used in this work, has been found to be both efficient and reliable for this application, and is also very tolerant of the adaptive learning parameters. For a complete description of the Rprop algorithm, see Riedmiller and Braun [8]. Fifteen years-worth of manually classified pressure data was divided randomly into three sections: training data (60%), validation data (20%) and test data (20%). To prevent over-fitting of the training data, the network configuration is saved when optimum performance on the validation data is reached. The statistically pure test data, that has not influenced the training procedure in any way, is then used to evaluate the network's performance.

4 Results

Figure 2 shows a graph of RMS error and the proportion of correctly classified patterns against cycles trained for a network with sixty-four hidden layer units. The network reaches a minimum RMS error on the validation data after 1000 epochs, at which time the network classifies 49.57% of the test patterns correctly. This compares favourably with the 43.6% scored by the rule based system currently used [5].

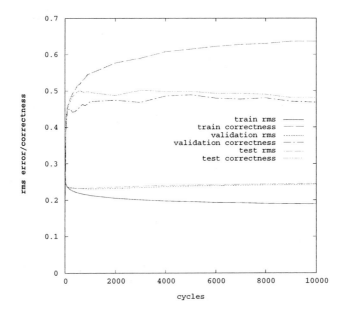

Fig. 2. Graph of RMS error against cycles trained for a multi-layer perceptron network with 64 hidden layer neurons.

Networks with 8, 16, 24, 32, 40, 48, 56 and 64 hidden units arranged in a single layer, were trained to determine the effect of network size on performance, measured by RMS error and the proportion of correctly classified patterns. In addition, a single layer perceptron network was trained to determine the extent to which the learning task is linearly separable. The results are depicted in figure 3. It is clear that the provision of hidden layer neurons immediately results in a substantial improvement in performance, but the improvement increases only slowly with increasing hidden layer size.

A confusion matrix for a neural Lamb classifier, for all available data from January 1973 to December 1988, is shown in figure 4. This provides a break down of the correctly classified patterns and false positive and false negative classifications for each of the 27 Lamb sub-types.

As noted in section 2, a given day may be left unclassified if the circulation pattern changed significantly during the day, as the neural classification is based on a single sea-level pressure chart, the classifier is unable to detect this. This accounts for a substantial portion of the false negative "unclassified" decisions. It can also be seen that the classifier is able to identify pure cyclonic and anti-cyclonic and pure directional types with a reasonable degree of accuracy, but performs less well on hybrid types. The wide dynamic range of the prior probab-

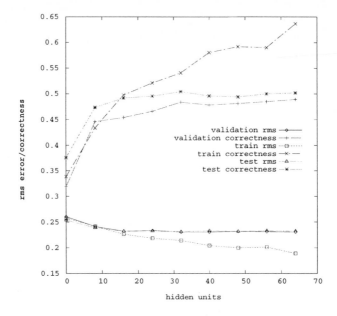

Fig. 3. Graph of RMS error and proportion of correctly classified patterns against hidden layer size, for training, validation and test data sets.

ilities for each class, is clearly a potential source of difficulty for any statistical classifier, including a neural network; for instance, westerly circulation patterns are particularly common due to the predisposition towards high pressure in the south and low pressure in the north, because of the temperature differential, whereas north-easterly patterns occur far less frequently. As a result, the overall error is minimised more efficiently by concentrating on the classes with the highest prior probability. As a result, the network correctly classifies 69.7% of the 726 westerly circulation patterns, but only 26% of the 61 north-easterly patterns.

5   Conclusions

In this initial study, a neural classifier has been constructed that is able to reproduce the Lamb catalogue more closely than the current rule based system by a modest, but significant amount. Both systems classify the prevailing circulation pattern according to the sea-level pressure at the same sixteen grid points centred on the U.K., since the subjective classification is also based on upper air patterns for each day, as well as the evolution of the circulation over the preceding days, this is clearly a simplified approach. We plan to take account of these other factors in future refinements of this work.

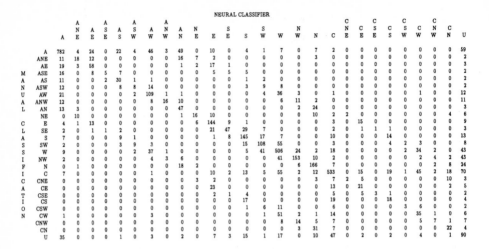

Fig. 4. Confusion matrix for neural weather pattern classifier on all available sea-level pressure data from 1973–1988.

References

1. T. D. Davies, S. R. Dorling, C. E. Pierce, R. J. Bathelmie, and G Farmer. The meteorological control on the anthropogenic ion content of precipitation at three sites in the UK: The utility of Lamb Weather Types. *International Journal of Climatology*, 11:795–807, 1991.

2. Hess and Brezowsky. Katalog der grosswetterlagen Europas. Technical Report Nr. 113, Bd. 15, 2, Berichte des Deuschen Wetterdienstes, 1969.

3. M. Hulme, K. R. Briffa, P. D. Jones, and C. A. Senior. Validation of GCM control simulations using indices of daily airflow types over the British Isles. *Climate Dynamics*, 9(2):95–105, 1993.

4. A. F. Jenkinson and F. P. Collison. An initial climatology of gales over the North Sea. Technical report, Synoptic Climatology Branch Memorandum No. 62, Meteorological Office, Bracknell, 1977.

5. P. D. Jones, M. Hulme, and K. R. Briffa. A comparison of Lamb circulation types with an objective classification scheme. *International Journal of Climatology*, 13:655–663, 1993.

6. P. D. Jones and P. M. Kelly. Principle Component Analysis of the Lamb catalogue of daily weather types: Part 1, annual frequencies. *International Journal of Climatology*, 2:147–159, 1982.

7. H. H. Lamb. British Isles weather types and a register of daily sequence of circulation patterns, 1861–1971. *Geophysical Memoirs*, 116, 1972.

8. M. Riedmiller and H. Braun. A direct adaptive method for faster backpropagation learning: The RPROP algorithm. In H. Ruspini, editor, *Proceedings of the IEEE International Conference on Neural Networks (ICNN)*, pages 586–591, San Francisco, 1993.

Two Gradient Descent Algorithms for Blind Signal Separation

H. H. Yang and S. Amari

Lab. for Information Representation, FRP, RIKEN
Hirosawa 2-1, Wako-shi, Saitama 351-01, JAPAN
E-mail: hhy@koala.riken.go.jp

Abstract. Two algorithms are derived based on the natural gradient of the mutual information of the linear transformed mixtures. These algorithms can be easily implemented on a neural network like system. Two performance functions are introduced based on the two approximation methods for evaluating the mutual information. These two functions depend only on the outputs and the de-mixing matrix. They are very useful in comparing the performance of different blind separation algorithms. The performance of the new algorithms is compared to that of some well known algorithms by using these performance functions. The new algorithms generally perform better because they minimize the mutual information directly. This is verified by the simulation results.

1   Introduction

The mutual information (MI) is one of the best contrast functions for designing blind separation algorithms because it has several invariant properties from the information geometrical point of view [1]. However, it is generally very difficult to obtain the exact function form of the MI since the probability density functions (pdfs) of the outputs are unknown. The Edgeworth expansion and the Gram-Charlier expansion were applied in [7] and [3] respectively to approximate the pdfs of the outputs in order to estimate the MI. The estimated MI is useful not only for deriving blind separation algorithms, but also for evaluating the performance of the algorithms. In the literature of blind separation, the performance of a blind separation algorithm is often measured by the cross-talking defined as the summation of all non-dominant elements in the product of the mixing matrix and the de-mixing matrix. However, the cross-talking can not be evaluated in practice since the mixing matrix is unknown. The MI is recommended to measure the performance of different blind separation algorithms.

The algorithms in [3] and in this paper are derived by applying the natural gradient descent method to minimize the estimated MI. The information-maximization approach[4] is another way to find the independent components. The output is first transformed by a non-linear function, and then the gradient descent algorithm is applied to maximize the entropy of the transformed output. Although this approach derives a concise algorithm for blind source separation, it does not directly minimize the dependence among the outputs except for the case in which the non-linear transforms happened to be the cumulative density functions of the unknown sources[5].

In this paper, we derive two blind separation algorithms based on two functions to approximate the MI. The performance of the two algorithms is compared to other existing algorithms including the info-max algorithm in [4].

2 Approximations of the MI

Let $s(t) = [s^1(t), \cdots, s^n(t)]^T$ be n unknown independent sources. Each source is stationary and has moments of any order with a zero mean. The model for the mixtures $x(t) = [x^1(t), \cdots, x^n(t)]^T$ is

$$x(t) = As(t)$$

where $A \in R^{n \times n}$ is an unknown non-singular mixing matrix.

To recover the original signals from the observations $x(t)$ we apply the following linear transform:

$$y(t) = Wx(t)$$

where $y(t) = [y^1(t), \cdots, y^n(t)]^T$ and $W = [w_k^a]_{n \times n} \in R^{n \times n}$ is a de-mixing matrix driven by a learning algorithm which has no access to both the sources and the mixing matrix.

The basic idea of the ICA [7] is to minimize the MI:

$$I(y|W) = \int p(y) \log \frac{p(y)}{\prod_{a=1}^{n} p_a(y^a)} dy = -H(y|W) + \sum_{a=1}^{n} H(y^a|W)$$

where $p_a(y^a)$ is the marginal pdf, $H(y|W) = -\int p(y) \log p(y) dy$, and $H(y^a|W) = -\int p_a(y^a) \log p_a(y^a) dy^a$ the marginal entropy.

It is easy to compute the joint entropy $H(y|W) = H(x) + \log |det(W)|$. But the computation of the marginal entropy $H(y^a|W)$ is difficult.

Applying the Gram-Charlier expansion to approximate the marginal pdf $p_a(y^a)$, we find the approximation of the marginal entropy in [3]:

$$H(y^a) \approx \frac{1}{2} \log(2\pi e) - \frac{(\kappa_3^a)^2}{2 \cdot 3!} - \frac{(\kappa_4^a)^2}{2 \cdot 4!} + \frac{3}{8}(\kappa_3^a)^2 \kappa_4^a + \frac{1}{16}(\kappa_4^a)^3 \tag{1}$$

where $\kappa_3^a = m_3^a$, $\kappa_4^a = m_4^a - 3$, and $m_k^a = E[(y^a)^k]$.

Let $F_1(\kappa_3^a, \kappa_4^a)$ denote the right hand side of the approximation (1), then

$$I(y|W) \approx -H(x) - \log |det(W)| + \sum_{a=1}^{n} F_1(\kappa_3^a, \kappa_4^a). \tag{2}$$

The Edgeworth expansion is used in [7] to approximate the pdf $p_a(y^a)$. The approximation of the marginal entropy is found to be

$$H(y^a) \approx \frac{1}{2} \log(2\pi e) - \frac{1}{2*3!}(\kappa_3^a)^2 - \frac{1}{2*4!}(\kappa_4^a)^2 - \frac{7}{48}(\kappa_3^a)^4 + \frac{1}{8}(\kappa_3^a)^2 \kappa_4^a. \tag{3}$$

But the cubic of the 4th order cumulant which is crucial for separation is not included in this approximation. We shall use the following approximation instead of (3) to derive the learning algorithm:

$$H(y^a|\boldsymbol{W}) \approx \frac{1}{2}\log(2\pi e) - \frac{1}{2\cdot 3!}(\kappa_3^a)^2 - \frac{1}{2\cdot 4!}(\kappa_4^a)^2 + \frac{1}{8}(\kappa_3^a)^2\kappa_4^a + \frac{1}{48}(\kappa_4^a)^3 \quad (4)$$

From this formula, we obtain another approximation for the MI:

$$I(\boldsymbol{y}|\boldsymbol{W}) \approx -H(\boldsymbol{x}) - \log|det(\boldsymbol{W})| + \sum_{a=1}^{n} F_2(\kappa_3^a, \kappa_4^a) \quad (5)$$

where $F_2(\kappa_3^a, \kappa_4^a)$ is defined by the right hand side of (4). Our blind separation algorithms are derived by minimizing the MI based on (2) and (5).

3 Algorithms

To minimize $I(\boldsymbol{y}|\boldsymbol{W})$, we use the following natural gradient descent algorithm:

$$\frac{d\boldsymbol{W}}{dt} = -\eta(t)\frac{\partial I(\boldsymbol{y}|\boldsymbol{W})}{\partial \boldsymbol{W}}\boldsymbol{W}^T\boldsymbol{W} \quad (6)$$

where $\eta(t)$ is a learning rate function.

From (2) and (6), we can derive the first algorithm:

$$\frac{d\boldsymbol{W}}{dt} = \eta(t)\{\boldsymbol{I} - (\boldsymbol{f}_1(\kappa_3,\kappa_4)\circ\boldsymbol{y}^2)\boldsymbol{y}^T - (\boldsymbol{g}_1(\kappa_3,\kappa_4)\circ\boldsymbol{y}^3)\boldsymbol{y}^T\}\boldsymbol{W} \quad (A1)$$

where $\circ$ denotes the Hadamard product of two matrices: $\boldsymbol{C}\circ\boldsymbol{D} = [c_{ij}d_{ij}]$,

$\boldsymbol{I}$ is an identity matrix, $\boldsymbol{y}^k = [(y^1)^k, \cdots, (y^n)^k]^T$ for $k = 2, 3$,

$\boldsymbol{f}_1(\kappa_3,\kappa_4) = [f_1(\kappa_3^1,\kappa_4^1), \cdots, f_1(\kappa_3^n,\kappa_4^n)]^T$, $f_1(y,z) = -\frac{1}{2}y + \frac{9}{4}yz$,

$\boldsymbol{g}_1(\kappa_3,\kappa_4) = [g_1(\kappa_3^1,\kappa_4^1), \cdots, g_1(\kappa_3^n,\kappa_4^n)]^T$, $g_1(y,z) = -\frac{1}{6}z + \frac{3}{2}y^2 + \frac{3}{4}z^2$.

In (A1), the cumulants κ_3^a and κ_4^a are driven by the following equations:

$$\frac{d\kappa_3^a}{dt} = -\mu(t)(\kappa_3^a - (y^a)^3), \quad \frac{d\kappa_4^a}{dt} = -\mu(t)(\kappa_4^a - (y^a)^4 + 3) \quad (7)$$

where $\mu(t)$ is a learning rate function.

Using (5) instead of (2), we have the second algorithm:

$$\frac{d\boldsymbol{W}}{dt} = \eta(t)\{\boldsymbol{I} - (\boldsymbol{f}_2(\kappa_3,\kappa_4)\circ\boldsymbol{y}^2)\boldsymbol{y}^T - (\boldsymbol{g}_2(\kappa_3,\kappa_4)\circ\boldsymbol{y}^3)\boldsymbol{y}^T\}\boldsymbol{W} \quad (A2)$$

where $\boldsymbol{f}_2(\cdot,\cdot)$ and $\boldsymbol{g}_2(\cdot,\cdot)$ are vector functions defined from $f_2(y,z) = -\frac{1}{2}y + \frac{3}{4}yz$ and $g_2(y,z) = -\frac{1}{6}z + \frac{1}{2}y^2 + \frac{1}{4}z^2$. The equation (7) is used again to trace the cumulants.

Both algorithms (A1) and (A2) can be rewritten as

$$\frac{dW}{dt} = \eta(t)\{I - h(y)y^T\}W \qquad\qquad (A)$$

where $h(y) = [h^1(y^1), \cdots, h^n(y^n)]^T$, $h^a(y^a) = f_i(\kappa_3, \kappa_4)(y^a)^2 + g_i(\kappa_3, \kappa_4)(y^a)^3$, for $i = 1$ or 2.

The info-max algorithm in [4] has the following form

$$\frac{dW}{dt} = \eta(t)\{(W^{-1})^T - f(y)x^T\} \qquad\qquad (8)$$

which is different from the algorithm (A). But if the natural gradient of the entropy is used, the algorithm (8) becomes

$$\frac{dW}{dt} = \eta(t)\{I - f(y)y^T\}W \qquad\qquad (B)$$

where the activation function $f(y)$ is determined by the sigmoid function used in transforming the outputs. For example, if the transform function is $\tanh(x)$, the activation function is $2\tanh(x)$. Note the algorithm (B) also works well for some other activation functions such as those proposed in [2, 4, 6, 8].

Both algorithms (A) and (B) have the same "equivariant" property as the algorithms in [6]. A significant difference between (A) and (B) is that the activation function employed in the former is data dependent while the activation functions employed in the later are data independent. The data dependence of the activation functions makes the algorithm (A) more adaptive. The tracking of κ_3^a and κ_4^a by the equation (7) is part of the algorithm (A). To compare the performance of the algorithm (A) and (B), we define

$$I_p(y|W) = -\log|det(W)| + \sum_{a=1}^{n} F_1(\kappa_3^a, \kappa_4^a).$$

It is the changing part of the MI when W is updated. By using this function, we compare the performance of the algorithm (A) to that of the algorithm (B).

4 Simulation

In the rest of this paper, we use (B1) and (B2) to refer the two versions of the algorithm (B) when the functions x^3 and $2\tanh(x)$ are used respectively. We choose four modulated signals and one uniformly distributed noise as the unknown sources in the simulation. The five sources are mixed by a 5×5 mixing matrix A whose elements are randomly chosen in $[-1, +1]$. The learning rate exponentially decreases to zero. The same learning rate is chosen for all algorithms in every simulation.

The simulation results are shown in Figure 1 and Figure 2. In each sub-figure of Figure 1, the curves for four outputs are shifted upwards from the zero level in order for a better illustration, and all the outputs shown there are within the same time window $[0.1, 0.2]$. In Figure 2, the total history of I_p for running (A1), (A2), (B1) and (B2) is displayed to compare their performance. From the simulation results shown in these figures, we have the following observations:

1. The algorithm (A) is faster in speed and better in quality than the algorithm (B) in separating the mixed sources.
2. The performance of blind separation algorithms can be measured objectively by the function $I_p(\boldsymbol{y}|\boldsymbol{W})$ by tracking the moments of the outputs.

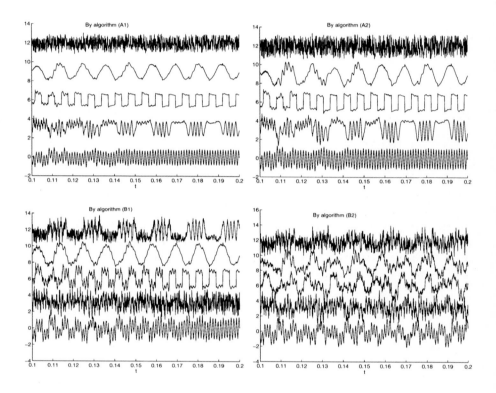

Fig. 1. The comparison of the separation by the algorithms (A1), (A2), (B1) and (B2).

5   Conclusion

In this paper, the performance function $I_p(\boldsymbol{y}|\boldsymbol{W})$ is introduced to evaluate the separation quality of blind separation algorithms. The algorithm (A) is derived based on the natural gradient descent algorithm to minimize the MI in the most efficient way. It is an on-line learning algorithm with the "equivariant" property. The activation functions used in this algorithm are data dependent. This data dependence makes the algorithm more adaptive. The simulation shows that the algorithm (A) can achieve the separation faster in speed and better in quality than the algorithm (B) with the two most popular activation functions x^3 and $2\tanh(x)$.

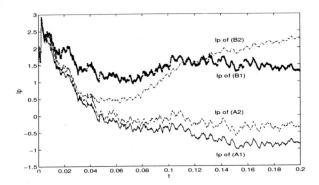

Fig. 2. The history of I_p in running the algorithms (A1), (A2), (B1) and (B2).

References

1. S. Amari. *Differential-Geometrical Methods in Statistics, Lecture Notes in Statistics vol.28.* Springer, 1985.
2. S. Amari, A. Cichocki, and H. H. Yang. Recurrent neural networks for blind separation of sources. In *Proceedings 1995 International Symposium on Nonlinear Theory and Applications*, volume I, pages 37–42, December 1995.
3. S. Amari, A. Cichocki, and H. H. Yang. A new learning algorithm for blind signal separation. In *Advances in Neural Information Processing Systems, 8, eds. David S. Touretzky, Michael C. Mozer and Michael E. Hasselmo, MIT Press: Cambridge, MA. (to appear)*, 1996.
4. A. J. Bell and T. J. Sejnowski. An information-maximisation approach to blind separation and blind deconvolution. *Neural Computation*, 7:1129–1159, 1995.
5. A. J. Bell and T. J. Sejnowski. Fast blind separation based on information theory. In *Proceedings 1995 International Symposium on Nonlinear Theory and Applications*, volume I, pages 43–47, December 1995.
6. J.-F. Cardoso and Beate Laheld. Equivariant adaptive source separation. *To appear in IEEE Trans. on Signal Processing*, 1996.
7. P. Comon. Independent component analysis, a new concept? *Signal Processing*, 36:287–314, 1994.
8. C. Jutten and J. Herault. Blind separation of sources, part i: An adaptive algorithm based on neuromimetic architecture. *Signal Processing*, 24:1–10, 1991.

Classification Rejection by Prediction

de LASSUS H.[1], DAIGREMONT Ph.[2], BADRAN F.[2], THIRIA S.[2,3],
LECACHEUX A.[1]

[1] Laboratoire ARPEGES, CNRS URA 1757,
Observatoire de Paris-Meudon - 92195 MEUDON cedex, France
[2] CEDRIC, Conservatoire National des Arts et Métiers,
292 rue Saint Martin - 75003 PARIS, France
[3] Laboratoire d'Océanographie et de Climatologie (LODYC), Université de Paris VI,
4 place Jussieu, T14 - 75005 PARIS, France

Abstract. We address the problem of autonomous decision making
in classification of radioastronomy spectrograms from spacecraft. It is
known that the assessment of the decision process can be divided into
acceptation of the classification, instant rejection of the current signal
classification, or rejection of the entire classifier model. We propose to
combine prediction and classification with a double architecture of Time
Delay Neural Network (TDNN) to optimize a decision minimizing the
false alarm risk. Results on real data from URAP experiment aboard
Ulysses spacecraft show that this scheme is tractable and effective.
Keywords: spectrogram classification, radioastronomy, neural network
classification, transient signal prediction.

1 Introduction

We address the problem of optimizing the decision process for classification of
spectrograms of mixed planetary radio signals with additive plasma noise. The
solution we suggest is based on neural prediction of the signal followed by a
neural classification. In this paradigm, the prediction is used as an assessment of
the classification decision. With this method, it is possible to punctually reject
the classifier's choice, or, even if necessary, to decide that the current classifier
becomes unreliable and has to be changed. The example we show belongs to low
frequency radio astronomy, but this approach seems applicable in many other
fields of interest where geophysical signals are under study.

2 Motivation

Classification of low frequency planetary radio bursts displayed in the time frequency plane is a task similar to the well known "cocktail party" problem: identify many sources emitting in the same time. The signals under study originate
from different parts of the solar system, and the distances, as well as the observation angles, are heterogeneous and changing fast. This leads to moving patterns
on the spectrogram while we want to classify the signal. Thus, as the sources

number may change, as well as their unstable patterns, it may happen that the classifier finds himself facing situations unseen before. In such situation, there is a need to assess the quality of the decision, and to evaluate the ability of the classifier to carry on his mission successfully.

When the classification has to be done on spectrogram representation of the signal, the TDNN architecture has proved to be a convenient choice [3] [2].

3   The rejection dilemma

Let d_1 be the decision of accepting a classifier's suggestion, and d_0 the rejection of this proposal. Let H_1 be the hypothesis that the signal belongs to the label suggested by the classifier, and H_0 hypothesis that it does not belong to this label. Then, we have [1]:

- choose d_1 when H_1: correct classification.
- choose d_1 when H_0: false alarm.
- choose d_0 when H_1: undue rejection.
- choose d_0 when H_0: justified rejection.

We would like to lessen the risk of $d_1 H_0$ without loosing too much on $d_1 H_1$. Many solutions have been suggested:

- use information from the classifier output distribution. It happens that neural classification of radio planetary signals yields similar distribution for $d_1 H_1$ and $d_1 H_0$, therefore, no solution is available using this technique.
- use parallel classifiers on correlated samples, or many different types of classifiers on the same sample and merge their results. These two methods are efficient.
- use expert knowledge to assess the classifier decision. Cellular automata are well suited for this task.

We propose here rejection by prediction: that is to predict the signal outside of the TDNN input. If this short term prediction is confirmed by futur samples, then the postponed decision is confirmed.

4   Rejection by Prediction

4.1   Punctual rejection

Principles. In order to minimize the classification risk $d_1 \ H_0$, we need a measure of quality to motivate our decision. We suggest to use in parallel two TDNNs sharing the same input: a window cut on the spectrogram. This window will be centered on the peak of a signal that has just occured. One TDNN will predict the signal expected to come just ouside the window in the very near future.The other TDNN will classify the signal with the information available from the window. Thus, the decision of the classifier TDNN will be kept for a while, waiting, for the result of the prediction. When the future predicted signal is there, it is confronted with the prediction. The distance from the prediction to the real signal yields the measure of quality we needed to confirm the decision of the classifier.

Autocorrelation Approach. The problem is to choose the right window and the suitable time step for the prediction. It is known [5] that the optimal input for our classifier is determined by the first zero crossings of the autocorrelation function of the signal. There is a useful lag between the zero crossing at $(t + \tau)$ and the peak at $(t = 0)$. If we restrict the input of the classifier to a size smaller than the distance between the zero crossings, it will be suboptimal, but experiments show that the performances on "reasonable" signals are still good. We decided to adjust the TDNN's input window to the flat part of the positive autocorrelation, and to predict the signal in its steep part, inside the zero crossings. Of course, the signals to be classified do not share the same autocorrelation functions, and therefore, the zero crossings are varying. Therefore, this method is restricted to signals sharing a certain scale of stationarity. When signals with very short term stationarities, are mixed with signals having longer stationarities, a choice has to be made: good results cannot be obtained simultaneously on the two classes of signals. We decided to adjust the size of the window and the time step for the signals for which other assessing methods where not efficient: smooth signal belonging to sources for which no rotation tables were available.

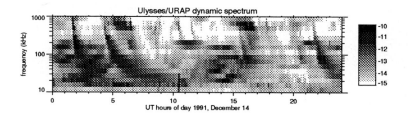

Fig. 1. Typical Ulysses Spectrogram with 4 different types of signals overlaping

Data Analysis. For our experiment with real data, we used eight months of data obtained by the URAP experiment aboard Ulysses spacecraft [4] (September 1991 - April 1992). In these data, radio bursts from the sun and jupiter are continuously present. Since Ulysses flew by around jupiter on Feb 8th, 1992 and changed its trajectory plane by jupiter gravity assist, the morphology of the highly directive radio emissions from jupiter changed abruptly. Radio spectra were acquired every 144 sec; they are made of 16 frequency channels logarithmically spaced from 10 to 1000 kHz. Signal intensities were normalized between (-1,+1). From September to February, four kinds of radio emissions are present, two smooth signals (solar type III, Jovian Nkom) and two bursty signals (Jovian

Hom and Bkom), then from February the 8th 1992 until april, the two bursty signals (jovian Hom and Bkom) disappear abruptly and a new bursty signal is present (Kom). The two smooth signals are present on the entire data set.

Experiment. We selected a learning set of 9609 events from September to December 1991. And a test set of 486 events collected during these months, but not the same days. The test set was used to stop learning early enough to keep the generalization properties of the TDNNs. A validation set of 951 events was then selected from January to February 8th, 1992.

We experimented our method on the 100Khz frequency, which lies at the center of the spectrogram, because it was there that the task was the most difficult with four overlapping signals. In fact, any other frequency could have been chosen. The optimal input window was calculated with an autocorrelation function estimator [5]. For the rejection, we decided to focus on the smooth type III signal, as we had already good cellular automata [2] rejections methods for the bursty ones.

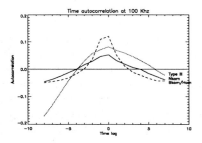 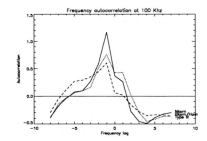

Fig. 2. autocorrelation of spectrogram time axis

Fig. 3. autocorrelation of spectrogram frequency axis

For the type III signal, zero crossings were $(-5, +7)$ in time and $(-6, +3)$ in frequency. So we chose a $(-5, +4)$ window in time and $(-6, +3)$ in frequency for classification and we predicted the signal at three time steps ie. $(+5, +7)$ in time and 100 Khz in frequency. The classifier is a $100 \times 5 \times 4 \times 3$ TDNN with logistic output function. the predictor is the same TDNN architecture but with an hyperbolictangent, output function. Therefore, the predictor yields an output between $(-1, +1)$ to extrapolate the normalized input function, while the classifier yields an output between $(0, 1)$, suitable for classification. A threshold is calculated from statistics (μ and σ) of the prediction error from the learning data set. A classified sample is rejected when its prediction error is higher than the threshold.

Results and Discussion. As expected, results on table 1 show that rejection by prediction is efficient on the smooth Type III signal. If the prediction error threshold is fixed at $\mu + \sigma$, then we have 60% of $d_1 H_0$ rejection at a cost of 25% of $d_1 H_1$ rejection. The Nkom signal which zero crossing is on the prediction lag does not respond significantly to this method. It yields 5% of $d_1 H_0$ rejection at a cost of 3% of $d_1 H_1$ rejection. Whereas the bursty Bkom/Hom signal cannot be assessed with the chosen input window. These results show that punctual rejection of $d_1 H_0$ is possible provided that the input window is optimized for the signal under study. We will see now, that this does not impede the possibility of model rejection whatever the signals.

Type III	$d1H0$	$d1H1$
$\mu + \sigma$	60	25
$\mu + 2\sigma$	0	2

Bkom/Hom	$d1H0$	$d1H1$
$\mu + \sigma$	2	5
$\mu + 2\sigma$	0	1

Nkom	$d1H0$	$d1H1$
$\mu + \sigma$	5	3
$\mu + 2\sigma$	1	0

Table 1. percentage of rejected decisions at different thresholds.

4.2 Model rejection

Principles. The idea is to use the TDNN predictor and the TDNN classifier jointly, to assess the ability of the classifier to pursue its task. That means that if the signals present in the learning set have disapeared, and have been replaced by new signals, it is clear that learning has to be resumed, and the classifier changed. The classifier alone gives no information on its ability to carry on its task. If the prediction error of the predictor TDNN is kept in memory, then its standard deviation, calculated on a moving window, gives information on the presence of a wave form that was not present in the learning data set. A suitable error bar calculated from the learning data will tell if the classifier has to be changed.

Experiment. For this experiment we chose the data from September 1991 to April 1992. On the 8th of February the Bkom/Hom disapears and is replaced by Kom, from this date, the classifier is unable to operate. This is obvious from the peak at time 8.10^4 on figure 5 wich represents classification error calculated for the period. Figure 4 shows the prediction error on the same period. The introduction of the new signal is accurately detected within few minutes. Spurious peaks after time 8.10^4 indicate clearly that the classifier is obsolete. Detailed analysis of the results indicate that the valleys in the prediction error are correlated with those of the classification error. They correspond to the presence of recognized signals (Type III/Nkom).

5 Conclusion

We have proposed a method to manage rejection in classification of spectrogram of low frequency radio astronomy signals. We have shown that a predictor TDNN

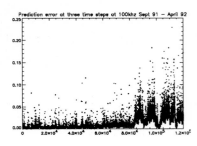
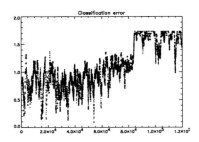

Fig. 4. Prediction error; spurious peaks start from Feb 8th, 1992

Fig. 5. Classification error; it abruptly increases after Feb 8th, 1992

and a classifier TDNN sharing the same inputs can be used to minimize the false alarm risk. Moreover, this scheme can be used to decide to change the classifier when the environment has evolved. This capacity is of prime importance for space radioastronomy detectors which have no directivity.

Now that we know that rejection is possible, there is a need for an algorithm enabling the learning of a new classifier with no further human intervention as soon as the situation has evolved. Among the tasks to fulfill to build a new learning data set, we will have to select recent reliable labelled samples belonging to classes still present. Our neural confirmer will yield us that small but secure number of labelled samples necessary to derive a new classifier. For this task, an efficiency of 60% of false alarm rejection that seemed somewhat unimpressive, will be of great value. Automatic derivation of reliable labelled samples for new classes of signal is the subject of our present research.

References

1. P. Y. ARQUES. *Décision en Traitement du Signal*. Collection Technique et Scientifique des Télécommunications. Masson, Paris, 1982.
2. H. DE LASSUS A. LECACHEUX S. THIRIA F. BADRAN. Neural Network Clusters and Cellular Automata for the Detection and Classification of Overlapping Transient Signals on Radio Astronomy Spectrograms from Spacecraft. In *International Symposium on Time-Frequency and Time-Scale Analysis*, Paris, France, june 1996. IEEE Signal Processing Society. Accepted paper.
3. K. J. LANG A. H. WAIBEL G. E. HINTON. A Time Delay Neural Network Architecture for Isolated Word Recognition. *Neural Networks*, 3:23–43, 1990.
4. STONE and al. The Unified Radio and Plasma Waves Investigation. *Astron. Astrophys. Suppl. Series*, 92:291–316, 1992.
5. GM. JENKINS DG. WATTS. *Spectral Analysis and its applications*. Holden-Day, San Francisco, holden-day edition, 1968.

Application of Radial Basis Function Neural Networks to Odour Sensing Using a Broad Specificity Array of Conducting Polymers

Dong-Hyun Lee, John S. Payne, Hyung-Gi Byun, Krishna C. Persaud

**Department of Instrumentation and Analytical Science,
UMIST, Po Box 88, Manchester, M60 1QD, UK.**
Email: KCPERSAUD@UMIST.AC.UK
Tel. +44 131 200 4892

ABSTRACT

Neural networks are increasingly being used to enhance the classification and recognition powers of data collected from sensory analysis. This paper details the effectiveness of an enhanced Radial Basis Function Neural Network combined with the outputs from an "Electronic Nose" using electrically conducting organic polymers as sensor materials. Robust performance and classification of data is demonstrated with data patterns from pure chemicals. The system has been used to monitor the changes of odour with age of dry foodstuffs.

INTRODUCTION

Odour perception within the animal kingdom is still not fully understood, and even less is known of odour recognition in the brain of the animal. It is generally accepted that a need exists for chemical sensing systems that mimic the action of biological olfaction in applications such as quality control of foods, environmental monitoring, raw material quality, and so on. Over the last decade Persaud and co-workers have developed an "Electronic Nose" which uses organic electrically conducting polymers based on aromatic or heteroaromatic compounds such as polypyrrole and polythiophene [1, 2]. Conducting polymer sensor arrays are being increasing reported for use in odour detection and identification [3, 4]. Although the understanding of chemical interaction with conducting polymers is still poor, it is believed that reversible characteristics of conformational and/or charge transfer take place between volatile odour chemical and polymer. Several advantages exist over other technologies such as little poisoning effects, rapid reversibility, use at room temperature (little power consumption and no breakdown of volatiles at the sensor surface from increased heating), rapid response absorption/desorption within seconds to most volatile chemicals, and a long sensor lifetime of several years.

The response to a volatile odour of each sensor is measured and is unique for each type of single or complex odour. With each conducting polymer sensor having a certain response character, an array of sensors with broad but different chemical specificities provides a measurement pattern of broad overlapping selectivity. These responses, or signals, are processed to produce a set of descriptors for the input, which can be identified as a "fingerprint" for an odour, and then saved into a database for further manipulation within statistical pattern recognition methods, cluster analysis, and artificial neural networks. The input data into these methods is a normalised pattern of responses of each sensor relative to the whole array. Miniature sensor arrays have now been realised containing thirty two different conducting polymers, and are now commercially available along with an integrated system containing to hold the sensor array while also keeping samples within a constant environment. Consequently, for pure solvents the patterns produced are concentration-independent; however, when monitoring a complex odour or chemical mixture the patterns are non-linearly concentration dependent. This creates difficulties for information processing.

The use of neural networks within artificial sensory analysis has been growing in momentum in recent years. The ability to recognise pattern characteristics from relatively small pieces of information has led to growing interest in the possible applications and development within sensory recognition.

A variety of pattern recognition techniques including neural networks may be applied to the classification of different odours, quantitative prediction and recognition of unknown gases and odours. Backpropagation, a model of multilayer perceptron networks, is probably the most widely used neural network paradigm. One disadvantage is a difficulty in classifying a previously unknown pattern that does not classified to any prototypes in the training set. We have investigated the characteristics of radial basis function (RBF) networks applied to odour classification problems. RBF networks train rapidly, usually orders of magnitude faster than backpropagation, while exhibiting none of backpropagation's training pathologies such as paralysis or local minima problems. A RBF network [5] is a two-layer network where the output units form a linear combination of the basis functions computed by hidden units. The basis functions in the hidden layer produce a localised response to the input and typically uses hidden layer neurones with Gaussian response functions, in which case the activation level O_j of hidden unit j is calculated by

$$O_j = \exp[-(X - W_j) \cdot (X - W_j) / 2\sigma_j^2]$$

where X is the input vector, W_j is the weight vector associated with hidden unit j and σ is the normalisation factor which represents a measure of the spread of the data. The outputs of the hidden unit lie between 0 and 1; the closer the input to the centre of the Gaussian, the larger the response of the node. The activation level O_j of an output unit is determined by

$$O_j = \sum W_{ji} O_i$$

where W_{ji} is the weight from hidden unit i to output unit j.

The training in a RBF network is done by finding the centres, widths, and the weights connecting hidden nodes to the output nodes. The performance of radial basis function classifiers are highly dependent on the choice of centres and width [6]. This has been a focus of our attention in order to optimise RBF networks for odour classification. For a minimum number of nodes, the selected centres should well represent the training data for acceptable classification. One of the optimum solutions is to combine a learning vector quantizer (LVQ) [7] algorithm with RBF which minimally affects training efficiency, the LVQ being used to find suitable cluster centres. The LVQ network is based on competitive learning and is used to both quantize input pattern vectors into reference or 'codebook' values and to use these reference values for classification. The class of input pattern x is found using:

$$\|m_c - x\| = \min_i \|m_i - x\|$$

where m_i is codebook vectors and m_c is the reference vector closest to the input. The codebook vectors are adjusted according to the update rule which shifts the winning unit of codebook vector towards the input vector when the class is correct and shifts away from input vector when the class is incorrectly classified. The distance measurement between one vector and neighbouring vectors is a useful means of deciding whether correct classification is possible. A suitable way to initialise the LVQ is to use the k-nearest-neighbour (k-NN) method as the vector should always remain inside the respective class domains.

RESULTS

In order to test the classifier based on RBF, the odour sensing system was used to collect representative patterns of pure chemicals as well as a mixture of methanol and ethanol in fixed ratios. Once validated, the network was then applied to a real world, complex odour measurement problem involving characterisation of the ageing process of a foodstuff.

For network training, four centres for each class were chosen from a total of 108 exemplar patterns. The trained system was then tested against 197 patterns of nine pure chemicals. Figure 1(a) shows the output of the network to training patterns where Figure 1(b) shows the performance to previously unseen data. For clarity, the output from the RBF network is presented graphically instead of class labels and error values. The x-axis in the figure depicts the target values (or the desired output) ordered linearly; the y-axis represents the sequencing of individual patterns. Any symbol on (or closer to) a target line means the pattern is recognised as the class which holds the target value. In all cases the normalisation factor σ was kept constant.

The network is also tested against input patterns obtained from a range of concentrations of chemicals. Four mixtures of methanol and ethanol in ratio of 1:1, 1:2, 1:4, 1:8 were trained with 4 centres for each class from 48 training patterns and the network was tested against previously unseen 132 patterns. The results are shown in Figure 2(a)(b). Figure 3(b) shows the output when untrained input data of mixture 1:4 is tested with a network previously trained with methanol, ethanol and three other previous mixtures (as shown in Figure 2(a)(b)). In this case, 4 centres for each class from 108 training patterns are trained and tested against 132 patterns. These data indicate a very robust performance of RBF network to previously unseen patterns.

Further studies were then carried out on real complex odour changes from one type of dry food material. Data was collected over a thirteen week period on a particular food sample in order to assess if ageing effects on foods could be measured by the conducting polymer array and identified using RBF networks. Controls of water, methanol, ethyl acetate and butyl acetate, were used to assess the reliability and stability over this period. These results indicated little sensor drift and a high reliability in sensor response. At the start of experimentation, part of the food sample was frozen at -20°C to minimise any ageing effects, and were called the control samples. The other part of the food sample was kept at room temperature within the original food container. Each week an amount of control sample and aged sample was analysed by the thirty two sensor array. In Figure 4(a) the RBF network was trained with control and aged samples from weeks 1, 3, 5, 7, 9, 11, and 13. The network was then tested in Figure 4(b) with previously unseen data of representing controls and aged data from weeks 2, 4, 8, 10 and 12. This network was trained with 4 centres for each class from a total number of 144 training patterns.

The radial basis function worked very well in the prediction of unknown unaged and aged data compared to known unaged and aged data. A trend over the thirteen week period of ageing is clearly seen. The capability to quantify complex odours has been demonstrated.

CONCLUSIONS

The enhancements to the RBF, involving LVQ and k-NN as part of the preliminary parameter choice of cluster centres, has produced a very robust and rapid classifier which has potential for complex odour measurements with conducting polymer sensor arrays.

ACKNOWLEDGEMENTS

This work was supported by EPSRC, BBSRC and AromaScan plc. UK.

REFERENCES

1. Pelosi, P. and Persaud, K.C., "Gas sensors: towards an artificial nose", In: Sensors and sensory systems for advanced robots, (Ed. Dario, P.), NATO ASI Series Vol. F43, Springer-Verlag, Berlin, 1988, p. 49-70.
2. Persaud, K.C. and Travers, P., "Multielement arrays for sensing volatile chemicals", Intelligent Instruments and Computers, July/August, Elsevier Press, 1991, p. 147-154.
3. Pisanelli, A-M., Qutob, A.A., Travers, P., Szyszko, S. and Persaud, K.C., "Applications of Multiarray Polymer Sensors to Food Industries", Life Chemistry Reports, Vol. 11, Harwood Academic Pubs., GmbH, 1994, p. 303-308.
4. Persaud, K.C., Khaffaf, S.M., Hobbs, P.J., Misselbrook, T.H. and Sneath, R.W., "Application of Conducting Polymer Odour Sensing Arrays to Agricultural Malodour Monitoring", In: International Conference Proceedings on Air Pollution from Agricultural Operations, Kansas City, Missouri, February 7-9, 1996, p. 249-253.
5. Moody, J., and Darken, C.J., "Fast learning networks of locally-tuned processing units", Neural Computation 1(2), 1989, p.281-294.
6. Musavi, M.T., Ahmed, W., Chan, K.H., Faris, K.B. and Hummels, D.M., "On the training of radial basis function classifiers", Neural Networks, Vol. 5, 1992, p595-63.
7. Kohonen, T., "Self-Organization and Associative Memory", Springer-Verlag, 3rd ed., 1989, 199-202.

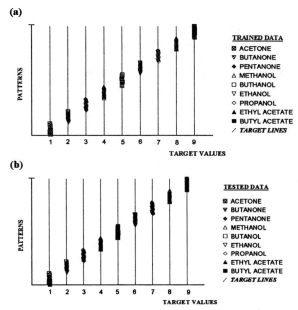

FIGURE 1. *Trained output (a) data and tested output (b) data of nine pure chemicals. Data in (b) had not previously been seen by the network during training.*

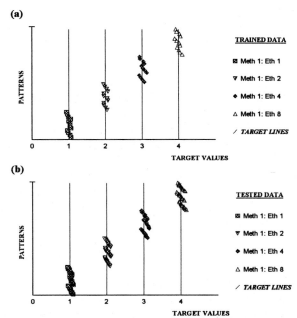

FIGURE 2. *Trained output (a) data and tested output (b) data of four methanol:ethanol mixtures in 1:1, 1:2, 1:4 and 1:8 ratios. The data indicate that the network is able to predict the composition of known mixtures.*

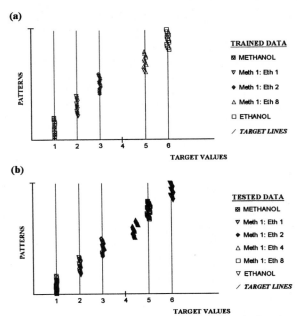

FIGURE 3. *Trained output (a) data of methanol, ethanol and methanol:ethanol mixtures in 1:1, 1:2 and 1:8 ratios, and tested output (b) of previously unseen data of methanol, ethanol and the mixtures in 1:1, 1:2, 1:4 and 1:8 ratios. The output of mixture in 1:4 ratio shows the potential of RBF for predicting concentrations.*

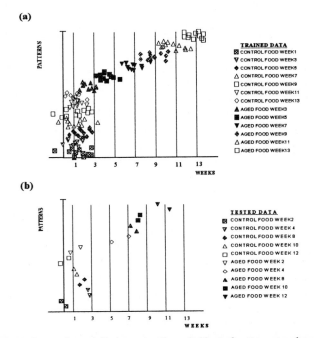

FIGURE 4. *Trained (a) and tested (b) data using the radial basis function network to analyse the ageing effects of complex odours from real food. The data in (b) were not previously seen by the network during training. They indicate that the system can be used to predict the age of the product.*

A Hybrid Object Recognition Architecture

Gunther Heidemann, Franz Kummert, Helge Ritter, Gerhard Sagerer

University of Bielefeld,
33501 Bielefeld,
Germany

Abstract. We present an architecture for 3D-object recognition based on the integration of neural and semantic networks. The architecture consists of mainly two components. A neural object recognition system generates object hypotheses, which are verified or rejected by a semantic network. Thus the advantages of both paradigms are combined: in the low level field adaptivity and the ability to learn from examples is realized by a neural network, whereas the high level analysis is performed by representing structured knowledge in a semantic network.

1 Introduction

One of the main reasons that make 3D-object recognition an extremely hard task in computer vision is that knowledge acquisition and representation has to cover the wide range from very low level (pixel) data up to a high level symbolic representation. On the one hand there are sensor data which are hard to describe by explicit models, but can be classified holistically by an artificial neural network (ANN), on the other hand the structure of objects often is too complex for a pure holistic recognition, but can be modeled explicitly in a semantic net. Therefore, it seems reasonable to combine the benefits of ANNs and semantic networks in a hybrid approach. Knowledge about the objects that can be well structured such as (in our case) shape is represented by a semantic net whereas the bridge to the pixel data is realized using a neural object recognition system that can be trained from examples.

2 The hybrid object recognition architecture

In our approach, the hybrid system performs object recognition in mainly three steps: 1. a low level preprocessing and search for regions of interest, 2. generation of object hypotheses by the neural recognition system, 3. knowledge based analysis and verification or rejection of the hypotheses by the semantic network.

In the low level part, first a segmentation for colors of the domain of the objects is performed. From this we get regions of interest which are the basis for both the neural and the semantic analysis. The semantic net operates on features of the regions such as eccentricity and compactness. Moreover, in the low level preprocessing the monochrome intensity image is transformed to an edge

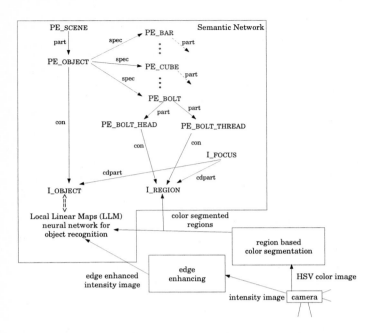

Fig. 1. The hybrid object recognition architecture. Only a part of the knowledge base is shown.

enhanced image by laplace filtering and a subsequent logarithmic transformation. From this image the features for the neural classification are extracted by Gabor filter kernels within the regions of interest.

The neural system then tries to classify the features extracted and generates up to three competing object hypotheses, combined with a judgment. By this means the search space of the semantic net can be directed, and search is started with the hypotheses with highest probability. The semantic net then tries to verify or reject the hypotheses by decomposition of the objects according to the knowledge base. In other words, it is the task of the ANN to have a "first look" at the scene and give an overview quickly, which is the starting point for a closer inspection by the semantic net.

2.1 The neural object recognition system

From the low level color segmentation the neural system gets the blob centers as "focus points". At each focus point, a feature vector is extracted by currently 16 Gabor filter kernels. The parameters of the Gabor filter kernels (location with respect to the focus point, width, wavevector and phase) were optimized to the classification task by a method outlined in [5, 1]. In short, the proposed algorithm optimizes the parameters of the filter kernels by (local) minimization of an energy function on the parameters, which is constructed such that the extracted feature vectors belonging to one type of object tend to cluster in feature space, whereas clusters belonging to different object types are separated as far as possible.

Classification of the feature vectors is performed by an ANN of the Local Linear Map (LLM) – type. The LLM network is related to the self-organizing map [2] and the GRBF approach. It can be trained to approximate a nonlinear function by a set of locally valid linear mappings, for details see e.g. [6]. For the classification task we use a "winner takes all" network. In this case, for a given input $\mathbf{x}$ only one node, the best match or "winner" node k, contributes to the output vector $\mathbf{y}$:

$$\mathbf{y} = \mathbf{w}_k^{out} + \mathbf{A}_k(\mathbf{x} - \mathbf{w}_k^{in}), \tag{1}$$

where $\mathbf{w}_k^{in}$ and $\mathbf{w}_k^{out}$ are the input and output weight vectors of node k, respectively. The input space has in our case as dimension the number of Gabor kernels n_G, the output space is $(n_O + 1)$ dimensional, this is the number of object classes n_O plus one as a rejection class. Therefore, $\mathbf{A}_k$ is a $(n_O + 1) \times n_G$ matrix associated with node k.

For the training, the output vector of training example α has the form

$$y_i^{(\alpha)}(l) = \delta_{il}, \quad \text{with} \quad i, l = 1 \ldots n_O + 1, \tag{2}$$

where l is the class of the object to be trained. When applying the network to an unknown input vector, the resulting class o_{res} is determined by

$$o_{res} = \arg \max_{i=1 \ldots n_O + 1} (y_i). \tag{3}$$

The main limitation of the neural recognition system is the lack of a universal rejection class. Up to now, only objects trained to be rejected will be classified correctly. However, the semantic analysis is able to reject completely unknown objects in most cases.

2.2 Knowledge based object recognition

The semantic network language ERNEST [3] builds the basis for knowledge representation and utilization. In contrast to other approaches like KL-ONE or PSN in ERNEST only three different types of nodes and three different types of links exist. *Concepts* represent classes of objects, events, or abstract conceptions having some common properties. In the context of image understanding an important step is the interpretation of the sensor signal in terms modeled in the knowledge base. The second node type, called *instance*, represents these extensions of a concept. It associates certain areas of the image with concepts of the knowledge base. In an intermediate state of processing instances may not be computable because certain prerequisites are missing. Nevertheless, the available information can be used to constrain an uninstantiated concept. This is done via the node type *modified concept*. As in all approaches to semantic networks the link *part* decomposes a concept into its natural components. Another link type is the *specialization* with a related inheritance mechanism by which a concept inherits all properties of the general one. For a clear distinction of knowledge of different levels of abstraction the link type *concrete* is introduced. Additionally, a concept is described by attributes representing numerical features and restrictions on these values according to the modeled term. Furthermore, relations defining constraints for attributes can be specified and must be satisfied for valid instances. The creation of modified concepts and instances constitutes the knowledge utilization in the semantic network. For the creation of instances, this process is based on the fact that the recognition of a complex object needs the detection of all its parts as a prerequisite. Since the results of an initial segmentation are not perfect, the definition of a concept is completed by a judgment function estimating the degree of correspondence of an image area to the term defined by the related concept. On the basis of these estimates and the inference rules an A*-like control algorithm is applied.

In the following the declarative knowledge base (see Fig. 1) and the processing strategy is described in some detail. Actually, the network consists of two levels of abstraction namely the image level (indicated by the prefix I_) and the level of perception (indicated by the prefix PE_). The concept I_FOCUS was motivated by the works of Moratz, see e.g. [4]. It mainly allows to focus on certain areas in the image to restrict the object recognition task. This focus can be established by an utterance or a gesture during the construction dialogue (actually not yet considered) or by the objects detected so far. This concept has two context–dependent parts namely I_REGION representing a color segmented region and I_OBJECT representing an object hypothesis calculated by the underlying LLM network. For every competing LLM hypothesis an instance I_OBJECT$^{(I)}$ is created which are stored in competing search tree nodes. Dependent on the object type detected by the LLM network the corresponding concept in the perceptual level is selected to verify the object hypothesis due to the structural knowledge stored in the semantic network. That means if an instance I_OBJECT$^{(I)}$ with type 'bolt head' exists then a modified concept PE_BOLT$^{(M)}$ is created with the concretization I_OBJECT$^{(I)}$. This link is inherited by the concept PE_OBJECT. In the next step, the control algorithm tries to detect the parts of a modified perceptual object as they are modeled in the semantic network. For our bolt example this yields in instances for 'bolt head' and 'bolt thread' which are concretized by one instance of I_REGION respectively. Currently, these instances are based on the regions detected by the preprocessing. But we are working on an object-dependant segmentation relying on inter-object comparisons. During the instantiation process restrictions for position, color and shape are propagated in a model–driven way. Additionally, the restrictions of the actual focus are taken into account. If a successful instance of a perceptual object is created then it is added as part of PE_SCENE which refers to all objects in the scene detected so far. After this step, the focus is adapted due to the newly detected object and the next object hypotheses are processed.

3 Results

The proposed architecture has been investigated so far in the scenario of SFB 360. The task is the recognition of a set of wooden toy pieces ("Baufix"), which are a "bar" with three, five or seven holes, a "felly", a "cube", a "rhomb-nut" a "tyre", a "socket", a "ring", and "bolts" with round or hexagon head. The bolts have four different lengths. The objects may be freely arranged within the

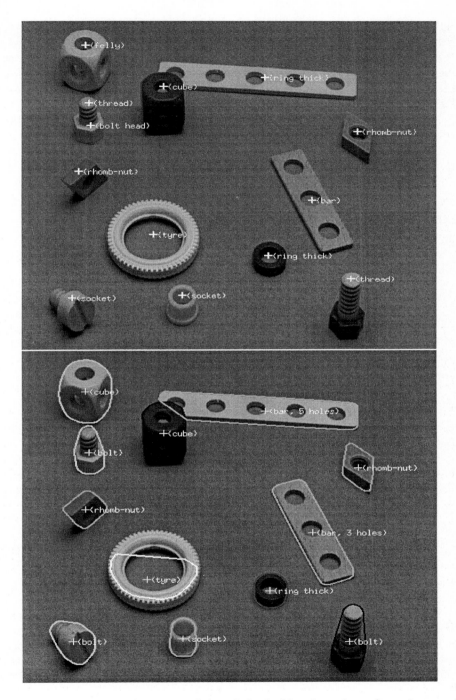

Fig. 2. Above: Wooden toys with best judged object hypotheses from the neural system, below: region boundaries and correct classification by the semantic analysis

range of a table from where the training images were taken as long as there is no occlusion, see Fig. 2. As a training set for the neural recognition system, 50 images of each part were used, for the bolts 200 images were used. On the training images the parts are arranged in different views and distances from the camera. By this way rotational invariance and scaling up to 30% were trained. For the LLM, 40 nodes approved to be the optimum.

The misclassifcation rate of the neural system is about 20%, it is reduced by the semantic analysis to about 10%.

4 Conclusion, outlook, and acknowledgement

We have presented a hybrid architecture for 3D-object recognition. Due to the hybrid architecture, knowledge acquistion becomes simple because using a semantic net we have the possibility of structuring, but avoid the difficulty of modeling knowledge about the sensor data explicitly by use of a neural network. By this means robustness and computational efficiency can be achieved.

Up to now the system is bound to a special geometric situation, because the ANN is trained only to a limited range of camera distance and angles. Adding an initialization phase, in which the semantic analyzer checks for camera distance and angle without help of the ANN, we want to get the parameters needed to choose a specialized ANN for the specific situation. After this initialization phase, the system could run as described here. This will be the aim of further investigation.

This work has been supported by the German Research Foundation (DFG) under SFB 360.

References

1. G. Heidemann and H. Ritter. A Neural 3-D Object Recognition Architecture Using Optimized Gabor Filters. In *Proceedings of 13th International Conference on Pattern Recognition, Vienna.* IEEE Computer Society Press, 1996. To appear.
2. T. Kohonen. Self-organization and associative memory. In *Springer Series in Information Sciences 8.* Springer Verlag Heidelberg, 1984.
3. F. Kummert, H. Niemann, R. Prechtel, and G. Sagerer. Control and Explanation in a Signal Understanding Environment. *Signal Processing, special issue on 'Intelligent Systems for Signal and Image Understanding',* 32:111–145, 1993.
4. R. Moratz, H.J. Eikmeyer, B. Hildebrandt, A. Knoll, F. Kummert, G. Rickheit, and G. Sagerer. Selective visual perception driven by cues from speech processing. In *7th Portuguese Conference on AI, EPIA95, Workshop on Applications of AI to Robotics and Vision Systems,* pages 63–72, Portugal, 1995. Trans Tech Pub. Ltd.
5. H. Ritter, G. Sagerer, G. Heidemann, and R. Moratz. Hybride Wissensrepräsentation: neuronale und semantische Netzwerke für die Bildanalyse. In *Arbeits- und Ergebnisbericht,* pages 27–65. Universität Bielefeld, SFB 360, 1995.
6. H.J. Ritter, T.M. Martinetz, and K.J. Schulten. *Neuronale Netze.* Addison-Wesley, München, 1992.

Robot Learning in Analog Neural Hardware[*]

A. Bühlmeier[1], G. Manteuffel[2], M. Rossmann[3] and K. Goser[3]

[1] Universität Bremen, FB-3, Postfach 330440, D-28334 Bremen, Germany
email: andreas@informatik.uni-bremen.de
[2] FBN, Wilhelm-Stahl-Allee 2, D-18196 Dummerstorf, Germany
[3] Universität Dortmund, LS Bauelemente der Elektrotechnik, D-44221 Dortmund,
Germany

Abstract. This paper describes a mobile robot that learned local maneuvers according to the principles of classical and operant conditioning, which were performed in an analog neural hardware implementation. The neurons were equipped with fixed gain and Hebbian inputs, each of which was low-pass filtered for short term memory effects. A wheelchair equipped with sonar and tactile sensors was used as a mobile robot that was able to steer autonomously through narrow doorways after learning an obstacle avoidance task. The system presented here performed operant conditioning in analog hardware controlling a physical mobile robot, which, to our knowledge, was not shown before.

1   Motivation

Emulating neural networks by analog electronic hardware rather than simulating on a sequential computer is a way to maintain some of the superior features of natural signal processing systems (Mead, 1990). However, analog neural networks did not result in too many practical applications yet, because of inaccuracies, e.g. offsets and variations on the chip, and other problems. For successful applications of analog learning hardware, we need a deeper understanding of how to enhance the system's robustness, e.g. by considering learning rules that do not rely on high numerical accuracy. To achieve this, it should be promising to investigate which mechanisms of animal learning can be applied to learning in the real world using analog hardware.

Conditioning is one example in animal learning that is relatively well understood and can be modeled on the basis of Hebbian learning, which is more appropriate for an analog implementation compared to, e.g., Backpropagation (Card *et al.*, 1994). The idea behind this paper is that the interconnection of 'analog' neurons can be used to construct robust adaptive controllers in hardware. Analog neurons as defined here possess the following properties: continuous input, output and weights, Hebbian learning, leaky integrator neuron model, all connections are used only unidirectionally.

The main advantages of this rather neuromorphic (Douglas *et al.*, 1995) approach may be summarized: i) true coincidence of signals is possible, ii) real-time

---

[*] This work was supported through the DFG and the Freie Hansestadt Bremen.

performance is inherent, i.e. response delays do not depend on the number of incoming signals, iii) physical properties of electronic components can be used to perform otherwise expensive computations, iv) compared to a standard software solution, higher robustness against hardware failures is achieved, v) high power efficiency, vi) similar constraints as in natural systems provide a better link to neuroscience. (See also Bühlmeier (1996) for more details and evidence.)

2 Experimental Setup

2.1 The Robot

An off-the-shelf electrically driven wheelchair (Meyra (Trademark) model 3.422) was used as an experimental platform (see figure 1). There were two driven wheels in the front and two steering wheels in the back. Steering this type of wheelchair is very similar to driving a car backwards. We added a frame of aluminum to attach bumpers, tactile and sonar sensors. The robot's width was 70cm and the length including bumpers amounted 127cm. Using a wheelchair instead of a standard holonomic [4] research robot provides more challenging problems of local navigation. In narrow places, for example, the wheelchair had to combine forward and backward movements to perform a desired turn which a holonomic robot can execute directly. On the other hand, in many outdoor applications experimental research robots would be probably unable to move at all. Analog neural circuits were mounted in a rack located on the arm rests. These circuits mimicked properties of natural neurons, especially adaptive Hebbian synapses and low pass filter characteristics; spiking was not implemented. The neuromimetic analog circuits ('Analog Neurons', see also german patent DPA Nr. 4137863, hold by Prof. G. Manteuffel) are based on earlier work (Manteuffel, 1992). Neurons were connected to other neurons, sensors and motor control.

2.2 The Sensors

Sonar measurements were acquired by microcontroller based circuits, which supplied an analog voltage that decreased with the distance between the sensor and an object. Sonar with a wide detection angle were used to provide 'near' signals, indicating obstacles closer than 60cm. Narrow angle sonar sensors were used as a 'far' signal detecting obstacles up to 100cm.

Two front and two back bumpers and tactile sensors on each side were used for collision detection. The measured steering angle was processed to obtain two different signals for left and right hand steering.

2.3 The Neuromimetic Circuits

One analog hardware neuron occupied one printed circuit board with a size of 100mm × 320mm. Using discrete components on the relatively large board enhances the flexibility to perform changes quite easily as compared to a VLSI

---

[4] A holonomic vehicle is able to perform a turn with any desired steering angle.

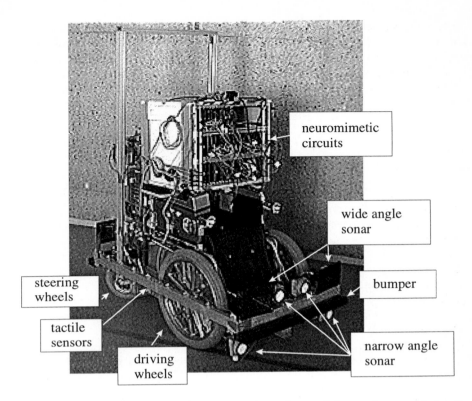

Fig. 1. A wheelchair equipped with sensors and a rack containing analog neural circuits served as an experimental platform.

design. The neuromimetic circuit's most important properties can be summarized as follows:

- **Inputs:** excitatory and inhibitory, i.e. inputs with a fixed gain of 0.5 and -0.5, respectively, and a Hebbian input with a variable gain up to 0.5. Each input is low pass filtered and can be shunted by an additional shunting input.
- **Learning:** heterosynaptic Hebbian learning with active and passive decay
- **Accuracy:** standard components were used, i.e. standard operational amplifiers, resistors and capacitors; weight storage was implemented by gold cap capacitors.
- **Weighting:** the modifiable gain of the Hebbian input was implemented by a highly nonlinear multiplier consisting of one (J-FET) transistor.
- **Transfer function:** The output was computed as the weighted sum of the input. Non-linearities are included by i) clipping the output to the range of 0 ... 9.5 volts, ii) amplifier saturation effects, iii) the ability to shunt inputs and iv) an adjustable offset at the output. No additional non-linearities (like a sigmoid function) were implemented.

Heterosynaptic hebbian learning as implemented here, requires that an input (in_i) is active (above a threshold θ_1) and the sum of all other weighted inputs $(\sum_{j \neq i} w_j in_j)$ is also greater than a second threshold (θ_2). Weights w saturate at an upper limit w_{max}. To prevent further weight increment after one weight reached the upper limit, learning is not possible after the sum of weights exceeded a threshold θ_3. For the i-th input in_i change of the modifiable weight w_i is:

$$
dw_i/dt = \begin{cases} K_1(w_{max} - w_i) & : \quad \text{if } in_i \geq \theta_1 \wedge \sum_{j \neq i} w_j in_j \geq \theta_2 \wedge \sum_j w_j < \theta_3 \\ -K_2 w_i & : \quad \text{if } in_i \geq \theta_1 \wedge \sum_{j \neq i} w_j in_j < \theta_2 \\ -K_3 w_i & : \quad \text{else} \end{cases}
$$

$$\tag{1}$$

where $\theta_1 = 2.35$ V, $\theta_2 = 5.9$ V, $\theta_3 = 5.35$ V, $1/K_1 \approx 33$ seconds, $1/K_2 \approx 400$ seconds, $1/K_3 \approx 26h$ and $w_{max} = 5$ V.

2.4 Classical and Operant Conditioning Network

Figure 2 depicts the structure that implemented classical and operant conditioning in the experiment described here (see Bühlmeier (1996) for full details). Each rectangle in this figure represents one or more neurons.

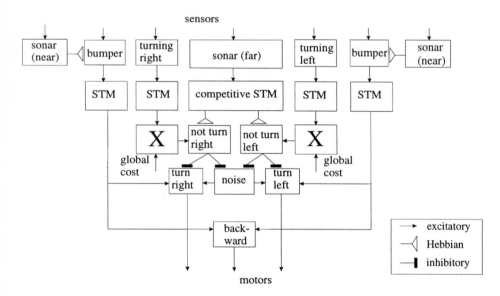

Fig. 2. A structure for classical and operant conditioning. Open triangles denote Hebbian, arrows fixed and thick horizontal lines inhibitory connections. Left and right hand side 'near' sonar signals were connected to both bumper neurons. The differences among 'far' sonar readings were enhanced in a competitive STM. All four STM signals of 'far' sonars were provided to both, the 'turn left' and the 'turn right' sonar. The 'X' depicts a multiplier, which was implemented with the help of shunting inputs. Each rectangle in this figure represents one or more neurons See text for details.

The naive robot's behavior was determined by reflexes (i.e. going backwards when a bumper was hit) and exploration (going forward with random steering). Reflexes constitute the basis for classical conditioning: the robot's bumper signal triggered a retreat movement (go backwards and turn away from the bump location). Since the retreat movement should be performed for a specific time, a short term memory (STM) was implemented that stores the bumper signal, and after association, sonar signals. Each STM circuit was realized by a positive feedback loop of two neurons. In the case of a collision, typically both, a bumper and one of the 'near' sonar signals were active at the same time and hence, the sonar signal of one side became associated to the bumper on the same side according to Hebbian learning. Hence, after a couple of collisions the robot will be able to anticipate the bumper signal, i.e. the robot will use the sonar signals to trigger the retreat movement before the bumper is touched.

In operant conditioning an unspecific reward or punishment is delivered. Typically, this reinforcement signal is assigned to the last action being performed and to the specific context. The principle of operant conditioning was applied here to adapt appropriate forward steering using 'far' sonar signals. In the implementation shown here the robot's task was to go forward and, hence, a signal indicating that the robot drove *backward* was chosen as an unspecific cost signal. The agent explored different actions and learned to avoid those actions in specific situations that resulted in 'cost' signals in these specific situations. For operant conditioning, the most important step is to obtain a specific signal to control which connections should be increased. In the model proposed here, the proprioceptive signal, indicating a specific action had been performed, was multiplied with the global cost signal (negative reinforcement) and was thus serving as a signal assessing that the specific action had not been appropriate in the last situation. The specific situation was indicated by a representation in a short term memory (STM). Time constants of the 'far' sonar STM and the proprioceptive STM were approximately 1.5s, so that signals in both STM were temporally related. Hebbian association of the specific situation to the specific action caused a suppression of this action in the specific situation.

Both, the 'turn left' and the 'turn right' neurons received fixed input (not shown in figure 2). Since the resulting steering angle depended on the difference of both direction inputs, the opposite steering angle of the suppressed direction was performed. Exploration signals were provided by a low frequency noise generator, which is depicted in the lower middle of figure 2. The influence of noise signals was suppressed when strong associations were formed to 'not turn right' and 'not turn left' units.

3 Experimental Results

As a result of the exploration signal, the naive robot went forward and randomly steered left or right until a collision forced the robot to go backwards. In this case, global cost and a proprioceptive signal were active simultaneously and triggered the 'not turn right' or 'not turn left' neuron, which associate sensor data from the short-term memory above. As a result, the robot learned which steering angles should be suppressed in specific situations.

In the case of an collision, the bumper signals became associated to 'near' sonar signals according to the principle of classical conditioning. The left side bumper was associated to the left 'near' sonar signal and vice versa. Both, the forward detour and the retreat movement were adapted in less than 10 minutes. Exploration of steering angles and the association of sonar sensor readings to unsuccessful steering commands enabled the robot to circumvent obstacles without going backward. Left 'far' sonar signals resulted in right hand steering and vice versa. Since the sonar connected to the forward steering possess a narrow detection angle of approximately 10°, the robot was able after learning to steer through a doorway which left only 12cm on each side.

Using the same hardware, a variety of other networks for the adaptation of feedback loops, velocity control, visual preprocessing, parallel learning with different learning rates and others were implemented and will be described by Bühlmeier (1996).

4   Conclusion

This paper described a network for classical and operant conditioning implemented by analog neural circuits. The wiring of reflexes, which lay the basis for classical conditioning, does not impose an additional effort in most cases, since protective circuits are often found in physical systems. For operant conditioning a reinforcement or cost signal has to be defined to enable an assessment of sensor data and to allow sensory motor couplings that result in minimized costs. Classical conditioning as performed here adapted obstacle avoidance and anticipated the cost signal that served as a reinforcement signal for the network's operant conditioning part. An analog VLSI implementation is planned and a digital implementation was performed by Rossmann et al. (1996).

References

BÜHLMEIER, A. 1996. *Analog Neural Networks in Autonomous Systems*. Ph.D. thesis, Universität Bremen, Germany. forthcoming.

CARD, H.C., DOLENKO, B.K., McNEILL, D.K., SCHNEIDER, C.R., & SCHNEIDER, R.S. 1994. Is VLSI Neural Learning Robust Against Circuit Limitations ? *Pages 1889–1893 of: Proc. 1994 IEEE International Conference on Neural Networks.*

DOUGLAS, R., MAHOWALD, M., & MEAD, C. 1995. Neuromorphic Analogue VLSI. *Ann. Rev. Neuroscience*, **18**, 255–281.

MANTEUFFEL, G. 1992. Neuronal Analog Models with Plastic Synapses for Neurobiological Teaching and Robotics. *Pages 155–162 of:* SCHUSTER, H.G. (ed), *Applications of Neural Networks*. VCH Verlagsgesellschaft.

MEAD, C. 1990. Neuromorphic Electronic Systems. *Proc. IEEE*, **78**, 1629–1636.

ROSSMANN, M., HESSE, B., GOSER, K., BÜHLMEIER, A., & MANTEUFFEL, G. 1996. Implementation of a Biologically Inspired Neuron-Model in FPGA. *Pages 322–329 of: Proceedings of MicroNeuro 96.*

Visual Gesture Recognition
by a Modular Neural System

Enno Littmann[1], Andrea Drees[2], and Helge Ritter[2]

[1] Abt. Neuroinformatik, Universität Ulm, Oberer Eselsberg,
D–89069 Ulm, FRG; enno@neuro.informatik.uni-ulm.de
[2] AG Neuroinformatik, Technische Fakultät, Universität Bielefeld,
D–33501 Bielefeld, FRG; andrea,helge@techfak.uni-bielefeld.de

Abstract. The visual recognition of human hand pointing gestures from stereo pairs of video camera images provides a very intuitive kind of man-machine interface. We show that a modular, neural network based system can solve this task in a realistic laboratory environment. Several neural networks account for image segmentation, estimation of hand location, estimation of 3D-pointing direction, and necessary transforms from image to world coordinates and vice versa. The functions of all network modules can be learned from data examples only, by exploiting various learning algorithms. We investigate the performance of such a system and dicuss the problem of operator-independent recognition.

1 Introduction

The fields of robotics and virtual reality currently face a rapid technological change. This challenges us to develop new and more powerful interfaces that allow humans to configure and control such devices. These interfaces should be intuitive for the human advisor and comfortable to use. Practical solutions so far require the human to wear a device that can transfer the necessary information. One typical example is the data glove [11, 10]. Clearly, in the long run solutions that are contactless will be much more desirable, and vision is one of the major modalities that appears especially suited for the realization of such solutions.

One such solution would be the guidance of robot pick-and-place actions by unconstrained human pointing gestures in a realistic laboratory environment — a still restricted but very important task in robot control. The input of target locations by pointing gestures provides a powerful, very intuitive and comfortable functionality for a vision-based man-machine interface for guiding robots and extends previous work that focused on the detection of hand location or the discrimination of a small, discrete number of hand gestures only [9, 1, 2, 6]. Besides two color cameras, no special device is necessary to evaluate the gesture of the human operator.

In the present paper, we focus on the visual recognition of human pointing gestures. We are particularly interested in the investigation of how to build a neural system for such a complex task from several neural modules. The development of advanced artificial neural systems challenges us with the task of finding architectures for the cooperation of multiple functional modules such that part of the structure of the overall system can be designed at a useful level of abstraction, but at the same time learning can be used to create or fine-tune the functionality of parts of the system on the basis of suitable training examples.

2 Image Processing

Environment: The recognition is performed in a complex laboratory environment. A test person is positioned at one side of a table that is covered with a black 10x10 grid on a yellow surface. This scenery is observed by two cameras yielding a frontal and a lateral image from an upper oblique viewpoint. The person points with one hand at one of the crossings of the grid. The cameras provide a stereo pair of color images of the scene. The task for the system is to determine from the camera images the 3D-pointing direction of the hand and the referenced 2D-location on the table surface.

Training Data: Two persons pointed at all 100 crosspoints of the 10x10 grid, thus yielding a total of 400 camera images consisting of 320x240 color pixels. For the hand segmentation, five images (pointing at the four corners and to the center) were segmented manually. One is used as training image, the others for the evaluation of the classification performance. For all 400 images, the actual hand center was determined manually. This information is used for the evaluation of the hand localization accuracy. The training set for the pointing recognition consists of 50 images per person and camera view arranged in a zigzag grid as indicated in fig. 6. The other 50 images serve as test set.

Segmentation and Localization: The segmentation of the human hand from the background scenery is performed by a LLM network consisting of 30 nodes. Currently, we train the LLM net on a single labeled image only. The training is based on the local color information. One training step consists of presenting one color pixel together with the corresponding label ("+1" for "hand" and "−1" for "background"). The number of training steps depends on the image size. In low resolution (40x30) we perform 10^4 steps ($\approx$ 8 epochs), in high resolution (320x240) 10^5 steps ($\approx$ 2.5 epochs).

After training the network can be used as a filter for the pixelwise segmentation of an image. For each image pixel, it yields a scalar value which can be interpreted as a probability value for this pixel to belong to the hand region. Fig. 2 (right) shows the segmentation result for a low resolution image, fig. 3 (left center) for a subframe of the high resolution image. Black dots indicate high probability for "hand". More details of this method can be found in [5].

The segmentation consists of two steps: (1) The images are segmented in *low* resolution and the hand position is extracted by calculating the centroid of the resulting activity image (e.g. fig. 2 (right)). (2) A small subframe around the estimated hand position (fig. 3 (left)) is processed in *high* resolution by another dedicated LLM network (fig. 3 (left center)). The centroid determined in the second image serves as reference point for the pointing recognition.

Pointing direction: The extraction of the pointing direction is based on a method proposed in [8]. It requires an image containing the hand only (fig. 3 (right center)) as provided by a threshold operation on the segmented subframe described above. This image is filtered by 36 Gabor masks arranged on a 3x3 grid with 4 directions per grid position and centered on the image. Filter kernels have a radius of 40 pixels, distance between the grid points is 20 pixels. The 36 filter responses (fig. 3 (right)) form the input vector for a LLM net.

Fig. 1. Stereo image pair. Scenery observed by camera A *(left)* and B *(right)*.

Fig. 2. Low resolution image of fig. 1 *(left)* and neural activity response *(right)*.

Fig. 3. Processing stages of the pointing direction recognition. The intensity subframe of the original image containing the hand *(left)* is weighted with its hand activity values. The resulting image *(left center)* is postprocessed *(right center)*. On this image the grid of Gabor masks yield a response shown *(right)*.

One major problem in recognizing human pointing gestures is the variability of these gestures and their measurement for the acquisition of reliable training information. Different persons follow different strategies where and how to point. Therefore, this information is gained indirectly. The person is told to point at a certain grid position with known world coordinates. From the camera images we extract the pixel positions of the hand center and transform them to world coordinates. Given these coordinates the angles of the vector describing the intended pointing gesture can be calculated trigonometrically. These angles form the target vector for the supervised training of a LLM network with 10 nodes.

Independent training and test set consist of 50 images each. 500 training steps are performed, i.e. 10 epochs. Each camera view and each person requires an own network. After training, the output of the net is used to calculate the point where the pointing vector intersects the table surface. For evaluation of the network performance we measure the Euclidian distance between this point and the actual grid point where the person intended to point at.

3 Results

An appropriate measure for segmentation performance is the relation of specifity — number of pixels *correctly* assigned to "hand" divided by the *total number* of pixels assigned to "hand" — and sensitivity — number of pixels *correctly assigned* to "hand" divided by the total number of pixels actually *belonging* to "hand". Fig. 4 shows such a curve for the LLM trained on the high resolution images of camera A applied to the test set images. The graph shows regions where specifity and sensitivity reach high values simultaneously which is equivalent to a good segmentation performance. For the fine segmentation, where we need this performance only on a subframe, the results are even more favorable. The localization accuracy is calculated by measuring the pixel distance between the centroid determined on the manually segmented image and the centroid of the activity image provided by the neural network. Table 1 gives quantitative results.

On the whole, the two-step cascade of LLM networks yields for *399 out of 400 images* an activity image precisely centered on the human hand. Only in one image, the first LLM net missed the hand completely, due to a second hand in the image that could be clearly seen in this view. This image was excluded from further processing and from the evaluation of the localization accuracy.

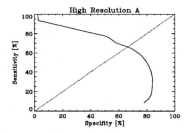

Fig. 4. Specifity/sensitivity curves for the high resolution images of person B on test images of camera A.

	Camera A		Camera B	
	Pixel deviation	NRMSE	Pixel deviation	NRMSE
Person A	0.8 ± 1.2	0.03 ± 0.06	0.8 ± 2.2	0.03 ± 0.09
Person B	1.3 ± 1.4	0.06 ± 0.11	2.2 ± 2.8	0.11 ± 0.21

Table 1. Estimation error of the hand localization on the test set. Absolute error in pixels and normalized error for both persons and both camera images.

Recognition of pointing direction: Fig. 5 shows the euclidean error MEE of the estimated target position as a function of the number of learning steps. "Manual" curves are obtained by centering the Gabor masks manually on the hand, "neuronal" curves by the neural segmentation. This allows us to study the influence of the error of the segmentation and localization steps on the pointing recognition. This influence is rather small. The MEE increases from 17 mm for the optimal method to 19 mm for the neural method, which is hardly visible in practice.

An interesting effect can be seen in fig. 6. The table grid points can be reconstructed according to the network output. If we apply the network to the trained person both training and test grid (fig. 6 (left, center)) can be reconstructed quite exactly. If we apply the network to another person, the error rises to $\approx$ 60 mm. However, the graph reveals a systematic deviation where the grid structure is still clearly visible (fig. 6 (right)). This deviation is probably caused by a different pointing strategy that might be compensated by calibration.

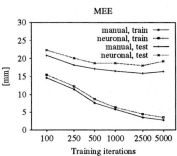

Fig. 5. The euclidean error of the estimated target point calculated using the network output depends on the preprocessing.

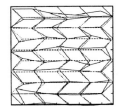 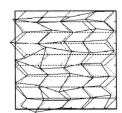 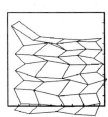

Fig. 6. The table grid points can be reconstructed according to the network output. The target grid is dotted. Reconstruction of training grid *(left)* and test grid *(center)* for the test person, and the reconstruction of the test grid for another person *(right)*.

4 Discussion

There are many practically important problems that admit an approach with a single neural network, trained to perform the desired function. Within the man-machine interface domain we have recently developed a number of single-network solutions to more resticted problems [7, 3, 5]. The majority of tasks, however, is too complex to be solved in such a simple fashion. Evidently, the next important transition step for the discipline of "neural engineering" will be the development of a capability for the systematic construction of comprehensive *neural systems* in which the functions of multiple neural networks become integrated in a modular way. To study some of the issues that are involved in the construction of neural systems, we have begun to integrate our single-network solutions into a larger system for the solution of the demanding and practically important vision task of recognizing human pointing gestures in video images of a complex laboratory environment. This integrated system allowed us for the first time to evaluate the interaction of the submodules in a complex setting.

The chosen pointing task has the favorable property that despite the complexity of the overall task system performance still can be measured in a single meaningful number (the accuracy of the inferred pointing direction) so that the comparison of alternatives becomes significantly facilitated. In our environment, we were able to achieve the rather impressive accuracy of only 2 cm deviation from the target point in a workspace of 50×50 cm for the test data of the same person. The implemented system thus demonstrates the solution of a rather complex recognition task with a system that for its function almost exclusively relies on neural components. This system has been further integrated in the SEE-EAGLE Environment where a human can guide robot pick-and-place actions by pointing gestures [4]. Further studies will aim at enhancing the robustness of the pointing recognition to varying lighting conditions and different persons.

Acknowledgements: We want to thank Th. Wengerek, J. Walter, and P. Ziemeck for hardware and software support. This work was supported by BMFT Grant No. ITN9104AO.

References

1. T. J. Darell and A. P. Pentland. Classifying hand gestures with a view-based distributed representation. In J. D. Cowan, G. Tesauro, and J. Alspector, editors, *Neural Information Processing Systems 6*, pages 945–952. Morgan Kaufman, 1994.

2. J. Davis and M. Shah. Recognizing hand gestures. In J.-O. Eklundh, editor, *Computer Vision — ECCV '94*, volume 800 of *Lecture Notes in Computer Science*, pages 331–340. Springer-Verlag, Berlin Heidelberg New York, 1994.

3. F. Kummert, E. Littmann, A. Meyering, S. Posch, H. Ritter, and G. Sagerer. Recognition of 3d-hand orientation from monocular color images by neural semantic networks. *Pattern Recognition and Image Analysis*, 3(3):311–316, 1993.

4. E. Littmann, A. Drees, and H. Ritter. Visual gesture-based robot guidance with a modular neural system. In D. Touretzky, M. Mozer, and M. Hasselmo, editors, *NIPS 8*. Morgan Kaufman Publishers, San Mateo, CA, 1996. To appear.

5. E. Littmann and H. Ritter. Neural and statistical methods for adaptive color segmentation — a comparison. In G. Sagerer, S. Posch, and F. Kummert, editors, *Mustererkennung 1995*, pages 84–93. Springer-Verlag, Heidelberg, 1995.

6. C. Maggioni. A novel device for using the hand as a human-computer interface. In *Proc. HCI'93 — Human Control Interface*, Loughborough, Great Britain, 1993.

7. A. Meyering and H. Ritter. Learning 3D hand postures from perspective pixel images. In I. Aleksander and J. Taylor, editors, *Artificial Neural Networks 2*, pages 821–824. Elsevier Science Publishers B.V., North Holland, 1992.

8. A. Meyering and H. Ritter. Learning 3D shape perception with local linear maps. In *Proc. of the IJCNN*, volume IV, pages 432–436, Baltimore, MD, 1992.

9. Steven J. Nowlan and John C. Platt. A convolutional neural network hand tracker. In *Neural Information Processing Systems 7*. Morgan Kaufman Publishers, 1995.

10. K. Väänänen and K. Böhm. Gesture driven interaction as a human factor in virtual environments – an approach with neural networks. In R. Earnshaw, M. Gigante, and H. Jones, editors, *Virtual reality systems*, pages 93–106. Academic Press, 1993.

11. T. G. Zimmermann, J. Lanier, C. Blanchard, S. Bryson, and Y. Harvill. A hand gesture interface device. In *Proc. CHI+GI*, pages 189–192, 1987.

Tracking and Learning Graphs on Image Sequences of Faces*

Thomas Maurer[1] and Christoph von der Malsburg[1,2]

[1] Ruhr-Universität Bochum, Institut für Neuroinformatik, 44780 Bochum, Germany
[2] University of Southern California, Dept. of Computer Science and
Section for Neurobiology, Los Angeles, USA

Abstract. We demonstrate a system capable of tracking, in real world image sequences, landmarks such as eyes, mouth, or chin on a face. In a first version knowledge previously collected about faces is used for finding the landmarks in the first frame. In a second version the system is able to track the face without any prior knowledge about faces and is thus applicable to other object classes.

1   Introduction

In the last decade there has been much development in the area of face recognition: Systems are being built to track a person in a scene, extract the head and recognize the face. Especially recognition of frontal views has become robust against changes of expression [1] and shows good performance even on large databases with hundreds of images [2, 3]. Indeed, the first face recognition systems have already left the laboratory and are being deployed under realistic conditions [4]. But these successful systems need (at least approximately) frontal face images. When the head is rotated appreciably ($> 30°$) from the frontal view in any direction, the performance of all systems decreases dramatically. Although different attempts have been made to overcome this problem by learning some sort of transformation from one pose or view to another [5, 6], none of those systems seems to have the capability of view-independent face recognition for practical applications in the near future. Nearly all systems work on static images taken from different views, and then apply time intensive methods like flow field computation (e.g., [5]) to find corresponding points in these images, or simply match corresponding points by hand. But as long as we don't have easy-to-use tools to feed a computer with the knowledge of how a face (or, for that matter, any object) looks like from all directions together with the information on point correspondence between views, there will be no chance to learn 3D representations. This is also true in the general case of 3D object recognition.

Consequently, if we want to achieve progress with 3D face and object recognition, we should provide computers with the same input that natural systems

---

* Supported by the German Federal Ministry of Science and Technology and by the US Army Research Lab.

have: with image sequences continuous not only in space but also in time. If image sequences are taken with sufficiently high frame rates, it should be possible to track the points over the whole sequence, there then being only small changes from frame to frame. In this way, the correspondence problem should be much easier to solve than in sparse static images. This is the strategy we will follow in this paper.

2   Brief description of the face finding system

As the basic visual feature we use a local image descriptor in the form of a vector called 'jet' [1]. Each component of a jet is the filter response of a Gabor wavelet of specific frequency and orientation, extracted at some point **x**. We typically use wavelets of four different frequencies and eight different orientations, for a total of 32 complex components. Such a jet describes the area surrounding **x**. A set of jets taken from different positions form a graph which describes an object in the image. To compare jets and graphs, similarity functions are defined: The normalized dot product of two jets yields their similarity, the sum over jet similarities being the similarity of graphs of identical geometry. Additionally, distortions of the grid can be taken into account to compute graph similarity [1]. Features and similarities defined in this way are robust against changes in illumination and contrast. In order to create an appropriate face graph for a gallery of 50–80 frontal faces of equal size, we place the nodes by hand at landmarks that we subjectively consider important to the task, e.g., at the center of the eyes or at the tip of the nose [2]. There are typically 20–40 nodes forming a face graph. Such a gallery constitutes the general face knowledge for frontal faces. For a new frontal face of approximately the same size these nodes of the graph can then be found automatically, even if this person is not in the gallery [2].

3   Tracking of face graphs

Given a sequence of images taken with a sufficient frame rate (>10 Hz) and showing the head of a human subject moving and rotating, the task is to track the landmarks as represented by the graph's nodes over the sequence. In order to get these nodes in the first frame as a starting point, we told the subjects to initially look straight into the camera. Then, the process for finding landmarks in frontal faces described above [2] can be used to place the graph on the first image automatically. The nodes of this graph were chosen *ad hoc* and are not necessarily optimal for tracking; how to find optimal nodes will be described in the next section.   Although we demonstrate our system on faces here, we don't want to be restricted to this case. Therefore, we avoid using any model knowledge or assumptions on the tracking device; only continuity in time (frame rate high enough) is needed. We do not make use of the relative movement of the nodes; they are tracked independently, and the edges of the graphs are for

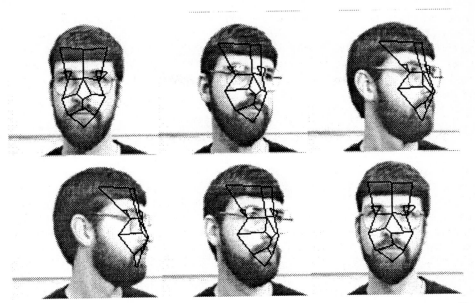

Fig. 1. Sequence of 50 frames of a face rotating from frontal view to right profile and back to frontal. Only frames 1, 10, 20, 30, 40, and 50 are shown, the nodes are placed and tracked completely automatically. The light spots on the glasses caused inaccuracies by shifting the eye nodes opposite to the movement direction, and some nodes (on hair and cheek) did not have enough local structure to track their position on the face properly.

illustration only. In this way we get a general tracking system, which can be further optimized for special cases if necessary.

To compute the displacement of a single node in two consecutive frames we use a method developed for disparity estimation in stereo images, based on [7] and [8]. The strong variation of the phases of the complex filter responses is used explicitly to compute the displacement with subpixel accuracy. Implemented as an iterative coarse-to-fine process, displacements up to half the wavelength of the kernel with the lowest frequency used can be computed (see [9] for details). Using jets with four different frequencies leads to a maximal displacement of 6–7 pixels. As already mentioned in the introduction, a much larger range would help only in the special case of a purely translational movement, in all other cases larger displacements are associated with greater changes in the image, and then the corresponding node position might not be found anyway. So in its simple form the tracking works as follows: For all nodes of the graph in frame n the displacement vectors with respect to frame $n + 1$ are computed, and then a graph is created with its nodes at these new corresponding positions in the new frame. But by doing so we would throw away all information from the preceding frames, and once an error occurs, this could never be repaired again. So we store all jets ever computed on the node positions and use them in the same way as

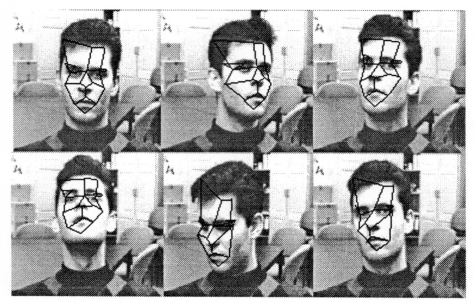

Fig. 2. Sequence of 50 frames of a person looking in different directions, with structured background. Only the frames 1, 15, 25, 30, 44, and 50 are shown to illustrate the trajectory of the head. Here the initial landmark finding misplaced some nodes on the left side as well as on the right eye. During tracking, the different viewing directions lead to varying shadows, resulting in inaccuracies especially in the eye region.

the general face knowledge for landmark finding: The displacement estimation is done using all jets already collected from frames 1 to n, and only the one with the highest confidence (i.e., jet similarity) is taken. This makes the tracking robust; Fig. 1 and 2 show graphs tracked over 50 frames without and with structured background. In Fig. 1 the tracking of the graph back from profile to frontal would not have been that accurate without this accumulation of knowledge.

4   Learning graphs

Given an image sequence, instead of starting with an initial graph in the first frame (placed automatically or, in the case of a new object, by hand), one can simply track *all* pixels from the first frame over a part of the sequence (more precisely, put a node on every pixel in the first frame and track all nodes) and somehow make use of this information. Of course we are only interested in nodes that could be tracked well, i.e., those that appear to be attached to the surface of the moving object. A first attempt could be to track all nodes forth and back on the image sequence and then choose those coming closest to their starting coordinates in the first frame. But this task is easier to fulfill by nodes in regions moving with lower velocity than by nodes in regions moving faster, and it is

Fig. 3. Graphs learned and tracked in three different sequences, the upper image being the start frame, the lower frame 30. While the graph has been learned only on the first 15 frames in the example on the left (it is the same sequence as shown in Fig. 1), it was tracked further to frame 50 looking exactly like frame 1. In the two examples on the right the graphs have been learned and tracked on the whole sequence of 30 frames. Only small inaccuracies can be found, e.g., the node on the chin in the middle example.

trivially fulfilled by nodes on the background. The path the nodes have taken during tracking must be taken into account as well. Also, if tracking errors occur, then mostly because the nodes could not follow, the movement of the object being too fast for them. Therefore we should preferably select nodes with larger movement. So our algorithm is as follows: We assume the object motion to have one major direction, which is the normal case for a short sequence. Now we track[3] all nodes from the first to the last frame, compute the average movement direction (movement vector averaged over all nodes, normalized to length 1), and compute the movement m_i of every node i made in this direction. Then we track all nodes back to the first frame on the same images and compute the deviations d_i of the end positions from the starting positions, which would ideally be zero. Now the nodes can be labeled with their cost values $c_i = d_i - m_i$, "good" nodes having low costs. To create a graph on the object consisting of nodes with minimal cost values, all nodes are sorted with respect to their cost and, starting with the best node, all nodes spatially too close to the good

---

[3] Note that we don't store the jets during tracking here, simply because there are so many. To find the positions in a new frame only the jets in the preceding frame are used. As long as the tracked nodes do not return to already seen regimes, this does not make much difference, because then the jets from the preceding frame are nearly always the most similar ones.

ones in the start frame are deleted (to avoid the jets having too much overlap, i.e., to keep redundancy small). Finally, the best N nodes are connected by a minimal spanning tree resulting in the graph optimal for tracking the object in this sequence. The number N of nodes is the only free parameter of the algorithm; we have chosen $N = 20$ in all examples (see Fig. 3). To summarize, this system can extract graphs of objects and track them completely automatically, without knowing anything *a priori* about those objects.

5   Conclusion and Outlook

We have demonstrated a system capable of tracking face graphs in natural image sequences. In a first version, predefined graphs are placed on the first frame automatically and then tracked over the sequence. In a second version, the graphs to be tracked are created automatically during the sequence. Continuity in time being the only constraint, the system is in no way restricted to faces. It can therefore serve as a basic tool to build 3D representations by collecting the information at the nodes tracked, e.g., to learn features that are robust with respect to variation of view point. On the other hand, to make the task of face tracking robust enough for everyday tasks, our system could be extended by incorporating a Kalman filter. It should not be too difficult to extract the shape- and motion parameters from the moving face graph, yielding a reliable pose estimator.

References

1. M. Lades, J.C. Vorbrüggen, J. Buhmann, J. Lange, C. von der Malsburg, R.P. Würtz, W. Konen, *Distortion Invariant Object Recognition in the Dynamic Link Architecture*, IEEE Trans. Comp., Vol. 42, No. 3, p. 300-311, 1993.
2. L. Wiskott, J.M. Fellous, N. Krüger, C. von der Malsburg, *Face Recognition and Gender Determination*, Proc. of the International Workshop on Automatic Face- and Gesture-Recognition (IWAFGR), p. 92, Zürich, 1995.
3. M. Turk & A. Pentland, *Eigenfaces for Recognition*, Journal of Cognitive Neuroscience, Vol. 3, No. 1, p. 71, 1991.
4. W. Konen & E. Schulze-Krüger, *ZN-Face: A system for access control using automated face recognition*, IWAFGR, p. 18, Zürich, 1995.
5. T. Poggio & D. Beymer, *Learning networks for face analysis and synthesis*, IWAFGR, p. 160, Zürich, 1995.
6. T. Maurer & C. von der Malsburg, *Learning Feature Transformations to Recognize Faces Rotated in Depth*, ICANN, Vol. 1, p. 353, Paris, 1995.
7. D.J. Fleet & A.D. Jepson, *Computation of component image velocity from local phase information*, Int. Journal of Computer Vision, Vol. 5, No. 1, p. 77, 1990.
8. W.M. Theimer & H.A. Mallot, *Phase-based binocular vergence control and depth reconstruction using active vision*, CVGIP: Image Understanding, Vol. 60, No. 3, p. 343, 1994.
9. L. Wiskott, *Labeled Graphs and Dynamic Link Matching for Face Recognition and Scene Analysis*, Verlag Harri Deutsch, Thun, Frankfurt a. Main, Reihe Physik, Vol. 53, 1995.

Neural Network Model Recalling Spatial Maps

Kunihiko Fukushima, Yoshio Yamaguchi and Masato Okada

Faculty of Engineering Science, Osaka University, Toyonaka, Osaka 560, Japan

Abstract. When driving through a place we have been before, we can recall and imagine the scenery that we shall see soon. Triggered by the newly recalled image, we can also recall other scenery further ahead of us. This paper offers a neural network model of such a recalling process. The model uses a correlation matrix memory for memorizing and recalling patterns. A correlation matrix memory by itself, however, does not accept shifts in location of stimulus patterns. In order to place stimulus patterns accurately at the location of one of the memorized patterns, we propose using the cross-correlation between the stimulus pattern and the "piled pattern".
A map of Europe is divided into a number of overlapping segments, and these segments are memorized in the proposed model. A map around Scotland is input to the model as the initial image. Triggered by the initial image, the model recalls maps of other parts of Europe sequentially up to Italy, for example.

1 Introduction

How does the brain memorize and recall spatial information of the external world? How is the spatial information decoded and represented in the brain? Neurophysiological and psychological experiments suggest that the spatial information of the external world is represented in the brain in a form of a topologically ordered spatial pattern like a map with an ego-centric coordinate system.

For example, this is suggested from observation of patients with a lesion in one cerebral hemisphere [1],[2]. A patient with a large lesion in the right cerebral hemisphere was requested to imagine himself looking at a famous building in a very familiar square in the city where he lived for many years. He reported several elements (buildings, streets, etc.) on his right side but none or very few on his left side, which is contralateral to the damaged cortex. When requested to imagine himself turned on his heels and to describe the square from the opposite perspective, he reported more elements on the right side of his mental image, which had previously been the left side of his mental image. This suggests that, when we are imagining a spatial map around us, the map is represented as a spatial activity pattern of neurons in some area of the brain, keeping the topology of the geographical features of the external world.

Let us consider a situation in which we are driving through a place we have been before. Even if our memory is ambiguous and we cannot draw a map from memory, we often can find the correct way to a destination when we actually go there. We can recall and imagine the scenery or geographical features that we cannot see yet but shall see soon. Triggered by the newly recalled image, we can also recall other scenery further ahead of us. We can thus imagine scenery of a wide area by a chain of recalling processes.

If the spatial information of the external world is represented in the brain as a topologically ordered spatial pattern like a map, the visual imagery recalled at one time with a certain resolution must be bounded in size because of the limited number of cells in the brain. Furthermore, spatial information memorized at one time is also limited because of the limited size of our visual field. However, we can combine two different scenes memorized at adjacent locations and create one unified image in our mind.

Even if we are brought to a place without being told where to go, a chain process of recalling spatial maps around us can start once we see a familiar scene. This suggests that the brain stores fragmentary maps without appending any information representing the locations of the maps.

This paper offers a neural network model of spatial memory of this type. The model stores spatial information in the form of fragmentary patterns like spatial maps. From the memory of fragmentary patterns, an image covering an infinitely wide area is retrieved by a continuous chain process of recalling.

2 The Model

Figure 1 illustrates the architecture of the proposed model. The model consists of a correlation matrix memory and many layers of cells. Each layer is a two-dimensional array of cells.

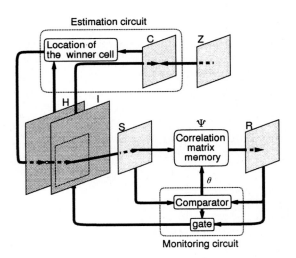

Fig. 1. Architecture of the proposed model.

In the computer simulation discussed later, a spatial pattern (or a line-drawing of a map) of a large area is divided into many pieces of fragmentary two-dimensional patterns ($N \times N$ in size) overlapping each other. These fragmentary patterns are then memorized in an associative memory circuit. Let $\boldsymbol{X}^{(k)}$ be the kth fragmentary pattern to be memorized, and $X^{(k)}(\boldsymbol{n})$ be the value (or brightness) of its pixel located at $\boldsymbol{n} \in P$. They are binary images, whose pixels take value 1 or 0. An upper bar, like $\bar{X}^{(k)}$, will be used to represent the average value of all pixels in a pattern.

The imagery layer $\boldsymbol{I}$, which represents the mental image in our mind, is somewhat larger $((2N-1) \times (2N-1))$ than the memorized fragmentary patterns. The mental image generated in the imagery layer is always represented in the form of a spatial pattern with an ego-centric coordinate system. When one moves, the mental image shifts opposite to the direction of body movement, so as to keep the body always at the center of the imagery layer. Since the mental image has only a finite size, a vacant region appears in the imagery layer after the shift. $\boldsymbol{I}$ is a binary pattern, but the values of the pixels in the vacant region in $\boldsymbol{I}$ are left undetermined.

In order to seamlessly fill the vacancy with a new pattern, the model searches the associative memory circuit Ψ for a pattern that tallies with the residual part of the mental image. If an appropriate pattern is found, it is united to the mental image to fill the vacancy in the imagery layer.

In the model, an auto-associative correlation matrix memory with recurrent connections is used to memorize and recall fragmentary patterns. As is well known, an auto-associative correlation matrix memory can recall a complete pattern from a portion of it, provided that a certain orthogonality condition among memorized patterns is satisfied. If K patterns have been memorized, the strength of the connections, namely, the elements of the correlation matrix Ψ, take the following values:

$$\psi(\boldsymbol{n}, \boldsymbol{m}) = \sum_{k=1}^{K} \{X^{(k)}(\boldsymbol{n}) - \bar{X}^{(k)}\} \{X^{(k)}(\boldsymbol{m}) - \bar{X}^{(k)}\}. \tag{1}$$

In the recalling phase, an area of $N \times N$ pixels, which has a vacant region, is cut out from $\boldsymbol{I}$ and is used as the stimulus $\boldsymbol{S}$ to the correlation matrix memory. Since a correlation matrix memory by itself does not accept shifts in location of stimulus patterns, it becomes necessary to place stimulus patterns accurately at the location of one of the memorized patterns. The estimation circuit in Fig. 1 determines where in layer $\boldsymbol{I}$ the stimulus pattern $\boldsymbol{S}$ should be cut out from.

At first, the estimation circuit calculates the cross-correlation function $C(\boldsymbol{\nu})$ between the stimulus pattern $\boldsymbol{S}$ and the "piled pattern" $\boldsymbol{Z}$. The "piled pattern" is made by a pixelwise sum of all patterns memorized in the correlation matrix, and is stored in the model together with the correlation matrix. That is,

$$Z(\boldsymbol{n}) = \sum_{k=1}^{K} X^{(k)}(\boldsymbol{n}). \tag{2}$$

The estimation circuit then searches for location $\boldsymbol{\nu}_0$ that maximizes $C(\boldsymbol{\nu})$. This becomes the estimated location of the center of $\boldsymbol{S}$ in layer $\boldsymbol{I}$, if it does not coincide with any of the estimated locations in the past, which are stored in layer $\boldsymbol{H}$. Incidentally, layer $\boldsymbol{H}$ stores the history of estimated locations, at which the centers of $\boldsymbol{S}$ have once been placed and recalling processes have been tried before. The response of layer $\boldsymbol{H}$ is made to shift in linkage with the shift of the mental image in layer $\boldsymbol{I}$, and is always expressed in the ego-centric coordinate system. This prevents the recall of the same patterns twice and also prohibits the repeated use of erroneously estimated locations.

Once $\boldsymbol{\nu}_0$ is determined by the estimation circuit, the area of $N \times N$ whose center is at $\boldsymbol{\nu}_0$ is cut out from $\boldsymbol{I}$ and is used as $\boldsymbol{S}$. $\boldsymbol{S}$ usually contains a missing region. The values of the pixels in this vacant region are replaced with the mean value of the pixels in the rest (non-vacant region) of $\boldsymbol{S}$.

Stimulus pattern $\boldsymbol{S}$ is fed to the auto-associative correlation matrix memory $\boldsymbol{\Psi}$. The output of the correlation matrix is calculated iteratively by

$$U_{t+1}(\boldsymbol{n}) = 1\left[\sum_{m \in P} \psi(\boldsymbol{n}, \boldsymbol{m})\{U_t(\boldsymbol{m}) - \bar{U}_t\} - \theta_t\right], \tag{3}$$

where $\boldsymbol{U}_t$ is the transient response of the cells of the correlation matrix at discrete time t, and the initial value $\boldsymbol{U}_0$ is set equal to the stimulus pattern $\boldsymbol{S}$. $1[x]$ is a threshold function, which takes value 1 or 0 depending on $x > 0$ or $x \leq 0$. Threshold θ_t is set so as to minimize the error, which is defined from the number of erroneously recalled pixels (or Hamming distance) in the non-vacant region of $\boldsymbol{S}$. The iterative calculation is repeated until the response reaches a steady state, in which we have $\boldsymbol{U}_{t+1} = \boldsymbol{U}_t$. The response at this steady state is the recalled output $\boldsymbol{R}$ from this auto-associative correlation matrix memory.

Although this recalling process sometimes fails by the error in estimation of $\boldsymbol{\nu}_0$, it usually does no harm because the model contains a monitoring circuit that detects the failure. The monitoring circuit compares $\boldsymbol{R}$ with $\boldsymbol{S}$ and measures the error rate ε, which is defined from the number of erroneously recalled pixels (or Hamming distance) in the non-vacant region of $\boldsymbol{S}$. If $\boldsymbol{R}$ gives ε smaller than a certain criterion value, the monitoring circuit judges that the recall is successful, and the recalled pattern $\boldsymbol{R}$ is united to the mental image to fill the vacancy of the imagery layer $\boldsymbol{I}$. If error rate ε is larger than the criterion, the monitoring circuit judges that the recall has failed, and discards the recalled pattern $\boldsymbol{R}$. The recalling process is repeated again after some period of time when the body has moved to another location.

In the computer simulation, the recalling process is started after every five pixels of movements of the body in the imagery layer $\boldsymbol{I}$, provided that more than 10% of the central $N \times N$ area of layer $\boldsymbol{I}$ is vacant.

3 Computer Simulation

In the computer simulation of the proposed model, a map of Europe (Fig. 2a), which consists of 240×240 pixels, is divided into many fragmentary patterns

(39×39 in size) overlapping each other as shown in Fig. 2b. Figure 2c displays these fragmentary patterns. These fragmentary patterns are then memorized in the associative memory circuit. At the same time, the piled pattern $\boldsymbol{Z}$, which is shown in Fig. 2d, is created and stored.

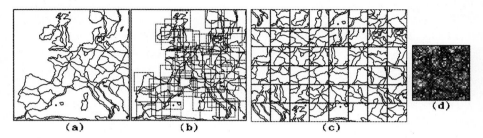

Fig. 2. Patterns memorized in associative memory.

We simulate a situation in which a person (or the model) catches a train at a city in Scotland and travels along a railway via Paris to Italy. A map, whose center is at the starting city, is first put at the central $N \times N$ area of the imagery layer $\boldsymbol{I}$. We will show that, triggered by the initial image, maps of other parts of Europe are recalled sequentially up to the south end of Italy.

In the model, the mental image generated in the imagery layer is always represented in an ego-centric coordinate system. When one moves, the mental image shifts in the direction opposite to the direction of body movement, so as to keep the body always at the center of the imagery layer. Since the process of tracing the railway is not the problem of interest in this paper, we simply use an arbitrary technique of line tracing here.

It should be noted here that the centers of the patterns memorized in the correlation matrix are not necessarily located on the railway to be traced. If they are all distributed on the railway, the problem becomes too easy and trivial, because the stimulus pattern need not be shifted for a correct recalling.

Figure 3 shows an example of the response of the model. A map centered at the starting city is first presented in the central area of the imagery layer. After a short excursion along a railway, we had an egocentric image $\boldsymbol{I}$ as shown in the figure. Now the cross-correlation $\boldsymbol{C}$ between $\boldsymbol{I}$ and $\boldsymbol{Z}$ is obtained. The maximum-output cell in $\boldsymbol{C}$ was located at $\boldsymbol{\nu}_0 = (-18, -10)$. The $N \times N$ area centered at $\boldsymbol{\nu}_0$ was cut out from $\boldsymbol{I}$, and $\boldsymbol{S}$ was produced. This stimulus $\boldsymbol{S}$ was input to the associative memory Ψ, and $\boldsymbol{R}$ was recalled. Error rate was $\varepsilon = 0.00$, and it was judged that the estimation of the location of the stimulus pattern $\boldsymbol{S}$ was correct. Therefore, $\boldsymbol{R}$ was shifted by $\boldsymbol{\nu}_0$ then united to $\boldsymbol{I}$. Thus, we had a new image $\boldsymbol{I}'$. Incidentally, $\boldsymbol{u}_1$ in the figure represents $\boldsymbol{U}_1$ before threshold operation, that is, the argument of the function in the left side of (3) after the first step of iterative calculation.

Figure 4 summarizes the patterns recalled during the trip from Scotland to

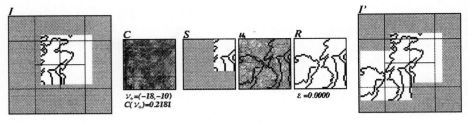

Fig. 3. Example of the model response.

Italy. Pattern <00> in the figure shows the map at the starting point of the trip, and the others are the patterns recalled sequentially during the trip. Below each pattern are shown the coordinates of the center of the pattern in the original map of Europe; the estimated relative location ν_0 from the center of I; and the number of pixels erroneously recalled, i.e., the Hamming distance between R and the pattern actually memorized in the correlation matrix.

It can be seen from this figure that, triggered by the initial image of Scotland, maps of other parts of Europe were recalled sequentially up to Italy without interruption. No error occurred in any of the recalled patterns throughout this trip. Although failures of recalling fragmentary patterns occurred sometimes by estimation errors of the location of S, they were all detected by the monitoring circuit, and erroneous uniting of irrelevant patterns was successfully prevented.

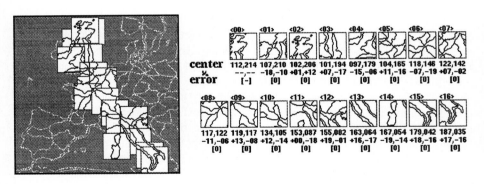

Fig. 4. Sequence of patterns obtained by a chain process of recalling during the trip from Scotland via Paris to Italy.

References

1. E. Bisiach, C. Luzzatti: "Unilateral neglect of representational space", *Cortex*, **14**, pp. 129–133 (1978).
2. C. Guariglia, A. Padovani, P. Pantano, L. Pizzamiglio: "Unilateral neglect restricted to visual imagery", *Nature*, **364**[6434], pp. 235–237 (July 1993).

Neural Field Dynamics for Motion Perception

M.A. Giese[1,2], G. Schöner[2], H.S. Hock[3]

[1] Institut für Neuroinformatik, Ruhr-Universität, D-44780, Bochum, Germany
[2] CNRS-CRNC, 31, Chemin Joseph Aiguier, F-13402 Marseille Cedex 20, France
[3] Dept. of Psychology, Florida Atlantic University, FL 33431 Boca Raton, USA

Abstract

In this paper we present a model for the perception of apparent motion based on dynamic neural fields. We show that both, the organization of percepts and their specific dynamical properties can be derived from the same perceptual dynamics. We show that neural fields adequately formalize this perceptual dynamics. This leads to a model that is sufficient in the sense that it reproduces correctly the perceptual organization of motion. We show that the model allows also to identify elements which are necessary for understanding perceptual organization together with the dynamic properties of the percepts (e.g. hysteresis and its link to switching rates). Such elements are an activation dynamics which is coupled to adaptation processes with different time-scales, and different mechanisms for perceptual switching. We show how the model can be linked quantitatively to a large set of psychophysical data for the motion quartet.

1 Introduction

Motion percepts have dynamic properties like temporal integration of relevant information [16] and visual inertia [4]. Moreover, motion perception shows essentially nonlinear properties like multi-stability and hysteresis [9,17]. Recently, these specific dynamic properties of motion percepts were measured in quantitative detail in psychophysical experiments [9,10]. In addition, it has been argued that cooperative neural networks can be used to solve the motion correspondence problem and allow to reproduce the formation of actual percepts [6,14]. This leads to the hypothesis that perceptual organization and the dynamic properties of percepts can be derived from a single *perceptual dynamics* signifying that the formation of motion percepts is based on a well defined dynamical process.

The aim of this paper is to evaluate this hypothesis based on psychophysical data using an adequate mathematical framework which allows to deal at the same time with a distributed representation of local information and dynamic self-organization phenomena. We propose the *dynamic neural field* as a suitable theoretical concept, in terms of which models can be formulated that can be related quantitatively to psychophysical data. We first show that the dynamic neural field model is *sufficient* to understand the perceptual organization of motion and its dynamical properties. Then we show how that model can be used to identify *necessary* components of the perceptual dynamics. This allows to construct a model that is *minimal* by containing only necessary components and nevertheless capturing the relevant dynamic phenomena of perceptual organization. We believe that this model reveals essential functional components of perceptual organization in general, and allows an analysis how these components interact dynamically during the formation of percepts.

2 Neural Field Model for the Perception of Apparent Motion

The model is based on the following basic concepts: (1) *Distributed neural representation of motion* in an abstract neural representation field: The percept can be characterized by a distribution of neural activation[1] u over a 4-dimensional perceptual space[2] (cf. [15]). It is parameterized by the retinal position of local motion (x, y) and the perceived motion vector in polar coordinates (v, ϕ). At each point of the activation field $u(\mathbf{z}; t) = u(x, y, v, \phi; t)$ the local activations u are thresholded by a sigmoidal function f. The value $f(u(\mathbf{z}; t))$ signifies the (gradual) presence of the corresponding motion specified by $\mathbf{z}$ ($f(u) \approx 1$: present, $f(u) \approx 0$: not present) in the percept. (2) *Neural dynamics:* Each point of the neural activation field is governed by a neural dynamics derived from a continuous neural network model by Amari [1]:

$$
\begin{aligned}
\tau \dot{u}(\mathbf{z}, t) &= -u(\mathbf{z}, t) + \int w(\mathbf{z} - \mathbf{z}') f(u(\mathbf{z}', t)) \, d\mathbf{z}' + \xi(\mathbf{z}, \mathbf{t}) \\
&\quad + S(\mathbf{z}, t) - h
\end{aligned}
\tag{1}
$$

We stress the following basic properties of the model: (1) The *time-scale* τ for the local neural dynamics is well-defined. (2) The threshold function, f, makes the system qualitatively *nonlinear.* (3) The neural field has recurrent *cooperative interactions* characterized by the convolution kernel $w(\mathbf{z}) = w(x, y, v, \phi)$. By choosing this interaction structure the space of stable dynamical patterns is strongly reduced, as compared, for instance, to general Hopfield-type (fully connected) network. The kernel also introduces a *topology* over the space of interactions which is crucial both, mathematically to generate stability properties of the neural field, and empirically to account for the topology of interaction effects in perceptual organization. The shape of the functional interaction kernel can be reconstructed from a set of psychophysical experiments. (4) *Stochastic* contributions, $\xi(\mathbf{z}, \mathbf{t})$, are not only necessary to account for fluctuations of the perceptual state, but also to prevent the field from relaxation to spurious (perceptually irrelevant) stable states[3]. For ambiguous stimulation, fluctuations force a perceptual decision. (5) The feed-forward *stimulus signal* $S(\mathbf{z}, t)$ may be viewed as output of a preceding stage of local motion detectors. It represents the capacity of the stimulus to specify the motion percept. Only stimulated local

---

[1]The neural activation should be interpreted as average activation of functional neural ensembles rather than as single neuron activity. A direct connection between neural field models and neurophysiological data can be achieved using adequate population coding techniques, c.f [11].

[2]To keep the model simple we restrict ourselves to the simplest case: 2-dimensional motion in the retinal plane.

[3]These states can be identified with local minima of the associated potential function.

motions ($S > 0$) participate in the cooperation process. (6) Finally, the constant parameter h fixes the perceptual threshold.

The percept results from two distinct contributions to the vector-field of the neural field dynamics: The *non-autonomous part* of equation (1) expresses the direct influence of the stimulus on the percept. The *autonomous part* restricts the space of possible stable solutions in a specific way which represents internal hypotheses of the nervous system about the structure of biologically meaningful visual stimuli. Examples are the smoothness of the motion-field (cf. [18]) or the Gestalt-law of "common fate".

3 Sufficiency of the Neural Organization Field

To evaluate if the proposed mathematical framework is sufficient to account for typical phenomena in motion perception we derived a set of critical properties of motion perception known from psychophysical experiments. (1) Localized motion stimuli lead to localized activation patterns (percepts) in the neural field. This reflects that coherently moving objects are represented by local motion percepts. (2) The model predicts the correctly organized percept for simple motion stimuli such as the motion quartet [9,10]. For the quartet this implies perceptual multi-stability and the suppression of potentially perceivable elementary motions. (3) Stable solutions of the model for random-dot stimuli correspond with solutions of the motion correspondence-problem. This shows that a one-layer neural dynamics is sufficient for the solution of the motion correspondence problem (cf. [8]). (4) Experimental results on *visual inertia* can be accounted for in *quantitative* detail (cf. figure 1). Inertia results from the continuous variation of the perceptual state over time with a well-defined time constant. (5) We were able to reproduce global cooperative effects (see figure 2 for an example, cf. [3]). We also reproduced the distance dependence of motion-entrainment/contrast effects, cf. [13]. (6) Presentation of ambiguous stimuli reveals dynamic stability properties of the model. We found multi-stability and hysteresis for motion-quartets and ambiguous random-dot stimuli (cf. [9] and [5]). (7) For symmetric ambiguous stimuli spontaneous symmetry-breaking occurs through fluctuations on an adequate intermediate time scale. Only one of the perceptual alternatives is stably observed over typical time scales [5,9]. (8) A quantitative account for the statistics of perceptual switching can be obtained for the motion quartet stimulus [10]. (9) The neural organization field reproduces the dependence of motion coherence and transparency on the angle difference of the component grating [12]. This is possible because the neural activation dynamics allows the co-existence of different motion percepts at the same position contrary to cooperative neural models which are based on a winner-takes-all mechanism. The highly cooperative perceptual organization dynamics shows a bifurcation as a function of the angular difference of the component gratings.

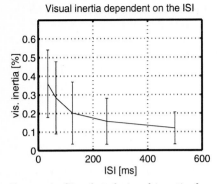

Visual inertia dependent on the ISI

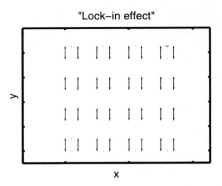

"Lock–in effect"

figure 1: Simulated visual inertia dependent on ISI corresponding to the results of Anstis & Ramachandran [4].

figure 2: Cooperative "lock-in" for multiple quartet-stimuli, cf. [3]. The length of the arrows corresponds to the perceptual strength of the local motion.

4 Necessary Components of the Perceptual Dynamics

By detailed *quantitative* comparison between the model and psychophysical results for *motion quartet* displays (figure 3) we were able to identify *necessary* components of the perceptual dynamics. The motion quartet is particularly well suited for quantitative assessment of the dynamic properties of biological percepts for two reasons: (1) Its temporal and geometrical structure can be quantified simply and completely. (2) The direct observation of switches between two distinct perceptual alternatives makes the dynamic properties of the percepts measurable.

Quartet stimuli produce local activated zones in the perceptual organization field (cf. figure 3). Integrating over each of the four (potentially) activated zones that represent the possible elementary motions[4] leads to a discrete network model with only four model neurons. This simplified model approximates the behavior of the complete neural field for the quartet stimulus. The interaction-matrix of the discrete model is *not* symmetric for quartet geometries other than square (aspect ratios other than 1.) That asymmetric case violates *detailed balance*[5] [7] and as a result the network has much more complicated stochastic behavior than Hopfield-networks.

The discrete approximation was fitted quantitatively to a large experimental data set [9,10]. This involved a mathematical and numerical analysis of the stable and unstable states, the stochastical dynamics for un-symmetric weight-matrices, the switching behavior and switching-time statistics. The quantitative

---

[4]Based on time-scale arguments we neglect the difference between forward and backward motion.

[5]This is equivalent to the absence of an equivalent classical potential for the dynamics

results reveal *necessary properties* of the perceptual dynamics: (1) The dynamics of the individual formal neurons in the discrete limit must be extended by two adaptation processes with very distinct time-scales. This can be mathematically expressed by introducing an adaptation dynamics for the thresholds h of the elementary motions. The biological processes which might be reflected by these time-scales might include the relatively fast intra-cortical inhibition ($\tau_1 \approx 600$ ms) and slower processes of neural fatigue ($\tau_2 > 60$ s). (2) Two different mechanisms of perceptual switching, stochastical fluctuations and adaptation, could be contrasted, and their relative contribution to perceptual switching have been determined. The quantitative results suggest that both mechanisms must substantially be involved in the formation of visual motion percepts.

Including these additional necessary components into the perceptual dynamics leads to a *minimal model* for the motion quartet that reproduces *quantitatively* a large set of experimental details: a) dependence of the stationary percept probabilities on the stimulus geometry (aspect ratio): The perceptual alternative with shorter motion matches occurs with higher probability. b) dynamic stability of the perceptual alternatives as measured with different methods: Displays with extreme aspect ratios are more stable that display with aspect ratio 1. c) stochastic switching behavior and switching-time statistics: More stable displays lead to less switches. The switching times vary over time. Switches can occur without time delay. d) reciprocal interdependence between adaptation and stability: stable percepts adapt more and adaptation changes the stability.

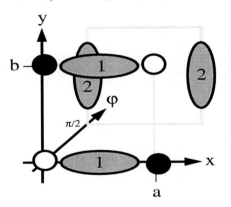

figure 3: The motion quartet consists of two pairs of dots flashing in alternating sequence (black dots: first frame; white dots: second frame; etc.). The geometry is determined by the side-length a and b. The percept is multi-stable and depends on the aspect ratio b/a. The stimulus causes local activated zones in the 3-dimensional projection of the motion organization field. Exclusively, either horizontal motion (zones 1) or vertical motion (zones 2) is seen.

5   Conclusions

We have evaluated the hypothesis that perceptual organization and dynamic properties of percepts both can be derived from a single perceptual dynamics. As adequate mathematical language to describe this perceptual dynamics we proposed a dynamic neural-field model which allows to treat distributed representation of information and dynamic self-organization phenoma within the same context. Our aim was to isolate the set of functional components which is sufficient and necessary to capture the dynamic aspects of perceptual organization in motion perception. We have shown that the neural field model is sufficient in the sense that it captures characteristic (dynamic) properties of motion perception (solution of the correspondence problem, dynamic stability, visual inertia, cooperative effects, perceptual switching and its statistics, transparency percepts). A detailed quantitative comparison of the model to experimental data revealed necessary features of the perceptual dynamics (topology of the cooperative interactions in the perceptual space, presence of adaptation and its relation to the neural activation, stochastical switching). We conclude that dynamic effects in motion perception and the perceptual organization are consequences of a single underlying perceptual dynamics. It was shown that this dynamics can be mathematically adequately described and analyzed and how its characteristic parameters can be determined in psychophysical experiments. This leads to the interpretation that motion perception must be interpreted as dynamical self-organization process. We have isolated functional components of this process and have analyzed how they interact. We believe also that we have developed a methodology which allows an analysis of dynamic perceptual organization processes in general.

Supported by the Studienstiftung des deutschen Volkes.

Literature:
[1] Amari, S.: *Biol. Cyb.*, **27**, 77-87 (1977).
[2] Arrowsmith D.K., Place C.M.: An introduction to dynamical systems. Cambride University Press (1990).
[3] Anstis, S., Ramachandran, V.S.: *Perception*, **14**, 135-143 (1985).
[4] Anstis, S., Ramachandran, V.S.: *Vis. Res.*, **27**(5), 755-764 (1987).
[5] Chang, J.J., Julesz B.: *Vis. Res.*, **24**(12), 1781-1788 (1990).
[6] Dawson, R.W.: *Psych. Rev.*, **98**(4), 569-603 (1991).
[7] Gardiner, C.W.: Handbook of stochastic methods. Springer (1989).
[8] Grossberg S, Mingolla E: *Perc Psychop*, **33**, 243-278 (1993).
[9] Hock, H.S., Kelso, J.A.S., Schöner, G.: *J. Exp. Psych.*, **19**(1), 63-80 (1993).
[10] Hock, H.S., Schöner, G., Hochstein, S.: *submitted*.
[11] Jancke D, Akhavan A C, Erlhagen W, Giese M A, Steinhage A, Dinse H, Schöner G: *this volume*.
[12] Kim J, Wilson HR: *Vis Res*, **33**, 2479-2489 (1993).
[13] Nawrot, M., Sekuler, R.: *Vis. Res.*, **30**, 1439-1451 (1990).
[14] Nowlan, S.J., Sejnowski, T.J.: *J. Opt. Soc. Am.*, (1994).
[15] Simoncelli, E.: MIT Technical Report no. 209 (1993).
[16] Snowden, R.J., Braddick: *Vis. Res.*, **29**(11), 1621-1630 (1989).
[17] Williams, D., Phillips, G., Sekuler, R.: *Nature*, **324**(20), 253-255 (1986).
[18] Yuille, A.L., Grzywacz, N.M.: *Nature*, **333**, 71-74 (1989).

Analytical Technique for Deriving Connectionist Representations of Symbol Structures

Robert E Callan

Southampton Institute, East Park Terrace, Southampton, UK, SO14 OYN
e-mail: callan_r@southampton-institute.ac.uk Tel: 44 (0)1703 319262

Dominic Palmer-Brown

Nottingham-Trent University, Burton Street, Nottingham, UK, NG1 4BU
e-mail: dpb@doc.ntu.ac.uk Tel: 44 (0)1159 418418 ext. 2867

Abstract

Recursive auto-associative memory (RAAM) has become established in the connectionist literature as a key development in the strive to develop connectionist representations of symbol structures. In this paper we present an important extension to an analytical version of RAAM, *simplified*RAAM ((S)RAAM). This development means that (S)RAAM models closely a RAAM in that a single constructor matrix is derived which can be applied recursively to construct connectionist representations of symbol structures. The derivation like RAAM exhibits a moving target effect but unlike RAAM the training is fast. A mathematical framework for (S)RAAM allows a clear statement to be made about generalisation characteristics and the conditions necessary for convergence.

1 Introduction

There has been much interest in representing and processing symbol structures using connectionist architectures (e.g., Hinton 1990, Smolensky 1990, Touretzky 1990). Connectionist representations of symbol structures have a number of desirable properties: they allow nearest neighbour judgements, they convey context (e.g., Chalmers 1992, Clark 1993, Sharkey and Jackson 1994) and are suited to holographic transformations (e.g., Blank *et al.* 1992, Chalmers 1990). A key feature of symbol structures is that a new whole can be composed from existing parts. One connectionist model that allows recursive composition of fixed valence trees (i.e., all trees have the same number of branches emerging from each node) is Recursive Auto-associative Memory (RAAM). Others have used RAAM representations for connectionist tasks (e.g., Chalmers 1990, Chrisman 1991, Sperduti and Starita 1993, Reilly 1992) but RAAMs can be troublesome to train and generalisation characteristics difficult to predict. Callan and Palmer-Brown (1995a) presented an analytical model of a RAAM without recursive composition and later (Callan and Palmer-Brown, 1995b) with recursive composition; the model is denoted as (S)RAAM - simplified RAAM. (S)RAAMs are important because the training is quick, generalisation is predictable and they are amenable to simpler mathematical analysis than RAAMs. In this paper we present a significant extension to the recursive (S)RAAM; this extension allows for a single mapping matrix that is applied recursively, it exhibits the moving target effect of a conventional RAAM and importantly conditions for convergence can be stated.

2 RAAM

A RAAM (devised by Pollack 1990) is a connectionist architecture that creates compact distributed representations of compositional symbolic structures. A RAAM is basically an auto-encoder network trained using backpropagation. Hidden unit activations can be stored on a stack to serve as input at a later stage; the hidden activations change during training and therefore some of the training patterns change thus leading to a moving target effect. The number of input fields (and therefore output fields) corresponds to the valence of the trees in the training data and the number of units within a field is usually the same as the number of hidden units. Fig. 1. shows an example RAAM to encode a binary tree.

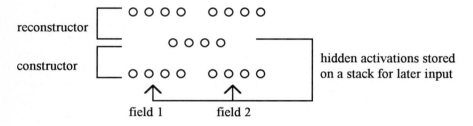

Fig. 1. A binary RAAM

Fig. 2 lists the sequence of training patterns to represent ((A B)(C D)). The terminals (A, B, C, D) are encoded for input as binary strings. Upon convergence, within some specified tolerance: $A = A^\sim$, $B = B^\sim$, $C = C^\sim$, $D = D^\sim$, $h_1 = h_1^\sim$, $h_2 = h_2^\sim$. Note that h_3 is a compressed representation for the whole tree.

input: field 1	input: field 2	hidden field	output: field 1	output: field 2
A	B	h_1	$A^\sim$	$B^\sim$
C	D	h_2	$C^\sim$	$D^\sim$
h_1	h_2	h_3	$h_1^\sim$	$h_2^\sim$

Fig. 2. Pattern presentations to represent ((A B)(C D))

3 (S)RAAM

(S)RAAM is an analytical approach to deriving RAAM style representations and therefore does not depend on slow network training. The first stage in deriving a (S)RAAM is to generate a bit representation for all trees (including subtrees). For example, to derive a (S)RAAM to represent the two parse trees:

((D (A N)) (V (P (D N))))
(D (A (A (A N))))

where D is determiner, A - adjective, P - preposition, V - verb, N - noun, the full training set would comprise:

Pattern ID	Pattern
r_1	((D (A N)) (V (P (D N))))
r_2	(D (A N))
r_3	(V (P (D N)))
r_4	(A N)
r_5	(P (D N))
r_6	(D N)
r_7	(D (A (A (A N))))
r_8	(A (A (A N)))
r_9	(A (A N))

The bit representations for each pattern are generated in a similar manner to Smolenky's Tensor Product role decomposition (1990) and each pattern is made the same length by padding with a NIL vector. Unlike a true role decomposition, the bit representations can be formed by concatenating the terminal codes where the order of terminals follows a breadth-first parse of the tree. If the terminals are encoded as 1-in-5 bit vectors and assuming 7 roles (the maximum number of terminals in any single pattern) then for the above set of trees a pattern matrix of size 9 x (7 * 5) would be generated.

Principal Component Analysis (PCA) is applied to the pattern matrix to provide a reduced description for each tree. A second matrix is then composed using the reduced descriptions:

Pattern ID	Pattern
r_1	((r_2)' (r_3)')
r_2	(D (r_4)')
r_3	(V (r_5)')
r_4	(A N)
r_5	(P (r_6)')
r_6	(D N)
r_7	(D (r_8)')
r_8	(A (r_9)')
r_9	(A (r_4)')

The superscript denotes the PCA reduced description. A second PCA analysis is performed on the newly defined matrix. If the number of eigenvalues from the second mapping is the same as from the first then no further PCA is performed: usually two PCA mappings are sufficient but occasionally further mappings may be required. A finite or small number of mappings is always sufficient since the trees are of finite length and many roles are shared. The eigenvectors from the second matrix define the constructor matrix, ψ. In order to use the mapping matrix recursively some of the patterns (internal nodes) have to change so that

$$\psi^{(r_1)} = \psi(\ \psi^{(r_2)}\ \ \psi^{(r_3)})), \text{ etc.}$$

The change in some patterns is the moving target effect; the reduced descriptions are allowed to change but the mapping matrix, ψ, remains fixed. The moving target phenomenon is implemented by iteratively applying the mapping: the pattern matrix is mapped with ψ and a new pattern matrix defined using the newly derived reduced descriptions and the process repeated until convergence. It should be emphasised that the process is very fast: what may take several thousand epochs with a conventional RAAM typically takes less than 30 epochs with (S)RAAM. To illustrate further the difference in speed, the representation of a set of sequences took 4000 epochs and five hours training for RAAM compared to 19 epochs and one and half minutes for (S)RAAM using a 486 machine; actual times quoted. It can also be shown that convergence will usually be guaranteed, the limiting case being where a subtree is a part of every other tree in the training set and the largest eignevalue of ψ is 0.5 or greater; this is an extreme example however.

The reconstructor is defined as: $(\psi)^{-1} = \left[(\psi)^T C^t \psi\right]^{-1} (\psi)^T C^t$ where t is used to denote stage of recursion, C^t denotes the covariance matrix defined from the pattern matrix once convergence has been achieved and T is the transpose.

4 Discussion and Conclusions

Kwasny and Kalman (1995) have proposed an interesting technique for developing RAAM style representations. Their solution is to convert a tree to a sequential list which allows a modified simple recurrent network (SRN) to be used for training. They report that their SRN technique is a more straightforward approach than RAAM for developing compressed representations. The SRN technique of Kwasny and Kalman still requires considerable time to train compared to (S)RAAM and further work needs to be performed to test whether their sequential mapping of a tree permits holographic transformations to be performed on the compressed representation. (S)RAAM compares well with other analytical techniques such as Plate's (1995) Holographic Reduced Representations (HRRs) and Smolensky's (1990) Tensor Products. With recursion (S)RAAM does not expand as rapidly as other matrix techniques such as Tensor Products; for example (S)RAAM can develop recursive compressed representations using three elements to represent all eight sequences of three bits. Compared to HRRs the operation of (S)RAAMs is less complicated in that an item memory is not required for clean up during retrieval. (S)RAAMs have been devised for a number of experiments with good results. A good example is a repeat of Chalmers (1990) work on sentence transformations from active to passive form. Chalmers used a RAAM to represent the sentences but had difficulty in attaining 100% generalisation. Repeating the experiment with (S)RAAM we have attained 100% generalisation over new structures. The generalisation properties of (S)RAAM are predictable in that we expect to be able to represent all trees that lie in the subspace spanned by the training set.

The suitability of the representations to nearest neighbour judgements is evident from the cluster diagram of Fig. 3 where trees have in the main been grouped according to phrase structure. This example is taken from Pollack (1990) where a number of parse trees are generated according to the following simple grammar:

S -> NP VP | NP V
NP -> D AP | D N | NP PP
PP -> P NP
VP -> V NP | V PP
AP -> A AP | A N

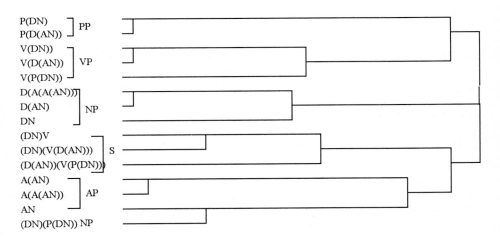

Fig. 3. Cluster diagram of (S)RAAM representations for syntactic parse trees.

In conclusion, we have satisfied one of Pollack's stated goals to analytically derive representations without reliance on slow network training. (S)RAAMs show good generalisation performance and the representations are suited to neural style processing. The mathematical framework and concept of (S)RAAM contributes to the study of recurrent style networks such as RAAM and to the development of symbol/connectionist integration which many view as desirable in order for high level cognitive tasks to be tackled.

References

Blank, D., Meedon, L., & Marshall, J., (1992). "Exploring the Symbolic/Subsymbolic Continuum: A Case Study of RAAM." In J. Dinsmore (Ed), *The Symbolic and Connectionist Paradigms - Closing the Gap* (pp. 113-148). Hillsdale, New Jersey: Lawrence Erlbaum Associates.

Callan, R., & Palmer-Brown, D., (1995a). "Deriving Symbolic Representations using a Simplification to Recursive Auto-Associative Memory." Vol II: 275-279. *Proc. World Congress on Neural Networks 95, Washington D.C.*

Callan, R., & Palmer-Brown, D., (1995b). "Analytical Derivation of a (S)RAAM." Vol 2: 485-490. *Proc. International Conference on Artificial Neural Networks (ICANN) 95, Paris.*

Chalmers, D., (1992). "Symbolic Computation and the Chinese Room." In J. Dinsmore (Ed), *The Symbolic and Connectionist Paradigms - Closing the Gap* (pp. 25-48). Hillsdale, New Jersey: Lawrence Erlbaum Associates.

Chalmers, D., (1990). "Syntactic Transformations on Distributed Representations." *Connection Science*, Vol 2, Nos 1& 2:53-62.

Chrisman, L., (1991). "Learning Recursive Distributed Representations for Holistic Computation." *Connection Science*, 3, No. 4:45-366.

Clark, A., (1993). *Associative Engines. Connectionsim, Concepts, and Representational Change.* A Bradford Book, MIT Press.

Hinton, G., (1990). "Mapping Part-Whole Hierarchies into Connectionist Networks." *Artificial Intelligence.*, 46:47-75.

Kwasny, S., Kalman, B., (1995). "Tail-recursive Distributed Representations and Simple Recurrent Networks". *Connection Science*, 7, No.1:61-80.

Plate, T., (1995). "Holographic Reduced Representations." *IEEE Transactions on Neural Networks*, 6:623-641.

Pollack, J., (1990). "Recursive Distributed Representations." *Artificial Intelligence*, 46:77-105.

Reilly, D., (1992). "Connectionist technique for on-line parsing." *Network*, 3:37-45.

Sharkey, N., & Jackson, S., (1994). "Three Horns of the Representational Trilemma." pp. 155-189. In Honavar, V., & Uhr, L. (Eds.), *Artificial Intelligence and Neural Networks. Steps toward Principled Integration.* London: Academic Press.

Smolensky, P., (1990). "Tensor Product Variable Binding and the Representation of Symbolic Structures in Connectionist Systems." *Artificial Intelligence*, 46:159-216.

Sperduti, A., & Starita, A., (1993). "An Example of Neural Code: Neural Trees Implemented by LRAAMs." *Conf. On Artificial Neural Networks and Genetic Algorithms*, pp. 33-39.

Touretzky, D., (1990). "BoltzCONS: Dynamic Symbol Structures in a Connectionist Network." *Artificial Intelligence*, 46:5-46.

Modeling Human Word Recognition with Sequences of Artificial Neurons

P. Wittenburg, D. van Kuijk, T. Dijkstra*

Peter.Wittenburg@mpi.nl
Max-Planck-Institute for Psycholinguistics, Nijmegen
*NICI, University of Nijmegen, Nijmegen
The Netherlands

Abstract. A new psycholinguistically motivated and neural network based model of human word recognition is presented. In contrast to earlier models it uses real speech as input. At the word layer acoustical and temporal information is stored by sequences of connected sensory neurons which pass on sensor potentials to a word neuron. In experiments with a small lexicon which includes groups of very similar word forms, the model meets high standards with respect to word recognition and simulates a number of well-known psycholinguistical effects.

1 Introduction

Human listeners can process speech seemingly effortless. Even degraded speech can be processed at a rate of about 2 to 3 words per second. Listeners can perform this amazing feat by relying not only on the signal (bottom-up information), but also on syntactic and semantic information extracted from the left-hand context (top-down information). However, even words spoken in isolation are recognized extremely well by human listeners, although performance is not always flawless.

Psycholinguists investigate the word recognition process in humans to better understand the mapping of the incoming acoustic/phonetic information onto the words stored in the mental lexicon. McQueen and Cutler [1] provide a broad overview of many of the known effects in this area. During the last few years, processing models have tried to integrate such psycholinguistic evidence concerning word recognition [2,3]. However, up till now these models have not used real speech input. Instead, they have used mock speech input, based on a phonetic transcription of words, to investigate relevant psychological issues (such as competition effects among word candidates). In doing so, human word recognition is reduced to a relatively simple string matching.

Mock input presumes a solution of how the speech signal is mapped onto abstract phonemic categories, and ignores all nuances in the human recognition process that depend on the signal in relation to the recognition task at hand. According to a phonemic transcription, a word sequence like /ship inquiry/ has the word /shipping/ fully embedded in it. However, in real speech the two spoken sequences /shipping/ and /ship inq.../ are acoustically/phonetically different in several details. Thus, a model based on real speech will to some extent behave differently from one that uses phonemes in the input sequence.

Therefore, the RAW-model (Real-speech model for Auditory Word recognition) was designed to serve as a starting point for a simulation lab which combines the use of real speech and the implementation of current psycholinguistic knowledge. The model intends to (a) adhere to the constraints defined by psycholinguists as much as possible, (b) use real speech as input, (c) store temporal patterns in a plausible way (relative to, e.g., TRACE [3]), and (d) allow later extensions to account for the use of prosodic information and improved attentional and incremental learning mechanisms.

In the design of the model we explicitly chose not to use current main-stream HMM-based or hybrid [see e.g. 4] techniques from the world of Automatic Speech Recognition (ASR). The rationale behind this choice was that these techniques do not provide a good basis for simulating psycholinguistic issues such as gradual lexicon expansion and active competition between words. Furthermore, these architectures are not open and flexible enough to allow easy introduction of extra knowledge sources like prosody. An overview of some major limitations of these systems and suggestions for new ideas can be found in [5] and [6].

2 The Architecture of RAW

Like earlier models and systems, RAW relies on a hierarchical approach. Besides a preprocessing step, RAW incorporates a phonemic and a word layer. The preprocessing of the speech signal results in a number of speech vectors which form the input to a phonemic map (p-map) yielding typical activity distributions for each vector. The p-map can be seen as a kind of spatio-temporal filter. Word neurons in the word map (w-map) sum up the activity distributions over time, leading to an activity distribution in the w-map as well. Competition between the word-neurons (which is suggested by psycholinguistic findings) can be simulated with the help of lateral inhibitory links.

2.1 Pre-lexical processing

In ASR mainly two techniques for preprocessing are used. The first is RASTA-mel-cepstra [7] which especially yields robustness against variations in channel characteristics, and the second is Bark-scaled filter bank preprocessing as suggested by Hermansky [8]. We used the second technique because it is simple, and because we did not have to cope with largely varying channel characteristics. The speech signal was multiplied by a 17.5 ms Hamming window with a stepsize of 8.75 ms. Next, the spectral representations obtained with a 512 point FFT were nonlinearly transformed to the Bark scale and finally multiplied with equidistant acoustical bandfilters [8]. These 16-dimensional spectral representations were further preprocessed by energy normalisation and noise filters.

The p-map is necessary for context-dependent decomposition of the highly modulated segmental information in the speech signal. The ultimate goal of the p-map is to generate, for every incoming speech vector, an activation distribution characterizing the speech segment represented by that vector in its acoustic/phonetic context. The scalar output of supervised trained MLPs [4], TDNNs [9], or RNNs [10] as used in hybrid ASR systems is too restricted for this goal. Instead we used a self-organizing feature map [11], which carries out a data and dimension reduction of the input space, at the same time preserving similarity relations between the input vectors. For example, in the trained map the activation peaks for different realisations of a /th/ will be almost identical in shape and position on the map. The activation peaks for /s/, which is acoustically very similar to /th/, will arise near the peaks for /th/, but it will have a different shape. For phonemes which are acoustically more different from /th/ the activation peaks will be clearly separable from the /th/-peaks. The implementation of the use of context in the p-map is currently under study. Since this paper focusses on the construction of the word map, the p-map and its characteristics will not further be discussed.

2.2 The word map

The word-map (w-map) uses the activity distributions in the p-map over time to store the sequential and the spatial representation of each word. The word neuron assemblies must accumulate matching information, but also block incoming activation when no input is expected. These neuron assemblies must further possess the capacity to store the inherent timing of spoken words and to store temporal order. This is done in RAW by defining for each word at least one sequence of sensor neurons connected via so-called gate signals (see figure 1).

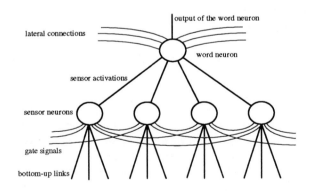

output of the word neuron

lateral connections

word neuron

sensor activations

sensor neurons

gate signals

bottom-up links

Figure 1 shows the assembly of neurons storing the pattern of one word. It exists of a sequence of sensor neurons which are connected to the p-map via excitatory connections. Bottom-up activation leads to sensor activations which are passed on by means of a special gating mechanism to the word neurons. Word neurons can compete with other word neurons via lateral links.

Every sensor neuron is sensitive to the activation pattern of only a certain spatial location in the p-map. The potential of the sensor neurons is computed as a quadratic form difference between the vector described by the afferent weights and that

$$q_{i,k}^{pot}(t) = \frac{1}{N} \sum_{n=0}^{N} e^{a1*(w_{i,k,n}-x_n(t))^2} \quad (1) \qquad q_{i,k}^{act}(t) = g_{i,k}^{act}(t)*q_{i,k}^{pot}(t) \quad (2)$$

described by the relevant activity distribution of the p-map neurons they are linked to (1), where a1 is a parameter <0, n is a neuron in the p-map, $w_{i,k,n}$ the link from a p-map location to sensory neuron k of word neuron I, x_n the activation of p-map neuron n, and N is the number of afferent links of sensor neuron k. The potential is transformed into the sensor activation by multiplying it with the corresponding gate signal (2). In doing so, a sensor activation will be high when the stored typical pattern appears in the p-map, given that the gate signal is high, i.e. if the pattern occurs at the expected moment relative to the preceding stored patterns.

The gate signal is the most important construct for representing the sequential structure of a word. At any moment during processing it represents the expectation value for the activation of the corresponding sensor neuron. Thus, it prevents the transfer of a high sensor potential, caused by a high local match, at a wrong moment in time. At the same time, the gate signal should be high when the previous input matches perfectly well with the stored pattern up to that moment. In this case the sensor potential should contribute to the global word activation. Equation 3 describes the complex dynamics of the gate potential:

$$g_{i,k}^{pot}(t+1) = g_{i,k}^{pot}(t) * (1 - dec1) + a2 * q_{i,k}^{act}(t) + a3 * \sum_{r=0}^{R} (\gamma_r * g_{i,k-r}^{act}(t)) \qquad (3)$$

with a2, a3, and dec1 parameters, γ a Gaussian function, and R the scope of the Gaussian. The recursive influence of the gate values g^{pot} effectuates a slow decrease of the gate signals. The second additive term describes the influence of the local match at the previous moment, and the third term represents the sum of earlier gate values weighted by a Gaussian function. This, and the dependency on the amount of match at the preceding sensor neurons, allows RAW to catch up with speaking rate variations, omissions, and other distortions of the speech signal. The gate signals are normalized variants of the gate potentials (4).

$$g_{i,k}^{act}(t) = \frac{2}{1 + e^{-g_{i,k}^{pot}(t)}} - 1 \qquad (4)$$

In fact, the gate signals have to fulfil a job which is similar to that carried out by dynamic programming algorithms in HMM-based systems. However, gate signals do not need backtracking and operate directly. The global potential of the word neuron is a summation of all sensor activations at any moment in time and the accumulated potential from earlier time steps (5) where dec2 and a4 are parameters, p_i the accumulated potential of word neuron i, and $q_{i,k}^{act}$ the contributions of the sensor neuron k. The final word activation is achieved by normalization (6).

$$p_i(t) = p_i(t-1) * (1 - dec2) + a4 * \sum_{k=0}^{K} q_{i,k}^{act}(t) \; ; \; y_i(t) = \frac{2}{1 + e^{-p_i(t)}} - 1 \qquad (5,6)$$

The afferent weights between the p-map and the sensory neurons are trained in two phases: a bootstrapping phase and a fine tuning phase. In the bootstrapping phase one well-articulated token is used to initialize the sequence of sensor neurons which results in an excellent match of this specific token. During the supervised fine tuning phase the weights are adapted such that they focus on the salient and discriminative aspects of the different variants produced. The strengths of the weights are correlated with the contribution of the singular connections to the sensor potential.

3 Simulation results

For the simulation results we constructed a lexicon with 32 word entries, introducing particular relationships between the words. A number of entries were chosen that were very similar in phonological form, e.g., words like /thanking/, /thinking/, and /sinking/. Furthermore, different types of embeddings of words in other words were introduced, such as /sister/ in /sisters/, and /tree/ in /treaties/. We analysed the behavior of the model with respect to a number of psycholinguistically relevant factors: Uniqueness Point (UP), Cohort Size, and Word Frequency. The uniqueness point is that point in the acoustic signal at which the word becomes unique with respect to all other entries in the lexicon. The cohort of a word is that number of words that is still consistent with the speech signal at a particular moment in time. At the UP the cohort size is reduced to one. All three factors affect the recognition process of humans.

The simulations showed a clear relation between the UP of the words in the lexicon and their point of recognition. This is consistent with findings in experiments with humans. With respect to the influence of the size of the cohort we also obtained consistent results. Words in a small cohort were recognized earlier than those in a large cohort, similar to human word recognition. The way the frequency effect was implemented in RAW, multiplying the contributions to the global word potential with a factor correlated with the word frequency, led to a clear frequency effect in recognition. Unfortunately, this implementation of word frequency also led to some side effects unknown for human subjects.

For a more detailed discussion of the simulation results and psycholinguistic findings we refer to another forthcoming paper [12].

4 Discussion and Conclusions

The RAW model shows a potential to simulate a number of psycholinguistic effects on word recognition and therefore to serve as a simulation and theorizing tool. The dynamics of the network provides a promising basis for further investigation. It performs better than TRACE [3] in that it may pick up early deviations in the pronunciation, for instance, the difference in the way the /I/ is spoken in /tree/ and /treaties/. While this seems to be a desirable feature, more information should be collected about how and when human listeners make use of such more subtle differences in the speech signal. The current implementation of word frequency in RAW resulted in some undesirable effects. Relating word frequency and bottom-up information to the same activation function may be a main reason for this. Other plausible architectural solutions which fit better with current psycholinguistic insights have to be implemented and tested.

The speech recognition component in RAW can be improved in two ways: (1) As already mentioned we are still looking for a better pre-lexical processing. The method has to take the acoustic context into account, like for example RNN do. On the other hand, the method should deliver a richer and more tunable output representation than that of RNN. A temporal solution will be the inclusion of the spectral and energy differences trained in separate maps. A further improvement of pre-lexical processing is expected by reducing the number of stored patterns with the help of a trace segmentation algorithm. This will give more weight to the dynamic parts of the signal. (2) The fine tuning phase of the training has to include attentional, i.e. discriminative mechanisms. In case of similar sound patterns of two words, the w-map has to be trained such that the stored patterns yield a maximal difference in activation. In case of strong differences between tokens of the same word, a new assembly of neurons will be used to store this variant. Fine tuning comes out on the one hand as modifying the representations of an existing neuron assembly or on the other hand as storing largely deviating patterns in separate neuron assemblies.

Of great importance is the inclusion of prosodic and syllabic information in psycholinguistic models. Syllable boundaries are possible moments of articulatory synchronisation. It has to be investigated in how far the gate signals can be optimized by using this information.

References:

[1] McQueen, J.M. & Cutler,A. (in press). Cognitive Processes in Speech Perception. In W.J. Hardcastle & J. Laver (Eds.), A Handbook of Phonetic Science. Oxford: Blackwell
[2] Norris, D.G. (1994). Shortlist: A connectionist model of continuous speech recognition. Cognition, 52(3), 1212-1232.
[3] McClelland, J. & Elman,J.L. (1986). The TRACE model of speech perception. Cognitive Psychology, 18, 1-86
[4] Morgan, N. & Bourlard, H. (1995). Connectionist Speech Recognition. Dordrecht:Kluwer.
[5] Bourlard, H. (1995). Towards Increasing Speech Recognition Error Rates. Proceedings Eurospeech95, Madrid
[6] Wittenburg, P., van Kuijk, D. & Behnke, K. (1995) Automatic and Human Speech Recognition Systems: a Comparison. Proceedings 3. SNN Symposium, Nijmegen, NL.
[7] Hermansky, H. & Morgan, N. (1994). RASTA processing of speech. IEEE Trans. Speech Audio, 2(4)
[8] Hermansky, H. (1990). Perceptional Linear Predictive Analysis of Speech. J. Acoust. Soc. Am. 87 (4). 1738-1752
[9] Waibel, A. et al. (1987). Phoneme Recognition Using Time-Delay Neural Networks. Technical Report TR-1-0006, ART Interpreting Telephony Research Laboratories.
[10] Wittenburg, P. & Couwenberg, R. (1991). Recurrent Neural Networks as Phoneme Spotters. In T. Kohonen et al. (Eds.), Artificial Neural Networks. Amsterdam: North Holland.
[11] Kohonen, T. (1989). Self-Organization and Associative Memory. Berlin: Springer Verlag.
[12] van Kuijk, D.,Wittenburg, P. & Dijkstra, T. (1996). A connectionist model for the simulation of human spoken-word recognition. Proceedings of the 6th Workshop Computers in Psychology 1996, Amsterdam.

A Connectionist Variation on Inheritance

Mikael Bodén

The Connectionist Research Group, Comp. Sci. Dept., University of Skövde, Sweden.
mikael.boden@ida.his.se

Abstract. A connectionist architecture is outlined which makes use of RAAM to generate representations for objects in inheritance networks and extended learning to make such representations context-sensitive. The architecture embodies inheritance quite differently by relying on associative similarities and regions in representational space. The model avoids many of the problems identified for traditional inheritance.

1 Introduction

Inheritance reasoning is concerned with *transfer* of properties or behaviours, sometimes referred to as *defaults*, between objects organised according to their class in *networks*. The obvious advantages are concerned with (i) avoidance of representing all knowledge explicitly by reasoning with defaults, and that (ii) inheritance generates reasonable assumptions when knowledge is incomplete. The mechanisms by which such transfer is handled are traditionally based on transitivity and measures derived from the network representation. Judging from the number of attempts to resolve situations that may occur in inheritance networks, reasoning with defaults is problematic [5]. A so-called *exception* contradicts some of the properties attached to one of its classes. Hence, an inference mechanism based solely on transitivity cannot be sound since it may return contradicting results for a member of the exception. Intuitively, an exception *cancels* inheritance of a particular property from its classes and thus an object should *prefer* to inherit the property of the nearest class. If a network contains redundant connections, a more general inheritance principle is required. A particular object should prefer to inherit the most *specific* properties. That is, whenever there is a conflict the property of the most specialized class should be preferred [5].

In Fig. 1 two inheritance networks are shown. The first contains two exceptions and is problematic if a first-order reasoner is used. The second network contains both exceptions, redundant links and multiple inheritance. To resolve such situations many have resorted to use representational and inferential measures to *preempt* some paths in the network [5].

Many of the attempts to formalise the notion of inheritance are indirectly constrained by accounting for the particular form of representation and the computational consequences of a particular formal model of inheritance. The model-freedom of connectionist systems relieves us from making such decisions. Connectionist systems *acquire* their own internal representation of objects through learning. Hence, representational inconsistencies, in a logical sense, can be accounted for and incorporated continuously. Appropriately configured connectionist systems *generalise* to account for data outside a training set. Hence, reasoning about objects, of which no knowledge is explicitly available, is possible.

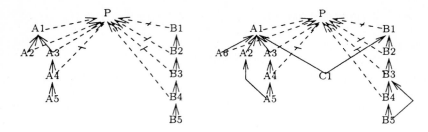

Fig. 1. Two inheritance networks that are problematic for inheritance reasoners. Inheritance-relations between objects (classes and members) are shown as solid arrows and property-relations as dashed arrows. Negative links are crossed.

2 Connectionist representation of inheritance networks

The knowledge contained in an inheritance network is symbolic in nature. What is required is a connectionist system that accepts as input knowledge in symbolic form. Symbols are arbitrary, static and discrete and thus don't exhibit features that make them suitable for associative and generalised processing in a connectionist network. The idea is to let the network develop its own input and output patterns for the associative task. Studies [2] show that Recursive Auto-Associative Memories (RAAMs) [3] produce patterns that reflect structural features of the inputs. The inputs in this case consist of tree-structures, e.g. *((a b) c)*, with bit patterns, e.g. $a=[1, 0, 0]$, $b=[0, 0, 1]$, $c=[0, 1, 0]$.

The three modules in the architecture are shown in Fig. 2. The strategy is as

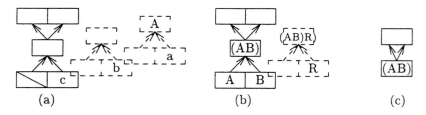

(a)   (b)   (c)

Fig. 2. The three modules, (a) the object encoder/decoder takes a tree of patterns to encode a representation for an object, (b) the assertion encoder/decoder takes, first, two objects to encode a representation for an assertion, then, a relation pattern which is combined with the assertion to yield a representation of an assertion with direction. The assertion and the assertion with direction are then used for training or testing (c) the associative module.

follows. A binary, fully-connected, RAAM with $n + m$ input and output units, and n hidden units is used. Select a unique m-bit pattern for each object in the inheritance network. To avoid introducing phoney relationships among these patterns orthogonal patterns are used. To generate a *representation* for an object A which is a B which is a C, a number steps need to be performed. First, set the first n units to an "end-of-tree" pattern. Then, set the remaining m input units to C's *bit pattern*, called c. Next, compute the outputs of the network. A RAAM is, as the name suggests, auto-associative which means that the target of the training is the input. Errors are calculated according to backpropagation. The next step is to copy the hidden activity to the first part of the input and introduce the bit pattern for B (called b) in the second. It was suggested by Pollack that the hidden activity can be regarded as a *reduced description* of the input. Another feedforward activation in the network would then result in a reduced description of the tree *((nil c) b)*. The hidden activity is yet again copied back to the first input slot and used together with the bit pattern selected for A (called a) to form (after another network activation) a reduced description for *(((nil c) b) a)*. Due to the auto-associative training the resulting reduced descriptions fulfill certain properties. Studies show that structural features such as tree depth and balance are present. Furthermore, the produced pattern is sensitive to the constituent patterns. The later a pattern is introduced the more influence it has on the resulting pattern. According to some analytical techniques, based on decision hyperplane analysis (cf. [4]), regions in hidden space of RAAM for trees with similar constituents can be identified. The regions are derived by determining on which sides of the decision hyperplanes, implemented by the output units of which the output activity is known, the hidden activity must fall. The method is described in full in [1]. One implication is that objects that belong to the same class occupy the same region. The mere existence of such regions entails that it should be possible to convey such from a sufficiently large set of samples (drawn from each region) to a second network which then could generalise to them.

A representation for each object in the inheritance networks is generated according to the same principle. The described encoding technique can be extended to cover multiple inheritance, in theory by imposing an order among the classes, in practice by iterating through each class as if they were transitively related. It should be pointed out that this is a problem since the order can be interpreted as imposing preference among the classes.

Properties, which are also objects, are attached to objects with a relation using a second RAAM. This RAAM takes two encoded object representations as input and generates a hidden pattern that is used to represent the expression without the relation, e.g. *(A B)*. This pattern is also used for creating a representation for an expression with relation by using it as input together with the relation representation, e.g. *((A B) R)*. The use of a second RAAM for encoding representations for expressions is justified by two aspects. First, RAAM generates patterns that are influenced by the inputs. Hence, patterns cluster whenever parts of inputs are shared. For example, *(A C)* ends up close to *(B C)*, especially if A's representation is similar to B's representation in the first place. Second, RAAM provides an explanation of the behaviour using the same technique as was suggested for the first RAAM. Hence, the existence of regions occupied by representations of one type, e.g. *((A B) R)* always ends up in the same region as all other expressions with relation R, and *(x B)* expressions always end up in the same region independently of the representation for x. Thus, it is possible to operate on the *form* of expressions, similar to a syntactic process.

Finally, these two types of representations are used as input and target patterns for an associative network. For the input it can be useful to think of the associative network as having the input-to-hidden weights in the second RAAM as an extra input layer. Thus, the input contains two object representations. The first layer of weights can then distinguish between regions for representations of a certain class. This goes for both the input slots. The remaining layer(s) of weights account(s) for regions imposed by the second RAAM.

The described model (called M1) was tested on the first of the two inheritance networks in Fig. 1. Orthogonal bit-patterns with 10 elements ($m = 10$) were selected to symbolize all objects in the network. The hidden layer of the RAAM was set to 3 ($n = 3$). Thus, the size of the slots in the second RAAM and the input and output layers in the associative network were set to 3. The associative network contained one hidden layer with 2 units. The above configuration was the best found for this task. Training in the first RAAM included all objects and their inheritance relations to other objects. Training in the second RAAM included all possible expressions that may be used in training or testing the associative network. Training in the associative network included all explicit relations in the inheritance network but excluded all inheritable relations. Backpropagation was used for all steps. All the networks were trained in parallel for 10000 cycles through the training set. This was sufficient to reach a stable state both in terms of generated patterns and weights in the networks. The experiment was repeated 5 times to ensure that the results were consistent and valid. Cluster analyses were performed on the generated patterns. The object representations tended to cluster with respect to the two branches. The second RAAM maintained the representational characteristics of the constituents and influenced the generated patterns accordingly. The expressions for which a relation can be inherited were used to test the performance of the system. Two ways of evaluating the results were used. First, the output patterns were compared using Euclidean distances from all of the previously encoded representations. Of the 5 runs only 2 were completely correct. For several reasons (cf. [4]), e.g. presence of discriminating weights, such measures are not telling the whole story. Instead the distances from each decision[1] in the hidden-to-output layer of the associative network were used. Inspection of the diagrams revealed that all of the 5 runs were successful. One such diagram is shown in Fig. 3. Considering the second inheritance network it is noted that M1 is incapable of using specificity when redundant links are present. This is due to separate encoding stages which are not influenced by the facts in the domain, e.g. the coding of $A5$ cannot take into account the fact that among its two immediate classes, $A4$ and $A2$, one ($A4$) has more specific information on P. This inability led to the idea of using feedback of influence.

3 Learning for establishing preferences

The first RAAM can be regarded in a restricted sense as another input layer to the second RAAM. Hence, learning errors for the associative network can be

---

[1] The distance from the switch (0.5 using the logistic function) of the output.

backpropagated through all of the modules. The addition of virtual feedback connectivity constitutes the M3 model.

M3 was tested on the two inheritance networks in Fig. 1. The setup was the same as for M1 with the minor change that a hidden layer in the associative network was no longer required and m was set to 12. Cluster analyses were performed on the different representations generated by the model. Now, object representations tended to cluster much more in terms of their intended function, e.g. *A4* was less close to *A3* since *A4* is an exception. Diagrams visualizing the distances to the output decision hyperplanes were evaluated. The results are shown in Fig. 3. The hyperplane diagrams indicate that the M3 model is capable

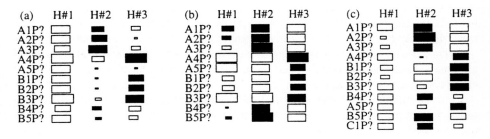

Fig. 3. Diagram for visualizing distances to output decision hyperplanes of the associative network for relevant inputs. (a) M1 on the first inheritance network, (b) M3 on the first inheritance network, and (c) M3 on the second inheritance network. The diagrams were all generated from the first runs which appeared to be quite representative. White boxes signify that the input activation is below the hyperplane and black is above. The size of the boxes indicates the magnitude of the distance.

of handling redundant links according to the specificity criterion. However, the ambiguous multiple inheritance situation exemplified by *C1* was handled arbitrarily. However, the presence of feedback gives M3 the potential of handling such situations through contextual information in the domain. Such information was not available here.

4 Conclusions

The experiments indicate that the model (with or without feedback) prefers more specific information if redundant links are disallowed. At a functional level this works similarly for most formal models of inheritance. The experiments with the model with feedback connectivity indicate that the model resolves exceptions in the presence of redundant links according to the specificity criterion.

A functional comparison between existing preemption strategies and the proposed alternatives was performed for the second inheritance network. In the case of *B5* both on-path and off-path preemption (cf. [5]; independently of them being skeptical or credulous) resulted in *P* which is the intuitively correct answer.

M3 produces also P although empirically validated. In the case of $A5$ on-path preemption is incapable of handling the redundant path through $A2$. Off-path preemption handles the added complexity of intermediate nodes and returns a path supporting $\neg P$. M3 produces also $\neg P$ which is the correct solution with respect to specificity. The $C1$ case is not handled by any of the reviewed methods. M3 has the potential of resolving similar situations using related (according to learning) information in the domain.

The experiments show that RAAM encoding is *sufficient* for successful operation. Feedback seems *necessary* to avoid the problems for M1 on domains with redundant or multiple links. The first RAAM is responsible for generating representations that obey the specificity constraint. The empirical study on the M1 model provides evidence that this is the case. The feedback present in M3 is responsible for resolving conflicts in the domain by providing task-specific context for the encoders. The empirical study supports this claim. Furthermore, the addition of feedback connectivity seems to enhance the systems ability to manage without hidden units in the associative module. The information needed for correct mapping is distributed throughout the modules. Hence, better generalisation capabilities are enforced.

The variation of the results of M3 on the multiple inheritance situation suggests, first, that in order to deal with such situations more experimental data may be needed before jumping to conclusions. Second, the ordering among classes imposed by the encoding technique has no obvious effect.

The connectionist system presented here embodies inheritance quite differently from classical approaches, by relying on facts about members of the same class (at the same level). For example, if B is an A and C is an A, the characteristics of C depends on facts about B. There is a certain representational coherence imposed by the encoders among members of the same class. The results presented here suggest that this might be what inheritance really is.

References

1. M. Bodén. *Representing and Reasoning with Nonmonotonic Inheritance Structures using Context-sensitive Connectionist Representations.* PhD thesis, University of Exeter, United Kingdom, 1996. In preparation.
2. M. Bodén and L. Niklasson. Features of distributed representations for tree-structures: A study of RAAM. In *Current Trends in Connectionism - Proceedings of the 1995 Swedish Conference on Connectionism,* pages 121–140, 1995. LEA.
3. J. B. Pollack. Recursive distributed representations. *Artificial Intelligence,* (46):77–105, 1990.
4. N. E. Sharkey and S. Jackson. An internal report for connectionists. In R. Sun and L. Bookman, editors, *Computational architectures integrating neural and symbolic processes,* pages 223–244. Kluwer Academic Press, 1994.
5. D. S. Touretzky, J. F. Horty, and R. H. Thomason. A clash of intuitions: The current state of nonmonotonic multiple inheritance systems. In *Proceedings of IJCAI 87,* 1987.

Mapping of Multilayer Perceptron Networks to Partial Tree Shape Parallel Neurocomputer

Pasi Kolinummi, Timo Hämäläinen, Harri Klapuri, and Kimmo Kaski
Electronics Laboratory, Tampere University of Technology
P.O. Box 692, FIN-33101 Tampere, Finland
pasiko@ele.tut.fi, timoh@ele.tut.fi, harrik@ele.tut.fi, kaski@ele.tut.fi

ABSTRACT

Mapping of multilayer perceptron networks are presented for a parallel neurocomputer system called PARNEU (PARtial tree shape NEUrocomputer). The partial tree shape architecture of the system offers implementation possibilities at several levels of parallelism for both execution and network training. An example mapping is given for error back-propagation learning utilizing combined neuron and weight parallel schemes. The time required in each step is estimated and compared to corresponding values found in bus, ring and complete tree architectures.

1. Introduction

Perceptron is the most commonly known artificial neuron model in the early days of artificial neural network research [1]. Since then, perceptrons have been used in a variety of applications involving classification or modelling of unknown functions. This is due to *multilayered perceptron networks* (MLPs), which have been proven to solve problems in any high dimensional data spaces [2]. Training of MLP networks can be done in several ways, but the *error back-propagation* (BP) algorithm [3] is still a highly popular method and can be applied successfully in many cases. Practical implementations of MLP networks can be planned most conveniently using parallel processing. Although each perceptron, or neuron, in the network introduces fairly simple computations, the great number of neurons and high connectivity between them makes sequential computation inefficient. Thus, implementations have been presented not only for general purpose parallel computers but also for *neurocomputer* systems, e.g. CNAPS [4], REMAP [5], RAP [6], SYNAPSE-1 [7], and TUTNC [8]. In addition, special integrated circuits have been used for implementations [9]. In this paper we present different mapping possibilities and an example implementation of MLP networks for PARNEU. In the next section we briefly introduce the PARNEU system and consider different mapping choices. A combined neuron and weight parallel implementation is presented step by step. In addition, the time required in each step is given.

2. PARNEU System

The PARNEU (PARtial tree shape NEUrocomputer) system consists of identical cards connected to a host computer, as depicted in Fig. 1. Both PCs and workstations can act as a host for convenient user interface. In each card there are four processing units (PUs) and a subtree of four nodes. Cards are connected such that subtrees together form the partial tree, in which global summation and comparison operations can be performed. In addition, each PU is connected to adjacent one via the ring bus. Thus, data can also be circulated between PUs in a systolic way. The global bus connects all cards to the host to allow efficient data broadcast operation.

The architecture offers linear expandability due to modular structure and many choices to implement artificial neural networks at several levels of parallelism, being *training session*, *training example*, *layer*, *neuron* and *weight* (synapse) parallelism [5]. For real time applications, neuron and weight parallelism are natural choices. It is also possible to mix several levels to achieve optimal performance. In the next section we consider the mapping of multilayer perceptron networks with back-propagation learning.

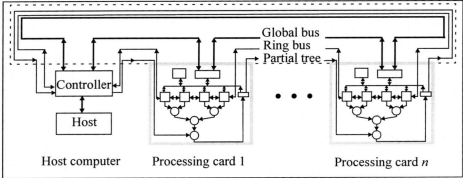

Fig. 1. Block diagram of partial tree shape neurocomputer.

3. Mapping

In this section, we first consider how neuron and weight parallelism can be utilized in PARNEU and then present an efficient combined mapping in more detail. The neuron parallel mapping is accomplished by assigning one or more neurons to each PU and computing weighted inputs and the output activation function locally. The input vector is most conveniently broadcast from the host to all PUs via the global bus. The host also collects neuron outputs and forms an output vector for that particular layer of neurons. The next neuron layer is processed by first broadcasting the output vector of the previous layer and repeating other steps.

In the weight parallel mapping, the computation of one neuron output is distributed to all PUs, as illustrated in Fig. 3. Due to this, the input vector is now written elementarily to PUs, such that the first PU gets the first element and so on. The task of the individual PU is simply to multiply its input element with a weight assigned to that neuron input. Now the partial tree can be utilized to sum up these weighted inputs to yield weighted sum of the neuron inputs. Activation function is applied in the host, which can also store the neuron output to a vector for further usage. The output of the next neuron in a layer is computed by changing the weights in each PU and repeating all the other steps.

To compare these mappings, the weight parallel mapping requires more communication during execution, while in the neuron parallel mapping communication is only needed in the input vector broadcast and neuron output collection. In the weight parallel mapping each PU needs only one input vector element and associated weights while in the neuron parallel mapping all the input vector elements are needed at a time. In both mappings, however, input vector elements or weighted sums have to be transferred quite frequently between PUs and the host. To minimize transfers, multilayer perceptron networks can be best mapped to PARNEU in combined neuron and weight parallel manner. In this scheme, the neuron parallelism is applied to the first neuron layer and the second layer is processed using weight parallelism. This is continued such that neuron and weight parallelism are alternated from one layer to another.

Consider next in more detail the combined mapping in both forward pass and training with back-propagation algorithm. For simplicity, we consider a two-layer network as an example. In the following computation, X is the number of input nodes, Y is the number of output nodes, H is the total number of hidden neurons and H^p is the number of hidden neurons in PU^p. In addition, Δpw refers to the previous weight change, η is the learning rate parameter, α is the momentum rate parameter and P is the number of PUs. We have used logistic activation function in which the values are obtained using a precomputed look-up table.

Forward pass

Step 1. The input vector $\underline{x} = [x_1, ..., x_X]$ is broadcast to PUs via the global bus.

Step 2. Each PU computes the outputs of its local hidden neurons without any data transfers between PUs or the host, as illustrated in Fig. 2. Specifically, PU^p computes $h_j^p = g(\text{net}_{1j}^p) = g\left(\sum_{i=1}^{X} w_{1ji}^p x_i\right)$, where $j \in \{1, H^p\}$ and $p \in \{1, P\}$.

Step 3. The output layer is processed in weight parallel manner. Since the neuron outputs of the previous layer are already available in PUs, no data transfers are needed to start the computation. During this step each PU multiplies a weight and the output of the hidden layer and sends it to the partial tree network, as illustrated in Fig. 3. Specifically, PU^p computes $\text{net}_{2i}^p = w_{2ij}^p \cdot h_j^p$, where $j \in \{1, H^p\}$, $i \in \{1, Y\}$ and $p \in \{1, P\}$. The tree adds up the partial results to form the weighted sum, $\text{net}_{2i} = \sum_{p=1}^{P} \text{net}_{2i}^p$ for the ith neuron in the output layer. The host computer reads the net_{2i} values, computes the activation function $y_i = g(\text{net}_{2i})$ for each neuron i and sends the result back to the PUs using the ring bus. This is done for each neuron in pipelined way to achieve good parallel efficiency.

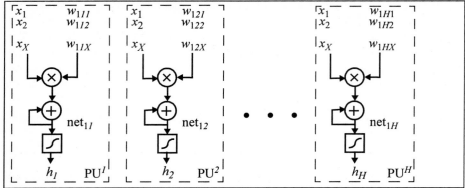

Fig. 2. Neuron parallel computation in step 2.

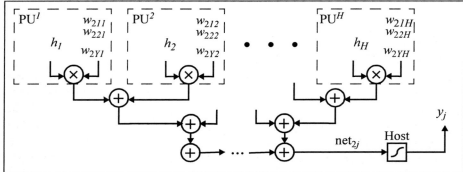

Fig. 3. Weight parallel computation in step 3.

Training pass

The error back-propagation learning consist of forward pass using known input-output pairs and backward pass for weights update. The backward pass is computationally quite similar to the forward pass, but the signal directions are reversed, error terms are used as inputs and activation functions are replaced with their gradients. The computation is mostly performed locally in PUs as in the forward pass. Steps one to three are similar with steps of the forward pass and are not repeated here.

Step 4. The desired output, or *target vector* $\underline{t} = [t_1, \ldots, t_Y]$ is broadcast to each PU. After that, PUs compute the sums of squared errors of their output neurons so that PU^p computes $E_s^p = \sum_{i=1}^{Y^p} (t_i^p - y_i^p)^2$, where $p \in \{1, P\}$. These partial sums are summed up in the tree network to yield the final squared error, which is used to determine whether the execution is stopped or continued.

Step 5. If the execution is continued, each PU computes error terms of their output neurons. The error term is a product of the output error and its gradient at this point. Thus, PU^p computes $e_{2j}^p = (t_j^p - y_j^p) \cdot y_j^p (1 - y_j^p)$, where $j \in \{1, Y^p\}$, Y^p is the number of output layer neurons in PU^p and $p \in \{1, P\}$. No data transfers are needed in this step, because all necessary values have already been stored in each PU.

Step 6. PUs compute the error terms for the hidden layer. First, all the output error terms are circulated to each PU via the ring bus. Then, PU^p computes

$$sum_j^p = \sum_{i=1}^{Y} e_{2i} \cdot w_{2ij}, \text{ where } j \in \{1, H^p\}.$$ PU^p has now both sum_j^p and h_j^p to compute the error of the hidden neurons: $e_{1j}^p = sum_j^p \cdot h_j^p(1 - h_j^p)$, where $j \in \{1, H^p\}$.

Step 7. PUs update the weights of the output neurons in parallel. At first, PU^p computes the change of the weight $\Delta pw_{2ji}^p = \alpha \cdot \Delta pw_{2ji}^p + \eta \cdot e_{2j}^p \cdot h_i^p$ and then updates the value $w_{2ji}^p = w_{2ji}^p + \Delta pw_{2ji}^p$, where $i \in \{1, H^p\}$, $j \in \{1, Y\}$ and $p \in \{1, P\}$.

Step 8. PUs update in parallel the weights of the hidden neurons as in step 7.

4. Performance issues

In order to estimate the execution performance, we consider next the *computational complexity*, or the time required in each step. As shown in Table 1, this depends on the input vector dimension X, the number of hidden neurons H and the number of outputs Y as well as the number of PUs P. The mapping implements good parallelism due to even distribution of neurons to PUs and pipelining of operations. In addition, the performance will increase when more PUs are added to the system as long as there are more than one neuron per PU (i.e. $H/P > 1$ in Tables 1 and 2). In an ideal situation, the term H/P becomes one and the total computational complexity shown in Table 1 becomes $O(X + Y + P)$, which demonstrates a linear dependence on the network size and PU count.

Table 1. Computational complexity in each step in back-propagation learning.

Step	Complexity	Step	Complexity	Step	Complexity
1	$O(X)$	4	$O(Y + P)$	7	$O\left(\frac{H}{P}Y\right)$
2	$O\left(\frac{H}{P}X\right)$	5	$O\left(\frac{Y}{P}\right)$	8	$O\left(\frac{H}{P}X\right)$
3	$O\left(\frac{H}{P}Y + P\right)$	6	$O\left(\frac{H}{P}Y\right)$	Σ	$O\left(\frac{H}{P}(X + Y) + P\right)$

In order to compare PARNEU with other architectures, the total computational complexities are also listed for *ring, complete tree* and *bus* architectures in Table 2. The same combined neuron and weight parallel mapping with BP learning is used in all cases. The time in each architecture is quite similar due to similar local operations in PUs, but differences occur in phases that require communication. As shown in Table 2, the time required in the ring and bus depends on the product of P and Y, which degrades the performance of large neural networks or systems with many PUs. In the tree the computational complexity depends on $P\log P$. In PARNEU there is no product terms in the best case. Thus, we can conclude that PARNEU combines the good features of bus, tree and ring architectures for efficient communication.

Table 2. Total computational complexity in BP learning for different architectures.

PARNEU	Ring	Tree	Bus
$O\left(\frac{H}{P}(X+Y)+P\right)$	$O\left(\frac{H}{P}(X+Y)+PY\right)$	$O\left(\frac{H}{P}(X+Y)+P\log P\right)$	$O\left(\frac{H}{P}(X+Y)+PY\right)$

5. Summary and Conclusions

Parallel mapping of perceptron networks has been presented for PARNEU. The flexible architecture of the system gives many choices for mappings. The combined neuron and weight parallel mapping optimizes computation and communication for best performance. Compared to other architectures, PARNEU features linear dependence on computational complexity. This means that good performance can be achieved also with large number of processing units or large neural network sizes.

References
[1]   F. Rosenblatt, "The Perceptron: A Probabilistic Model for Information Storage and Organization in the Brain", *Psychical Review*, Vol. 65, 1958, pp. 386-408.
[2]   S. Haykin, *Neural Networks: A Comprehensive Foundation*, Macmillan College Publishing Company, NY, USA, 1994.
[3]   D. Rumelhart, G. Hinton and R. Williams, "Learning Internal Representations by Error Propagation", *Parallel Distributed Processing: Explorations in the Microstructure of Cognition*, Vol. 1, MIT Press, USA, 1986, pp. 318-362.
[4]   D. Hammerström, "A Highly Parallel Digital Architecture for Neural Network Emulation", *VLSI for Artificial Intelligence and Neural Networks*, J. G. Delgado-Frias and W. R. Moore (Eds.), Plenum Publishing Company, New York, USA, 1990.
[5]   T. Nordström and B. Svensson, "Designing and Using Massively Parallel Computers for Artificial Neural Networks", *Journal of Parallel and Distributed Computing*, Vol. 14, No. 3, March 1992, pp. 260-285.
[6]   N. Morgan, J. Beck, J. Kohn, J. Bilmes, E. Allman and J. Beer, "The Ring Array Processor: A Multiprocessing Peripheral for Connectionist Applications", *Journal of Parallel and Distributed Computing*, Vol. 14, No. 3, March 1992, pp. 248-259.
[7]   U. Ramacher, "SYNAPSE - A Neurocomputer That Synthesizes Neural Algorithms on a Parallel Systolic Engine", *Journal of Parallel and Distributed Computing*, Vol. 14, No. 3, March 1992, pp. 306-318
[8] .T. Hämäläinen, J. Saarinen and K. Kaski, "TUTNC: A General Purpose Parallel Computer for Neural Network Computations", *Microprocessors and Microsystems*, Vol. 19, No. 8, 1995, pp. 447-465.
[9]   J. Tomberg and K. Kaski, "Pulse Density Modulation Technique in VLSI Implementations of Neural Network Algorithms", *IEEE Journal of Solid-state Circuits*, Vol. 25, No. 5, October 1990, pp. 1277-1286.

Linearly Expandable Partial Tree Shape Architecture for Parallel Neurocomputer

Timo Hämäläinen, Pasi Kolinummi and Kimmo Kaski
Electronics Laboratory, Tampere University of Technology
P.O. Box 692, FIN-33101 Tampere, Finland
timoh@ele.tut.fi, pasiko@ele.tut.fi, kaski@ele.tut.fi

ABSTRACT

The architecture of linearly expandable partial tree shape neurocomputer (PARNEU) is presented. The system is designed for efficient, general-purpose artificial neural network computations utilizing parallel processing. Linear expandability is due to modular architecture, which combines bus, ring and tree topologies. Mappings of algorithms are presented for Hopfield and perceptron networks, Sparse Distributed Memory, and Self-Organizing Map. Performance is discussed with figures of computational complexity.

1. Introduction

The need of efficient computing hardware has emerged abreast with exploding number of real-time and embedded applications of Artificial Neural Networks (ANNs) [1]. The large connectivity and well localized subcomputations in most ANN algorithms make them very suitable for parallel implementations. *General purpose parallel computers* like Connection Machine [2] have been used successfully in ANN computations, but it would be more cost-effective and even more efficient to use special parallel systems. *Dedicated neurocomputers* typically offer the highest performance and can be best utilized in specific, restricted applications. On the other hand, several different neural algorithms can be executed in *general purpose neurocomputers*, which can also be used for data pre- or post-processing in companion with the ANN algorithm. Examples of such systems are REMAP [3], RAP [4], CNAPS [5], SYN-APSE-I [6], and MUSIC [7]. A common approach in these systems is to connect a number of *processing units* (PUs) to a *communication network*, whose topology may be a bus, a mesh, a hypercube, a torus, a tree or a ring. PUs may be realized with either application specific integrated circuits or some commercial processors. Very often a conventional computer, PC or workstation, is used to host the system for convenient software development and user interface.

Previously we have studied and implemented a *complete tree shape architecture* in a system called TUTNC (Tampere University of Technology Neural Computer) [8]. Based on good experiences with TUTNC, we now present a new architecture called PARNEU (PARtial tree shape NEUrocomputer) for practical expandability and more flexibility. In next sections we describe the new architecture in detail, and give example mappings of Hopfield networks, multilayer perceptron networks, Sparse Distributed Memory, and Self-Organizing Map. Performance is estimated using computational complexity.

2. Architecture

The architecture of PARNEU is depicted in Fig. 1. As can be seen, the system consists of identical cards, which are connected to each other and to a host computer with three buses. In each card there are four PUs and a subtree of four nodes. The *Global Bus* is used for efficient data broadcast, write and read transfers between the host and the cards, while the *Chain Bus* is dedicated for communication between PUs. The *Tree Bus* connects subtrees together for a larger tree. The tree has leaves only on its left side and can thus be called a *partial tree*. Each node of the tree can perform summation and comparison operations to the data sent by PUs or upper nodes in the tree, enabling efficient global operations. All the buses are pipelined, which ensures controlled operation and improved performance.

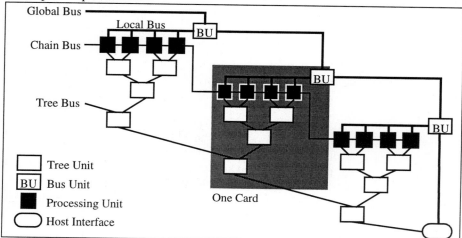

Fig. 1. Architecture of the PARNEU system.

An important feature is, that the system can be linearly expanded. Thus, a fully functional system can be realized with only one card, and more performance can be introduced by adding cards. Since the architecture combines the bus, tree and ring topologies, communication delays can be minimized by choosing the most reasonable communication path for each phase of execution. In addition, the architecture offers many implementation choices for different neural networks.

Next we discuss one basic card depicted in Fig. 2. In each card, there is a *Local Bus* common for four PUs and the subtree composed of Tree Units (TUs). An input from the previous card is included to the bottom-most TU to allow the expansion of the tree. The output of this TU can be routed either to the next card or to the local bus. In this way, the tree can be utilized as stand-alone to allow independent operation of cards. A Bus Unit (BU) implements one pipeline stage and a bidirectional connection between the local and global buses. The Chain Bus transfer data in serial mode between adjacent PUs and it can be completed to a ring. In addition, a shared memory (RAM) is included for each card for temporary storage of large data sets.

The internal structure of one PU is depicted in Fig. 3. The main parts are dual-ported memory (RAM), a processor (CPU) and communication links. A preliminary choice for the PU is Analog Device's ADSP-21060-SHARC DSP [9], which contains on-chip 512 kilobytes of dual ported memory and six communication links, i.e. the whole PU can be implemented with this chip. Similar features can also be found in many other commercial DSP processors [9]. One TU (Fig. 3) contains an adder and comparator as well as a multiplexer to route the result down to the next TU or local bus. One large Field Programmable Gate Array (FPGA) chip is used to implement the subtree for compact realization.

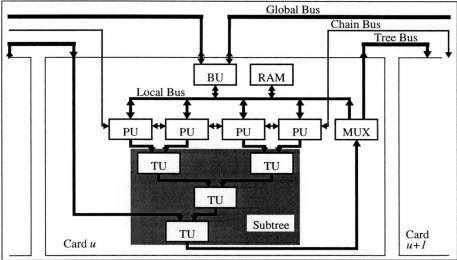

Fig. 2. Block diagram of one card for PARNEU.

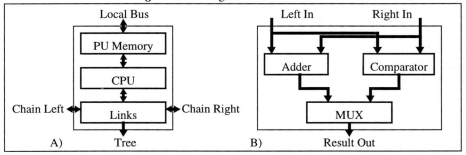

Fig 3. Block diagram of A) Processing Unit and B) Tree Unit.

3. Mapping of Algorithms

The PARNEU architecture can be utilized at several levels of parallelism involved in ANNs, being *network, training set, layer, neuron, weight* and *bit parallelism* [3]. A single card is a parallel system itself, and thus all these levels of parallelism can also be applied in it. In addition, it is possible to mix several levels, e. g. neuron parallelism applied within each card, and all cards together implement layer parallelism. As examples, we consider a weight parallel mapping of Hopfield networks as well as a combined neuron and weight parallel mapping of multilayer perceptron networks (MLP) and Sparse Distributed Memory (SDM). Furthermore, we discuss a neuron parallel mapping for Self-Organizing Maps (SOM). For each mapping, the execution phases are summarized in Tables 1 and 2 with estimated values of *computational complexity*, which describes the increase of the execution time with respect to network parameters.

Recurrent networks, such as the Hopfield network, use their output y at iteration step t as a new input for the next step $t + 1$. Such networks can be mapped in weight parallel manner by computing the weighted inputs $y_j(t)w_{ji}$ for a neuron i simultaneously in all PUs, and performing the summation in the adder tree, as illustrated in Fig. 4. The chain bus can then be used to transfer the sum results back to PUs. Each PU thresholds its input sum_j, which results in the new input value. Since the adder tree and the chain bus are pipelined, a new input element $y_j(t + 1)$ is produced in a system clock cycle.

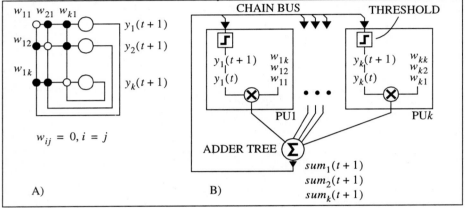

Fig. 4. Hopfield A) network B) weight parallel implementation.

Multilayer perceptron networks can be mapped in a similar manner utilizing weight parallelism. However, the time for communication during execution can be reduced by applying the combined mapping. In this scheme, the first neuron layer (a) is processed in neuron parallel way, and the second (b) using weight parallelism. This is continued such that both are alternated from one layer to another. In the neuron parallel phase, each PUj computes locally the output h_j of its neuron using input vector $\underline{x} = [x_1...x_X]$ broadcast by the host and associated weights $w_{ij}^a, i \in \{1, X\}, j \in \{1, H\}$. After that, PUs first compute the weighted inputs $h_j w_{ji}^b, i \in \{1, Y\}$ in weight parallel phase and the tree is used to sum up the input terms. The host can then perform thresholding to yield one output element y_k. This is repeated, until each neuron in the layer is processed. The host broadcasts $\underline{y} = [y_1...y_Y]$ to all PUs for the next neuron parallel processing.

Sparse distributed memory can be thought as a fully connected, three-layered ANN or a special associative memory. Using the latter interpretation, the mapping can be implemented by delivering the address matrix row-wise to PUs, such that each PU has one or more rows. The data memory matrix is partitioned columnwise to PUs. Execution starts by first broadcasting the input address vector $\underline{a} = [a_1...a_k]$ to the PUs, each computing the Hamming distance d_j between their hard addresses h_j and the input address. After that, each PU looks for close hard addresses to be selected for further phases. Then, PUs exchange indices of selected rows (s_j) using the Chain Bus.

Table 1. Complexity in each step for Hopfield and perceptron networks.

Step in Hopfield	Complexity	Step in MLP	Complexity
		Broadcast x	$O(X)$
$p_i = y_i(t)w_{ij}$	$O\left(\dfrac{k}{P}\right)$	$h_j = f\left(\displaystyle\sum_{i=1}^{X} x_i w_{ij}^a\right)$	$O(X)$
$sum_i(t+1) = \displaystyle\sum p_i$	$O(k)$	$p_i = h_j w_{ji}^b$	$O\left(\dfrac{H}{P}Y\right)$
write sum_i	$O(k)$	$net_i = \displaystyle\sum p_i$	$O(Y)$
$y_i(t+1) = f(sum_i)$	$O(1)$	$y_i = f(net_i)$	$O(Y)$
Total	$O(k)$	Total	$O\left(X + \dfrac{H}{P}Y\right)$

Table 2. Complexity in each step for SDM and SOM.

Step in SDM	Complexity	Step in SOM	Complexity
Broadcast a	$O(k)$		
$s_j = f\left(\displaystyle\sum_{i=1}^{k}(a_i \oplus h_{ji})\right)$	$O(k)$	Broadcast x	$O(k)$
Deliver s_j to all PUs	$O(P)$	$d_j = \displaystyle\sum_{i=1}^{k}(x_i - w_{ji})^2$	$O\left(k\dfrac{n}{P}\right)$
Broadcast x (Write)	$O(k)$	Find BMU	$O\left(\dfrac{n}{P}\right)$
$m_{ji} = m_{ji} + x_{ji} \& s_j$ (Write)	$O(R)$	Find global winner $min(d_j)$	$O(P)$
$y_{ji} = f\left(\displaystyle\sum_{j=1}^{R}(m_{ji}\&s_j)\right)$ (Read)	$O(R)$	Broadcast index of $min(d_j)$	$O(1)$
Collect y_{ji} to host (Read)	$O(P)$	$w_{ji} = w_{ji} + \alpha N_j(x_i - w_{ji})$	$O\left(k\dfrac{n}{P}\right)$
Total	$O(P + k + R)$	Total	$O\left(k\dfrac{n}{P} + P\right)$

Writing to the memory is performed by first broadcasting an input data vector $\underline{x} = [x_1 \ldots x_k]$ to PUs, which pick up only those elements corresponding to their data matrix columns. After that, PUs update hard data locations m_j in parallel. In the read operation, each PU sums up their columns of selected rows to form a sum vector. This vector is then thresholded elementwise to yield final output vector elements for this particular PU. The host reads elements from all PUs and forms the final output vector $\underline{y} = [y_1 \ldots y_k]$.

Our final example is Kohonen's self-organizing map, which can be parallelized in a neuron parallel manner. The n-neuron map is partitioned to PUs such that there is one or more neurons per PU. The input vector $\underline{x} \in \{1, k\}$ is broadcast to PUs, which computes a local best matching unit (BMU). The tree is now used to compare these BMUs to find the global winner. The index of the winner is broadcast from the host to PUs, which can now compute the neigborhood N_j of the winner and perform weight updates to neigborhood neurons.

4. Conclusions and further work

The architecture of PARNEU offers wide flexibility and many choices for mappings of ANN algorithms. Recurrent networks can be executed efficiently using weight parallel scheme, while combined neuron and weight parallel mapping is very suitable for perceptron networks. With SDM, combined row- and columnwise mapping balances execution in PUs in both addressing and read or write operations. Neuron parallel mapping is suitable for self-organizing maps. Figures of computational complexity show, that each algorithm can be well parallelized. In order to verify estimates, a prototype system is being built during 1996 for detailed performance measurements of given mappings.

Acknowledgements

This research work was supported by the Academy of Finland and Graduate School Programme on Electronics, Telecommunications and Automation.

References

[1] S. Haykin, *Neural Networks: A Comprehensive Foundation*, Macmillan College Publishing Company, NY, USA, 1994.

[2] L. Trucker, and C. Robertson, "Architecture and Applications of the Connection Machine", *IEEE Computer*, August 1988, pp. 26-38.

[3] T. Nordström and B. Svensson, *"Designing and Using Massively Parallel Computers for Artificial Neural Networks"*, *Journal of Parallel and Distributed Computing*, Vol. 14, No. 3, March 1992, pp. 260-285.

[4] N. Morgan, J. Beck, J. Kohn, J. Bilmes, E. Allman and J. Beer, "The Ring Array Processor: A Multiprocessing Peripheral for Connectionist Applications", *Journal of Parallel and Distributed Computing*, Vol. 14, No. 3, March 1992, pp. 248-259.

[5] D. Hammerström, "A Highly Parallel Digital Architecture for Neural Network Emulation", *VLSI for Artificial Intelligence and Neural Networks*, J. G. Delgado-Frias and W. R. Moore (Eds.), Plenum Publishing Company, New York, USA, 1990.

[6] U. Ramacher, "SYNAPSE - A Neurocomputer That Synthesizes Neural Algorithms on a Parallel Systolic Engine", *Journal of Parallel and Distributed Computing,* Vol. 14, No. 3, March 1992, pp. 306-318.

[7] U. Müller, A. Gunzinger and W. Guggenbühl, "Fast Neural Net Simulation with a DSP Processor Array", *IEEE Transactions on Neural Networks*, Vol. 6, No. 1, January 1995, pp. 203-213.

[8] T. Hämäläinen, J. Saarinen and K. Kaski, "TUTNC: A General Purpose Parallel Computer for Neural Network Computations", *Microprocessors and Microsystems*, Vol. 19, No. 8, 1995, pp. 447-465.

[9] M. Levy and J. Leonard, "EDN's DSP-Chip Directory" *EDN Magazine*, May 1995, pp. 40-95.

A High-Speed Scalable CMOS Current-Mode Winner-Take-All Network

Andreas Demosthenous[1], John Taylor[1] and Sean Smedley[2]

[1]Department of Electronic and Electrical Engineering, University College, Torrington Place, London, WC1E 7JE, UK
[2]Data Conversion Systems, Dirac House, St Johns Innovation Park, Cowley Road, Cambridge, CB4 4WS, UK

Abstract. A CMOS modular high-speed current-mode 2-input Winner-Take-All (2-WTA) circuit for use in VLSI tree-structure WTA networks is described. The classification speed of the design is not input pattern dependent, but is a function of the value of the largest current input only. Simulations show that the new circuit can resolve input currents differing by less than 1µA with a negligible loss of operating speed. Detailed simulations and preliminary measured results of a single WTA cell and of a complete 8-input tree WTA network are presented.

1 Introduction

The *Winner-Take-All* (WTA) network is an important component of artificial neural systems [1] and also a fundamental building block of fuzzy logic systems [2] where it is known as the *Max operator*. In general, the function of the WTA network is to select and identify the largest variable from a specified set of M competing variables and inhibit the remaining $(M-1)$ variables. Several recent publications ([3] - [6]) describe analogue WTA designs that are very compact and economical in terms of power dissipation. The advantages of the different approaches are outlined in [6] where an analogue WTA network based on an asynchronous feedforward *tree* is described. Since only local device matching is required in this structure, the tree WTA network lends itself more readily to scaling for a large number of inputs than designs based on *current conveyors* [3], [4]. The tree WTA network presented in [6] consists of $(M-1)$, 2-input current winner-take-all cells (2-WTAs), and has $\log_2 M$ layers, each layer having half the number of 2-WTAs of that immediately above it. Each 2-WTA identifies the larger of its two input currents and produces an output current which is a replica of this local 'winner'. The output from the final 2-WTA is thus a copy of the largest input (i.e., the global 'winner') current. After the completion of this *forward pass*, the states of the 2-WTAs enable the identity of the winner to be determined by means of a *backward pass* involving some combinatorial logic circuitry.

The BiCMOS 2-WTA described in [6] is based on a current sensitive bridge arrangement which subtracts the input currents (I_1, I_2) from each other and develops a difference current, $|I_1 - I_2|$. The circuit changes state when unequal input currents are applied and $|I_1 - I_2|$ is non-zero. Additional circuitry steers the 'winning' current to the output of the 2-WTA cell to be passed on to the next level in the tree. This arrangement suffers from several disadvantages, however:

(a)   In [6] it was reported that the speed with which the bridge changes state depends not only on I_1 and I_2 but also on $|I_1 - I_2|$. Consequently, the classification speed of the complete WTA network is pattern dependent, leading to serious loss of classification speed for certain applied patterns.

(b)   Since the input *pnp* mirrors (Q1-6) are of the emitter follower augmented (EFA) type, a potential for high frequency instability exists [7]. The resulting increased settling time can reduce the classification speed of the WTA network.

(c)   The BiCMOS circuit requires more devices and hence consumes more static power than the new CMOS 2-WTA. In addition, since more stages of current copying are required in the BiCMOS design, there is a greater potential for inaccuracy in replicating the winning input current at each stage in the tree.

These problems are addressed and minimised in the CMOS 2-WTA circuit described in this paper.

2   Circuit Description

1. Forward pass circuitry (FPC): The forward pass is implemented by the circuit shown in Fig. 1. Its function is to identify the larger of the two input currents I_1 and I_2 and to produce a copy of this local 'winner' at the output, I_o. The input currents I_1 and I_2 are applied to a pair of NMOS cascode current mirrors M1 - M4 and M5 - M8 respectively. In addition to generating replicas of the currents I_1 and I_2, these current mirrors realise voltages V_1 and V_2 at the drains of M3 and M7. These voltages are applied to the gates of transistors M9 and M10, which operate competitively as a current steering latch. So, for example, if I_1 is greater than I_2, then V_1 is greater than V_2 and the drain current of M9 is greater than that of M10. Since the drains of M9 and M10 are cross-coupled to I_2 and I_1 respectively, the fraction of I_2 flowing in M5 and M6 decreases, as more is directed into M9. Finally, (virtually) all of I_2 flows to ground through M9, the drain current of M7 decreases to zero, and I_1 'wins' the competition. At this point, I_1 is copied to the output, I_o, by means of M2 and M4 and the PMOS cascode current mirror M13 - M16. Finally, at the completion of the WTA process, the latches are reset globally by means of the transistors M11 and M12 and by the RESET line. Note that the RESET line should be high when the input currents are set up, pulsed low to start the competition and then set high again to reset the latches.

The operating speed of the proposed circuit is essentially limited by the rise times of the three cascode current mirrors (M1 - M4, M5 - M8 and M13 - M16) and these depend on the individual input currents I_1 and I_2 and *not* on the difference between them. In the case of the BiCMOS design [6], as noted earlier, a change of state of the current sensitive bridge is driven by a current which is the difference between the input currents. The response time of the BiCMOS circuit is therefore a function of $|I_1 - I_2|$, leading to the undesirable pattern dependence reported in [6]. Due to the current subtraction, the operating speed will also tend to be highly sensitive to device parameter variations, which is not the case in the proposed CMOS circuit.

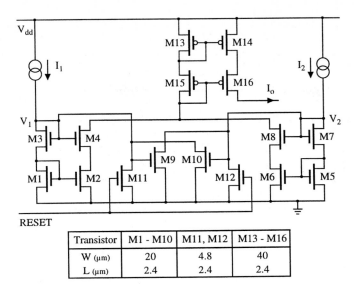

Fig. 1. Forward pass circuitry and transistor dimensions

Transistor	M1 - M10	M11, M12	M13 - M16
W (μm)	20	4.8	40
L (μm)	2.4	2.4	2.4

The potential high frequency instability problems associated with the EFA current mirrors employed in the BiCMOS circuit are avoided in the proposed CMOS design by using cascode current mirrors, which are overdamped systems. Although the use of cascode mirrors generally leads to sub-optimal high frequency performance [8], we show in section 3 below that the new circuit achieves a higher operating speed than the BiCMOS circuit in [6].

2. Backward pass circuitry (BPC): At the completion of the forward pass, the value of the winning input current $I_{o(max)}$ is known. Also, the voltages V_1 and V_2 are set high or low for each 2-WTA cell, corresponding to logic levels $\{1, 0\}$. The purpose of the backward pass is to identify the origin of $I_{o(max)}$ and to represent the point of origin by a binary encoded word of length $\log_2 M$. The backward pass phase is initiated when the ENABLE line in the last cell of the tree goes from logical 0 to 1 (this function is supplied externally for the purposes of this paper) and proceeds in the manner described in [6]. The logic circuitry is shown schematically in Fig. 2 together with the relevant truth tables. The NAND gates and INVERTERS are standard CMOS units and transistor level details are not shown.

M1 and M2 realise a tri-state buffer, which drives the BUS lines representing the binary output codes, which are interconnected in the manner described in [6]. An additional transistor, M3, is required for each row of the tree WTA network all connected to the ENABLE line by means of a single buffer. This arrangement sets the state of the tri-state BUS at ground when ENABLE is low, i.e., for the duration of the forward pass.

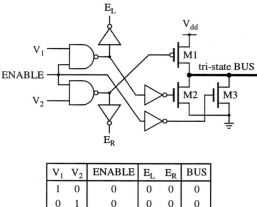

V_1	V_2	ENABLE	E_L	E_R	BUS
1	0	0	0	0	0
0	1	0	0	0	0
1	0	1	1	0	0
0	1	1	0	1	1

Fig. 2. Logic circuitry and tri-state buffer

3 Simulated and Measured Results

A CMOS 2-WTA circuit prototype of the type described above has been designed and fabricated using the MIETEC 2.4µm double-poly, double-metal analogue CMOS process with 5V supplies and with HSPICE parameters supplied by EUROCHIP [9]. The simulations described below are based on an extracted version of the complete circuit and hence include all layout parasitics.

1. FPC Results: Two sets of simulations were carried out to determine the properties of the forward pass circuit. In both cases, the circuit was triggered by a negative-going step being applied to the RESET line. Firstly, I_1 was held constant at 5µA while I_2 was increased from 10µA to 100µA in steps of 10µA. We define the forward pass *classification period*, τ_f, as the period between the point where the RESET pulse falls to 50% of its initial value and the output current reaching 90% of its final value. The results of this simulation are given in Fig. 3.

In the second simulation, $(I_2 - I_1)$ was held constant at 10µA while I_2 was once again varied from 10µA to 100µA in steps of 10µA. These results are indistinguishable from the first set, i.e., τ_f depends only on the value of the largest input current, I_2 in this case, and not on $|I_1 - I_2|$ as in the BiCMOS 2-WTA described in [6]. Further simulations revealed that the circuit can differentiate between currents differing by less than 1µA with only a small loss of operating speed. Hence, for a WTA network with $\log_2 M$ rows the total classification period for the forward pass is given by:

$$T_f = \tau_{f(\max)} . \log_2 M \qquad (1)$$

where $\tau_{f(max)}$ is the classification period corresponding to the largest (winning) input current. Figure 3 also includes preliminary measured results for the 2-WTA chip described above. The measured classification periods are very close to and slightly smaller than the HSPICE simulations.

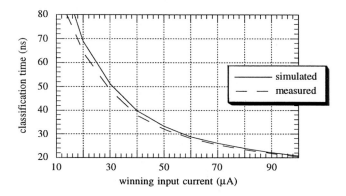

Fig. 3. 2-WTA cell performance curves

An HSPICE simulation of the response of a WTA network with 8 inputs (i.e., 3 rows) is given in Fig. 4. The three curves represent the outputs of the 2-WTA cells receiving the largest input current, which is 40μA in this case. The inter-row delays in Fig. 4 agree closely with the values predicted by the curves in Fig. 3.

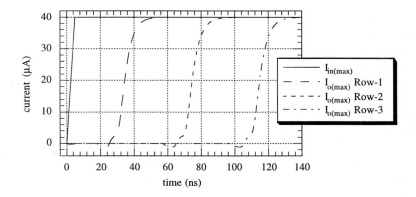

Fig. 4. Transient response of a 3 row asynchronous tree WTA network

The active area of the CMOS 2-WTA cell is $80\mu m \times 140\mu m$, compared to $100\mu m \times 140\mu m$ for the equivalent unit realised in BiCMOS technology [6]. The steady state-state power dissipation for the new circuit with $I_1 = 10\mu A$ and $I_2 = 15\mu A$ is 150μW compared to 450μW for the BiCMOS circuit operated under the same conditions. These results emphasise the benefits of the new CMOS circuit in terms of size and power consumption. Further benefits could be obtained by the use of a more advanced process with shorter gate length transistors.

2. BPC Results: HSPICE analysis of the arrangement shown in Fig. 2 gave a delay, τ_e, of 2ns from the setting of the ENABLE line to the point where E_L or E_R reach 90% of V_{dd}. For the purposes of simulation, the tri-state BUS was loaded with a capacitance of 2pF and the additional delay required to set the tri-state BUS, high (90% V_{dd}) or low (10% V_{dd}) was 3ns. Using (1), for a WTA network with $\log_2 M$ rows, the *total classification period*, T_s (in nanoseconds) is given by:

$$T_s = \left\{ (\tau_e + \tau_{f(\max)}) \cdot \log_2 M \right\} + 3 \tag{2}$$

As an example, consider the design of a binary tree WTA network with $M = 1024$ inputs, and where the maximum input current is 100μA. From Fig. 3, $\tau_{f(\max)} = 20$ns and so T_s is 223ns. The classification period for a BiCMOS tree WTA network with $M = 1024$ was quoted in [6] as 233ns. However, this measurement *did not include the effects of layout parasitics*. The fact that the new design, including all parasitics, is still slightly faster than the BiCMOS design indicates the advantages of the approach presented in this paper. This improvement is obtained in addition to the gains in size and power consumption described earlier in this section.

References

1. R. Lippmann, "An introduction to computing with neural nets," *IEEE ASSP Mag.*, vol. 4, pp. 4-22, Apr 1987.
2. J. Ramirez-Angulo, "Building blocks for fuzzy processors," *IEEE Circuits & Devices Mag.*, vol. 10, pp. 48-50, Jul. 1994.
3. D. Grant, J. Taylor, and P. Houselander, "Design, implementation and evaluation of a high-speed integrated Hamming neural classifier," *IEEE J. Solid-State Circuits*", vol. 29, pp. 1154-1157, Sep. 1994.
4. J. Choi and B. Sheu, "A high-precision VLSI winner-take-all circuit for self-organising neural networks," *IEEE J. Solid-State Circuits*, vol. 28, pp. 576- 584, May 1993.
5. U. Çilingiroglu, "A charge-based neural Hamming classifier," *IEEE J. Solid-State Circuits*, vol. 28, pp. 59-67, Jan. 1993.
6. S. Smedley, J. Taylor, and M. Wilby, "A scalable high-speed current-mode winner-take-all network for VLSI neural applications," *IEEE Trans. Circuits and Systems-Part I*, vol. 42, pp. 289-291, May 1995.
7. K. Laker and W. Sansen, *Design of Analog Integrated Circuits and Systems*. USA: McGraw-Hill, 1994, pp. 385-386.
8. G. Di Cataldo, G. Palumbo and S. Stivala, "New CMOS current mirrors with improved high-frequency response," *Int. J. Circuit Theory and Applications*, pp. 443-450, 1993.
9. C. Das, "MIETEC 2.4 μm CMOS MPC - Electrical parameters," EUROCHIP Service Org. (RAL), Didcot, UK, Doc. MIE/F/02, Jan. 1993.

An Architectural Study of a Massively Parallel Processor for Convolution-Type Operations in Complex Vision Tasks

Martin Franz   René Schüffny

Technische Universität Dresden,
Institut für Grundlagen der Elektrotechnik und Elektronik,
Mommsenstraße 13, 01069 Dresden, Germany
E – Mail: schueffn@iee.et.tu-dresden.de

Abstract. Complex vision tasks, e.g., face recognition or wavelet based image compression, impose severe demands on computational resources to meet the real-time requirements of the applications. Clearly, the bottleneck in computation can be identified in the first processing steps, where basic features are computed from full size images, as motion cues and Gabor or wavelet transform coefficients. This paper presents an architectural study of a vision processor, which was particulary designed to overcome this bottleneck.

1   Introduction

The computational requirements of complex vision tasks based on real-time image sequences impose severe demands on the processing hardware. The primary computational steps performed on the full size image frames represent a bottleneck in processing. State-of-the-art processor hardware cannot cope with the computational complexity imposed by spatio-temporal convolution, gabor and wavelet transformation, motion estimation by block matching, and recurrent neural fields in real-time. To meet these requirements for realistic image size, e.g. 512 × 512 pixels, a dedicated massively parallel processor architecture was investigated [7]. Our processor architecture is derived from the real-time computational and functional requirements of several representative complex vision applications, e.g. object recognition based on jets computed from Gabor transform coefficients [4], 3D-vision for autonomous vehicles [2], wavelet transformation and vector quantization for video conference image compression [5], block matching based motion estimation for automotive applications, and recurrent neural fields [3]. However, the architecture provides sufficient generality to be widely applicable in machine vision.

In the following section, an outline of the supported algorithms will be given. Then the issue of reduced accuracy will be addressed and the processor architecture will be presented. Finally, verification of the processor architecture will be described and verilog simulation results discussed.

2 Image Processing Algorithms

2.1 Gaussian Pyramid

The Gaussian Pyramid provides a multiscale image representation, which can be exploited for a reduced complexity hierarchical processing, e.g. in object recognition and tracking. Assuming a separable Gaussian filter mask with five coefficients and an image size of $N \times N$ pixels, $10\,N^2$ multiplications and $8\,N^2$ additions are required for the first pyramid level. For each following level, the computational requirement is reduced by a factor of four.

2.2 Wavelet Transformation

The wavelet transformation [1] is employed for preprocessing in image compression because of its salient entropy reduction properties. Also, the wavelet decomposition can serve to compute features, e.g. texture analysis. From the computational point of view, the forward and backward wavelet transformation are based on one-dimensional convolution. For image processing the one-dimensional convolution is applied in both image dimensions. In applications 4 to 20 filter coefficients are used. Assuming a maximum number of 20 coefficients and an input image size of $N \times N$ pixels, the following amount of operations is required:

Operation	Multiplications	Additions
2 Convolutions with 20 coefficients, $N \times N$ pixels	$40\,N^2$	$39\,N^2$
4 Convolutions with 20 coefficients, $N \times \frac{N}{2}$ pixels	$40\,N^2$	$39\,N^2$
$\sum$	$80\,N^2$	$78\,N^2$

2.3 Gabor Transformation

This transformation is based on the two-dimensional complex Gabor function according to [2], which mainly is determined by orientation ϑ and wavenumber k. The Gabor transformation is defined by convolution operator G on an image $I(x, y)$ as follows:

$$(G_{k,\vartheta} I)(x_0, y_0) = \sum_{\epsilon_x} \sum_{\epsilon_y} g_{k,\vartheta}(\epsilon_x, \epsilon_y)\, I(x_0 + \epsilon_x, y_0 + \epsilon_y) \tag{1}$$

The results are usually stored as column vectors, denoted as *jets* $J(\mathbf{x_0})$

$$J(x_0, y_0) = \{(G_{k,\vartheta} I)(x_0, y_0) \mid k \in \mathbb{K}, \vartheta \in \Theta\}, \tag{2}$$

where $\mathbb{K}$ is a set of different wavenumbers and Θ a set of different orientations. The jet $J(x_0, y_0)$ contains informations about the local wavenumber spectrum with respect to different orientations at the jet position in an image. A collection of jets thus provides a good representation of image features, which can serve for robust object recognition [4] and vergence control in active 3D- vision applications [2]. For a complete Gabor transformation we have to convolve the whole picture with 40 masks (8 different orientations ϑ and 5 different wavenumbers k) containing 11×11 complex coefficients. Assuming an input image size of $N \times N$ pixels, the following amount of operations is required:

Operation	Multiplications	Additions
40 Convolutions with 242 coefficients, $N \times N$ pixels	$9680\,N^2$	$9640\,N^2$

2.4 Neural fields

Also, the computation of neural fields according to [3, 6] and of cellular neural networks by recurrent convolution with arbitrary neighbourhood radius are supported and require the following amount of operations:

Operation	Multiplications	Additions
$N \times N$ Neural Field with 7×7 weights, one iteration	$49\,N^2$	$50\,N^2$

3 Processor Design

In this chapter we derive a processor architecture for image processing based on the requirements of the previously described algorithms. We focused our work on the optimum implementation of the Gabor transformation, due to its complexity and the enormous amount of required computations.

3.1 Computational Requirements

All algorithms are implemented using fixed-point number representation to avoid area consuming floating-point arithmetic blocks. Thus, the processor is designed for fixed-point computation. According to simulation results, the following accuracies were determined for the algorithms:

Algorithms	Cache Resolution	Coefficient/ Weight Resolution	Result Resolution
Gaussian Filtering	8 bit	8 bit	8 bit
Wavelet transformation	8 bit	16 bit	8 bit
Gabor transformation	8 bit	32 bit	32 bit
Neural fields [3, 6]	32 bit	32 bit	32 bit

We concentrated on an optimization of the processor architecture with respect to the acceleration of convolutions. In our architecture the convolution computation for a pixel position is sequentially carried out by a single processing element, but 64 pixel positions are concurrently processed by 64 available processing units. Figure 1 shows the corresponding processor unit architecture.

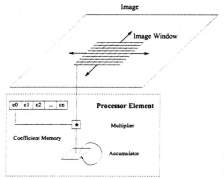

Fig. 1.: Calculation of Convolution Operations

An on-chip image cache with a size of 8×8 pixels stores a window of the input image, which can be shifted in two dimensions. Each memory cell corresponds to a processor element (PE) containing multiplier and adder. In figure 1 the shifter cache and one of the 64 PEs is shown. At each clock cycle a coefficient is multiplied with the gray value of the corresponding pixel position and the result is added to the accumulator. Then, the window is shifted one pixel in the input image. Only eight new pixels have to be read from image memory to the cache for the next 64 operations. After the multiplication of the last coefficient, the convolution for 64 pixels is completed. Our architecture achieves a minimum I/O-bandwidth along with an optimum computation rate, provided that convolution type locality can be assumed for the computation.

3.2 Data Path

Figure 2 shows the general structure of the SIMD-array image processor, which contains 64 PEs, a command controller, an I/O controller and a central processor unit, which controls the program flow and computes special numerical operations, e.g. division. The shifter cache is not explicitly shown here, as each of its cells is considered as integrated in the corresponding slice. For efficient minimum or maximum computation, as required for matching algorithms, a 64-input comparator tree is implemented. The internal structure of the PEs is shown in figure 3. The coefficient memory contains 72 16-bit registers. Coefficient resolutions of 8, 16 and 32 bit are feasible. For storage of a 32-bit coefficient two registers are required. The shifter cache is implemented with four levels, allowing the concurrent storage and access to four different

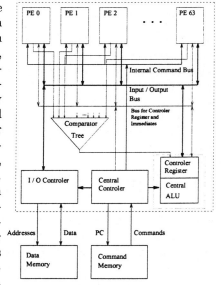

Fig. 2.: Processor Architecture

images. The pixel values can be represented as 8 bit unsigned or 16 bit signed integers. The multiplier is configurable dependent on pixel and coefficient representation, e.g. for 8×32 bit and 16×16 bit multiplications in one clock cycle, or 16×32 bit in two cycles and 32×32 bit in four clock cycles. The partial products can be added to or subtracted from the accumulator content. Finally, the accumulator with a resolution of 64 bit can be normalized by right shifts.

The normalised result is moved to result memory, which consists of 52 16-bit registers. The result will usually be transported from result memory over the I/O-bus (64-bit Bus D) to the external memory, but it can be used as input for following calculations, too. The architecture uses three 16-bit busses (A,B,C), which allow operand access in the shifter cache levels and in the coefficient or result memories. Coefficient as well as pixel value precisions can be configured. The I/O-bus is connected to the coefficient memory in order to provide an initial loading of mask

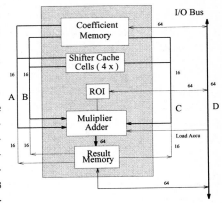

Fig. 3.: Processor Element

coefficients. The abbreviation ROI stands for "region of interest" and means a bit array, which is used for local masking of numerical and logical operations, so that computations are limited to significant image sections. This bit array can be loaded from the external memory via I/O-bus.

4 Processor Verification

4.1 Modelling in Verilog-XL

The processor architecture was completely implemented in Verilog-XL. A symbolic assembler for the generation of machine code was developed to facilitate processor programming and testing. The fixed instruction length of the RISC type processor is 32 bit. The set of 51 processor commands contains 22 complex arithmetic commands, 29 instructions for data transfers, 12 commands for controlling program flow, and 18 initialization instructions. As example for a complex command the *MAC-and-shift* instruction `mula.r.r 32x16 shr sl8 rf127 rs8 roi23` is given.

4.2 Simulation Results

For a verification of the processor model the Gaussian filtering, Wavelet and Gabor transformations where realized in assembler code. Because of the long simulation time the input picture was limited to 64 × 64 pixels. The cycle diagrams are shown in figure 4. A Gaussian filter mask with five coefficients was

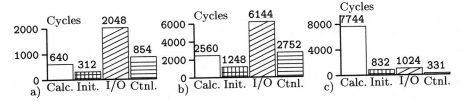

Fig. 4. Cycle Diagrams for a) Gaussian filtering, b) Wavelet Transformation, c) Gabor Transformation

used for simulation. For Wavelet transformation the Haar Wavelet was chosen, which leads to two coefficients. This results in a very small amount of calculations in comparison to the Gabor transformation, which requires at least 16 masks with 11×11 coefficients. Figure 4a and 4b show, that the ratio between I/O operations and calculations is not satisfying in these cases, where the processor mainly performs I/O operations. On the contrary, for Gabor transformation the amount of I/O transfers is negligible in comparison to the number of calculations. An optimum acceleration is achieved for applications imposing a large amount of computations. Assuming a clock cycle of 100 MHz and a 512×512 input image, the following processor performance is obtained[1]:

	Gaussian Filtering	Wavelet Transformation	Gabor Transformation
Performance	405 frames/s	123 frames/s	10 frames/s

5 Conclusion

In this work, a massively parallel SIMD-array processor architecture with 64 processing elements was proposed and simulated at behavioral-level, that provides an optimum performance for convolution type operations. Our processor represents a real-time computation platform for a number of significant front-end real-world vision tasks [1, 2, 3, 4, 5]. The VERILOG simulation of our processor proved the feasibility and the correct functionality of the implementation. A performance improvement for algorithms with small masks could be achieved in a future revision by doubling the I/O-bus size.

References

1. Daubechis, I.: *Ten Lectures on Wavelets*, CBMS-NSF Regional Conf. Series in Appl. Math., Vol. 61, Society for Industrial and Appl. Math., Philadelphia, 1992
2. Theimer, W. M.; Mallot, H. A.: *Binocular Vergence Control and Depth Reconstruction Using a Phase Method*, ICANN Proceedings 1992, pp. 517-520, Elsevier Science Publishers B.V. 1992
3. Seelen, W.v., *A Neural Architecture for Autonomous Visually Guided Robots – Results of the NAMOS Project*, *Fortschrittsberichte VDI, Nr. 388, 1995*
4. Wiskott, L.; Malsburg, C. v.: *A Neural System for the Recognition of Partially Occluded Objects in Cluttered Scenes: A Pilot Study*, IEEE Transactions on Pattern Recognition and Artificial Intelligence, Vol. 7, No. 4, 1993
5. Buhmann, J.; Kühnel, H.; *Vector Quantization with complexity Costs*, IEEE Transactions on Information Theory, Vol. 39, No. 4, 1993
6. Chua, L. O. and Yang, L.: *Cellular Neural Networks: Theory and Applications*, IEEE Transactions on Circuits and Systems, Vol. 35, No. 10, 1988, pp 1257 – 1290
7. Franz, M.: *Entwurf eines Faltungsprozessors*, TU Dresden, Institut für Grundlagen der Elektrotechnik und Elektronik, Diploma Thesis, 1996

---

[1] The processing speed of 10 frames/s, achieved for Gabor transformation, is a pessimistic estimation, as in the application of face recognition the jet computation is only carried out for a small number of pixel feature positions.

FPGA Implementation of an Adaptable-Size Neural Network

Andrés Pérez-Uribe and Eduardo Sanchez

Logic Systems Laboratory
Swiss Federal Institute of Technology
CH–1015 Lausanne, Switzerland
{Andres.Perez,Eduardo.Sanchez}@di.epfl.ch

Abstract. Artificial neural networks achieve fast parallel processing via massively parallel non-linear computational elements. Most neural network models base their ability to adapt to problems on changing the strength of the interconnections between computational elements according to a given learning algorithm. However, constrained interconnection structures may limit such ability. Field programmable hardware devices allow the implementation of neural networks with in-circuit structure adaptation. This paper describes an FPGA implementation of the *FAST* (Flexible Adaptable-Size Topology) architecture, a neural network that dynamically changes its size. Since initial experiments indicated a good performance on pattern clustering tasks, we have applied our dynamic-structure FAST neural network to an image segmentation and recognition problem.

1 Introduction

Artificial neural network models offer an attractive paradigm: learning to solve problems from examples. Most neural network models base their ability to adapt to problems on changing the strength of their interconnections according to a learning algorithm. However, the lack of knowledge in determining the appropriate number of layers, the number of neurons per layer, and how they will be interconnected, limits such ability. The so-called *ontogenic neural networks* [4] try to overcome this problem by offering the possibility of dynamically modifying their topology. Other potential advantages of ontogenic neural networks are improved generalization and better implementation optimization (for size and/or execution speed). Neural networks with dynamic topologies for both supervised and unsupervised learning have been investigated by others. The Carpenter and Grossberg's ART model (Adaptive Resonance Theory) [2], the Grow And Represent (GAR) [1] model and Fritzke's Growing Cell Structures [5] are ontogenic competitive learning neural networks, and have inspired our work. While software implementations on conventional von Neumann machines are very useful for investigating the capabilities of neural network models, a hardware implementation is essential to fully profit from their inherent parallelism, as well as for real-time processing in real-world problems. Recent progress in semiconductor devices enable us to use programmable hardware devices such as Field

Programmable Gate Arrays (FPGA) [9] and the Field Programmable Interconnection Circuits (FPIC) which allow users to reconfigure their internal circuit connections and node logic functionalities. Such devices afford rapid prototyping and inexpensive neural network implementations. The reprogrammability of field programmable hardware devices enables the implementation of neural networks with in-circuit structure adaptation. This paper describes an FPGA implementation of the *FAST* architecture (Flexible Adaptable-Size Topology), a neural network that dynamically adapts its size. This paper is organized as follows: the learning and changing topology algorithm is described in section 2, section 3 is dedicated to presenting the FPGA implementation, section 4 deals with a color segmentation application, and finally,section 5 presents some conclusions and future extensions of this work.

2 Algorithm

The **FAST** neural network is an unsupervised learning network with **F**lexible **A**daptable-**S**ize **T**opology. The network is feedforward, fully connected, and consists of two layers, an input layer and an output one. Essentially, the aim of the network is to cluster or categorize the input data. Clusters must be determines by the network itself based on correlations of the inputs (unsupervised learning). The network's size increases by adding a new neuron in the output layer when a *sufficiently distinct* input vector is encountered and decreases by deleting an operational neuron through the application of probabilistic deactivation. Each neuron j maintains an n-dimensional reference vector, W_j, and a threshold, T_j, both of which determine its *sensitivity region*, i.e. the input vectors to which it is maximally "sensitive". At the startup, the network consists only of an input layer, with no output neurons. Input patterns are then presented and the network evolves through application of the FAST algorithm, which is driven by three processes: (1) learning, (2) incremental growth, and (3) pruning:

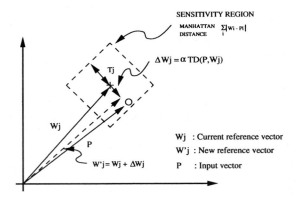

Fig. 1. Learning in FAST

1. *Learning.* The learning mechanism adapts the neuronal reference vectors; as each input vector P is presented to the network, the distance, $D(P, W_j)$, between it and every reference vector W_j is computed. If $D(P, W_j) < T_j$ (the threshold of neuron j), W_j is updated as $W'_j = W_j + \alpha T_j D(P, W_j)$, where W'_j is the new reference vector, and α is a learning parameter in the range [0,1] (Figure 1). In our implementation, the Manhattan distance $D(P, W_j) = \sum_i |P_i - W_{ji}|$ is used as a measure of similarity between the reference vectors and the current i-dimensional input. It defines the (diamond-shaped) sensitivity region of neuron j (Figure 1). Activation of neuron j also entails a linear decrease in its threshold value: $T'_j = T_j - \gamma(T_j - T_{min})$, where T'_j is the new threshold, γ is a gain parameter in the range [0,1], and T_{min} is the minimal threshold. This threshold modification decreases the sensitivity region size for neurons in high density areas of the vector space, and vice versa.

2. *Incremental growth.* When an input vector P does not lie within the sensitivity region of any currently operational neuron, a new neuron is added, with its reference vector set to P, and the threshold set to an initial value *Tini.*

3. *Pruning.* The network decreases in size, i.e. output neurons are deleted, through a pruning process. The probability of an operational neuron being deleted, Pr_j, increases in direct proportion to the overlap between its sensitivity region and the regions of its neighbors. The overlap between the sensitivity regions of several neurons is estimated by computing the frequency of activation of the overlapping neurons with a same input vector. Therefore, the probability of deletion of a neuron increases linearly with slope η towards a maximum $1-Pr_{min}$ every time an input vector activates two or more neurons. Once a deletion has occurred the probability Pr_j of the units involved is reset.

3 FPGA Implementation

The FAST neural network architecture was implemented on a custom machine based on programmable logic devices which afford rapid prototyping and easy reconfiguration, and are relatively inexpensive. It is composed of a 68331 microcontroller, four Xilinx XC4013-6 FPGA chips [9], and four globally-connected FPID devices (Crossbar switches I-Cube 320) [6]. We used two of the Xilinx XC4013 FPGAs and two of the I-Cube 320 FPIDs to implement four two-dimensional FAST neurons, the network's sequencer, and a bank of I/O mapping registers. The sequencer is implemented in one of the Xilinx XC4013 chips along with two of the FAST neurons. The other two FAST neurons are implemented in the second XC4013 chip along with the I/O mapping registers. The sequencer, the FAST neuron architecture, and the I/O mapping register bank will now be briefly discussed.

The Fast neuron

A FAST neuron is composed of three blocks: Manhattan distance computation, learning (i.e. modification of reference vector and threshold), and

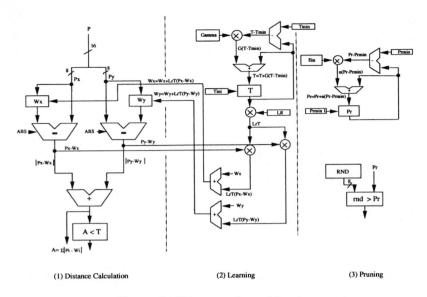

Fig. 2. FAST neuron's architecture

pruning (Figure 2). The system currently supports 8-bit computation and two dimensional vectors. Each neuron includes nine 8-bit adders and a single 8-bit shift-add multiplier, so the additions occur in parallel but the multiplications involved in the learning phase are executed sequentially. The maximal number of multiplications during learning phase is five and each requires 8 clock cycles. To implement the pruning process, each neuron includes a random number generator, consisting of an 8-cell heterogeneous, one-dimensional cellular automata implementing rules 90 and 150 of Wolfram's classification [8]. It has a period of 255 and is implemented using a small number of components: 8 D flip-flops, 5 two-input XOR gates, and 3 three-input XOR gates.

The sequencer

The sequencer is a finite state machine (FSM) which handles the addition and deletion of neuronal units in the output layer and synchronizes the neuron's operation during Manhattan distance calculations, weight and threshold updates (learning) and probabilistic unit deactivations (pruning).

I/O mapping register bank

The hardware device is connected to a host computer which is used to generate input vectors for the system and to display its status. It communicates serially with the 68331 microcontroller who reads/writes signals from/to the I/O mapping registers. It consists of ten 8-bit registers used to map every input and output of the neural network. A polling subroutine runs on the 68331 microcontroller in order to generate the read/write control signals and to receive/send data when the host computer requires it.

The maximal XC4013's pin-to-pin delay is 122.1ns, and the I-CUBE's pin-to-pin delay is 12ns. The processing time per input vector is up to 57 clock cycles depending on the number of multiplications during learning, thus enabling the introduction of a new input vector every $8\mu s$, given the current hardware. Our tests had a slower rate of inputs because of the serial link used.

Fig. 3. van Gogh's *Sunflowers* in a 61-color image and FAST-segmented image

4 Example

The algorithm described in the previous sections has been applied to a color learning and recognition problem in digital images. Given a digital image from a real scene, the problem is clustering image pixels by chromatic similarity properties. Similar color pixels belong to a so-called *Chromatic Class*. By identifying chromatic classes in an image and assigning a code to each class, we can do color image segmentation and recognition. In image processing this is called a *Learning Process* [3].

In our experiments we have considered the R-G-B components of the 294x353 pixels of a 61-color image of van Gogh's *Sunflowers*. Thus, every pixel is represented in a 3D space. Then, a color space transformation is used for represent the pixels in a 2D space called I2-I3 : I2 = R - B, I3 = G - 0.5 (R + B) [7]. This 2D space is more suitable for color representation than the R-G-B because coordinates in the former are less correlated than in the latter.

These bidimensional pixel coordinates are randomly presented to the FAST neural network. The resulting segmented image is shown in Figure 3: four chromatic clusters are identified (one per neuron), and objects in the image are consequently easily recognized in image analysis tasks. Also, we have compared software simulations of our algorithm with another neural network learning algorithm which combines the thresholding in I2-I3 space and the fuzzy Kohonen clustering nets approach [3]. While in this learning algorithm a histogram is first obtained to determine the number of clusters and their shapes, ours determines rhomboid clusters dynamically. We are still experimenting with this model and current results are very encouraging.

5 Conclusions

We have used high-density programmable logic devices (FPGAs) and field programmable interconnection devices (FPID) to successfully implement an adaptable-size neural network architecture. Our system offers the advantages of any neural network hardware implementation : parallel computation and operation at very high speeds, enabling real-time processing. The algorithm also offers implicit fault tolerance. Our initial experiments with this recently completed system indicate that high performance can be achieved on pattern clustering tasks. As an example, we have applied our FAST neural network to an image segmentation and recognition problem. It has been found that the results obtained with this methodology closely resemble those obtained with other neural algorithms which are not suitable for hardware implementation. We have currently begun the design of a new system, based on the principles presented in this paper, with a larger number of neurons. New FPGA architectures with valuable features such as partial reconfiguration should lead us to more complex adaptable-topology neural networks.

References

1. A. Alpaydin: *Neural Models of Incremental Supervised and Unsupervised Learning* Thesis 863, Swiss Federal Institute of Technology, Lausanne, 1990.
2. G. Carpenter and S. Grossberg: "Self-Organization of stable category recognition codes for analog pattern" *Applied Optics* Vol 26, 1987
3. D. Benitez Diaz and J. Garcma Quesada: "Learning Algorithm with Gaussian Membership Function for Fuzzy RBF Neural Networks" *From Natural to Artificial Neural Computation* Springer Verlag 1995.
4. E. Fiesler: "Comparative Bibliography of Ontogenic Neural Networks" *Proceedings of the International Conference on Artificial Neural Networks (ICANN'94)*
5. B.Fritzke: "Growing Cell Structures - a self-organizing network in k dimensions" *Artificial Neural Networks 2* 1992
6. "I-CUBE. The FPID Family Data Sheet" *I-CUBE,Inc* May, 1994
7. Y. Ohta: "Knowledge-Based Interpretation of Outdoor Natural Color Scenes" *Pitman Advanced Publishing Program* 1995
8. W. Pries and A. Thanailakis and H. C. Card: "Group Properties of Cellular Automata and VLSI Applications" *IEEE Transactions on Computers* December, 1986
9. S.M. Trimberger: *Field-Programmable Gate Array Technology* Kluwer Academic Publishers, Boston, 1994.

Extraction of Coherent Information from Non-Overlapping Receptive Fields

Dario Floreano

AREA Science Park, Cognitive Technology Laboratory, Padriciano 99, I-34012
Trieste, Italy

Abstract. It has been suggested that long-range lateral connections in the cortex play a contextual role in that they modulate the gain of the response to primary receptive field input. In the first part of this paper I show that a network with a set of such pre-wired connections has a short-term dynamics that enhances and stabilizes coherent information defined across multiple, non-overlapping receptive fields. In the second part, I suggest a simple Hebbian rule that can develop the required pattern of synaptic strengths and describe two simulations where the networks discover information that is defined only by its coherence across receptive fields.

1 A contextual role for long-range lateral connections

Neurophysiological and pharmacological experiments have indicated that long-range lateral connections in the cortex have a modulatory effect on the postsynaptic cell [3]. Their action could be described by a mechanism that increases or decreases the gain of the cell response to the receptive field input, but cannot alter the feature that is transmitted by the receptive field of the cell. Since these cortico-cortical connections extend from 200μm up to 6mm, they connect groups of neurons with non-overlapping receptive fields. Each group is internally organized in a fairly similar fashion as a local circuit of neurons with largely-overlapping receptive fields and inhibitory interneurons, such as the orientation-selective hypercolumns in the visual cortex. In this paper such a local circuit will be called a *processor*. Modulatory long-range connections between processors could serve various purposes. For example, they could synchronize activity of processors responding to similar features distributed across the input surface and make them stand out from background activity. Since they affect postsynaptic activity, they could also play a role in the development of the primary receptive fields of the individual processors that they connect.

A continuous activation function has been proposed [5] that combines driving signals from the primary receptive fields with modulatory signals from lateral processing units . The output of the processor is a bipolar value $\{-1, +1\}$ whose probability is given by filtering the activation strength through a sigmoid function. The activation function intends to capture the main biological properties of the interaction between the two types of signals: *a)* the sign of the output is determined solely by the primary receptive fields, *b)* the strength of the activation is increased when the sign of the integrated modulatory signal agrees with the sign of the integrated driving signal, and *c)* it is reduced when the sign of the integrated modulatory signal is in contrast with the sign of the integrated driving signal; *d)* the activation function is equivalent to a sigmoid function when the modulatory signal is nil.

$$\mathcal{A}(d_i, m_i) = (1/2)d_i(1 + \exp(2d_i m_i)) \tag{1}$$

where $d_i = \sum_{j=1}^{n} x_j w_{ij}^d$ is the integrated driving signal from the receptive field components x_j and $m_i = \sum_{k=1, k \neq i}^{n} y_k w_{ik}^m$ is the integrated modulatory signal from the other processors y_k. The output probability $P(y_i = 1)$ is then computed by passing $\mathcal{A}$ through a sigmoid function, or through a tanh function for a mean-field approximation.

The modulatory connections provide *contextual guidance* in that they modulate the postsynaptic activity according to what is computed by the other processors. It has also been suggested that contextual – or modulatory – connections might be used to extract information that is coherent across the receptive fields of several processors. The goal of each processor is then the maximization of the three-way mutual information between its own output, the receptive field input, and the contextual field input. Learning rules can be derived by performing gradient ascent on this objective function [5]. Similar goals have been explored also with other learning schemes and architectures [1, 7].

Using this approach, section 2 shows that a network of such processors with pre-wired contextual connections can dynamically enhance and stabilize a noisy signal. Section 3 suggests a simple Hebbian rule that can provide the required functionality.

2   Enhancement and stabilization of noisy signals

Consider a set of 100 processors organized as a square matrix (Figure 1, left).

Each processor receives a single primary receptive field input from a corresponding element in the input surface where a subset of 4 x 4 elements have a value of 0.6 with added random noise from a uniform distribution in the range ±0.3 and all the remaining elements have a random value in the ±0.6 range. Only the processors that receive a primary signal from the central 4 x 4 input elements are mutually linked by contextual connections which have a synaptic value of 1. The output of each processor is updated for several iterations: at each iteration the values of the input elements are calculated anew. Note that intial output activity, before contextual modulation, is less than input (Figure 1, step 1). After a single iteration the processors within the 4 x 4 inner matrix become fully active (Figure 1, step 2) and maintain the activity level despite strong fluctuations at the input level by sustaining each other through the contextual signals (Figure 1, step 3). Similar behaviours are observed when the input surface consists of ambiguous figures analogous to the Necker cube or to the Rubin vase and inhibitory contextual connections are added between processors connected

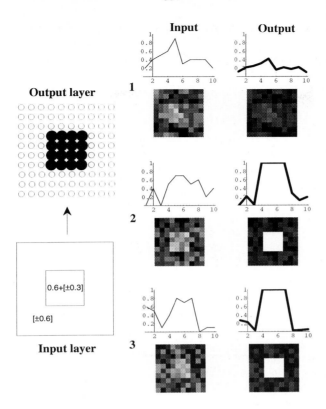

Fig. 1. Left: Architecture of the network. Each processor receives a single primary receptive field input from a corresponding element in the input surface. Processors drawn in bold face are mutually linked by excitatory contextual connections. **Right**: Input and output activity (the absolute value of tanh $[\mathcal{A}(d_i, m_i)]$) for three iterations; at each iteration the input elements and the processor outputs are computed anew. Step 1 shows intial output activity before the contribution of the contextual signal. For each iteration are shown the complete matrices of input and output activity levels (white and black meaning respectively fully active and completely inactive) and graphs correponding to a slice through the middle row.

to different input subgroups: in this case the network can shift between alternative interpretations of the image and maintain them for a number of iterations [2, 6].

The dynamic properties of this model are potentially useful in multi-stage levels of processing. Appropriate configurations of lateral connections can segregate and enhance a noisy pattern from the background according to its spatial and/or temporal coherence, thus facilitate processing at successive stages. This functionality is also similar to the grouping properties of synchronizing connections [6].

3   Adaptation of receptive and contextual fields

This section introduces a local Hebbian rule that develops the pattern of synaptic configuration assumed in the model above. Since learning is done on-line, the output of each processor is computed using the mean field approximation

$$y_i = \tanh\left[\mathcal{A}(d_i, m_i)\right] \tag{2}$$

Both the receptive field weights and the contextual field weights are simultaneously adapted by the same learning rule. The main idea is that synaptic weights should be adapted when postsynaptic activity is boosted by the contextual signal, and gradually decay when postsynaptic activity is dampened by the contextual signal. A single binary variable M_i describes the contextual effect on processor i after L iterations of activity ($L = 3$ in the experiments described below)

$$M_i = \begin{cases} 1 \text{ if } |y_i^t| - |y_i^{t-L}| \geq 0 \\ 0 \text{ otherwise} \end{cases} \tag{3}$$

Synaptic modification occurs on a slower time scale: all the weights in the network (receptive field and contextual field weights) are simultaneously updated after computation of short-term dynamics according to the same learning rule

$$w_{i,j}^t = w_{i,j}^{t-1} + \eta \underbrace{y_i \left(x_j \overline{y_i x_j} - y_i w_{i,j}^{t-1} \right) M_i}_{\text{adaptation}} - \underbrace{(1 - M_i) \overline{y_i x_j} w_{i,j}^{t-1}}_{\text{decay}} \tag{4}$$

where η is the learning rate, x_j is the preynaptic activity, and $\overline{y_i x_j}$ is a moving average of pre - and postsynaptic activity over a restricted time window (5 learning cycles in the experiments described below). The adaptation component of learning is similar to Oja's learning rule for extraction of the first principal component [4] and makes sure that the synaptic weight vectors tend to length 1. Both the adaptation and the decay phase depend on the average correlation between pre- and postsynaptic activity $\overline{y_i x_j}$: the stronger the correlation, the faster the weights are adapted if boosting of postsynaptic activity has occurred ($M_i = 1$) and the faster they decay to zero if dampening of postsynaptic activity has occurred ($M_i = 0$).

Consider a network with two processors that have non-overlapping receptive fields and are mutually linked by contextual connections. The primary input to each processor is a random vector of three elements that can take bipolar values {-1, +1} with equal probability; however, the sign of the first input element is correlated with the sign of the same input element in the other processor. Both receptive field weights w_{ij}^d and contextual field weights w_{ij}^m are initialized to random values in the ± 0.001 range, and the learning rate η is set to 0.1. Both the processors learn to signal the sign of the first element in their own input. This is reflected by the pattern of connection strengths (Figure 2, **a**). Only the synaptic weights corresponding to the input element correlated across processors are strengthened. The contextual connections between the two processors are also strengthened at the same time. The small fluctuations of the remaining connections reflect temporary weak correlations between the other input elements and, although they are too small to affect the response of the processors, they can be easily reduced by widening the time window over which $\overline{y_i x_j}$ is computed. If the processors do not share correlated information in their input or such information ceases to exist over time, the contextual connections would gradually decay to zero and each processor would transmit the most informative component of its own receptive field. The same algorithm has also been used to extract higher order information defined only across processors. Each receptive field input can be visualized as a 2 x 2 square matrix whose entries r_{ij} take bipolar values {-1, +1} with equal probability (Figure 2, **b**, left side). Therefore, within each of the two streams of processing all possible inputs occur with equal probability. The higher-order input variable correlated across the two streams is the sign of the "horizontal edge" $E_H = \left(\sum_j r_{1j} - \sum_j r_{2j} \right)$ which is the sign of the difference between the sums of the two row components [5]. Both processors learn to signal the sign of the horizontal edge in less than 450 training cycles (all the training parameters are the same employed in the previous experiments). The final receptive field weights reflect the structure of the variable correlated across processors (Figure 2, **b**, right side).

Summarising, the model outlined in this paper is capable of extracting coherent information defined across non-overlapping receptive fields, and of developing the appropriate pattern of contextual connections. Since the learning rule includes a normalizing factor, the postsynaptic activation is never saturated and the modulatory signals can generate the short-term dynamics described in section 2.

References

1. S. Becker and G. E. Hinton. Self-organizing neural network that discovers surfaces in random-dot stereograms. *Nature*, vol. 355:pp. 161–163, 1992.
2. D. Floreano, J. Kay, and W. A. Phillips. A computational theory of learning visual features via contextual guidance. European Conference on Visual Perception, August, Tübingen, 1995.
3. J. A. Hirsch and C. D. Gilbert. Synaptic physiology of horizontal connections in cat's visual cortex. *Journal of Neuroscience*, vol. 1:pp. 1800–1809, 1991.
4. E. Oja. A simplified neuron model as a principal component analyzer. *Journal of Mathematical Biology*, vol. 15:pp. 267–273, 1982.
5. W. A. Phillips, J. Kay, and D. M. Smyth. The discovery of structure by multi-stream networks of local processors with contextual guidance. *Network*, vol. 6:pp. 225–246, 1995.
6. W. A. Phillips and W. Singer. In search of common foundations for cortical computation. Submitted to *Brain and Behavioural Science*, 1995.
7. J. Schmidhuber and D. Prelinger. Discovering predictable classifications. *Neural Computation*, vol. 4:pp. 625–635, 1993.

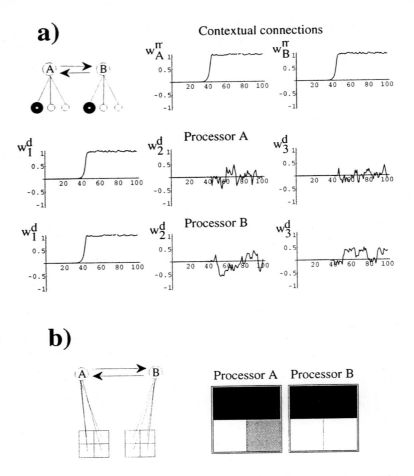

Fig. 2. Extraction of coherent information when there is no structure within the receptive field input of the processors. **a:** Each processor receives primary input from
three random bipolar units, but the sign of the first unit (evidentiated in bold face)
is correlated across receptive fields. Development of synaptic strengths during learning
(data are plotted every 10 cycles): both contextual weights and receptive field weights
learn at the same time to signal the sign of the correlated variable. **b:** Architecture and
final receptive field weights of a network that has discovered the sign of the horizontal
edge variable defined only across processors. Black is inhibitory and white is excitatory
(the gray level stands for a small excitatory weight).

Cortico-Tectal Interactions in the Cat Visual System

Michael Brecht and Andreas K. Engel

Max-Planck-Institut für Hirnforschung
Deutschordenstraße 46, 60528 Frankfurt, Germany

Introduction

The optic tectum is the major visual processing center in lower vertebrates. However, in some mammalian lines such as carnivores and primates an enormous expansion of neocortical processing centers has taken place and in these species the thalamocortical system has become the major target of the retinofugal projection. In these mammals anatomical studies have demonstrated massive corticotectal connections. From both the evolutionary and the anatomical pattern it is to be expected that in cats and monkeys there should be a strong cortical influence on superior colliculus (SC) function. Therefore, numerous physiological studies have addressed the issue of corticotectal interactions. Almost all of these studies have focused on how either SC or cortex contribute to the receptive-field (RF) properties of neurons in the respective other structure. Following this line of research, corticotectal neurons have been identified and their RFs classified, and the consequences of collicular or cortical inactivation on RFs in the respective other structure have been analyzed. This work has shown that cortical input contributes to the general responsiveness, direction selectivity and binocularity of collicular cells and similar conclusions have been drawn about the tectal impact on cortical RFs (Chalupa, 1984).

Meanwhile, theoretical studies have argued that neural coding might not be understood completely by describing how many spikes a cell emits in response to stimulation (i. e., by analysis of average firing rates). Rather, it has been suggested that assemblies of cells code for stimuli (Hebb, 1949; Braitenberg, 1978) and that temporal relationships in the activity of neural populations might define such assemblies (von der Malsburg, 1986). Recent experimental studies support the notion that spike timing matters (Vaadia et al., 1995), and a large number of recent results from the visual system support the hypothesis that assemblies of synchronized neurons are used for the representation of visual stimuli (Singer and Gray, 1995).

The concept of temporal coding in cortical information processing poses new questions for the study of corticotectal interactions: Is spike timing in the SC affected by the cortical input, i. e., is there a possibility that the SC reads out cortical temporal information? What type of interaction is reflected by temporal correlations between cortex and SC? Do corticotectal correlations resemble corticocortical long-range interactions or do they differ in their dynamics? We sought to address these and related questions by applying cross-correlation analysis to responses that were simultaneously recorded from SC and a variety of cortical areas. Specifically, we have investigated visual cortical areas 17, 18, PMLS, PLLS and the superficial collicular layers. These structures are known to be connected by massive corticotectal connections and by indirect tectocortical feedback via the thalamus.

Methods
　　All data were recorded from adult cats in the anaesthetized preparation according to a standard protocol (for details, see Engel et al., 1991a, 1991b, 1991c). Multi-unit activity was recorded from the SC and from visual cortical areas 17, 18, PMLS or PLLS with arrays of two to five microelectrodes in each structure. The electrode signals were amplified, band-pass filtered and fed through a Schmitt-trigger to obtain TTL-pulses which signaled spike timing. Receptive fields were plotted onto a tangent screen, and the ocular dominance, orientation preference as well as the direction preference of each multi-unit cluster was assessed with hand-held stimuli. For quantitative measurements visual stimuli were projected onto the tangent screen using a computer-controlled optical bench. Usually, moving bars were applied as stimuli. Cells with overlapping or close receptive fields were activated with a single bar, whereas units with distant receptive fields were activated with two coherently moving bars. For all responses peristimulus time histograms as well as auto- and crosscorrelation functions were computed. To quantitatively assess the correlogram modulation, generalized Gabor functions were fitted to the data as described previously (König, 1994).

Results
　　Before addressing the question of corticotectal correlations a few comments about the temporal correlations observed within cortex and SC seem appropriate. In accordance with earlier studies (reviewed in Singer and Gray, 1995) temporal interactions in visual cortical areas occurred with 0-time lags and were often accompanied by oscillatory components with frequencies between 20-80 Hz. In the SC we also frequently encountered temporal correlations with 0-time shifts. Oscillatory modulation of correlograms was common. These oscillations occurred in a broad frequency range of 5-80 Hz and the frequencies evoked tended to be quite variable from one stimulus presentation to the next. Generally, correlogram peaks for SC responses were broader than for cortical recordings and the dominant oscillation frequencies were lower.
　　In many instances we have observed temporal correlations between responses in areas 17, 18, PMLS, PLLS and the SC. The corresponding crosscorrelograms computed for corticotectal responses resembled in shape the correlation patterns observed within the SC (Fig. 1): most peaks were broad (>20 ms) and most oscillatory side peaks indicated low frequencies (5-30 Hz), but occasionally also narrow peaks and high frequency coupling (>50 Hz) were seen. Correlations occurred both in epochs where cortical and tectal units were active in different frequency bands as well as in episodes where the correlated cells were active at similar frequencies. With respect to their interactions with the tectum, we observed no major differences between the cortical areas studied.
　　Corticotectal interactions differed in three respects from intracortical and intratectal correlations: (i) They occurred less frequently than interactions within either cortex or colliculus. (ii) Corticotectal correlations tended to be weaker than the cortical or collicular correlations. (iii) With very few exceptions, corticotectal interactions did not occur at 0-time shifts. Rather, systematic phase shifts were observed, with the cortical units leading up to 25 ms over collicular responses (Fig. 1).
　　The incidence of corticotectal correlations depended on the retinotopic position of RFs of the respective units. Correlations were observed significantly

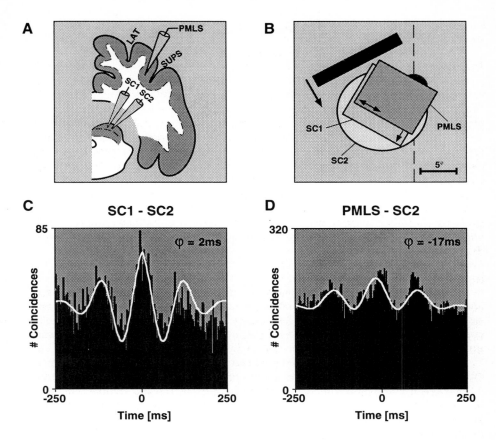

Fig. 1: Corticotectal interactions in the cat. (A) Data are from a measurement where we recorded simultaneously from two sites in the SC (SC1, SC2) and the extrastriate area PMLS. LAT, lateral sulcus; SUPS, suprasylvian sulcus. (B) Schematic plot of the receptive fields for all three recording sites. All fields were overlapping and, hence, the cells were activated with a single moving light bar. (C) Crosscorrelogram for the interaction between the two tectal cell groups. Note that the interaction occurs at a frequency in the alpha range (around 10 Hz) and with only a small phase shift (φ). (D) Crosscorrelogram for the responses obtained at SC2 and at the PMLS recording site. The correlation occurs in a similar frequency band, but shows a pronounced phase shift of about 17 ms. The continuous line superimposed to the correlograms represents the generalized Gabor function that was fitted to the data.

more often if RFs were overlapping as compared to pairs of recordings with non-overlapping fields.

Discussion

This study demonstrates that neurons in both striate and extrastriate visual cortex can show temporal correlations with cells located in the superficial layers of the SC. These interactions occur with a millisecond precision and are most pronounced between populations of neurons at retinotopically corresponding sites.

In our sample of corticotectal interactions the cortical activity usually preceded the tectal discharges by several milliseconds. These systematic time shifts distinguish corticotectal interactions from intracortical long-range interactions, since cortical intra- and interareal synchronization has consistently been found to occur with 0-time shifts (Engel et al., 1991a, 1991b; Eckhorn, 1992). Presumably, these differences are related to the anatomical differences between corticocortical and the corticotectal connections. Whereas corticocortical connections are reciprocal, the corticotectal connection is asymmetric because cortical cells project monosynaptically onto SC-cells, but receive disynaptic (and probably less massive) tectocortical feedback via the thalamus. In hierarchical concepts of visual processing the SC is often placed downstream from cortex and is thought to carry out its functions under cortical control. The result that cortical activity is leading in phase over tectal activity may be viewed as consistent with this idea. The fact that corticotectal correlations tend to be weaker than intracortical or intracollicular correlations is not unexpected given that the connectivity within cortex and tectum is much richer than the connections between these structures.

Our data indicate that, with respect to precision and frequency range, corticotectal interactions are more similar to collicular correlations than to the interactions found within cortex. This finding argues against the intuitive view that cortical correlation patterns might simply be imposed onto the SC. These aspects of corticotectal dynamics require further investigation.

Our data demonstrate for the first time that the corticotectal pathway is not only involved in setting up certain RF properties in SC, but also transmits temporal information. For visual cortex it has been suggested that temporal coding mediates basic processing steps such as scene segmentation (Engel et al., 1991c; Singer and Gray, 1995). The possibility to establish specific and temporally precise correlations across the corticotectal pathway might be a crucial prerequisite to transmit such information to the SC. Thus, SC neurons might be able to "read" relations within cortical populations which are defined by synchronized firing. It is generally assumed that population coding plays a major role in collicular sensorimotor transformations such as saccade generation (McIlwain, 1991). In future studies we will address the question whether temporal codes are involved in these SC functions.

Acknowledgements

The work described here is supported by the Minna-James-Heineman Foundation and by the Heisenberg Program of the Deutsche Forschungsgemeinschaft. We thank Wolf Singer for continuous support and many inspiring discussions.

References

Braitenberg V (1978) Cell assemblies in the cerebral cortex. In *Lecture Notes in Biomathematics 21, Theoretical approaches in complex systems.* Eds Heim R, Palm G, pp 171-188. Berlin, Springer.

Chalupa LM (1984) Visual physiology of the mammalian superior colliculus. In *Comparative Neurology of the Optic Tectum.* Ed Vanegas H, pp 775-818. New York, Plenum.

Eckhorn R, Schanze T, Brosch M, Salem W, Bauer R (1992) Stimulus-specific synchronizations in cat visual cortex: multiple microelectrode and correlation studies from several cortical areas. In *Induced rhythms in the brain.* Eds Basar E, Bullock TH, pp 47-82. Boston, Birkhäuser.

Engel AK, König P, Kreiter AK, Singer W (1991a) Interhemispheric synchronization of oscillatory neuronal responses in cat visual cortex. *Science* 252, 1177-1179.

Engel AK, Kreiter AK, König P, Singer W (1991b) Synchronization of oscillatory neuronal responses between striate and extrastriate visual cortical areas of the cat. *Proc. Natl. Acad. Sci. USA* 88, 6048-6052.

Engel AK, König P, Singer W (1991c) Direct physiological evidence for scene segmentation by temporal coding. *Proc. Natl. Acad. Sci. USA* 88, 9136-9140.

Hebb DO (1949) *The organization of behavior.* New York, Wiley.

König P (1994) A method for the quantification of synchrony and oscillatory properties of neuronal activity. *J. Neurosci. Meth.* 54, 31-37.

McIlwain JT (1991) Distributed spatial coding in the superior colliculus: A review. *Vis. Neurosci.* 6, 3-13.

Singer W, Gray CM (1995) Visual feature integration and the temporal correlation hypothesis. *Annu. Rev. Neurosci.* 18, 555-586.

Vaadia E, Haalman I, Abeles M, Bergman H, Prut Y, Slovin H, Aertsen A (1995) Dynamics of neuronal interactions in monkey cortex in relation to behavioral events. *Nature* 373, 515-518.

von der Malsburg C (1986) Am I thinking assemblies? In *Brain theory.* Eds Palm G, Aertsen A, pp 161-176. Berlin, Springer.

Detecting and Measuring
Higher Order Synchronization Among Neurons:
A Bayesian Approach

Laura Martignon
Max Planck Institute for
Psych. Research
Center for Adaptive Behavior
and Cognition
80802 Munich, Germany
laura@mpipf-
muenchen.mpg.de

Kathryn Laskey
Department of Systems
Engineering
George Mason
University
Fairfax, VA 22030
klaskey@gmu.edu

Arne Schwarz
Max Planck Institute for
Psych. Research
Center for Adaptive Behavior
and Cognition
80802 Munich, Germany
a.schwarz@mpipf-
muenchen.mpg.de

Eilon Vaadia
Dept. of Physiology,
Hadassah Med. School
The Hebrew University
of Jerusalem
Jerusalem 91010, Israel
eilon@hbf.huji.ac.il

Abstract

A Bayesian approach to modeling and inferring patterns of synchronous activation in a group of neurons. A major objective of the research is to provide statistical tools for detecting changes in synchronization patterns. Our framework is not restricted to the case of correlated pairs, but generalizes the Boltzmann machine model to allow for higher order interactions. A Markov Chain Monte Carlo Model Composition (MC^3) algorithm is applied in order to search over connectivity structures and uses Laplace's method to approximate their posterior probabilities. Performance of the method was first tested on synthetic data. The method was then applied to data obtained on multi-unit recordings of six neurons in the visual cortex of a rhesus monkey in two different attentional states. The obtained results indicate that the interaction structure predicted by the data is richer than just a set of synchronous pairs. They also confirmed the experimenter's conjecture that different attentional states were associated with different interaction structures.

Keywords: Nonhierarchical loglinear models, Markov Chain Monte Carlo Model Composition, Laplace's Method, Neural Networks

1. Introduction

Simultaneous activation of two or more neurons is denoted by Aertsen and Grün the *unitary event* for information processing in the brain [1]. Since the time of Hebb's [2] fundamental thesis it has been conjectured that complex information processing in the brain arises from the collective interaction of groups of neurons. Experimental advances in the last decade enable the direct study of firing patterns of groups of neurons. Abeles and his coworkers have developed and applied new measurement methods, and have reported coincidences occurring with a much higher probability than chance ([3]; [4]; [5]; [6]). It is of great concern to neuroscientists to be able to detect with precision the size and firing internsity of each synchronous cluster of neurons. Martignon, et al. [10] developed a family of models capable of representing higher order interactions. Yet the frequentist approach presented in [10] is subject to well-known difficulties, especially those associated with multiple simultaneous hypothesis tests. The methods reported here were developed to overcome these problems.

This paper presents a family of loglinear models capable of capturing interactions of all orders. An algorithm is presented for learning both structure and parameters in a unified Bayesian framework. Each model structure specifies a set of clusters of nodes, and structure-specific parameters represent the directions and strengths of interactions among them. The Bayesian learning algorithm gives high posterior probability to models that are consistent with the data. An advantage of the Bayesian approach is the ease of interpretation of the results of analysis. Results include a probability, given the observations, that a set of neurons fires simultyaneously, and a posterior probability distribution for the strength of the interaction, conditional on its occurrence.

2. The Family of Models for Interactions

Let L be a set of nodes labeled 1 through k. At a given time each node may be in state 1 (active) or 0 (inactive). The state of node i at a given time is represented by a random variable X_i. The lowercase letter x_i is used to denote the state of X_i. The state of the set L is denoted by the random vector $\mathbf{X} = (X_1, \ldots, X_k)$, which can be in one of 2^k configurations. The actual configuration at a given time is denoted by $x = (x_1, \ldots, x_k)$.

A cluster of nodes exhibits a positive (or excitatory) interaction when all nodes in the cluster tend to be active simultaneously and a negative (inhibitory) interaction when simultaneous activity tends not to occur. Our objective is to introduce a mathematical model that allows us to define excitatory and inhibitory interactions in rigorous terms, to identify clusters exhibiting interactions, and to estimate the strength of interaction among nodes in each given cluster.

Let the probability of occurrence of configuration x in a given context be denoted by $p(x)$. The context refers to the situation in which the activity occurs; e.g., the task the animal is performing, the point in the task at which the measurement is taken, etc. If nodes act independently, the probability distribution $p(x)$ is equal to the product of marginal probabilities for the individual neurons:

$$p(x) = p(x_1) \cdots p(x_k). \tag{1}$$

A system of neurons is said to exhibit *interaction* if their joint distribution cannot be characterized by (1).

To formalize these ideas, let X be a set of clusters of nodes, where each cluster x is a subset of L; the empty set Δ is required to belong to X. Define the random variable θ to have the value 1 if all nodes in cluster x are active and 0 if any node in cluster x is inactive. That is, T_x is the product of the X_i for all $i \in$ x. The distribution $p(x)$ is assumed to be strictly positive, i.e., all configurations are possible. Thus, its logarithm is well-defined and can be expanded in terms of the components x_i for $i=1,\ldots N$. Let h(x) be the natural logarithm of the probability $p(x)$ of configuration x. We say that the nodes in set L exhibit *interaction structure* X with *interaction parameters* $\{q_X\}_{X \in X}$ if $h(x)$ can be written:

$$h(x) = \log p(x) = \sum_{\xi \in \Xi} \theta_\xi t_\xi \;. \tag{2}$$

A cluster x is said to exhibit excitatory interaction if $q_X > 0$ and inhibitory interaction if $q_X < 0$. From (2) it follows that when an excitatory interaction $q_X > 0$ occurs, the probability of simultaneous firing of the neurons in x is greater than what would be the case if $x \in X$ and all other q_X for $x \neq \Delta$ remained the same.

Let **q** denote the vector of interaction parameters for a structure X. To emphasize dependence of the configuration probability on **q** and X, we adopt the notation $p(x|\mathbf{q},X) = \exp\{h(x|\mathbf{q},X)\}$ for the configuration probabilities and their natural logarithms.

We are concerned with two issues: identifying the interaction structure X and estimating the interaction strength vector **q**. Our approach is Bayesian. Information about $p(x)$ prior to observing any data is represented by a joint probability distribution called the *prior distribution* over X and the q's. Observations are used to update this probability distribution to obtain a *posterior distribution* over structures and parameters. The posterior probability of a cluster x can be interpreted as the probability that the r nodes in cluster x exhibit a degree-r interaction. The posterior distribution for q_X represents structure-specific information about the magnitude of the interaction. The mean or mode of the posterior distribution can be used as a point estimate of the interaction strength; the standard deviation of the posterior distribution reflects remaining uncertainty about the interaction strength.

Initial information about q_X is expressed as a prior probability distribution $g(\mathbf{q}|X)$. Let $x_1, \ldots, x_N$ be a sample of N independent observations from the distribution $p(x|\mathbf{q},X)$. The joint probability of the observed set of configurations, $p(x_1|\theta,\Xi) \cdots p(x_N,\theta,\Xi)$, viewed as a function of **q**, is called the *likelihood function* for **q**. From (2), it follows that the likelihood function can be written

$$L(\mathbf{q}) \quad = \quad p(x_1|\theta,\Xi) \cdots p(x_N|\theta,\Xi)$$

$$= \quad \exp\left\{\sum_i h(x_i|\theta)\right\} \quad = \quad \exp\left\{N \sum_{\xi \in \Xi} \theta_\xi \bar{t}_\xi\right\}, \tag{3}$$

where $\bar{t}_\zeta = \frac{1}{N}\sum t_{\xi i}$ is the frequency of simultaneous activation of the nodes in cluster x. The expected value of $\bar{T}_\zeta$ is the probability that $T_{\mathbf{H}} = 1$, and is often referred to as the marginal function for cluster x. The observed value $\bar{t}_\zeta$ is referred to as the marginal frequency for cluster x.

The posterior distribution for **q** given the sample $x_1, \ldots, x_N$ is given by:

$$g(\theta|\Xi, x_1, \ldots, x_N) = K p(x_1|\theta,\Xi) \cdots p(x_N|\theta,\Xi) g(\theta|\Xi) \tag{4}$$

where the constant of proportionality K is chosen so that (4) integrates to 1:
A structure X indicates which q_X are nonzero. The prior distribution $g(q|X)$ expresses prior information about the nonzero q_X. We assume that the nonzero q_X are normally distributed with zero mean and standard deviation $s=2$. That is, we are assuming that the q_X are symmetrically distributed about zero (i.e., excitatory and inhibitory interactions are equally likely) and that most q_X lie between -4 and 4. The standard deviation $s=2$ is based on previous experience applying this class of models [10] to other data.
The posterior distribution $g(\theta|\Xi, x_1, \ldots, x_N)$ cannot be obtained in closed form. Thus approximation is necessary. Options for approximating the posterior distribution include analytical or sampling-based methods. Because we must estimate posterior distributions for a large number of structures, Monte Carlo methods are infeasibly slow. We therefore use an analytical approximation method. We approximate $g(\theta|\Xi, x_1, \ldots, x_N)$ as the standard large-sample normal approximation. The approximate posterior mean is given by the mode $\tilde{\theta}$ of the posterior distribution. The posterior mode $\tilde{\theta}$ can be obtained by using Newton's method to maximize the logarithm of the joint mass/density function:

$$\tilde{\theta} = \arg\max_\theta \left\{ \log\left(p(x_1 | \theta, \Xi) \cdots p(x_N \theta, \Xi) g(\theta|\Xi) \right) \right\} \tag{5}$$

The approximate posterior covariance is equal to the inverse of the Hessian matrix of second derivatives (which can be obtained in closed form).

3. Posterior Probabilities for Structures

To perform inference with multiple structures, it is natural to assign a prior distribution for **q** that is a *mixture* of distributions of the form (7). That is, a structure probability p_X and a prior distribution $g(q|X)$ is specified for each of the interaction structures under consideration. The structure-specific parameter distribution $g(q|X)$ was described above in Section 3. To obtain the prior distribution for structures, we used discussions with neuroscientists and experience fitting similar models to other data sets. We expected all q_X for single-element x to be nonzero. We expected most other q_X to be zero [27]. The prior distribution we used assumed that the q_X were zero or nonzero independently, and each q_X with $|x|>1$ had prior probability of .1 of being nonzero. The results of the analysis were insensitive to the (nonzero) prior probability of including singleton x.
The posterior distribution of **q** given a sample of observations is also a mixture distribution:

$g(q|x_1, \ldots, x_N) = \sum_\Xi \pi_\Xi^* g(\theta|\Xi, x_1, \ldots, x_N)$ The ratio of posterior probabilities for two structures X_1 and X_2 is given by:

$$\frac{\pi_{\Xi_1}^*}{\pi_{\Xi_2}^*} = \frac{p(\Xi_1|x_1, \ldots, x_N)}{p(\Xi_2|x_1, \ldots, x_N)} = \frac{p(x_1, \ldots, x_N|\Xi_1)}{p(x_1, \ldots, x_N|\Xi_2)} \frac{\pi_{\Xi_1}}{\pi_{\Xi_2}} \tag{6}$$

Thus, the posterior odds ratio is obtained by multiplying the prior odds ratio by the ratio of marginal probabilities of the observations under each of the two structures. This ratio of marginal probabilities is called the Bayes factor [19]. The marginal probability of the observations under structure X is obtained by integrating **q'** out of the joint conditional density:

$$p(x_1, \ldots x_N | \Xi) = \int p(x_1 | \theta', \Xi) \cdots p(x_N | \theta', \Xi) g(\theta' | \Xi, \tau, \eta) d\theta' \qquad (7)$$

We assumed in Section 3 that the joint mass/density function (and hence the posterior density function for **q**) was highly peaked about its maximum and approximately normally distributed. The marginal probability $p(x_1, \ldots x_N | \Xi)$ is approximated by integrating this normal density:

$$\tilde{p}(x_1, \ldots x_N | \Xi) = (2\pi)^{d/2} |\tilde{\Sigma}|^{1/2} p(x_1 | \tilde{\theta}, \Xi) \cdots p(x_N \tilde{\theta}, \Xi) g(\tilde{\theta} | \Xi) \qquad (8)$$

This approximation (12) is known as Laplace's approximation [19]; [28].
The posterior expected value of **q** is a weighted average of the conditional posterior expected values. As noted in Section 3, these expected values are not available in closed form. Using the conditional MAP estimate $\tilde{\theta}_\Xi$ to approximate the conditional posterior expected value, we obtain the point estimate

$$\tilde{\theta} = \sum_\Xi \pi_\Xi^* \tilde{\theta}_\Xi \ . \qquad (9)$$

In (13), the value $\tilde{\theta}_\xi = 0$ is used for structures not containing x. Also of interest is an is an estimate of the value of q_X given that $q_X \neq 0$, which is obtained by dividing $\tilde{\theta}_\xi$ by the sum of the p_X for which $x \oe X$. Equation (8) can be used to approximate the posterior covariance matrix $\tilde{\Sigma}_\Xi$ for **q** conditional on structure X. Again, the unconditional covariance can be approximated by a weighted average:

$$\tilde{\Sigma} = \sum_\Xi \pi_\Xi^* \tilde{\Sigma}_\Xi \ . \qquad (10)$$

(Zero variance is assumed for q_X when $x \oe X$.) The standard deviation of q_X conditional on $x \oe X$ can be obtained by dividing the standard deviation $\hat{\sigma}_\xi$ estimated from (14) by the sum of the p_X for which $x \oe X$.

Computing the posterior distribution (9) requires computing, for each possible structure X, the posterior probability of X and the conditional distribution of **q** given X. This is clearly infeasible: for k nodes there are 2^k activation clusters, and therefore 2^{2^k} possible structures. It is therefore necessary to approximate (9) by sampling a subset of structures. We used a Markov Chain Monte Carlo Model Composition (MC[3]) algorithm [13] to sample structures. A Markov chain on structures was constructed such that it converges to an equilibrium distribution in which the probability of being at structure X is equal to the Laplace approximation to the structure posterior probability π_Ξ^*. We used a Metropolis-Hastings sampling scheme, defined as follows. When the process is at at structure X, a new structure X' is sampled by either adding or deleting a single cluster. The cluster x' to add or delete is chosen by a probability distribution r(x'|X). The selected cluster x' is then either accepted or rejected, with the probability of acceptance given by:

$$\min\left\{1, \frac{p(x_1,\ldots x_N|\Xi')\pi_\Xi\cdot\rho(\xi'|\Xi)}{p(x_1,\ldots x_N|\Xi)\pi_\Xi\rho(\xi'|\Xi')}\right\}$$

4. Application of the Methods

We applied our models to data from an experiment performed in the lab of the fourth author. which spiking events among 6-16 of neurons were analyzed through multi-unit recordings in the cortex of a Rhesus monkey.The monkeys were trained to localize a source of light and, after a delay, to touch the target from which the light blink was presented. At the beginning of each trial the monkeys touched a "ready-key", then the central ready light was turned on. Three to six seconds later, a visual cue was given in the form of a 200-ms light blink coming from either the left or the right. Then, after a delay of 1 to 32 seconds, the color of the ready light changed from red to orange and the monkeys had to release the ready key and touch the target from which the cue was given.

The spiking events (in the 1 millisecond range) of each neuron were encoded as a sequence of zeros and ones, and the activity of the whole group was described as a sequence of configurations or vectors of these binary states. Since the method presented in this paper does not take into account temporal correlation or nonstationarity, the fouth author provided data corresponding to stationary segments of the trials, which are also those presenting the least temporal correlation, corresponding to intervals of 2000 milliseconds around the ready-signal. He adjoined these 94 segments and formed a data-set of 188,000 milliseconds. The data were then binned in time windows of 40 milliseconds. The frequencies of configurations of zeros and ones in these windows are the data used for analysis in this paper. We selected a subset of six of the neurons for which data were recorded.

Cluster ξ	Posterior Probability of ξ (Frequency)	Posterior Probability of ξ (Best 100 models)	MAP estimate $\bar\theta_\xi$	Standard Deviation of θ_ξ
1	1.00	1.00	-1.52	0.06
2	1.00	1.00	-1.73	0.07
3	1.00	1.00	-3.13	0.15
4	1.00	1.00	-0.82	0.06
5	1.00	1.00	-2.76	0.10
6	1.00	1.00	-0.83	0.06
4,6	0.98	1.00	0.49	0.11
2,3,4,5	0.33	0.36	2.34	0.67
3,4,6	0.22	0.15	0.80	0.27
2,3,4,5,6	0.16	0.13	2.53	0.88
1,4,5,6	0.16	0.10	6.67	1.19
3,4	0.12	0.08	0.63	0.23

Table 1:   Results for Pre-Ready Signal Data
Effects with Posterior Probability > 0.1

The data analysis confirms the prior expectation that not many interactions would be present [8]. There was a high probability second-order interaction in each data set $\xi_{4,6}$ in the pre-ready data and $\xi_{3,4}$ in the post-ready data. In the pre-ready data, a fourth-order interaction $\xi_{2,3,4,5}$had posterior probability about 1/3 (this represents approximately three times the prior probability of 1) The tables show a few other interactions with posterior probability larger than their prior probability

Cluster ξ	Posterior Probability of ξ (Frequency)	Posterior Probability of ξ (Best 100 models)	MAP estimate $\bar{\theta}_\xi$	Standard Deviation of θ_ξ
1	1.00	1.00	-1.03	0.06
2	1.00	1.00	-2.54	0.10
3	1.00	1.00	-3.86	0.24
4	1.00	1.00	-0.40	0.05
5	1.00	1.00	-3.06	0.12
6	1.00	1.00	-0.50	0.05
3,4	0.86	0.94	1.00	0.27
2,5	0.25	0.18	0.98	0.34
1,4,5,6	0.18	0.13	1.06	0.36
1,4,6	0.15	0.08	0.38	0.13

Table 2: Results for Post-Ready Signal Data:
Effects with Posterior Probability > 0.1

References

1. Grün, S., A. Aertsen, E. Vaadia, and A. Riehle. "Behavior-Related Unitary Events in Cortical Activity." Learning and Memory: 23rd Gottingen Neurobiology Conference (poster presentation), 1995.

2. Hebb, D. *The Organization of Behavior*. New York: Wiley, 1949.

3. Abeles, M. *Corticonics: Neural Circuits of the Cerebral Cortex*. Cambridge: Cambridge University Press, 1991.

4. Abeles, M., H. Bergman, E. Margalit, and E. Vaadia. "Spatiotemporal Firing Patterns in the Frontal Cortex of Behaving Monkeys." *Journal of Neurophysiology* 70, no. 4 (1993): pp. 1629-1638

5. Abeles, M., and G. Gerstein. "Detecting Spatiotemporal Firing Patterns Among Simultaneously Recorded Single Neurons." *Journal of Neurophysiology* 60 (1988): 904-924.

6. Abeles, M., E. Vaadia, Y. Prut, I. Haalman, and H. Slovin. "Dynamics of Neuronal Interactions in the Frontal Cortex of Behaving Monkeys." *Concepts in Neuroscience* 4, no. 2 (1993): pp. 131-158

7. Martignon, L., H. Hasseln, S. Grün, A. Aertsen, and G. Palm. "Detecting the Interactions in a Set of Neurons: A Markov Field Approach." *Biological Cybernetics* (1995):

8. Braitenberg, V., and A. Schüz. "Anatomy of the Cortex: Statistics and Geometry." In *Studies in Brain Function. Springer-Verlag* (1994).

9. Kass, Robert E., and Adrian E. Raftery. "Bayes Factors." *Journal of the American Statistical Association* 90, no. 430 (1995): 773-795.

10. Madigan, D., and J. York. "Bayesian Graphical Models for Discrete Data," Technical Report 239, Department of Statistics, University of Washington, 1993.

11. Tierney, L., and J. B. Kadane. "Accurate Approximations for Posterior Moments and Marginal Densities." *Journal of the American Statistical Association* 81 (1986): pp. 82-86.

Analyzing the Formation of Structure in High-Dimensional Self-Organizing Maps Reveals Differences to Feature Map Models

Maximilian Riesenhuber[1], Hans-Ulrich Bauer[2] and Theo Geisel[2]

[1] Center for Biological & Computational Learning and Department of
Brain & Cognitive Sciences, MIT, Cambridge, MA 02142, USA
[2] Max-Planck-Institut für Strömungsforschung, Göttingen and SFB 185
"Nichtlineare Dynamik", Univ. Frankfurt, 60054 Frankfurt, Germany

Abstract. We present a method for calculating phase diagrams for the high-dimensional variant of the Self-Organizing Map (SOM). The method requires only an ansatz for the tesselation of the data space induced by the map, not for the explicit state of the map. Using this method we analyze two recently proposed models for the development of orientation and ocular dominance column maps. The phase transition condition for the orientation map turns out to be of different form than of the corresponding low-dimensional map.

1 Introduction

The high-dimensional self-organizing map (SOM) and the low-dimensional self-organizing feature (SOFM) map have been used to model a variety of self-organizational paradigms in the brain and in technical applications [1]. In both variants stimuli from an input space are mapped to a lattice of output elements (neurons), each characterized by a position $\mathbf{r}$ in the lattice plus a receptive field $\mathbf{w_r}$. A stimulus $\mathbf{v}$ is mapped onto that neuron $\mathbf{s}$ whose receptive field $\mathbf{w_s}$ matches $\mathbf{v}$ best. This amounts to a winner-take-all rule, i.e. a strong lateral nonlinearity.

In the SOM (as well as in previous [2] or more recent [3] formulations of map formation processes) stimuli are normalized patterns of activity in a high-dimensional space (eg. images on a discretized retina). The map results as a stationary state of a self-organization process, which successively changes all receptive fields $\mathbf{w_r}$,

$$\Delta\mathbf{w_r} = \epsilon h_{\mathbf{rs}}(\mathbf{v} - \mathbf{w_r}), \qquad h_{\mathbf{rs}} = e^{-\frac{||\mathbf{r}-\mathbf{s}||^2}{2\sigma^2}},\qquad (1)$$

following the presentation of stimuli $\mathbf{v}$. ϵ controls the size of learning steps. The neighborhood function $h_{\mathbf{rs}}$ enforces neighboring neurons to align their receptive fields, thereby imposing the property of topography on the SOM. The best-matching neuron $\mathbf{s}$ for a particular stimulus is determined by

$$\mathbf{s} = \arg\max_{\mathbf{r}}(\mathbf{w_r} \cdot \mathbf{v}). \qquad (2)$$

In the low-dimensional SOFM the full stimulus distribution $\mathbf{v}$ and the receptive field distributions $\mathbf{w_r}$ are replaced by (low-dimensional) features $\tilde{\mathbf{v}}$ and $\tilde{\mathbf{w}}_{\mathbf{r}}$ (eg. center of gravity-coordinates) which can be extracted from $\mathbf{v}$ and $\mathbf{w_r}$ by application of a linear operator R,

$$\tilde{\mathbf{v}} = R(\mathbf{v}), \qquad \tilde{\mathbf{w}}_{\mathbf{r}} = R(\mathbf{w_r}). \tag{3}$$

Application of the operator R to the learning rule (1) yields (exploiting the linearity of R and using (3)) the SOFM learning rule

$$R(\Delta\mathbf{w_r}) = R(\epsilon h_{\mathbf{rs}}(\mathbf{v} - \mathbf{w_r})) = \epsilon h_{\mathbf{rs}}(\tilde{\mathbf{v}} - \tilde{\mathbf{w}}_{\mathbf{r}}). \tag{4}$$

So the features $\tilde{\mathbf{w}}_{\mathbf{r}}$ obey dynamics of the same form as the dynamics for the full receptive field parameters $\mathbf{w_r}$.

However, as the dot product is not a useful measure in the feature space, the mapping rule (2) has to be changed for the SOFM. Here, the best-matching neuron $\mathbf{s}$ is determined as the neuron whose receptive field has the smallest Euclidean distance to the stimulus.

These different mapping rules would yield identical results when operating on the same vectors $\mathbf{w}$ and $\mathbf{v}$ only if the vectors were square normalized (whereas in the SOM, they are sum-normalized). Even more important, the distance measures operate on vectors in different spaces which can yield different best-matching neurons even if the normalization issue is ignored (illustrated in Fig. 1). Therefore, the two versions of the algorithm can possibly deliver different structure formation results when applied to analogous systems.

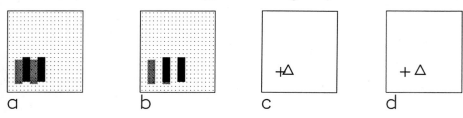

Figure 1: SOMs (**a,b**) and SOFMs (**c,d**) can yield different winner neurons when stimulated with analogous stimuli. Consider a square input space, discretized by a square lattice for the SOM case. The stimulus is a double bar, as indicated as the (combined) gray regions. Two exemplary neurons have receptive fields indicated by the (combined) black regions in (a) and (b), resp. The stimulus-receptive field overlap is larger in **b**, so this neuron will be the winner. In (c,d), stimuli and receptive fields are represented by their centers of gravity (crosses and triangles, resp.). Now, the neuron depicted in **c** (and **a**) is the winner.

2 Evaluating tesselations of data sets

So far, a mathematical analysis of SOM map development process has only been achieved for the feature map variant [9] but not for the high-dimensional

SOMs. Here, we propose to make an ansatz for plausible states of the map, to evaluate a distortion function for these states and to determine the state of lowest distortion. Since such an ansatz in terms of the explicit weight vectors $\mathbf{w_r}$ is difficult, if not impossible to make, we consider a new distortion measure

$$E_\mathbf{v} = \sum_\mathbf{r} \sum_\mathbf{r'} \sum_{\mathbf{v'} \in \Omega_{\mathbf{r'}}} \sum_{\mathbf{v} \in \Omega_\mathbf{r}} (\mathbf{v'} - \mathbf{v})^2 e^{\left(-\frac{||\mathbf{r}-\mathbf{r'}||^2}{2\sigma^2}\right)}, \qquad (5)$$

which requires only an ansatz for the stimulus space tesselations as given by the $\Omega_\mathbf{r}$ ($\Omega_\mathbf{r}$ denotes the set of stimuli $\mathbf{v}$ which are mapped onto node $\mathbf{r}$. By Eq. (2 and depends on the $\mathbf{w_r}$ in an implicit way). Under quite general assumptions, the minima of $E_\mathbf{v}$ coincide with those of the naive "energy function" for the SOM-model [7] in the limit of $\sigma \to 0$, and the deviations are small otherwise. A great advantage of this new distortion measure lies in the fact that the $\Omega_\mathbf{r}$-ansatz necessary for an evaluation of Eq. (5) is comparatively easy to make (see following examples). A more detailed account of this method can be found in forthcoming publications [5, 6].

3 Results

Using Eq. (5), we now analyze two SOM-models for the development of visual maps, one for the development of orientation maps, in which the emergence of oriented receptive fields is contingent upon the presence of oriented stimuli, and one for the development of ocular dominance maps. Finally we make a brief reference to a third model in which oriented receptive fields arise from non-oriented stimuli, a symmetry breaking phenomenon which cannot be described in feature map approximation but only in high-dimensional SOMs.

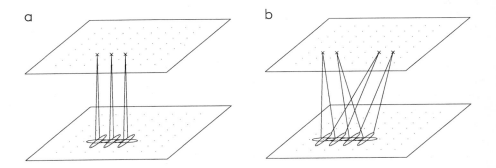

Figure 2: SOM-model for the formation of orientation maps. Oriented stimuli in a retinal layer are mapped to a sheet of cortical neurons. **a:** Stimuli with different orientations, but located at the same retinal position, are mapped to the same cortical neuron: non-oriented receptive fields. **b:** Same orientation, different retinal positions: oriented receptive fields.

3.1 SOM Model for the Development of Orientation Maps

The first model is concerned with the development of orientation maps. Here the map input space is a two-dimensional sheet of retinal input channels, discretized as a $N \times N$-dimensional grid. Using ellipsoidal Gaussian activity distributions as stimuli (minor axis σ_1, major axis $\sigma_2 > \sigma_1$), simulations of this model led to maps with orientation preference for substantially elongated stimuli [4]. Using rather circular stimuli ($\sigma_2 \approx \sigma_1$) maps with neurons of no orientation preference were also observed [8]. To simplify calculations we restrict the number of stimulus orientations to two (horizontal and vertical), and the possible stimulus centers to the retinal channels (see Fig. 2).

For this restricted stimulus set, we have twice as many stimuli as neurons so the Voronoy sets $\Omega_{\mathbf{r}}$ contain two stimuli on average. What are sensible stimulus space tesselations which correspond to maps with or without orientation preference of the individual neurons? It is a sensible ansatz to assume that the non-oriented map is characterized by a tesselation of the stimulus set such that stimuli of both orientations, but centered at the retinal location, go to one neuron. In contrast we assume for the maps with orientation preference to have stimuli of the same orientation, but located at neighboring retinal positions in the sets $\Omega_{\mathbf{r}}$. For these different tesselations we can evaluate and compare the resp. values of our distortion measure. We obtain, in the limit of $\sigma_{1,2} \gg \sigma \gg 1$

$$\sigma_{2,crit} \approx \sigma_1 + \sqrt{3}\sigma. \tag{6}$$

The condition (6) for the break of symmetry from non-oriented to oriented receptive fields is very well corroborated by numerical simulations using the reduced stimulus set (Fig. 2), as well as by simulations with the full stimulus set (all orientations and positions). The additive relation between σ_1 and $\sigma_{2,crit}$ deviates from the multiplicative relation found for a corresponding SOFM [9].

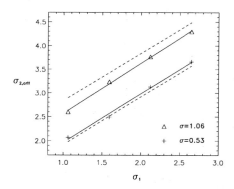

Figure 3: Critical value $\sigma_{2,crit}$ for the occurrence of an orientation map, as a function of σ_1, for two exemplary values of the map neighborhood function width σ. Symbols indicate the results of simulations of SOMs, the solid line is a fit to these four points, resp. The dashed lines show the corresponding analytic results (6).

This high-dimensional SOM-model for orientation map formation also formed the basis for the simulations in an accompanying paper which provides an explanation for the recently observed similarity of orientation maps which developed under uncorrelated stimulation conditions [10]

3.2   SOM-Model for the Development of Ocular Dominance Maps

Second, we analyze a SOM-model for the development of ocular dominance (OD) maps [11]. Two retinal layers for the two eyes are mapped to a cortical output layer. The stimuli are small patches of activity, located at the same (random) position in both input layers, with an amplitude ratio of $1 : c$ or $c : 1$. For $c \approx 1$, the weight vectors $\mathbf{w_r}$ develop symmetrically in both retinae: each $\Omega_\mathbf{r}$ contains two stimuli of opposite ocularity, but centered at the same retinal position (solution type **a**). With decreasing c, a symmetry breaking transition takes place, and neurons develop a preference for one or the other retina (ocular dominance). The $\Omega_\mathbf{r}$ now contain two stimuli of the same ocularity, but centered at different (neighboring) positions. Depending on the clustering pattern of same-ocularity neurons in the map, different types of solutions can be distinguished. Here we consider the following arrangements: a chequerboard-type alternation (type **b**), or bands of length N in one direction, and widths 1 (**c**), 2 (**d**), or $N/2$ (**e**) in the orthogonal direction. Type **e** corresponds to a degenerate solution, which can be found in models [13], but not in the visual cortex.

Evaluating the distortions $E_\mathbf{v}$ for the different tesselations, we obtain the phase diagram depicted in Fig. 3a. For decreasing values of c, the preferred solution changes from no ocular dominance, via solutions with increasing band width, to the degenerate type **e** solution with only two OD bands. This transition scenario coincides with the findings of a recent neuroanatomical experiment [12]. Simulations corroborated the analytic results very well (Fig. 3b).

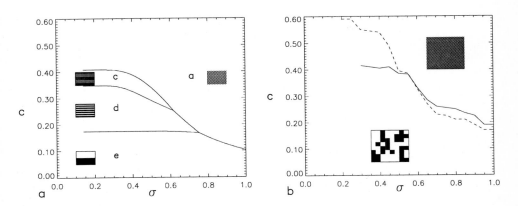

Figure 4: Phase diagrams for a SOM-model for the development of ocular dominance maps. **a**: analytical solution. **b**: numerical solution. Solid line: multiplicative normalization Eq. (1), dashed line: subtractive normalization.

3.3 Development of oriented receptive fields from non-oriented stimuli

Finally we briefly describe results of a recent project involving an SOM model for the development of orientation maps based on a competition of circular symmetric on-center and off-center cell inputs. The ability or non-ability of SOM-based models to generate a) the receptive fields of individual neurons in addition to the map layout, and b) oriented receptive fields from non-oriented stimuli [3] has been at the center of a recent debate about the merits and disadvantages of different map formation models. A formation of receptive field shapes without prior parametrization of receptive field properties cannot be achieved in feature map approximation but only in the high-dimensional SOM variant. We here report that our above-described method allowed us to calculate the regime of stimulus and map parameters for the On-center- Off-center-cell competition problem. In subsequent simulations we then obtained maps with oriented receptive fields [14]. With regard to the above-mentioned debate we can now state that SOM models exhibit on a qualitative level the same pattern formation properties as a the competing correlation-based models, plus the effect of an increase in ocular dominance stripe width upon less correlated stimulation as observed in a recent experiment. An assessment of more quantitative properties of high-dimensional SOM maps, as well as investigation of a combined orientation and ocular dominance map formation will be adressed in future work. We expect that our new method of analysis will provide a valuable guiding line in these investigations.

References

1. T. Kohonen, *Self-Organizing Maps*, Springer, Berlin (1995).
2. C. von der Malsburg, Kybernetik **14**, 85 (1973); Biol. Cyb. **32**, 49 (1979).
3. K. D. Miller, J. Neurosci. **14**, 409 (1994).
4. K. Obermayer, H. Ritter, K. Schulten, Proc. Nat. Acad. Sci. USA **87**, 8345 (1990).
5. H.-U. Bauer, M. Riesenhuber, T. Geisel, submitted to Phys. Rev. Letters (1995).
6. M. Riesenhuber, H.-U. Bauer, T. Geisel, submitted to Biol. Cyb. (1995).
7. E. Erwin, K. Obermayer, K. Schulten, Biol. Cyb. **67**, 47 (1992).
8. K. Obermayer, *Adaptive Neuronale Netze und ihre Anwendung als Modelle der Entwicklung kortikaler Karten*, infix Verlag, Sankt Augustin (1993).
9. K. Obermayer, G. G. Blasdel, K. Schulten, Phys. Rev. A **45**, 7568 (1992).
10. K. Pawelzik, H.-U. Bauer, F. Wolf, T. Geisel, Proc. ICANN 96, this volume (1996).
11. G. J. Goodhill, Biol. Cyb. **69**, 109 (1993).
12. S. Löwel, J. Neurosci. **14**, 7451 (1994).
13. G. J. Goodhill, D. J. Willshaw, Network **1**, 41 (1990); P. Dayan, Neur. Comp. **5**, 392 (1993).
14. M. Riesenhuber, H.-U. Bauer, T. Geisel, submitted to CNS 96, Boston (1996).

Precise Restoration of Cortical Orientation Maps Explained by Hebbian Dynamics of Geniculocortical Connections

K.Pawelzik, H.–U. Bauer, F. Wolf, and T. Geisel

MPI für Strömungsforschung, Göttingen, and
SFB 185 Nichtlineare Dynamik, Universität Frankfurt, Germany.
e-mail: {klaus,bauer,fred,geisel}@chaos.uni-frankfurt.de

Abstract. It is widely assumed that experience dependence and variability of the individual layout of cortical columns are the hallmarks of activity–dependent self–organization during development. Recent optical imaging studies of visual cortical development seem to indicate a lack of such intinsic variability in visual cortical development. These studies showed that orientation maps which were forced to form independently for the left and right eye exhibited a high degree of similarity in area 18 of cat visual cortex [7, 4]. It has been argued that this result must be viewed are evidence for an innate predetermination of orientation preference if orientation preference is determined by the pattern of geniculocortical connections.

Here we point out that the observed phenomenon can in fact be explained by the activity–dependent development of geniculocortical connections if geometric constraints and retinotopic organization of area 18 are taken into acount. In particular we argue that symmetries, which would lead to a strong variability of the emerging orientation map can be broken by boundary effects and interactions between the orientation preference map and the retinotopic organization. As a consequence independently developing orientation maps should exhibit the same layout in area 18, but not in area 17. Simulations of the formation of orientation columns in narrow areas indeed produce uniquely defined orientation preference maps irrespective of the particular set of stimuli driving the development.

1 Confinement of Columnar Patterns in Area 18

It has long been suggested by theoretical studies that the activity–dependent refinement of geniculocortical connections might follow a dynamics analogous to symmetry breaking and pattern–formation in physical systems (see e.g.[10]). One expects that such a dynamics typically results in a pronounced variability of the resulting columnar patterns. In contrast to this expectation recent studies of Bonhoeffer and coworkers show that orientation maps which were forced to form independently for the left and right eye exhibited a high degree of similarity in area 18 of cat visual cortex [7, 4].

To explain the observations of Bonhoeffer and coworkers within the framework of dynamic self–organization it must be noted that orientation maps in area 18 are probably subject to specific geometric constraints. In cats area 18 forms a narrow belt, which lines the much larger area 17 laterally and measures only about $3mm$ in its mediolateral extension [15]. The characteristic wavelength of orientation columns in area 18 is even larger than in area 17 and measures approximatelly $1.3mm$ [9]. In contrast to the situation encountered in striate cortex the lateral extension of area 18 therefore is in the same order as the characteristic wavelength of its orientation map.

Several lines of evidence suggest that areal borders are an important factor determining the layout of cortical orientation maps. It has been observed in several species that iso–orientation domains tend to be aligned orthogonal to areal boundaries [8]. This effect is also visible in the data of Gödecke and Bonhoeffer were domains activated by stimuli of one orientation form bands of mediolateral extension, perpendicular to the borders of area 17 and 19 (see Figs.1 and 2 of [4]). Alignment relative to external borders is generally expected in pattern–forming systems confined to a small spatial domain [3]. With respect to the activity–dependent formation of neuronal connection patterns comparable effects have been demonstrated in Hebbian modells for the formation of cortical maps[5, 1]. While activity-dependent development is generally asumed to result in a large variability of individual patterns, boundary effects and spatial confinement therefore could dramatically reduce the number of stationary patterns, i.e. the actual variability of the system. If orientation maps in area 18 in fact develop according to a constrained, pattern–forming dynamics, the observed precise restoration of orientation maps therefore would indicate that this dynamics displays only one or a few very similar stationary solutions or that only one is reached from typical initial conditions.

2   Basic Symmetries of Orientation Maps

It is instructive to state this scenario more formally. For simplicity the orientation map is described by a complex scalar field $z(\mathbf{x})$ [14] and we assume that its emergence during development can be described by a dynamics

$$\partial_t z(\mathbf{x}) = \mathbf{F}[\mathbf{z}(\mathbf{x})] \tag{1}$$

where $F[\cdot]$ is a nonlinear operator. To model the development of area 18 we consider a model area of finite width and infinite length and assume that the stationary solutions of Eq.(1) are subject to bondary conditions which force iso–orientation contours $arg(z(\mathbf{x})) = \mathbf{const.}$ to intersect the lateral borders at right angles. Quite generally Eq.(1) will possess a discrete set of stationary solutions $z_\mu(\mathbf{x})$, each of which is twofold degenerate. On the one hand translations of a stationary solution $z'_\mu(\mathbf{x}) = \mathbf{z}_\mu(\mathbf{x} + \mathbf{de_y})$ must also be stationary. On the other hand orientation shifts $z'_\mu(\mathbf{x}) = \mathbf{e}^{i\phi}\mathbf{z}_\mu(\mathbf{x})$ must also form stationary patterns. The second degeneracy is linked to the assumption that orientation preference is determined by the pattern of geniculocortical connections, which implies that the spatial arrangement of iso–orientation domains or its relation

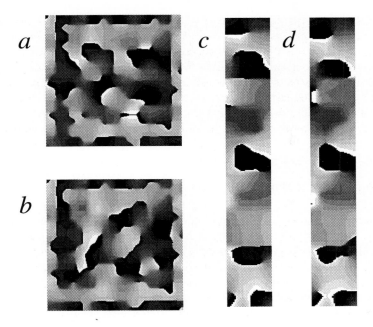

Fig. 1. Results of simulations of a simple model for the self–organization of orientation maps in a quadratic (a,b) and a strongly elongated target area (c,d). Pictures show spatial patterns of preferred orientations. Orientation angles are displayed in grey scale. Due to the cyclic nature of orientation black and white indicate similar orientations. Starting from identical initial conditions and using identical model parameters, the two self–organization processes differed only in the sequence of randomly chosen stimuli. In the first case the maps exhibit differing layouts (mean correlation 0.32 ± 0.07, evaluated in the center region for three pairs of maps), in the second case the maps are virtually identical (mean correlation 0.89 ± 0.06, evaluated for three pairs of maps). The maps were generated using Kohonen's Self-Organizing Map algorithm, with 16×16 (retinal) input channels mapped to 24×24 map neurons (a,b), and 16×8 input channels mapped to 36×6 map neurons (c,d), open boundary conditions. Stimuli were activity distributions of elliptic Gaussian shape, length of halfaxes $\sigma_1 = 1$, $\sigma_2 = 5$. Orientation and location of the stimuli were chosen by a numerical pseudo–random procedure, with different random sequences for (a) and (b), and (c) and (d), resp. . Further parameters of the algorithm were the adaptation step size $\epsilon = 5 \times 10^{-5}$, the width of the neighborhood function $\sigma = 1$, and the number of adaptation steps $N = 3 \times 10^6$. The maps were intialized with pairwise identical, retinotopic connectivity patterns.

to the retinotopic organization is not tightly related to the actual values of orientation preferences displayed. Instead orientation preference might be changed while the arrangement of iso–orientation domains is kept constant. The observed restoration however leads to a precise reestablishment of particular pattern of orientation preferences and not only to the reemergence of a particular pattern of iso–orientation domains. To expain this observation by activity–dependent mechanisms it therefore is necessary to show that both symmetries can actually be broken in a more realistic model for the development of geniculocortical connections.

3 Precise Restoration by Hebbian Learning

To investigate this posibility we studied an idealized model for the development of afferent receptive fields under the influence of pattern vision [12] in cortical areas of different shape. For a detailed analysis of the model see Riesenhuber et al., this volume. The emergence of orientation maps was simulated using sequences of random stimuli applied to a single model retina. In this system a reverse suture experiment should be compared to a pair of simulations starting from identical initial conditions but using different random sequences of stimuli. The system exhibited qualitatively different behaviour depending on the chosen geometry of the cortical area. In areas with large extension in both spatial dimensions, the maps displayed only a general structural similarity (Fig.1a,b). In narrow and elongated areas, however, simulations utilizing different sequences of random stimuli led to the emergence of virtually identical orientation maps (Fig.1c,d). These simulations also reproduced the tendency of iso–orientation domains to intersect areal borders at right angles. As this tendency has been widely observed in the visual cortex [6, 9, 13] and is also visible in the data of Gödecke and Bonhoeffer (see Figs.1 and 2 of [4]) the latter behaviour strongly suggests that the model studied embodies a mechanism equivalent to the real interaction of areal borders and orientation columns in the visual cortex. We therefore conclude that precise restoration of orientation maps in cat area 18 is in fact expected, if orientation maps arise by activity–dependent refinement of geniculocortical connections.

4 Discussion

If external boundary conditions actually determine the development of area 18, the results of a reverse–suture experiment might be qualitatively different in a larger area where most columns are distant from areal borders. This is in fact suggested by two studies of the effects of reverse–suture in area 17 [2, 11] which are largely inconsistent with the results obtained in area 18. These studies indicated that neurons which regain input from a previously sutured eye exhibit independent orientation preferences in the two eyes. It therefore might turn out that the phenomenon discovered by Bohoeffer and coworkers does not represent evidence for an innate predetermination of orientation preference but indicates that the activity–dependent refinement of geniculocortical connections participates in a spatially coherent dynamics.

References

1. Hans-Ulrich Bauer. Development of oriented ocular dominance bands as a consequence of areal geometry. *Neural Comp.*, 7:36–50, 1995.
2. Colin Blakemore and R.C. Van Sluyters. Reversal of the physiological effects of monocular deprivation in kittens: further evidence for a sensitive periode. *J. Physiol.*, 237:195–216, 1974.
3. M.C. Cross and P.C. Hohenberg. Pattern-formation outside of equilibrium. *Rev. Mod. Phys.*, 65(3):850–1112, 1993.
4. Imke Gödecke and Tobias Bonhoeffer. Development of identical orientation maps for two eyes without common visual experience. *Nature*, 379:251–254, 1996.
5. Geoffrey J. Goodhill and David Willshaw. Elastic net model of ocular dominance: Overall stripe pattern and monocular deprivation. *Neural Comp.*, 6:615–621, 1994.
6. A. Humphrey, L.C. Skeen, and T.T Norton. Topographic organization of the orientation column system in the striate cortex of the tree shrew (*tupaia glis*) II: deoxyglucose mapping. *J. Comp. Neurol.*, 192:544–566, 1980.
7. Dae-Shik Kim and Tobias Bonhoeffer. Reverse occlusion leads to a precise restoration of orientation preference maps in visual cortex. *Nature*, 370:370–372, 1994.
8. Simon LeVay and Sacha B. Nelson. Columnar organization of the visual cortex. In *Vision and Visual Dysfunction*, chapter 11, pages 266–315. Macmillan, Houndamills, 1991.
9. Siegrid Löwel, Brian Freeman, and Wolf Singer. Topographic organization of the orientation column system in large flat–mounts of the cat visual cortex: A 2-deoxyglucose study. *J. Comp. Neurol.*, 255:401–415, 1987.
10. Kenneth D. Miller. Receptive fields and maps in the visual cortex: Models of ocular dominance and orientation columns. In E. Domany, J.L. van Hemmen, and K. Schulten, editors, *Models of Neural Networks III*, volume 14. Springer–Verlag, NY, 1995.
11. J. Anthony Movshon. Reversal of the physiological effects of monocular deprivation in the kitten's visual cortex. *J. Physiol.*, 261:125–174, 1976.
12. K. Obermayer, H. Ritter, and K. Schulten. A principle for the formation of cortical feature maps. *Proc. Natl. Acad. Sci. USA*, 87:8345–8349, 1990.
13. C. Redies, M. Diksic, and H. Riml. Functional organization in the ferret visual cortex: A double–lable 2–deoxyglucose study. *J. Neurosci.*, 10:2791–2803, 1990.
14. N.V. Swindale. Iso–orientation domains and their relationship with cytochrome oxidase patches. In D. Rose and V.G. Dobson, editors, *Models of the Visual Cortex*, chapter 47, pages 452–461. Wiley, 1985.
15. R.J. Tusa, A.C. Rosenquist, and L.A. Palmer. Retinotopic organization of areas 18 and 19 in the cat. *J. Comp. Neurol.*, 185:657–678, 1979.

Qu. What about growth of cortex + effect a boundaries?

Nature pap, in press

bw effecting sensitive to threshold setting etc

Modification of Kohonen's SOFM to Simulate Cortical Plasticity Induced by Coactivation Input Patterns

Marianne Andres, Oliver Schlüter, Friederike Spengler, Hubert R. Dinse

Inst. f. Neuroinformatik, Ruhr-Universität Bochum, 44780 Bochum, Germany
e-mail: marianne@neuroinformatik.ruhr-uni-bochum.de

Abstract

We present a modification of Kohonens SOFMin order to simulate cortical plasticity induced by coactivation patterns. This is accomplished by introducing a probabilistic mode of stimulus presentation and by substituting the winner-takes-all mechanism by selecting the winner from a set of best matching neurons.

1   Introduction

Post-ontogenetic cortical plasticity represents a powerful capacity to adjust to changing conditions in the environment of an adult organism. The need for adjustment can be due to central or peripheral lesions or injuries or to the selective and repeated use or disuse of sensori-motor-functions. Such reorganizational changes are observable at the level of receptive fields of single cells and of representational maps. It is assumed that the temporal coincidence of events is a necessary prerequisite to evoke changes of synaptic excitability [6]. The probability of the sensory inputs constitute the basic mechanisms which governs the form of reorganization. From a theoretical point of view, the concept of selforganizing feature maps (SOFM) is a highly effective and an often used approach to deal with the implications of the input statistics in terms of topographic maps [4]. However, the idea of plasticity of dynamically maintained cortical maps and receptive fields implies continuous adaptational processes throughout life which do not allow a convergence to a final state. Moreover, experimentally induced cortical reorganizations are limited by the fact that stimulation that lacks the aspect of input coactivation do not lead to substantial plastic changes. In contrast, the conventional feature map approach is highly sensitive to a single point stimulation paradigm leading to complete deformation of the topography. We therefore extended the KOHONEN feature map approach in two ways: 1) by introducing a probabilistic mode of stimulus presentation, and 2) by substituting the winner-takes-all mechanism by selecting the winner from a set of best matching neurons.

2   Cortical plasticity induced by an associative pairing protocol

We studied plastic reorganization of receptive fields (RF) and representational maps in the rat hindpaw representation in the primary somatosensory cortex [2]. Plasticity was induced by an associative pairing of tactile stimulation (paired peripheral tactile stimulation, PPTS). The required temporal coincidence of inputs is generated by coactivation of two simultaneously applied tactile stimuli at two non-overlapping RFs. To test the impact of lacking coincidence, a single point stimulation (SPPS) is applied to a single receptive field. In order to explore the potential perceptual consequences of PPTS-induced short-term plastic processes, we studied tactile spatial two-point discrimination performance in humans [2]. PPTS results in

- selective cortical reorganization after 6 to 12 hours of PPTS,
- enlargement of RFs by integration of the stimulated skin sites,
- emergence of new RFs in cortical zones of former non-somatic responsiveness,
- enlargement of the cortical representation of the stimulated skin,
- reversibility after 12 hours after terminating the PPTS protocol,
- improvement in the human spatial discrimination performance.

SPPS resulted in only slight reorganizational changes.

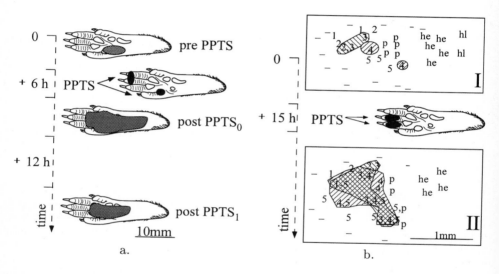

Figure 1: PPTS-induced cortical plasticity. a. Reversible receptive field enlargement after 6 hours of PPTS. b. Cortical map reorganization induced by 15 hours of PPTS. I: pre-PPTS cortical map; II: post-PPTS cortical map.

3   Description of the discrete model

We simulate the somatosensory representation of a rat paw of size $12mm \times 30mm$. The simulated paw contains a set $\mathcal{M} := \{(x_i, y_i) \mid (x_i, y_i) \in \mathbb{R}^2 \cap [0, 12] \times [0, 30], i = 0, \cdots, M - 1\}$ of $M = 400$ receptors with positions (x_i, y_i) which are uniformly distributed. The activity distribution of the tactile stimuli $v^s = (v_0^s, \cdots, v_{M-1}^s)$ is a 2-dimensional Gaussian function over an ellipse with the center $(x_s, y_s) \in \mathcal{M}$:

$$v_i^s = \frac{1}{2\pi ab} exp \left(-\frac{1}{2} \left[\frac{(x_i - x_s)^2}{a^2} + \frac{(y_i - y_s)^2}{b^2} \right] \right), \quad i = 0, \cdots, M - 1 \qquad (1)$$

In this simulation we take a radial symmetric Gaussian function with a mean size of the variance $a^2 = b^2$. Therefore $\mathcal{V} := \{v^s \mid s = 0, \cdots, M-1\}$ defines *the basic stimulus set* $\mathcal{V}$ which is the set of the training samples used for selforganizing the somatosensory map [7].

The 2-dimensional map consists of $N \times N$ ($N = 40$) neurons, each neuron $r_i = (r_i^0, r_i^1)$ of the grid receives input from all receptors of the simulated paw. The learning process is similar to the Hebbian learning in the KOHONEN algorithm [5, 8]. Instead of adapting at time t the neuron with the best match to the sample $v^s(t) \in \mathcal{V}$ and its neighborhood, the winner neuron r_g is selected randomly from the set $\mathcal{S}$ of the first K best matching neurons r_j ($j = 0, \cdots, K-1$):

$$\mathcal{S} = \{r_j \in N \times N \mid \|v^s(t) - w_{r_0}(t)\| \le \|v^s(t) - w_{r_1}(t)\| \le \cdots \le \|v^s(t) - w_{r_{K-1}}(t)\|\} \tag{2}$$

The weights $w_{r_g}(t)$ of a such selected winner neuron and the weights $w_{r_j}(t)$ of its neighbors are adapted by a Hebbian learning rule:

$$w_{r_j}(t+1) = w_{r_j}(t) + \epsilon(t) h_{r_g r_j}(t)(v(t) - w_{r_j}(t)) \tag{3}$$

The coupling function $h_{r_g r_j}(t)$ is a 2-dimensional Gaussian function located at the winner neuron $r_g = (r_g^0, r_g^1)$ on the map with variance $\sigma^2(t)$.

$$h_{r_g r_j}(t) = \frac{1}{2\pi\sigma^2(t)} exp\left(-\frac{1}{2}\left[\frac{(r_j^0 - r_g^0)^2}{\sigma^2(t)} + \frac{(r_j^1 - r_g^1)^2}{\sigma^2(t)}\right]\right) \tag{4}$$

During the simulation for the ontogenetic map $\sigma^2(t)$ decreases linearly with time t to a constant value of 1.5 and the learning rate $\epsilon(t)$ decreases proportional to $1/t$ to a constant value of 0.05 to maintain plastic properties.

4   Simulation of post-ontogenetic plasticity

In order to provide a framework for the post-ontogenetic plastic changes, an ontogenetic map is selforganized from the *basic stimulus set*. The grid in fig. 2.Ia shows the topographic representation of the paw together with a typical RF of the neuron marked in fig. 2.Ib and Ic. The representation of the regions of the digits and the palm is depicted in the map in fig. 2.Ib. Fig. 2.Ic shows the greyvalue coded RF-size of each neuron in the map. As expected, the neurons in the central region of the map have roughly homogeneous RF-sizes, but at the border regions they have smaller RFs which is an effect of the limited spatial extend of the simulated paw. The ontogenetic map (pre) serves as reference map for all simulations of post-ontogenetic plasticity.

Simultaneous paired stimulation (PS) v^p – which is analogous to experimental PPTS – is composed of two stimuli $v^s, v^{s'} \in \mathcal{V}$ with distance $\|(x_s, y_s) - (x_{s'}, y_{s'})\|$ on the simulated paw by $v^p = v^s + v^{s'}$. The paired stimulation is presented at timesteps $t = n \cdot \tau_1$, $n \in \mathbb{N}$, $\tau_1 = const$. Strong deformation and loss of the paw topography is prevented by presenting randomly choosen stimuli from the *basic stimulus set* at the remaining timesteps (see fig. 2.II). This might correspond to the spontaneous neural activity and randomly occurring sensory input.

In order to test the effect of temporal coupling, we introduce a paired stimulation with a time delay τ_2. The stimuli v^s and $v^{s'}$ are presented at time

$t = (2n - 1) \cdot \tau_2$ and $t = 2n \cdot \tau_2$, respectively. During the other timesteps the system learns randomly choosen stimuli from the *basic stimulus set* (see fig. 3.I).

To compare the different effects of simultaneous PS (fig. 2.II), PS with time delay (fig. 3I) and single point stimulation (fig. 3II), we consider the changes in RF-size and cortical magnification:

stimulus type	RF-size	cortical magnification
simultaneous PS	increasing	increase and attraction of the reorganizational foci
PS with time delay	slightly decreasing	small increase at the two foci
single point stimulation	slightly decreasing	small increase at the focus

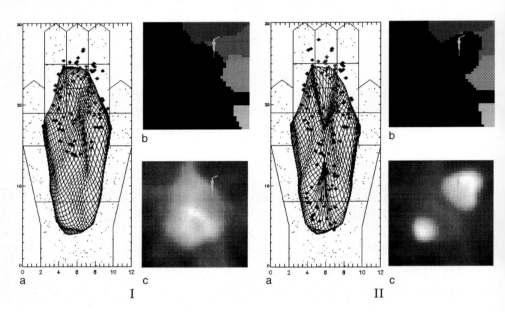

Figure 2: I. Selforganizing ontogenetic map (pre). II. After simultaneous PS. a. Topographic representation, b. cortical representation of the digits and the palm, c. RF-sizes of the neurons in the map (light $\rightarrow$ dark $\equiv$ large $\rightarrow$ small RFs).

In order to analyze the distance dependence of the effects in the simultaneous PS, we vary the distance of the stimulation centers. The distance-dependent changes of the RF-sizes are depicted in fig. 4. For short distances, stimulation occurs in slightly overlapping RFs. The neurons with large RFs tend to align towards the location of stimulated neurons. The stimulation in non-overlapping, but neighboring RFs, shows plastic reorganization by increasing RF-sizes and emergence of two separated regions representing the stimulated locations. Stimulation in non-overlapping RFs, which are separated by a mean RF-size, leads to further increasing RF-sizes and to the further separation of the representation of the stimulated locations.

The aspect of reversibility (fig. 5) is modelled by switching from the paired stimulation mode to random stimulation from the *basic stimulus set* [1].

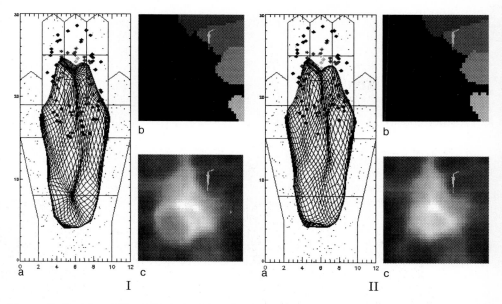

Figure 3: I. After PS with time delay II. After single point stimulation. a., b., c. as in the previous figure.

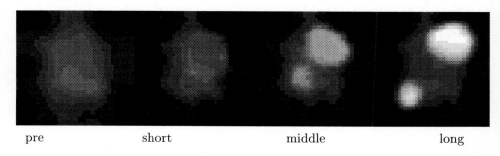

pre   short   middle   long

Figure 4: Distance-dependent RF-size distribution of the neurons in the maps. Pre-map compared with maps after simultaneous PS at short, middle and long interstimulus distance.

5 Discussion

Our attempt to modify the SOFM allowed a successful quantitative (fig. 5) simulation of reorganizational changes observed experimentally in rat somatosensory cortex induced by an associative pairing protocol that utilizes coactivation input patterns. In fig. 5 absolute RF-sizes and their plastic changes induced by the paired stimulation protocol are shown for the experimental and the simulation data, which illustrates the high degree of correspondence between both approaches.

Modifications included a probabilistic mode of stimulus presentation, which results in a stabilization of the topography when only a single point stimulation is

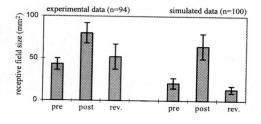

Figure 5:
Comparison of reversibel RF reorganization of experimental and simulated data.

used, but increases the sensitivity to coactivation input patterns. In addition, we introduced the aspect of fuzziness that was implemented by substituting the 'winner takes all' mechanism by selecting the winner from a set of best matched neurons. These changes lead to a quasi-parallel behavior of the SOFM. Both changes were crucial for the treatment of coactivation patterns typical in the paired stimulation protocol and thus for the simulation of realistic cortical plasticity. It is known from experimental data that the receptors on the rat paws are inhomogenously distributed. Similarily, the RF-size within the cortical paw representation is unequally distributed. In order to be able to differentiate effects due to plastic reorganization and those related to the topographic aspects of receptor density and receptive field size, we assumed a homogenous distribution of both parameters. It is possible by introducing biological based gradients to analyze the impact of inhomogeneity on the type of reorganizational changes in the simulations.

The impact of plastic receptive field changes as induced by the associative pairing on a perceptual level are analyzed in respect to the increased spatial discrimination performance using the coarse coding approach [3].

References

[1] M. Andres, O. Schlüter, F. Spengler, and H. R. Dinse. A model of fast and reversible representational plasticity using Kohonen mapping. In *Artificial Neural Networks, Proceedings of ICANN 94*, pages 306 – 309. Springer-Verlag, 1994.

[2] H.R. Dinse, B. Godde, and F. Spengler. Short-term plasticity of topographic organization of somatosensory cortex and improvement of spatial discrimination performance induced by an associative pairing of tactile stimulation. Technical Report 95-01, Institut für Neuroinformatik, Ruhr-Univ. Bochum, 1995.

[3] C.W. Eurich, H.R. Dinse, U. Dicke, B. Godde, and H. Schwegler. A population model for the increase in spatial discrimination performance induced by an associate pairing of tactile stimulation in humans and rats. In *ICANN '96. International Conference on Artificial Neural Networks*, page this meeting, Bochum, 1996.

[4] F. Joublin, F. Spengler, S. Wacquant, and H.R. Dinse. A columnar model of somatosensory reorganizational plasticity based on Hebbian and non-Hebbian learning rules. *Biological Cybernetics*, page in press, 1996.

[5] T. Kohonen. Analysis of a simple self–organizing process. *Biol. Cyb.*, 44:135–140, 1982.

[6] M.M. Merzenich and K. Sameshima. Cortical plasticity and memory. *Current Opinion Neurobiol.*, 3:187–196, 1993.

[7] K. Obermayer, H. Ritter, and K. Schulten. Large–scale simulations of self–organizing neural networks on parallel computers: applications to biological modelling. *Parallel Computing*, 14:381–404, 1990.

[8] H. Ritter, T. Martinetz, and K. Schulten. *Neuronale Netze*. Addison-Wesly, Germany, 1990.

Cortical Map Development Driven by Spontaneous Retinal Activity Waves

Christian Piepenbrock[1], Helge Ritter[2], and Klaus Obermayer[1]

[1] Technische Universität Berlin, Informatik, Berlin, Germany
[2] Universität Bielefeld, Technische Fakultät, Bielefeld, Germany

Abstract. Correlation-based learning (CBL) models have been quite successful when applied to the activity driven development of cortical orientation selectivity (OR) or ocular dominance (OD) maps. We ask under which conditions waves of spontaneous activity as have been observed in the developing retina may provide the proper activity to drive the CBL process. We demonstrate that the currently assumed statistics of spontaneous activity waves may drive OD and OR development only, if (*i*) the visual pathways interact before they reach the cortex, and (*ii*) if the CBL model involves nonlinear mechanisms or constraints to couple OD and OR development. Finally, simulated cortical maps are examined.

Introduction

Correlation-based learning (CBL) describes the development of simple-cell receptive fields and cortical maps in the primary visual cortex, including orientation selectivity (OR) and ocular dominance (OD) [5, 3, 4, 2]. Visual signals are projected from the retinas of the two eyes through the LGN to simple cells in the visual cortex. Specific response properties of these cells include sensitivity to (*i*) stimuli from a certain location on the retina (topographic projection), (*ii*) stimuli from one of the two eyes (OD), and (*iii*) stimuli of a certain orientation (OR). There is evidence that these are properties of the geniculo-cortical projection and that OD and OR develop very early driven by neuronal activity [7, 9]. A simple model suggests that cells exhibit OR because they develop receptive fields with alternating light and dark sensitive subfields.

Previous approaches have assumed filtered white noise activity from the retina as input for development [4, 8]. This is based on the idea that before eye opening retinal ganglion cells receive spontaneous excitation within their center-surround receptive fields from other neurons in the retina.

However, there has been little evidence for such activity patterns in the retina. On the contrary, in the ferret retina it has been possible to record structured waves of spontaneous activity for periods of many days. Long before eye opening such waves of excitation frequently move across the retina in arbitrary directions [10]. It has been argued [4] that these waves may account for OD but cannot explain OR development since they are too wide to provide learning stimuli for simple cell receptive fields. In this paper we show that this does not necessarily hold true and use a model for such waves as realistic input patterns

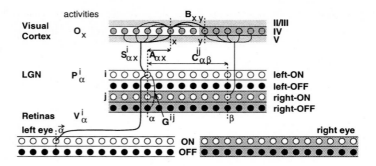

Fig. 1. The CBL model.

to a CBL model for the development of OD and OR. In particular we demonstrate that the width of waves is not critical as long as the ON- and OFF-center ganglion cells differ in their activity in some way.

The Model

Neurons are arranged in layers (see fig. 1) where all locations $(\alpha, \beta, \mathbf{x}, \mathbf{y})$ are given in retinotopic coordinates. We consider four populations of retinal ganglion cells, which we will call left-eye-ON, left-eye-OFF, right-eye-ON, and right-eye-OFF $(i = 1, \ldots, 4)$. Activity in the retina V_α^i is projected via the lateral geniculate nucleus (LGN) to the primary visual cortex layer IVc. CBL models make the basic assumptions that (i) the formation of OD and OR depends on retinal electrical activity V_α^i, (ii) OD and OR are linked to properties of the geniculo-cortical projection $S_{\alpha,\mathbf{x}}^i$, and (iii) OR emerges as a result of competition between ON-/OFF-pathways in a similar way as OD emerges as a result of inter-ocular competition.

Activity P_α^i in the LGN and $O_\mathbf{x}$ in V1 depends on the retinal activity V_α^i (we use a connectionist neuron model with activation function φ):

$$P_\alpha^i(t + 1) = \varphi\Big(V_\alpha^i(t) + \sum_j G^{ij} P_\alpha^j(t)\Big) \tag{1}$$

$$O_\mathbf{x}(t + 1) = \varphi\Big(\sum_{i,\alpha} P_\alpha^i(t) S_{\alpha,\mathbf{x}}^i(t) + \sum_\mathbf{y} B_{\mathbf{x},\mathbf{y}} O_\mathbf{y}(t)\Big) \tag{2}$$

where G^{ij} are interactions between different layers in the LGN and $B_{\mathbf{x},\mathbf{y}}$ are lateral connections within V1. OD and OR should then emerge as a result of Hebbian learning

$$\frac{d}{dt} S_{\alpha,\mathbf{x}}^i(t) = \eta A_{\alpha,\mathbf{x}} P_{\alpha,\mathbf{x}}^i(t) O_\mathbf{x}(t) - \gamma(\mathbf{x}, t) S_{\alpha,\mathbf{x}}^i(t), \quad S_{\alpha,\mathbf{x}}^i(t) \geq 0 \tag{3}$$

which happens in the geniculo-cortical projection. The arborization function $A_{\alpha,\mathbf{x}}$ is localized and enforces a topographic mapping from the LGN to the cortex. The synaptic connection strengths $S^i_{\boldsymbol{\alpha},\mathbf{x}}$ can assume only positive values and a multiplicative constraint (implemented as a time dependent $\gamma(\mathbf{x},t)$) limits the total synaptic strength of each cortical neuron. Linearization and adiabatic elimination of the neuron activities lead to the CBL model

$$\frac{d}{dt}S^i_{\boldsymbol{\alpha},\mathbf{x}}(t) = \eta A_{\boldsymbol{\alpha},\mathbf{x}} \sum_{j,\boldsymbol{\beta},\mathbf{y}} I_{\mathbf{x},\mathbf{y}} C^{ij}_{\boldsymbol{\alpha},\boldsymbol{\beta}} S^j_{\boldsymbol{\beta},\mathbf{y}}(t) - \gamma(\mathbf{x},t)S^i_{\boldsymbol{\alpha},\mathbf{x}}(t), \qquad (4)$$

where $C^{ij}_{\boldsymbol{\alpha},\boldsymbol{\beta}} = \sum_l \left((1-G)^{-2}\right)_{il}\langle V^l_{\boldsymbol{\alpha}} V^j_{\boldsymbol{\beta}}\rangle$ and $I_{\mathbf{x},\mathbf{y}} = \sum_{n=0}^{\infty} (B^n)_{\mathbf{x},\mathbf{y}}$ (5)

are the LGN input correlation function and the effective cortical interaction function, respectively.

Analysis I: Role of Symmetry

In the following we make two simplifying assumptions to derive conditions for biologically realistic development on these functions. First, we assume that the neuronal layers have a homogeneous structure and that correlations and connectivity patterns only depend on differences between coordinates. Secondly, we assume that the activity patterns in different LGN layers (left and right eye, ON- and OFF-center) are not identical but have identical statistical properties, i.e.

$$C = (C^{ij}) = \begin{bmatrix} C^1 & C^2 & C^3 & C^4 \\ C^2 & C^1 & C^4 & C^3 \\ C^3 & C^4 & C^1 & C^2 \\ C^4 & C^3 & C^2 & C^1 \end{bmatrix} \quad \text{and} \quad G = (G^{ij}) = \begin{bmatrix} g_1 & g_2 & g_3 & g_4 \\ g_2 & g_1 & g_4 & g_3 \\ g_3 & g_4 & g_1 & g_2 \\ g_4 & g_3 & g_2 & g_1 \end{bmatrix} \qquad (6)$$

As a result we can easily diagonalize the C^{ij} matrix and recover the following eigenvalues and eigenvectors:

property	left eye	right eye[3]	eigenvalue	eigenvector	left ON	left OFF	right ON	right OFF
unspecific			$\lambda^1_{\mathbf{w}\mathbf{k}}=\eta \hat{I}_{\mathbf{k}}(\hat{C}^1_{\mathbf{w}} + \hat{C}^2_{\mathbf{w}} + \hat{C}^3_{\mathbf{w}} + \hat{C}^4_{\mathbf{w}})$	$\nu_1 = [$	1	1	1	$1]^T$
OD			$\lambda^2_{\mathbf{w}\mathbf{k}}=\eta \hat{I}_{\mathbf{k}}(\hat{C}^1_{\mathbf{w}} + \hat{C}^2_{\mathbf{w}} - \hat{C}^3_{\mathbf{w}} - \hat{C}^4_{\mathbf{w}})$	$\nu_2 = [$	1	1	-1	$-1]^T$
OR			$\lambda^3_{\mathbf{w}\mathbf{k}}=\eta \hat{I}_{\mathbf{k}}(\hat{C}^1_{\mathbf{w}} - \hat{C}^2_{\mathbf{w}} + \hat{C}^3_{\mathbf{w}} - \hat{C}^4_{\mathbf{w}})$	$\nu_3 = [$	1	-1	1	$-1]^T$
OR (neg.)			$\lambda^4_{\mathbf{w}\mathbf{k}}=\eta \hat{I}_{\mathbf{k}}(\hat{C}^1_{\mathbf{w}} - \hat{C}^2_{\mathbf{w}} - \hat{C}^3_{\mathbf{w}} + \hat{C}^4_{\mathbf{w}})$	$\nu_4 = [$	1	-1	-1	$1]^T$

The development of the synaptic connection strengths will be dominated by the eigenvector with the largest eigenvalue. Biologically realistic OD develops

---

[3] The images show typical cortical receptive fields for each mode for an arbor function with diameter 13 grid points (for λ^1, λ^2 the sum and for λ^3, λ^4 the difference of the ON and OFF afferents is shown).

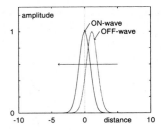 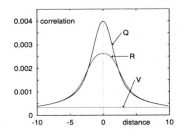

Fig. 2. (left) Model waves of spontaneous retinal activity. ON-wave $= V_\alpha^1$ and OFF-wave $= V_\alpha^2$ (for the left eye). **(right) Correlation functions.** $Q_{\alpha-\beta} = \langle V_\alpha^1 V_\beta^1 \rangle$ and $R_{\alpha-\beta} = \langle V_\alpha^1 V_\beta^2 \rangle$ and $(V)^2 = \langle V_\alpha^1 V_\beta^3 \rangle = \langle V_\alpha^1 V_\beta^4 \rangle$.

if $\lambda_{\mathbf{w},\mathbf{k}}^2$ is the maximum eigenvalue for $\mathbf{w} \approx 0$ and $\mathbf{k} \approx \mathbf{k}_{OD}$ where $\mathbf{k}_{OD}$ is the typical spatial frequency of OD bands. For biologically realistic OR development the maximum eigenvalue has to be $\lambda_{\mathbf{w},\mathbf{k}}^3$ for $\mathbf{w} \approx \mathbf{w}_{OR}$ and $\mathbf{k} \approx \mathbf{k}_{OR}$, the typical spatial frequencies of receptive subfields and of cortical orientation maps, respectively. From the eigenvalue decomposition we conclude that the two properties develop independently of each other and a concurrent development of OR and OD requires nonlinear mechanisms to link the modes.

Analysis II: Input Correlations

Now we address the question whether waves of spontaneous retinal activity as have been observed in the ferret retina suffice to generate OR and OD. The necessary condition within our model framework is that the OD and OR eigenvalues, hence the LGN correlation functions, have the right structure to meet the conditions from the last section. Waves of spontaneous activity have been observed in the ferret retina long before birth. They slowly move across the retinas in arbitrary directions and lead to correlated firing of neighboring ganglion cells as well as uncorrelated firing patterns between both retinas. We model these waves by Gaussians and compute their correlation functions according to eqn. (5) as shown in figure 2. ON- and OFF-center ganglion cells are experimentally not distinguishable at this developmental stage. Thus we may allow for a model in which the activity of retinal ON- and OFF-cells is approximately identical except that "OFF-waves" may follow (or precede) "ON-waves" with a small delay (see figure 2). This results in a correlation function R between the ON- and OFF-populations which is somewhat wider than the intra-population correlation function Q. All correlation functions are rotationally symmetric.

If we assume that the four visual input pathways stay completely segregated until they reach the cortex ($G^{ij} = 0$) at most OD development is possible. The reason is that the eigenvalues for the OD and unspecific mode are degenerate (except for the frequency $\mathbf{w} = 0$) and larger than the OR eigenvalues. As a result OR cannot develop in this scenario and even OD may emerge only if a nonlinearity breaks the degeneracy in favor of the OD mode.

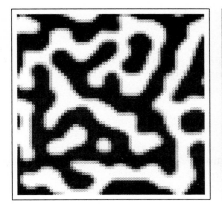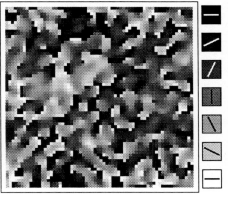

Fig. 3. Simulation results. OD and OR simulation using the same retinal input. The simulation was run with four input layers and a rectangular A with a diameter of 13 grid points. Retina, LGN and cortex were of size $64 * 64$ with periodic boundary conditions; the retinal activity waves had a width of 2.4 grid points and a distance of 1.2 between the waves of different cell types; $I_{\mathbf{x}} = \exp(-\mathbf{x}^2/4) - 1/9 \exp(-\mathbf{x}^2/36) + \delta_{\mathbf{x}}$. **(left) Cortical OD map.** Parameters were $g_2 = -0.15$, $g_3 = -0.15$, and $g_4 = -0.15$. Cortical cells with input dominated by left or right eye afferents are shown in white or black, respectively. **(right) Cortical OR map.** Parameters were $g_2 = -0.45$, $g_3 = 0$, and $g_4 = -0.02$. Orientation preference is coded by grey value.

The situation changes completely if we include interactions between different LGN layers in our model. We suggest a competitive network configuration with inhibitory interactions between layers (as has been found physiologically [6, Wörgötter, pers. communication]) and self-excitation. As a result the intra-LGN coupling parameters G^{ij} scale the eigenvalues such that any one of the four modes might govern development. This allows OD as well as OR development. Furthermore, OR development depends on the Fourier transform of $Q - R$. Consequently, (i) OR can only develop if there is a difference between "ON-waves" and "OFF-waves" and (ii) the difference between "ON-waves" and "OFF-waves" is more significant for development than the actual shape of the waves.

Figure 3 shows some simulation results of cortical map development driven by spontaneous retinal waves. In each simulation either an ocular dominance or an orientation map develops for the same input signals depending only on the parameters of the intra-LGN interactions.

OR and OD decouple and a joint development is possible only if they develop with about equal strength (i.e. the corresponding eigenvalues are almost degenerate). Furthermore, nonlinear mechanisms are necessary to couple the development of both properties. Erwin and Miller [1], e.g. suggest clipping of synaptic weights at zero and at some maximum value. Such mechanisms "freeze" the synaptic connection patterns that emerge during the initial development when a number of modes grow roughly equally strong. This creates a non-vanishing parameter region in which a joint development of OR and OD is possible. In contrast, there is a sharp phase transition between OD and OR development in a purely linear model.

Acknowledgements. This research was funded in part by HFSPO (RG 98/94B) and DFG (OB 10212-1). Computing time (CM5 parallel computer) was made available by NCSA and HLRZ Jülich.

References

1. E. Erwin and K.D. Miller. Modeling joint development of ocular dominance and orientation in primary visual cortex. *Proc. of the CNS*, 1995.
2. E. Erwin, K. Obermayer, and K. Schulten. Models of orientation and ocular dominance columns in the visual cortex: A critical comparison. *Neur. Comp.*, 7:425–468, 1995.
3. K.D. Miller. Correlation-based models of neural development. In M.A. Gluck and D.E. Rumelhart, editors, *Neuroscience and Connectionist Theory*, pages 267–353. Lawrence Erlbaum Associates, 1990.
4. K.D. Miller. A model for the development of simple cell receptive fields and the ordered arrangements of orientation columns through activity-dependent competition between ON- and OFF-center inputs. *J. Neurosci.*, 14:409–441, 1994.
5. K.D. Miller, J. B. Keller, and M. P. Stryker. Ocular dominance column development: Analysis and simulation. *Science*, 245:605–615, 1989.
6. H. C. Pape and U. T. Eysel. Bionocular interactions in the lateral geniculate nucleus of the cat: Gabaergic inhibition reduced by dominant afferent activity. *Exp. Brain Research*, 61:265–271, 1986.
7. P. Rakic. Prenatal genesis of connections subserving ocular dominance in the rhesus monkey. *Nature*, 261:467–471, 1976.
8. M. Stetter, A. Muller, and E. W. Lang. Neural network model for the coordinated formation of orientation preference and orientation selectivity maps. *Phys. Rev. E*, 50:4167–4181, 1994.
9. T. N. Wiesel and D. H. Hubel. Ordered arrangement of orientation columns in monkeys lacking visual experience. *J. Comp. Neurol.*, 158:307–318, 1974.
10. R. O. L. Wong, M. Meister, and C. J. Shatz. Transient period of correlated bursting activity during development of the mammalian retina. *Neuron*, 11:923–938, 1993.

Simplifying neural networks for controlling walking by exploiting physical properties

Holk Cruse, Christian Bartling, Jeffrey Dean, Thomas Kindermann, Josef Schmitz, Michael Schumm, Hendrik Wagner

Fac. of Biology, University of Bielefeld, Postfach 100131, D-33501 Bielefeld, Germany

Abstract

A network for controlling a six-legged, insect-like walking system is proposed. The network contains internal recurrent connections, but important recurrent connections utilize the loop through the environment. This approach leads to a subnet for controlling the three joints of a leg during its swing which is arguably the simplest possible solution. The task for the stance subnet appears more difficult because the movements of a larger and varying number of joints (9-18: three for each leg in stance) have to be controlled such that each leg contributes efficiently to support and propulsion and legs do not work at cross purposes. Already inherently non-linear, four factors further complicate this task: 1) the combination of legs in stance varies continuously, 2) during curve walking, legs must move at different speeds, 3) on compliant substrates, the speed of the individual leg may vary unpredicatably, and 4) the geometry of the system may vary through growth and injury or due to non-rigid suspension of the joints. We show that an extremely decentralized, simple network copes with all these problems by exploiting the physical properties of the system.

Introduction

In the present article, we shall discuss a motor system for a complex behavior in which hard physical interactions play an essential role. We will describe how the system generates adaptable motor rhythms using a decentralized control system. Furthermore, we will present a model illustrating how the loop through the world can be exploited to drastically simplify the structure of the network controller and to solve otherwise intractable computational problems. The results are based on experimental investigation of the behavior of the biological control system underlying the locomotion of six-legged arthropods in general and the stick insect in particular.

The step cycle of the walking leg can be divided into two functional states. During stance, the leg is on the ground, supports the body and, in the forward walking animal, moves backwards with respect to the body. During swing, the leg is lifted off the ground and moved in the direction of walking to where it can begin a new stance. The relaxation oscillator making up the step pattern generator is assumed to consist of two agents--a swing network and a stance network controlling the movement of the leg during swing and stance, respectively (Fig. 1b). The transition between swing and stance is controlled by a third agent, the selector network (Fig. 1b). The swing network and the stance network are always active, but the selector network determines which of the two agents controls the motor output. The selector net consists of a two-layer feedforward net with positive feedback connections in the second layer. These positive-feedback connections serve to stabilize the ongoing activity, namely stance or swing.

Control of the swing and the stance movement

The task of finding a network that produces a swing movement seems to be easier than finding a network to control the stance movement because a leg in swing is mechanically uncoupled from the environment and therefore, due to its small mass, essentially uncoupled from the movement of the other legs. The geometry of the leg is shown in Fig. 1a. The coxa-femur and femur-tibia joints, the two distal joints, are simple hinge joints with one degree of freedom corresponding to elevation and to extension of the tarsus, respectively. The subcoxal joint is more complex, but most of its movement is in a rostrocaudal direction around the nearly vertical axis. Thus, the control network must have at least three output channels, one for each leg joint. As has been shown by Cruse et al. (1995), a simple, two-layer feedforward net with three output units and six input units can produce movements which closely resemble the swing movements observed in walking stick insects. In the simulation, the three outputs of this net, interpreted as the angular velocities $d\alpha/dt$, $d\beta/dt$, and $d\gamma/dt$, are fed into an integrator (not shown in Fig. 1b), which corresponds to the leg itself in the animal, to obtain the joint angles. The actual angles are measured and fed back into the net. In this net the feedback loop does not correspond to an internal neuronal channel, but passes via the periphery, i.e., the leg and its motion. The swing net, using only 8 or 9 nonzero weights, which most probably represents the simplest possible network, is able to generalize over a considerable range of untrained situations. Furthermore, the swing net is remarkably tolerant with respect to external disturbances. The learned trajectories represent a kind of attractor to which the disturbed trajectory returns. This compensation for disturbances occurs because the system does not compute explicit trajectories, but simply exploits the physical properties of the world.

In natural situations, the task of controlling the movements of all the legs on the ground involves several problems. It is not enough simply to specify a movement for each leg on its own: the mechanical coupling through the substrate means that efficient locomotion requires coordinated movement of all the joints of all the legs in contact with the substrate, that is, a total of 18 joints when all legs of an insect are on the ground. However, the number and combination of mechanically coupled joints varies from one moment to the next, depending on which legs are lifted. The task is quite nonlinear, particularly when the rotational axes of the joints are not orthogonal, as is often the case in insects, particularly for the basal leg joint. A further complication arises when the animal negotiates a curve and the different legs have to move at different speeds. Further complexities arise in more complex, natural walking situations, making solution difficult even with high computational power. These occur, for example, when an animal or a machine walks on a slippery surface or on a compliant substrate, such as the leaves and twigs encountered by stick insects. Any flexibility in the suspension of the joints further increases the degrees of freedom that must be considered and the complexity of the com-

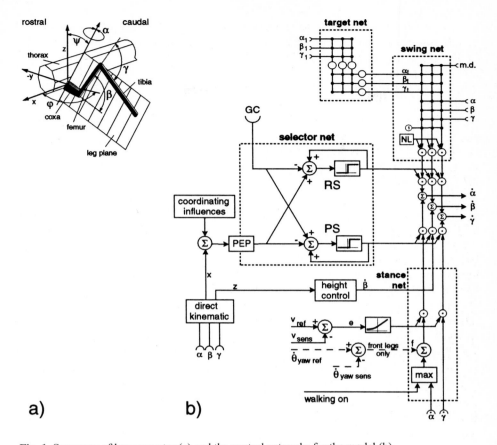

Fig. 1. Summary of leg geometry (a) and the control networks for the model (b).

putation. Further problems for an exact, analytical solution occur when the length of leg segments changes during growth or their shape changes through injury. In such cases, knowledge of the geometrical situation is incomplete, making an explicit calculation difficult, if not impossible.

Despite the evident complexity of these tasks, they are mastered even by insects with their „simple" nervous systems. Therefore, there has to be a solution that is fast enough that on-line computation is possible even for slow neuronal systems. How can this be done? To solve the particular problem at hand, we propose a central controller for the stance movement with distributed control in the form of local positive feedback. Compared to earlier versions (Cruse et al. 1995), this change permits the stance net to be radically simplified. The positive feedback occurs at the level of single joints: the position signal of each is fed back to control the motor output of the same joint (Fig. 1b, stance net). How does this system work? Let us assume that any one of the joints is moved actively. Then, because of the mechanical connections, all other joints begin to move passively, but in exactly the proper way. Thus, the movement direction and speed of each joint does not have to be computed because this information is already provided by the physics. The positive feedback then transforms this passive movement into an active movement.

There are, however, several problems to be solved. The first is that positive feedback using the raw position signal would lead to unpredictable changes in movement speed, not the nearly constant walking speed which is usually desired. This problem can be solved by introducing a kind of band-pass filter into the feedback loop. The effect is to make the feedback proportional to the angular velocity of joint movement, not the angular position. In the simulation, this is done by feeding back a signal proportional to the angular change over the preceding time interval, i.e., the current angle minus the angle at the previous time step. The second problem is that using positive feedback for all three leg joints leads to unpredictable changes in body height, even in a computer simulation neglecting gravity. A physical system, of course, would be pulled downward by gravity and the positive feedback would accelerate this movement. In summary, a system with positive feedback at all joints does not maintain a constant body height even in the absence of gravity; it is even less able to do so under the influence of gravity. Such control, of course, is an essential characteristic of an efficient walking system. To solve this problem we assume that during walking positive feedback is provided for the α joints and the γ joints, (Fig. 1), but not for the β joints. The β joint is the major determinant of the separation between leg insertion and substrate, which determines body height. Thus, in the control scheme for walking proposed here, the β joint remains under classical negative feedback control as in the standing animal. In this way, it is possible to solve all the problems mentioned above in an easy and computationally simple way.

Preliminary tests showed that continuous walks are possible but several problems occur. First, the system did not walk straight but moved in a slight curve as a result of a tendency to yaw (rotate around the vertical body axis). To prevent this yawing, a supervisory system was introduced which, in a simple way, simulates optomotor mechanisms for course stabilisation that are well-known from insects and have also been applied in robotics (Franceschini et al. 1992). This supervisory system uses information on the rate of yaw ($d\theta_{yaw\ sens}/dt$, Fig. 1b, stance net), such as could be provided by visual movement detectors. It is based on negative feedback of the deviation between the desired turning rate and the actual change in heading over the last time step. The error signal controls additional impulses to the α joints of the front legs which have magnitudes proportional to the deviation and opposite signs for the right and left sides. With this addition and $d\theta_{yaw\ ref}/dt$ set to zero, the system moves straight (not shown) with small, side-to-side oscillations in heading such as can be observed in walking insects. To simulate curve walking, the reference value is given a small positive or negative bias to determine curvature direction and magnitude. Fig. 2 shows an example of a left turn ($d\theta_{yaw\ ref}/dt \neq 0$).

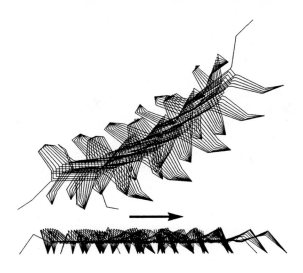

Fig. 2. Simulated walk by the basic six-legged system with negative feedback applied to all six β joints and positive feedback to all α and γ joints as shown in Fig. 1b. Movement direction is from left to right (arrow). Leg positions are illustrated only during stance and only for every second time interval in the simulation. Each leg makes about five steps. Upper part: top view, lower part: side view.

A second problem inherent in using positive feedback is the following. Let us assume that a stationary insect is pulled backward by gravity or by a brief tug from an experimentor. With positive feedback control as described, the insect should then continue to walk backwards even after the initial pull ends. This has never been observed. Therefore, we assume that a supervisory system exists which is not only responsible for switching on and off the entire walking system, but also specifies walking direction (normally forward). This influence is represented by applying a small, positive input value (Fig. 1b, walking on) which replaces the sensory signal if it is larger than the latter (the box "max" in Fig. 1b, stance net). Finally, we have to address the question of how walking speed is determined in such a positive feedback controller. Again, we assume a central value which represents the desired walking speed v_{ref}. This is compared with the actual speed, which could be measured by visual inputs or by monitoring leg movement. This error signal is subject to a nonlinear transformation $\{y = (1+e)$ for $e \geq = 0$ and $y = 1/(1-e)$ for $e < 0\}$ and then multiplied with the channels providing the positive feedback for all α and γ joints of all six legs (Fig. 1b, stance net). No examples for different walking speeds are given. Sometimes, the following unexpected behavior was observed. In the course of a long, apparently stable walk, the system stumbled for as yet undetermined reasons and fell to the ground, but in all cases, the system stood up by itself and resumed proper walking. This happened even when the fall placed the six legs in an extremely disordered arrangement. This means that the simple solution proposed here also eliminates the need for a special supervisory system to rearrange leg positions after such an emergency.

Discussion

Six individual leg controllers are coupled by a small number of connections forming a recurrent network (Cruse et al. 1995). Together with the ever present external recurrent connections - during walking the legs are mechanically coupled via the substrate - this recurrent system produces a proper spatio-temporal pattern corresponding to different gait patterns found in insects. One subsystem, the swing net, consists of an extremely simple feedforward net which exploits the recurrent information via the sensorimotor system as the leg´s behavior is influenced by the environment to generate swing movements in time. The system does not compute explicit trajectories; instead, it computes changes in time based on the actual and the desired final joint angles. Thus, the control system exploits the physical properties of the leg. This organization permits an extremely simple network and responds in an adaptive way to external disturbances.

A more severe problem has to be solved by the stance system, the second subsystem of the leg controller. It appears that solving this complex task requires quite a high level of "motor intelligence". Nevertheless, a complicated, centralized control system is not required. On the contrary, we have shown in simulations that an extremely decentralized solution copes with all these problems and, at the same time, allows a very simple structure for the local controllers. No neuronal connections among the joint controllers, even among those of the same leg, are necessary during stance. The necessary constraints are simply provided by the physical coupling of the joints. Thus, the system controlling the joint movements of the legs during stance is not only "intelligent" in terms of its behavioral properties, but also in terms of simplicity of construction: the most difficult problems are solved by the most decentralized control structure. This simplification is possible because the physical properties of the system and its interaction with the world are exploited to replace an abstract, explicit computation. Thus, "the world" is used as "its own best model" (Brooks 1991). Due to its extremely decentralized organization and the simple type of calculation required, this method allows very fast computation.

References

Brooks,R.A. (1991): Intelligence without reason. IJCAI-91, Sydney, Australia, 569-595

Cruse,H., Bartling,Ch., Cymbalyuk,G., Dean,J., Dreifert,M. (1995): A modular artificial neural net for controlling a six-legged walking system. Biol. Cybern. 72, 421-430

Franceschini,N., Pichon,J.M., Blanes,C. (1992): From insect vision to robot vision. Phil.Trans.R.Soc. 337,283-294

Saccade Control Through the Collicular Motor Map: Two-Dimensional Neural Field Model

Andreas Schierwagen and Herrad Werner

Institut für Informatik, Universität Leipzig, FRG

Abstract. In this paper principles of neural computing underlying saccadic eye movement control are applied to formulate a *mapped neural field model* of the spatio-temporal dynamics in the motor map of the superior colliculus. A key feature of this model is the assumption that dynamic error coding in the motor map might be realized without efference copy feedback but only through the nonlinear, spatio-temporal dynamics of neural populations organized in mapped neural fields. The observed activity pattern suggests a particular, space-variant scheme of lateral interconnections within the motor map.

1 Introduction

A major role in gaze control plays the superior colliculus (SC), a sensorimotor transformation center in the mammalian brain stem. Whereas the upper SC layers contain a retinotopic map of the visual hemisphere, the deeper layers host a motor map of rapid eye movements (saccades) [Sparks and Mays, 1990]. That is, the location of active neurons in the deep layers represents the vector of motor error, i.e. the amplitude and direction of the saccade to reach the target. The range of possible saccades prior to which a given neuron discharges defines its movement field (MF) while the neuron is firing for a specific but large range of movements. The topography of the motor map is such that horizontal motor error is encoded mainly along the rostro-caudal axis, and vertical motor error along the medio-lateral axis. The visual and the motor map seems to be in spatial register [Grantyn, 1988; Hilbig and Schierwagen, 1994]. Activity in the visual map then could be conveyed to corresponding areas in the motor map, thus moving the eye from its current fixation point to the target that activates the visual cells.

Experimental studies [Munoz *et al.*, 1991] showed that in the cat's SC during a saccade the activity hill travels through the motor map from its initial location towards the fixation zone (a hill-shift effect). The instantaneous hill location on the map specifies the remaining motor error (dynamic error coding). The saccade is terminated when the traveling activity hill reaches the fixation zone where it ceases. Recent investigations demonstrated that in the monkeys' SC the hill-shift effect is also present although the situation seems to be more complex [Munoz and Wurtz, 1995]. These studies suggest how the spatial representation of activity on the motor map is transformed into the temporal code (frequency and duration of discharge) required by motoneurons: by continuous, dynamic collicular control of the eye movement. Thus, the trajectory of activity on the SC motor map seems to spatially encode the instantaneous eye movement motor error signal.

The neural mechanisms by which the collicular motor map might realize this dynamic motor error coding are currently under debate. Most distributed models of saccade generation used fast efference copy feedback for continuously updating the dynamic motor error (for review, see [Wurtz and Optican, 1994]). In this paper we employ an alternative assumption which is both structurally and computationally simpler to show that collicular dynamic error coding might be realized without feedback but only through the nonlinear, spatio-temporal population dynamics of the collicular neurons. It turns out that a particular type of lateral interconnections within the motor map is favored by the observed activity pattern.

2   Neural Field Model of SC

The neurobiological findings suggest to consider the SC as a neural system the layers of which may be characterized by anatomical and physiological parameters remaining more or less constant within the layer but varying between layers. A particular useful approach for studying the dynamic behavior of neural populations in layers is represented by continuous neural field models [Milton et al., 1995]. Thus, we modeled the motor map as a 2-dimensional neural field.

Let $u(\mathbf{x}, t)$ be the average membrane potential of neurons located at position $\mathbf{x} = (x, y)$ at time t. The average activity (firing rate) of neurons at $\mathbf{x}$ at t is given by the sigmoid-shaped nonlinearity $f[u(\mathbf{x}, t)]$, and the average strength of synaptic connections from neurons at position $\mathbf{x}'$ to those at position $\mathbf{x}$ by $w(\mathbf{x}, \mathbf{x}')$. For homogeneous fields $w(\mathbf{x}, \mathbf{x}') = w(|\mathbf{x} - \mathbf{x}'|)$ holds. With u_0 the global threshold of the field and $s(\mathbf{x}, t)$ the intensity of applied stimulus from the outside of the field to the neurons at position $\mathbf{x}$, the neural field equation reads

$$\tau \frac{\partial u(\mathbf{x}, t)}{\partial t} = -u(\mathbf{x}, t) + \iint_{\mathcal{R}^2} w(\mathbf{x}, \mathbf{x}') f[u(\mathbf{x}', t)] d\mathbf{x}' - u_0 + s(\mathbf{x}, t), \qquad (1)$$

which is the 2-dimensional generalization of Amari's equation [Amari, 1977].

For fields of lateral-inhibition type, excitatory connections dominate for proximal neurons and inhibitory connections dominate at greater distances, described e.g. by a radially symmetrical weighting function of on-center off-surround type which can be modeled by a difference of Gaussians,

$$w(\mathbf{x} - \mathbf{x}') = g_e \cdot \exp\left(-\left(\frac{\mathbf{x} - \mathbf{x}'}{\sigma_e}\right)^2\right) - g_i \cdot exp\left(-\left(\frac{\mathbf{x} - \mathbf{x}'}{\sigma_i}\right)^2\right) \qquad (2)$$

where g_e and σ_e are the height and width of the excitatory center and g_i and σ_i are the corresponding values for the inhibitory surround.

A categorization of the dynamics of 1-dimensional fields has been provided (see [Amari, 1977] for details). In the case of 2-dimensional neural fields, a similar categorization of the dynamics can be given but the results are more complex. Neural wave patterns include standing, traveling and rotating activity waves, depending on the relative spatial extent of excitatory and inhibitory connectivity. We note that 1-dimensional trajectories of activity hills as observed in the motor map of SC are not found in *continuous* 2-dimensional neural fields.

3 Model Structure and Simulations

In Amari's theory of neural fields, space-invariance or homogeneity of the weighting function is essential. However, there is evidence that, e.g. in the internal cortical processing, space-variance occurs [Mallot *et al.*, 1990]. Likewise, it has been shown that MFs of collicular neurons typically have a skewed (asymmetrical) sensitivity profile [Ottes *et al.*, 1986] which suggests that the strengths of the lateral connections depend on both input and output site rather than on their mere difference.

Space-variant couplings are also suggested in the present context by the analyses of specific *active media* (AM), i.e. fields of FitzHugh-Nagumo (FHN) neurons. The FHN model holds a prominent position in studies of AM because of its simplicity and qualitative correspondence with more complicated models [Schierwagen, 1989]. A particularly important result from investigations of 2-dimensional FHN fields is the following: a *discrete* field model with space-variant, asymmetrical weighting functions exhibits dynamic behavior which corresponds to 1-dimensional trajectories of activity hills. This means, these hills fail to spread laterally to excite the adjacent field regions, if the asymmetry of the weighting function is strong enough [Keener, 1988].

Based on these theoretical results we modeled the collicular motor map by a discrete version of a *mapped neural field* of the Amari type. That is, the field model, Eqn. (1), was endowed with rotationally asymmetric, space-variant weighting functions $w(\mathbf{x}, \mathbf{x}')$ described as a difference of 2-dimensional Gaussian functions. The weighting function $w(\mathbf{x}, \mathbf{x}')$ is given in normal form by

$$
\begin{aligned}
w(\mathbf{x}, 0) = \quad & g_e \cdot \exp\left(-\left(\frac{x}{\sigma_{e_x}}\right)^2 - \left(\frac{y}{\sigma_{e_y}}\right)^2\right) \\
& - g_i \cdot \exp\left(-\left(\frac{x}{\sigma_{i_x}}\right)^2 - \left(\frac{y}{\sigma_{i_y}}\right)^2\right)
\end{aligned}
\tag{3}
$$

where $\mathbf{x} = (x, y)$. g_e, σ_{e_x} and σ_{e_y} are the height, x-axis width and y-axis width of the excitatory center, and g_i, σ_{i_x} and σ_{i_y} are the corresponding values for the inhibitory surround. In addition, the major semi-axes of the elliptical isoefficacy contours, described by σ_{e_x} and σ_{i_x}, differed for the half-planes $x < 0$ and $x > 0$ which was characterized by the compression factor, κ.

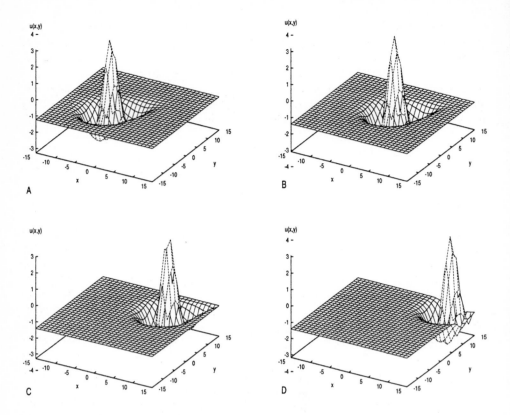

Fig. 1. Simulation of the hill-shift effect with the mapped neural field model. The SC motor map was spatially represented by a 30 × 30 grid. Eqn. (1) was integrated employing the Euler scheme with a time step of $\Delta t = 0.1$. The 2-dimensional, spatial integral in Eqn. (1) was computed by repeated 1-dimensional integration, using the extended trapezoidal rule with stepsize $\Delta x = \Delta y = 1.0$. (**A**)-(**D**) Example of simulated activity pattern in the neural field model. Starting from initial location $(-5, -5)$, the activity hill moves over the field which is shown at different, consecutive times $t = 2$ (**A**), $t = 9$ (**B**), $t = 16$ (**C**), $t = 23$ (**D**). Movement direction is defined by the steeper decaying flank of the weighting function, $w(x, y)$. Parameters used in the simulations are $\tau = 1$, $u_0 = 1.4$, $s(\mathbf{x}, t) = 0$, $g_e = 3.6$, $g_i = 4.0$, $\sigma_{e_x} = 2.8$, $\sigma_{e_y} = 1.4$, $\sigma_{i_x} = 5.7$, $\sigma_{i_y} = 2.8$, $\kappa = 0.5$.

The 2-dimensional motor map was modeled by a 30 × 30 grid. All the neurons of the grid had a step transfer function f. The map origin corresponding to the fovea region was fixed at point $(15, 15)$ (Fig. 1). Space-variant processing in the model has been realized through the asymmetrical center and surround structure of the weighting function described by Eqn.(3) and through the radial organization of the weighting functions, in accordance with the radially organized asymmetry of MFs found in collicular neurons [Ottes et al., 1986].

As the initial state we chose a localized excitation on the grid that eventually developed into a hill of activity on the mapped field. The localized excitation had cubic shape and covered a 2×2 area within which the field potential was set to $u = 1.0$. We run 30 simulations, placing the activity hill in randomly selected locations of the field.

The simulated pattern of activity in the neural field model showed a clear anisotropy, resulting from the space-variant coupling of the neurons. The direction of movement was determined by the orientation of the steeper decaying flank of the weighting function defined by κ. In the example shown in Fig. 1, the initial location of the activity hill was at $(-5, -5)$. The asymmetrical, radially oriented weighting functions caused the hill to move towards the origin $(15, 15)$ of the map coordinate system where it stopped and finally decayed (Fig. 1A-D).

4 Conclusions

In this paper we proposed a model of the spatio-temporal dynamics in the motor map of the SC. In particular, we studied which qualitative behaviors of the collicular neurons can be obtained if the model is based on some high-level computing principles governing saccadic eye movement control, i.e. neural mapping, population coding and space-variant processing in layered systems. A model type that is especially appropriate to incorporate these principles is provided by neural fields.

In generalizing Amari's concept of homogeneous, space-invariant fields, we modeled the collicular motor map as a 2-dimensional neural field endowed with space-variant weighting functions. We presented evidence for the assumption that the lateral connectivity within the collicular motor map might be inhomogeneous. Therefore, we employed an asymmetrical, radially oriented coupling scheme within the collicular motor map. In simulations with the field model, we could reproduce the hill-shift effect observed experimentally in the collicular motor map, i.e. our model exhibited 1-dimensional trajectories of activity hills.

Altogether, our study presented evidence for the assumption that collicular dynamic error coding can be realized without efference copy feedback. A recent model study on the hill-flattening effect found in burst neurons of the monkey's SC arrived at similar conclusions [Massone and Khoshaba, 1995].

Acknowledgment
Supported in part by Deutsche Forschungsgemeinschaft, Grant No. Schi 333/4-1.

References

[Amari, 1977] S. Amari. Dynamics of pattern formation in lateral-inhibition type neural fields. *Biol. Cybern.*, 27:77–87, 1977.

[Grantyn, 1988] R. Grantyn. Gaze control through superior colliculus: Structure and function. In J. Büttner-Ennever, editor, *Neuroanatomy of the Oculomotor System*, pages 273–333. Elsevier, New York, 1988.

[Hilbig and Schierwagen, 1994] H. Hilbig and A. Schierwagen. Interlayer neurons in the superior colliculus of the rat - a tracer study using DiI/Di-ASP. *NeuroReport*, 5:477–480, 1994.

[Keener, 1988] J. P. Keener. On the formation of circulating patterns of excitation in anisotropic excitable media. *J. Math. Biol.*, 26:41–56, 1988.

[Mallot et al., 1990] H. A. Mallot, W. von Seelen, and F. Giannakopoulos. Neural mapping and space-variant processing. *Neural Networks*, 3:245–263, 1990.

[Massone and Khoshaba, 1995] L.L.E. Massone and T. Khoshaba. Local dynamic interactions in the collicular motor map: A neural network model. *Network*, 6:1–18, 1995.

[Milton et al., 1995] J. Milton, T. Mundel, U. an der Heiden, J.-P. Spire, and J. Cowan. Traveling activity waves. In M.A. Arbib, editor, *The Handbook of Brain Theory and Neural Networks*, pages 994–997. MIT Press, Cambridge, MA, 1995.

[Munoz and Wurtz, 1995] D. P. Munoz and R. H. Wurtz. Saccade-related activity in monkey superior colliculus. II. Spread of activity during saccades. *Journal of Neurophysiology*, 73:2334–2348, 1995.

[Munoz et al., 1991] D. P. Munoz, D. Pellison, and D. Guitton. Movement of neural activity on the superior colliculus motor map during gaze shifts. *Science*, 251:1358–1360, 1991.

[Ottes et al., 1986] F. P. Ottes, J. A. M. Van Gisbergen, and J. J. Eggermont. Visuomotor fields of the superior colliculus: a quantitative model. *Vision Res.*, 26:857–873, 1986.

[Schierwagen, 1989] A. K. Schierwagen. The continuum approach as applied to wave phenomena in physiological systems. In J. Engelbrecht, editor, *Nonlinear Waves in Active Media*, pages 185–191. Springer-Verlag, Berlin Heidelberg New York Tokyo, 1989.

[Sparks and Mays, 1990] D. L. Sparks and L. E. Mays. Signal transformations required for the generation of saccadic eye movements. *Annu. Rev. Neurosci.*, 13:309–336, 1990.

[Wurtz and Optican, 1994] R.H. Wurtz and L.M. Optican. Superior colliculus cell types and models of saccade generation. *Curr.Opin.Neurobiol.*, 4, 1994.

Plasticity of Neocortical Synapses Enables Transitions Between Rate and Temporal Coding

Misha V. Tsodyks and Henry Markram

Department of Neurobiology, Weizmann Institute of Science, Rehovot 76100, Israel.

Abstract. The transmission across neocortical synapses changes dynamically as a function of presynaptic activity [1]. A switch in the manner in which a complex signal, such as a burst of presynaptic action potentials, is transmitted between two neocortical layer 5 pyramidal neurons was observed after coactivation of both neurons. The switch involved a redistribution of synaptic efficacy during the burst such that the synapses transmitted more effectively only the first action potential in the burst. A computational analysis reveals that this modification in dynamically changing transmission enables pyramidal neurons to extract a rich array of dynamic features of ongoing activity in network of pyramidal neurons, such as the onset and amplitude of abrupt synchronized frequency transitions in groups of presynaptic neurons, the size of the group of neurons involved and the degree of synchrony. These synapses transmit information about dynamic features by causing transient increases of postsynaptic current which have characteristic amplitudes and durations. At the same time, the ability of synapses to signal the sustained level of presynaptic activity is limited to a narrow range of low frequencies, which become even narrower after synaptic modification.

1 Dynamic transmission in pyramidal synapses

A marked feature of synaptic contacts between pyramidal neurons, which is typically overlooked in cortical neural network modeling, is fast depression of postsynaptic responses evoked by consecutive spikes in a burst [1]. Modifications in synaptic responses to bursts of presynaptic spikes following simultaneous activity in two layer V pyramidal neurons (referred to as synaptic learning in the present study) was investigated [2]. It was found that synaptic learning was not associated with an overall increase in the synaptic response to a burst, but with a redistribution of the synaptic efficacy. In particular, while the amplitude of the response to the first spike was increased, the responses to the second and few subsequent spikes were more often depressed. Overall, the changes in the amplitude of responses were confined to the first few spikes in a burst (usually $4-5$), after which the responses were unaffected. A characteristic case is shown in Fig. 1A,B.

1.1 Synaptic model

In the present study, we explore the computational significance of these striking synaptic properties for signaling between pyramidal neocortical neurons. For this purpose we chose a minimal model to simulate the synaptic behavior which ignores a presynaptic factor and considers a three-state kinetic scheme for postsynaptic receptor desensitization [3] (Fig. 1C). In this scheme, influx of the neurotransmitter into the synaptic cleft instantaneously opens some fraction of receptors, which then desensitize with a time constant of a few milliseconds and recover with a time constant of about 1 second. The crucial factor in this model is an overall amount of neurotransmitter released by each spike, which in turn determines the fraction of receptors, r, opened. Synaptic modification of the type reported [2] are incorporated in the model through an increase of the average fraction of receptors opened by a spike, r. This could result either from presynaptic factors, such as an increase in probability of release [4, 5, 6] and/or number of release sites [7], or from the postsynaptic factors, such as an increase in the affinity of the receptors to neurotransmitter. If the fraction of receptors opened by a spike is sufficiently large, the subsequent spike, even when releasing more neurotransmitter, can produce a weaker effect because of receptor desensitization. This model captures the main properties of plasticity in these synapses (Fig. 1).

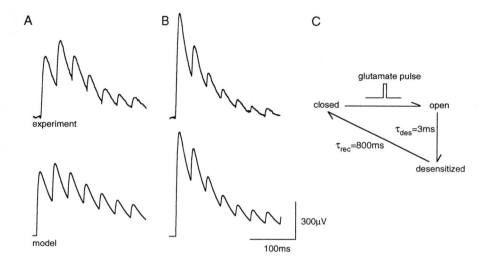

Fig. 1. A minimal model of the synapse between neocortical layer V pyramidal neurons. Postsynaptic potential due to a burst of 7 presynaptic spikes at a frequency of 23Hz (A) before and (B) after synaptic learning. Upper traces - experiment, lower traces - model. (C) Kinetic scheme used to compute the postsynaptic current. To calculate the postsynaptic potential, we incorporated the typical passive membrane mechanism with the membrane time constant of 50 msec and input resistance $100\,M\Omega$. Parameters: $E = 250$ pA, $\tau_{des} = 3$ msec, $\tau_{rec} = 800$ msec. $r = 0.35$ before learning, $r = 0.67$ after learning.

Kinetic equations corresponding to this scheme read:

$$\frac{dC}{dt} = \frac{D}{\tau_{rec}}$$

$$\frac{dO}{dt} = -\frac{O}{\tau_{des}}$$
$$D = 1 - C - O, \tag{1}$$

where C, O, D are the fraction of receptors in the corresponding state. τ_{des} and τ_{rec} are the time constants of receptor desensitization and recovery, respectively. As the spike arrives, fraction r of receptors available in closed state opens instantaneously. The net postsynaptic current generated by the synapse is taken to be proportional to a fraction of receptors in the open state, O.

The model equations allow iterative expressions for the successive amplitudes of excitatory postsynaptic current (EPSC) produced by a burst of presynaptic spikes.

$$EPSC_{n+1} = EPSC_n(1-r)e^{-\Delta t/\tau_{rec}} + Er(1 - e^{-\Delta t/\tau_{rec}}), \tag{2}$$

where $EPSC_n$ is the postsynaptic current for nth ($n = 0, 1, 2...$) spike in a burst, and Δt is the time interval between nth and $(n+1)$th spike, which is assumed to be larger then the time constant of receptor desensitization. E is a maximal $EPSC$ which would have been produced by opening all of the receptors. The $EPSC$ for the first spike equals Er. If presynaptic spikes arrive at equal time intervals between each other, then the plateau level of $EPSCs$ is reached after several spikes, which equals

$$EPSC_{plateau} = E(1/r - 1 + \frac{1}{1 - e^{-\Delta t/\tau_{rec}}})^{-1}. \tag{3}$$

2 Coding of population activity

2.1 Firing rate coding

Two important observations can be made about the expression (3). First, in the broad range of input frequencies where $\tau_{des} << \Delta t << r\tau_{rec}$, the $EPSC_{plateau}$ is inversely proportional to the frequency of spikes in the presynaptic burst; $EPSC_{plateau} \approx E\frac{\Delta t}{\tau_{rec}}$. Second, in this frequency range, the amplitude of the plateau does not depend on the parameter r and therefore is not sensitive to long-term synaptic learning, consistent with experimental results. The first point leads to an unexpected conclusion when considering a neuron which receives input from an ensemble of other pyramidal neurons. Since the average $EPSC$ produced by each neuron is proportional to the product of the presynaptic frequency and the evoked $EPSC$ of an individual spike, the frequency dependence cancels out in the result. Thus, after several initial spikes, the average synaptic current will hardly depend on the firing rates of presynaptic neurons, except when their frequencies are low (boundary is proportional to $\frac{1}{r\tau_{rec}}$). In other words, at sufficiently high frequencies, these synapses transmit information to the postsynaptic neuron, which, when integrated with that of other synapses, becomes binary, in the sense that it only indicates whether the presynaptic neuron is active or quiescent. We emphasize, however, that the region of frequencies for which analog signaling is possible can become substantial if synapses operate in a very 'untrained' state, characterized by low values of r, e.g. due to a high failure rate (Fig. 2).

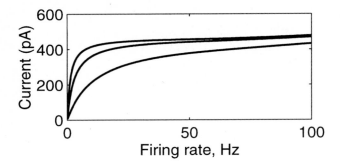

Fig. 2. The stationary postsynaptic current as a function of sustained presynaptic frequency for a population of 500 neurons for three different states of the synapses. Analytical curve obtained from Eq. (3) under the assumption that all the neurons have the same rate. Lower curve - very 'untrained' synapses with $r = 0.1$. Middle curve - $r = 0.37$, upper curve - $r = 0.67$. In this study we assume linear summation of the EPSCs from different presynaptic neurons, and do not consider the implications of postsynaptic transformations of the current produced by activation of voltage-dependent conductances.

2.2 Temporal coding

An increase in firing rates of presynaptic neurons, however, does not go completely unnoticed - it leads to a *transient* increase in the synaptic current when the rates change abruptly and in synchrony (Fig. 2A). Moreover, the amplitude of the current transient increases linearly as a function of the change in presynaptic firing rate and the rate of this change. Thus, during these transitions, the signal transmitted to the neuron is *analog*.

Neurons are able to detect changes in the pattern of presynaptic activity even if the mean frequency of the presynaptic population remains unchanged (see fourth transient in Fig. 3B).

Sharp current transients have been shown to be very effective in driving a neocortical neuron [8]. A postsynaptic neuron could therefore encode these dynamic features in its spiking activity and thereby convey them to other targets. To confirm this possibility, we injected the current obtained in simulations into pyramidal neurons (Fig. 3). We found that these neurons reliably detect the sharp transitions in the presynaptic populations. Synaptic learning facilitates this detection, by increasing the amplitude of the transients, provided that the transition occurs from a low initial rate.

2.3 Switch between rate and temporal coding

In order to further elucidate the striking redistribution of synaptic efficacy observed in neocortical pyramidal synapses, we compared it with another kind of synaptic modification, which is usually assumed in models of learning - uniform increase in synaptic strength. We consider the synapses in an untrained state ($r = 0.1$), and increased it's strength, given by parameter E, two-fold. The resulting simulated synaptic current and recorded response of pyramidal neuron upon injecting this current, is shown in the lower panel of Fig. 3. One can see that the

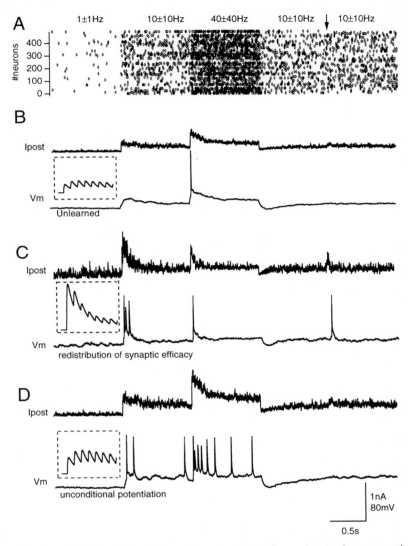

Fig. 3. Simulated effect of abrupt transitions in the discharge rates of presynaptic neurons on the postsynaptic current. On top - raster plot of spikes for a population of 500 presynaptic neurons (only the spikes of each 10th neuron are shown). The spike train for each neuron is modeled as an independent Poisson process. The rates of individual neurons are distributed randomly around the mean value with the spread of the rates which equals the mean. At the first transition, the mean rate of presynaptic neurons instantaneously increases from 1 Hz (spontaneous activity) to 10 Hz. At the second transition, the rates of all neurons are instantaneously increased 4-fold. At the third transition, the rates are decreased back to the mean of 10 Hz. Finally, at the last transition the rates remain the same on average, but individual neurons exchange their rates, so that the slowest firing neurons become the fastest, and vice versa. Below - the resulting simulated postsynaptic current, together with the response of a pyramidal neuron to which the simulated synaptic current was injected. First two panels correspond to synapses with different values of r - 0.1 and 0.68 from top to bottom. Lowest panel - the synapses are in the untrained state ($r = 0.1$) with a two fold strength increase, so that $E = 500pA$.

response of the neuron changed dramatically compared to the one shown in the previous panel, and reflects mostly the firing rate of the presynaptic population.

3 Conclusion

The presented analysis suggests that the trained state of synaptic contacts between neocortical pyramidal neurons determines the nature of the signal these synapses transmit to their targets about the activity of the presynaptic population. If the synapses are untrained, the postsynaptic current encodes the average sustained firing rate of presynaptic neurons. In the trained state, the same synapses are able to transmit a rich array of dynamics features of presynaptic population activity, such as rapid synchronized changes in activity of neuronal populations, by means of transients in the postsynaptic current. Their ability to signal the sustained activity level is limited to low rates.

The fact that the form of synaptic learning observed experimentally [2] selectively facilitates the dynamic component of interactions between pyramidal neurons, provides some support to the ideas of a neuronal code based on synchronicity [10, 11, 12] or coincidences of spikes [13, 14]. It is still possible that other forms of learning, such as an uniform increase in synaptic strength or a creation of new synaptic connections [15], could affect stationary states of neocortical networks.

The present study suggests that dynamic synaptic properties are crucial for the computational abilities of neocortical neuronal ensembles. It is therefore imperative to study the synaptic properties for other classes of neocortical neurons [1] in order to reveal the computational sequences occurring in cortical networks.

References

1. Thomson, A.M. & Deuchars, J.J. *TINS* **17**, 119-126 (1994).
2. Markram, H., *Nature*, submitted (1996).
3. Destexhe, A., Mainen, Z.F. & Sejnowski, T.J. *J. Comput. Neurosci.* **1**, 195-230 (1994).
4. Muller, D. & Lynch, G. *Proc. Natl. Acad. Sci. USA* **85**, 9346-50 (1988).
5. Kauer, J.A., Malenka, R.C. & Nicoll, R.A. *Neuron* **1**, 911-917 (1988).
6. Bekkers, J. M. & Stevens, C.F. *Nature* **346**, 724-729 (1990).
7. Schulz, P.E., Cook, E.P. & Johnston, D. *J. Physiol. Paris* **89**, 3-9 (1995).
8. Mainen, Z.F. & Sejnowski, T.J. *Science* **268**, 1503-1506 (1995).
9. Miyashita, Y. & Chang, H.S. *Nature* **331**, 68 (1988).
10. von der Malsburg, C. & Schneider, W. *Biol. Cybern.* **2**, 29-40 (1986).
11. Singer, W. *Int. Rev. of Neurobiol.* **37**, 153-183 (1994).
12. Hansel, D. & Sompolinsky, H. *J. Comp. Neurosci.*, in press (1996).
13. Abeles, M. *Corticonics*, (Cambridge Univ Press, NY 1991).
14. Hopfield, J.J. *Nature* **376**, 33-36 (1995).
15. Liao, D.Z., Hessler, N.A. & Malinow, R. *Nature* **375**, 400-404 (1995).

Controlling the Speed of Synfire Chains

Thomas Wennekers and Günther Palm

University of Ulm, Department of Neural Information Processing,
D-89069 Ulm, Germany

Abstract. This paper deals with the propagation velocity of synfire chain activation in locally connected networks of artificial spiking neurons. Analytical expressions for the propagation speed are derived taking into account form and range of local connectivity, explicitly modelled synaptic potentials, transmission delays and axonal conduction velocities. Wave velocities particularly depend on the level of external input to the network indicating that synfire chain propagation in real networks should also be controllable by appropriate inputs. The results are numerically tested for a network consisting of 'integrate-and-fire' neurons.

1 Introduction

The concept of synfire chains has been introduced by Abeles [1] in order to explain precisely correlated spike events observable in the cortex on surprisingly large time-scales of up to hundreds of milliseconds. The main idea is that highly specific spatio-temporal firing patterns occur in the brain in such a way that synchronously firing pools of neurons iteratively excite other pools, whereby a chain of activation evolves and propagates through the network. Experiments with trained awake monkeys show that the occurrence of synfire chain activation in frontal cortices is clearly correlated with behaviorally relevant events.[2] This observation relates the synfire chain idea to cognitive processes, although the explicit relation is still a matter of discussion. It is furthermore reasonable to believe that spatio-temporal excitation patterns similar to synfire chains also underly the neuronal generation of (elementary) movements. Since those can be executed at variable speed it seems promising to investigate neural network models capable of generating patterns of spike activity at a controllable speed.

Synfire chain models have been analyzed for their memory capacity, the length of chains or the fusion of several chains in a single network [4, 5, 7]. Propagation velocities of activation have only recently been studied - numerically by Arnoldi and Brauer [4] and mathematically, using integro-differential equations, by Arndt et al. [3].

Here we present a model of synfire chain activation consisting of a one-dimensional chain of locally connected *spiking neurons* with 'integrate-and-fire'-like dynamics. We derive a general exact but implicit formula for the propagation speed of travelling pulses as well as explicit approximative solutions for special cases. Background input can be used to control wave propagation speed, which also depends on the form of postsynaptic potentials and the coupling width. The theoretical results are checked numerically in section 3.

2 Theory

Since in the original synfire chain model [1] all neurons in one synfire 'node' fire synchronously, each node may be represented by a single neuron. This leads to a one-dimensional model, which we assume to be infinitely extended. The velocity of pulse-like activation patterns can then be derived similar to the work of Idiart and Abbott [6] who analyzed the propagation of excitation in extended cortical tissue. In [6] a merely macroscopic point of view is taken, but the mathematical method is largely independent of the particularly chosen single unit model. In the sequel we apply it to spiking neurons. We require that the state of a neuron is given by a single variable y representing its membrane potential. A pulse-like output z is released whenever y reaches a given threshold ϑ. After firing some mechanism should prevent the cell from firing immediately again for a certain time, but the details of implementation of such refractoriness do not matter at this point. The above requirements are satisfied by many types of artificial spiking neurons, in particular the well known 'integrate and fire'-model. If a spike generated by cell x' arrives at another cell x it evokes a post-synaptic potential (PSP) at x of a given form $g(t)$ and a height $K(x, x')$. PSP's of several cells are supposed to sum linearly. Then we can write for the time-evolution of cell x:

$$y(x,t) = I(x,t) + \int_{-\infty}^{\infty} K(x-x') \int_{t_0}^{t} g(t-t')z\left(x',t'-\frac{|x-x'|}{c}\right) dt'dx' \ . \quad (1)$$

For each neuron x the temporal output pattern of spikes $z(x,t)$ in (1) has to be computed selfconsistently from $y(x,t)$ according to the threshold mechanism described above. $I(x,t)$ is some external input into the network and the double integral in (1) represents neuronal interactions. It has the form of a spatio-temporal convolution, describing the linear summation of PSP's. Note that the spatial coupling kernel K is chosen shift-invariant, $K(x, x') = K(x - x')$, though not necessarily isotropic, e.g. $K(x) = K(-x)$. Furthermore, we have included a finite propagation velocity c of signals in (1). This velocity represents the finite speed of action potentials along axonal fibres and should not be confused with the speed of wave propagation computed below.

Equation (1) supplemented with a suitable refractory mechanism describes the time-evolution of general excitation modes of the network. Now, we set $I(x,t) = I(< \vartheta)$ and consider the special case that a solitary wave of spikes propagates to the right along the chain with velocity v. Thus we impose

$$z(x,t) = \delta\left(t - \frac{x}{v}\right) \ , \quad (2)$$

where δ is the Dirac pulse and the time origin is chosen such that the wave arrives at $x = 0$ at time $t = 0$. But the cell at $x = 0$ *can* only fire if it reaches threshold. Therefore selfconsistency requires $y(0,0) = \vartheta$. Then, inserting (2) into (1) and evaluating the integrals for $x = t = 0$ leads to

$$\vartheta = I + \int_{0}^{\infty} K(x)g(\alpha x)dx \quad \text{with} \quad \alpha = \frac{1}{v} - \frac{1}{c} \ . \quad (3)$$

This is an implicit equation determining the unknown parameter α and therefore the wave velocity v. Note, that c, the axonal conduction velocity, is contained in (3) only through the parameter α. If we have solved (3) for a particular choice of K, g, I and ϑ we can easily compute v for every c.

Equation (3) has a simple physical meaning: assume for the moment that the spike train travels with an arbitrary fixed speed. Then it evokes a compound PSP at $x = 0$ which for suitable K and g will be an unimodal function of time added to the background I. The compound PSP depends on the propagation speed: it is larger, when the wave travels faster, since then individual PSP's superimpose more synchronously and the compound PSP is less spread out. Now, by construction, the integral in (3) gives the value of the PSP at time $t = 0$, therefore (3) determines exactly *that* speed (respectively α) for which the membrane potential intersects the threshold at $t = 0$. (Strictly speaking there are generically two intersections: one on the leading and one on the trailing edge of the PSP. Of course, only the first is physically meaningful.)

Note that wave velocities derived from (3) will depend on I. Background input acts as a predepolarization; hence membrane potentials are nearer to the firing threshold and the compound PSP reaches threshold more quickly. Furthermore, since typically the integral in (3) will be bounded for arbitrary α, there exists a minimum I for which (3) can have solutions. The biological meaning again is simple: if we inhibit cortical tissue sufficiently strong, then it will show no reverberating activation since the overall recurrent excitatory efficacy alone becomes too small.

We proceed by specifying particular choices for g. Typical synaptic interaction functions are of the form

$$g_n(t) = \frac{1}{n!\tau^{n+1}}(t - \Delta)^n e^{-(t-\Delta)/\tau}\Theta(t - \Delta) \, , n = 0, 1, \cdots . \tag{4}$$

Here, $\Theta(x)$ is the Heavyside step function, which we introduced to enforce causality. $\Delta \geq 0$ takes finite synaptic transmission delays into account, and $\tau > 0$ defines the time-scale of the synaptic response. Using (4) we can compute an approximative solution of (3). To this end, we assume, that K is localized in space and the wave velocity is large, e.g. α small, which will be particularly the case when the cells are near threshold. Furthermore we assume $\Delta \ll \tau$. For fixed n and $k < n$ all (right handed) derivatives $g_n^{(k)}(\Delta+)$ vanish. Then we can expand g in (3) to lowest non-vanishing order yielding the following condition for α:

$$\kappa := \vartheta - I = g_n^{(n)}(\Delta+) \int_0^\infty K(x)(\alpha x - \Delta)^n \Theta(\alpha x - \Delta)dx \, . \tag{5}$$

Furthermore we choose the rectangular coupling kernel

$$K(x) = K_0\Theta(a - x)\Theta(a + x) \, , \tag{6}$$

where K_0 and a measure strength and range of the synaptic interactions. The choice is somewhat arbitrary and in particular applications one may solve (5) or

(3) numerically with other choices. Here, the simple form of (6) together with (5) allows us finally to compute explicit formulas for the wave velocity

$$n = 0: \quad \alpha = \frac{1}{v} - \frac{1}{c} = \frac{\Delta}{a}\left(1 - \frac{\tau\kappa}{aK_0}\right)^{-1} \tag{7}$$

$$n = 1: \quad \alpha = \frac{\Delta}{a} + x\left(1 + \sqrt{1 + 2\frac{\Delta}{ax}}\right), \text{ with } x = \frac{\tau^2\kappa}{a^2 K_0}. \tag{8}$$

3 Simulations

We tested the results of the previous section by simulating a long discrete chain of $N = 512$ 'integrate and fire' neurons. Beside implementing the time-evolution of membrane potentials $y(i, t), i = 1, 2, \cdots N$, given by Eqn. (1), the following refractory mechanism was chosen: after reaching the threshold value $\vartheta = 1$ the membrane potential of neuron i was reset to zero and exponentially relaxed back to the input value I. This was effectively done by adding $-\exp(-(t - t_f)/T)$ to $y(i, t)$ after a spike at $t = t_f$ with a time constant T long enough to prevent cells from firing more than once during each simulated wave. The PSP-function and coupling kernel were chosen as in (4) and (6) with $n = 0$ and $n = 1; \tau = 10.0, \Delta = 1.0, K_0 = 0.1$ and $a = 50$. The axonal conduction velocity was $c = 300.0$. (Here and below, all velocities are measured in units of cells/time.) Waves were evoked by forcing a sufficient number of cells at one end of the chain to fire.

Figure 1 combines results from theory and simulations. Figure 1a displays the dependency of the wave velocity on $\kappa/K_0 = (\vartheta - I)/K_0$ for $n = 0$ and $n = 1$. $\kappa := \vartheta - I$ measures the distance between the external (subthreshold) Input and the firing threshold of cells. In Fig. 1a we have scaled κ by K_0 since only the fraction κ/K_0 enters in (5), (7), and (8). Thus, if we change the synaptic strength K_0 the curves in Fig. 1a still remain valid.

The solid lines represent exact solutions derived from (3) by numerical root-finding (bisectioning). Dashed lines display the approximations (7) and (8), which fit the exact curves quite well. Asterisks indicate simulation results.

First observe that the velocities for $n = 1$ are appreciable smaller than for $n = 0$. This, because g jumps for $n = 0$ discontinuously to its maximum value as soon as the transmission delay Δ is over, whereas the PSP's for $n = 1$ grow slowly and reach a stronger influence on succeeding cells much later; thereby delaying the wave. Figure 1a shows further, that the velocity increases with increasing input or coupling strength as it should be expected, and that there exists a critical input necessary for the stable development of synfire chains (dotted vertical lines). This is significantly higher for $n = 1$ because then PSP's are broad and relatively small; therefore it is harder for them to reach threshold.

If I tends to ϑ Eqns. (7) and (8) both give a limiting α of .02. For $c = \infty$ this would yield a maximum velocity of $v = 50$. The value becomes somewhat lowered (to ≈ 42) due to the finite value of c (cf. the equation for α in (3)). Since biological conduction velocities are large the maximum cortical wave speed will

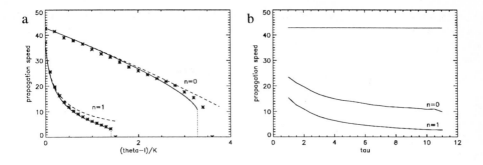

Fig. 1. Wave velocities for different parameter variations. Solid: exact solutions; dashed: approximations; asterisks: simulation results. (See text for details.)

probably be mainly determined by synaptic transmission delays as long as only localized pools of cells are considered.

The PSP-functions (4) are characterized by two time-constants Δ and τ, representing the synaptic transmission delay and time scale of PSP variations. Combining Δ and τ with the range a of the coupling kernel yields two characteristic velocities $a/\Delta = 50$ and $a/\tau = 5$. The first, as mentioned above, is an upper bound for the highest possible speed. It can be shown that the second gives the order of the minimum speed if $\Delta \ll \tau$ and $a/\Delta \ll c$. Then these two velocities approximately determine the range of possible velocity variations. Since rise-times of real synaptic PSP's are probably not much larger than synaptic transmission delays [1] it is opportune to ask for the range of speed-modulation when $\Delta \approx \tau$. This is considered in Fig.1b, where τ has been varied for fixed $\Delta = 1.0$. Displayed are the maximum speed of 42 (which is independent of τ, cf. (7) and (8)) and the minimum speeds derived from (3) by root-finding. As can be seen the possible range of variation only weakly depends on τ.

Wave velocities also depend on the coupling width a. Changing a leaves the qualitative results intact, but absolute values and ranges of speed variation may change. Furthermore, in discrete chains with small a more complex dynamical phenomena can occur (e.g. discontinuities and hysteresis in $v(I)$); lack of space prevents us from going into details. Nearest neighbor couplings $a = 1$, as in the original synfire chain model [1], together with $n = 0$ present a somewhat singular case, insofar as then firing times between nodes are always Δ independent of I. However, any finite slope of g abolishes this effect.

4 Discussion

Under general conditions we have derived the propagation speed of pulse-like waves in networks of spiking neurons. The velocities show a marked dependency on the level of external input. By this mechanism wave speeds should also be controllable in real cortices. However, our results present new problems concerning both of the initially mentioned phenomena, synfire activity in frontal cortex and movement control.

The synfire idea has originally been developed to explain highly precise correlations of spike events, occuring with an accuracy of less than 3ms for temporal delays between participating spikes of up to hundreds of milliseconds. Changing velocities of synfire activity contradict such precise timing and the repetition of exact spike patterns. Only the constancy of background activation over long time intervals could explain such a pattern. This possibility is supported by [2].

The opposite reasoning applies to movement control. Looking at Fig. 1b reveals that the speed can be modulated roughly by a factor of 2-5, otherwise time-constants have to be chosen in an unreasonable way. This means, it is not very likely that the proposed mechanism can account for the whole dynamical range of speeds with which subjects are able to conduct specific movements. Nonetheless this mechanism might still help to support fine-tuning of submodules in a more complex architecture of movement control.

Let us finally mention, that the proposed mechanism also occurs in other architectures than our abstract model (cf.[4, 3]). Small random variations in single cell or PSP-parameters ($\tau, \vartheta, \Delta, K$) can be considered theoretically yielding averaged forms of Eqn. (3) (see also [1], chapter 7). Similar calculations furthermore apply to replay speeds in associative memories which store pattern sequences.

Acknowledgement

This work has been supported by DFG, Pa 268/8-1.

References

1. Abeles, M.: Corticonics: neural circuits of the cerebral cortex. Cambridge University Press, Cambridge UK, 1991
2. Abeles, M., Bergman, H., Gat, I. Meilijson, I., Seidemann, E., Tishby, N., and Vaadia, E.: Cortical Activity Flips Among Quasi Stationary States. PNAS **92** (1995) 8616–8620
3. Arndt, M., Erlhagen, W., and Aertsen, A.: Propagation of Synfire Activity in Cortical Networks - a Dynamical Systems Approach. In: Lappen, B. and Gielen, S.: Neural Networks: Artificial Intelligence and Industrial Applications. Proceedings of the Third Annual SNN Symposium on Neural Networks. Springer, Berlin, 1995
4. Arnoldi, H.M.R., Brauer, W.: Synchronization without oscillatory neurons. Biol.Cybern. **74** (1996) 209–223
5. Bienenstock, E.: A model of the neocortex. Network **6** (1995) 179–224
6. Idiart, M.A.P., Abbott, L.F.: Propagation of excitation in neural network models. Network **4** (1993) 285–294
7. Palm, G.: On the internal structure of cell assemblies. In Aertsen, A. (ed) Brain Theorie, pp 261-271. Elsevier Science Publishers, Amsterdam, 1993

Temporal Compositional Processing by a DSOM Hierarchical Model

Claudio M. Privitera and Lokendra Shastri

International Computer Science Institute
1947 Center St. - Suite 600 - Berkeley, California 94704-1198

Abstract. Any intelligent system, whether human or robotic, must be capable of dealing with patterns over time. Temporal pattern processing can be achieved if the system has a short-term memory capacity (STM) so that different representations can be maintained for some time. In this work we propose a neural model wherein STM is realized by leaky integrators in a self-organizing system. The model exhibits *compositionality*, that is, it has the ability to extract and construct progressively complex and structured associations in an hierarchical manner, starting with basic and primitive (temporal) elements.

1 Compositionality and temporal pattern processing

Many cognitive processes must be *compositional*, that is, they must be capable of extracting and constructing progressively complex and structured representations in a hierarchical manner, starting with simple primitive elements [1]. For example, visual recognition of complex objects and events requires the rapid assembly of diverse information about the subcomponents of the object or event. Another phenomena where compositionality plays an ubiquitous and critical role is language where a sequence of acoustic input is mapped to a complex and high dimensional description of events and states. A central requirement of compositionality and hierarchical processing is the maintenance and propagation of dynamic *binding* between appropriate components of a multi-level and distributed representation [5].

Recent neurological data highlights the potential role of the temporal structure of neural activity in the expression and propagation of dynamic bindings. A particularly promising hypothesis is that all information pertaining to a single entity is bound together as a result of the synchronous firing of the various nodes encoding information about this entity [7]. Thus compositionality, binding, and temporal pattern processing are interrelated and seem to play an essential role in cognition.

In this work, compositionality is achieved by a hierarchical model of Self-Organizing Maps (SOM) and the temporal processing ability is achieved by exploiting neurons capable of supporting short-term memory (STM). Each map in the hierarchy (which we refer to as Dynamic SOM — DSOM) represents a specific abstraction of information expressed in maps that are lower in the hierarchy. The unification among different maps is realized by a synchronization of the activity of appropriate nodes across these maps.

Due to space limit, this paper present only a condensed description of the whole model. A complete description can be found in [2].

2 Dynamic Self-Organization computational Maps

Temporal processing requires a STM capacity in order to represent different temporal components over a window of time. We use an extended form of SOM that exhibits the necessary STM property. Unlike SOM, the extended model, D-SOM, uses a dynamic version of a classical neuron model.

In the D-SOM model, the activation of an element in the map is represented by a membrane potential $P_i(t)$ which in turn is a function of the input activation function $u_i(\epsilon(t)))$ which measures the degree of match between the element's weights vector and the temporal input vector $\epsilon(t)$ (for example, $u_i(\epsilon(t))$ could be the softmax function). $P_i(t)$ takes also into account the dynamic properties of the biological membrane of a cell which are approximated by a generic RC circuit.

Note that the input activation function $u_i(\cdot)$ of a map element is a function of time $u_i(\epsilon(t)))$. Thus the membrane potential of an element is correlated with the temporal evolution of its input and the element can not only analyze a static input vector ϵ, it can also perform a temporal integration of the function $u_i(\epsilon(t))$ (hence the name, Leaky Integrator element (see for example [4])). If this temporal integration depends on the occurrence of certain spatio-temporal features/subevents in the evolving temporal input, then the activation of the element indicates the occurrence of an external event composed of these subevents. Thus one can view each element as a recognizer of a complex event.

In view of the above, each element is composed of a set of sub-elements — classically called *taps* [3] — each of which represents different critical points in the temporal evolution of the input vector. Specifically, if an element recognizes an event composed of m temporal features, it consists of m distinct taps. If the input vector is k-dimensional, $\epsilon(t) \in \Re^K$ then, each tap is characterized by a weight vector $\pi \in \Re^K$. At each time instance, the input vector $\epsilon(t)$ is processed by all the taps and the membrane potential of the element is evaluated on the basis of the tap outputs:

$$\frac{dP_i}{dt} = -\frac{P_i(t)}{\tau_i} + \sum_{j \in Act_i(t)} u_{ij}(\epsilon(t)) \qquad (1)$$

where, $u_{ij}((\epsilon(t))$ is the activation function of the $j - th$ taps of the $i - th$ neuron and τ_i is the time-constant of the neuron.

The term $Act_i(t)$ represents the set of active taps of the $i - th$ neuron at a generic time instant. As soon as the activation function of a tap exceeds a prefixed threshold it is inhibited and excluded from the set of active taps. Different elements may have different activation strategies for taps. In some cases, all taps may be active initially. In other cases, taps may become active one at a time, and hence, respond to temporal features occurring in a specific order. This

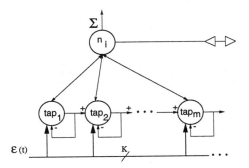

Fig. 1. The general structure of partitioned leaky integrator neuron. The taps can be organized by an input layer of neurons connected each other in a way representing the activation policy of the model: once a generic tap matches the input, it inhibits itself and sequentially enables the next tap. The integrating neuron n_i keeps memory of all the sequence of matching.

is achieved by connecting the taps in a sequence as shown in Figure 1. Notice that taps are represented by neurons, $tap_1, \ldots tap_m$, that read a common input vector. A tap can either be in an enabled state, wherein it actively processes its input, or it can be in a disabled state. Initially only the first tap is enabled. When an enabled tap receive a matching input signal, it crosses its threshold and fires, enables its successor tap, and at the same time inhibits itself thereby entering a disabled state.

The neuron $n_i - th$ integrates the tap outputs and maintains a continuous potential according to Eq. (1). This neuron eventually fires if all the taps have been activated by the temporal input. The time constant τ_i governs the delay limit between two sequential temporal features recognizing and the STM of the integrating neuron.

The learning of tap weights can be done in a traditional Hebbian manner exploiting for each iteration an unique vector $\pi \in \Re^{K \times m}$. In [3] an alternative learning method has been presented where the taps are learned in real time using a sequence of input vectors. In [8] another method is proposed where taps are pre-fixed and the learning process only effects the connections between taps and neuron.

3 A visual example: multi-classification of input trajectories

The general behavior of the model can be studied in the context of a visual task where the system must analyze the motion of multiple points in a two dimensional space and recognize the occurrence of certain coordinated patterns of movement. An example of such a coordinated pattern of movement is shown in Fig. 3b where two points are moving along diverging and converging paths.

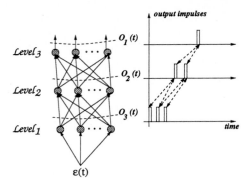

Fig. 2. A schematization of a three-level architecture: for each level, the sequence of firing neurons represents the input impulses sequence $O_i(t)$ to the upper level. Binding is represented by *phasing* relationship between the input and the output sequence of impulses. This is a bidirectional relationship.

This task involves three levels of representation: the position of the two points in the visual scene at time t, the relationship between the local trajectories of the two points at the same time and the evolution of this relationship over time. Consequently, three levels are defined in the DSOM (Figure 2 shows a schematization of the architecture). A brief description of the system follows.

Figure 3 shows in details the third level of the architecture after the learning process. At this level, the map consists of to a 10×10 grid of elements. Each element is composed of four distinct taps, each representing a particular combination of the two input trajectories. More specifically, each tap is a two-dimensional vector and represents the mean and the difference of the angles of the two input trajectories. For example, two moving points that follow the *zigzag* pattern shown in Fig. 3b are characterized by a set of taps that respond maximally if the angular mean remains constant and the sign of the angular difference alternates.

The first level is composed of elements with a non-symmetric two-dimensional Gaussian filter whose tilt represents the angle of the input trajectory. The input scene consists of one or more dots moving in a two-dimensional space. All movements are sampled at various time interval and for each interval the most active elements are allowed to fire. A global competitive mechanism among the elements of a map (see for example [6]) together with an absolute refractory period (ARP) [5] [2] prevents elements that represent different trajectories from firing at the same time.

The map at level two analyses all the possible pairs of impulses from the first level which represent different trajectories, and produces output impulses representing the mean and the difference of the corresponding angles. Finally, the elements at the third level are trained to recognize certain patterns of global coordination in the two trajectories. Since the patterns being learned involve four segments (see Fig. 3a) the elements in this level have four taps. All the maps are trained separately using the softmax learning rule.

Figure 3.b shows a test pattern consisting of a *zig-zaging* five-stroke sequence. This pattern forms the input to the first level. The arrows mark the instants at which the trajectories were sampled. Figure 3.c shows four temporal snapshot of the activity level of the third level. The small x's identify the elements: the gray circles around the elements represent activation levels. A black circle indicates that an element has achieved a firing threshold. It is worth observing that the two elements at level three corresponding to a *zig-zag*-like pattern fire at $t = 182$ and $t = 242$. These are the times when the system has completed an observation of the appropriate patterns.

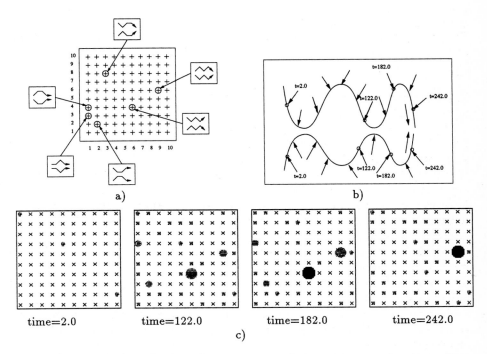

a)

b)

time=2.0   time=122.0   time=182.0   time=242.0

c)

Fig. 3. a) DSOM after the learning process: each input temporal pattern (two points moving with a specific correspondence) is associated with the best responding neuron. b) the test input pattern: the bold arrows represent the direction of the two movements and the small arrows the sampling instants. c) four different snapshot of the third level neuronal activity during observation of the two input movements.

4 Conclusions

The ability of DSOM to process temporal patterns has been used to implement a compositional processing model where primitive descriptions of time varying situations can be processed in a hierarchical manner to produce progressively abstract and structured descriptions. The model represents information as a sequence of impulses. It can analyze an input consisting of time varying sequence of impulses and generate an output sequence of impulses, wherein each impulse represents the occurrence of certain patterns in the input sequence.

The temporal properties of the map elements enable the system to maintain correct bindings between the input and output impulses: in fact, even though each element detects an input temporal pattern at the end of its occurrence, the firing of each tap in the element is synchronized with the occurrence of the appropriate sub-event that makes up the input temporal pattern.

The proposed paradigm has been tested in general situations where multiple moving dots in an input visual image generate different classes of events at each level in the hierarchy. In future work, we plan to investigate the use of DSOM structures to address additional problems involving compositionality that arise in language, handwriting recognition, and rapid reasoning.

References

1. E. Bienenstock and S. Geman. Compositionality in neural systems. In M.A. Arbib, editor, *The Handbook of Brain Theory and Neural Networks*, pages 223–226. The Mit Press, 1995.
2. C. M. Privitera and L. Shastri. A DSOM hierarchic model for reflex processing: a visual tracking example. Technical Report 96-010, International Computer Science Institute, 1996.
3. C.M. Privitera and P. Morasso. The analysis of continuous temporal sequences by a map of sequential leaky integrators. In *Proceedings of IEEE International Conference on Neural Networks*, volume 5, pages 3127–3130, Orlando, Florida, 1994.
4. M. Reiss and J.G. Taylor. Storing temporal sequences. *Neural Networks*, 4:773–787, 1991.
5. L. Shastri and V. Ajjanagadde. From simple associations to systematic reasoning: A connectionist representation of rules, variables and dynamic bindings using temporal synchrony. *Behavioral and brain sciences*, 16:417–494, 1993.
6. C. von der Malsburg. Self-organization of orientation sensitive cells in the striata cortex. *Kibernetik*, 14:85–100, 1973.
7. C. von der Malsburg. Am i thinking assemblies? *Brain Theory*, 1986.
8. D.L. Wang and M.A. Arbib. Timing and chunking in processing temporal order. *IEEE Trans. on System, Man, and Cybernetics*, 23(4):993–1009, 1993.

A Genetic Model and the Hopfield Networks

A. Bertoni, P. Campadelli, M. Carpentieri, G. Grossi

Dipartimento di Scienze dell'Informazione
Università degli Studi di Milano
via Comelico, 39 - 20135 Milano Italy

Abstract. In this paper a genetic model is presented and the dynamics in the thermodynamic limit is derived. Analogies and differences with neural networks are discussed and attractors of the genetic model are characterized as equilibria points of Hopfield's networks. The neural network and the genetic system are experimentally compared as approximate algorithms for the MAX-CUT problem.

Keywords: Hopfield's networks, genetic algorithms, optimization.

1 Introduction

Genetic algorithms are probabilistic search algorithms inspired by mechanisms of natural selection. They have received considerable attention because of their applications to several fields as optimization, adaptive control and others [5], [10], [11], [12]. By simplifying natural laws, genetic algorithms simulate reproductive processes over a population of individuals in arbitrary environment.

In this paper we present a model that preserves the main properties of classical genetic algorithms but it is simpler to analyze. We give a formal description through a homogeneous Markov chain [3], [6], [13] and derive the equations of the deterministic dynamics in the thermodynamic limit, that is for large populations.

We discuss the case of quadratic fitness functions; this case is particularly interesting in combinatorial optimization since the problem of maximizing quadratic functions is **NP**-hard and many natural problems can be polynomially reduced to it [4].

Hopfield's networks are a useful tool for the qualitative analysis of the genetic system; in fact we prove that the genetic system attractors in $\{0, 1\}^l$ are the equilibrium points of the a discrete Hopfield's network whose energy is the fitness function. As far as the system trajectories is concerned, we derive an approximate solution of the equations which recalls the dynamics of neural networks with long-time connections [8]. We discuss analogies and differences and note that simulation results show that the genetic system has good stabilizing properties.

The performances of the genetic algorithm and of discrete Hopfield's networks have been experimentally compared for the MAX-CUT problem; MAX-CUT is one of the **NP**-hard combinatorial problem for which the networks are a good approximation algorithm [1]. Preliminary results show that the performances are comparable, though the network behaves slightly better.

2 The Genetic Model

The genetic model we propose is described by a Markov chain which depends on a fitness function $f : \{0,1\}^l \to \mathbf{N}$. The model preserves the classical principle that genetic material of individuals is recombined with a probability proportional to their fitness and depends on similarities among chromosomes. The states of the chain are populations, that is vectors $\Phi = (\Phi_1, \ldots, \Phi_n)$ of n binary words of length l, denoted by $\Phi_i = \Phi_{i1} \cdots \Phi_{il} \in \{0,1\}^l$ $(1 \leq i \leq n)$.

Let Φ be a fixed population; the fitness f_Φ of Φ is defined by

$$f_\Phi = \sum_{i=1}^{n} f(\Phi_i)$$

while the contribution $f_{k\Phi}$ of the words with 1 in position k to the fitness of Φ is

$$f_{k\Phi} = \sum_{i=1}^{n} \Phi_{ik} f(\Phi_i) \ (1 \leq k \leq l).$$

The Markov chain is then described by the stochastic matrix $(A_{\Phi\Phi'})$, where $A_{\Phi\Phi'}$ is the probability of generating the population Φ', and Φ'_{ik} $(1 \leq i \leq n, 1 \leq k \leq l)$ are the results of n independent Bernoulli trials each with probability $p_{k\Phi} = \frac{f_{k\Phi}}{f_\Phi}$ of obtaining 1 (if $f_\Phi = 0$ we set $p_{k\Phi} = \sum_{i=1}^{n} \frac{\Phi_{ik}}{n}$). Notice that our genetic model differs from that ones found in the literature, where new populations are generated by selecting chromosomes in pairs, each with a probability proportional to its fitness, and, then, by swapping the sequences of genes starting from a random position.

The genetic model is simulated by the following algorithm:

Input: $\Phi :=$ initial population Φ_0;

      while (not term cond) **do** $\begin{cases} \text{generate } \Phi' \text{ with probability } A_{\Phi\Phi'}; \\ \Phi := \Phi'; \end{cases}$

Output: $f(\Phi)$.

For $n \to \infty$ the stochastic model described above can be approximated by a deterministic one. Let μ_Φ be the probability distribution over $\{0,1\}^l$ such that:

$$\mu_\Phi(x_1, \ldots, x_l) = \prod_{k=1}^{l} g_k(x_k),$$

where $g_k(1) = p_{k\Phi}$ and $g_k(0) = 1 - p_{k\Phi}$. Consider the random variables f and f_k over the probability space $\langle \{0,1\}^l, \mu_\Phi \rangle$, where $f_k(x_1, \ldots, x_l) = x_k f(x_1, \ldots, x_l)$, and denote their expectations by $E_{\mu_\Phi}(f)$ and $E_{\mu_\Phi}(f_k)$. Let M be the maximum value that the random variable f can assume. By applying techniques on uniform convergence of sample mean of i.i.d. bounded random variables to their expectation, we are able to prove the following theorem.

Theorem 1. *Let* $n \geq \frac{36}{\epsilon^2} \log \frac{l}{\delta} \left(\frac{M}{E_{\mu_\Phi}(f)} \right)^2$. *If at time* t *the system is in the state* Φ, *then the state* Φ' *at time* $t+1$ *is such that for all* k

$$\left| p_{k\Phi'} - \frac{E_{\mu_\Phi}(f_k)}{E_{\mu_\Phi}(f)} \right| \leq \epsilon$$

with probability at least $1 - \delta$.

Consider now pseudo boolean quadratic fitness functions, of the following form:

$$f(x_1, \ldots, x_l) = \sum_{i,j} w_{ij} x_i x_j + \sum_i \lambda_i x_i, \tag{1}$$

where, without loss of generality, $w_{ij} = w_{ji}$ and $w_{ii} = 0$. As a consequence of Theorem 1 and by (1) we can conclude that with high probability

$$p_{k\Phi'} \approx p_{k\Phi} \frac{\sum_{i \neq k} p_{i\Phi} \left(\sum_{j \neq k} w_{ij} p_{j\Phi} + \lambda_i \right) + 2 \sum_i w_{ki} p_{i\Phi} + \lambda_k}{\sum_{i,j} w_{ij} p_{i\Phi} p_{j\Phi} + \sum_i \lambda_i p_{i\Phi}}.$$

Therefore, for $n \to \infty$ the stochastic genetic system can be approximated by the iterative deterministic system whose states are vectors $(p_1, \ldots, p_l) \in [0,1]^l$ and whose dynamics is described by the equation

$$p_k(t+1) = p_k(t) \frac{\sum_{i \neq k} p_i(t) \left(\sum_{j \neq k} w_{ij} p_j(t) + \lambda_i \right) + 2 \sum_i w_{ki} p_i(t) + \lambda_k}{\sum_{i,j} w_{ij} p_i(t) p_j(t) + \sum_i \lambda_i p_i(t)}, \tag{2}$$

where $1 \leq k \leq l$.

3 Analysis and Comparison with Neural Models

Equation (2) can be rewritten in the form:

$$\Delta p_k = p_k (1 - p_k) \frac{b_k(p)}{f(p)}, \tag{3}$$

where

$$f(p) = \sum_{i,j} w_{ij} p_i p_j + \sum_i \lambda_i p_i \quad \text{and} \quad b_k = 2 \sum_i w_{ki} p_i + \lambda_k.$$

The comparison of this dynamic system with neural algorithms used to solve difficult combinatorial optimization problems is interesting. The following theorem relates the qualitative behaviour of the iterative system described by equation (2) with the behaviour of discrete Hopfield's networks with sequential updating and the fitness $f(x_1, \ldots, x_l)$ as energy function.

Theorem 2. *The fitness function* $f : [0,1]^l \to N$ *of the iterative system (2) has its maximum value on elements of* $\{0,1\}^l$. *Every element* $x \in \{0,1\}^l$ *is an equilibrium point for the iterative system; if* $b_k(x) \neq 0$ *for all* k. *then* $x \in \{0,1\}^l$ *is a stable attractor for the system if and only if it is an equilibrium point for the Hopfield's network.*

As a consequence, the global maximum of the fitness is an attractor of the genetic system. Moreover, the network can be used as an algorithm to find the attractors.

As far as the behaviour of the trajectories is concerned, the following considerations show analogies and differences between neural models and the genetic one. Equation (3) can be rewritten as:

$$\frac{\Delta p_k}{p_k(1 - p_k)} = \frac{b_k}{f},$$

from it we obtain:

$$\Delta(\lg \frac{p_k}{1 - p_k}) \approx \frac{b_k}{f}$$

and

$$p_k(t+1) \approx \sigma_a \left(\sum_j w_{kj} \left(\frac{b_j(p(t))}{f(p(t))} + \cdots + \frac{b_j(p(0))}{f(p(0))} \right) \right), \tag{4}$$

where σ_a is the sigmoidal function $\sigma_a(x) = \frac{Ae^x}{1+Ae^x}$ and $A = \frac{p_k(0)}{1-p_k(0)}$.

Equation (4) recall networks with so called "long-time connections" [8]. In such networks the inputs $h_k(t)$ of the unit k at time t is given by:

$$h_k(t) = \sum_j (w_{kj}^S S_j(t) + w_{kj}^L \overline{S}_j(t)),$$

where w_{kj}^S and w_{kj}^L are respectively the short-time and the long-time connections, $S_j(t)$ is the state of unit j at time t and $\overline{S}_j(t)$ is a "memory trace" of the past values of S_j defined by:

$$\overline{S}_j(t) = \sum_{k=0}^{t} s(t-k)S_j(k).$$

It is required that $\sum_{k=0}^{t} s(k) = 1$. In equation (4) the short-time connections are missing, long-time connections are symmetric and the kernel s is given by $s(k) = \frac{1}{f(t-k)}$; unlike the long-time connections model, $s(k)$ explicitly depends here on the dynamics and $\sum_{k=0}^{t} s(k)$ is different from 1. The symmetry of the connections can explain the good stabilization properties observed in simulations experiments.

4 Simulation Results

To compare the performance of discrete Hopfield's networks with that of the deterministic iterative model described by equation (3) we have chosen the problem MAX-CUT, i.e. that of partitioning the nodes of an arbitrary graph G into two subsets A and $\overline{A}$ in such a way that the number $\omega(A)$ of edges with one endpoint in A and the other in $\overline{A}$ is maximum. The decision version of MAX-CUT is well known to be **NP**-complete, therefore polynomial time algorithms can find only approximate solutions, unless $\mathbf{P} = \mathbf{NP}$.

Given an instance of MAX-CUT, that is an arbitrary undirected graph G, the corresponding energy function of a Hopfield's network is $E = \sum_{i,j} w_{ij} x_i (1 - x_j)$, where (w_{ij}) is the adjacency matrix of G. Moreover, the *relative error* $\epsilon(G, A)$ of an approximate solution A for the instance G is defined by $\epsilon(G, A) = \frac{\omega^* - \omega(A)}{\omega^*}$, where ω^* is the cost of an optimal solution.

The relative error of the neural algorithm is at most 0.5. and only recently a polynomial time approximation algorithm, based on semidefinite programming, as been proposed with smaller relative error [7]. Nevertheless this new algorithm has a complex structure and it is computationally expensive. As for lower bounds, recent results on the approximability of **NP**-hard optimization problems states that for the MAX-CUT problem there exists $\epsilon > 0$ such that any polynomial time approximation algorithm exhibits a relative error at least ϵ.

Some preliminary results obtained by the network and by the iterative system with the same fitness function are presented in Table 1. Both algorithms have been tested on graphs of different dimensions (30 and 50 nodes). For each dimension 100 random graphs with probability 1/4 and 1/7 have been generated. The mean size of the cuts found by the networks and by the iterative systems are reported in columns 2 and 3 respectively of the table 1.

Algorithm	G_{30}	G_{50}
Neural	79.76	209.43
Genetic	79.42	208.63

Table 1. Mean size of cuts found by the two algorithms.

We observe that the results are comparable; the network performs slightly better with 90% confidence level.

References

1. M. A. Alberti, A. Bertoni, P. Campadelli, G. Grossi, R. Posenato, *A Neural Algorithm for MAX-2SAT: Performance Analysis and Circuit Implementation*, to be published on Neural Networks.

2. G. Ausiello, P. Crescenzi, M. Protasi, *Approximate Solution of NP Optimization Problems*, Theoretical Computer Science, Vol. 150, 1995.

3. A. E. Eiben, E. H. L. Aarts, K. M. van Hee, *Global Convergence of Genetic Algorithms: a Markov Chain Analysis*, Parallel Problem Solving from Nature, 1st Workshop PPSN I, Berlin, 1991.

4. M. R. Garey, D. S. Johnson, *Computers and Intractability. A Guide to the Theory of NP-Completeness*, W. H. Freeman & Co., San Francisco, 1979.

5. D. E. Goldberg, *Genetic Algorithms in Search, Optimization and Machine Learning*, Addison-Wesley, Reading, MA, 1989.

6. D. E. Goldberg, P. Segrest, *Finite Markov Chain Analysis of Genetic Algorithms*, Genetic Algorithms and their Aplication: Proceedings of the Second International Conference on Genetic Algorithms, pp. 1-8, 1987.

7. M. Goemans, D. Williamson, *.878 approximation algorithms for the max cut and max 2sat*, Proceedings of the twenty-sixth annual symposium on the theory of computing, pag. 422, 1994.

8. J. Hertz, A. Krogh, R. J. Palmer, *Introduction to the Theory of Neural Computation*, Addison-Wesley, 1991.

9. J. J. Hopfield, *Neural networks and physical systems with emergent collective computational abilities*, Proceedings of the National Academy of Sciences 79, pp. 2554 − 2558, 1982.

10. J. H. Holland, *Induction, Processes of Inference, Learning and Discovery*, MIT Press, Cambridge, 1989.

11. J. H. Holland, *Adaptation in Natural and Artificial Systems*, MIT Press, Cambridge, 1992.

12. J. R. Koza, *Genetic Programming*, MIT Press, Cambridge, 1992.

13. A. E. Nix, M. D. Vose, *Modelling Genetic Algorithms with Markov Chains*, Annals of Mathematics and Artificial Intelligence, Vol. 5, pp. 79-88, 1992.

Desaturating Coefficient for Projection Learning Rule

Dmitry O. Gorodnichy*

Dept. of Computing Science, University of Alberta, Edmonton, AB, Canada T6G 2H1
Email: dmitri@cs.ualberta.ca

Abstract. A Hopfield-like neural network designed with projection learning rule is considered. The relationship between the weight values and the number of prototypes is obtained. A coefficient of self-connection reduction, termed the desaturating coefficient, is introduced and the technique which allows the network to exhibit complete error correction for learning ratios up to 75% is suggested. The paper presents experimental data and provides theoretical background explaining the results.

1 Introduction

Since 1982, when Hopfield introduced a neural network (NN) based on the spin glass analogy, the question of its capacity and error correction capability has been discussed in earnest. The first result obtained was that for the Hebbian (outer-product) learning rule the memory capacity of the network can not exceed 14% of the number of neurons (e.g. see [2]). Then, in 1985 Personnaz *et al.* introduced a new learning mechanism called Projection Learning Rule, for which in 1986 they [7] showed that the network exhibits error correction until M becomes of the order $N/2$ (M is the number of prototypes, N – the number of neurons).

This paper makes a further contribution to the investigation of the behaviour of the NN designed with the Projection Learning Rule. The relationship between the weight coefficients and the number of prototypes is studied. It is shown that introducing the coefficient of self-connection reduction, termed the desaturating coefficient, improves considerably the retrieval capability of the network. More specifically, Section 2 gives the description of the model under the consideration. The formulas for weights coefficients as functions of M are obtained in Section 3, and the desaturating coefficient is introduced in Section 4. Finally, Section 5 shows the results obtained by simulation of the model.

---

* This research was mainly done in Cybernetics Center of Ukrainian Academy of Sciences, Kiev, Ukraine

2 Presentation of the Model

2.1 The Neural Network Model

Consider a dynamical system of N totally interconnected binary *neurons*, the dynamics of which is determined by the *synchronous Update Rule*

$$\mathbf{Y}(t+1) = F[\mathbf{CY}(t)], \qquad \text{until} \quad \mathbf{Y}(t+1) = \mathbf{Y}(t) \equiv \mathbf{Y}(t^*) \qquad (1)$$

where $\mathbf{Y}(t) \in \Re^N$, $Y_i(t) \in \{+1, -1\}$, $i = 1..N$, is a *state of the network* at time t. $F[x]$ is the "hard" (binary) threshold function.

The output of this system, $\mathbf{Y}(t^*)$, totally depends upon two values: 1) the input state vector $\mathbf{Y}(0)$, and 2) the parameters of the systems, which are $N \cdot N$ coefficients of the *weight matrix* $\mathbf{C}$.

We want the system to exhibit information-retrieval property, which means, that having a distorted (noisy) version of a prototype as an input, we want the network to output the original, retrieved from noise, prototype. Here the main two problems arise:

– How to adjust, or to calculate, the weight matrix $\mathbf{C}$, so that the network will exhibit information-retrieval property with respect to a given set of prototypes $\mathbf{V}^1, ..., \mathbf{V}^M$?

– How many prototypes M can be retrieved by the network of size N?

The first problem can be reformulated as following: Find such a matrix $\mathbf{C}$ that the following requirements are satisfied:

R1: No iteration occurs, if $\mathbf{V}$ is given as initial state[2], i.e. $\mathbf{V} = F[\mathbf{CV}]$.
R2: For $\mathbf{Y}(t) \neq \mathbf{V}$: $\mathbf{Y}(t+1) \neq \mathbf{Y}(t)$.
R3: At some iteration t^*: $\mathbf{Y}(t^*+1) \equiv F[\mathbf{CY}(t^*)] = \mathbf{V}^m$.

Here, we say that vector $\mathbf{Y}(t)$ has fallen in the attractor basin of the vector $\mathbf{V}^m$. The larger the attractor basin radius ($Hattr$), the more prototype attractivity; and conversely, when $Hattr$ is less than unity, a NN stops exhibiting complete error correction.

2.2 Projection Learning Rule

The decision for the above presented problem was suggested by Personnaz *et al.* in 1985. Considering the system of $M \cdot N$ equations $(\mathbf{CV})_i = \alpha_i^m V_i$, $\alpha_i^m > 0$ $(i = 1..N, m = 1..M)$, which follows from the requirement R1, and imposing $\alpha_i^m = \alpha = 1$, they obtained the solution

$$\mathbf{C} = \mathbf{VV}^+ \qquad (2)$$

where $\mathbf{V} = (\mathbf{V}^1, .., \mathbf{V}^M)$ is the matrix made of column prototype vectors, $\mathbf{V}^+ \doteq \lim_{\delta \to 0}(\mathbf{V}^T\mathbf{V} + \delta^2 \cdot \mathbf{I})^{-1}\mathbf{V}^T$ is the *pseudoinverse* of matrix $\mathbf{V}$. $\mathbf{I}$ is the identity matrix.

The matrix defined by eqn (2) is the orthogonal projection matrix in the subspace spanned by prototype vectors $\{\mathbf{V}^m\}$. Therefore, the obtained rule is

---

[2] Hereafter $\mathbf{V}$ will designate a prototype vector, while $\mathbf{Y}$ – any state vector.

also termed the *Projection Learning Rule* (ProjLR). It can be noticed, that in the case of orthogonal prototypes ProjLR reduces exactly to the classical Hebbian rule: $\mathbf{C} = \frac{1}{N}\mathbf{V}\mathbf{V}^T$.

2.3   The Evolution of the Network

In [2] Personnaz *et al.* showed that, if the weight matrix is a projection matrix (i.e. calculated by eqn (2)), then the following conditions are true:

(C1) The energy of the network, defined as $E(\mathbf{Y(t)}) \doteq -\frac{1}{2}\mathbf{Y(t)}^T\mathbf{S(t)}$, is decreasing during the evolution of the network. This means, that the network evolves, until it reaches either global or local minimum of the energy in the state space. Global minimum corresponds to a prototype state and is given by $E(\mathbf{V}) = -\frac{1}{2}\mathbf{V}^T\mathbf{C}\mathbf{V} = -\frac{N}{2}$. Local one corresponds to a spurious stable state; the greater the number of prototypes M, the greater the number of spurious minima.

(C2) The attraction radius of the network is given by $Hattr = \frac{N}{2M}$, which means that, if M becomes of the order $N/2$, then the attractivity of prototypes falls sharply, and the network stops exhibiting error correction.

From these it follows, that in order to build a network which will be able to retrieve more than $N/2$ prototypes, one has to modify the weight matrix (i.e. to deviate from eqn (2)). But, of course, the modification should be done in such a way that it will not change the location of the global minima in the state space.

3   Evaluation of Weights

Starting from the following properties of a projection matrix (see [1]):

P1: It is idempotent, i.e. $\mathbf{C} = \mathbf{C}^2$.

P2: It is symmetric, i.e. $\mathbf{C} = \mathbf{C}^T$.

P3: It has exactly M eigen-values $\lambda_i = 1$ and $N - M$ eigen-values $\lambda_i = 0$.

let us evaluate the weights C_{ij} finding $\overline{C_{ij}}$, where term $\overline{X_i}$ designates the arithmetical mean of values X_i.

From P3 we have for characteristic polynomial of matrix $\mathbf{C}$ the equality: $det(\mathbf{C} - \lambda\mathbf{E}) = \lambda^{N-M}(1-\lambda)^M$. Hence, the trace $tr(\mathbf{C}) \doteq \sum_{i=1}^{N} C_{ii} = M$, and

$$\overline{C_{ii}} = \frac{M}{N} \tag{3}$$

<u>Note</u>: From P1 and P2 it follows that $C_{ii} = \sum_{j=1}^{N} C_{ij} \cdot C_{ji} = C_{ii}^2 + \sum_{j=1, j\neq i}^{N} C_{ij}^2$, and that $0 \leq C_{ii} \leq 1$. Using this inequality and eqn (3), we obtain that the larger M, the closer C_{ii} to $\overline{C_{ii}}$ (i.e. $\lim_{M\to N} C_{ii} = \overline{C_{ii}} = \frac{M}{N}$).

Then, from P1 and P2 we can write $M = \sum_{i=1}^{N} C_{ii} = \sum_{i=1}^{N}\sum_{j=1}^{N} C_{ij}^2 = \sum_{i=1}^{N} C_{ii}^2 + \sum_{i=1}^{N}\sum_{j=1, j\neq i}^{N} C_{ij}^2$. Using eqn (3) and assuming that $\overline{C_{ij}}^2 = \overline{C_{ij}^2}$, which is true for large M (see note above), for $N \gg 1$ we obtain

$$\overline{C_{ij}^2} = \frac{M - N\frac{M^2}{N^2}}{N(N-1)} = \frac{M(N-M)}{N^3} \tag{4}$$

Note: From the note above it follows that the obtained estimates (eqns (3) and (4)) are especially precise for large M, which is the case in question in this paper.

4 Desaturating Coefficient

As can be seen from eqns (3) and (4) the ratio $R \doteq \frac{\overline{C_{ii}}}{|C_{ij}|} = \sqrt{\frac{MN}{N-M}} = \sqrt{\frac{N^2}{N-M} - N}$
is an increasing function of M. Looking at this ratio R we can judge how the network is saturated. That is, given a network with already calculated weights and a distorted version of a prototype as an input, we can predict how good the prototype retrieval will be. To do this we can simply calculate the ratio R and find M corresponding to this R — the greater M, the worse error correction. For example, if R is greater than $R(M = \frac{N}{2}) = \frac{1/2}{1/(2\sqrt{N})} = \frac{1}{4}\sqrt{N}$, then, according to Sec. 2.3, we can expect poor error correction.

Such an understanding of the problem inspires to modify the learning rule in the following way: After all weights have been calculated by eqn (3), the operation of *self-connection reduction* is applied

$$C'_{ii} := D \cdot C_{ii}, \quad 0 < D < 1 \tag{5}$$

It can be easily shown that changing diagonal terms does not influence the location of the minima of the energy function in the state space[3], but changing the weights will change the evolution of the network in time.

First, by deviating from the projection matrix, we make condition (C1) not to be true. That is, now the energy is not necessarily decreasing all the time during the evolution. The result of this will be the network becoming less prone to getting trapped in local minima. But this may also result in occurrence of cycles and even in the divergence of the network.

Second, as it follows from above, by decreasing diagonal weights, we decrease the ratio R. The decreased value of R will correspond to a network, which is trained with a smaller number of prototypes (say, $M' = M - K$). That is, we force the network to behave as if it were learnt with a smaller number of prototypes. By other words, we *desaturate* the network. This is why the coefficient of self-connection reduction D is termed the *desaturating coefficient*. It will be demonstrated in the subsequent section that, by appropriately choosing the coefficient D, any network (assigned by its prototype family, numbers N and M) can be desaturated, i.e. it can be transferred into a state, in which it will exhibit better error-correction.

Here, of course, we have to fully realize, that the newly created matrix will be an approximation of the original one; and the more we deviate from the original matrix (i.e. the smaller the desaturating coefficient D is), the more the behaviour of the new network differs from that of the original network. Thus, the main interest of the research is to find such an optimal *desaturating coefficient*

---

[3] $E'(\mathbf{Y}) \equiv -\frac{1}{2}\mathbf{Y}^T\mathbf{C}'\mathbf{Y} = -\frac{1}{2}\mathbf{Y}^T\mathbf{C}\mathbf{Y} - \frac{1}{2}\mathbf{Y}^T(1-D)\mathbf{I}\mathbf{Y} = E(\mathbf{Y}) - \frac{1}{2}(1-D)N$, where $(1-D)N$ is a constant value.

D that will allow the network to escape local minima, but will not bring about the divergence of the network.

The idea of investigating the influence of self-connections on the performance of a network is not new. This has been thoroughly studied for the Hebbian learning rule. In particular, it was observed that the existence of self-connections indicates an increased recall capacity of the network but, at the same time, the greater these self-connections, the less sensitive the network and the greater the number of spurious states [4,5,8]. As can be seen, these observations agree with the reasoning made above.

5 Simulation Results

In the experiments the program NEUTRAM, designed for simulation of neural networks onto transputers, was used. For the description of the program see [3].

In the training stage a NN of size N is learnt on a set of M prototypes. Prototypes are ordinary letter patterns, i.e. they are not supposed to be orthogonal and can be partially linearly dependent. After all weights have been calculated by eqn (2), self-connection weights are decreased by eqn (5).

In the stage of retrieval, the noisy versions of prototypes are applied. Noise is created by randomly inverting $H0$ pixels of a pattern. Starting from these states, the NN evolutionizes according to eqn (1), until it converges or a cycle occurs. The difference H between the final state of the network and the corresponding prototype is recorded. Its average over M patterns $\overline{H}$ is then calculated. The number of iterations It, their average $\overline{It}$, and the number of cycles occured Cyc are recorded as well. If for some pattern from the set the network diverges, i.e. on every successive iteration the state of the network goes away from the prototype state, then the data are marked as "diverging".

Such a procedure is carried out for different values of noise $H0$, various (over ten) implementations of each noise, for different prototype sets and for different learning ratios M/N (from 0.3 to 0.8). The objective of the experiments is to observe the dependency of the performance of the network upon the desaturating coefficient D. Tables 1 and 2 show such a dependency. The results presented in the tables correspond to $N = 100$. But the experiments were carried out also for N=256 and 600, and the relationship among the investigated parameters was the same in all these cases.

Table 1 shows the dependency of $\overline{H}$ and $\overline{It}$ upon the coefficient D for different learning ratios. For each learning ratio the table shows the data for maximal value of noise $H0$, for which it is possible by appropriately choosing the desaturating coefficient to achieve complete error correction, i.e. there exists such D that yields H equal to zero.

Refer to the table and consider, for example, the data obtained for the set of 50 patterns ($M = 50$). Conventional approach (i.e. without reduction of self-connections) yields little error correction: It decreases noise from $H0 = 9$ to $H = 5.12$ in average, the number of iterations rarely exceeds one ($\overline{It} = 1.08$). Reducing self-connections (i.e. decreasing the value of D) makes the network

Table 1: The influence of the desaturating coefficient D on the performance of the network. The number of neurons N=100

D	HO=20 M=30 $\overline{H}$	$\overline{It}$	HO=13 M=40 $\overline{H}$	$\overline{It}$	HO=9 M=50 $\overline{H}$	$\overline{It}$	HO=5 M=60 $\overline{H}$	$\overline{It}$	HO=3 M=70 $\overline{H}$	$\overline{It}$	HO=2 M=75 $\overline{H}$	$\overline{It}$
.01	0.50	4.47(2)		div		div		div		div		div
.05	0.00	4.17	1.87	6.28(9)		div		div		div		div
.10	0.10	3.53	0.72	5.03(5)	1.62	4.84(12)		div	0.04	1.81(1)	0.03	1.64(1)
.15	0.00	3.00	0.12	4.67(1)	0.08	3.48	0.11	2.13	0.00	1.59	0.32	1.44
.20	0.00	3.10	0.27	3.90(1)	0.12	2.84	0.10	2.05	0.42	1.30	1.20	0.40
.30	0.16	2.83	0.35	3.27	0.12	2.62	0.23	2.33	1.57	0.80	1.60	0.40
.40	0.19	2.90	1.02	3.06	0.76	2.54	1.28	1.27	2.88	0.11	2.00	0.00
.50	0.20	3.07	1.62	2.85	2.06	1.72	1.53	1.15	3.00	0.00	2.00	0.00
.80	2.60	2.83	4.12	2.10	4.62	1.22	2.26	1.07	3.00	0.00	2.00	0.00
1.0	3.16	2.83	5.20	1.65	5.12	1.08	2.45	1.17	3.00	0.00	2.00	0.00

The performance of the NN is represented by $\overline{H}$ and $\overline{It}$ (the first and the second number in each column, respectively). The number of cycles (Cyc) if any is written in parenthesis. "div" denotes the divergence of the network. Each column corresponds to a particular number M of prototypes in learning set. For each M the table shows the data for maximal value of noise $H0$, for which it is possible by appropriately choosing the desaturating coefficient to achieve complete error-correction. The most appropriate cases are underlined as well as the cases of the conventional approach ($D = 1$). Note the improvement of error correction.

more "plastic" – the number of iterations increases and the retrieval improves. This is the result of the fact, that the network succeeds to escape some local minima, which is due to the changing of the evolution of the network energy in time – the energy is not always decreasing now (see Sec. 4). Finally, for $D = 0.15$ we observe $\overline{H} = 0.08$, which means that all of 50 presented noisy patterns but four were retrieved completely (with $H = 0$).

Further decreasing of D (down to 0.1) makes the network unstable at the global minimum itself. For 12 of 50 patterns a cycle occurs — the network makes oscillations around a global minimum, which is observed as the blinking of two pixels in the pattern. This is because the energy, while being decreasing near a global minimum, is not decreasing at the global minimum itself. When $D = 0.05$ or less the network "goes away" from the minima — it diverges.

I would like to emphasize that for $M = 50$ conventional approach does not allow a network to exhibit complete error correction even for small amount of noise, while reducing self-connections yields successful retrieval of patterns with up to 9% of noise.

Table 1 shows the values of H and It that are obtained for one implementation of noise and prototype set. For other noise implementations and prototype sets the data observed may be very different, as some patterns are more similar to each other than others, and it does matter what pixels are inverted. Table 2 shows the influence of the coefficient D on the performance of the network for two different prototype sets and for four different noises. The parameters fixed are the value of noise ($H0 = 5$) and the number of prototypes ($M = 60$). As can be seen from the table the situation is the same: the introduction of the

Table 2: The influence of the coefficient D on the performance of the network.
For different noise implementations and prototype sets.
The number of neurons N=100. The number of patterns M=60. Noise H0=5

D	1 $\overline{H}$ $\overline{It}$	2 $\overline{H}$ $\overline{It}$ 1st set of prototypes	3 $\overline{H}$ $\overline{It}$	4 $\overline{H}$ $\overline{It}$	1 $\overline{H}$ $\overline{It}$	2 $\overline{H}$ $\overline{It}$ 2nd set of prototypes	3 $\overline{H}$ $\overline{It}$	4 $\overline{H}$ $\overline{It}$
.10	div	div	div	div	div	div	div	1.23 3.5(14)
.15	0.11 2.13	0.28 2.7(1)	0.31 2.18	0.95 2.7(12)	1.50 3.7(8)	1.66 4.5(6)	0.05 2.70	0.63 2.9(3)
.20	0.10 2.05	0.60 2.25	0.58 2.09	0.21 3.0(1)	1.35 2.5(1)	0.85 2.10	0.40 2.13	0.05 2.33
.30	0.23 2.33	0.61 2.20	0.71 2.08	0.43 2.23	1.55 1.63	0.33 2.05	1.46 1.18	1.26 1.68
.40	1.28 1.27	2.45 1.13	2.01 1.48	0.56 1.98	1.43 1.32	0.73 2.03	2.01 1.10	0.90 1.47
.50	1.53 1.15	2.85 1.05	2.23 1.42	1.13 1.58	1.98 1.00	1.41 1.73	2.25 1.21	1.25 1.33
.60	1.98 1.15	3.11 1.00	2.61 1.38	1.63 1.17	1.91 1.17	2.31 1.27	2.41 1.28	1.48 1.10
.80	2.26 1.07	3.53 1.00	4.10 0.68	2.48 1.30	3.05 0.95	3.53 0.85	3.08 1.25	1.80 1.13
1.0	2.45 1.17	3.80 1.00	4.86 0.10	3.61 1.05	3.90 0.60	4.25 0.58	3.76 0.75	2.53 0.90

Each prototype set was examined with four different noise implementations. Each noise implementation is represented by a column in the table. The recorded parameters are the same as in Table 1 ($\overline{H}$, $\overline{It}$, Cyc – in parenthesis). The data corresponding to the best and the worst error-correction are underlined. Note how decreasing of D improves the performance of the network.

desaturating coefficient does improve the performance of the NN.

More particularly, the results observed from Tables 1 and 2 are:

- the larger reduction of self-connection (i.e. the less D), the greater number of iterations;
- while D is greater than 0.2, the reduction of self-connection always results in improving error correction;
- if D becomes less then 0.2, than cycles may occur, and the large reduction ($D < 0.1$) often results in the divergence of the NN;
- the most appropriate value of the desaturating coefficient D lies between 0.1 and 0.2, provided that iterations terminated if a cycle occurs;

It is seen that even for the case of $M \geq 0.7N$, for which conventional approach ($D = 1$) does not show any error correction at all, the introduction of the desaturating coefficient allows the network to exhibit error correction. As an example, for some prototype set (which is far from being orthogonal) it is achieved complete error correction[4] (i.e. $H = 0$) even for $M = 0.75N$.

It is worth mentioning that the results observed were theoretically predicted (Sec. 4), and that the experiments show the consistency of the reasoning and provide insights for choosing the optimal value for the desaturating coefficient.

6 Conclusion

A new technique, based on reducing self-connections in the projection weight matrix, is suggested. The theory, which inspires the technique, is presented. It

---

[4] More exactly, we have $\overline{H} = 0.03$, which means that complete error-correction ($H = 0$) is achieved for all of 75 patterns except four, for which $H = 1$.

is explained, why the coefficient of self-connection reduction D is termed the desaturating coefficient. It is shown, that the proposed technique improves considerably the retrieval capability of the network, the improvement being especially noticeable for $0.1 < D < 0.2$. As a specific result, it is demonstrated that even for the great number of prototypes ($M = 0.75N$) it is still possible for the network to exhibit error correction. Further theoretical analysis of the results observed is under the investigation.

Acknowledgements: The author expresses his gratitude to A.M. Reznik for stimulating discussions.

References

1. Albert A. *Regression and Moore-Penrose pseudoinverse,* Academic New-York, 1972
2. Amit D.J., Gutfreund H. and Somplolinsky H. *Spin Glass Model of Neural Networks,* Phys. Rev.– A. (1985),32,p.1007-1018.
3. Gorodnichy D.O., Reznik A.M. *NEUTRAM - A Transputer Based Neural Network Simulator,* Proc. of Second Intern. Conf. on Software for Multiprocessors and Supercomputers Theory, Practice, Experience (SMS TPE'94), 136-142, Moscow, Russia, 1994
4. Gindi G.R., Gmitro A.F. and Parthasarathy K. *Hopfield model associative memory with non-zero diagonal terms in memory matrix,* Applied optics(1988), 27, 129-134
5. Marom E. *Associated Memory Neural Networks with Concatenated Vectors and Nonzero Diagonal Terms,* Neural Networks (1990), Vol.3, N3, 311-318
6. Personnaz I.,Guyon I.,Dreyfus G. *Information storage and retrieval in spin-glass like neural networks,* J.Physique Lett.(1985),46, L359
7. Personnaz I.,Guyon I.,Dreyfus G. *Collective computational properties of neural networks: New learning mechanisms,* Phys. Rev.– A. (1986),34,N5,p.4217-4228.
8. Yanai H. and Sawada Y. *Associative memory network composed of neurons with Hysteretic Property,* Neural Networks (1990), Vol.3, N2, 223-228.

Getting More Information out of SDM

Gunnar Sjödin

RWCP[1] Neuro SICS[2] Laboratory
Box 1263, S-164 28 Kista, Sweden

Abstract. A more efficient way of reading the SDM memory is presented. This is accomplished by using implicit information, hitherto not utilized, to find the information-carrying units and thus removing unnecessary noise when reading the memory.

1 Introduction

1.1 The SDM Model

The Kanerva Sparse Distributed Memory (SDM) is feed-forward net with a set of M locations (hidden units) of U-dimensional vectors of integers. The memory is addressed within a binary address space of dimension N, where $M \ll 2^N$.

A picture of the general structure is given in figure 1. It should serve as a reference when reading this paper. A good introduction to the SDM model is given in [4]. For other relevant literature on SDM see [1], [2], [5], [6], [8].

1.2 Activation with Address X

In the basic SDM model each location is given a randomly generated location address and those locations within a certain Hamming distance from X are activated. In general an activation mechanism should have the robustness property that similar addresses activate roughly the same set of locations.

1.3 Storing in Memory (Weight Adjustment)

When a U-dimensional binary vector W is stored "at" an address X, a mechanism activates a set of memory locations and W is added to their contents after first transforming its 0s to -1s.

An index of a datum-vector or a location will henceforth be called a *position*. It is convenient (cf. figure 1) to have a special memory location in which each datum is stored (i.e., it is activated by every storage operation). This location would then hold $\mathcal{W}$, which is the sum of the T stored data vectors. Furthermore, we extend each location with an extra position. When storing, the value of the

---

[1] Real World Computing Partnership
[2] Swedish Institute of Computer Science

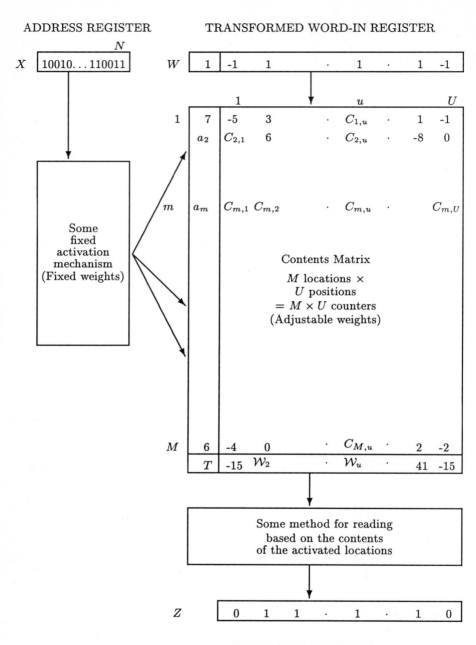

Fig. 1. SDM

corresponding position in the datum vector is always set to 1, i.e., the value of this position in each activated location is incremented by 1. Thus a location m will, in this extra position, hold a_m, which is the number of data vectors stored in location m. In particular, the extra position in the extra location would contain the total number of stored vectors, T. Using probabilistic methods, all the information obtained by using the extra location and the extra position can be obtained without them with probability of error converging to 0 with the size of the memory increasing. This is shown in [7] and [10]. Thus, no extra information has been added, but the architecture described here is good for efficiency.

2   Reading the Memory

Suppose that we read the memory at an address X close to an address X_0, where a datum W has been stored. We will then activate, by some mechanism, a certain set of locations, $\mathcal{H}$. Let $S = (S_1, \ldots, S_U)$ be the vector sum of the data vectors in $\mathcal{H}$. In the basic models of SDM, the output vector Z is calculated by thresholding S at 0, i.e.,

$$Z_u = 0 \text{ if } S_u < 0$$
$$Z_u = 1 \text{ if } S_u \geq 0$$

This is based on the assumption that, for each datum-vector position u, 1 and -1 have equal probability when storing a datum and thus also when reading a datum. This is usually not so. Even if the probability of a 1 is $\frac{1}{2}$, the actual outcome (i.e., the proportion of 1s) will differ in each position and it is possible to get better results even in this situation by conditioning the reading on the actual outcome of the proportions of 1s in each datum position. Furthermore, these proportions usually vary from position to position and as a rule it is very hard to say anything about them beforehand.

As we will show in this paper, this reading procedure can be considerably improved. It involves discarding activated locations that do not contain the desired data vector and adjusting the threshold to the bias in the data, but it is independent of the activation mechanism.

Let the probability of storing a 1 in position u be denoted by p_u and let $q_u = 1 - p_u, \nu_u = p_u - q_u$. These constants may all be estimated through

$$\nu_u = \frac{\mathcal{W}_u}{T} \tag{1}$$

Let $\mathbf{h} = \{m_1, \ldots, m_h\}$ be the set of locations in $\mathcal{H}$ where W has been stored. If ε is the proportional bit error in the address, and we use the Jaeckel/Karlsson activation mechanism (cf. [1], [5]) then the expected value of h is roughly $M^{\frac{1}{3}(1-\frac{2\varepsilon}{\log 2})}$ (cf. [10] page 29). Let $k = \sum_{m \in \mathcal{H}} a_m$ be the total number of summands in S. Note that k is simply the value of the extra position in S. Thus,

$$S = hW + Y \tag{2}$$

where Y is the sum of $c = k - h$ "noise" vectors. Hence, h and c are known if we read at the correct address, i.e., if $X = X_0$. As will be shown in section 3, h and c can be estimated with arbitrarily small risk of getting an incorrect value. Now, suppose that the h-locations are known. Then, using a Bayesian approach, it may be shown that the optimal criterion for choosing a 1 in position u is given by

$$p_u \prod_{j=1}^{h} q_u (1 + \frac{C_{m_j,u}}{a_{m_j}}) \geq q_u \prod_{j=1}^{h} p_u (1 - \frac{C_{m_j,u}}{a_{m_j}}) \tag{3}$$

where $C_{m,u}$ denotes the value of position u in location m.

3   Finding the Information-carrying Locations

It follows from (2), making the appropriate independence assumptions, that the variance of S_u is proportional to h. This makes it natural to introduce

$$G_2 = \frac{1}{U} \sum_{1}^{U} (S_u - k\nu_u)^2 \tag{4}$$

Let $\omega = \sum_{u=1}^{U} 4 p_u q_u$, i.e., the mean of the variances of the stochastic $\{-1, 1\}$ variables in position u. Then

$$\begin{cases} h + c = k \\ h^2 + c = \frac{EG_2}{\omega} \end{cases}$$

Thus the integer value of

$$\frac{1}{2} + \sqrt{\frac{1}{4} + \frac{G_2}{\omega}} - k \tag{5}$$

(or 0 if the expression is not real) is an estimate of h. This method of determining h is arbitrarily safe for big memories if $U \approx M^{\frac{2}{3}(1 + \frac{2\varepsilon}{\log 2})} + \delta$, where ε is the proportional bit error in the address and $\delta > 0$. This result is compatible with the values in the SDM interpretation of the cerebellum where $U \approx M^{0.7}$ (cf. [3]).

Center the variables $S_u, C_{m,u}$ by

$$\widetilde{S}_u = S - k\nu_u \tag{6}$$

$$\widetilde{C}_{m,u} = C_{m,u} - a_m \nu_u \tag{7}$$

By considering the mean values and variances of linear combinations of $\widetilde{S}_u^2$, $\widetilde{S}_u \widetilde{C}_{m,u}$ and $\widetilde{C}_{m,u}^2$, we are led to picking the h-locations to be the h locations in $\mathcal{H}$ with the largest

$$\mathbf{Q} = \frac{1}{U} \sum_{u=1}^{U} (\widetilde{S}_u - \widetilde{C}_{m,u}) \widetilde{C}_{m,u} \tag{8}$$

We have,

$$EQ = \begin{cases} 0 & , m \in \mathcal{H} - \mathbf{h} \\ (h-1)\omega & , m \in \mathbf{h} \end{cases}$$

The following proposition roughly says that if the memory is big and $U \approx M^{\frac{1}{3}}$ then the above procedure of finding the **h**-locations is safe for big memories.

Proposition 1. *Let $\delta > 0$ and let $\xi < \frac{\delta}{2}$. Then there is an $M_0 = M_0(\delta, \xi)$ and an $\alpha = \alpha(\delta, \xi) > 0$ such that if*

$$h \geq M^{\frac{1}{3}-\xi}, \ \omega \geq \delta \tag{9}$$

$$M \geq U \geq M^{\frac{1}{3}+\delta}, \ M \geq M_0 \tag{10}$$

then

$$P\,(\textit{we find the correct h-locations}) \geq 1 - e^{-M^{\alpha}} \tag{11}$$

The proof of this proposition, based on large deviation theory, is given in [10]. The restriction $U \leq M$ is required purely for technical reasons. Of course the proposition is valid without it. However, we are interested in making U small so the restriction does not matter. Note that it is easy to prove the proposition if we simply replace **Q** with its normal approximation. However, this is a dangerous procedure far out in the tails of the distribution. By considering the simple case when all $p_u = \frac{1}{2}$ it is seen that the exponent $\frac{1}{3} + \delta$ cannot be improved to an exponent less than $\frac{1}{3}$. The requirement on the size of U is much weaker here than when finding the h. In fact, by a minor adjustment in the proof of proposition 1, it can be shown that even if we don't know the value of h, but we know that the inequalities (9) are satisfied, we may still find the **h**-locations by taking the locations in $\mathcal{H}$ with

$$\mathbf{Q} > \frac{\omega}{2} M^{\frac{1}{3}-\xi}$$

This procedure is exponentially safe in the sense of (11) if $U = M^a$, where $a \approx \frac{1}{3}$. By counting the number of elements in **h**, we also get h, which we need for the reading methods in section 2.

4   Conclusion and Comments

The methods of this paper allow us to reduce the noise of the SDM memory by reading only from the information-carrying locations. The distortion caused by a bias towards a 0 or a 1 in the noise can be eliminated. It may be argued that the information-carrying locations, the **h**-locations, can easily be found using the "tagged memories" of [8]. This construction stores symbols, in extra positions of each location, indicating what has been stored in the location. However, it seems improbable that this construction is faithful to the structure of the cerebellum and therefore may not be well suited for solving other problems involving efforts to mimic human behaviour.

The efficiency of the methods in this paper has been tested. Below is an example obtained with data in which the probability of a 0 is 0.6. This bias towards 0 would not be detected by the basic SDM model. The reading addresses of the example are at Hamming distances of 5 from the addresses used for storing. This means that many of the activated locations will not contain information but just noise. Note that this noise is kept in the basic model. The other parameters of the example are:

$$N = 64, U = 3000, M = 8000$$
Number of stored data vectors, $T = 200$

When reading 20 of the stored data, i.e., in total 60000 bits, the number of bit-errors was 3 with the methods as above, i.e., first finding the h-locations, and then using the optimal threshold; the basic method of thresholding the sum of all activated locations at 0 gave 902 bit-errors.

If we read at an address X that is close to an address X_0 at which we have stored a datum, we would like to move X toward the unknown X_0 and read again, in order to capture more of the information-carrying locations. This problem is addressed in Sjödin [9].

References

1. Jaeckel, L.A. An alternative design for a Sparse Distributed Memory. Technical Report TR-89.28, RIACS, 1989.
2. Kanerva, P. *Sparse Distributed Memory*. The MIT Press, 1988.
3. Kanerva, P. Associative-memory models of the cerebellum. In I. Alexander and J. Taylor, editors, *Artificial Neural Networks, 2, Proceedings ICANN '92*. Elsevier, 1992.
4. Kanerva, P. Sparse distributed memory and related models. In Mohamad H. Hassoun, editor, *Associative Neural Memories*. Oxford University Press, 1993.
5. Karlsson, R. A fast activation mechanism for the Kanerva SDM memory, 1995. SICS Research Report R95:10, Swedish Institute of Computer Science.
6. Kristoferson, J. Best probabilities of activation and performance comparisons for several designs of Kanerva's SDM (Sparse Distributed Memory), 1995. SICS Research Report R95:09, Swedish Institute of Computer Science.
7. Kristoferson, J. Some comments on the information stored in sparse distributed memory, 1995. SICS Research Report R95:11, Swedish Institute of Computer Science.
8. Rogers, D. Using data tagging to improve the performance of Kanerva's sparse distributed memory. Technical Report TR-88.1, RIACS, 1988.
9. Sjödin, G. Convergence and new operations in SDM, 1995. SICS Research Report R95:13, Swedish Institute of Computer Science.
10. Sjödin, G. Improving the capacity of SDM, 1995. SICS Research Report R95:12, Swedish Institute of Computer Science.

Using a General Purpose Meta Neural Network to Adapt a Parameter of the Quickpropagation Learning Rule

Colin McCormack,
Dept. of Computer Science,
University College Cork,
Cork, Ireland.
E-mail: colin@odyssey.ucc.ie

Abstract

This paper proposes a refinement of an application independent method of automating learning rule parameter selection which uses a form of supervisor neural network, known as a Meta Neural Network, to alter the value of a learning rule parameter during training. The Meta Neural Network is trained using data generated by observing the training of a neural network and recording the effects of the selection of various parameter values. The Meta Neural Network is then combined with a normal learning rule to augment its performance. This paper investigates the combination of training sets for different Meta Neural Networks in order to improve the performance of a Meta Neural Network system. Experiments are undertaken to see how this method performs by using it to adapt a global parameter of the Quickpropagation learning rule.

1 Introduction

Despite the development of more efficient learning rules it remains necessary to manually select appropriate learning rule parameter values in order to achieve an acceptable solution. The learning rule parameters which yield a performance of highest quality (where quality can be defined as the speed of convergence and the accuracy of the resultant network) are usually unique for each problem with no effective method of judging what parameter value is suitable for which problem. This is a significant shortcoming in the area of learning rule usage as selection of an inappropriate parameter can have a marked effect on the performance of most learning rules [1]. Even learning rules which have been shown to be tolerant to arbitrary selection of initial learning rule parameters, such as RPROP [2], are inhibited by inappropriate parameter selection [2]. In most learning rules parameters are initialised with general values which are suggested by the rules authors. These general values are those which have been found to deliver the best results on average and may not be the ideal parameter values for every problem domain they are used for. A learning rule is therefore constrained in its performance by a non-deterministic process of selecting a parameter value

This paper investigates a method of parameter adaptation which involves the use of a separate neural network (called a Meta Neural Network) to select appropriate values for the ε parameter of the Quickpropagation [3] learning rule. We look at the results obtained when a standard Quickpropagation rule and a Meta Neural Network are applied to four benchmark problems.

1.1 Quickpropagation

Quickpropagation (Quickprop) [3] is a widely used adaptive learning rule and has been shown to be one of the most effective [1]. There is only one global parameter making a significant contribution to the result, the ε parameter.

Quickpropagation uses a set of heuristics to optimise Backpropagation, the condition where ε is used is when the sign for the current slope and previous slope for the weight is the same. In this situation the update rule for each weight is stated to be: $\Delta w(t) = \dfrac{S(t)}{S(t-1)-S(t)} . \Delta w(t-1) + \varepsilon . S(t)$ where S(t) and S(t-1) are the current and previous values of the summed gradient information (slope), $\delta E / \delta w$, over all the patterns in the training set. The value of ε is split by dividing it by the number of inputs to each unit which helps ε stay in a useful range, ε is usually set to 0.01.

2 Meta Neural Network Description

2.1 MNN Methodology

The scheme for creating a MNN to aid a conventional neural network is composed of three stages. In the first stage data for training the MNN is created, in the second stage the MNN is trained and in the third stage the MNN is used to guide a conventional learning rule.

In stage 1 a backtracking system was set up which allowed a learning algorithm to see the results of the selection of the next learning rule parameter value (i.e. the value of ε for Quickpropagation). This method is known as backtracking since it allows the learning rule to backtrack from a parameter value choice that does not lead to a short term decrease in error value. At each epoch a potential parameter value is evaluated with the value leading to the greatest reduction in the training set error being retained and the training process continued. The value of the parameter is limited to six values, in the range 0.1 to 0.000001. At each epoch the parameter value is allowed to increase or decrease by one step (where a step is a factor of 10) in its range or remain at the current value, these three values are then evaluated. After the evaluation the results are used to augment a set S with the current network slope, the previous network slope and the action (increment/maintain/decrement) which produced the best value of the parameter (i.e. the value of the parameter which caused the largest reduction in error).

In stage 2 the set S is used as a training set for a MNN, where the inputs are: current network slope, previous network slope and the output is a single value which indicated whether the value of the parameter increases, decreases or remains the same.

In stage 3 at each epoch the learning rule passes the value of the current network slope and the previous network slope to the MNN which suggests an increase/decrease or no change in the value of the parameter.

2.2 Benchmark Problem Description

Four benchmark problems: Thyroid, Building, Heart and Credit Card are used to evaluate the effectiveness of MNN's. Three of these problems are also used to produce data for the MNN's which are in turn trained and evaluated. The problems are taken from a comprehensive study of neural network benchmarks [4] and have seen wide use in AI and neural network literature. The problems are referred to in [4] as 'Thyroid 1', 'Building 1', 'Heart 1' and 'Card 1'.

3 Experiments

Earlier work using individual MNNs [5,6] to adapt a learning rules parameter value indicated that the approach was successful but not always guaranteed to produce a consistent result. In an attempt to improve the operation of a MNN based system this paper proposes a system which combines the training sets for individual MNNs into a large training set which forms a general purpose MNN known as a combination MNN.

In the set of experiments undertaken groups of ten networks were trained using the standard Quickprop method to gauge the average performance. Additional standard Quickprop learning algorithms were trained using 'optimal' parameter values derived from previous experimentation. Each MNN is trained using a set S derived from backtracking on a particular problem. Networks which used a learning rule aided by a MNN were trained for comparison. For all the experiments the initial learning rule and network parameters were fixed apart from the initial weight set which was random.

3.1 Result Evaluation

In the series of experiments undertaken 50% of the problems total available examples are allocated for the training set, 25% for the validation set and 25% for the test set. The error measure, E, used was the squared error percentage [4], this was derived from the normalisation of the mean squared error to reduce its dependence on the number of coefficients in the problem representation and on the range of output values used: $E = 100.\dfrac{o_{max} - o_{min}}{N.P} \sum\limits_{p=1}^{P} \sum\limits_{i=1}^{N} (o_{pi} - tr_{pi})^2$ where o_{min} and o_{max} are the minimum and maximum values of the output coefficients used in the problem, N is the number of output nodes of the network, P is the number of patterns in the data set, o is the network output and tr is the target value.

Training progress P [4] is measured after a training strip of length k, which is a sequence of k epochs numbered n+1,...,n+k where n is divisible by k:

$$P_k(t) = 1000.\left(\dfrac{\sum_{t' \in t-k+1...t} E_{tr}(t')}{k.\min_{t' \in t-k+1...t} E_{tr}(t')} - 1 \right)$$

, where E_{tr} is the training set error. In the experiments detailed in this paper k=5. In the experiments performed in this paper training is halted when the progress P drops below 0.1. Since the goal of this work is to contribute towards a method whereby neural networks can be used with little or no initialisation or intervention the errors reported are those obtained at the cessation of training as opposed to the minimum error obtained during training.

4 Results

Each MNN is trained using a set S derived from backtracking on a particular problem. The networks trained using these MNN to suggest parameter values are known as 'Thyroid MNN' (i.e. the MNN was trained using results obtained from training on the Thyroid problem), 'Building MNN' and 'Heart MNN'.

Networks were trained for the normal Quickprop parameter ($\varepsilon = 0.01$), the parameter for ε which was found to be optimal from trial and error experiments, three MNNs and a combination MNN. This trial and error approach is one of the most popular means of ascertaining the most appropriate learning rule parameters. The 'optimal' ε values were 0.01 for the Building problem, 0.001 for the Thyroid problem, 0.1 for the Heart problem and 0.01 for the Card problem. No 'optimal' results are shown for the Building or Card problems since these problems 'optimal' ε value is equal to the normal ε value.

The average of the errors at the cessation of training for the training set (Train), validation set (Valid) and the test set (Test) are presented in figures 1-4. The standard deviation for the validation set is illustrated in the form of an error bar. This particular standard deviation was chosen because it was usually the most significant and gave a good idea of the consistency of the learning rule used. The average number of epochs taken to reach cessation of the training process is included in the figures as a column. The keys used are 'Normal QP' for Quickpropagation using the normal parameter, 'Optimal QP' for Quickpropagation using the 'optimal' parameter, 'Thyroid MNN', 'Building MNN' and 'Heart MNN' for the MNN using the individual training sets and 'Combi MNN' for the MNN using a combination of the three individual MNNs training sets.

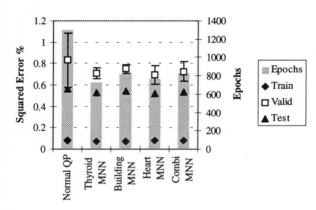

Figure 1: Results for the Building problem

For the Building problem (figure 1, left) the results for the MNNs in general are quite good, improving on the performance of the normal Quickpropagation parameter both for average errors and training speed. The errors for the combination of MNN training sets are about the median of the performance of the other MNNs. The standard deviation for the combination MNN is not as low as that for some of the other MNNs but is very encouraging in comparison to normal Quickpropagation

The results for the Thyroid problem (figure 2, right) indicate that not only do MNNs improve on the performance obtained using normal and optimal Quickpropagation but the combination MNN returns a performance average to those of the other MNN while at the same time producing a standard deviation lower than that of most of the MNNs.

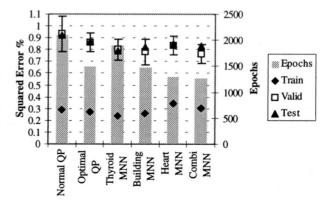

Figure 2: Results for the Thyroid problem.

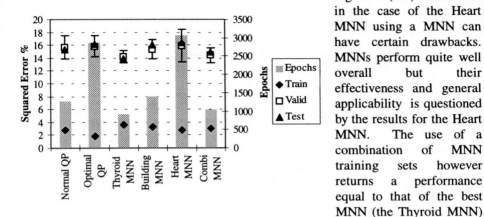

Figure 3 (left) shows that in the case of the Heart MNN using a MNN can have certain drawbacks. MNNs perform quite well overall but their effectiveness and general applicability is questioned by the results for the Heart MNN. The use of a combination of MNN training sets however returns a performance equal to that of the best MNN (the Thyroid MNN) and returns a very low

Figure 3: Results for the Heart problem.

standard deviation indicating it is a consistent performer.

For the Credit Card problem (figure 4, right), the MNN system does not produce a considerable improvement over the normal ε parameter. However not only does a combination of MNNs improve slightly on the normal parameter but it returns an extremely low standard deviation. This shows that even for problems where the use of

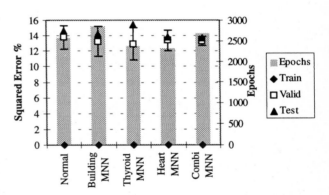

Figure 4: Results for the Credit Card problem.

a MNN does not present a significant benefit the combination of MNNs will result in the best solution.

5 Conclusion

A Meta Neural Network is a means of acquiring and using information about the learning mechanism of a neural network. It is a method of combining neural networks to improve the quality of the solution by using a neural network as an extension to a learning rule.

In this paper we have seen that a MNN trained using knowledge derived from an unrelated problem domain can be used to outperform a learning rule which has had the 'optimal' parameter values for that problem domain preselected by trial and error. The advantage of a MNN scheme is that a MNN's effectiveness on a problem is independent of the MNN's original training problem and its associated architecture. MNN's can thus be trained once and used successfully on different problem domains. The disadvantage of a MNN scheme is that MNNs which apparently work quite well for some problems can return disappointing results for others. Particularly disappointing is the fact that a MNN trained on a particular problem domain will not always perform well on that domain. This is because the adaptation strategy acquired by the MNNs is not always appropriate for the circumstances in which it is applied. Unfortunately there is no way of determining a MNNs appropriateness. The aim of a combined MNN scheme was to try to build a parameter adaptation scheme which contained all of the advantages of a MNN based scheme without being encumbered by any of the disadvantages of that scheme. A combination of MNN training sets stabilised the performance obtained using a MNN without affecting the quality of the solution. We also saw that while a MNN based scheme can be used to select a parameter as an alternative to tuning the learning rule the three MNNs trained seem to be suitable for specific problems. This leads to the situation where instead of having to select the best learning rule parameter by trial and error we must now select the most appropriate MNN. The use of a system of combining the training sets for individual MNNs overcomes that problem and shows that MNNs can now be implemented without the need to intervene in the choice of MNN.

This paper contributes towards a learning method which requires a minimum of initialisation or intervention. The method of parameter adaptation presented in this paper reduces the amount of information needed to use a learning rule without reducing the quality of the solution.

References

[1] Schiffmann, W. Optimisation of the Backpropagation Algorithm for training multilayer perceptrons. *Technical report, Dept of Physics, University of Kobelnz, Germany.*ftp://archive.cis.ohio-state.edu/pub/neuroprose/schiff.bp_speedup.ps.Z (1993)

[2] Riedmiller, M. Advanced supervised learning in multi-layered perceptrons: From Backpropagation to adaptive learning algorithms. *Computer Standards and Interfaces.* vol 16 part 3. 265-278. (1994)

[3] Fahlman, S. Faster-Learning variations on Back-Propagation: An Empirical Study. *Proceedings of the 1988 Connectionist Models Summer School.* Morgan Kaufmann. (1988)

[4] Prechelt, L. PROBEN1: A set of neural network benchmark problems and benchmarking rules. *Technical Report 21/94, Dept. of Informatics, University of Karlsruhe, Germany.* ftp://ftp.ira.uka.de/pub/neuron/proben1.tar.gz. (1994)

[5] McCormack, C. A study of the Adaptation of Learning Rule Parameters using a Meta Neural Network. *13th European Meeting on Systems and Cybernetic Research, Vienna.* Vol 2, 1043-1048. (1996)

[6] McCormack, C. Parameter Adaptation using a Meta Neural Network. *Proceedings of the World Conference on Neural Networks, Washington D.C.,* Vol 3, 147-150. (1995)

Unsupervised Learning of the Minor Subspace

Fa-Long Luo and Rolf Unbehauen
Lehrstuhl für Allgemeine und Theoretische Elektrotechnik
Universität Erlangen-Nürnberg, Cauerstraße 7
91058, Erlangen, Germany

Abstract

This paper proposes a learning algorithm which extracts adaptively the subspace (the minor subspace) spanned by the eigenvectors corresponding to the smallest eigenvalues of the autocorrelation matrix of an input signal. We will show both analytically and by simulation results that the vectors provided by the proposed algorithm are guaranteed to converge to the minor subspace of the input signal. For more practical applications, we also generalize the proposed algorithm to the complex-valued case.

1. Introduction

The subspace spanned by the eigenvectors corresponding to the smallest eigenvalues of the autocorrelation matrix of an input signal is defined as the minor subspace. The minor subspace usually represents the statistics of the additive noise, and hence it is also referred to as the noise subspace in many fields of signal processing such as frequency estimation, bearing estimation, digital beamforming, moving target indication and clutter cancellation [1]. The adaptive estimation of the minor subspace is a primary requirement in these fields. The minor subspace is the counterpart of the principal subspace (the subspace spanned by the eigenvectors corresponding to the largest eigenvalues of the autocorrelation matrix). Many effective principal subspace analysis (PSA) approaches have been recently proposed and extensively analyzed (for reviews, see [1]-[2]). Unfortunately, these PSA approaches can not be generalized to the minor subspace analysis (MSA) with simply reversing the sign [2]. A few significant methods related to the MSA have been reported such as the minor component analysis (MCA) algorithm developed by Oja [2] and the orthogonal eigensubspace analysis algorithm proposed in [3]. The MCA algorithm in [2] requires the assumption that the smallest eigenvalue of the autocorrelation matrix of input signals is less than unity. Moreover, as Reference [4] pointed out, the proof of the convergence of the continuous-time differential equations corresponding to the algorithm given in [2] is unacceptable mainly because the stationary values of the lower eigenvectors are in effect assumed for the proof of the convergence of the higher eigenvectors, as a result, the convergence of this algorithm remains an unsolved problem. A rigorous convergence analysis of the algorithm proposed in [3] has been given, but this algorithm can only provide the eigensubspace corresponding to the repeated minimum eigenvalue of the autocorrelation matrix. The main purpose of this paper is to develop an alternative MSA algorithm and investigate its properties. We will show both analytically and by simulation results that this algorithm could extract adaptively the minor subspace of the autocorrelation matrix of input signals without the above mentioned conditions on the eigenvalues.

2. The Proposed MSA Algorithm

Let us consider a linear single-layer network with inputs $\boldsymbol{X}(t) = [x_1(t), x_2(t), \ldots, x_N(t)]^T$ and outputs $\boldsymbol{Z}(t) = [z_1(t), z_2(t), \ldots, z_M(t)]^T$

$$z_j(t) = \sum_{i=1}^{N} w_{ij}(t) x_i(t) = \boldsymbol{W}_j^T(t) \boldsymbol{X}(t) \tag{1}$$

where $\boldsymbol{W}_j(t) = [w_{1j}(t), w_{2j}(t), \ldots, w_{Nj}(t)]^T$ is the connection weight vector.
The proposed MSA algorithm is

$$\boldsymbol{W}_j(t+1) = \boldsymbol{W}_j(t) - \gamma(t) z_j(t) \{ \boldsymbol{W}_j^T(t) \boldsymbol{W}_j(t) [\boldsymbol{X}(t) - \sum_{i=1}^{j-1} z_i(t) \boldsymbol{W}_i(t)]$$

$$- z_j(t) \boldsymbol{W}_j(t) + \sum_{i=1}^{j-1} z_i(t) \boldsymbol{W}_j^T(t) \boldsymbol{W}_i(t) \boldsymbol{W}_j(t) \} \tag{2}$$

$$(\text{for } j = 1, 2, \ldots, M)$$

where $\gamma(t)$ is a positive scalar gain parameter.

According to the stochastic approximation theory in References [5-6] and further explanations in References [2],[7], it can be shown that the asymptotic limits of the above discrete learning algorithm can be solved by applying the corresponding continuous-time differential equations

$$\frac{d\boldsymbol{W}_j(t)}{dt} = -z_j(t) \{ \boldsymbol{W}_j^T(t) \boldsymbol{W}_j(t) [\boldsymbol{X}(t) - \sum_{i=1}^{j-1} z_i(t) \boldsymbol{W}_i(t)]$$

$$- z_j(t) \boldsymbol{W}_j(t) + \sum_{i=1}^{j-1} z_i(t) \boldsymbol{W}_j^T(t) \boldsymbol{W}_i(t) \boldsymbol{W}_j(t) \} \tag{3}$$

and the averaging differential equations

$$\frac{d\boldsymbol{W}_j(t)}{dt} = \boldsymbol{W}_j^T(t) [\boldsymbol{I} - \sum_{i=1}^{j-1} \boldsymbol{W}_i(t) \boldsymbol{W}_i^T(t)] \boldsymbol{R} \boldsymbol{W}_j(t) \boldsymbol{W}_j(t)$$

$$- \boldsymbol{W}_j^T(t) \boldsymbol{W}_j(t) [\boldsymbol{I} - \sum_{i=1}^{j-1} \boldsymbol{W}_i(t) \boldsymbol{W}_i^T(t)] \boldsymbol{R} \boldsymbol{W}_j(t) \tag{4}$$

where $\boldsymbol{R} = E[\boldsymbol{X}(t) \boldsymbol{X}^T(t)]$ is the autocorrelation matrix of the input vector $\boldsymbol{X}(t)$ with the eigenvalues $0 < \lambda_1 \leq \lambda_2 \leq \ldots \lambda_N$ and the corresponding orthonormal eigenvectors $\boldsymbol{S}_1, \boldsymbol{S}_2, \ldots, \boldsymbol{S}_N$. The matrix $\boldsymbol{I}$ is the identity matrix.

If we set

$$R_j = R - \sum_{i=1}^{j-1} W_i(t)W_i^T(t)R, \tag{5}$$

$$\psi_j(t) = W_j^T(t)W_j(t), \tag{6}$$

$$\phi_j(t) = W_j^T(t)R_jW_j(t) \tag{7}$$

then (4) can be written as

$$\frac{dW_j(t)}{dt} = \phi_j(t)W_j(t) - \psi_j(t)R_jW_j(t) \tag{8}$$

Concerning the dynamics of (8), we have the following theorems:

Theorem 1

The norm of each connection weight vector $W_j(t)$ is invariant during the time evolution and is equal to the norm of the initial state $W_j(0)$, that is:

$$\| W_j(t) \|^2 = \| W_j(0) \|^2, \quad t \geq 0 \tag{9}$$

$$(\text{for } j = 1, 2, \dots, M).$$

The proof of Theorem 1 reads as follows:

Multiplying (8) by $W_j^T(t)$ on the left yields

$$\frac{d \| W_j(t) \|^2}{dt} = 2\{\phi_j(t)W_j^T(t)W_j(t) - \psi_j(t)W_j^T(t)R_jW_j(t)\}. \tag{10}$$

Substituting (6) and (7) into (10) gives

$$\frac{d \| W_j(t) \|^2}{dt} = 2\{\phi_j(t)\psi_j(t) - \psi_j(t)\phi_j(t)\} = 0 \tag{11}$$

which shows that (9) holds and concludes the proof of Theorem 1.

If we set

$$W_j(t) = \sum_{i=1}^{N} y_{ij}(t)S_i \tag{12}$$

$$(\text{for } j = 1, 2, \dots, M)$$

then Theorem 1 shows

$$0 \leq y_{ij}^2(t) \leq \psi_j(0), \quad t \geq 0 \tag{13}$$

$$(\text{for } i = 1, 2, \dots, N; j = 1, 2, \dots, M).$$

Theorem 2

If the initial values of the weight vector satisfy $W_j^T(0)S_j \neq 0$ (for $j = 1, 2, \dots, M$), we have

$$\lim_{t \to \infty} W_j(t) = \lim_{t \to \infty} \sum_{i=1}^{j} y_{ij}(t)S_i \tag{14}$$

$$(\text{for } j = 1, 2, \ldots, M).$$

For limit of space, the details of the proof of Theorem 2 can be referred to [1].

Theorem 1 and Theorem 2 guarantee that the continuous-time differential equations corresponding to the proposed MSA algorithm will converge to the minor subspace of the autocorrelation matrix of the input signal with the norm of the initial weight vector. Based on the above, we can also obtain a modified MSA algorithm as

$$\boldsymbol{W}_j(t+1) = \boldsymbol{W}_j(t) - \gamma(t)z_j(t)\{\boldsymbol{W}_j^T(0)\boldsymbol{W}_j(0)[\boldsymbol{X}(t) - \sum_{i=1}^{j-1} z_i(t)\boldsymbol{W}_i(t)]$$

$$-z_j(t)\boldsymbol{W}_j(t) + \sum_{i=1}^{j-1} z_i(t)\boldsymbol{W}_j^T(t)\boldsymbol{W}_i(t)\boldsymbol{W}_j(t)\} \quad (15)$$

$$(\text{for } j = 1, 2, \ldots, M).$$

Compared with (2), the algorithm (15) has less computational complexity because $\boldsymbol{W}_j^T(t)\boldsymbol{W}_j(t)$ has been replaced by $\boldsymbol{W}_j^T(0)\boldsymbol{W}_j(0)$ which is a constant during the learning phase. Note that the above theorems still hold for (15).

3. Generalization to the Complex-Valued Case

For more practical applications, it is desirable to generalize the proposed MSA algorithm to the case where the input vector $\boldsymbol{X}(t)$ takes complex values.

In the complex-valued case, (1) becomes

$$z_j(t) = \boldsymbol{W}_j^H(t)\boldsymbol{X}(t) \quad (16)$$

$$(\text{for } j = 1, 2, \ldots, M).$$

Let us use the notation

$$\boldsymbol{z}_{jc}(t) = \begin{pmatrix} z_{jr}(t) \\ z_{ji}(t) \end{pmatrix}, \quad \boldsymbol{W}_{jc}(t) = \begin{pmatrix} \boldsymbol{W}_{jr}(t) \\ \boldsymbol{W}_{ji}(t) \end{pmatrix}.$$

Then, we have

$$z_{jr}(t) = \boldsymbol{W}_{jr}^T(t)\boldsymbol{X}_r(t) + \boldsymbol{W}_{ji}^T(t)\boldsymbol{X}_i(t) \quad (17)$$

$$z_{ji}(t) = -\boldsymbol{W}_{jr}^T(t)\boldsymbol{X}_i(t) + \boldsymbol{W}_{ji}^T(t)\boldsymbol{X}_r(t) \quad (18)$$

where $\boldsymbol{X}_r(t)$, $\boldsymbol{W}_{jr}(t)$, $z_{jr}(t)$ and $\boldsymbol{X}_i(t)$, $\boldsymbol{W}_{ji}(t)$, $z_{ji}(t)$ are the real and imaginary parts of the variables $\boldsymbol{X}(t)$, $\boldsymbol{W}_j(t)$ $z_j(t)$, respectively. Moreover,

$$\boldsymbol{z}_{jc}(t) = \boldsymbol{X}_c^T(t)\boldsymbol{W}_{jc}(t) \quad (19)$$

where

$$\boldsymbol{X}_c(t) = \begin{pmatrix} \boldsymbol{X}_r(t) & -\boldsymbol{X}_i(t) \\ \boldsymbol{X}_i(t) & \boldsymbol{X}_r(t) \end{pmatrix}. \quad (20)$$

Then from (2) we have the algorithm to update the weights as

$$
\begin{aligned}
\boldsymbol{W}_{jc}(t+1) \;=\; & \boldsymbol{W}_{jc}(t) - \gamma(t)\{\boldsymbol{W}_{jc}^T(t)\boldsymbol{W}_{jc}(t)[\boldsymbol{X}_c(t)\boldsymbol{z}_{jc}(t) \\
& - \sum_{i=1}^{j-1} \boldsymbol{z}_{jc}^T(t)\boldsymbol{z}_{ic}(t)\boldsymbol{W}_{ic}(t)] - \boldsymbol{z}_{jc}^T(t)\boldsymbol{z}_{jc}(t)\boldsymbol{W}_{jc}(t) \\
& + \sum_{i=1}^{j-1} \boldsymbol{z}_{jc}^T(t)\boldsymbol{z}_{ic}(t)\boldsymbol{W}_{jc}^T(t)\boldsymbol{W}_{ic}(t)\boldsymbol{W}_{jc}(t)\}
\end{aligned}
\tag{21}
$$

$$(j = 1, 2, \ldots\ldots, M).$$

According to Theorem 2, we know that the weight vectors provided by (21) will converge to the minor subspace of the matrix $\boldsymbol{R}_c = E[\boldsymbol{X}_c(t)\boldsymbol{X}_c^T(t)]$. Furthermore, it is easy to prove that $E[\boldsymbol{X}_c(t)\boldsymbol{X}_c^T(t)] = \begin{pmatrix} \boldsymbol{R}_r & -\boldsymbol{R}_i \\ \boldsymbol{R}_i & \boldsymbol{R}_r \end{pmatrix}$ where $\boldsymbol{R}_r$ and $\boldsymbol{R}_i$ are the real and imaginary parts of the autocorrelation matrix $\boldsymbol{R} = E[\boldsymbol{X}(t) \cdot \boldsymbol{X}^H(t)]$ of the complex-valued input signal $\boldsymbol{X}(t)$, respectively. This shows that the algorithm (21) can provide the desired minor subspace.

4. Simulation Results

We have simulated the proposed MSA algorithm. Two examples are given in this paper. In these simulations, we generate a random vector $\boldsymbol{X}(t)$ with the autocorrelation matrix $\boldsymbol{R}$ and take it as the input signals. The first example is with distinct eigenvalues, the second example is with non-distinct eigenvalues. For simplicity, in the simulations we let $\gamma(t)$ be 0.001 and unchanged during the learning phase. A more sophisticated and better selection of $\gamma(t)$ could be made according to the Robbins-Monro procedures [7]. $\boldsymbol{W}_j(f)$ and $\boldsymbol{W}_j(0)$ (for $j = 1, 2, \ldots, M$) are the weight vectors in the steady state and the initial state, respectively. $\lambda_j(f)$ are the values computed by $\boldsymbol{W}_j(f)$, $\lambda_j(a)$ (for $j = 1, 2, \ldots, M$) are the exact eigenvalues of the matrix $\boldsymbol{R}$. These simulation results demonstrate the accuracy of the above analyses and the effectiveness of the proposed algorithm.

Example 1

$$
\boldsymbol{R} = \begin{pmatrix}
2.7262 & 1.1921 & 1.0135 & -3.7443 \\
1.1921 & 5.7048 & 4.3103 & -3.0969 \\
1.0135 & 4.3103 & 5.1518 & -2.6711 \\
-3.7443 & -3.0969 & -2.6711 & 9.4544
\end{pmatrix}
$$

$$\boldsymbol{W}_1(f) = \begin{pmatrix} 0.9104 & 0.0183 & 0.0287 & 0.4206 \end{pmatrix}^T$$

$$\boldsymbol{W}_2(f) = \begin{pmatrix} 0.0093 & 0.7058 & -0.7271 & 0.0351 \end{pmatrix}^T$$

$$\boldsymbol{W}_1(0) = \begin{pmatrix} 1.0000 & 0.0000 & 0.0000 & 0.0000 \end{pmatrix}^T$$

$$\boldsymbol{W}_2(0) = \begin{pmatrix} 0.0000 & 1.0000 & 0.0000 & 0.0000 \end{pmatrix}^T$$

$$\lambda_1(f) = 1.0486, \quad \lambda_2(f) = 1.1047$$
$$\lambda_1(a) = 1.0484, \quad \lambda_2(a) = 1.1048$$

Example 2

$$R = \begin{pmatrix} 2.5678 & 1.1626 & 0.9302 & -3.5269 \\ 1.1626 & 6.1180 & 4.3627 & -3.4317 \\ 0.9302 & 4.3627 & 4.7218 & -2.7970 \\ -3.5269 & -3.4317 & -2.7970 & 9.0905 \end{pmatrix}$$

$$W_1(f) = \begin{pmatrix} 0.9059 & 0.0403 & 0.0444 & 0.4261 \end{pmatrix}^T$$
$$W_2(f) = \begin{pmatrix} 0.0471 & 0.6701 & -0.7577 & 0.0457 \end{pmatrix}^T$$
$$W_1(0) = \begin{pmatrix} 1.0000 & 0.0000 & 0.0000 & 0.0000 \end{pmatrix}^T$$
$$W_2(0) = \begin{pmatrix} 0.0000 & 1.0000 & 0.0000 & 0.0000 \end{pmatrix}^T$$
$$\lambda_1(f) = 1.0001, \quad \lambda_2(f) = 1.0003$$
$$\lambda_1(a) = 1.0000, \quad \lambda_2(a) = 1.0000$$

5. Acknowledgements

This paper is partially supported by the German Research Foundation.

References

[1] F.-L. Luo and R.Unbehauen, *Applied Neural Networks for Signal Processing*, Cambridge University Press, 1996.

[2] E.Oja, "Principal components, minor components and linear neural networks," Neural Networks, Vol.5, 1992, pp.927-935.

[3] G.Mathew, V.U.Reddy, "Orthogonal Eigensubspace estimation using neural networks," IEEE Trans. on SP, Vol.42, No.7, 1994. pp.1803-1811.

[4] H. Chen and R.W.Liu., "An on-line unsupervised learning machine for adaptive extraction," IEEE Trans. on CAS,II, Vol.41, 1994, pp.87-98.

[5] L.Ljung, "Analysis of recursive stochastic algorithm," IEEE Trans. on AC, Vol.22, 1977, pp.551-575.

[6] H.J.Kushner and D.S.Clark, *Stochastic Approximation Methods for Constrained and Unstrained Systems*, Springer-Verlag, 1978.

[7] L.Xu, E. Oja and C.Y.Suen, "Modified Hebbian learning for curve and surface fitting," Neural Networks, Vol.5, 1992, pp.441-457.

A Unification of Genetic Algorithms, Neural Networks and Fuzzy Logic: The GANNFL Approach

Martin Schmidt, University of Aarhus, Denmark, Email: marsch@daimi.aau.dk,
msc@bording.dk, WWW: http://www.daimi.aau.dk/~marsch, Fax: +45 89423225

Abstract: The GANNFL Approach uses a steady state Genetic Algorithm (GA) to build and train hybrid classifiers which are combinations of Neural Networks (NN's) and Fuzzy Logic (FL). This novel approach finds both the architecture, the types of the hidden units, the types of the output units, and all weights ! The designed modular and tight GA encoding together with the GA fitness function lets the GA develop high performance hybrid classifiers, which consist of NN parts and FL parts co-operating tightly within the same architecture. By analysing the behaviour of the GA it will be investigated whether there is evidence for preferring NN classifiers or FL classifiers. Further, the importance of the types of the hidden units is investigated. Parameter reduction is a very important issue according to the theory of the VC-dimension and Ockham's Razor. Hence, the GANNFL Approach also focuses on parameter reduction, which is achieved by automatically pruning unnecessary weights and units. The GANNFL Approach was tested on 5 well known classification problems: The artificial, noisy monks3 problem and 4 problems representing difficult real-world problems that contain missing, noisy, misclassified, and *few* data: The cancer, card, diabetes and glass problems. The results are compared to public available results found by other NN approaches, the sGANN approach (simple GA to train NN's), and the ssGAFL approach (steady state GA to train FL classifiers). In every case the GANNFL Approach found a better or comparable result than the best other approach !

Keywords: Genetic Algorithm, Fuzzy Logic, Neural Networks, uncertain real-world data.

1. Introduction

Many researchers doubt that one effectively can use Genetic Algorithms (GA's) to train Neural Networks (NN's) or use GA's to train Fuzzy Logic Classifiers (FLC's). Nevertheless it has been shown several times that this is possible. Only recently it was shown that one successfully can train high performance NN's and FLC's using GA's [2,3,4]. Note though, that these approaches only used one classifier style, i.e. NN style or FL style. Furthermore, these approaches used only one fixed type of hidden units. The GANNFL Approach continues this research line by stating the following: What really matters when training classifiers is: The exact combination of NN parts and FL parts, the training objective, the number of free parameters and the types of the hidden units ! In order to investigate whether this holds, we designed a classifier that can express both NN style and FL style within the same architecture, where the GA chooses the combination of styles that gives the best performance. Furthermore, the GA chooses the type of every hidden unit and finds each weight, and the GA can prune any hidden unit and any connection. We also designed a new GA fitness function together with a modular and tight GA encoding. After describing the hybrid NNFL classifiers (section 2), we will describe the GA chromosomes (section 3) and the GA fitness function (section 4), followed by other GA related descriptions (section 5). Finally we will show all results (section 6) and give overall conclusions (section 7).

2. Description of the Hybrid NNFL Classifiers

The NNFL classifiers consists of 3 layers: The usual input layer, and the special hidden layer and output layer which can contain different types of hidden units and different types of output units (see figure 1).

We will now explain figure 1:

- Every connection between an input unit k and a hidden unit j has two parameters $A_{k,j}$ and $B_{k,j}$: /* T_j = type of hidden unit j */

$$A_{k,j} = \begin{cases} weight & \text{if } T_j = SIGMOID \\ center\ of\ the\ gaussian & \text{if } T_j = GAUSSIAN \end{cases}$$

$$B_{k,j} = \begin{cases} not\ used & \text{if } T_j = SIGMOID \\ width\ of\ the\ gaussian & \text{if } T_j = GAUSSIAN \end{cases}$$

- Each hidden unit can be one of two types: A sigmoid unit (global decision boundary, a hyperplane) or a gaussian unit (local decision boundary, a gaussian cluster). Every hidden unit j has three parameters T_j , S_j and $BIAS_j$:

$$T_j \in \{ NOT\_USED,\ SIGMOID,\ GAUSSIAN \} \quad \text{/* type of hidden unit j */}$$

$$S_j = \begin{cases} steepness\ of\ sigmoid & \text{if } T_j = SIGMOID \\ not\ used & \text{if } T_j = GAUSSIAN \end{cases}$$

$$BIAS_j = \begin{cases} bias\ weight & \text{if } T_j = SIGMOID \\ not\ used & \text{if } T_j = GAUSSIAN \end{cases}$$

- Connections between a hidden unit j and an output unit i has the parameter $h_{j,i}$:

  $h_{j,i}$ = weight from hidden unit j to output unit i

- Each output unit i has two parameters t_i and $bias_i$:

$$t_i \in \{ NN\_STYLE,\ FL\_STYLE \} \quad \text{/* type of output unit i */}$$

$$bias_i = \begin{cases} bias\ weight & \text{if } t_i = NN\_STYLE \\ not\ used & \text{if } t_i = FL\_STYLE \end{cases}$$

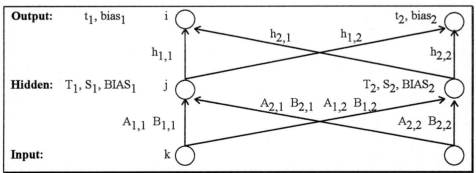

Figure 1: Explanation of the NNFL architecture

We will now explain how the NNFL classifiers operate:
1. First present some data to the input layer.
2. Calculate the value V_j of each hidden unit j : /* usual sigmoid or gaussian */

$$V_j = \begin{cases} \dfrac{1}{1+\exp\left(-2*S_j*\left(\left(\sum\limits_{k=1}^{no.\ input} A_{k,j}*input_k\right) - BIAS_j\right)\right)} & if\ T_j = SIGMOID \\[2em] \prod\limits_{k=1}^{no.\ input} \exp\left(-\left(\dfrac{A_{k,j} - input_k}{0.1*B_{k,j}}\right)^2\right) & if\ T_j = GAUSSIAN \end{cases}$$

3. Calculate the value Out_i of each output unit i:

   /* usual NN style or FL "centre of gravity defuzzification" */

$$Out_i = \begin{cases} \sum\limits_{j=1}^{no.\ hidden} h_{j,i}*V_j - bias_i & if\ t_i = NN\_STYLE \\[2em] \dfrac{\sum\limits_{j=1}^{no.\ hidden} h_{j,i}*V_j}{\sum\limits_{j=1}^{no.\ hidden} V_j} & if\ t_i = FL\_STYLE \end{cases}$$

3. Description of the GA Chromosomes

The idea with the designed encoding is to have a modular and hence a tight encoding which enables the GA to find good building blocks. Hence, we defined the following encoding for each NNFL classifier:

H_1	H_2	...	$H_{no.\ hidden}$	O_1	O_2	...	$O_{no.\ output}$

where each hidden unit H_j consisted of:

T_j	$A_{1,j}$	$B_{1,j}$	$A_{2,j}$	$B_{2,j}$	...	$A_{no.\ input,j}$	$B_{no.\ input,j}$	$BIAS_j$	S_j

and where each output unit O_i consisted of:

t_i	$h_{1,i}$	$h_{2,i}$	...	$h_{no.\ hidden,i}$	$bias_i$

All parameters are encoded as floating point values and are initialised randomly in the range [-1.0, +1.0]. Since we used a fixed length encoding, the maximal size of the NNFL architecture was fixed and could only be reduced (using pruning).

4. Description of the GA Fitness Function

The used GA fitness function is based on earlier work [2,3] but has been redesigned and extended partly based on research results found in [4].

- *Fitness $_i$* = #*correct* - f_0 + f_1 - f_2 - f_3 + f_4 /* ranked according to importance */
1. #correct = the number of correctly classified training patterns

2. f_0 = number of used hidden units (hidden units where $T_j \neq$ NOT_USED).

3. $f_1 = \alpha_1 * \sum\limits_{n=1}^{no.\ data} 2^\beta$, where $\alpha_1 = 1 / ($ no. data $* 2^{no.\ output})$ and

   β = no. output units with a smaller value than the wanted winning unit [1]

4. $f_2 = \alpha_2 *$ no. used connections between the input layer and the hidden layer

   $\alpha_2 = 1/($ no. input units $*$ no. hidden units $)$

5. $f_3 = \alpha_3 *$ no. used connections between the hidden layer and the output layer

   $\alpha_3 = 1/($ no. hidden units $*$ no. output units $)$

6. $f_4 = \alpha_4 * \sum\limits_{n=1}^{no.\ data} \begin{cases} 0.1*(max_1 - max_2) & \textit{if } pattern_n \textit{ is learned correctly} \\ -\left((max_1 - wanted) + \sqrt{(max_1 - wanted)} \right) & \textit{else} \end{cases}$

   $\alpha_4 = 0.1 / ($ no. data $)$, max_1 (max_2) is the value of the largest (second largest) output unit for $pattern_n$

5. Description of Other GA Related Issues

The used GA operates like the following: For each generation the GA selects two parents, generates two offspring, and replaces two individuals with the two generated offspring. The GA stops after G=500*N/2 generations without improvement (see also [4]). The following describes the GA we used:

- Steady State GA with a population size N, such that the population takes up 16 MB of memory.
- Binary tournament selection to select two parents at a time (see also [4]).
- Triple tournament replacement to select two individuals to be replaced (see [4]).
- Usual two-point crossover (2X).
- With a probability of 50% swap 2 randomly chosen hidden units and all corresponding connections (recommended in [4]). The swapping of hidden units allows the GA to combine two good hidden units, which initially are at the same position within the GA chromosome.
- Choose one gene on every generated offspring and vary its value slightly.
- Heavy mutation on every generated offspring: With at probability of 5% make a small "variation" (see above) m=0.05*L times, where L=chromosome length.
- Stochastic hillclimbing on every generated offspring: With a probability P, make a non-negative one-point mutation, i.e. make 1 mutation (a small mutation like above) and accept the mutation if the fitness is not decreased. Probability P is increased dynamically if the GA does not find an improved individual.

6. Results of the Tests

We will now present all results found by the new GANNFL approach, the ssGAFL approach [4], the sGANN [2,3] approach, and results found by other NN techniques

---

[1] We work with winner-takes-all classifiers, hence we only have one wanted winning node.

[1,5]. For each of the 5 problems we made 5 runs to collect statistics. Further, we used a maximal number of 10 hidden units in all tests.

The following abbreviations will be used: IN = no. input units, OUT = no. output units, N = population size, L = chromosome length (no. floating point parameters), <error> = average (avg.) generalisation error, dev = standard deviation of the generalisation error, best = best found generalisation error, <sig> = avg. no. sigmoid units, <gaus> = avg. no. gaussian units, <NN> = avg. no. NN units, <FL> = avg. no. FL units, and finally "-" = not published.

Monks3: IN=17, OUT=2, N=5322, L=3944, no. train data=122, no. test data=432

Technique	<error>	dev	best	<sig>	<gaus>	<NN>	<FL>
GANNFL	3.8	2.1	0.0	1.2	0.0	1.4	0.6
ssGAFL	5.0	1.7	2.8				
sGANN	5.0	0.9	3.5				
BP weight decay [5]	-	-	2.8				
Cascade Correlation [5]	-	-	2.8				
BP [5]	-	-	6.9				

GANNFL performs extremely well using ≅ 1 sigmoid and mostly NN style units.

Cancer: IN=9, OUT=2, N=8962, L=234, no. train data=350, no. test data=174

Technique	<error>	dev	best	<sig>	<gaus>	<NN>	<FL>
GANNFL	1.4	0.3	0.8	1.0	0.2	1.5	0.5
BP 1 hidden layer [1]	1.4	0.5	-				
sGANN	2.3	1.2	0.6				
ssGAFL	2.7	1.1	1.3				

Again the GANNFL used ≅ 1 sigmoid unit and used mostly NN style output units.

Card: IN=51, OUT=2, N=1952, L=1074, no. train data=345, no. test data=172

Technique	<error>	dev	best	<sig>	<gaus>	<NN>	<FL>
GANNFL	13.9	0.6	13.1	2.2	0.0	0.8	1.2
ssGAFL	14.0	0.8	13.1				
BP 1 hidden layer [1]	14.0	1.0	-				
sGANN	14.8	1.2	12.8				

Using ≅ 2 sigmoid units together with a *combination* of NN and FL units were best.

Diabetes: IN=8, OUT=2, N=9799, L=214, no. train data=384, no. test data=192

Technique	<error>	dev	best	<sig>	<gaus>	<NN>	<FL>
GANNFL	23.7	1.2	22.3	0.6	1.6	1.2	0.8
sGANN	24.1	1.5	22.4				
BP 1 hidden layer [1]	24.1	1.9	-				
ssGAFL	24.6	0.6	23.6				

Again a *combination* of NN and FL were best, but 1-2 *gaussian* were necessary.

Glass: IN=9, OUT=6, N=7436, L=282, no. train data=107, no. test data=53

Technique	<error>	dev	best	<sig>	<gaus>	<NN>	<FL>
GANNFL	29.8	2.7	26.4	0.0	5.0	1.2	4.8
ssGAFL	30.8	3.1	26.4				
BP 1 hidden layer [1]	32.7	5.3	-				
sGANN	33.2	4.5	28.3				

For the glass problem the GANNFL Approach avoided sigmoid units and used only gaussian units. Further, the GANNFL Approach used $\cong 1$ NN unit and $\cong 5$ FL units.

It is obvious that the GANNFL Approach outperforms all other approaches, which is achieved by adapting the types of the hidden units and the types of the output units to the problem at hand.

7. Overall Conclusions

The described GANNFL Approach outperforms all other approaches over a broad range of difficult real-world problems. The results strongly indicate that one can find superior classifiers using a GA to train NN's and FLC's.

We conclude that the success of the GANNFL Approach is based on the following:

- The GA fitness function is based on "#correct", i.e. the number of correctly learned training patterns.
- We used a modular and tight encoding.
- Swapping of whole hidden units is preferable.
- Pruning is naturally implemented in the GANNFL Approach.
- A simple stochastic hillclimber gives the "last finish".
- It is important to combine different types of hidden units.
- One should allow different types of classifiers within the same architecture.

Acknowledgement
I would like to thank professor Brian Mayoh for commenting drafts of this article.

References
[1] L. Prechelt, "Proben1 - A Set of Neural Network Benchmark Problems and Benchmarking Rules", TR 21/94 (1994), anonymous ftp: /pub/papers/techreports/1994/1994-19.ps.Z on ftp.ira.uka.de.
[2] Martin Schmidt and Thomas Stidsen, "Using GA to train NN using sharing and pruning", SCAI'95, anon. ftp: pub/empl/marsch/SCAI95.ps.Z on ftp.daimi.aau.dk
[3] Martin Schmidt and Thomas Stidsen, "GA to train NN's using sharing and pruning: Global GA search combined with local BP search", ADT'95, anon. ftp: pub/empl/marsch/ADT95.ps.Z on ftp.daimi.aau.dk, also in John Wiley's book "Neural Networks and their Applications" (chapter 8).
[4] Martin Schmidt, "Genetic Algorithm to train Fuzzy Logic Rulebases and Neural Networks", FLAMOC'96, anon. ftp: pub/empl/marsch/FLAMOC96.ps.Z on ftp.daimi.aau.dk
[5] Wnek, Sarma, Wahab and Michalski, "Comparison learning paradigms via diagrammatic Visualization", TR (1990), George Mason University

Active Learning of the Generalized High-Low-Game

Martina Hasenjäger and Helge Ritter

Universität Bielefeld, Technische Fakultät,
Postfach 10 01 31, D-33501 Bielefeld, Germany
e-mail: {pmhasenj, helge}@techfak.uni-bielefeld.de

Abstract. In this paper, we study the performance of active learning with the query algorithm *Query by Committee* (QBC), which selects a new query such that it approximately maximizes the expected information gain. As target functions, we introduce a generalization of the *High-Low-Game*, for which we derive a theoretically optimal query sequence. This allows us to compare the performance of a QBC-learner with an information-optimal active learner. Simulations show that an active learner that selects queries with QBC rapidly converges against a learner trained with theoretically optimal queries.

1 Introduction

In supervised learning, the common approach is to provide the learner with randomly selected training data. This may lead to unnecessarily large, redundant training sets.

In order to solve this problem, a different paradigm for data selection has emerged during the last few years: *active learning*. Here, the learner is given the power to *actively* select new training samples. That is, it is allowed to ask questions about the target function it is supposed to learn.

We study an approach to active learning that is based on principles of information theory and which selects a query that yields the maximum expected information gain. In particular, we consider a query algorithm called *Query by Committee* (QBC) [2, 1] which implements an approximation of this optimal query strategy in a convenient way. QBC employs an ensemble or committee of learners to estimate the information gain of a query. A query is selected in such a way that the committee's estimation of the expected information gain that can be obtained by the answer is maximized.

As target functions for our investigations, we introduce an extension of the well-known *High-Low-Game* [2, 3]. We show that for this *High-Low-Game of order N* a query sequence can be derived analytically that is optimal in the sense of maximization of the information gain. This allows us to compare the exact optimum that can be achieved by this approach to the approximation of the optimal query strategy conducted by QBC. As we will see, QBC converges rapidly against the optimal solution as the number of the members of the committee increases.

2 The Query Algorithm

QBC [2] was proposed for incremental query learning of binary classification tasks, which means that the training set is built up one example at a time.

The query algorithm proceeds as follows. Set up a committee of $2k$ learners. Find an input vector that is classified as an example of class $c = 0$ by k members of the committee, and as an example for class $c = 1$ by the other k. Ask for the correct classification of this input vector, add this new example to the training set, train the committee on the enlarged training set and repeat.

This procedure implements an approximation of the query selection criterion of maximum expected information gain in the following way. In terms of weight vectors, a query/answer pair will provide maximal information gain about the target function, if it restricts the version space (i. e. the set of weight vectors that are consistent with the training data) as much as possible. In the case of binary classification, this is achieved by bipartitioning the version space after each query. Since the committee of learners serves as a sample of the version space, an approximate bipartitioning of the version space can be achieved with a query that splits the committee into two opposite camps of the same size. Thus a query that leads to maximal disagreement among the committee members can be expected to be highly instructive.

Here some remarks are appropriate. The existence of such a "bipartitioning" query can not be guaranteed. In such cases this requirement should be replaced by a sufficiently weakened condition. The success of QBC depends crucially on the properties of the training algorithm. For optimal performance of the query algorithm, the training algorithm should ensure a uniform distribution of the committee members throughout the training process, otherwise the committee looses its property of being a representative sample of the version space. The size of this sample, i. e. the number of the committee members, has a decisive influence on the quality of a query. In this paper, we mainly focus on the performance of QBC in dependence on the size of the committee.

3 The High-Low-Game of Order N

For our studies, we choose a generalization of the *High-Low-Game*, in which the learning task requires to identify the N locations where a $\{0,1\}$-valued function on the interval $[l, r]$ changes. (In the original High-Low-Game we have $[l, r] = [0, 1]$ and $N = 1$.) We will denote such a generalized High-Low-Game as $HLG(N, l, r)$, with the underlying function $f_{\mathbf{s}} : [l, r] \subset \mathcal{R} \to \{0, 1\}$, with parameters $\mathbf{s} = (s_1, \ldots, s_N) \in [l, r]$, defined as follows:

- if $N = 0$: $f_{\mathbf{s}}(x) = 1$ $\forall x \in [l, r]$
- if N is even and $N > 0$:

$$f_{\mathbf{s}}(x) = \begin{cases} 1 & \text{if} \quad s_{2i} \le x < s_{2i+1} \quad i = 0, \ldots, N/2 \\ 0 & \text{if} \quad s_{2i+1} < x \le s_{2i+2} \quad i = 0, \ldots, N/2 - 1 \end{cases}$$

– if N is odd and $N > 0$:

$$f_\mathbf{s}(x) = \begin{cases} 1 & \text{if} \quad s_{2i+1} \leq x < s_{2i+2} \quad i = 0, \ldots, (N-1)/2 \\ 0 & \text{if} \quad s_{2i} < x \leq s_{2i+1} \quad i = 0, \ldots, (N-1)/2 \end{cases}$$

where $s_0 = l$ and $s_{N+1} = r$. The components of the parameter vector $\mathbf{s}$ denote the points in the input space where the function value changes. Fig. 1 shows an example of a $HLG(N, l, r)$.

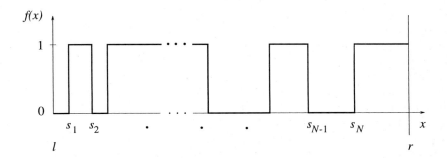

Fig. 1. Example of a High-Low-Game of order N in the interval $[l, r]$, denoted as $HLG(N, l, r)$.

We are interested in an expression for the information content of a query for such a target function.

To this end, we make the assumption of uniform and independent distributions for the components of the parameter vector $\mathbf{s}$, whose determination is the goal of the queries. Each query leads to a partition of the parameter space into two sets of parameter vectors which cause an answer $f_\mathbf{s}(q) = 0$ and $f_\mathbf{s}(q) = 1$, respectively. The probabilities $P^0(q)$ and $P^1(q)$ for the occurrence of either answer can be estimated by the volume of the corresponding part of the parameter space:

$$P^i(q) = \frac{V_N^i(q; l, r)}{\hat{V}_N(l, r)} \quad , \qquad i = 0, 1 \tag{1}$$

where $V_N^i(q; l, r)$, $i = 0, 1$, denotes the partial volume of the parameter space of a $HLG(N, l, r)$ that is occupied by parameter values $\mathbf{s}$ that cause the answer $f_\mathbf{s}(q) = i$ to query q. $\hat{V}_N(l, r)$ denotes the total volume $V_N^0(q; l, r) + V_N^1(q; l, r)$. $\hat{V}_N(l, r)$ is given by

$$\hat{V}_N(l, r) = \frac{1}{N!}(r - l)^N. \tag{2}$$

The information content of a query q is given by the entropy $S(q)$ for the pair of probabilities (1), i. e.

$$S(q) = - \sum_{i=0,1} P^i(q) \log P^i(q) \quad .$$ (3)

The optimal query q^* in the sense of maximizing the information gain of a query is that one which maximizes (3).

In order to derive an expression for $V_N^i(q; l, r)$, we observe that a new query/answer pair reduces the original HLG to two HLGs of lower order. To make this reduction clearer, we adopt the following notation.

Let $\omega = (q, i)$ denote a query/answer pair with $(q, i) \in [0, 1] \times \{0, 1\}$; let $F_N^{(2)}(\omega, \omega')$ denote the partial volume of a $HLG(N, q, q')$ with $f_{\mathbf{s}}(q) = i$ and $f_{\mathbf{s}}(q') = i'$; let $F_N^{(3)}(\omega, \omega', \omega'')$ denote the partial volume of a $HLG(N, q, q'')$ that is bound by a query $\omega' \in [q, q'']$; and let $F_N^{(k)}(\omega^{(1)}, \ldots, \omega^{(k)})$ denote the partial volume of a $HLG(N, q^{(1)}, q^{(k)})$ that is bound by $(k - 2)$ queries in $[q^{(1)}, q^{(k)}]$. $F_N^{(2)}$ and $F_N^{(3)}$ play an important role in the reduction of $F_N^{(k)}$ to $F_N^{(k-1)}$. The consideration of a new query ω' leads to a decomposition of a HLG of order N into two HLGs of order n and n', respectively, i.e. the HLG with N parameters s_i in the interval $[q, q'']$ is broken up into two HLGs, one with n parameters in the interval $[q, q']$, the other one with n' parameters in the interval $[q', q'']$. All we know about n and n' is $n + n' = N$. That is, the parameter volume consists of a sum of volume products for each decomposition of N in n and n'.

$$F_N^{(3)}(\omega, \omega', \omega'') = \sum_{n=0}^{N} F_n^{(2)}(\omega, \omega') F_{N-n}^{(2)}(\omega', \omega'') \quad .$$ (4)

Each factor $F_n^{(2)}(\omega, \omega')$ can be expressed in terms of $\hat{V}_n(q, q')$:

$$F_n^{(2)}((q, i), (q', i')) = \begin{cases} \hat{V}_n(q, q') & \text{if } \{(i \neq i' \text{ and } n \text{ is odd}) \text{ or} \\ & \qquad (i = i' \text{ and } n \text{ is even})\} \\ 0 & \text{otherwise} \end{cases}$$ (5)

It can be shown by induction that in the case of $(k - 2)$ queries, $k \geq 3$, (4) generalizes to

$$F_N^{(k)}\left(\omega^{(1)}, \omega^{(2)}, \ldots, \omega^{(k)}\right) =$$

$$\sum_{\substack{n_1 \ldots n_{k-1} \\ n_1 + \ldots + n_{k-1} = N}} F_{n_1}^{(2)}\left(\omega^{(1)}, \omega^{(2)}\right) F_{n_2}^{(2)}\left(\omega^{(2)}, \omega^{(3)}\right) \ldots F_{n_{k-1}}^{(2)}\left(\omega^{(k-1)}, \omega^{(k)}\right)$$ (6)

This means, the sum runs over all $\binom{N+k-2}{k-2}$ decompositions of N into $(k - 1)$ summands $n_i \geq 0$. Many of these terms vanish because of condition (5). Now we can utilize the F_N to express the value of the partial volumes V_N^i, $i = 0, 1$.

The partial volume $V_N^i(q; l, r)$ of the parameter space of a HLG of order N, that is consistent with the answer $f_{\mathbf{s}}(q) = i$ to the $(k+1)$-th query q, $q^j < q < q^{j+1}$, is finally given by

$$V_N^i(q; l, r) = F_N^{(k+1)} \left(\omega^{(1)}, \ldots, \omega^{(j)}, (q, i), \omega^{(j+1)}, \ldots, \omega^{(k)} \right) \quad . \tag{7}$$

Equation (7), together with the recursion (6) and (1-3) form the essential elements of an algorithm to determine an optimal query q^* at each query step.

4 Results

In this section, we compare the course of the average generalization error for a HLG of order $N = 2$ in the interval $[0, 1]$ of learners trained with a) optimal training data maximizing the information gain of each query which are determined numerically according to Sect. 3 , b) training data selected by the approximation of QBC with various sizes of the committee, c) randomly selected training data. For the learner we used threshold nets with one hidden layer; only the thresholds of the hidden layer were adaptable. This leads to a learner that is able to realize the target function perfectly. In the training process, a new weight vector was selected at random from the version space, i. e. a weight vector was drawn at random from the weight space and accepted as new weight vector, if it was consistent with the training data. This was done to ensure a uniform distribution of the committee members in the version space in the case of QBC.

The results of our simulations are shown in Fig. 2. Here the average test error of 20 runs of the simulations is plotted versus the number of training samples available. In comparison to a passive learner (curve E), the use of an active learner (curves A to C) leads to a considerable acceleration of the learning progress. This is even true for the smallest possible committee with only two members (curve A). In fact, this rough approximation of the optimal strategy already contributes a large part to the improvement that can be obtained by active learning. As the number of the committee members increases (see curve B for a committee with 8 members and curve C for a committee with 64 members), the QBC-learner approaches the optimal course of the generalization error (curve D).

5 Summary

In this paper, we studied active learning with the query algorithm *Query by Committee* (QBC), which uses a committee of learners to estimate the expected information gain of a query and selects the one with the maximum expected information gain. The performance of such a QBC-learner is compared with both an information-optimal active learner and a passive learner using random sampling.

We introduced the *High-Low-Game of order N* for which we derived an optimal query sequence and showed that the generalization error of an active learner that selects the training samples with the aid of QBC rapidly approaches the generalization error of a learner that was trained with the optimal sequence of

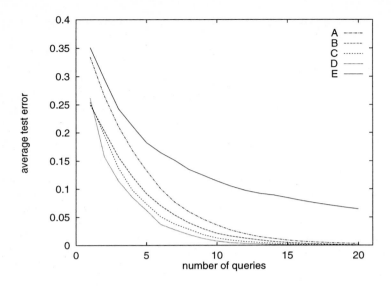

Fig. 2. HLG of order 2 learned by a committee of threshold nets: A – 2 committee members, B – 8 committee members, C – 64 committee members, D – theory, E – random samples

queries as the number of committee members increases.

The comparison of the performance of the two active learners with the passive learner underscores the capability of active learning to achieve much faster convergence and to offer an attractive approach to make learning also more feasible from a practical viewpoint.

References

1. Freund, Y., Seung, H. S., Shamir, E., and Tishby, N.: Information, prediction, and query by committee. In Neural Information Processing Systems 5, ed. by S. J. Hanson, J. D. Cowan, and C. L. Giles (Morgan Kaufmann, San Mateo, CA 1993) pp. 483–490
2. Seung, H. S., Opper, M., and Sompolinsky, H.: Query by committee. In Proc. of the 5th Annual ACM Workshop on Computational Learning Theory (ACM Pr., New York, NY 1992) pp. 287–294
3. Sollich, P.: Query construction, entropy and generalization in neural network models. Phys. Rev. E, **49** (1994) 4637–4651

Optimality of Pocket Algorithm

Marco Muselli

Istituto per i Circuiti Elettronici
Consiglio Nazionale delle Ricerche
via De Marini, 6 - 16149 Genova, Italy
Email: muselli@ice.ge.cnr.it

Abstract. Many constructive methods use the pocket algorithm as a basic component in the training of multilayer perceptrons. This is mainly due to the good properties of the pocket algorithm confirmed by a proper convergence theorem which asserts its optimality.

Unfortunately the original proof holds vacuously and does not ensure the asymptotical achievement of an optimal weight vector in a general situation. This inadequacy can be overcome by a different approach that leads to the desired result.

Moreover, a modified version of this learning method, called pocket algorithm with ratchet, is shown to obtain an optimal configuration within a finite number of iterations independently of the given training set.

1 Introduction

Besides the minimization of the error scored on a given training set an effective learning technique for supervised neural networks must pursue another important aim: the optimization of the number of weights in the final configuration. In fact, the reduction of the number of parameters in a multilayer perceptron lowers the probability of overfitting and increases its generalization ability in the employment phase [1].

Unlike the back-propagation algorithm which works on a fixed architecture and is therefore unable to optimize the resulting configuration, constructive methods generate the desired neural network by consecutively adding units to the hidden layers only when necessary [2, 3, 4]. This allows to obtain in a direct way the optimal or near-optimal number of weights needed to solve a given problem.

In most cases the execution of a constructive method requires the repeated application of a training algorithm for single neuron which provides proper values for the weights of the unit that is going to be added to the network. Such an algorithm must have good convergence properties which lead to a high compactness of the final configuration. For this aim the pocket algorithm [5] is often used, since a favorable theorem ensures the achievement of an optimal weight vector when the execution time increases indefinitely.

A careful examination of the proposed proof [5] shows that the assertion of the convergence theorem holds vacuously and the achievement of an optimal configuration is not generally ensured. The present paper has therefore the aim of extending previous results in order to dissipate any doubt on the asymptotical optimality of the pocket algorithm. Moreover, a new theorem shows that the version with ratchet is finite-optimal, i.e. it reaches the desired configuration within a finite number of iterations.

For sake of brevity all the proofs are only outlined leaving details to a future publication [7].

2   Optimality definitions

Let $\mathcal{N}(\mathbf{w})$ denote a single threshold neuron which input weights and bias are contained in the vector $\mathbf{w}$. The output is binary and can assume values in $\{0, 1\}$. Furthermore, let $\varepsilon(\mathbf{w})$ be the cumulative error (total number of unsatisfied input-output pairs) scored by $\mathcal{N}(\mathbf{w})$ on a given training set S containing a finite number s of samples and denote with $\varepsilon_{\min}$ the minimum of $\varepsilon(\mathbf{w})$. A learning algorithm for the neuron $\mathcal{N}(\mathbf{w})$ provides at every iteration t a weight vector $\mathbf{w}(t)$ which depends on the particular rule chosen for minimizing the cumulative error ε.

Thus we can introduce the following definitions where the probability P is defined over the possible sequences $\mathbf{w}(t)$ generated by the given procedure:

Definition 1. A learning algorithm for the neuron $\mathcal{N}(\mathbf{w})$ is called *(asymptotically) optimal* if

$$\lim_{t \to +\infty} P\left(\varepsilon(\mathbf{w}(t)) - \varepsilon_{\min} < \eta\right) = 1 \quad \text{for every } \eta > 0 \tag{1}$$

independently of the given training set S.

Definition 2. A learning algorithm for the neuron $\mathcal{N}(\mathbf{w})$ is called *finite-optimal* if there exists $\bar{t}$ such that

$$P\left(\varepsilon(\mathbf{w}(t)) = \varepsilon_{\min} , \text{ for every } t \geq \bar{t}\right) = 1 \tag{2}$$

independently of the given training set S.

To reach an optimal configuration the pocket algorithm repeatedly executes perceptron learning whose basic iteration is formed by the following two steps:

1. a sample in the training set S is randomly chosen with uniform probability.
2. if the current threshold neuron $\mathcal{N}(\mathbf{v})$ does not satisfy this sample its weight vector $\mathbf{v}$ is modified in such a way to provide the correct output.

It can be shown that perceptron learning is finite-optimal when the training set S is linearly separable [8]. In this case there exists a threshold neuron that correctly satisfies all the samples contained in the training set S.

In the opposite case the perceptron algorithm cycles indefinitely through feasible weight vectors without providing any (deterministic or probabilistic) information about their optimality. However, we can observe that a neuron $\mathcal{N}(\mathbf{v})$ which satisfies a high number of samples of the training set S has a small probability of being modified at step 2.

This consideration motivates the procedure followed by the pocket algorithm: it holds in memory at every iteration t the weight vector $\mathbf{w}(t)$ (called *pocket vector*) which has remained unchanged for the greatest number of iterations during perceptron learning. The corresponding neuron $\mathcal{N}(\mathbf{w}(t))$ forms the output of the training.

Before approaching the convergence theorems for the pocket algorithm we need to introduce some notations which will be used in the relative proofs. If the training set S contains a finite number s of samples, we can always subdivide the feasible weight vectors for the considered threshold neuron in $s + 1$ (eventually empty) sets W_m, with $m = 0, 1, \ldots, s$. Each of them contains the configurations that satisfy exactly m samples of the given training set and consequently $\varepsilon(\mathbf{w}) = s - m$ for every $\mathbf{w} \in W_m$.

If r is the number of input patterns in S correctly classified by an optimal neuron, then we have $W_m = \emptyset$ for $m > r$ and $\varepsilon_{\min} = s - r$. Thus, the set W_r contains all the optimal weight vectors for the given problem.

Now, let $S(\mathbf{w})$ denote the subset of samples of the training set S which are satisfied by the neuron $\mathcal{N}(\mathbf{w})$. For the perceptron convergence theorem [6, 8], if samples in $S(\mathbf{w})$ are repeatedly chosen, then the weight vector $\mathbf{w}$ will be generated in a finite number of iterations, starting from any initial configuration.

Furthermore, a corollary of the perceptron cycling theorem [8] ensures that the number of different weight vectors which can be generated by the perceptron algorithm is always finite if the input patterns of the training set have integer or rational components. Then, there exists a maximum number τ of iterations required for reaching a configuration $\mathbf{w}$ starting from any initial weight vector, assuming that only samples in $S(\mathbf{w})$ are chosen in the perceptron learning algorithm.

3   Convergence theorems for the pocket algorithm

The existence of a convergence theorem for the pocket algorithm [5] which ensures the achievement of an optimal configuration when the number of iterations increases indefinitely has been an important motivation for the employment of this learning procedure. If we use the definition introduced in the previous section this theorem asserts that the pocket algorithm is optimal (1) when the training set S contains a finite number of samples and all their inputs are integer or rational.

Nevertheless, a correct interpretation of the corresponding proof requires the introduction of some additional quantities which are used in the mathematical formulation. As a matter of fact the convergence relations are characterized by the number N of generations (visits) of an optimal weight vector instead of the corresponding number t of iterations. The value of N does not take into account the permanences obtained by a given weight vector after each generation.

Then, let MN be the greatest number of generations scored by a non-optimal configuration during the same training time t; it can be shown that the ratio M remains finite when the number t of iterations increases indefinitely. In fact, as we noted in the previous section, the set of different weight vectors generated by the pocket algorithm is finite if the inputs in the training set are integer of rational.

Moreover, we can observe that the configurations reached by perceptron learning form a Markov chain since the generation of a new weight vector only depends on the previous one and on the randomly chosen sample of the training set S. Now, the theory of Markov chains [9] ensures that for every pair of configurations the ratio between the number of generations remains constant when the learning time increases indefinitely. Since the number of these pairs is finite, the maximum M of the ratio above always exists.

With these premises we can analyze the original proof for the pocket convergence theorem: it asserts that for every $\sigma \in (0,1)$ the pocket vector $\mathbf{w}(t)$ is optimal with probability greater than σ if the number k of permanences scored by this vector is contained in the following interval:

$$
I(\sigma) = \left[\frac{\log\left(1 - {}^M\sqrt[N]{\sigma}\right)}{\log p_m}, \frac{\log\left(1 - \sqrt[N]{1 - \sigma}\right)}{\log p_r} \right) \tag{3}
$$

for $N \geq \overline{N}$. The value of $\overline{N}$ depends on σ although an explicit relation is not given. The probability p_m in the expression for $I(\sigma)$ is related to the non-optimal weight vector $\mathbf{v}$ which has obtained the greatest number of permanences at iteration t ($\mathbf{v} \in W_m$ with $m < r$).

The validity of the original proof is then limited by the constraint (3) on the number k of permanences obtained by the pocket vector. In fact general conditions on the feasible values for k during the training of a threshold neuron are not given.

On the other hand, by following definition 1 we obtain that the pocket algorithm is asymptotically optimal only if the convergence to the optimal configuration occurs independently of the particular samples chosen at every iteration. Thus, the proof of the pocket convergence theorem must hold for any value assumed by the number of permanences of the pocket vector or by other learning dependent quantities.

Consequently, the pocket convergence theorem has only been shown to hold vacuously and the convergence to an optimal weight vector is not ensured in a general situation. A theoretical proof which presents the desired range of applicability can be obtained by a different approach: consider a binary experiment (such as a coin toss) with two possible outcomes, one of which has probability p and is called *success*. The other outcome has then probability $q = 1 - p$ of occurring and is named *failure*.

Let $Q_k^t(p)$ denote the probability that in t trials of this experiment there does not exist a run of consecutive successes having length k. The complexity of the exact expression for $Q_k^t(p)$ [10] prevent us from directly using it in the mathematical equations; if we denote with $\lfloor x \rfloor$ the truncation of x, good approximations for $Q_k^t(p)$ are given by:

Lemma 3. *If* $1 \leq k \leq t$ *and* $0 \leq p \leq 1$ *the following inequalities hold:*

$$\left(1 - p^k\right)^{t-k+1} \leq Q_k^t(p) \leq \left(1 - p^k\right)^{\lfloor t/k \rfloor} \qquad (4)$$

This lemma, whose technical proof is omitted for sake of brevity, gives a mathematical basis to obtain the asymptotical optimality of pocket algorithm.

Theorem 4. (pocket convergence theorem) *The pocket algorithm is optimal if the training set is finite and contains only input patterns with integer or rational components.*

Proof (outline). Consider t iterations of the pocket algorithm and denote with $\mathbf{v}^*$ and $\mathbf{v}$ the optimal and non-optimal weight vectors respectively which have obtained the greatest number of permanences during the training. Let a_t and b_t be these numbers of permanences.

From the procedure followed by the pocket algorithm we have that the current pocket vector $\mathbf{w}(t)$ is equal to $\mathbf{v}^*$ or $\mathbf{v}$ according to the values of a_t and b_t; in particular, if $a_t > b_t$ the saved weight vector is optimal. Consequently we obtain for $0 < \eta < 1$:

$$P\left(\varepsilon\left(\mathbf{w}(t)\right) - \varepsilon_{\min} < \eta\right) \geq 1 - \sum_{k=0}^{t} P(a_t \leq k | b_t = k) P(b_t = k) \qquad (5)$$

But the training set is finite and contains only input patterns with integer or rational components; thus, if the algorithm consecutively chooses $k + \tau + 1$ samples belonging to $S(\mathbf{v}^*)$ we obtain $k + 1$ permanences of the weight vector $\mathbf{v}^*$. Some mathematical passages involving elementary probabilistic concepts and the application of lemma 4 yield the following inequalities:

$$P(a_t \leq k | b_t = k) \leq \min\left(1, \left(1 - p_r^{k+\tau+1}\right)^{\frac{t}{k+\tau+1}-3}\right) \qquad (6)$$

$$P(b_t = k) \leq 1 - Q_k^t(p_m) \leq 1 - \left(1 - p_m^k\right)^t \qquad (7)$$

where $p_r = r/s$ is the probability of choosing a sample in $S(\mathbf{v}^*)$ and $p_m = m/s$ $(m < r)$ is the probability of having a permanence for $\mathbf{v}$.

By substituting (6) and (7) in (5) we obtain an upper bound for the probability of error at iteration t (note that $0 < p_m < p_r < 1$)

$$P\left(\varepsilon\left(\mathbf{w}(t)\right) - \varepsilon_{\min} \geq \eta\right) \leq \sum_{k=0}^{t} \min\left(1, \left(1 - p_r^{k+\tau+1}\right)^{\frac{t}{k+\tau+1}-3}\right) \left(1 - \left(1 - p_m^k\right)^t\right) \qquad (8)$$

Thus, the optimality of the pocket algorithm is directly shown if the right side of (8) vanishes when the number t of iterations increases indefinitely. This can be verified by breaking the summation above in three contributions having values of the integer k belonging to the intervals $[0, \lfloor t_1 \rfloor]$, $[\lfloor t_1 \rfloor + 1, \lfloor t_2 \rfloor]$, $[\lfloor t_2 \rfloor + 1, t]$, where

$$t_1 = -\alpha \frac{\log t}{\log p_r} \quad , \quad t_2 = t^{(1-\alpha)/2} \quad , \quad \alpha = \frac{\log p_r}{\log \sqrt{p_r p_m}} \qquad \blacksquare$$

Unfortunately, the cumulative error $\varepsilon(\mathbf{w}(t))$ associated with the pocket vector $\mathbf{w}(t)$ does not decrease monotonically with the number t of iterations. To eliminate this undesired effect a modified version of the learning method, called pocket algorithm with ratchet, has been proposed. It computes the corresponding error $\varepsilon(\mathbf{w}(t))$ at every possible saving of a new pocket vector $\mathbf{w}(t)$ and maintains the previous configuration when the number of misclassified samples would increase.

In this way the cumulative error ε decreases monotonically when a new pocket vector is saved and the stability of the algorithm increases. The theoretical properties of the pocket algorithm with ratchet are supported by the following:

Theorem 5. *The pocket algorithm with ratchet is finite-optimal if the training set is finite and contains only input patterns with integer or rational components.*

Proof. From the procedure followed by the pocket algorithm with ratchet we obtain that the cumulative error $\varepsilon(\mathbf{w}(t))$ associated with the pocket vector $\mathbf{w}(t)$ decreases monotonically towards a minimum $\varepsilon(\mathbf{w}(\bar{t}))$ corresponding to the last saved configuration. We have thus $\mathbf{w}(t) = \mathbf{w}(\bar{t})$ for every $t \geq \bar{t}$.

Let $\bar{k}$ denote the number of permanences scored by $\mathbf{w}(\bar{t})$; we have

$$P\left(\varepsilon\left(\mathbf{w}(t)\right) = \varepsilon_{\min} \text{ , for every } t \geq \bar{t}\right) = P\left(\mathbf{w}(t) \notin W_r \text{ , for every } t \geq \bar{t}\right) \leq$$
$$\leq \lim_{t \to +\infty} P\left(\mathbf{w}(t) \notin W_r\right) \leq \lim_{t \to +\infty} Q^t_{\bar{k}+\tau+1}(p_r) \leq 0$$

having used the upper bound in (4). ∎

References

1. HERTZ, J., KROGH, A., AND PALMER, R. G. *Introduction to the Theory of Neural Computation.* Redwood City, CA: Addison-Wesley, 1991.
2. MÉZARD, M., AND NADAL, J.-P. Learning in feedforward layered networks: The tiling algorithm. *Journal of Physics A* **22** (1989), 2191–2203.
3. FREAN, M. The upstart algorithm: A method for constructing and training feedforward neural networks. *Neural Computation* **2** (1990), 198–209.
4. MUSELLI, M. On sequential construction of binary neural networks. *IEEE Transactions on Neural Networks* **6** (1995), 678–690.
5. GALLANT, S. I. Perceptron-based learning algorithms. *IEEE Transactions on Neural Networks* **1** (1990), 179–191.
6. ROSENBLATT, F. *Principles of Neurodynamics.* Washington, DC: Spartan Press, 1961.
7. MUSELLI, M. On convergence properties of pocket algorithm. Submitted for publication on *IEEE Transactions on Neural Networks*.
8. MINSKY, M., AND PAPERT, S. *Perceptrons: An Introduction to Computational Geometry.* Cambridge, MA: MIT Press, 1969.
9. NUMMELIN, E. *General Irreducible Markov Chains and Non-Negative Operators.* New York: Cambridge University Press, 1984.
10. GODBOLE, A. P. Specific formulae for some success run distributions. *Statistics & Probability Letters* **10** (1990), 119–124.

Improvements and Extensions to the Constructive Algorithm CARVE

Steven Young* and Tom Downs
Intelligent Machines Laboratory
Department of Electrical and Computer Engineering
The University of Queensland
Brisbane, Australia, 4072

1. Introduction

The CARVE algorithm, which was presented in [1], extends the sequential learning algorithm of [2] from binary inputs to the real valued input case. The algorithm constructs a feedforward network with a single hidden layer of threshold units, beginning with an empty hidden layer and adding units one at a time. A threshold unit implements a hyperplane in the input domain and the aim is to find a hyperplane that separates a set of points of one class only from the remaining training examples. The set of points that are separated by the hyperplane (the *carve set*) is removed from the training set. The next unit to be added to the network aims to separate another set of points of one class only, but now only from the reduced training set.

In the method described in [1] the search for carve sets belonging to a given class commenced with the construction of the convex hull of the set of points making up all other classes. The rationale for this was that only points outside this convex hull are separable from the other classes. The methods that were used for constructing convex hulls had computational complexity that was exponential in the input dimension and this restricted the applicability of the algorithm to data sets of moderate size and dimension. It is the purpose of this paper to present a new method whose complexity is polynomial in the input dimension and further to present an extension of the algorithm to accommodate higher order threshold units. The new version of CARVE is described in the next section and the extension to higher order units is detailed in section 3. A set of results using relatively high dimension data and using higher order units is given in section 4. Some concluding remarks are presented in section 5.

2. A Polynomial Time Version of CARVE

The CARVE algorithm described in [1] employs a technique known as the beneath-beyond method to generate a data structure that gives a full description of each required convex hull. Inspection of the set of hyperplanes that touch the convex hull then allows the largest carve set to be found. This version of the CARVE

---

* Now with Department of Experimental Psychology, University of Oxford, OX1,3UDUK

algorithm has time complexity which is exponential in the input dimension and thus its applicability is restricted to problems of only moderate size . A more widely applicable polynomial time algorithm can be derived by ceasing to search for all the boundaries of the convex hull and instead attempting to find large carve sets. This we have done using the gift-wrapping method [3], as a basis.

Our procedure for finding large carve sets, which is a slightly modified version of part of the gift-wrapping method, is a two-step process as follows:

Step 1: Find a hyperplane that passes through a boundary point (or boundary points) of the convex hull. This is done by choosing a random direction and determining the hyperplane normal to this direction which passes through a boundary point (or points) of the hull and which has the remainder of the hull set in one half space. We call this the *initial hyperplane*, and the boundary point (or points) the hyperplane passes through is called the *boundary set*.

Step 2: Choose another random direction and determine another hyperplane, normal to that direction, which passes through a point (or points) of the boundary set and which has the remainder of the hull set in one half space. Rotate the initial hyperplane around its intersection with the second hyperplane (in both of the available directions) until it touches another point (or points) in the hull set. Determine and record the sets of points encountered by the hyperplane during its rotation - these constitute a potential carve set.

In order to build up a profile of potential carve sets, Steps 1 and 2 are repeated a prescribed number of times. The number of times we implement Step 1, which determines an initial hyperplane, is denoted by NInit and the number of times we implement Step 2 for each initial hyperplane is denoted by NRot. Thus there is a total of NInit x NRot hyperplane rotations involved in the process of determining a carve set. And after these rotations have been carried out, our selection criterion chooses the carve set that constitutes the largest proportion of a class of points that can be separated.

It is straightforward to show [4] that the worst-case time complexity of the CARVE algorithm is $O(n^2(\log n + d) \times NInit \times NRot)$, where n is the number of training examples and d, as above, is the dimension of the set of training examples. The values of NInit and NRot are selected by the user of the algorithm.

3.   Higher-order CARVE

The CARVE algorithm described in the previous section employs linear threshold units to provide the hyperplanes which separate sets of points of the same class from the remaining points. We will now briefly discuss the implementation of a higher-order variant of the basic algorithm.   The fact that higher-order units implement more general decision surfaces than hyperplanes implies that hidden units generated by a higher-order version of the algorithm should generally be capable of finding larger carve sets than those generated by the basic algorithm.

Extension to the higher-order case is quite straightforward in principle. Note that higher-order threshold functions consist of weighted sums of polynomial terms in the input variables, i.e. weighted sums of $x_1, x_2, x_3, ...,x_1x_2, x_1x_3, ..., x_1^2, x_2^2, ...,$ etc. If we construct a space using the full set of these polynomial terms as coordinates, then the threshold function will define a hyperplane in that space. It follows that if the training data are transformed into this space, we can consider convex hulls of the transformed data points and apply the convex-hull boundary searching methods that we use in the basic algorithm. Hyperplanes in the space of polynomial terms of course correspond to higher-order surfaces in the original input space.

4.   Experimental Results

For any classification task over a finite set of training examples CARVE will generate a single hidden layer threshold network which correctly classifies the training set. Note that because the CARVE procedure depends upon randomly-generated hyperplanes, multiple applications of the procedure to the same learning task will normally lead to the production of different networks, and, in particular, to the production of networks of different sizes. Thus, we have run multiple trials on each learning task in order to obtain some statistics on the sizes of network generated. In addition, we have experimented with the values of the parameters NInit and NRot in order to assess their effect upon network size.

The first learning task below concerns the learning of random Boolean functions and we use this simply as a vehicle for comparing the size of network produced by our new version of CARVE with network size obtained using other well-known techniques. For this first learning problem, we employ only first-order CARVE (ie we use linear threshold units only). The other two learning tasks that we consider are employed to demonstrate the performance of CARVE when it is applied to problems that are more easily solved when the improvements and extensions described in this paper are incorporated into our algorithm. Thus, we demonstrate performance of the higher-order version of CARVE on application to the two-spiral problem and we also demonstrate performance on a well known problem that involves input data of high dimension.

4.1   Random Boolean Functions

A random Boolean function of dimension n consists of a training set containing all 2^n Boolean vectors with each training vector being randomly assigned to one of two classes with equal probability. We applied CARVE to the random

n	CARVE <100>	Sequential [1] <100>	Regular [5] <200>	Tiling [6] <100>	Upstart [7] <25>
4	2.40 ± 0.69			3.9	3
5	3.73 ± 0.58				4.5
6	5.88 ± 0.67	7.28 ± 0.82	15.8 ± 2.2	16.19	8
7	9.47 ± 0.74				16
8	16.23 ± 0.86	18.3 ± 0.69		56.98	33

Table 1: Comparison of network size of constructive algorithms on random Boolean functions.

Boolean function task, and the average network sizes it generated over 100 trials are given, along with standard deviation, in Table 1. These network sizes were obtained after experimenting with values of (NInit, NRot). For $n = 4$, 5 and 6, the network sizes given in Table 1 were obtained with values of (NInit, NRot) less than 50, but the network sizes given for $n = 7$, 8 were obtained with (NInit, NRot) = (70,100). Table 1 also contains results reported for four other constructive algorithms that have been applied to this task. The value in the angle braces at the top of each column is the number of trials over which the network size is averaged.

4.2   Two Spirals

The two spirals problem is a classification task which is highly nonlinearly separable. Single hidden layer networks trained by backpropagation generally fail to produce solutions to this problem and constructive algorithms have been more successful. Application of the first order CARVE algorithm to the two spirals task generates single hidden layer threshold network solutions. A typical example uses (NInit, NRot) = (10, 10), for which the average network size obtained (over 100 trials) was 34.34 ± 4.14, with a minimum network size of 25 hidden units and maximum size of 43 hidden units. This range of network sizes obtained for the two spirals problem

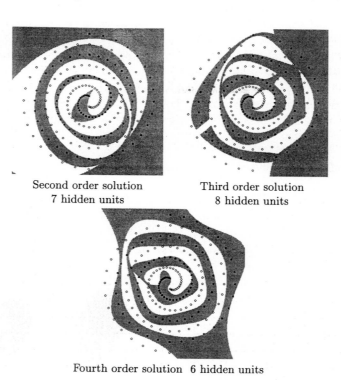

Second order solution   Third order solution
7 hidden units   8 hidden units

Fourth order solution 6 hidden units

Figure 1: Higher order solutions to two spirals.

improves upon the 50 unit single hidden layer network solution reported by Baum and Lang for their query learning scheme [8].

The solutions obtained using first order CARVE are similar in appearance to those presented in [1]. They do not approximate the spiral regions very accurately and better solutions can be obtained by using CARVE with higher order threshold units. The smallest second order solutions had 7 hidden units and were obtained consistently for (NInit, NRot) values above (35,35). The average size for third order solutions with (NInit, NRot) = (100,100) was 10.0, with minimal network size having 8 hidden units. (Our results with second order units imply that third-order solutions with seven and possibly fewer hidden units should exist, but no such solution was found by CARVE). The minimal size obtained using fourth order solutions was 6 hidden units, but networks of this size were not obtained consistently. For values of (NInit, NRot) larger than (50,50) the average size of network solutions was 7 hidden units. Examples of these higher order solutions are given in Fig 1. The smoothness of higher order polynomial threshold surfaces seems well suited to the two spirals task.

4.3 Sonar Data

Gorman and Sejnowski's sonar returns data [9] consists of 208 training examples classified into two classes (Mines/Rocks). Each example consists of 60 real valued variables. Gorman and Sejnowski conducted two experiments to test generalization performance on the sonar data. The first experiment uses cross validation, splitting the 208 training examples into 13 sets of 16 examples each. Each set, in turn, is left out of the training set and is used as the test set. 10 trials on each training set are performed, and test set performance is measured. This experiment ignores the fact that the training examples come from sonar returns with different aspect angle. The second experiment takes this fact into account and splits the data set into equal sized training and test sets with examples from each aspect angle. The generalization performance over 10 trials is averaged. We carried out Gorman and

Network Size	Training set perf. (%)	Test Set perf (%)	Network Size	Training set perf (%)	Test Set perf (%)
CARVE (25,26)			CARVE (100, 100)		
6.9 ± 0.5	100	81.2 ± 10.1	3.2 ± 0.4	100	79.5 ± 3.8
CARVE 2nd Order (25,25)			CARVE 2nd Order (50, 45)		
8.1 ± 0.6	100	76.9 ± 9.7	4.5 ± 0.5	100	78.7 ± 4.5
BP			BP		
0	89.4 ± 2.1	77.1 ± 8.3	0	79.3 ± 3.4	73.1 ± 4.8
2	96.5 ± 0.7	81.9 ± 6.2	2	96.2 ± 2.2	85.7 ± 6.3
3	98.8 ± 0.4	82.0 ± 7.3	3	98.1 ± 1.5	87.6 ± 3.0
6	99.7 ± 0.2	83.5 ± 5.6	6	99.4 ± 0.9	89.3 ± 2.4
12	99.8 ± 0.1	84.7 ± 5.7	12	99.8 ± 0.6	90.4 ± 1.8
24	99.8 ± 0.1	84.5 ± 5.7	24	100	89.2 ± 1.4

Table 2: Sonar Data Experiment 1: Aspect Angle Independent Series

Table 3: Sonar Data Experiment 2 : Aspect Angle Dependent Series

Sejnowski's experiments using CARVE instead of backpropagation and the results are presented in Tables 2 & 3.

For experiment 2, since a single run was only for 10 trials (rather than the 130 trials in experiment 1) we ran experiments for higher values of (NInit, NRot) to find the minimal network size that CARVE can achieve on the task. A difficulty faced by the second order algorithm is that the 60-dimensional input vector for the sonar data becomes a second order vector with 1891 terms so that it is more difficult for the second order algorithm to find smaller network solutions in the same time as the first order algorithm.

5.　　Concluding Remarks

Our experimental results demonstrate that first order CARVE produces smaller networks than those reported for other constructive algorithms on the random Boolean functions problem. This in particular, demonstrates that our method is able to find better CARVE sets than the sequential learning method in [2]. The results obtained for higher-order CARVE on application to the two spirals problem show a much better fit to the problem than those obtained using first-order CARVE. For the sonar data, however, the generalization performance is quite poor. A likely reason for this is that CARVE always produces networks that correctly classify all the training data, implying the possibility of overfitting. This possibility is greater in the higher order case where the expressive power of the network is increased; this is reflected in the results.

References
[1] S Young and T Downs, "CARVE - A constructive algorithm for real valued examples," Proceedings ICANN'94, Sorrento, Italy, pp 785-788.
[2] M Marchand, M Golea and P Rujan, "A convergence theorem for sequential learning in two-layer perceptrons," Europhysics Letters, Vol 11, pp 487-492, 1990.
[3] G Swart, "Finding the convex hull facet by facet," Journal of Algorithms, Vol 6, pp17-48, 1985.
[4] S Young, "Constructive neural network training algorithms for pattern classification using computational geometry techniques," PhD Thesis, University of Queensland, 1995.
[5] R Rujan and M Marchand, "Learning by Minimizing Resources in Neural Networks", Complex Systems, 3, pp. 229-241, 1989.
[6] M Mezard and J-P Nadal, "Learning in feedforward layered networks: the tiling algorithm", Journal of Physics A; Mathematical and General, 22, pp 2191-2203, 1989
[7] M Frean, "The Upstart Algorithm: A Method for Constructing and Training Feedforward Neural Networks", Neural Computation, 2, pp 198-209, 1990
[8] E B Baum and K J Lang, "Constructing hidden units using examples and queries," Advances in Neural Information Processing Systems," Vol 2, pp 904-910, 1990.
[9] R P Gorman and T Sejnowski, "Analysis of hidden units in a layered network trained to classify sonar targets," Neural Networks, Vol 1, pp75-89, 1988.

Annealed RNN Learning
of Finite State Automata

Ken-ichi Arai and Ryohei Nakano

NTT Communication Science Laboratories
2, Hikaridai, Seika-cho, Soraku-gun, Kyoto 619-02 Japan
E-mail: {ken,nakano}@cslab.kecl.ntt.jp
Tel: +81 774 95 1825

Abstract. In recurrent neural network (RNN) learning of finite state automata (FSA), we discuss how a neuro gain (β) influences the stability of the state representation and the performance of the learning. We formally show that the existence of the critical neuro gain (β_0) : any β larger than β_0 makes an RNN maintain the stable representation of states of an acquired FSA. Considering the existence of β_0 and avoidance of local minima, we propose a new RNN learning method with the scheduling of β, called an annealed RNN learning. Our experiments show that the annealed RNN learning went beyond than a constant β learning.

1 Introduction

Recently, there have been many studies on training *recurrent neural networks* (RNNs) to behave like automata or to recognize languages ([6][5][3]). RNNs have been shown to be able to learn *finite state automata* (FSA) and also Push Down Automata (PDA). The studies have shown the powerful capabilities of RNNs to approximate large classes of nonlinear systems.

The orbits of a trained RNN form clusters, where the orbits mean series of the activation values of the state units; the clusters correspond to states of an FSA. Thus, automata can be extracted from values of state units using vector quantization techniques.

We will say that an RNN has a capability to maintain the stable representation of states if for test data which are much longer than training data clusters are distinguishable and the RNN behaves like an FSA that the RNN has acquired for training data. However, most of successfully trained RNNs do not have the capability to maintain the stable representation of states, namely, they lack generalization capabilities for data with the long length. Additionally, there is the serious and general problem associated with the gradient descent method, so-called, the local minimum. It leads an RNN to learn a different FSA from an underlying FSA of training data.

Throughout this paper, we discuss discrete time, continuous space, first-order RNNs. After the brief introduction of an FSA, an RNN and how to train an RNN, in section 3 we show there exists a critical neuro gain (β_0) in learning an FSA and propose an annealed method. In section 4, we show how the annealed learning worked in the experiments.

2 Background

2.1 Finite State Automaton (FSA)

We introduce a finite state automaton (FSA), employing a Moore type definition. A Moore machine $\mathcal{M}$ is a five-tuple

$$\mathcal{M} = < S, I, O, f, g >,$$

where S is a set of states, I is a set of input symbols, O is a set of output symbols, $f : S \times I \to S$ is a state transition function and $g : S \to O$ is an output function. We fix the initial state $s_0 \in S$. A sequence of input symbols is called a *word*.

2.2 Recurrent Neural Network (RNN)

We employ a recurrent neural network (RNN) called the Elman net [1]. An input layer L_I contains n_I input units, including a bias unit whose activation is always set to 1. A state layer L_S has n_S state units, each of which has recurrent connections. An output layer L_O contains n_O output units.

Each unit activation evolves with dynamics defined by the equations

$$h_i(t) = \sum w_{ij}^S S_j(t) + \sum w_{ij}^I \xi_j(t) \tag{1}$$
$$S_i(t+1) = \sigma(h_i(t)), \tag{2}$$

where $h_i(t)$ is an input summation at unit i at time t, $S_i(t)$ is an activation of unit i at time t, w_{ij}^S and w_{ij}^I are connection weights to a unit i from a state unit j and an input unit j, respectively, $\xi_j(t)$ is an input to an input unit i at time t, and σ is a sigmoid function, $\sigma(x) = (1 + \tanh(x))/2$.

We can rewrite the above equations with similar forms to an FSA.

$$\mathbf{S}(t+1) = \mathbf{f}(\mathbf{S}(t), \xi(t)), \tag{3}$$
$$\mathbf{O}(t+1) = \mathbf{g}(\mathbf{S}(t)), \tag{4}$$

where $\mathbf{S} \in (0,1)^{n_S}$, $\xi \in (0,1)^{n_I}$, $\mathbf{O} \in (0,1)^{n_O}$, $\mathbf{f} : (0,1)^{n_S} \times (0,1)^{n_I} \mapsto (0,1)^{n_S}$, and $\mathbf{g} : (0,1)^{n_S} \mapsto (0,1)^{n_O}$. We call $\mathbf{f}$ and $\mathbf{g}$ *neuro-maps*.

2.3 Training procedure

Using training data, a set of words and their corresponding sequences of output symbols, we train an RNN to make the network mimic the behaviour of an FSA. Each symbol appearing in the training data is translated into numerical values for input or output units. We train an RNN by a gradient decent method, using Back Propagation Through Time (BPTT)[7]. The error can be defined as $E = \sum_{t=1}^{L} \sum_i (S_i(t) - \zeta_i(t))^2$, where ζ_i denotes the desired output value and L is a length of data. Thus, we have

$$\Delta w_{ij} = -\eta \frac{\partial E}{\partial w_{ij}} = -\eta \sum_{t=1}^{L} z_i(t) S_j(t-1) \sigma'(h_i(t-1)), \tag{5}$$

where $z_i(t)$ can be calculated backwards in time by the following equation with the final condition $z_i(L) = S_i(L) - \zeta_i(L)$.

$$z_i(t) = S_j(t) - \zeta_j(t-1) + \sum_j z_j(t+1)w_{ji}\sigma'(h_j(t)) \qquad (6)$$

3 Annealed RNN Learning

We focus on a *neuro gain* β which appears in the modified activation function, $S_i(t+1) = \sigma(\beta h_i(t))$. Additionally, $\mathbf{f}_\beta$ and $\mathbf{g}_\beta$ denote neuro-maps $\mathbf{f}$ and $\mathbf{g}$ with the above defined β.

3.1 Existence of critical β_0

We extend the theorem of a stable fixed point for neuro-map [2] to the theorem of stable orbits for neuro-maps with various inputs.

Let $\mathbf{v} = (v_1, \ldots, v_{nS})$ be a vertex, where $v_i \in \{0,1\}$ and V be a set of all vertices, i.e. $V = \{0,1\}^{nS}$. For a given connection matrix W and input data ξ^α corresponding to a symbol α, each $\sum_j w_{ij}^S S_j + \sum_j w_{ij}^I \xi_j^\alpha = 0$ forms a hyperplane H_i^α and we define a domain $(D_\mathbf{v})$ and a partition $(P_{\mathbf{v},\alpha})$ as follows.

$$D_\mathbf{v} = \{ \mathbf{S} \mid 0 \leq S_i < \tfrac{1}{2} \ if \ v_i = 0,$$
$$\tfrac{1}{2} < S_i \leq 1 \quad if \quad v_i = 1, \quad for \quad i = 1, \ldots, n_S\}, \qquad (7)$$
$$P_{\mathbf{v},\alpha} = (\mathbf{f}_\beta^\alpha)^{-1}(D_\mathbf{v}) \ , \qquad (8)$$

where $(\mathbf{f}_\beta^\alpha)^{-1}$ denotes the inverse of $\mathbf{f}_\beta^\alpha$ and $\mathbf{f}_\beta^\alpha$ means $\mathbf{f}_\beta(\mathbf{S}, \xi^\alpha)$. Note that the hyperplanes $\{H_i^\alpha\}$ separate $\Re^{nS}$ into partitions $\{P_{\mathbf{v},\alpha}\}$. Let $\mathcal{F} : \{V \times I\} \mapsto V$ be a vertex transition function. If there exists a word w such that a vertex $\mathbf{v}_q$ can be reached from $\mathbf{v}_p$ by iterative operations of $\mathcal{F}$, then $\mathbf{v}_p$ and $\mathbf{v}_q$ are called *connected*. Let $V_c \subseteq V$ be a set of connected vertices. We define the neighborhood $(N_\mathbf{v})$ of a vertex $\mathbf{v}$, guaranteeing the valid transition for any input α_i,

$$N_\mathbf{v} = D_\mathbf{v} \cap P_{\mathcal{F}(\mathbf{v},\alpha_1),\alpha_1} \cap \ldots \cap P_{\mathcal{F}(\mathbf{v},\alpha_m),\alpha_m}. \qquad (9)$$

Note that for any $\mathbf{S} \in N_\mathbf{v}$, $\mathbf{f}(\mathbf{S}, \xi^{\alpha_i})$ is in $D_{\mathcal{F}(\mathbf{v},\alpha_i)}$, $i = 1, \ldots, m$.

Theorem If $\mathbf{v} \in N_\mathbf{v}$ for all $\mathbf{v} \in V_c$, then there exists a finite β_0 such that for any $\beta > \beta_0$

$$\mathbf{f}_\beta^{\alpha_n} \circ \cdots \circ \mathbf{f}_\beta^{\alpha_2} \circ \mathbf{f}_\beta^{\alpha_1}(D_{\mathbf{v}_0}) \subset D_{\mathbf{v}_n}, \quad where \ \mathbf{v}_n = \mathcal{F}^{\alpha_n} \circ \cdots \circ \mathcal{F}^{\alpha_1}(\mathbf{v}_0) \qquad (10)$$

for any $\mathbf{v}_0 \in V_c$ and any word $\alpha_1\alpha_2 \ldots \alpha_n \in I^*$ having an arbitrary length n. Here, $\mathcal{F}^\alpha(\mathbf{v})$ means $\mathcal{F}(\mathbf{v}, \alpha)$. The value β_0 is called the *critical neuro gain*.

Although a formal proof is omitted due to the space limitation, the intuitive explanation is given below. Suppose that a vertex $\mathbf{v}_0$ is mapped to $\mathbf{v}_1$, $\mathbf{v}_1$ is mapped to $\mathbf{v}_2$, ... for a word $\alpha_1\alpha_2 \ldots$ by $\mathcal{F}$. Then any $\mathbf{S}_0 \in N_{\mathbf{v}_0}$ is mapped to $\mathbf{S}_1 \in D_{\mathbf{v}_1}$ by $\mathbf{f}_\beta$. But if $\beta > \beta_0$, then $\mathbf{S}_1$ is close enough to $\mathbf{v}_1$, in the neighborhood $N_{\mathbf{v}_1}$. Similarly $\mathbf{S}_1$ is mapped to $\mathbf{S}_2 \in N_{\mathbf{v}_2}$. For any word, repetitions of such operation make any orbit be always in the neighborhood of vertices.

Thus we see that for any $\beta(> \beta_0)$ the orbits are robust for small noise and all orbits are stably located in the vicinities of vertices. In such situations, an RNN gets the stable representation of states and decreases error.

3.2 Annealed Learning Method

We propose a new learning method with the scheduling of β [8], called *annealed RNN learning method*. Although an RNN can acquire the stable representation for a large β, we should not start a training with a large β. Because an error surface forms an exact and fine landscape and has many local minima. If β is small, then an error surface is relatively smooth and has a small number of local minima. When β increases a little, we can assume that a new global minimum is close to the previous one. Thus, it is expected that a learning traces a global minimum while β increases slowly. Once an RNN gets a correct FSA, we can rapidly increase β as theorem indicates even if the representation is unstable. Considering the above, we propose a scheduling that increases β slowly at the beginning of learning and gradually goes faster corresponding to the ratio of right recognition.

4 Experiments

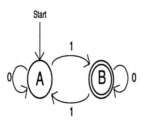

Fig. 1. Original FSA

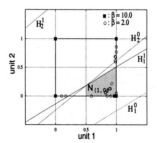

Fig. 2. Activation Space

We consider learning an odd parity automaton shown in Fig. 1, using the RNN and the learning algorithm stated in section 2. The structure of the network is as follows: $n_I = 2$, $n_S = 2$ and $n_O = 1$. Hence, we can examine full activations on two-dimensional space.

Figure 2 shows an example of the internal representation of a trained RNN for $\beta = 2.0$ and 10.0. Hyperplanes are also shown in the same figure. It is shown that the orbits for $\beta = 2.0$ form some indistinct clusters. However, in this example, each neighborhood $N_{\mathbf{v}}$ includes the vertex $\mathbf{v}$; for example, the shaded portion represents $N_{(1,0)}$ and it includes the vertex $(1,0)$. On the other hand, the orbits for $\beta = 10.0$ are localized in the vicinities of vertices. In this case, the clusters are clearly divided, realizing a stable representation of the states as the theorem indicates. Thus, for $\beta(> \beta_0)$ an RNN completely recognizes an FSA acquired by training data of the short length.

Figure 3 shows an error surface for $\beta = 1.0$. The error surface has gentle slopes and there is only one local minimum. Figure 4 shows an error surface for $\beta = 10.0$ and the error surface forms a rugged stair-like landscape. Thus, if we start a training with a large β, then an RNN is often trapped in local minima. Figure 5 shows a typical incorrect FSA extracted from an RNN trapped in local minima. On the other hand, Fig. 6 shows a typical correct FSA extracted from an RNN that learned by the annealed RNN method. We can find that state A in Fig. 1 splits into state A, C and D in Fig. 6, which is called *state-splitting* [4]. But the state-splitting is not seen in Fig. 5.

Figure 7 shows how an average Mean Squared Error (MSE) of 100 samples changes, comparing β scheduling and non-scheduling. We evaluate MSEs using the same large β $(> \beta_0)$ to treat both fairly. The best fixed β is 4.1 in this case. We employ annealing scheduling of $\beta = \frac{3.3}{1.5-r}$, where r is the ratio of correct recognition. Compared with the constant β learning, the learning with β scheduling worked better. Although the best constant β sharply worked at the beginning of learning, in the end, the β scheduling provided a better recognizer than any constant β.

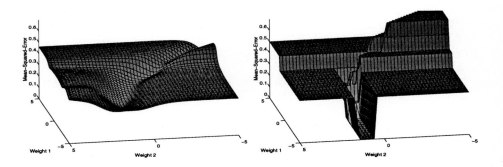

Fig. 3. Error Surface ($\beta = 1.0$)   **Fig. 4.** Error Surface ($\beta = 10.0$)

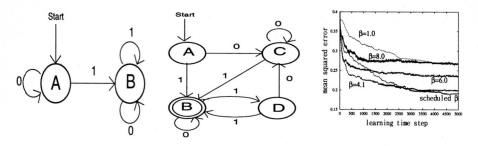

Fig. 5. Incorrect FSA   **Fig. 6.** Correct FSA

Fig. 7. Mean Squared Error

5   Conclusion

In RNN learning of FSA, focussing on a neuro gain, we show that there exists the critical neuro gain (β_0) such that for any $\beta(> \beta_0)$ an RNN gets the stable representation of states of an FSA and propose a new annealed RNN learning to avoid local minima. The experiments using the small example show that the annealed RNN learning makes an RNN better recognizer than a fixed β learning. We plan to evaluate the proposed learning using a larger example.

References

1. J. Hertz, A. Krogh and R.G. Palmer, *Introduction to the Theory of Neural Computation*, Addison-Wesley (1991)
2. Leong K.L., Fixed Point Analysis for Discrete-Time Recurrent Neural Networks, *Proc. IJCNN* **IV** (1992) 134–139
3. P. Tiňo, B.G. Horne, C.L. Giles and P.C. Collingwood, Finite State Machines and Recurrent Neural Networks - Automata and Dynamical System Approaches, *UMIACS-TR-95-1, CS-TR-3396* (1995)
4. M.W. Goudreau, C.L. Giles, S.T. Chakradhar and D. Chen, First-Order Versus Second-Order Single-Layer Recurrent Neural Networks, *IEEE Neural Networks*, Vol. **5** No. 3 (1994) 511–111
5. P. Manolios and R. Fanelli, First-Order Recurrent Neural Networks and Deterministic Finite State Automata, *Neural Computation*, Vol. **6** (1994) 1155–1173
6. S. Das, C.L. Giles and G.Z. Sun, Using Prior Knowledge in a NNPDA to Learn Context-Free Languages, *Advances in Neural Information Processing Systems*, Vol. **5** (1993) 65–71.
7. D.E. Rumelhart, G.E. Hinton and R.J. Williams, Learning Internal Representations by Error Propagation, in *Parallel Distributed Processing*, D.E. Rumelhart and J.L. McClelland (1986) Vol. **1** 318–362
8. N. Ueda and R. Nakano, Deterministic Annealing Variant of the EM Algorithm, *Advances in Neural Information Processing Systems*, Vol. **7** (1995) 545–552.

A Hierarchical Learning Rule for Independent Component Analysis

Bernd Freisleben[1] and Claudia Hagen[2]

[1]Department of Electrical Engineering & Computer Science, University of Siegen
Hölderlinstr. 3, D–57068 Siegen, Germany
E-Mail: freisleb@informatik.uni-siegen.de

[2]Department of Computer Science, University of Darmstadt
Julius-Reiber-Str. 17, D–64293 Darmstadt, Germany
E-Mail: hagen@iti.informatik.th-darmstadt.de

Abstract. In this paper, a two–layer neural network is presented that organizes itself to perform *Independent Component Analysis* (ICA). A hierarchical, nonlinear learning rule is proposed which allows to extract the unknown independent source signals out of a linear mixture. The convergence behaviour of the network is analyzed mathematically.

1 Introduction

Independent Component Analysis (ICA) [4] is a novel signal processing technique for extracting the individual, unobservable source signals out of a linear noisy mixture of them. ICA relies on the assumption of the statistical independence of the source signals, since only in this case the existence of a separation matrix is guaranteed.

Using information-theoretic considerations, Comon [4] derived an adaptive non-neural algorithm for estimating the separation matrix. Another algorithm based on a stochastic gradient descent of a simple objective function for measuring the statistical independence has been developed by Delfosse and Loubaton [6]. Recently, several neural network approaches have been proposed [1, 2, 3, 5, 8, 12] in which recurrent networks with nonlinear functions are used. In [10, 14], it was shown that the introduction of a nonlinear function in a well known neural learning algorithm for *Principal Component Analysis* (PCA) [7] performs one step of ICA, the separation of the source signals, if the input vector and the distribution of the source signals meet some conditions, in order to deal with the higher order moments required in ICA.

In this paper, a new ICA learning rule for a two-layer network is presented. It is hierarchical in the sense that the output units extract the independent components one after the other, similar to the Sanger's linear PCA learning rule [11]. The proposed learning rule is based on the contrast function used in the non-neural ICA algorithm described in [6].

The paper is organized as follows. Section 2 briefly reviews the definition of the ICA basis and criteria for measuring the degree of statistical independence.

In section 3, the new learning rule is presented. In section 4, a mathematical analysis of the convergence behaviour of the network is given. Section 5 concludes the paper and outlines areas for further research.

2 Independent Component Analysis

In this section the general model of ICA is described.

Assume that there are n source signals $s_1(t), \ldots, s_n(t)$ at discrete time $t = 1, 2, \ldots$ which meet the following conditions:

(i) $s_i(t), i = 1, 2, \ldots n, t = 1, 2, \ldots$ are the values of a stationary scalar-valued random process at time t, having unit variance and zero mean.

(ii) $s_i(t), i = 1, 2, \ldots, n, t = 1, 2, \ldots$ are statistically independent, i.e. the joint probability density is the product of the marginal densities:

$$p(s_1(t), \ldots, s_n(t)) = \prod_{i=1}^{n} p_i(s_i(t)) \qquad (1)$$

(iii) $s_i(t), i = 1, 2, \ldots, n, t = 1, 2, \ldots$ have non-Gaussian probability density functions, except at most one which may be a Gaussian.

(iv) $s_i(t), i = 1, 2, \ldots, n, t = 1, 2, \ldots$ are not observable.

The strongest condition is (ii), but it is realistic if the source signals come from different physical devices. The condition (i) can always be achieved by normalization. Condition (iii) reflects the fact that a mixture of Gaussians can not be separated. In many practical situations, the source signals are either *sub-Gaussians* or *super-Gaussians*, i.e. distributions with densities flatter or sharper than that of Gaussians, respectively. Speech and audio signals are typically super-Gaussians, while image signals are often sub-Gaussians [10, 14].

The ICA model can be described as follows: Assume that a linear noisy mixture of source signals $s_i(t), i = 1, 2, \ldots n, t = 1, 2, \ldots$ at time t satisfying the conditions above is given:

$$\mathbf{x}(t) = \mathbf{M}\,\mathbf{s}(t) + \mathbf{n}(t) \qquad (2)$$

where $\mathbf{s}(t)$ is the n-dimensional source signal vector, $\mathbf{x}(t)$ is the m-dimensional observed signal vector, $m \geq n$, $\mathbf{M}$ is the unknown constant $m \times n$ mixing matrix and $\mathbf{n}(t)$ is a noise vector. The pair $(\mathbf{H}, \mathbf{D})$ is called the ICA of the random vector $\mathbf{x}$ if the following conditions are met:

1. The covariance matrix $\mathbf{C}_{xx} = E(\mathbf{x}\mathbf{x}^T)$ can be decomposed into

$$\mathbf{C}_{xx} = \mathbf{H}\,\mathbf{D}^2\,\mathbf{H}^T \qquad (3)$$

where $\mathbf{H}$ is a $m \times n$ matrix with full column rank n, and $\mathbf{D}$ is a $n \times n$ diagonal matrix, consisting of the positive roots of the eigenvalues of $\mathbf{C}_{xx}$.

2. The input vector $\mathbf{x}$ is related to $\mathbf{H}$ through:

$$\mathbf{x}(t) = \mathbf{H}\,\mathbf{y}(t), \tag{4}$$

where the unknown random vector $\mathbf{y} \in \mathbf{R}^n$ has as its covariance matrix $E(\mathbf{y}\mathbf{y}^T) = \mathbf{D}^2$ and has a maximal degree of statistical independence, which can be measured by a so called *contrast function*.

The ICA $(\mathbf{H}, \mathbf{D})$ can be uniquely determined under the following additional conditions:

1. The diagonal matrix $\mathbf{D}$ has its elements in decreasing order.
2. The column vectors of the ICA matrix have unit norm.
3. The entry of the largest element in each column of $\mathbf{H}$ is positive.

A contrast function $\mathcal{F}$ is a real valued function defined on a set $\mathcal{X}$ of random vectors and has to satisfy three conditions:

1. $\mathcal{F}(\mathbf{x})$ depends only on the probability distribution of $\mathbf{x}$ for all $\mathbf{x} \in \mathcal{X}$.
2. The value of $\mathcal{F}(\mathbf{x})$ is invariant while scaling the vector $\mathbf{x}$ with a diagonal matrix.
3. If the components of $\mathbf{x}$ are independent non-Gaussians, $\mathcal{F}(\mathbf{x}) \geq \mathcal{F}(\mathbf{A}\,\mathbf{x})$ for any invertible matrix $\mathbf{A}$ and equal if $\mathbf{A}$ is a diagonal matrix.

In several approaches [3, 4, 6, 10, 14], the contrast function is assumed to depend on the fourth-order cumulants of the components x_i of the vector $\mathbf{x}$. An exact definition of the fourth-order cumulants $c_4(x_i)$ can be found in [9] and an approximation of them in [13]. In [3, 4, 6, 10], different contrast functions are defined in order to approximate the fourth-order cumulants on one hand and to have a formula as simple as possible on the other hand.

The first task of the ICA according to [4, 6] is to estimate the separation matrix $\mathbf{B}$ with $\mathbf{y}(t) = \mathbf{B}\mathbf{x}(t)$ in order to estimate the source signals by the components of vector $\mathbf{y}$. Assuming that the observed signal vector $\mathbf{x}$ has the identity matrix as its covariance matrix $\mathbf{C}_{xx}$, the search for the separation matrix can be reduced to the search on the set of all orthogonal $m \times n$ matrices [6]. The vectors meeting this condition are called *whitened* [10]. In the remainder of this paper, we assume that the signal vector to be separated is whitened. This can simply be obtained by appropriately preprocessing the signal vector $\mathbf{x}(t)$. Several algorithms, both neural and non-neural, for performing such a *prewhitening* are presented in [10, 14]. For example, all PCA algorithms perform this task by computing the diagonal matrix $\mathbf{D}$ of the roots of the n first eigenvalues of the covariance matrix $E(\mathbf{x}\mathbf{x}^T) = \mathbf{F}\,\mathbf{D}^2\,\mathbf{F}^T$ with $\mathbf{F}$ being the orthogonal matrix of the eigenvectors in its columns. The *whitened* signal vector is obtained by transforming and projecting it according to:

$$\tilde{\mathbf{x}}(t) = \mathbf{D}^{-1}\,\mathbf{F}^T\,\mathbf{x}(t) \tag{5}$$

For the sake of simplicity, we denote this prewhitened signal vector again with $\mathbf{x}(t)$ and assume that its dimension coincides with the number n of the source signals.

3   A New ICA Learning Rule

The neural network proposed in this paper consists of an input layer of n input units and one output layer of n output units. Both layers are fully connected to each other. The connection weights are the n-dimensional vectors $\mathbf{w}_i, i = 1, \ldots, n$ which form the columns of the $n \times n$ weight matrix $\mathbf{W}$. The output $y_i, i = 1, \ldots, n$ of the i-th output unit is

$$y_i(t) = \mathbf{w}_i(t)^T \mathbf{x}(t) \quad \text{or} \quad \mathbf{y}(t) = \mathbf{W}(t)^T \mathbf{x}(t) \tag{6}$$

where $\mathbf{x} = (x_1, \cdots, x_n)^T$ is the input vector and $\mathbf{y} = (y_1, \cdots, y_n)^T$ is the output vector.

The input vector $\mathbf{x}$ is assumed to be a linear noisy mixture of unknown source signals, already preprocessed, such that the covariance matrix $E(\mathbf{x}\mathbf{x}^T) = \mathbf{I}$. Similar to other approaches [6, 10, 14], in which particular nonlinear functions are used to allow the separation of either entirely sub-Gaussian or super-Gaussian source signals, we replace condition (iii) of the general ICA model presented in the previous section by assuming that all source signals are sub-Gaussians.

The weight vectors $\mathbf{w}_i, i = 1, 2, \ldots, n$ are updated according to the learning rule

$$\tilde{\mathbf{w}}_i(t+1) = \mathbf{w}_i(t) - \alpha(t) \left[(\mathbf{w}_i(t)^T \mathbf{x}(t))^3 \left(\mathbf{I} - \sum_{j=1}^{i-1} \mathbf{w}_j(t) \mathbf{w}_j(t)^T \right) \mathbf{x}(t) \right] \tag{7}$$

$$\mathbf{w}_i(t+1) = \tilde{\mathbf{w}}_i(t+1)/|\tilde{\mathbf{w}}_i(t+1)| \tag{8}$$

This learning rule is similar to the *robust PCA learning rule* proposed in [10]. The difference is that in our proposal the learning rule is a gradient descent algorithm and it is hierarchical in the sense that output unit i receives contributions from all units j with $j < i$, i.e. the learning rule is different for each of the output units, as in the PCA rule proposed by Sanger [11]. Furthermore, our learning rule is based on an explicit normalization step to keep the weight vectors on the unit sphere. In fact, the authors of [10] indicated that their robust PCA learning rule is not stable when using the nonlinear function $g(y) = y^3$ and a negative learning rate, such that a gradient descent is performed. The authors therefore had to use a different nonlinear function, $g(y) = \tanh y$, to be able to separate sub-Gaussians.

4   Mathematical Analysis

In this section we will show that the weight matrix $\mathbf{W}^T$, adapted using the learning rule given by equations (7), (8) converges to the separation matrix $\mathbf{B}$. The simplest way to prove convergence of a stochastic algorithm is to find it being a gradient of a cost function. If the contrast function $\mathcal{K}$ derived in [6] is taken, a gradient algorithm simply yields the proposed learning rule. The contrast function $\mathcal{K}$ is defined on the unit sphere $\mathcal{S}$ of n-dimensional unit norm vectors:

$$\mathcal{K}(\mathbf{w}) = -E\{(\mathbf{w}^T\mathbf{x})^4\}/4 \qquad (9)$$

Assuming that all source signals $s_i(t), i = 1, \ldots, n$ are sub-Gaussians, the function $-\mathcal{K}$ can be minimized instead of maximizing $\mathcal{K}$ with respect to $\mathbf{w}$. In this case, it was proven in [6] that the vectors $(\pm\mathbf{b}_k)_{k=1,\cdots,n}$, i.e. the rows of the separation matrix $\mathbf{B}$, coincide with the arguments of the local minima of $-\mathcal{K}$ on the unit sphere $\mathcal{S}$. Consequently, a gradient descent algorithm will find one of these local minima. The gradient descent of $-\mathcal{K}$ yields:

$$\mathbf{w}(t+1) = \mathbf{w}(t) - \alpha(t)\frac{\partial(-\mathcal{K})}{\partial\mathbf{w}} \qquad (10)$$

$$= \mathbf{w}(t) - \alpha(t) \cdot (\mathbf{w}(t)^T\mathbf{x}(t))^3\mathbf{x}(t) \qquad (11)$$

This is exactly the learning rule for the first weight vector $\mathbf{w}_1$. In addition, the vector has to be normalized after each step in order to keep it on the unit sphere $\mathcal{S}$. Therefore, we can conclude that the first weight vector $\mathbf{w}_1$ converges to one of the vectors $\pm\mathbf{b}_{k_0}$ for some $k_0 \in \{1, \cdots, n\}$, but we cannot predict to which particular vector $\pm\mathbf{b}_{k_0}$ the weight vector $\mathbf{w}_1$ will converge.

After the first weight vector $\mathbf{b}_{k_0}$ has been found by the minimization of $-\mathcal{K}$, we follow an approach different from the one proposed in [6] in order to find all other local minima $\mathbf{b}_j, j \neq k_0$. The adaptive but non-neural algorithm of [6] is based on the parametrization of a unit vector using rotation matrices. Since we know that the separation matrix $\mathbf{B}$ is orthogonal, we introduce the additional condition

$$\mathbf{w}_i^T\mathbf{w}_j = 0, \quad \forall i \neq j \qquad (12)$$

This condition, combined with the minimization of the function $-\mathcal{K}$, leads to the Lagrange function

$$L(\mathbf{w}_i, \lambda_{i1}, \cdots, \lambda_{i,i-1}) = E\{(\mathbf{w}_i^T\mathbf{x})^4\}/4 + \sum_{j=1}^{i-1}\mathbf{w}_i^T\mathbf{w}_j\,\lambda_{ij} \qquad (13)$$

with λ_{ij} being the Lagrange multipliers.

The optimal values λ_{ij}^* are found by setting the gradient of (13) to zero

$$\lambda_{ij}^* = -E\left(\mathbf{w}_i^T\mathbf{y}\right)^3 \cdot \mathbf{w}_j^T\mathbf{y} \qquad (14)$$

and a gradient descent yields the proposed learning rule:

$$\mathbf{w}_i(t+1) = \mathbf{w}_i(t) - \alpha(t)\left[(\mathbf{w}_i(t)^T\mathbf{x}(t))^3\left(\mathbf{I} - \sum_{j=1}^{i-1}\mathbf{w}_j(t)\mathbf{w}_j(t)^T\right)\mathbf{x}(t)\right] \qquad (15)$$

5 Conclusions

In this paper, we have presented a self-organizing neural network which can be used to separate independent source signals out of a linear mixture of these signals. A new ICA learning rule based on a nonlinear function has been introduced which is hierarchical in the sense that the output units extract the independent components one after the other. Assuming that the input vectors are whitened and the source signals are all sub-Gaussians, it was shown that the columns of the weight matrix $\mathbf{W}$ converge to the rows of the separation matrix $\mathbf{B}$. The task of estimating the ICA matrix itself can be performed by both neural and non-neural algorithms [10, 14].

There are several areas for future research, such as investigating the suitability of using other PCA network architectures or nonlinear learning rules for performing ICA, and the use of the proposed ICA learning rule in signal separation applications.

References

1. S. Amari, A. Cichocki, H. H. Yang. Recurrent Neural Networks For Blind Separation of Sources. In *Proceedings of the 1995 International Symposium on Nonlinear Theory and Applications*, Vol. I, pp. 37-42, 1995.
2. S. Amari, A. Cichocki, H. H. Yang. A New Learning Algorithm for Blind Signal Separation. In *Advances in Neural Information Processing Systems 1995*, Vol. 8, MIT Press, 1996.
3. G. Burel. Blind Separation of Sources: A Nonlinear Algorithm. In *Neural Networks*, Vol. 5, pp. 937-947, 1992
4. P. Comon. Independent Component Analysis: A New Concept. In *Signal Processing*, Vol. 36, pp. 287-314, 1994.
5. P. Comon, C. Jutten, J.Herault. Blind Separation of Sources, Part II: Problem Statements. In *Signal Processing*, Vol. 24, pp. 11-20, 1991.
6. N. Delfosse, P. Loubaton. Adaptive Blind Separation of Independent Sources: A Deflation Approach. In *Signal Processing*, Vol. 45, pp. 59-83, 1995.
7. Jolliffe, I.T. *Principal Component Analysis*. Springer-Verlag, New York, 1986.
8. C. Jutten, J.Herault. Blind Separation of Sources, Part I: An Adaptive Algorithm Based on a Neuromimetic Architecture. In *Signal Processing*, Vol. 24, pp. 1-10, 1991.
9. M. Kendall, A. Stuart. In *The Advanced Theory of Statistics*, Vol. 1, 1977.
10. J. Karhunen, E. Oja, L. Wang, R. Vigario, J. Joutsensalo. A Class of Neural Networks for Independent Component Analysis. Technical Report A 28, Helsinki University, Finland, 1995.
11. Sanger, T.D. Optimal Unsupervised Learning in a Single-Layer Linear Feedforward Neural Network. *Neural Networks*, 2:459–473, 1989.
12. E. Sorouchyari. Blind Separation of Sources, Part III: Stability Analysis. In *Signal Processing*, Vol. 24, pp. 21-29, 1991.
13. D. L. Wallace. Assymptotic Approximations to Distributions. In *Annual Mathematical Statistics*, Vol. 29, pp. 635-654, 1958.
14. L. Wang, J. Karhunen, E. Oja. A Bigradient Optimization Approach for Robust PCA, MCA and Source Separation. In *Proceedings of the 1995 IEEE International Conference on Neural Networks*, Perth, Australia, pp. 1684-1689, 1995.

Improving Neural Network Training Based on Jacobian Rank Deficiency

Guian Zhou and Jennie Si

Department of Electrical Engineering
Arizona State University
Tempe, AZ 85287-5706

Abstract. Analysis and experimental results obtained in [1] have revealed that many network training problems are ill-conditioned and may not be solved efficiently by the Gauss-Newton method. The Levenberg-Marquardt algorithm has been used successfully in solving nonlinear least squares problems, however only for reasonable size problems due to its significant computation and memory complexities within each iteration. In the present paper we develop a new algorithm in the form of a modified Gauss-Newton which on one hand takes advantage of the Jacobian rank deficiency to reduce computation and memory complexities, and on the other hand, still has similar features to the Levenberg-Marquardt algorithm with better convergence properties than first order methods.

1 Introduction

Let $\mathbf{e} = (e_1, e_2, \cdots, e_m)^T$ represent the error between training target and network output, $\mathbf{w} = (w_1, w_2, \cdots, w_n)^T$ be a network weight vector, $J = [j_{ij}]_{m \times n}$ be the Jacobian matrix with $j_{ij} = \frac{\partial e_i}{\partial w_j}$. The Gauss-Newton algorithm for updating the weight vector at the kth iteration is given by [2]

$$(J^T J)\Delta\mathbf{w} = -J^T \mathbf{e}. \tag{1}$$

However, it was observed in [1] that the Jacobian matrix is commonly rank deficient. Experiments have revealed that the rank of a deficient Jacobian matrix is about $60 \sim 80\%$ of the size of the Jacobian on average, and it can reach as low as 20%.

The Levenberg-Marquardt algorithm has been shown to be an efficient modification to the Gauss-Newton method and significantly outperforms the conjugate gradient and the back-propagation with variable learning rate in terms of training time and accuracy [3, 4]. However the computation and memory requirements of the algorithm are about $O(n^3)$ and $O(n^2)$, respectively within each iteration.

---

* Supported in part by NSF under grant ECS-9553202, by EPRI under grant RP8015-03, and by Motorola.

2   A New Algorithm Based on Jacobian Rank Deficiency

The objective of training a feedforward network is to associate input-output training pairs $\{(x_{(1)}, t_{(1)}), (x_{(2)}, t_{(2)}), \cdots, (x_{(m)}, t_{(m)})\}$ by properly adjusting the weights $\mathbf{w}$ in the network such that an error measure $E(\mathbf{w})$ as follows is minimized.

$$E(\mathbf{w}) = \frac{1}{2} \sum_{i=1}^{m} e_i^2 = \frac{1}{2} \sum_{i=1}^{m} (y_{(i)} - t_{(i)})^2, \tag{2}$$

where $y_{(i)}$ represents the network output when the input is $x_{(i)}$.

Let $\mathbf{w}^{(k)}$ denote the network weight vector at the kth iteration. A new weight vector $\mathbf{w}^{(k+1)}$ is obtained by

$$\mathbf{w}^{(k+1)} = \mathbf{w}^{(k)} + \Delta\mathbf{w}, \tag{3}$$

where $\Delta\mathbf{w}$ represents the weight change at the kth iteration.

When the Gauss-Newton update rule is employed, the weight change $\Delta\mathbf{w}$ is computed from

$$(J^T J)\Delta\mathbf{w} = -J^T\mathbf{e}. \tag{4}$$

When J is rank deficient, which is often the case, (4) is not applicable. Based on the diagonal pivoting triangular factorization [5, Chap. 5], a new procedure is devised as follows to obtain $\Delta\mathbf{w}$.

Let $P_1 \in \mathcal{R}^{n \times n}$ be a pivoting matrix. The matrix $J^T J$ can be pivoted and factorized into

$$(JP_1)^T (JP_1) = L^T DL, \tag{5}$$

where $L \in \mathcal{R}^{n \times n}$ is an upper triangular matrix with unit diagonal elements, $D \in \mathcal{R}^{n \times n}$ is a diagonal matrix of the form

$$D = diag(d_1, d_2, \cdots, d_r, 0, \cdots, 0), \quad (d_i \geq d_j, \quad \text{for} \quad i < j)$$

where r is the rank of J.

Let $\Delta\tilde{\mathbf{w}} = P_1^{-1}\Delta\mathbf{w}$, $\mathbf{b} = (L^T)^{-1}(JP_1)^T\mathbf{e}$. Left multiplying P_1^T on both sides of (4) and then substitute JP_1 in (5) and $\Delta\mathbf{w}$ and $\mathbf{b}$ to (4), we have

$$(L^T DL)\Delta\tilde{\mathbf{w}} = -L^T\mathbf{b}, \tag{6}$$

equivalently,

$$DL\Delta\tilde{\mathbf{w}} = -\mathbf{b}. \tag{7}$$

As D is not full rank, we represent D, L, $\Delta\tilde{\mathbf{w}}$ and $\mathbf{b}$ in appropriate block matrix forms as

$$D = \begin{bmatrix} D_1 \\ & 0 \end{bmatrix}, \quad L = \begin{bmatrix} L_1 & L_2 \\ & L_3 \end{bmatrix}, \quad \Delta\tilde{\mathbf{w}} = \begin{bmatrix} \Delta\tilde{\mathbf{w}}_1 \\ \Delta\tilde{\mathbf{w}}_2 \end{bmatrix} \quad \text{and} \quad \mathbf{b} = \begin{bmatrix} \mathbf{b}_1 \\ \mathbf{b}_2 \end{bmatrix} \tag{8}$$

where $D_1 \in \mathcal{R}^{r \times r}$ is a non-zero diagonal matrix, $L_1 \in \mathcal{R}^{r \times r}$ is an upper triangular matrix with unit diagonal elements, and other sub-matrices and sub-vectors have proper dimensions.

(7) then becomes

$$D_1(L_1\Delta\tilde{\mathbf{w}}_1 + L_2\Delta\tilde{\mathbf{w}}_2) = -\mathbf{b}_1. \tag{9}$$

Let $\Delta\tilde{\mathbf{w}}_2 = 0$, we obtain that

$$\Delta\tilde{\mathbf{w}}_1 = -L_1^{-1}D_1^{-1}\mathbf{b}_1. \tag{10}$$

However, it is almost impossible to obtain a converging result by searching along the direction of (10) with adjustable step length in our network training experiments. One of the major problems was observed to be near orthogonal searching directions between (10) and the gradient $J^T\mathbf{e}$.

To improve upon this situation, we modify (10) by adding an rth order symmetric positive definite matrix S as follows

$$\Delta\tilde{\mathbf{w}}_1 = -(L_1^T D_1 L_1 + \mu_1 S)^{-1}L_1^T\mathbf{b}_1, \tag{11}$$

where μ_1 is a positive trial-and-error parameter.

Let $S = L_1^T L_1$. (11) then becomes

$$\begin{aligned}\Delta\tilde{\mathbf{w}}_1 &= -(L_1^T D_1 L_1 + \mu_1 L_1^T L_1)^{-1}L_1^T\mathbf{b}_1 \\ &= -L_1^{-1}(D_1 + \mu_1 I_1)^{-1}\mathbf{b}_1,\end{aligned} \tag{12}$$

where I_1 is an identity matrix with proper dimension.

When (12) is used in training the network weights, μ_1 is decreased after each successful step and increased only if a step increases the training error. This update rule provides an advantage in the sense that if a given μ_1 within an iteration causes an increase in the training error, one only has to solve the equation one more time for $\Delta\tilde{\mathbf{w}}_1$ for an increased μ_1 by backward substitution with a computational complexity of only $O(r^2/2)$.

Experiments have revealed that the updating formula (12) with $\Delta\tilde{\mathbf{w}}_2 = 0$ generally has a satisfactory convergence property. However, this formula is sometimes not able to converge to an ideal accuracy, in particular when the rank of the Jacobian is very low.

Notice also that (9) is not balanced if $\Delta\tilde{\mathbf{w}}_1$ is computed with (12) where μ_1 is not equal to 0. In fact, (9) reduces to

$$D_1 L_2 \Delta\tilde{\mathbf{w}}_2 = -\mu_1(D_1 + \mu_1 I_1)^{-1}\mathbf{b}_1. \tag{13}$$

Perform column pivoting orthogonal decomposition [5, Chap. 9] on $D_1 L_2$, i.e.,

$$D_1 L_2 P_2 = QR \tag{14}$$

where $P_2 \in \mathcal{R}^{(n-r)\times(n-r)}$ is a column pivoting matrix, $Q \in \mathcal{R}^{r\times(n-r)}$ is a unit orthogonal matrix, $R \in \mathcal{R}^{(n-r)\times(n-r)}$ is an upper triangular matrix with non-negative decreasing diagonal elements.

Let $\Delta\check{\mathbf{w}}_2 = P_2^{-1}\Delta\tilde{\mathbf{w}}_2$. (13) then becomes

$$R\Delta\check{\mathbf{w}}_2 = -\mathbf{h} \tag{15}$$

where $\mathbf{h} = \mu_1 Q^{-1}(D_1 + \mu_1 I_1)^{-1}\mathbf{b}_1$.

If R is not full rank, then R, $\Delta\check{\mathbf{w}}_2$ and $\mathbf{h}$ can be divided into appropriate block submatrices like L, $\Delta\tilde{\mathbf{w}}$ and $\mathbf{b}$ in (8), respectively. A similar form to (9) can be obtained as

$$R_1\Delta\check{\mathbf{w}}_{21} + R_2\Delta\check{\mathbf{w}}_{22} = -\mathbf{h}_1 \qquad (16)$$

where R_1 and R_2 are the block submatrices of R, R_1 has the same rank as R, $\Delta\check{\mathbf{w}}_{21}$ and $\Delta\check{\mathbf{w}}_{22}$ are the block sub-vectors of $\Delta\check{\mathbf{w}}_2$, and $\mathbf{h}_1$ is a sub-vector of $\mathbf{h}$.

Solving $\Delta\check{\mathbf{w}}_{21}$ directly from (16) is usually not satisfactory for improving convergence. Similar to the procedure we used to reach (12) from (9), we obtain

$$\Delta\check{\mathbf{w}}_{21} = -R_1^{-1}(D_2 + \mu_2 I_2)^{-1}D_2\mathbf{h}_1 \qquad (17)$$
$$\Delta\check{\mathbf{w}}_{22} = 0$$

where D_2 is a diagonal matrix satisfying $D_2^{1/2} = diag\{R_1\}$, μ_2 is a parameter like μ_1 and I_2 is an identity matrix with proper dimension.

Note that an improved convergence is observed due to a larger search domain resulted from updating more weights by (17) in addition to (12). However, this in turn introduces more computational cost. In practice, (17) is used as little as possible if (12) alone can deliver satisfactory results.

An approximate comparison on memory and computation complexities between the Levenberg-Marquardt algorithm and the new algorithm consisting of (12) and (17) is summarized in Table 2.1.

Table 2.1 An approximate comparison on complexities between the two algorithms

	the new algorithm		the Levenberg-Marquardt	
Memory	$r^2/2 \sim rn\text{-}r^2/2$		n^2	
Multiplications	$m(r^2/2 \sim rn\text{-}r^2/2)$	for $(JP_1)^T JP_1$	$mn^2/2$ for $J^T J$	
	$r^3/6 \sim nr^2/2\text{-}r^3/3$	for L_1, L_2	$n^3/6$	for first factorization
	$0 \sim r(n\text{-}r)^2\text{-}(n\text{-}r)^3/3$ for R_1, R_2		$n^3/6$	for second factorization

3 Applications

In the following examples, if (12) is used alone for updating the weights, we refer to this procedure as the Level One Decomposition (LOD), and the Level Two Decomposition (LTD) refers to the procedure where both (12) and (17) are used for updating the weights. The performance of the algorithm is evaluated quantitatively in terms of training time, training accuracy, convergence properties, *etc.* As a basis of comparison, the Levenberg-Marquardt (LM) algorithm with the same parameters as the LOD and the LTD is used to train the same problems.

Example 1. A ten-step ahead prediction. The system which is also considered in [6] satisfies the following equation,

$$y(i+1) = \frac{y(i)y(i-1)y(i-2)u(i-1)(y(i-2)-1)+u(i)}{1+y^2(i-1)+y^2(i-2)}$$

A two-layer feedforward network $\mathcal{N}_{[14\times20\times1]}$ was trained to predict ten-step ahead in the process. The ith training pattern is formulated as the target output

$t_i = y(i+10)$ and the input $x_i = (y(i), y(i-1), y(i-2), u(i+9), \cdots, u(i-1))^T$, where $u(i)$ was generated from a uniform distribution on [-1, 1]. A total of 700 training samples were used for training the network until $E(\mathbf{w})$ was less than the error goal of 0.2. 20 experimental runs were carried out for 20 different arbitrary initial weights. Table 3.1 summarizes simulation results for the 19 successful runs out of 20.

Table 3.1 Simulation results for a ten-step ahead prediction

	Initial error	Iterations	μ trials	Rank	Final error	Time(s)
LOD	540.81	22.05	46.47	228.42	≤ 0.2	21,339
LTD	535.49	21.37	45.00	231.00	≤ 0.2	22,962
LM	535.49	23.89	48.11	321.00	≤ 0.2	37,332

As can be seen from Table 3.1 that due to the fact that the LOD and the LTD are operating on a lower rank matrix than the LM, the computation time and memory requirements for the LOD and the LTD are much lower than the LM to achieve the same accuracy.

Example 2. A distillation process identification. In this example, a 26-tray methannol-isopropanol distillation column is considered. A mathematical model which describes the dynamic process in each tray of the column similar to the one in [7, Chap. 3] was used to generate data. The system inputs are generated from Gaussian distributions whose mean and covariance are 3.3 and 1/9, respectively.

A feedforward network $\mathcal{N}_{[20 \times 20 \times 1]}$ is used as the identification model. The ith training pattern is given as: the target output $t^i = y(i+1)$, and the input $x^i = (y(i), \cdots, y(i-9), u(i+1), u(i), \cdots, u(i-8))^T$. A total of 900 training patterns were used in the example. 20 experimental runs were carried out for 20 different arbitrary initial weights. Simulation results are tabulated in Table 3.2.

Table 3.2 Simulation Results for a distillation process

	Initial error	Iterations	μ trials	Rank	Final error	Time(s)
LOD	1787.4	20.85	42.70	177.25	≤ 0.04	23,631
LTD	1787.4	20.30	41.20	177.10	≤ 0.04	26,880
LM	1787.4	13.35	27.50	441.00	≤ 0.04	54,697

Similar observations to *Examples 1* still hold in example 2.

Example 3. Convergence comparison among algorithms. Our experience reveals that the LOD can obtain satisfactory convergence results with much less computation and memory complexities than the LM. However, we notice that the algorithm sometimes falls down to a local minimum. Here we adopt the example (Demolm1) which was used in Neural Network Toolboxs in Matlab to make a convergence comparison among the LOD, the LTD and the LM.

A network $\mathcal{N}_{[1 \times 10 \times 1]}$ was used. 30 experimental runs were carried out for 30 different arbitrary initial weights and the network was trained until the error $E(\mathbf{w})$ was less than the desired goals. Table 3.3 summarizes the number of runs which converged to different accuracies for the three algorithms.

Table 3.3   Numbers of runs converging to different accuracies

	$\leq 10^{-1}$	$\leq 10^{-2}$	$\leq 10^{-3}$	$\leq 10^{-4}$	$\leq 10^{-5}$
LOD	30	30	22	0	0
LTD	30	30	29	0	0
LM	30	30	30	1	0

As can be seen from Table 3.3, the LTD has similar convergence properties to the LM. Comparing all three algorithms, the LM takes least number of iterations, the next is the LTD, and then the LOD. However, overall the LOD converges fastest with the least memory and computation complexities, and then the LTD and the LM when accuracy requirement is not extremely high.

4   Conclusions

The Gauss-Newton algorithm is usually not applicable to neural network training problems since the Jacobian matrix is commonly rank deficient. In this paper, a modified Gauss-Newton algorithm based on Jacobian rank deficiency is proposed for training neural networks.

The two versions of the algorithm have two obvious advantages: firstly, it requires at most half of the memory as the Levenberg-Marquardt, which means the algorithm practically applicable to training larger size networks; secondly, the algorithm generally needs only one triangular factorization on a sub-matrix of $J^T J$ whereas the Levenberg-Marquardt needs at least two factorizations on the whole $J^T J$ with an identity matrix, which makes the overall training speed of the new algorithm about two times as fast as the Levenberg-Marquardt according to the examples given in the paper.

References

1. S. Saarinen, R.B. Bramley, and G. Cybenko, "The Numerical Solution of Neural Network Training Problems", CRSD Report No. 1089. Center for Supercomputing Research and Development, University of Illinois, Urbana, 1991.
2. J.E. Dennis, and R.B. Schnabel, *Numerical Methods for Unconstrained Optimization and Nonlinear Equations*. Prentice Hall, Englewood Cliffs, NJ, 1983.
3. M.T. Hagen, and *al el.*, "Training Feedforward Networks with the Marquardt Algorithm", *IEEE Trans. on Neural Networks*, vol.5, No.6, 1994, pp.989-993.
4. R. Battiti, "First- and Second-Order Methods for Learning: Between Steepest Descent and Newton's Method", *Neural Computation*, No.4, 1992, pp.141-166.
5. J.J Dongarra, C.B. Moler, J.R. Bunch, and G.W. Stewart, *LINPACK: Users' Guide*. SIAM Philadelphia, 1979.
6. K.S. Narendra, and K. Parthasarathy, "Identification and Control of Dynamical Systems Using Neural Networks", *IEEE Trans. on Neural Networks*, vol.1, No.1, 1990, pp.4-27.
7. W.L. Luyben, *Process Modeling, Simulation and Control for Chemical Engineering*, McGraw-Hill, 1990.

Neural Networks for Exact Constrained Optimization

József Bíró[1] Edith Halász[1] Tibor Trón[1] Miklós Boda[2] Gábor Privitzky[1]

[1] Technical University of Budapest
Department of Telecommunications and Telematics
H-1111, Sztoczek u. 2., Budapest, Hungary
[2] Ericsson Telecommunication Systems Laboratories

Abstract. The paper is concerned with analog neural networks which can solve nonlinear constrained optimization tasks using the penalty function approach. The neural model developed can be regarded as asymptotically exact dynamic solver in a sense that the equilibrium state represents a solution which can be *arbitrarily* close to that of the original constrained optimization task. Although it is a quite natural requirement, generally it can be fulfilled only with *arbitrarily* large penalty multipliers. The neural network presented overcomes this problem with the use of special nonlinearities, that is it can produce solutions arbitrarily close to the exact one at *finite* penalty multipliers. Stability analyses validating the usefulness of the model are also outlined.

1   Introduction

Hardware approaches of solving optimization tasks have long been of wide interest. Pyne [1] presented a circuit for linear programming as early as in 1956. Recently, there is an advent of this topic thanks to artificial neural networks of which one of the largest application areas is optimization. First, Hopfield and Tank [2] proposed an analog one-layer feed-back 'neural' circuit for solving combinatorial problems. Since then a wide spectrum of Hopfield-like neural networks has appeared for improving the performance of the original system. Tank and Hopfield also designed a two-layer linear programming neural network [3] which was extended (and corrected) for nonlinear programming by Kennedy and Chua [4].

The Kennedy & Chua's network, which is the basis of our works is a continuously operating gradient system and based on the penalty function method in solving nonlinear programming problems. It means that violations of constraints are penalized by adding penalty terms to the objective function to be optimized. In other words, the original constrained task is transformed to optimize an unconstrained objective function. The main drawback of this approach is that the solutions of the constrained and unconstrained tasks may not coincide at finite penalty controlling variables (multipliers) in case of smooth penalty function. It implies that the equilibrium state of pure analog gradient systems like the Kennedy & Chua's network represents only an approximation of the solution of the original problem and provides no bound for the resulted deviation [4].

Exact penalty functions may be used to overcome the difficulties mentioned above. This functions should be non-differentiable at the boundaries of the feasibility region determined by the constraints. In Section 3.1 we present a neural model which relies on efficient approximation of an exact penalty function. In this case, the penalty multipliers should be above *finite* lower bounds.

An interesting and utilizable property of the neural model is that the non-linearities assigned to the penalty functions can be time-varying while stability is preserved. In Section 3.2 the related analysis are performed. In addition, this feature can be utilized to explore which of the constraints influence the quality of the optimum and which don't. In other words, a quick sensitivity analysis can be performed. This also makes it possible to identify the bottleneck constraints and those which provide the highest reward in terms of solution quality for relaxing them. This scenario is also outlined in Section 3.2.

2 The Underlying Optimization Theory

First, let us consider the following constrained optimization task:

$$\text{Minimize } f(x), \text{ subject to } g_i(x) \leq 0, \ h_j(x) = 0, \ i = 1..p, \ j = 1..r, \ r < n \quad (1)$$

where $x \in \mathbf{R}^n$, $f : \mathbf{R}^n \to \mathbf{R}$ and g_i, $h_j : \mathbf{R}^n \to \mathbf{R}$ are scalar valued functions of n variables. We also assume that all the functions are continuously differentiable. This task can be transformed to the minimization of an unconstrained function $f(x)+cP(x)+dQ(x)$ where $P(x)$ and $Q(x)$ are continuously differentiable penalty functions for inequality and equality constraints, respectively. $P(x)$ $(Q(x))$ takes positive value if any of the inequality (equality) constraints is violated, otherwise takes zero. The penalty controlling parameters (multipliers) c and d are always positive.

The following theorem makes the relation between optimizers of the above-mentioned constrained and unconstrained tasks clear.

Theorem 1. *Let c_k and d_k be increasing sequences of penalty multipliers and x_k the sequence of the corresponding local minimizer of the unconstrained objective function $f(x) + c_k P(x) + d_k Q(x)$. If $k \to \infty$ then $x_k \to x^*$, where x^* is a local minimizer of (1).*

It implies that the coincidence of the two solution sets can generally be provided only with infinitely large controlling parameters which is a quite impractical issue. To overcome this difficulty exact penalty functions can be used which are differentiable except on the sets $\{x | g_i(x) = 0, \ i = 1..p\}$ and $\{x | h_j(x) = 0, \ j = 1..r\}$, respectively. In this case, the coincidence can be guaranteed at finite penalty multipliers [7].

Hereafter, we treat penalization in a distributed manner suitable for neural implementation [6], i.e. every constraint has an own penalty function and multiplier. It follows that the unconstrained minimization can be written as

$$\text{Minimize } f(x) + \sum_{i=1}^{p} c_i P_i(g_i(x)) + \sum_{j=1}^{r} d_j Q_j(h_j(x)) \quad (2)$$

If P_i and Q_j are exact penalty functions the solution of (2) will also be solution of (1) at *finite* parameters c_i and d_j. For the next theorem let's assume that P_i, Q_j are convex exact penalty functions and there exist the limits $\lim_{y \to 0, y > 0} P_i(y) = M_i$, $\lim_{y \to 0, y > 0} Q_j(y) = N_j = -\lim_{y \to 0, y < 0} Q_j(y)$.

Theorem 2. *If f, g_i are convex and h_j are affine differentiable functions and the task (1) is solvable then if $\forall(i, j)$, $c_i > u_i^* / M_i$ and $d_j > |v_j^* / N_j|$ the solution sets of problem (1) and (2) are identical. u_i^*, v_j^* are the Lagrangian multipliers for a local minimizer of (1).*

The proof is omitted, interested readers can construct it on the basis of Theorem 9.3.1 [8]. It is also worth noting that the convexity assumptions can be relaxed and substituted by certain second order sufficiency conditions for a constrained local minimizer of problem (1) [9], in this way Theorem 2 holds for a much larger set of optimization tasks than convex optimization.

3 Asymptotically Exact Dynamic Solvers

3.1 Approximation of Exact Penalty Functions

In this section, we introduce a neural network which can be regarded as *asymptotically exact* dynamic solver. The performance of this model is based on an efficient smooth approximation of simple penalty functions: $P_i(g_i(x)) = g_i(x)$ if $g_i(x) > 0$, otherwise $P_i = 0$; $Q_j(h_j(x)) = |h_j(x)|$. Let

$$\tilde{P}_i(x, T) = T \ln(\cosh(g_i(x)/T)) \text{ if } g_i(x) > 0 \text{ otherwise } 0$$
$$\tilde{Q}_j(x, T) = T \ln(\cosh(h_j(x)/T)). \tag{3}$$

where T is a positve constant. It can easily be shown that $\lim_{T \to 0} \tilde{P}_i(x, T) = P_i(x)$ and $\lim_{T \to 0} \tilde{Q}_j(x, T) = Q_j(x)$.

Now let's go on with introducing the neural model. The system consists of three types of neurons. CON_I and CON_E neurons are responsible for constraint fulfillment while DEC neurons produce the decision variables x_k, $k = 1..n$. It means that altogether $n + p + r$ neurons are in the network. CON_I and CON_E are characterized by Ω_i and Λ_j, respectively:

$$\Omega_i\left(g_i(x), T\right) = \tanh\left(g_i(x)/T\right) \text{ if } g_i(x) > 0, \text{ otherwise } 0$$
$$\Lambda_j\left(h_j(x), T\right) = \tanh\left(h_j(x)/T\right) \tag{4}$$

The decision variables can be obtained as outputs of DEC neurons through a nonlinear transformation, that is $x_k = \Theta(z_k)$ where z_k, $k = 1..n$ are state variables to be specified later and Θ is strictly monotone increasing function saturated by finite $X_{max} = \Theta(\infty)$ and $-X_{max} = \Theta(-\infty)$. X_{max} should be as large so that the feasibility set $\mathcal{F}$ determined by the constraints be a subset of the hypercube set $\mathcal{S}_H = [-X_{max}, X_{max}]^n$. It requires the boundedness assumption for $\mathcal{F}$, however, it is not restrictive for most practical applications.

The operation of the whole system can be characterized such that it starts operating at time $t = 0$ with $x(0) = 0$ and then $x(t)$ evolves according to the following set of differential equations:

$$\frac{dz_k}{dt} = -\frac{\partial f(x)}{\partial x_k} - \sum_{i=1}^{p} c_i \Omega_i(g_i(x), T) \frac{\partial g_i(x)}{\partial x_k} - \sum_{j=1}^{p} d_j \Lambda_j(h_j(x), T) \frac{\partial h_j(x)}{\partial x_k} \quad (5)$$

Since $\int \tanh(y/T) dy = T \ln(\cosh(y/T))$ the equation above can be written in the concise form:

$$\frac{dz_k}{dt} = -\nabla_x \varphi(x, c, d, T) = -\nabla_x \left(f(x) + \sum_{i=1}^{n} c_i \tilde{P}_i(x, T) + \sum_{j=1}^{r} d_j \tilde{Q}_j(x, T) \right) \quad (6)$$

It follows that the neural network is a gradient system and $\varphi(x, c, d, T)$ is a qualified Lyapunov function provided that all functions in (1) are bounded over $\mathcal{S}_H$ and f is bounded on the whole space. (Hereafter, c and d are vectors comprising c_i's and d_j's, respectively). In other words, the equilibrium state of the neural network represents a local minimum of $\varphi(x, c, d, T)$. It immediately implies that this local optimum can be arbitrarily close to the solution set of problem (1) if c_i and d_j are above *finite* lower bounds and T is arbitrarily small positive number (see Theorem 2). Therefore, this model is referred to as *asymptotically exact* dynamic solver.

3.2 A Time-varying Neural Network Model

An interesting and utilizable property of the neural network presented in the previous section is that it preserves the stability properties if parameter T varies in time. For simplicity and clarity let's assume that $T(t)$ is strictly monotone decreasing function of time such that $\forall t$, $\lim_{t \to \infty} T(t) = \varepsilon > 0$ and $\lim_{t \to \infty} \frac{dT(t)}{dt} = 0$. Furthermore, we define the following functions:

$$K_i(y, T(t)) = \frac{\partial \tilde{P}_i(y, T(t))}{\partial T(t)} , \quad L_j(y, T(t)) = \frac{\partial \tilde{Q}_j(y, T(t))}{\partial T(t)} \quad (7)$$

The neuron characteristics and the governing equations are the same as in Section 3.1 except the time-varying nature of $T(t)$.

Regarding network stability and solution optimality the next theorem summarizes the main results.

Theorem 3. *The neural network with penalty functions and time-varying parameter $T(t)$ defined above is asymptotically stable in Lyapunov sense and the equilibrium state represents a local minimizer of $\varphi(x, c, d, \varepsilon)$*

Proof: Since the assumed boundedness of g_i, h_j and $\dot{T}(t)$ there exist $\hat{K}_i$ and $\hat{L}_j$ such that $\forall t$, $\hat{K}_i < \inf_{x \in \mathcal{S}_H} K_i(g_i(x), T(t))$ and $\hat{L}_j < \inf_{x \in \mathcal{S}_H} L_j(h_j(x), T)$. Let's consider the function:

$$L(t) = f(x) + \sum_{i=1}^{p} c_i \left(\tilde{P}_i \big(g_i(x), T(t) \big) - T(t) \hat{K}_i \right) + \sum_{j=1}^{r} d_j \left(\hat{Q}_j \big(h_j(x), T(t) \big) - T(t) \hat{L}_j \right)$$
$$(8)$$

First, we show that this is a Lyapunov function of the system. Taking the time derivative of $L(t)$ it is obtained that

$$\frac{dL(t)}{dt} = \sum_{k=1}^{n} \frac{\partial f(x)}{\partial x_k} \frac{dx_k}{dt} + \sum_{i=1}^{p} c_i \Omega_i \big(g_i(x), T(t) \big) \sum_{k=1}^{n} \frac{\partial g_i(x)}{\partial x_k} \frac{dx_k}{dt}$$

$$+ \sum_{j=1}^{r} d_j \Lambda_j \big(h_j(x), T(t) \big) \sum_{k=1}^{n} \frac{\partial h_j(x)}{\partial x_k} \frac{dx_k}{dt} + \sum_{i=1}^{p} c_i \frac{dT(t)}{dt} \left(K_i(g_i(x), T(t)) - \hat{K}_i \right)$$

$$+ \sum_{j=1}^{r} d_j \frac{dT(t)}{dt} \left(L_j(h_j(x), T(t)) - \hat{L}_j \right) \tag{9}$$

Substituting the governing equation (5) into (9) we get the following more concise form for $\dot{L}(t)$.

$$-\sum_{k=1}^{n} \frac{dx_k}{dt} \frac{dz_k}{dt} + \sum_{i=1}^{p} c_i \frac{dT(t)}{dt} \left(K_i(g_i, T) - \hat{K}_i \right) + \sum_{j=1}^{r} d_j \frac{dT(t)}{dt} \left(L_j(h_j, T) - \hat{L}_j \right) \tag{10}$$

The first term is non-positive because Θ is strictly monotone increasing transfer function between z_k and x_k. The second and third summations are negative due to the definition of $\hat{K}_i$, $\hat{L}_j$ and the strictly monotone decrease of $T(t)$. Thus, $\dot{L}(t) < 0$ during the operation of the network. Since $L(t)$ is also bounded from below, the neural network is asymptotically stable according to the Lyapunov's theorem.

Due to the time-variation of $T(t)$, $L(t)$ may have time-varying local optima being followed by the decision variables x. However, the network finally will converge to a local minimizer of $L(t = \infty)$. Since $L(\infty)$ differs from $\varphi(x, c, d, \varepsilon)$ only in a constant, eventually the decision variables converge to a minimizer of $\varphi(x, c, d, \varepsilon)$. $\square$

Remark: Similar stability properties can be obtained if different $T_m(t)$'s ($m = 1..p + r$) are allowed to use in the nonlinearities Ω_i and Λ_j. Moreover, among $T_m(t)$'s there can also be strictly monotone increasing functions (tending to a finite value) or constants. It can be shown that in these cases the network remains stable and converges to a local minimizer of $\varphi(x, c, d, T(t = \infty))$ where $T(t)$ is now a vector comprising $T_m(t)$'s.

Discussion

The main practical implication of the neural network presented above and of Theorem 3 is that the time-varying nature of $T(t)$ does not destroy the basic capabilities of the neural network. This makes it possible to quickly identify the bottleneck constraints and those which provide the highest reward for relaxing them, that is, to perform a quick sensitivity analysis, as follows.

If the increase of a parameter T_m results in better optimum value, it means that the corresponding constraint was a bottleneck constraint, in the sense that its relaxation allows to have a better optimum. Otherwise it is not an essential constraint. Moreover, the $T_m(t)$, which results in the highest improvement in the optimum in the above sense, identifies the constraint which provides the highest return if we invest in relaxing it. This scenario has significance, for example, in telecommunication network dimensioning. An ongoing research is proceeding based on a similar concept [10].

Another practical issue of the time-varying nonlinearities is that they allow to tune the network during its operation in order to refine the quality of the solution, meanwhile the stability is kept, large oscillations or chaotic behaviours don't appear.

Acknowledgement

The authors are grateful to Ericsson Telecommunication Systems Laboratories for the continuous support.

References

1. Pyne, I.B.: Linear Programming on an Analog Computer. *Trans. Am. Ins. Ele. Eng.* **75** (1956) 139–143
2. Hopfield, J.J., Tank, D.W.: 'Neural' Computation on Decisions Optimization Problems. *Biological Cybern.* **52** (1985) 141–152
3. Tank, D.W., Hopfield, J.J.: Simple 'Neural' Optimization Networks: An A/D Converter, Signal Decision Circuit and a Linear Programming Circuit. *IEEE Trans. on Circuits and Systems.* **33** (1986) 533–541
4. Kennedy, M.P., Chua, L.O.: Neural Networks for Nonlinear Programming. *IEEE Trans. on Circuits and Systems.* **35** (1988) 554–562
5. Kennedy, M.P., Chua, L.O.: Unifying the Tank and Hopfield Linear Programming Network and the Canonical Nonlinear Programming Circuit of Chua and Lin. *IEEE Trans. on Circuits and Systems.* **34** (1987) 210–214
6. Bíró, J., Koronkai, Z., Brandt, H., Trón, T., Henk, T., Faragó, A.: Efficient Extensions of Nonlinear Programming Neural Networks. In *International Conference on Artificial Neural Networks.* **2** (1995) 407–411
7. Bertsekas, D.P.: Necessary and Sufficient Conditions for a Penalty Method to be Exact. *Mathematical Programming*, (1975) 87–99
8. Bazaraa, M.S., Sherali, H.D., Shetty, C.M.: *Nonlinear Programming–Theory and Algorithms.* John Wiley & Sons, Inc. 1993.
9. Charalambous, C.: A Lower Bound of the Controlling Parameters of the Exact Penalty Functions. *Mathematical Programming.* **7** (1979) 131–147
10. Faragó, A., Bíró, J., Henk, T., Boda, M.: Analog Neural Optimization for ATM Resource Management. submitted to *IEEE JSAC*, May 1995.

Capacity of Structured Multilayer Networks with Shared Weights

Sabine Kröner[1] and Reinhard Moratz[2]

[1] Technische Informatik I, TU Hamburg-Harburg, D-21071 Hamburg
[2] AG Angewandte Informatik, Universität Bielefeld, Postfach 100131, D-33501 Bielefeld, Germany

Abstract. The capacity or Vapnik-Chervonenkis dimension of a feedforward neural architecture is the maximum number of input patterns that can be mapped correctly to fixed arbitrary outputs. So far it is known that the upper bound for the capacity of two-layer feedforward architectures with independent weights depends on the number of connections in the neural architecture [1].

In this paper we focus on the capacity of multilayer feedforward networks structured by shared weights. We show that these structured architectures can be transformed into equivalent conventional multilayer feedforward architectures. Known estimations for the capacity are extended to achieve upper bounds for the capacity of these general multi-layer feedforward architectures. As a result an upper bound for the capacity of structured architectures is derived that increases with the number of independent network parameters. This means that weight sharing in a fixed neural architecture leads to a significant reduction of the upper bound of the capacity.

1   Introduction

Structured multi-layer feedforward networks gain more and more importance in speech- and image processing applications. Their characteristic is that a-priori knowledge about the task to be performed is already built into their architecture by use of nodes with shared weight vectors. Examples are time delay neural networks [10] and networks for invariant pattern recognition [4, 5].

One problem in the training of neural networks is the estimation of the number of training samples needed to achieve good generalization. In [1] is shown that for feedforward neural architectures this number is correlated with the capacity or Vapnik-Chervonenkis dimension of the architecture. So far an upper bound for the capacity has been derived for two-layer feedforward neural architectures with independent weights: it depends with $\mathcal{O}(\frac{w}{a} \cdot \ln \frac{q}{a})$ on the number w of connections in the neural architecture with q nodes and a output elements.

In this paper we focus on the calculation of upper bounds for the capacity of structured multi-layer feedforward neural architectures. First we give some definitions and introduce a new general terminology for the description of structured neural networks. In section 3 we apply this terminology on structured feedforward architectures first with one layer then with multiple layers. We show that they can be transformed into equivalent conventional multi-layer feedforward architectures, and derive upper bounds for the capacity of the structured architectures. At the end we comment the results.

2 Definitions

A *layered feedforward network architecture* $\mathcal{N}_{e,a}^r$ is a directed acyclic graph with a sequence of e input nodes, $r-1$ ($r \in \mathbb{N}$) intermediate (*hidden*) layers of nodes, and a final output layer with a nodes. Every node is connected only to nodes in the next layer.

To every node k with indegree $n \in \mathbb{N}$ a triplet $(\boldsymbol{w}_k, s_k, f_k)$ is assigned, consisting of a weight vector $\boldsymbol{w}_k \in \mathbb{R}^n$, a threshold value $s_k \in \mathbb{R}$, and an activation function $f_k : \mathbb{R} \to \{0,1\}$. The activation y_k of the node for an input vector $\boldsymbol{x} \in \mathbb{R}^n$ is computed in the common way: $y_k = f_k(\boldsymbol{w}_k \cdot \boldsymbol{x})$. For all input nodes (indegree 1) the weight vectors and the activation functions are fixed: $\boldsymbol{w} := (1), f := \mathrm{Id}$. The activation function for all other nodes is the hard limiter function, and without loss of generality we choose $s = 0$ for the threshold values of all nodes. An architecture $\mathcal{N}_{e,a}^r$ with given triplets $(\boldsymbol{w}, s, f)$ for all nodes we define as a *net* $N_{e,a}^r$. With the net itself a function $F : \mathbb{R}^e \mapsto \{0,1\}^a$ is associated.

Let an architecture $\mathcal{N}_{e,a}^r$ be given. A set of $m \in \mathbb{N}$ input vectors $\boldsymbol{x}_l \in \mathbb{R}^e$ ($l = 1, \ldots, m$) arranged as m rows in a $(m \times e)$-matrix S is denoted *input matrix* for $\mathcal{N}_{e,a}^r$. An input-matrix S is mapped to an $(m \times a)$-*output-matrix* T by a net $N_{e,a}^r$.

Let S be a fixed input-matrix S for $\mathcal{N}_{e,a}^r$. All nets $N_{e,a}^r$ that map S on the same output-matrix T are grouped in a *net class* of $\mathcal{N}_{e,a}^r$ related to S. $\Delta(S)$ is the number of net classes of $\mathcal{N}_{e,a}^r$ related to S. The *growth function* $g(m)$ of an architecture $\mathcal{N}_{e,a}^r$ with m input vectors is the maximum number of net classes over all $(m \times e)$-input matrices S.

Now we consider the nodes of the architecture $\mathcal{N}_{e,a}^r$ within one layer (except the input layer) with the same indegree $d \in \mathbb{N}$. All nodes k whose components of their weight vectors $\boldsymbol{w}_k \in \mathbb{R}^d$ can be permuted through a permutation $\pi_k : \mathbb{R}^d \to \mathbb{R}^d$ so that $\pi_k(\boldsymbol{w}_k) = \boldsymbol{w}$ $\forall k$ for some vector $\boldsymbol{w} \in \mathbb{R}^d$, are elements of the same *node class* $K_{\boldsymbol{w}}$. We call an architecture $\mathcal{N}_{e,a}^r$ *structured* if at least one node class has more than one element. Then the architecture with b node classes $K_{\boldsymbol{w}_i}$ ($i = 1, \ldots, b$) is denoted $\mathcal{N}_{e,a}^r(K_{\boldsymbol{w}_1}, \ldots, K_{\boldsymbol{w}_b})$.

The *Vapnik-Chervonenkis dimension* d_{VC} [9] of a feedforward architecture is defined by $d_{VC} := \sup\{m \in \mathbb{N} \mid g(m) = 2^{ma}\}$. Let $Q := \left\{m \in \mathbb{N} \mid \frac{g(m)}{2^{ma}} \geq \frac{1}{2}\right\}$. Then $c := \sup Q$ for $Q \neq \emptyset$, or $c := 0$ for $Q = \emptyset$, is an upper bound for the Vapnik-Chervonenkis dimension and also defined as capacity in [2, 7].

3 Upper bounds for the capacity

In this section is shown how structured architectures can be transformed into conventional architectures with independent weights. The upper bounds for the capacity of these conventional architectures then are applied to the structured architectures.

A basic transformation needed in the following derivations is the transformation of structured one-layer architectures $\mathcal{N}(K_{\boldsymbol{w}_1}, \ldots, K_{\boldsymbol{w}_b})$ with input nodes of outdegree ≥ 1 and input vectors $\boldsymbol{x}_l$ into structured one-layer architectures $\mathcal{N}'(K_{\boldsymbol{w}_1}, \ldots, K_{\boldsymbol{w}_b})$ with input nodes of outdegree 1 only and dependent input vectors $\boldsymbol{x}_l'$ $(l = 1, \ldots, m)$: Every input node with outdegree $z > 1$ is replaced by z copies of that input node. The outgoing edges of the input node are assigned to the copies in such a way that every copy has outdegree 1. The elements of the input vectors are duplicated in the same way. By permuting the input nodes and the corresponding components of the input vectors we get the architecture $\mathcal{N}''(K_{\boldsymbol{w}_1}, \ldots, K_{\boldsymbol{w}_b})$ without any intersecting edges.

3.1   Structured one-layer architectures

I) First we focus on structured one-layer architectures $\mathcal{N}_{e,a}^1(K_{\boldsymbol{w}})$ with a set $I := \{u_1, \ldots, u_e\}$ of e input nodes and the output layer $K := \{k_1, \ldots, k_a\}$. Let $K_{\boldsymbol{w}} := K$ be the only node class. All nodes in $K_{\boldsymbol{w}} = K$ have the same indegree $d \in \mathbb{N}$.

Theorem 1. *Let a structured one-layer architecture $\mathcal{N}_{e,a}^1(K_{\boldsymbol{w}})$ with only one node class $K_{\boldsymbol{w}} = K$ be given. Suppose $d \in \mathbb{N}$ as the indegree of all nodes in $K_{\boldsymbol{w}}$. The number of input nodes is $e \leq a \cdot d$. For m input vectors of length e*

$$C(m \cdot a, d) := 2 \cdot \sum_{i=0}^{d-1} \binom{m \cdot a - 1}{i}$$

is an upper bound for the growth function $g(m)$ of the structured one-layer architecture $\mathcal{N}_{e,a}^1(K_{\boldsymbol{w}})$.

Proof. At first we examine structured one-layer architectures with outdegree 1 for every input node, equivalent to architectures $\mathcal{N}_{e,a}^1(K_{\boldsymbol{w}})$ with $e = a \cdot d$ input nodes. By permuting the input nodes and the corresponding components of the input vectors we get the architecture $\mathcal{N}''(K_{\boldsymbol{w}})$. Without loss of generality we consider the permutation π of the node class $K_{\boldsymbol{w}}$ as the identity function. Thus we have $\boldsymbol{w} = \boldsymbol{w}(k_1) = \ldots = \boldsymbol{w}(k_a) \in \mathbb{R}^d$ for the a weight vectors (cf. Figure 1 a)). For m fixed input vectors $\boldsymbol{x}_l := (\boldsymbol{x}_l^1, \ldots, \boldsymbol{x}_l^a) \in \mathbb{R}^{ad}$ $(\boldsymbol{x}_l^i \in \mathbb{R}^d, l = 1, \ldots, m, i = 1, \ldots, a)$ let S be an $(m \times a \cdot d)$-input matrix for $\mathcal{N}_{e,a}^1(K_{\boldsymbol{w}})$:

$$S := \begin{pmatrix} \boldsymbol{x}_1 \\ \vdots \\ \boldsymbol{x}_m \end{pmatrix} = \begin{pmatrix} \boldsymbol{x}_1^1 & \cdots & \boldsymbol{x}_1^a \\ \vdots & \ddots & \vdots \\ \boldsymbol{x}_m^1 & \cdots & \boldsymbol{x}_m^a \end{pmatrix} .$$

A given weight vector $\boldsymbol{w}_1 \in \mathbb{R}^d$ defines a function $F_1 : \mathbb{R}^{ad} \to \{0,1\}^a$ or a net N_1 respectively. Let $\boldsymbol{w}_2$ be a weight vector that defines a function F_2 (a net N_2) different to F_1 on the input matrix S. Thus these two nets are elements of different net classes of $\mathcal{N}_{e,a}^1(K_{\boldsymbol{w}})$ related to the input matrix S.

a) Structured architecture $\mathcal{N}''(K_w)$ with $w(k_i) = w \in \mathbb{R}^d$, $i = 1 \ldots a$ b) Architecture $\mathcal{N}^1_{d,1}$

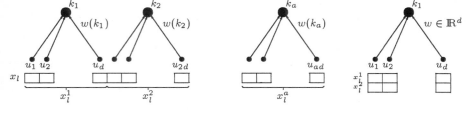

Fig. 1. a) Structured architecture $\mathcal{N}(K_w)''$ with an input vector x_l ($l \in \{1, \ldots, m\}$). b) Architecture $\mathcal{N}^1_{d,1}$ with the corresponding a input vectors $x^1_l, \ldots, x^a_l$.

Now we consider the one-layer architecture $\mathcal{N}^1_{d,1}$, consisting of a single node with indegree d. By rearranging the m rows of the input matrix S for $\mathcal{N}''(K_w)$ to one column of the $m \cdot a$ input vectors $x^i_l \in \mathbb{R}^d$ ($l = 1, \ldots, m$, $i = 1, \ldots, a$) we derive the $(m \cdot a \times d)$-input matrix $\widetilde{S}$ for $\mathcal{N}^1_{d,1}$ (cf. Figure 1 b)):

$$\widetilde{S} := \begin{pmatrix} x^1_1 \\ x^2_1 \\ \vdots \\ x^a_m \end{pmatrix}. \tag{1}$$

On $\mathcal{N}^1_{d,1}$ the weight vector w_1 (w_2) defines a function $\widetilde{F_1} : \mathbb{R}^d \to \{0,1\}$ ($\widetilde{F_2} : \mathbb{R}^d \to \{0,1\}$) or a net $\widetilde{N_1}$ ($\widetilde{N_2}$) respectively. Because of $F_1(x_s) \neq F_2(x_s)$ for at least one input vector x_s ($s \in \{1, \ldots, m\}$) and definition (1) the nets $\widetilde{N_1}$ and $\widetilde{N_2}$ are elements of different net classes of $\mathcal{N}^1_{d,1}$ related to the input matrix $\widetilde{S}$.

Summarizing we get: if two nets of the architecture $\mathcal{N}^1_{e,a}(K_w)$ are different related to any input matrix S we can define an input matrix $\widetilde{S}$ for $\mathcal{N}^1_{d,1}$ by (1), so that the corresponding nets are different, too. For the number of net classes this yields

$$\Delta(S) \leq \Delta(\widetilde{S}). \tag{2}$$

With the results of [7] the growth function of the architecture $\mathcal{N}^1_{d,1}$ is given by $C(m \cdot a, d)$. From (2) also follows that this is an upper bound for the growth function of the structured one-layer architecture $\mathcal{N}''(K_w)$ or $\mathcal{N}^1_{ad,a}(K_w)$ respectively: $g(m) \leq C(m \cdot a, d)$. The inequation $g(m) \geq C(m \cdot a, d)$ can easily be verified in a similar way, so it yields $g(m) = C(m \cdot a, d)$ for the growth function of structured one-layer architectures $\mathcal{N}^1_{e,a}(K_w)$ with outdegree 1 for every input node.

Now we consider structured one-layer architectures $\mathcal{N}_{e,a}^1(K_{\boldsymbol{w}})$ with outdegree $z > 1$ for some input nodes. These architectures can be transformed into structured one-layer architectures $\mathcal{N}''(K_{\boldsymbol{w}})$ with $e = a \cdot d$ input nodes all with outdegree 1. But the input vectors of the input matrix for the transformed architecture $\mathcal{N}''(K_{\boldsymbol{w}})$ cannot be chosen totally independent. Thus $C(m \cdot a, d)$ is an upper bound for the growth function of structured one-layer architectures $\mathcal{N}_{e,a}^1(K_{\boldsymbol{w}})$ with exactly one node class $K_{\boldsymbol{w}} = K$. $\qquad\square$

Remark. With [7] we find $\frac{2 \cdot d}{a}$ as an upper bound for the capacity of structured one-layer architectures $\mathcal{N}_{e,a}^1(K_{\boldsymbol{w}})$ with exactly one node class $K_{\boldsymbol{w}}$.

II) Second we focus on structured one-layer architectures $\mathcal{N}_{e,a}^1(K_{\boldsymbol{w}_1}, \ldots, K_{\boldsymbol{w}_b})$ with b $(2 \leq b < a)$ node classes $K_{\boldsymbol{w}_1}, \ldots, K_{\boldsymbol{w}_b}$. These classes form the set K of the a output nodes: $K = K_{\boldsymbol{w}_1} \dot{\cup} \ldots \dot{\cup} K_{\boldsymbol{w}_b}$.

Theorem 2. *Assume a structured one-layer architecture $\mathcal{N}_{e,a}^1(K_{\boldsymbol{w}_1}, \ldots, K_{\boldsymbol{w}_b})$ with $e \leq \sum_{i=1}^b \alpha_i \cdot d_i$ input nodes, $a = \sum_{i=1}^b \alpha_i$ output nodes, and $b \in \mathbb{N}$ ($2 \leq b \leq a$) node classes $K_{\boldsymbol{w}_i}$ $(i = 1, \ldots, b)$. Let $\alpha_i := |K_{\boldsymbol{w}_i}|$ be the sizes of the node classes $K_{\boldsymbol{w}_i}$, and d_i the indegrees of the nodes in $K_{\boldsymbol{w}_i}$ $(i = 1, \ldots, b)$. For m input vectors the product*

$$\prod_{i=1}^b C(m \cdot \alpha_i, d_i)$$

is an upper bound for the growth function of $\mathcal{N}_{e,a}^1(K_{\boldsymbol{w}_1}, \ldots, K_{\boldsymbol{w}_b})$.

Proof. First we examine structured one-layer architectures $\mathcal{N}_{e,a}^1(K_{\boldsymbol{w}_1}, \ldots, K_{\boldsymbol{w}_b})$ with $e = \sum_{i=1}^b \alpha_i \cdot d_i$ input nodes (all with outdegree 1) and an $(m \times e)$-input matrix S. The a nodes in the output layer K are permuted so that we get the ordered sequence $K = \{K_{\boldsymbol{w}_1}, \ldots, K_{\boldsymbol{w}_b}\}$ with $K_{\boldsymbol{w}_i} := \{k_1^i, \ldots, k_{\alpha_i}^i\}$ $(i = 1, \ldots, b)$. The input nodes and the corresponding components of the input vectors are permuted as in the Proof of Theorem 1: so no edges are intersecting and $\boldsymbol{w}(k_1^i) = \boldsymbol{w}(k_2^i) = \ldots = \boldsymbol{w}(k_{\alpha_i}^i)$ for $i = 1, \ldots, b$. These permutations generate b structured one-layer sub architectures $\mathcal{N}_{e_i, \alpha_i}^1(K_{\boldsymbol{w}_i})$ with $e_i := \alpha_i \cdot d_i$ input nodes, α_i output nodes and $(m \times e_i)$-sub input matrices S^i $(i = 1, \ldots, b)$ (cf. Figure 2).

With Theorem 1 we get $g_i(m) \leq C(m \cdot \alpha_i, d_i)$ for the growth functions $g_i(m)$ of the sub architectures $\mathcal{N}_{e_i, \alpha_i}^1(K_{\boldsymbol{w}_i})$. For the determination of the growth function $g(m)$ of $\mathcal{N}_{e,a}^1(K_{\boldsymbol{w}_1}, \ldots, K_{\boldsymbol{w}_b})$ the b input matrices S^i for the sub architectures can be chosen independently. Thus we get $g(m) = \prod_{i=1}^b g_i(m) \leq \prod_{i=1}^b C(m \cdot \alpha_i, d_i)$.

Now we examine structured one-layer architectures $\mathcal{N}_{e,a}^1(K_{\boldsymbol{w}_1}, \ldots, K_{\boldsymbol{w}_b})$ with outdegree ≥ 1 for some input nodes, equivalent to $e < \sum_{i=1}^b \alpha_i \cdot d_i$ input nodes. These architectures are transformed into structured one-layer architectures with $e = \sum_{i=1}^b \alpha_i \cdot d_i$ input nodes, each with outdegree 1 (cf. proof of Theorem 1) which then are transformed as in the beginning of this proof. As we have

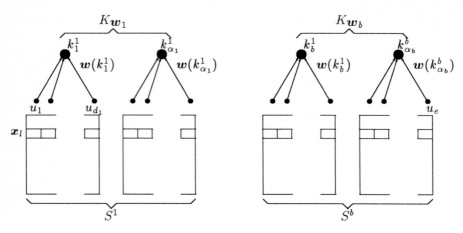

Fig. 2. b sub architectures $\mathcal{N}^1_{e_i,\alpha_i}(K\boldsymbol{w}_i)$ and the corresponding sub input matrices S^i ($i = 1, \ldots, b$) of a structured one-layer architecture $\mathcal{N}^1_{e,a}(K\boldsymbol{w}_1, \ldots, K\boldsymbol{w}_b)$ with b node classes $K\boldsymbol{w}_i$ and $e = \sum_{i=1}^{b} \alpha_i \cdot d_i$ input nodes.

seen in the proof of Theorem 1 there are dependences between some elements of the input vectors since some components of the input vectors are identical. So $\prod_{i=1}^{b} C(m \cdot \alpha_i, d_i)$ is an upper bound for the growth function of structured one-layer architectures $\mathcal{N}^1_{e,a}(K\boldsymbol{w}_1, \ldots, K\boldsymbol{w}_b)$ with $e < \sum_{i=1}^{b} \alpha_i \cdot d_i$ input nodes, too. $\qquad\square$

Theorem 3. *Given a structured one-layer architecture $\mathcal{N}^1_{e,a}(K\boldsymbol{w}_1, \ldots, K\boldsymbol{w}_b)$ with $b \in \mathbb{N}$ ($2 \leq b \leq a$) node classes $K\boldsymbol{w}_i$ ($i = 1, \ldots, b$), maximum indegree $\widehat{d} \geq 2$ for all nodes, maximum size $\widehat{\alpha} := |K\boldsymbol{w}_i|$ of the node classes, and $t := \frac{\widehat{\alpha} \cdot b}{a} \geq 2$, then for the capacity we get*

$$c = \mathcal{O}\left(\frac{b \cdot \widehat{d}}{a} \cdot \ln t \right) .$$

Proof Sketch. With the above definitions and Theorem 2 we get for the growth function $g(m)$ of the architecture $\mathcal{N}^1_{e,a}(K\boldsymbol{w}_1, \ldots, K\boldsymbol{w}_b)$:

$$g(m) \leq \prod_{i=1}^{b} C(m \cdot \alpha_i, d_i) \leq \prod_{i=1}^{b} C(m \cdot \widehat{\alpha}, \widehat{d}) = C(m \cdot \widehat{\alpha}, \widehat{d})^b .$$

This yields an upper bound for the capacity: $c \leq \sup \left\{ m \in \mathbb{N} \;\middle|\; \frac{C(m \cdot \widehat{\alpha}, \widehat{d})^b}{2^{ma}} \geq \frac{1}{2} \right\}$.

With some estimations and $const := \frac{2 + 2 \cdot \ln(2)}{(\ln(2))^2}$ it follows:

$$m \leq const \cdot \frac{t \cdot \widehat{d}}{\widehat{\alpha}} \cdot \ln(t) .$$

For details and further information see [8]. $\qquad\square$

3.2 Structured multi-layer architectures

Consider a structured r-layer architecture with e input nodes, a_j nodes in the hidden layers H^j ($j = 1, \ldots, r - 1$) and a nodes in the output layer K. Let the layers H^j be the disjoint union of the $b_j \leq a_j$ node classes $H_{\boldsymbol{w}_1^j}, \ldots, H_{\boldsymbol{w}_{b_j}^j}$ and the output layer the disjoint union of the node classes $K_{\boldsymbol{w}_i}$ ($i = 1, \ldots, b$). The number of node classes is $\sum_{j=1}^{r-1} b_j + b =: \beta$. The structured architecture is denoted by $\mathcal{N}_{e,a}^r(H_{\boldsymbol{w}_1^1}, \ldots, K_{\boldsymbol{w}_b})$.

A structured r-layer feedforward architecture $\mathcal{N}_{e,a}^r(H_{\boldsymbol{w}_1^1}, \ldots, K_{\boldsymbol{w}_b})$ can be regarded as a combination of r structured 1-layer feedforward architectures since the output matrices of each layer are the input matrices for the following layer. Thus we get an upper bound for the growth function $g(m)$ of $\mathcal{N}_{e,a}^r(H_{\boldsymbol{w}_1^1}, \ldots, K_{\boldsymbol{w}_b})$ by multiplying the growth functions of the r structured 1-layer architectures (refer to Theorem 2):

$$g(m) \leq \prod_{j=1}^{r-1} \left(\prod_{i=1}^{b_j} C(m \cdot \alpha_i^j, d_i^j) \right) \cdot \prod_{i=1}^{b} C(m \cdot \alpha_i, d_i) . \tag{3}$$

With the maximum size $\widehat{\alpha} := \max\left\{ \alpha_1, \ldots, \alpha_b, \alpha_1^1, \ldots, \alpha_{b_{r-1}}^{r-1} \right\}$ of the β node classes, and the maximum indegree $\widehat{d}$ of all nodes of the architecture $\mathcal{N}_{e,a}^r(H_{\boldsymbol{w}_1^1}, \ldots, K_{\boldsymbol{w}_b})$, $C(m \cdot \widehat{\alpha}, \widehat{d})^\beta$ is an upper bound for (3).

Theorem 4. *Let $\mathcal{N}_{e,a}^r(H_{\boldsymbol{w}_1^1}, \ldots, K_{\boldsymbol{w}_b})$ be a structured r-layer feedforward architecture with $\beta \geq 2$ node classes $H_{\boldsymbol{w}_1^1}, \ldots, K_{\boldsymbol{w}_b}$, $\widehat{d} \geq 2$ the maximum indegree of all nodes, $\widehat{\alpha}$ the maximum size of all β node classes, and $\widehat{t} := \frac{\widehat{\alpha} \cdot \beta}{a} \geq 2$. For the capacity of $\mathcal{N}_{e,a}^r(H_{\boldsymbol{w}_1^1}, \ldots, K_{\boldsymbol{w}_b})$ we get*

$$c = \mathcal{O}\left(\frac{\beta \cdot \widehat{d}}{a} \cdot \ln \widehat{t} \right) .$$

Proof. Analogous to the proof of Theorem 3. $\qquad\square$

An architecture $\mathcal{N}_{e,a}^r(H_{\boldsymbol{w}_1^1}, \ldots, K_{\boldsymbol{w}_b})$ with $\sum_{j=1}^{r-1} b_j + b = \sum_{j=1}^{r-1} a_j + a$ node classes is equivalent to an architecture $\mathcal{N}_{e,a}^r$ in which every node class has size 1. Thus the above upper bounds for the capacity hold good for conventional r-layer feedforward architectures with e input and a output nodes, too.

4 Conclusion

In this paper we determined upper bounds for the capacity of structured multi-layer feedforward neural architectures. By transforming architectures with shared weight vectors into equivalent conventional feedforward architectures and the extension of the definitions of the growth function and the capacity to multi-layer feedforward architectures we could give estimations for the upper bounds of the capacity of structured multi-layer architectures. These upper bounds depend with $\mathcal{O}(\frac{p}{a} \cdot \ln \widehat{t})$ on the number p of free parameters in the structured neural architecture with maximum size $\widehat{\alpha}$ of the β node classes, $\widehat{t} := \frac{\widehat{\alpha} \cdot \beta}{a} \geq 2$, and a nodes in the output layer. So weight sharing in a fixed neural architecture leads to a reduction of the upper bound of the capacity. The amount of the reduction increases with the extent of the weight sharing. With $\widehat{\alpha} = 1$ the upper bounds hold good for conventional feedforward networks with independent weights, too.

It is known that the generalization ability of a feedforward neural architecture improves within certain limits with a reduction of the capacity for a fixed number of training samples. As a consequence of our results a better generalization ability can be derived for structured neural architectures compared to the same unstructured ones. This is a theoretic justification for the generalization ability of structured neural architectures observed in experiments [5].

Further investigations will focus on an improvement of the upper bounds for the capacity, on the determination of capacity bounds for special structured architectures, and on the derivation of capacity bounds for structured architectures of nodes with continuous transfer functions [3, 6].

References

1. E. B. Baum, D. Haussler: *What Size Net gives Valid Generalization?*, Advances in Neural Information Processing Systems, D. Touretzky, Ed., Morgan Kaufmann, (1989).
2. T. M. Cover: *Geometrical and Statistical Properties of Systems of Linear Inequalities with Applications in Pattern Recognition*, IEEE Transactions on Electronic Computers, Vol. 14, 326-334, (1965).
3. P. Koiran, E.D. Sontag: *Neural Networks with Quadratic VC Dimension*, Neuro-COLT Technical Report Series, NC-TR-95-044, London, (1995)
4. S. Kröner, R. Moratz, H. Burkhardt: *An adaptive invariant transform using neural network techniques*, in Proceedings of EUSIPCO 94, 7th European Signal Processing Conf., Holt et al. (Ed.), Vol. III, 1489-1491, Edinburgh, (1994).
5. Y. le Cun: *Generalization and Network Design Strategies*, Connectionism in Perspective, R. Pfeiffer, Z. Schreter, F. Fogelman-Soulié, L. Steels (Eds.), Elsevier Science Publishers B.V. 143-155, North-Holland, (1989).
6. W. Maass: *Vapnik-Chervonenkis Dimension of Neural Nets*, Preprint, Technische Universität Graz, (1994).
7. G. J. Mitchison, R. M. Durbin: *Bounds on the Learning Capacity of Some Multi-Layer Networks*, Biological Cybernetics, Vol.60, No. 5, 345-356, (1989).
8. P. Rieper: *Zur Speicherfähigkeit vorwärtsgerichteter Architekturen künstlicher neuronaler Netze mit gekoppelten Knoten*, Diplomarbeit, Universität Hamburg, (1994).
9. V. Vapnik: *Estimation of Dependences Based on Empirical Data*, Springer-Verlag, Berlin, (1982).
10. A. Waibel: *Modular Construction of Time-Delay Neural Networks for Speech Recognition*, Neural Computation, Vol.1, 39-46, (1989).

Optimal Weight Decay in a Perceptron

Siegfried Bös

Lab for Information Representation, RIKEN
Hirosawa 2–1, Wako–shi, Saitama, 351–01, Japan
Tel: +81–48-, phone: -467–9625, fax: -462–9881
email: boes@zoo.riken.go.jp

Abstract. Weight decay was proposed to reduce overfitting as it often appears in the learning tasks of artificial neural networks. In this paper weight decay is applied to a well defined model system based on a single layer perceptron, which exhibits strong overfitting. Since the optimal non-overfitting solution is known for this system, we can compare the effect of the weight decay with this solution. A strategy to find the optimal weight decay strength is proposed, which leads to the optimal solution for any number of examples.

1 The Model

Overfitting is a problem in neural network learning which can reduce the performance drastically. Simply defined, overfitting means that the networks learns too many details from the examples and neglects its generalization ability. In this paper we study how weight decay, see [3], can help to avoid overfitting.

In previous works [1] and [2], we have established a model system, which can be handled analytically, but shows already many of the characteristics of general feedforward learning. The system is based on a continuous single–layer perceptron, which has one layer of N variable weights W_i between its input units and the output unit. To compute the output z, a continuous function g is applied on the weighted sum h of the inputs x_i, i.e.

$$z = g(h), \qquad \text{with} \qquad h = \frac{1}{\sqrt{N}} \sum_{i=1}^{N} W_i \, x_i \,. \tag{1}$$

Learning is done in a *supervised* fashion. That means one uses example inputs $\mathbf{x}^\mu$ (with $\mu = 1, \ldots, P$) from the input space I, for which the correct outputs z_*^μ are known. For theoretical purposes we assume that the learning task is provided by another network, which we call *teacher network*.

The mean squared error is used to measure the difference between the outputs of the teacher and student. Training tries to minimize the *training error* E_{T}, which is the mean error over the set of examples, and the performance is measured by the *generalization error* E_{G}, which is averaged over all possible inputs, i.e.

$$E_{\mathrm{T}} := \frac{1}{2P} \sum_{\mu=1}^{P} [z_*^\mu(\mathbf{x}^\mu) - z^\mu(\mathbf{x}^\mu)]^2 \,, \qquad E_{\mathrm{G}} := \frac{1}{2} < [z_*(\mathbf{x}) - z(\mathbf{x})]^2 >_{\{\mathbf{x} \in I\}} \,. \tag{2}$$

The concept of the teacher network allows an efficient monitoring of the training process. Suitable quantities for the monitoring of the training process are the normalized *order parameters*,

$$q := \sqrt{\frac{1}{N}\sum_{i=1}^{N}(W_i)^2} = ||\mathbf{W}||, \qquad r := \frac{1}{||\mathbf{W}||}\frac{1}{N}\sum_{i=1}^{N} W_i^* W_i, \qquad (3)$$

in which the '*' denotes the variables belonging to the teacher net. Both have obvious meanings, q is the norm of the student's weights and r is the cosine of the angle between the two weight vectors of teacher and student.

In this paper we will study how weight decay can help to reduce overfitting, which appears usually in unrealizable learning tasks. We implement the unrealizability simply by different choices of the output function for teacher and student, i.e. $g_*(h) = \tanh(\gamma h)$ and $g(h) = h$. We can assume $||\mathbf{W}^*|| = 1$, another norm can be taken into account by a change of the gain γ.

The generalization error (2) can be expressed in terms of these order parameters. We assume uniformally distributed random inputs $\mathbf{x}$, with $< x_i >= 0$ and $< (x_i)^2 >= 1$. Then the weighted sums h_* and h are correlated Gaussians. The average over the inputs $\mathbf{x}$ can be transformed in a Gaussian integral, with the uncorrelated variables $\tilde{h}_*$ and $\tilde{h}$,

$$E_\text{G}(r,q) = \frac{1}{2} < \left\{ g_*\left[\gamma\tilde{h}_*\right] - g\left[q(r\tilde{h}_* + \sqrt{1-r^2}\,\tilde{h})\right]\right\}^2 >_{\tilde{h}_*,\tilde{h}}, \qquad (4)$$

where $< \ldots >_{\tilde{h}}$ denotes the following integral,

$$< F(\tilde{h}) >_{\tilde{h}} := \int_{-\infty}^{\infty} \frac{d\tilde{h}}{\sqrt{2\pi}}\, \exp\left(-\frac{\tilde{h}^2}{2}\right) F(\tilde{h}).$$

The two constants, $G(\gamma)$ and $H(\gamma)$,

$$G(\gamma) := < g_*^2(\gamma\tilde{h}_*) >_{\tilde{h}_*}, \qquad H(\gamma) := < g_*(\gamma\tilde{h}_*)\,\tilde{h}_* >_{\tilde{h}_*}, \qquad (5)$$

summarize the dependence on the teacher. They can also be used to describe other learning tasks with a linear student network. Therefore our theory is not restricted to this special task, but can be applied to other tasks, like the noisy teacher (affects only G and H).

We start with a recapitulation of the results without weight decay, see [1]. Later we will see that these results are very important for the comprehension of the case with weight decay.

2 Without Weight Decay

Now we briefly discuss the results of the case without weight decay. The learning is done by gradient descent, $\Delta W_i = -\eta\frac{\partial E_\text{T}}{\partial W_i}$

A good theoretical description can be achieved by methods from *statistical mechanics*. This implies the thermodynamic limit, i.e. the number of weights $N \to \infty$. We assume that the number of examples P becomes also infinite, but the fraction $\alpha = \frac{P}{N}$ remains finite. Already systems of quite moderate size, such as $N \geq 100$, are well described by this theory. A general introduction to the calculation of the free energy with the *replica method* can be found [3]. For more details especially about this problem consult [1].

The free energy f describes the behavior of the system. In the case of the linear student network the free energy can be expressed analytically as $f(r, q, Q)$. The stationarity conditions, $\frac{\partial f}{\partial r} = 0$ and $\frac{\partial f}{\partial q} = 0$, define the values of the order parameters, i.e. $r(\alpha, a)$ and $q(\alpha, a)$. They depend on the parameters G and H, which describe the learning task, and a parameter,

$$a := 1 + \frac{1}{\beta(Q^2 - q^2)}, \tag{6}$$

which expresses the temperature dependence, since $\beta = T^{-1}$.

Both errors E_G and E_T are fully determined by the values of α and a. The parameter a relates the quality of the training, measured by $E_T(\alpha, a)$, to the generalization performance of the system $E_G(\alpha, a)$. *Exhaustive training* corresponds to the absolute minimum of the training error. In the thermodynamic theory this is the zero–temperature limit. The resulting generalization error shows strong overfitting around $\alpha = 1$ (see upper solid line in Fig. 1). Asymptotically, i.e. for $\alpha \gg 1$, it shows the optimal convergence rate,

$$E_G^{\text{exh}}(\alpha > 1) = \frac{G - H^2}{2} \frac{\alpha}{\alpha - 1}, \qquad E_T^{\text{exh}}(\alpha > 1) = \frac{G - H^2}{2} \frac{\alpha - 1}{\alpha}. \tag{7}$$

But in the intermediate regime better solutions can be found. In [1] it was shown that a certain finite training error can have positive effects on the generalization ability. The parameter a can be optimized leading to

$$E_G^{\text{opt}}(\alpha) = \frac{1}{2} \left(G - \frac{\alpha}{a^{\text{opt}}(\alpha)} H^2 \right), \tag{8}$$

with

$$a^{\text{opt}}(\alpha) := c + \sqrt{c^2 - \alpha}, \quad \text{and} \quad c = \frac{1}{2} \left(\alpha + \frac{G}{H^2} \right). \tag{9}$$

The resulting curve for the generalization error is shown in Fig. 1 as the lower solid line and exhibits no overfitting.

3 With Weight Decay

Now we want to examine whether weight decay can help to reach the optimal generalization for all values of α. Weight decay can be expressed by a penalty term in the training energy, i.e.

$$E_T = E_T + \frac{\lambda}{2N} \sum_{i=1}^{N} (W_i)^2, \tag{10}$$

where the parameter λ determines the relative strength of the weight decay. This changes the gradient descent learning in such a way that the additional term decreases the amount of the weights.

Now we have to recapitulate the statistical mechanics approach from the last section with the additional weight decay term. The calculation follows the same guidelines as without weight decay. As result we find that only one term is added to the free energy,

$$-\beta\tilde{f}(r,q,Q) = -\beta f(r,q,Q) - \frac{\alpha\beta\lambda Q^2}{2} . \tag{11}$$

Since this additional term is independent of q and r, the corresponding order parameter equations remain unchanged. Only the equation for Q is affected. This has serious implications because it means, that a system with weight decay can be transformed in an equivalent system without weight decay at another temperature.

The determination of Q, by $\frac{\partial f}{\partial Q} = 0$, leads to the following relation between a and λ,

$$b^2\alpha\lambda + b(\alpha\lambda + \alpha - 1) - 1 = 0, \quad \text{with} \quad b := \beta(Q^2 - q^2) = (a-1)^{-1} . \tag{12}$$

The problem is now already more or less solved. We can directly use the solution of the case without weight decay, if we rescale the temperature.

First, we will assume that the weight decay strength is fixed. This resembles the situation, when no further knowledge about the system is available and one has to make a more or less well educated guess about the weight decay strength. We resolve the equation (12) to receive a as a function of λ,

$$a(\lambda) = \frac{1}{b_{1,2}(\lambda)} + 1, \qquad b_{1,2}(\lambda) = \frac{1 - \alpha - \alpha\lambda \pm \sqrt{(1 - \alpha - \alpha\lambda)^2 + 4\alpha\lambda}}{2\alpha\lambda} . \tag{13}$$

Only the solution with the 'plus'–sign, i.e. b_1, is a relevant solution. Then we insert $a(\lambda)$, G, and H in the order parameter equations for r and q, which yields the behavior of the errors E_G and E_T.

Fig. 1 shows the generalization error as a function of the loading rate $\alpha = P/N$ for different fixed values of λ. The curves are the theoretical predictions and the points are simulated results. We can see that the overfitting is already reasonably reduced, but the optimal curve is reached only once for each choice of λ, see also [4]. So the weight decay strength should be chosen more accurately.

Now we want to determine the optimal value for the weight decay strength for each α. As a starting point we look at the generalization error as a function of λ for different choices of α, see Fig. 2. Again it can be seen that there is always one optimal choice for λ. If the number of examples becomes large the weight decay becomes less and less important, i.e. $\lambda^{\text{opt}}(\alpha \to \infty) = 0$.

To find the optimal value for the weight decay strength, we use the relation (12) between a and λ, and get λ as a function of a,

$$\lambda(a) = \frac{1 + (1-\alpha)b}{\alpha b(b+1)} = \frac{1}{\alpha}\frac{a-1}{a}(a - \alpha). \tag{14}$$

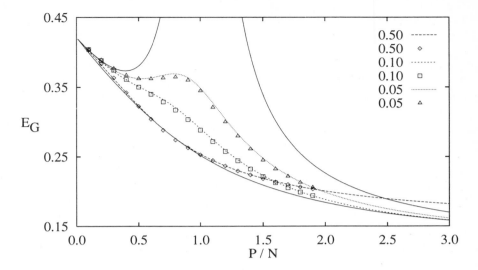

Fig. 1. The generalization error as a function of the loading rate $\alpha = P/N$ for different weight decay factors $\lambda = 0.5$, 0.1, and 0.05. Lines are calculated by the theory and points are simulation results ($N = 200$, 100 independent runs, gain $\gamma = 5$).

If we insert $a^{\mathrm{opt}}(\alpha)$ from eq. (9) we receive $\lambda^{\mathrm{opt}}(\alpha)$. It is surprising that the expression can be simplified extremely. After applying some algebra we find

$$\lambda^{\mathrm{opt}}(\alpha) = \left(\frac{G}{H^2} - 1 \right) \frac{1}{\alpha}. \tag{15}$$

This simple dependence of the optimal weight decay strength on α allows a practical training strategy, which should reach the optimal curve for all values of α. Only the constant prefactor has to be determined. From a theoretical point of view the equations above can be exploited. The initial learning error, i.e. $E_G(0) = G/2$, yields G. The value of H can be calculated from $E_T^{\mathrm{exh}}(\alpha)$, if we use expression (7). More practically the prefactor can be determined for a smaller system by the use of a cross–validation scheme. Based on the knowledge of the prefactor the optimal weight decay strength $\lambda^{\mathrm{opt}}(\alpha)$ is given for arbitrary system sizes.

4 Summary

In this paper we have studied the influence of weight decay on the generalization ability of a single–layer perceptron. A task with strong overfitting was investigated carefully. We could show that an optimally chosen weight decay strength can avoid the overfitting totally. A possible strategy to find this optimal weight decay strength has been proposed also.

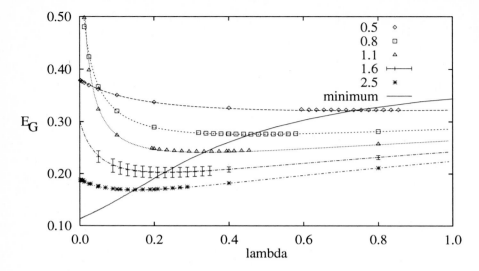

Fig. 2. The generalization error as a function of the weight decay strength λ for different $\alpha = 0.5, 0.8, 1.1, 1.6, 2.5$. Lines are calculated by the theory and points are simulation results ($N = 200, \gamma = 5$). For each α there is exactly one optimal weight decay strength $\lambda^{\mathrm{opt}}(\alpha)$, indicated by the solid line.

Furthermore, it is a nice application of the theory of learning with errors, which was introduced in [1]. In [2] we had shown how this theory can be applied to nonlinear systems. It would be interesting to extend it to more practical learning tasks.

Acknowledgment: We thank Xiao Yan SU and Herbert Wiklicky for comments on the manuscript.

References

1. S. Bös (1995), 'Avoiding overfitting by finite temperature learning and cross–validation', in *International Conference on Artificial Neural Networks 95 (ICANN'95)*, edited by EC2 & Cie, Vol.2, p.111–116.
2. S. Bös (1996), 'A realizable learning task which exhibits overfitting', in *Advances in Neural Information Processing Systems 8 (NIPS*95)*, editors D. Touretzky, M. Mozer, and M. Hasselmo, MIT Press, Cambridge MA, in press.
3. J. Hertz, A. Krogh, and R.G. Palmer (1991), *Introduction to the Theory of Neural Computation*, Addison–Wesley, Reading.
4. A. Krogh, and J. Hertz (1992), 'A simple weight decay can improve generalization', in *Advances in Neural Information Processing Systems 4*, editors J.E. Moody, S.J. Hanson and R.J. Lippmann, Kaufmann, San Mateo CA, p.950–957.

Bayesian Regularization
in Constructive Neural Networks

Tin-Yau Kwok (jamesk@cs.ust.hk)   Dit-Yan Yeung (dyyeung@cs.ust.hk)

Department of Computer Science
Hong Kong University of Science and Technology
Clear Water Bay, Kowloon
Hong Kong

Abstract. In this paper, we study the incorporation of Bayesian regularization into constructive neural networks. The degree of regularization is automatically controlled in the Bayesian inference framework and hence does not require manual setting. Simulation shows that regularization, with input training using a full Bayesian approach, produces networks with better generalization performance and lower susceptibility to over-fitting as the network size increases. Regularization with input training under MacKay's evidence framework, however, does not produce significant improvement on the problems tested.

1 Introduction

Multi-layer feedforward networks have been popularly used in many pattern classification and regression problems. While standard back-propagation performs gradient descent in the weight space of a network with fixed topology, *constructive* procedures [4] start with a small network and then grow additional hidden units and weights until a satisfactory solution is found. A particularly successful subclass of constructive procedures [3, 6] proceeds in a layer-by-layer manner. First, the weights feeding into the new hidden unit are trained (*input training*) by optimizing an objective function. They are then kept constant and the weights connecting the hidden units to the outputs are trained (*output training*). In so doing, only one layer of weights needs to be optimized at a time. There is never any need to back-propagate the error signals and hence is much faster. A number of objective functions have been proposed for the input training phase. Some [2, 7, 8] have simple computational forms and also ensure asymptotic convergence of the procedure to the target function. Constructive procedures using these objective functions are thus particularly interesting.

Regularization can be used to improve generalization performance. The basic idea is to encourage smoother network mappings by adding a penalty term to the error function being optimized, with the balance controlled by a regularization parameter. Early success of incorporating regularization into constructive procedures has been reported in [5]. *Optimal brain damage* (OBD) is used after training to prune away connections to the new hidden unit with low saliency. Application to the two-spirals classification problem reduces the number of connections by about one third without impairing performance. However, the main problem of this method is that the regularization parameter has to be set manually. Moreover, the relative contribution of regularization and OBD to this improvement is unclear.

Recently, various researchers [1, 9] have incorporated Bayesian methods into neural network learning. Regularization can then be accomplished by using appropriate priors that favor small network weights. For the inference part, MacKay [9, 10] introduced an evidence framework, while Buntine and Weigend [1] advocated a full Bayesian approach[1]. Because of the lack of space, these will not be reviewed here. Both methods have also been used in pruning procedures, but not in constructive procedures. The main difficulty in applying this to constructive procedures is that while the inference has to be probabilistic in nature, not all objective functions used in the input training of constructive procedures have proper probabilistic meanings.

2 Incorporating Bayesian Regularization

2.1 Input Training with Regularization

Estimating the Residual Error Without loss of generality, assume in the sequel that there is only one output unit. Consider the L^2 space of all square integrable functions. Let $\|\cdot\|$ be its norm, f be the true function to be learned, f_n be the function implemented by a network with n hidden units directly connected to the output unit, $e_n \equiv f - f_n$ be the corresponding residual error function, and Γ be the set containing all hidden unit functions that can possibly be implemented. The following proposition can be proved:

Proposition 1. *For a fixed $g \in \Gamma$ ($\|g\| \neq 0$), the expression $\|f - (f_n + cg)\|$ achieves its minimum iff $c = c^* = \langle e_n, g \rangle / \|g\|^2$, and*

$$\|f - (f_n + c^*g)\|^2 = \|e_n\|^2 - \langle e_n, g \rangle^2 / \|g\|^2. \tag{1}$$

Hence, if output training is not performed, Proposition 1 suggests that the objective function to be optimized during input training should be:

$$\langle e_n, g \rangle^2 / \|g\|^2. \tag{2}$$

In practice, (2) can only be calculated when the exact functional form of e_n is available. This is obviously impossible as the true f is unknown. A consistent estimate of (2) based on the training set is $(\frac{1}{N} \sum_p E_p H_p)^2 / \frac{1}{N} \sum_p H_p^2$, where N is the number of training patterns, H_p is the activation of the new hidden unit for pattern p, and E_p is the corresponding residual error before this new hidden

---

[1] The approach advocated in [1] takes a Gaussian approximation at the posterior mode, and hence is not a "full' Bayesian approach in the strict sense. However, it is still more in line with the Bayesian methodology than the evidence framework [9] because the hyperparameters are integrated out instead of being optimized.

unit is added. Dropping the constant (with respect to a fixed training set) factor $1/N$, we obtain the following objective function [8]:

$$S = (\sum_p E_p H_p)^2 / \sum_p H_p^2. \tag{3}$$

In [8], we demonstrated experimentally that S in (3) compares favorably with other objective functions used for input training. Moreover, if we ignore the decrease in residual error as a result of output training, then, from (1), we have the following approximation for the residual error:

$$E_{D,n+1} \equiv \frac{1}{2} \sum_p E_{p,n+1}^2 \simeq \frac{1}{2} \left(\sum_p E_{p,n}^2 - S \right) = E_{D,n} - \frac{S}{2}, \tag{4}$$

where, for clarity, we have added the subscript n to denote that the values correspond to a network with n hidden units.

Plugging in the Evidence Framework Plugging in this estimate for $E_{D,n+1}$, the objective function to be minimized in the level 1 inference is:

$$M_{n+1}(\mathbf{w}) \equiv \beta E_{D,n+1} + \alpha E_W \simeq \beta E_{D,n} - \frac{\beta S}{2} + \alpha E_W, \tag{5}$$

where α, β are the hyperparameters, and $E_W = \frac{1}{2} \sum w_i^2$ when Gaussian prior is used for the weights. As β is fixed in level 1, we can alternatively use:

$$\tilde{M}(\mathbf{w}) = -\frac{\beta S}{2} + \alpha E_W.$$

Minimizing $\tilde{M}$ is thus approximately the same as minimizing M_{n+1}. After this, level 2 inference is used to find optimal values for α, β and then levels 1 and 2 are iterated. The best candidate hidden unit in input training is the one that minimizes $\tilde{M}$. Alternatively, one may treat each candidate hidden unit as a model in explaining the residual error and use level 3 inference to compute its evidence. The one that has maximum evidence will be selected.

Plugging in the Full Bayesian Approach In the full Bayesian approach, the unknown hyperparameters are integrated out instead of being optimized as in the evidence framework. If we denote the prior on the weights by $p(\mathbf{w})$, the objective function to be minimized in input training is [1]:

$$C_{n+1}(\mathbf{w}) = \frac{N}{2} \log(2E_{D,n+1}) - \log p(\mathbf{w}) \simeq \frac{N}{2} \log(2E_{D,n} - S) - \log p(\mathbf{w}),$$

using again the approximation in (4). The best hidden unit is the one that maximizes the approximated size of the local area under the posterior weight distribution, given by:

$$C(\mathbf{w}) - \frac{k}{2} \log(2\pi) + \frac{1}{2} \log \det I(\mathbf{w}),$$

where k is the number of weights associated with the candidate hidden unit and $I(\mathbf{w})$ is the Fisher information matrix, or equivalently the hessian.

2.2 Output Training with Regularization

Output layer weights are regularized by using level 2 in the evidence framework to set the optimal regularization parameters. In this case, as all the weights of the hidden units are frozen, E_D can be computed exactly. Computing the hessian $A \equiv \beta \nabla^2 E_D + \alpha \nabla^2 E_W$ is also simple when the output unit is linear.

3 Simulation

The problems used here are the regression functions described in [6]: (1) simple interaction function, (2) radial function, (3) harmonic function, (4) additive function, and (5) complicated interaction function. Noisy training sets with 225 points are generated from the uniform distribution $U[0, 1]^2$, with iid Gaussian noise of mean zero and variance 0.25^2 added. The test set, of size 10000, is generated from a regularly spaced grid on $[0, 1]^2$. A total of 25 trials, each using an independently generated training set, have been performed.

The hidden units have cascade connections as in the cascade-correlation architecture [3]. The output unit is linear and a maximum of 40 sigmoidal hidden units can be generated. Instead of having the same hyperparameter for all weights, we group the input-to-hidden weights into one class and the cascade weights into another, with each class controlled by its own hyperparameter. Biases to the hidden units are not regularized. Moreover, the noise level is assumed to be unknown and so β also has to be estimated when the evidence framework is used.

Boxplots for the best attainable testing (generalization) errors, among 40 networks ranging from 1 to 40 hidden units, are shown in Figure 1. Regularization with input training using the full Bayesian approach produces networks with lower testing errors when compared to those with no regularization. However, when the evidence framework is used instead, with the selection of the best hidden unit done by either level 2 or level 3, regularization does not show significant improvement. Figure 2 shows a typical plot of the testing error against the number of hidden units installed in the network. The phenomenon of over-fitting is apparent when the number of hidden units is large. This problem is, again, not prominent when regularization with input training using the full Bayesian approach is adopted.

4 Conclusion

In this paper, we studied the incorporation of Bayesian regularization into constructive procedures. The degree of regularization can be automatically controlled. Networks with lower testing errors and lower susceptibility to the problem of over-fitting are obtained when regularization with input training using a full Bayesian approach is used. However, regularization with input training using the evidence framework does not produce significant improvement. This phenomenon will be further investigated in the future.

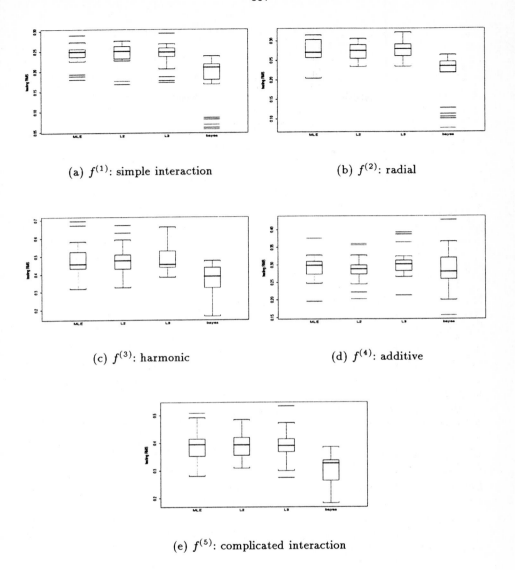

(a) $f^{(1)}$: simple interaction

(b) $f^{(2)}$: radial

(c) $f^{(3)}$: harmonic

(d) $f^{(4)}$: additive

(e) $f^{(5)}$: complicated interaction

Fig. 1. Testing (generalization) performance for networks with cascade connections. Here, MLE refers to the use of no regularization, L2 for using level 2 inference in input training, L3 for using level 3, and bayes for the full Bayesian approach. The horizontal line in the interior of each box is located at the median. The box height is equal to the interquartile distance (IQD). The dotted lines at the top and bottom of the box extend to the extreme values or a distance 1.5 × IQD from the center, whichever is less. Data points which fall outside these are indicated by horizontal lines.

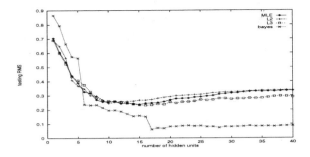

Fig. 2. A typical plot showing the testing (generalization) error against the number of hidden units.

The proposed incorporation of Bayesian regularization can also be used for other types of network. For example, it may be used to penalize the complicated hidden units in projection pursuit learning networks [6]. Moreover, if one uses a pool of candidate hidden units with varying complexity, Bayesian regularization provides an impartial way to select among these units and avoids the pitfall that more complicated candidates are capable of producing smaller training errors.

References

1. W.L. Buntine and A.S. Weigend. Bayesian back-propagation. *Complex Systems*, 5:603–643, 1991.
2. G.P. Drago and S. Ridella. Convergence properties of cascade correlation in function approximation. *Neural Computing & Applications*, 2:142–147, 1994.
3. S.E. Fahlman and C. Lebiere. The cascade-correlation learning architecture. In D.S. Touretzky, editor, *Advances in Neural Information Processing Systems 2*, pages 524–532. Morgan Kaufmann, Los Altos CA, 1990.
4. E. Fiesler. Comparative bibliography of ontogenic neural networks. In *Proceedings of the International Conference on Artificial Neural Networks*, volume 1, pages 793–796, Sorrento, Italy, May 1994.
5. L.K. Hansen and M.W. Pedersen. Controlled growth of cascade correlation nets. In *Proceedings of the International Conference on Artificial Neural Networks*, volume 1, pages 797–800, Sorrento, Italy, May 1994.
6. J.N. Hwang, S.R. Lay, M. Maechler, D. Martin, and J. Schimert. Regression modeling in back-propagation and projection pursuit learning. *IEEE Transactions on Neural Networks*, 5(3):342–353, May 1994.
7. V. Kůrková and B. Beliczynski. Incremental approximation by one-hidden-layer neural networks. In *Proceedings of the International Conference on Artificial Neural Networks*, volume 1, pages 505–510, Paris, France, October 1995.
8. T.Y. Kwok and D.Y. Yeung. Objective functions for training new hidden units in constructive neural networks, 1995. Submitted.
9. D.J.C. MacKay. Bayesian interpolation. *Neural Computation*, 4(3):415–447, May 1992.
10. D.J.C. MacKay. A practical Bayesian framework for backpropagation networks. *Neural Computation*, 4(3):448–472, May 1992.

A Nonlinear Discriminant Algorithm for Data Projection and Feature Extraction

C. Santa Cruz, J. Dorronsoro

Department of Computer Engineering and Instituto de Ingeniería del Conocimiento,
Universidad Autónoma de Madrid, 28049 Madrid, Spain

Abstract. A nonlinear supervised feature extraction algorithm that directly combines Fisher's criterion function with a preliminary non linear projection of vectors in pattern space will be described. After some computational details are given, a comparison with Fisher's linear method will be made over a concrete example.

1 Introduction

Several techniques have been devised for dimensionality reduction and pattern recognition, the best known certainly being Fisher's discriminant analysis [1]. Fisher's method performs a linear projection from pattern space into projection space that maximizes interclass distances against individual class variances. The resulting technique is very elegant, lends itself naturally to powerful statistical interpretation and has been successfully applied in a number of problems. Its linear nature however has also an important drawback: classification is done in projection space by a linear procedure. It will be thus successful only if pattern class projections are linearly separable, which often is not the case.

Neural methods have been proposed as nonlinear alternatives for pattern classification. Several approaches to both supervised and non-supervised classification can be followed by neural techniques [4], but for supervised pattern recognition multi-layer perceptrons (MLPs) are actually the tools of choice. In fact, if an appropriate labeling scheme is used, a linear MLP without hidden layers performs essentially as Fisher's method [3]. Moreover, when a single hidden layer networks with sigmoidal activations in the hidden units is added, it can be seen [5] that the internal hidden layer representation performs a non linear discriminant analysis in the sense that it maximizes a certain network discriminant function.

In that analysis, network training is done through conventional neural techniques; the interpretation of the network discriminant function does not depend on them, but rather on the concrete coding schemes to be used. In the approach proposed here, we won't concern ourselves with specific codings, but rather with a training procedure for nonlinear MLPs directly related to Fisher's method. More precisely, we will use directly Fisher's criterion function on the outputs of

---

[0] Both authors partially supported by grant TIC 95–965 of Spain's CICyT.

the hidden layer. That is, the final weight set will maximize the ratio of interclass distances to class variances for the hidden layer outputs.

Of course, this approach cannot rely on the usual network training schemes. However, several methods have been proposed recently for network training by means of layer by layer optimization [2]. In them, the network transfer function is considered as the composition of several partial intermediate functions that capture layer communication. Global network optimization is then performed by successive optimization of these intermediate functions, starting at the output layer and proceeding backwards, keeping fixed the weights of the rightmost layers.

Since Fisher's criterion function on hidden layer outputs is used for weight adjustment, a major concern in this approach could be training time. Fisher's criterion optimization require rather costly eigenvalue and eigenvector computations for output weight estimation (although they will require just one single pass). Moreover, given the rather complicated global criterion function we obtain, the availability of precise gradient information for hidden weight computation is not clear beforehand. That information is necessary to get good convergence rates in numerical optimization, because of the inherent errors of purely numerical gradient estimations. However, given the relatively simple final network structure, gradient computation for hidden weight estimation can be analytically computed. In the following section we will describe the concrete implementation, and we will finally give some numerical examples.

2   The learning algorithm

As mentioned before, the architecture to be used is that of a single hidden layer, fully connected perceptron, with sigmoidal activity functions in the hidden units and the identity function on the output units. There are $D+1$ input units, D for sample parameters and an extra one always with value one for bias effects; there are also H hidden units and $C-1$ output units. The resulting weight and bias sets will be denoted as $\mathbf{W} = (\mathbf{W}^I, \mathbf{W}^H)$, with $\mathbf{W}^I$ and $\mathbf{W}^H$ being respectively input and hidden layer weights. For a particular weight set, the network then computes a transfer function which we will denote by $\mathbf{F}(\mathbf{X}, \mathbf{W})$, and that can be seen as the combination of the transfer functions of two simpler input-output nets. That is, $\mathbf{F}(\mathbf{X}, \mathbf{W}) = \mathbf{H}(\mathbf{Y}, \mathbf{W}^H) = \mathbf{H}(\mathbf{G}(\mathbf{X}, \mathbf{W}^I), \mathbf{W}^H) = \mathbf{G}_H(\mathbf{X}, \mathbf{W}^I)$, where $\mathbf{H}(\mathbf{Y}, \mathbf{W}^H)$ reflects hidden weights influence in network output and $\mathbf{G}_H(\mathbf{X}, \mathbf{W}^I)$ that of input layer weights.

Learning proceeds performing layer by layer optimization: once a particular weight set $\mathbf{W}_k$ has been obtained, $\mathbf{W}_{k+1}$ is computed in a two step fashion. First we compute the hidden weights $\mathbf{W}_{k+1}^H$ by performing what essentially is multiple C class discriminant analysis taking here as pattern vectors the hidden unit outputs. Once these $\mathbf{W}^H$ have been obtained and the optimal partial transfer function $\mathbf{H}$ at this point defined, we proceed to compute the corresponding optimal weights $\mathbf{W}^I$. Recall that this technique has already been presented in the literature for usual sum of squared errors minimization. However we are using here a new, more complex network criterion function and the concrete optimization procedures to be used are totally different. We will next analyze them.

2.1 Optimization of hidden layer weights $\mathbf{W}^H$

As mentioned before, this essentially corresponds to performing linear discriminant analysis (LDA) on the hidden layer outputs. In accordance with Fisher's traditional procedure (see [1]), we will minimize as a function of the output weight matrix the following determinant based criterion function

$$J(\mathbf{W}^H) = \frac{|(\mathbf{W}^H)^t S_W \mathbf{W}^H|}{|(\mathbf{W}^H)^t S_B \mathbf{W}^H|} = \frac{|\tilde{S}_W|}{|\tilde{S}_B|} \tag{1}$$

(several other choices, as replacing determinants by matrix traces could also be used). Here we assume that the training set class i has N_i patterns, with a total number of patterns of N. Now, the within–class scatter matrix S_W is defined as $S_W = \sum_{i=1}^{C} \sum_{j=1}^{N_i} (\mathbf{out}_{ij} - \mathbf{m}_i)(\mathbf{out}_{ij} - \mathbf{m}_i)^t$, where $\mathbf{m}_i$ is the mean of class i pattern values at the hidden units, which we denote by $\mathbf{out}_{ij}$, $j = 1, \ldots, N_i$ (i.e., the hidden unit vector representation of the j–th element of class i). Also, S_B denotes the between class scatter matrix, $S_B = \sum_{i=1}^{C} N_i(\mathbf{m}_i - \mathbf{m})(\mathbf{m}_i - \mathbf{m})^t$, where $\mathbf{m}$ is the total sample mean at the hidden layer.

As it is well known (see [1], p. 120), the columns of the optimal $\mathbf{W}^H$ matrix are the eigenvector of the $C - 1$ largest eigenvalues of the equation $S_B W = \lambda S_W W$.

2.2 Optimization of the input layer weights $\mathbf{W}^I$

We now regard our criterion function (1) as a non–linear function of input weights, that is $J = J(\mathbf{W}^I)$. To minimize it we have chosen a quasi–Newton method, which is well known to be of second order. For it to be so, however, precise gradient estimation of J is required. To compute it, we consider the hidden layer values of each pattern vector as intermediate variables, that is, $J = J(out_{ij}^h)$, $i = 1, \ldots, C$, $j = 1, \ldots, N_i$, $h = 1, \ldots, H$. We then have:

$$\frac{\partial J}{\partial w_{kl}^I} = \sum_{i=1}^{C} \sum_{j=1}^{N_i} \frac{\partial J}{\partial out_{ij}^l} f'(act_{ij}^l) x_{ij}^k =$$

$$= \sum_{i=1}^{C} \sum_{j=1}^{N_i} \frac{|\tilde{S}_B| \frac{\partial |\tilde{S}_W|}{\partial out_{ij}^l} - |\tilde{S}_W| \frac{\partial |\tilde{S}_B|}{\partial out_{ij}^l}}{|\tilde{S}_B|^2} f'(act_{ij}^l) x_{ij}^k \tag{2}$$

Therefore, the derivatives of the $(C - 1) \times (C - 1)$ determinants $|\tilde{S}_B|$ and $|\tilde{S}_W|$ have to be determined, which in turn can be computed in terms of the component-wise partials $\frac{\partial (S_B)_{kh}}{\partial out_{ij}^l}$ and $\frac{\partial (S_W)_{kh}}{\partial out_{ij}^l}$ of the corresponding hidden layer covariance matrices. Taking into account that now we assume the weights $\mathbf{W}^H$ to be constant, it follows that

$$\frac{\partial \tilde{S}_B}{\partial out_{ij}^l} = (\mathbf{W}^H)^t \frac{\partial S_B}{\partial out_{ij}^l} \mathbf{W}^H \qquad \frac{\partial \tilde{S}_W}{\partial out_{ij}^l} = (\mathbf{W}^H)^t \frac{\partial S_W}{\partial out_{ij}^l} \mathbf{W}^H \qquad (3)$$

and it can be seen that, for instance,

$$\frac{\partial (S_B)_{kh}}{\partial out_{ij}^l} = \delta_{lk} \left[(m_i^h - m^h) - \sum_{s=1}^{C} \frac{N_s}{N}(m_s^h - m^h) \right] +$$

$$+ \delta_{hl} \left[(m_i^k - m^k) - \sum_{s=1}^{C} \frac{N_s}{N}(m_s^k - m^k) \right] \qquad (4)$$

where δ_{mn} denotes Kronecker's delta matrix, and m^h and m_i^h the components of global and class means. A similar equation is obtained for the derivatives of S_W. It is worth mentioning that big savings are possible when computing these equations. For instance, Kroneckers's delta ensures that for a fixed hidden unit index l, all but one of the rows and columns of the partial derivative matrix are zero. Also, the sums in equation (4) do not depend on the j index, that is, they are class dependent rather than pattern dependent. Similar savings are also possible for the $\frac{\partial (S_B)_{kh}}{\partial out_{ij}^l}$ partials. Finally, let's recall that once the optimal $\mathbf{W}^I$ have been computed through the quasi–Newton method, we go back to adjusting the $\mathbf{W}^H$ weights, and so on. This iterative procedure goes on until an appropriate termination test is met.

3 An example: handwritten character identification

As an example we will consider a dimensionality reduction application of these procedures to the problem of handwritten number recognition. For the sake of simplicity, only characters 3, 5 and 6 will be discussed (observe that 3 and 5, and 5 and 6 are prone to mixed identifications); the projection will thus take place in a two dimensional space and will be easily visualized. The bilevel images are scanned at 200 dpi, normalized to a matrix of 32×32 pixels, and encoded using the 40 most significant coefficients of their cosine transform. Therefore, dimensionality reduction will go from a 40 dimensional space to a bidimensional one.

Two independent sets of data will be used, one set to train the network and the other to test it. The training set is formed by 400 characters per class while the test set consist of 250 character per class. Notice that these sizes are very small for any OCR application; the interest here however lies in a comparison of nonlinear discrimination (NLD) as proposed here against Fisher's method.

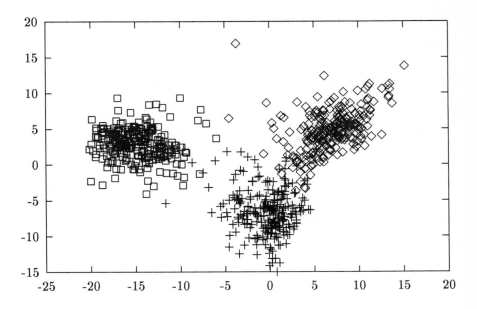

Fig. 1. Two dimensional test set projection by Fisher's method (Symbols: ◇ = character 3, + = character 5, □ = character 6).

Figure 1 shows the projection of the test set given by Fisher's method; figure 2 illustrates the same result for a 50 hidden unit network.

It is seen that projection boundaries defined in figure 2 (almost linear in fact), are much better than in figure 1: the projections of characters 5 and 6 are rather far apart, and only two handwritten fives and one three are slightly mixed. These figures and their training set counterparts (not shown here), in which NLD character projections are extremely concentrated, make clear that NLD has much stronger separation properties than Fisher's method.

4 Conclusions

We have presented a nonlinear classification method related to classical Fisher's analysis, but with much higher discriminant power. Although the potential application for pattern recognition of our procedure is clear, no comparisons have been made with other nonlinear discriminants, such as MLPs.

Clearly, its more complicated criterion function requires a higher computational effort (at least if that function is defined in terms of determinants). However it does not depend on global sample characteristics, such as squared error sums, but rather on individual class parameters. An advantage of our method is thus clear: it will be more robust to unequal or even greatly different training class sizes.

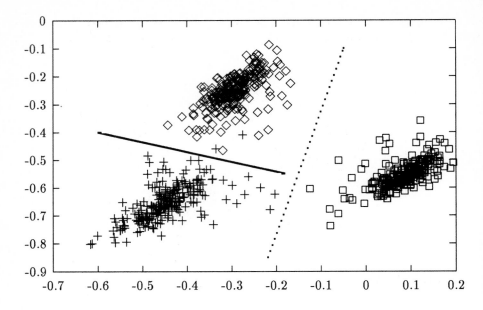

Fig. 2. Two dimensional test set NLD projection (Symbols as in figure 1).

Furthermore, instead of relying on arbitrarily chosen target codings as desired network outputs, NLD automatically determines output class groupings. Its training should therefore be also more robust when class labeling errors result in already misclassified training patterns. Preliminary experiments in these directions (to be reported elsewhere) are very encouraging.

References

1. R.O. Duda, P.E. Hart, "Pattern classification and scene analysis", Wiley, 1973.
2. S. Ergezinger, E. Thomsen "An Accelerated Learning Algorithm for Multilayer Perceptrons: Optimization Layer by Layer", IEEE Trans. on Neural Networks **6** (1995), 31–42.
3. P. Gallinari, S. Thiria, F. Badran, F. Fogelman–Soulie, "On the relations between discriminant analysis and multilayer perceptrons", Neural Networks **4** (1990), 349–360.
4. J. Mao and A.K. Jain, "Artificial Neural Networks for Feature Extraction and Multivariate Data Projection", IEEE Trans. on Neural Networks **6** (1995), 296–317.
5. A.R. Webb, D. Lowe "The optimized internal representation of multilayer classifier networks performs nonlinear discriminant analysis", Neural Networks **3** (1990), 367–375.

A Modified Spreading Algorithm for Autoassociation in Weightless Neural Networks

Christopher Browne, Joel de L. Pereira Castro Jnr, Amos Storkey
Imperial College of Science Technology and Medicine,
Exhibition Road- London - SW7 2BT
United Kingdom

Abstract. This paper describes a problem with the conventional Hamming metric spreading algorithm often employed in weightless neural networks. The algorithm can cause incorrect classifications in some cases where a section of the neuron input vector is noise. The conditions under which such error occurs are described and a modified spreading algorithm proposed to overcome this problem has its validity demonstrated theoretically and tested in two practical applications.

1 Background

The general neural unit (GNU) has been introduced as a building brick for neural systems [2]. This consists of K neural nodes of the weightless G-RAM variety [2]. In simulation the virtual VRAM is preferred as it has considerably lower memory requirements. The VRAM only stores the trained patterns [6]. G-RAMs are flexible neural devices which require a generalising algorithm and output resolution to be specified. The output resolution is usually three: 0, 1 and the u state. The u state has an equal probability of generating a 0 or a 1 at, the device output. The G-RAM starts with all its locations in the u state. Training writes a 1 into some G-RAM locations and a 0 into others. This depends upon the location addressed by the training patterns. The trained patterns create the subsequent system outputs. The generalisation is usually based on the Hamming metric. This means that during the spreading phase the trained patterns will be written to untrained locations which are most similar to them in Hamming distance. A simple example is to consider a system, just trained on the black, all 1's pattern and the white, all 0's pattern. The black pattern is trained to output a 1, the white a 0. During spreading all locations which have a majority of 1's will have a 1 written to them. All locations with a majority of 0's will have a 0 written to them. The even length patterns which have equal numbers of 1's and 0's will be left in the u state. The Hamming metric spreading has proved successful in many applications [5][7]. However limitations have been foreseen and other spreading algorithms proposed [3].

During research the authors have independently discovered a problem with the existing methods of spreading. The problem arises when an area of the node input vector becomes noise. In these cases the system output can become noisy, or even erroneous. This paper examines the theory behind a proposed change to the spreading algorithm to resolve this problem [4]. The paper is organised as follows. Firstly the problem and the proposed solution are described in a general way. Secondly a mathematical formulation is developed which enables the demonstration of the validity of the proposed solution. Lastly the authors illustrate this solution as applied to their research studies.

2 Problem and Solution

Consider a GNU where the output state has been split into two distinct fields, region (i) and region (ii). Both fields are five bits wide. The states A and B, chosen to better illustrate the point rather than from formal design considerations, are encoded by the patterns shown in table 1. During the operation region (i) is set to random noise. The state is then recalled from region (ii). This characterises a simple associative memory task. However there are patterns which can cause the incorrect association to be made. An example is 000111111. This is a Hamming distance of four from state A and a hamming distance of 2 from state B. Thus state B will be recalled. Examining the correct associative relationship reveals this to be an error. Region (ii), which is the partial pattern the full pattern is to be retrieved from, is identical to the region (ii) of state A, therefore state A should be recalled.

Now consider the same recall, but with region (i) of the input excluded from the Hamming distance calculation. The input 0000111111 is now a Hamming distance of zero from state A and a Hamming distance of one from state B. Thus state A is recalled. The proposed change to the spreading algorithm identifies sections of the input vector to each node which have become noise and eliminates them from the Hamming distance calculation. These areas do not contain any information concerning the required state transition. They merely disrupt this transition.

	Region (i)					Region (ii)				
State A	1	1	1	1	1	1	1	1	1	1
State B	0	0	0	0	0	1	1	1	1	0

Table 1. *A possible training set which could lead to incorrect recall under noise.*

3 Theory

This section develops a theoretical basis to describe the problem. It then demonstrates the validity of the proposed solution.

Take a situation with two arbitrary trained states, A and B, and an input state C with region (ii) identical to state A and region (i) randomly chosen (from a uniform distribution). How many of the possible input vectors C will be recalled as the trained state A?

For ease of calculation, let the bits of state A be represented by $\xi_i \in \{-1, 1\}$, those of state B be denoted by $\eta_i \in \{-1, 1\}$ and those of state C be denoted $x_i \in \{-1, 1\}$. In all these cases $i \in \{1, 2, ..., n\}$ where n is the length of the patterns. The region boundary occurs at m. In other words a bit is in region (i) if $i \in \{1, 2, ..., m\}$ and region (ii) if $i \in \{m, m+1, ..., n\}$.

For correct recall, as illustrated in the example x_i must be closer in Hamming distance to ξ_i than to η_i. i.e.

$$\frac{1}{2}\left(\sum_{i=1}^{n}x_i\xi_i - n\right) > \frac{1}{2}\left(\sum_{i=1}^{n}x_i\eta_i - n\right) \Rightarrow \sum_{i=1}^{n}x_i\xi_i > \sum_{i=1}^{n}x_i\eta_i \Rightarrow \sum_{i=1}^{n}(\xi_i - \eta_i)x_i > 0$$

$$\Rightarrow \sum_{i=1}^{m}(\xi_i - \eta_i)x_i + \sum_{i=m+1}^{n}(\xi_i - \eta_i)x_i > 0$$

Now the second sum in the equation is known:

$$\sum_{i=m+1}^{n}(\xi_i - \eta_i)x_i = 2(n - m - q)$$

where q is the total number of places where $\xi_i = \eta_i$, $i \in \{m, m+1, .m+2, ..., n\}$.

The requirement is now:

$$\sum_{i=1}^{m}(\xi_i - \eta_i)x_i > 2(q + m - n)$$

Define $S_i = (\xi_i - \eta_i)/2$ and

$$y_i = \begin{cases} x_i & S_i \neq -1 \\ -x_i & S_i = -1 \end{cases}$$

The above constraint now becomes

$$\sum_{i=1}^{m}S_i^2 y_i > (q + m - n)$$

Let y_i^* denote the projection of y_i onto a p-vector, formed by re-indexing the set $\{ y_i^* \mid i \in \{1, 2, ..., m\}, S_i \neq 0 \}$, Here p is the number of values of i $\in \{1, 2, ..., m\}$ for which $S_i \neq 0$. Then

$$\sum_{i=1}^{p}y_i^* > (q + m - n)$$

must be satisfied. Note that this is only a constraint on p out of m of the bits of the pattern x_i The other bits were unconstrained. Let r denote the number of values of i for which $y_i^* = 1$. Then

$$\sum_{i=1}^{p}y_i^* = r - (p - r) = 2r - p$$

This leads to the requirement $2r - p > (q + m - n)$ which implies that

$$\boxed{r > (q + m - n + p)/2}$$

Here n is the total vector length, m is the length of the region (ii), q is the number of bits in region (ii) where state A = state B, p is the number of bits in region (i) where state A $\neq$ state B and r is the number of palces in region 1 where state A $\neq$ state B AND state C = sate A.

The above calculation shows the constraints on the bits of state C which will ensure that it is closest in Hamming distance to state A. From this it is necessary to calculate the number of the possible patterns of state C which will be correctly recalled. So the total number of vectors y^* for which the above constraint holds is:

$$\sum_{r=r^*}^{p}{}^{p}C_r; \ r^* = \begin{cases} 0 & \text{if} \quad (q + m - n + p)/2 < 0 \\ (q + m - n + p)/2 & \text{if} \quad (q + m - n + p)/2 \geq 0 \text{ and is even} \\ (q + m - n + p + 1)/2 & \text{if} \quad (q + m - n + p)/2 \geq 0 \text{ and is odd} \end{cases}$$

So the total number of combinations of the vector x for which the constraint holds is:

(including the contribution from the unsconstrained bits) $2^{m-p} \sum\limits_{r=r*}^{p} \dfrac{p!}{(p-r)!r!}$

It is possible to convert this figure to a probability. If one of the possible patterns that state C could take is chosen at random, then the probability of it being closest in Hamming distance to A is given by the number of possible states which are closest to A divided by the total number of possible states. Therefore

P(state C closest to state A)$= 2^{m-p} \sum\limits_{r=r*}^{p} \dfrac{p!}{(p-r)!r!}$ divided by $2^{m} = 2^{-p} \sum\limits_{r=r*}^{p} \dfrac{p!}{(p-r)!r!}$

4 Discussion of theoretical result

This section attempts to relate this result to physical properties of the neural network, using the example above. In the example the incorrect state is recalled when a region of the state is set to noise. The theory predicts this can arise when the inequality $r > (q + m - n + p)/2$. The terms are defined above. In the example the values are; $r = 1$; $q = 4$; $m = 5$; $n = 10$; $p = 5$. So substituting into the above inequation gives $1 > (4 + 5 - 10 + 5)/2$

The inequality is not true, thus the condition does not hold and the incorrect pattern is recalled. The theory then goes on to develop an expression for the probability of a noise pattern causing incorrect recall. The applications this type of system are used for usually employ a random pattern. Thus this expression can be used for assessing the chance of an incorrect classification. The modified spreading algorithm will produce the correct classification under normal conditions. The conditions for classification are the usual conditions which apply to spreading algorithms using the Hamming metric.

5 Implementation

The modified spreading algorithm can be easily implemented on an existing GNU simulation. This is achieved by modifying the spreading algorithm function. The function should first scan through each region in the input vector in turn. These are region (i) and region (ii) in the example. The Hamming distance between the corresponding regions of the input pattern and the trained patterns is calculated. If there is a trained pattern region within a specified Hamming distance of the input region then that region is judged to be a learned state. The spreading algorithm is then set to include that region. Conversely if no learned pattern is found within the specified Hamming distance then that region is assumed to be noise and excluded from the final spreading calculation. The result is that the system only spreads over the areas if its input vectors which contain information. However it does this without any loss of information. If the spreading algorithm were altered to just to reject a region of the entire input vector then any information there would be lost. A pathological example of this would be a system where the external input region is ignored. This would cause the system to loose its ability to respond to an external, controlling input.

6 Application

The modified spreading technique has been successfully used in two different pieces of research by the authors. The first study relates to acquiring semantic power in a weightless neural network [4]. The second is a study of iconic prediction using a weightless neural network [1].

 The first study was carried out as part of an investigation of the GNU capabilities to cope with semantic representations. A Knowledge Base (KB) was created in semantic network terms using the GNU. Each of the 80 facts con trained in the KB were encoded by partitioning the trained patterns (facts) in three semantic regions. The validity of the representation was tested by posing queries (noisy versions of trained patterns) to the net. Altogether 80 queries were posed for 100 different GNU configurations (corresponding to systematic variations in the numbers of input and feed-back connections). Errors obtained when using the conventional Hamming distance algorithm motivated the development, implementation and testing of a modified spreading algorithm (see section 2). For a detailed discussion of the design, implementation and extended comparative analysis of the experiments using both spreading techniques see [4]. The results were as follows. Semantically erroneous patterns were retrieved more often when the conventional algorithm was used. A considerable improvement in performance was indicated by the higher Recall values (number of retrieved correct answers over the number of recovered answers) obtained when using the new spreading algorithm.

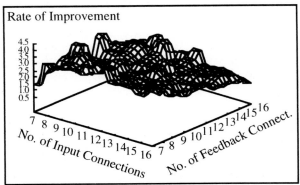

Figure 1. *Relative rate of recall of correct semantic patterns in a semantic neural network using the original Hamming metric based and modified spreading algorithmsover a range of varying-connectivity configurations.*

 Fig.1 shows this relative improvement in performance through the comparison of the Recall values obtained using both spreading algorithms for a range of configurations. It also shows that sparsely connected configurations yielded better results. This last result is specially relevant to reinforce the feasibility of the new spreading algorithm in the development of practical applications with low computational cost.

 In an iconic prediction experiment a probabilistic weightless neural network was trained on one hundred and forty four state transitions [1]. A set of five experimental runs was per formed on the system both with the simple Hamming distance algorithm and the modified spreading algorithm. Each run consisted of setting the system controlling input to a fixed noise pattern and the system state to one of the learned states. The system was then recalled for twenty time steps. Two measurements were taken. The first was the highest number of incorrect state transitions made in a run. The second was the average number of incorrect transitions over the five runs. An incorrect state transition is defined as a transition which does not result in a system state within a Hamming distance of forty of a learned state. The Hamming distance of forty was arbitrarily chosen. It is about the level of distortion where the states become impossible do visually discriminate between. The results from these runs are in table 2. The results show that the modified spreading algorithm performs consider ably better in this task.

	Highest no. of incorrect transitions	Average no. of incorrect transitions
Original spreading	11	7
Modified spreading	6	2.6

Table 2. *Comparison of performance of an iconic prediction system using the original Hamming metric based and modified spreading algorithms.*

7 Conclusion

The conditions under which the Hamming metric based spreading algorithm can fail in a weightless neural network have been derived theoretically. A modified spreading algorithm proposed to overcome the described problem has had its validity mathematically demonstrated. The performance enhancement achieved by the modified algorithm has been demonstrated in two different practical applications.

References

[1] C. Browne and I. Aleksander, Digital general neural units with controlled transition probabilities Electronics Letters, Vol. 32, No. 9, pp 824-825, 1996.

[2] I. Aleksander, Neural Systems Engineering: towards a unified design discipline? IEE J. Computation and Control, Vol.6, pp. 259-265, 1990.

[3] I. Aleksander, T. J. W. Clarke, and Braga A. P. Binary neural systems: A unified approach to their analysis and design.

[4] J. D. L. P. Castro. Acquiring Semantic Power in "Weightless Neural Network. In 1995 IEEE International Conference on Neural Networks Proceedings, pages 674-679.

[5] N. J. Sales and R. G. Evans. An approach to solving the symbol grounding problem: Neural networks for object naming and retrieval. In Proc. CMC-95, 1995.

[6] P. Ntourntoufis. Aspects of the theory of weightless artificial neural networks. PhD thesis, Imperial College of Science, Technology and Medicine, 1994.

[7] R Evans. Reconstructing quatisation-distorted images with a Neural State Machine. 1994.

Analysis of Multi-Fluorescence Signals Using a Modified Self-Organizing Feature Map

Stefan Schünemann[1], Bernd Michaelis[1], Walter Schubert[2]

Otto-von-Guericke University Magdeburg
[1] Institute for Measurement and Electronics
[2] Institute for Medical Neurobiology
P.O. Box 4120, D-39016 Magdeburg, Germany

Abstract. This paper introduces an algorithm for the multi-sensory integration of signals from the fluorescence microscopy. For the cluster analysis a Self-Organizing Feature Map (SOFM) is used. One basic property of these artificial neural nets is the smoothing of the input vectors and thus a certain insensitivity to clusters of low feature density. While classifying clusters of highly different feature density this property is undesirable. A modification of the learning algorithm of the SOFM, which makes a reproduction of low feature density clusters on a SOFM possible, is described.

1 Introduction

Series of microscope recordings, i.e. multiple signals of the same region of interest with different object markings and stimulating frequencies, are of special interest in biomedical research.

The quantitative analysis of higher-level combination patterns of human antigens in the cellular immune system, i.e. the detection of existing combination patterns under normal and pathological conditions, is the goal of advanced fluorescence microscopy. This microscopy is a repetitive method which provides a tool for the localization of random numbers of different receptor epitopes at cell surfaces [1].

The developed algorithm for the multi-sensory integration contains the following central parts: signal acquisition, preprocessing and cluster analysis. The different fluorescence markings of the cells and the acquisition of the multiple signals occur sequentially. An important task of the preprocessing is the detection of cells in the sensor signals. The cluster analysis yields the basic biological information.

The main demand on the analysis algorithm is that the cluster analysis in the high-dimensional feature space shall be executed without the previous definition of classes, borderlines and without a-priori knowledge of the number, position and form of the feature clusters. Self-Organizing Feature Maps (SOFM) satisfy these requirements. The smoothing of input vectors is a basic property of the reproduction of the feature space on the SOFM and leads to a relative insensitivity to clusters of low feature density [2]. In the case of the combination patterns of antigens the clusters of low feature density represent significant qualities of the distribution [3].

In the following a modified learning algorithm for SOFM, which makes a separate reproduction of low feature density clusters on a SOFM possible, will be introduced.

2 Overview

Fig. 1 shows the strategy for multi-sensory integration for the analysis of higher-level combinatorical receptor patterns in the cellular immune system.

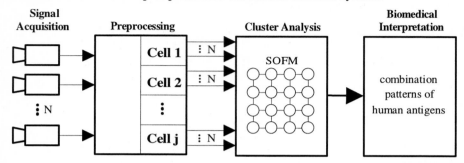

Fig. 1: Multi-sensory integration for analysis of the cellular immune system

The *Multi-Signal Aquisition* occurs by means of a fluorescence microscope and CCD-camera. The cells of the immune system are marked sequentially with N different receptor epitopes and the fluorescence signal is acquired.

The first task of the *Signal Preprocessing* is the normalization of the N sensor signals. Therefore systematic errors, caused by changes in the optical refraction for different spectral channels, and small mechanical and thermal movements during the signal acquisition, are corrected in the fluorescence signals. In addition the signal preprocessing executes a detection of cells in the sensor signals. For the signal description a feature vector V for each cell is created, which contains the spatially averaged cell signals representing the actual molecular combination pattern.

The components of extracted feature vectors span a N-dimensional feature space $V \in \mathbf{R}^N$. The distinct concentrations of antigens in the cells produce different clusters of feature vectors. The Self-Organizing Feature Maps (SOFM) provides an efficient method for *Cluster Analysis*, because it facilitates a reduction of high-dimensional feature spaces onto two-dimensional arrays of representative weight vectors [4, 5, 6]. A modification of the learning algorithm for SOFM supports a separate reproduction of low feature density clusters on a SOFM [7].

3 Cluster Analysis with Modified SOFM

In the training phase of the SOFM the weight vectors $W(x,y)$ are trained by means of the learning set for a suitable representation of the feature space. Each neuron calculates its Euclidean distance to the presented feature vector V:

$$d_v^w(x, y) = \sqrt{\sum_{k=1}^{N} \left[V_k - W_k(x, y) \right]^2} \tag{1}$$

The weight vectors of the winner neuron n_{xy} (the neuron featuring the minimal distance), and its coupled neighbors $n_{x'y'}$ are updated according to :

$$W^{i+1}(x, y) = W^i(x, y) + \alpha_i \, \varphi_i(x, y) \, [V - W^i(x, y)] \tag{2}$$

in the direction of the feature vector. The learning rate α_i determines the magnitude of the update and the neighborhood function φ_i describes the effective training region of the SOFM at learning step i.

The main idea of the modification is the definition of a neighborhood function φ_i^* in Eq. 3, which in contrast to the basic algorithm more strongly differentiates the degree of coupling of the neighboring neurons with the winner neuron.

$$\varphi_i^*(x, y) = \begin{cases} 1 & , \; n_{xy} : d_v^w(x, y) = \min. \\[2ex] \dfrac{\beta}{2 \, D_{x'y'}^{xy}}, & n_{x'y'} : r_i \geq D_{x'y'}^{xy} \\[2ex] 0 & , \; \text{else} \end{cases} \tag{3}$$

If the neuron is the winner neuron n_{xy} then simply results $\varphi^*=1$. The coupled neighborhood neurons $n_{x'y'}$ will be evaluated with the relation $1/(2{*}D)$ expanded by the multiplicative term β. Here, the neighborhood distance D is defined as the Vector distance between the winner and neighborhood neurons and the radius r_i characterizes the coupled neighborhood neurons [7]. Fig. 2 exemplifies the function φ^* for the coupled neighborhood neurons. The term β controls the update in the direction of clusters of low feature density in dependence on partial spaces. An idea concerning the initialization of the SOFM will be used for the description of partial spaces which utilizes the principal components analysis [6, 4] and defines a hyper-ellipsoid γ with the eigenvalues λ_k of the feature distribution as its half-axes:

$$\gamma = \frac{V_1'^2}{\lambda_1} + \frac{V_2'^2}{\lambda_2} + \cdots + \frac{V_N'^2}{\lambda_N} \tag{4}$$

where V' are the feature vectors transformed using the eigenvectors (Fig. 3). The partial space with low feature density is located outside the hyper-ellipsoid $\gamma{>}1$.

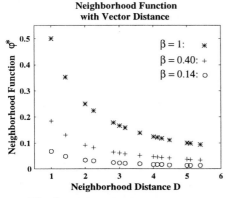

Fig. 2: Neighborhood function φ^*

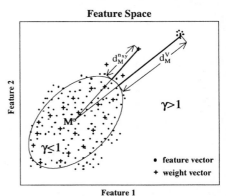

Fig. 3: Partial spaces and distances

In the following the effects of β in the distinct training phases [7] will be described:

During the initialization phase of the training and during all other phases within the hyper-ellipsoid $\gamma\leq1$ the term β is assigned the value $\beta=1$.

In the *Initialization Phase* the inital order of the SOFM is established. The relation $1/(2*D)$ causes a non-linear variation of the neuron's weights in the direction of the feature vector. There is no differentiation of the partial spaces $\gamma\leq1$ and $\gamma>1$ during this stage of training.

The *Orientation Phase* serves the further adaptation of the neuron weights to the feature clusters within the hyper-ellipsoid $\gamma\leq1$ and suppresses a contraction of the map into the ellipsoid at $\gamma>1$. The intention for this phase is the development of neurons outside the hyper-ellipsoid $\gamma>1$. If the winner neuron is pulled away from the ellipsoid by a feature vector V then the term β is assigned the value $\beta=1$. But, the weights of neighborhood neurons with $\gamma>1$ and whose winner neuron is changed towards the hyper-ellipsoid, experience an additional limitation of their adaptability through β. This way, it is possible to train some neurons in the direction of low feature density. For a measure of limitation, the length of the straight $d(n_{x'y'},M)$ from the weight of the neighborhood neuron $n_{x'y'}$ to the point of penetration into the ellipsoid in the direction of the center of gravity M of the distribution is used. Fig. 3 schematizes the relations in the feature space. The length $d(n_{x'y'},M)$ is normalized with the maximum distance $d(V,max)$ of the feature vectors. For the described case the empirically found function β is definied as:

$$\beta = \frac{1}{e^{10\frac{d_M^{n_{x'y'}}}{d_{\max}^V}}} \tag{5}$$

In the third phase of the training process, called *Convergence Phase*, the neuron weights converge for clusters within $\gamma\leq1$. Outside the hyper-ellipsoid $\gamma>1$ individual neurons will be further updated in the direction of clusters of low feature density. They also converge and thus represent reference vectors. All coupled neighborhood neurons, which are found outside $\gamma>1$ experience an additional limitation independent of the position of the winner neuron in $\mathbf{R}^N$ according to Eq.(5).

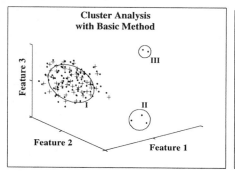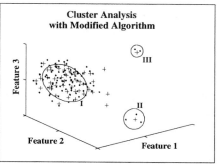

Fig. 4: Feature (•) and weight (+) vectors for the classification of an example distribution

Fig. 4 shows the result of a training of the SOFM in a three-dimensional feature space for the classification of an example distribution, side-by-side for both the basic method and the modified algorithm. From the figure it is evident that using the basic method, the weight vectors have been pulled into the direction of the ellipsoid, i.e. into the main cluster I. In contrast to the modified algorithm the clusters II and III are suppressed as separate clusters.

4 Application of the Algorithm

The performance of the described algorithm for the multi-sensory integration, i.e. the multi-signal aquisition, signal preprocessing and cluster analysis, for a reduced four-dimensional feature space for several combination patterns of human antigens: CD7,CD38,CD45 and CD71 in the blood as an example distribution are presented in Fig. 5 [3]. Normally in this application the feature space is 15 or higher dimensional.

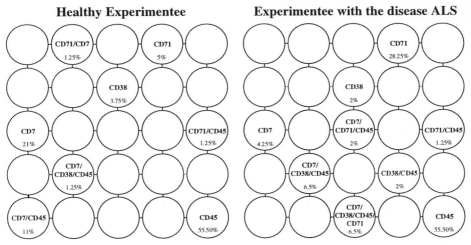

Fig. 5: SOFM for several combination patterns of human antigens

After the aquisition of the multiple signals from marked cells they experience a shift correction to rid them from system errors. All sensor signals are background subtracted and edge enhanced. Furthermore, a contrast stretching unifies the dynamic range of the different signals. For the detection of the cells in the blood, which approximate circles in the sensor signals, a modified Hough-Transformation-Technique is used [8]. The extracted feature vectors for a healthy and an experimentee suffering from sporadic Amyotrophic Lateral Sclerosis (ALS) are used to train a SOFM with 5*5 neurons. In the Fig. 5 separate recall phases of the normal and pathological conditions are shown.

The distinct frequencies shown in percentages and the different existing combination patterns for healthy and experimentees suffering from ALS are recognizable from Fig 5. Each existing combination of the fluorochrome labeled human antigens, i.e. each cluster of the used distribution, is stored in a separate neuron while maintaining the topographics in $\mathbf{R}^N$. The neuron weights of the SOFM are reference vectors of the desired clusters and make a biological interpretation possible.

In the case of an adequate number of training sets with representive feature vectors the cluster analysis, i.e. the determination of the number and position of the clusters, need not be repeated. Provided the measurement conditions for the fluorescence microscopy remain unchanged, the training phase is omitted. Small deviations could be corrected by means of system calibration.

5 Conclusion

The advantage of the introduced algorithm for the multi-sensory integration is that the detection and classification of molecular combination patterns of human antigens from multi-fluorescence signals is possible without previous definition of classes and borderlines. The algorithm is applied to the detection of human antigens in the blood, muscular tissue and organs. The described method is also suitable for other high-dimensional classification problems in biology and medicine with multi-markings of objects. The used algorithm of cluster analysis with SOFM is universally applicable.

The expansion of the cluster analysis to cluster-dependent multi-layer structures of SOFM in connection with the definition of several hyper-ellipsoids seems reasonable for a further improvement.

This work was supported by the DFG/BMBF grant (Innovationskolleg 15/A1).

References

1. Schubert, W.: Lymphocyte antigen Leu19 as a molecular marker of regeneration in human skeletal muscle, Proc. Natl. Acad. Sc USA 86, pp. 307-311, 1989
2. Obermayr, K.; et.al.: Statistical-Mechanical Analysis of Self-Organization and Pattern Formation During the Development of Visual Maps, Physical Review A, Vol. **45** (10), pp. 7568-7589, 1992
3. Rethfeldt, Ch.; et.al.: Multi-Dimensional Cluster Analysis of Higher-Level Differentiation at Cell-Surfaces, Proc. 1. Kongress der Neurowissenschaftlichen Gesellschaft, p. 170, Spektrum Akademischer Verlag, Berlin 1996
4. Kohonen, T.: Self-Organizing Maps, Springer Series in Information Sciences, Springer, New York 1995
5. McInerney, M.; Dhawan, A.: Training the Self-Organizing Feature Map using Hybrids of Genetic and Kohonen methods, Proc. The 1994 IEEE Int. Conf. on Neural Networks, pp. 641-644, Orlando 1994
6. Bertsch, H.; et.al.: Das selbstlernende System der topologischen Merkmalskarte zur Klassifikation und Bildsegmentierung, Proc. 10. Symposium Deutsche Arbeitsgemein-schaft für Mustererkennung, pp. 298-303, Zurich 1988
7. Schünemann, St.; Michaelis, B.: A Self-Organizing Map for Analysis of High-Dimensional Feature Spaces with Clusters of Highly Differing Feature Density, Proc. 4th European Symposium on Artificial Neural Networks, pp. 79-84, Bruges 1996
8. Sklansky, J.: On the Hough Technique for Curve Detection, IEEE Trans. on Comp. **27** (10), pp. 923-926, 1978

Visualizing Similarities in High Dimensional Input Spaces with a Growing and Splitting Neural Network

Monika Köhle, Dieter Merkl

Institute of Software Technology, Vienna University of Technology
Resselgasse 3/188, A-1040 Wien, Austria
{monika, dieter}@ifs.tuwien.ac.at

Abstract. The recognition of similarities in high dimensional input spaces and their visualization in low dimensional output spaces is a highly demanding application area for unsupervised artificial neural networks. Some of the problems inherent to structuring high dimensional input data may be shown with an application such as text document classification. One of the most prominent ways to represent text documents is by means of keywords extracted from the full-text of the documents and thus, document collections represent high dimensional input spaces by nature. We use a growing and splitting neural network as the underlying model for document classification. The apparent advantage of such a neural network is the adaptive network architecture that develops according to the specific requirements of the actual input space. As a consequence, the class structure of the input data becomes visible due to the separation of units into disconnected areas. The results from a growing and splitting neural network are further contrasted to the more conventional neural network approach to classification by means of self-organizing maps.

1. Introduction

Real world data is often represented in high dimensional spaces and hence the inherent similarities are hard to recognize and to illustrate. This fact makes it a challenging task to build tools that are able to visualize these similarities. By nature, visualization requires a mapping process from the high dimensional input space to a low dimensional output space. One well-known approach to dimension reduction is the self-organizing map [6]. This model preserves the structure of input data as faithfully as possible and results can be visualized on a 2D topographic map, thus indicating the similarities between input vectors in terms of the distance between the respective units. However, it does not represent cluster boundaries explicitly, these rather have to be drawn by hand. Only recently, variations on the basic idea of self-organizing maps with a special focus on adaptive architectures have been suggested [2], [4]. These architectures incrementally grow and split according to the specific requirements of the actual input space, thus producing isolated clusters of related items. This paper presents an investigation in the applicability of one of those models to a real world problem domain, namely the classification of text documents.

The remainder of the paper is organized as follows. In the next section we present a brief review of the network dynamics of Growing Cell Structures, a growing and splitting artificial neural network. In Section 3 we present the general framework for text document classification. Section 4 contains a description of experimental results from our application. These results are further compared to self-organizing maps. Finally, our conclusions are presented in Section 5.

2. Growing Cell Structures

For the experiments we used a neural network implemented in the spirit of Growing Cell Structures, GCS, as described in [4]. This model may be regarded as a variation on Kohonen's self-organizing maps [6]. In its basic form, GCS consist of a two-dimensional output space where the units are arranged in form of triangles. Each of these output units owns a weight vector which is of the same dimension as the input data. The learning process of GCS is a repetition of input vector presentation and weight vector adaptation. More formally, during the first step of the learning process the unit c with the smallest distance between its weight vector w_c and the current input vector x is selected. This very unit is further referred to as the best-matching unit, the winner in short. The selection process as such may be implemented by using the Euclidean distance measure as indicated in expression (1) where O denotes the set of units in the output space.

$$c: \|x-w_c\| \le \|x-w_i\|; \ \forall i \in O \tag{1}$$

The second step of the learning process is the adaptation of weight vectors in order to enable an improved representation of the input space. This adaptation is performed with the weight vector of the winner and the weight vectors of its directly neighboring units. Denoting the best-matching unit with c and the set of its neighboring units with N_c, the equations of weight vector adaptation may be written as given in expressions (2) and (3) with t denoting the current learning iteration. The terms ε_c and ε_n represent learning-rates in the range of [0, 1] for the winner and its neighbors. Notice that the definition of a neighborhood function as with self-organizing maps is omitted since the adaptation is restricted to direct neighbors only. As another difference to self-organizing maps the learning-rates are fixed throughout the entire learning process.

$$w_c(t+1) = w_c(t) + \varepsilon_c \cdot (x-w_c) \tag{2}$$
$$w_n(t+1) = w_n(t) + \varepsilon_n \cdot (x-w_n); \ \forall n \in N_c \tag{3}$$

Finally, the third step of the learning process constitutes the major difference to self-organizing maps. In addition to the weight vector, each unit owns a so-called signal counter variable τ. This variable represents an indication of how often a specific unit has been selected as the winner during the learning process. This information is further used to adapt the architecture of the artificial neural network to the specific requirements of the input space. We will return to this adaptation below. At the moment we will just provide the formulae for signal counter adjustment in equations (4) and (5). In this expressions c again refers to the best-matching unit. α is a fixed rate of signal counter reduction for each unit that is not the winner at the current learning iteration t.

$$\tau_c(t+1) = \tau_c(t) + 1 \tag{4}$$
$$\tau_i(t+1) = \tau_i(t) - \alpha \cdot \tau_i(t); \ i \ne c \tag{5}$$

Apart from the adaptation of weight vectors as described above the GCS learning process consists of an adaptation of the overall architecture of the artificial neural network. Pragmatically speaking, units are inserted in those regions of the output space that represent large portions of the input data whereas units are removed if they do not contribute sufficiently to input data representation. Additionally, due to the deletion of units, the output space of the GCS may split into several disconnected areas each of which representing a set of highly similar input data. This adaptation of the architecture is performed repeatedly after a fixed number of input presentations. Just to give the exact figure, the results presented below are based on an adaptation of the network structure each two epochs of input presentation.

Starting with the insertion of new units, in a first step the unit that served most often as the winner is selected. The selection is based in the signal counter variable as given in expressions (6) and (7). We denote the unit having the highest relative signal counter value with q.

$$h_i = \tau_i / \Sigma_j \tau_j \tag{6}$$
$$q: h_q \ge h_i; \ \forall i \in O \tag{7}$$

Subsequently, the neighboring unit r of q with the most dissimilar weight vector is to be determined as given in expression (8). In this formula N_q again represents the set of neighboring units of q.

$$r: \|w_r-w_q\| \ge \|w_p-w_q\|; \ \forall p \in N_q \tag{8}$$

An additional unit s is now added to the artificial neural network in between units q and r and its weight vector is initially set to the mean of the two existing weight vectors, i.e. $w_s = 1/2 \cdot (w_q + w_r)$.

Finally, the signal counter variables τ in the neighborhood N_s of the newly inserted unit s have to be adjusted. This adjustment represents an approximation to a hypothetical situation where unit s would have been existing throughout the learning process so far. The necessary operations are given in expressions (9) through (11). With card(N_p) we refer to the cardinality of set N_p. The term F_p represents an approximation of the size of the decision region covered by the units of N_p. Furthermore, t refers to the situation before and $t+1$ after the insertion of unit s. In a last step the signal counter variable of unit s is initialized with the sum of changes at existing units.

$$F_p = 1 \,/\, (\mathrm{card}(N_p) \cdot \Sigma_i \, \|w_p{-}w_i\|; \; \forall i \in N_p \tag{9}$$

$$\Delta\tau_p = ((F_p(t{+}1) - F_p(t)) \,/\, F_p(t)) \cdot \tau_p(t); \; \forall p \in N_s \tag{10}$$

$$\tau_s = -\Sigma_p \, \Delta\tau_p; \; \forall p \in N_s \tag{11}$$

Units that do not contribute sufficiently to input data representation are removed from the artificial neural network. Again, the contribution of a particular unit is measured by means of its relative signal counter value and the decision region the unit belongs to, i.e. $p_i = h_i \,/\, F_i; \; \forall i \in O$.

The final deletion is guided by a simple threshold logic. Thus, all units j with contribution p_j below a certain threshold η are removed, i.e. $p_j{<}\eta$. After this removal the artificial neural network has to be examined whether the triangular structure of neighboring units is violated. Incomplete triangles have to be trimmed.

3. A note on software library organization

In order to assess the capability of GCS we selected text document classification in the specific setting of software reuse as a real world application. Text document classification is perfectly well suited for such an experiment since the various inputs are represented in a high dimensional space. In general, the purpose of classification is to uncover semantic similarities between different text documents. Since natural language processing is still far from understanding arbitrarily complex documents, the various documents first have to be mapped on some representation language in order to be comparable. Still one of the most widely used representation languages is single-term indexing. Roughly speaking, the text documents are represented by the words they are built of. In order to achieve a better document representation, words that appear either too often or too rarely within the entire document collection are excluded from the representation language. Additional improvement is possible by means of prefix and suffix stripping, thus by reducing the words to radicals. As a result, each text document is represented by vectors of equal dimension where each component of such a vector corresponds to a particular word of the representation language. Each component may either have a value of one indicating that this specific word was extracted from the document at hand or a value of zero if this word is not contained in the document. This representation is known as the vector space model of information retrieval [10].

Text document classification is of vital importance for promoting software reuse since effective reuse of existing software components relies first on the existence of a library that contains these components and second, on an organization of the library that facilitates locating and retrieving the needed software component [3]. In other words, the stored software components ought to be arranged in a way that resembles their functional similarity as closely as possible. Such an arrangement may be found relying on the textual descriptions of the various components as contained in the software manual. In parentheses, however, we have to note that other approaches to software library organization relying on semantic networks have been proposed in literature [1], [9]. Yet, these approaches suffer from the requirement of heavy manual interference. Contrary to that, information retrieval based approaches count for highly automated software library organization.

Our approach to software library organization relies on automatically extracted keywords from the manual of the software components and a vector based component representation [7]. These vectors are further used as the input data for an artificial neural network which performs the task of library organization. So far we used the self-organizing map [7] and ART [8] for library organization.

4. Experimental results

In this section we provide a comparison of software library organization by means of GCS and self-organizing maps. We include the latter model since we described the former one as a variation on self-organizing maps and thus, the results can be contrasted with the original. More precisely, we compare the artificial neural networks in two case studies. In the first one, the experimental software library contains a number of commands from the MS-DOS operating system. Each of these commands is represented by a 37-dimensional feature vector containing the words that occur in the description of the command when invoking help. The whole test set comprises 36 MS-DOS commands. We use this example since the assessment is fairly easy due to the fact that the behavior of the various commands are well-known and an intuitive judgment of the result is possible.

A result from structuring operating system commands by means of the GCS model is presented in Figure 1. In order to ease comparison we structured the result according to the target of the different commands. As can be seen easily, GCS is successful in arranging the commands according to their functionality. However, the relationship between complementary commands, i.e. commands that undo the effect of each other, is by no means obvious from the GCS classification. As examples of complementary commands consider backup and restore or mirror and recover. It certainly is one of the self-organizing map's strengths to uncover this sort of relationship as can be seen in Figure 2. Finally, we want to direct the attention to Figure 1 (d). The cluster in the upper right contains a set of at first sight rather unrelated commands. On a second look, however, we find that each of these commands is used to display information on the screen and thus, the arrangement of these commands is not as random as it seemed.

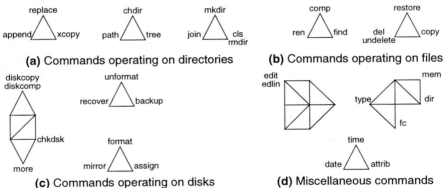

(a) Commands operating on directories **(b)** Commands operating on files

(c) Commands operating on disks **(d)** Miscellaneous commands

$\varepsilon_c = 0.095; \varepsilon_n = 0.01; \alpha = 0.0095; \eta = 0.015$

Figure 1. GCS classification of related MS-DOS commands

One of the advantages of GCS in comparison to self-organizing maps is the explicit representation of cluster boundaries thanks to the adaptive architecture of this artificial neural network. The difference is obvious when examining Figure 2 depicting the classification result from a self-organizing map. In case you are familiar with the behavior of the operating system commands it is easy to recognize a similar arrangement as with GCS. However, if you are not familiar with the input data and thus have to rely on the visualization as provided in Figure 2 you might have some problems in finding the borderline between commands such as time, date and diskcopy, format since they seem to be comparably similar.

more	.	format	.	diskcopy	.	time	.	date	.	mem
		chkdsk		diskcomp						
.	mirror	.	assign	.	unformat	.		date	.	type
recover					undelete					
.		mkdir		.	del	.		find	.	fc
ren		rmdir								comp
attrib		cls		restore	copy	xcopy	append	replace	path	tree
edit		edlin		backup	.	dir	chdir	.		join

Figure 2. SOM classification of related MS-DOS commands

As a second case study we used a more realistic example of a reusable software library. In particular, we used the NIH class library, NIHCL. The NIHCL comprises a set of classes implemented in the C++ language and includes generally useful data types, numbers of container classes, and provides the facilities to store arbitrarily complex data structures on disk. Each of the classes is described by a set of keywords extracted from the respective parts of the NIHCL manual [5]. Just to give the exact figure, each class is represented by using a 489 dimensional feature vector.

In Figure 3 we present a result from using GCS to classify the NIH classes. As with the MS-DOS example above, it is straight forward to find a justification for the GCS classification result. A typical self-organizing map result is depicted in Figure 4 for convenient comparison. In this particular example the difficulties in determining cluster boundaries with the self-organizing map becomes obvious. It is hardly possible to draw a line between classes performing file I/O and container classes without profound knowledge of the class library. With GCS this task is much easier since the respective classes are separated in disconnected parts of the output space.

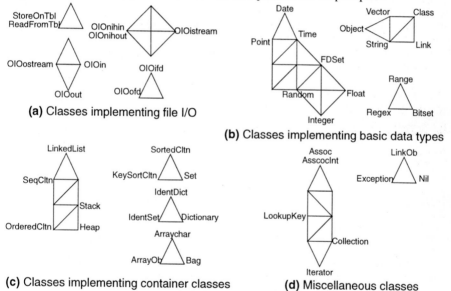

(a) Classes implementing file I/O

(b) Classes implementing basic data types

(c) Classes implementing container classes **(d)** Miscellaneous classes

$\varepsilon_c = 0.08; \varepsilon_n = 0.01; \alpha = 0.002; \eta = 0.015$

Figure 3. GCS classification of related NIH classes

However, in the GCS result the close relation between the container classes allowing keyed access to their elements gets lost. In particular, we refer to the classes Dictionary, IdentDict, and KeySortCltn. In the self-organizing map their close relationship is uncovered in the left part of the result depicted in Figure 4. Additionally, the classes that actually implement the keyed access, i.e. Assoc, AssocInt, and LookupKey, are mapped onto the very same region of the map. Finally, we want to stress the fact that there is nothing like a free lunch. The GCS

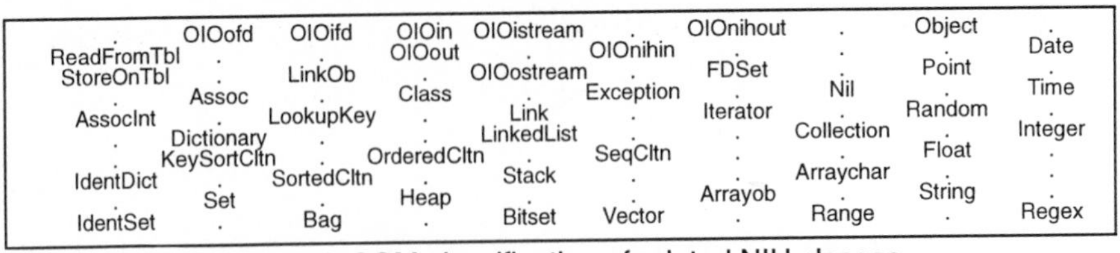

ReadFromTbl	OIOofd	OIOifd	OIOin OIOout	OIOistream	OIOnihin	OIOnihout		Object	Date
StoreOnTbl	Assoc	LinkOb	Class	OIOostream	Exception	FDSet	Nil	Point	Time
AssocInt	Dictionary	LookupKey		Link LinkedList		Iterator	Collection	Random	Integer
IdentDict	KeySortCltn	SortedCltn	OrderedCltn	Stack	SeqCltn		Arraychar	Float	
IdentSet	Set	Bag	Heap	Bitset	Vector	Arrayob	Range	String	Regex

Figure 4. SOM classification of related NIH classes

results seem to be advantageous in the sense that cluster boundaries become apparent due to the adaptive architecture of this artificial neural network model. Additionally, the input data is represented in a highly compact form with respect to the required amount of output units. However, in our experiments it turned out that GCS are much more susceptible to variations in the initial parameter settings than the self-organizing map. We have to note that with a wide variety of parameter values the GCS was unable to produce semantically meaningful classification results. On the other hand, the self-organizing map proved to be more stable. More precisely, the results of the self-organizing map are largely the same irrespective of the actual learning-rate and neighborhood-function.

5. Conclusion

We addressed the recognition and visualization of similarities in high dimensional input spaces by means of a growing and splitting artificial neural network. As a reference problem area we selected text document classification. One of the most prominent ways to represent text documents is by means of keywords extracted from the full-text of the documents. Hence, document collections represent high dimensional input spaces by nature. Our experiments indicated that a semantically meaningful classification of text documents is achievable with a growing and splitting neural network. Moreover, the various classes become apparent and cluster boundaries are shown explicitly due to the adaptive architecture of the neural network model. These results have been further contrasted to self-organizing maps. The major disadvantages of the growing and splitting neural network as compared with self-organizing maps, however, are first the loss of intercluster relations and second, the high sensibility regarding the actual choice of the learning parameters. In this sense, the self-organizing map has proved to be an extremely robust model that is by far not comparably susceptible to variations in its learning parameters.

References

[1] R. Adams. An experiment in software retrieval. *Proc European Software Eng Conf.* 1993.

[2] J. Blackmore and R. Miikkulainen. Visualizing high-dimensional structure with the incremental grid growing neural network. *Proc Int'l Conf on Machine Learning.* 1995.

[3] W. B. Frakes and S. Isoda. Success factors of systematic reuse. *IEEE Software* 11(5). 1994.

[4] B. Fritzke. Growing Cell Structures: A self-organizing neural network for unsupervised and supervised learning. *Neural Networks* 7(9). 1994.

[5] K. E. Gorlen. *NIH Class Library Reference Manual* (Revision 3.10). National Institutes of Health. Bethesda, MD. 1990.

[6] T. Kohonen. *Self-Organizing Maps.* Berlin: Springer-Verlag. 1995.

[7] D. Merkl, A M. Tjoa, and G. Kappel. Learning the semantic similarity of reusable software components. *Proc IEEE Int'l Conf on Software Reuse* (ICSR'94). 1994.

[8] D. Merkl. Content-based software classification by self-organization. *Proc IEEE Int'l Conf on Neural Networks* (ICNN'95). 1995.

[9] E. Ostertag, J. Hendler, R. Prieto-Diaz, and C. Braun. Computing similarity in a reuse library system: An AI-based approach: *ACM Trans on Software Eng and Meth* 1(3). 1993.

[10] H. R. Turtle and W. B. Croft. A comparison of text retrieval models. *Computer Journal* 35(3). 1992.

A Neural Lexical Post-Processor for Improved Neural Predictive Word Recognition

S. Garcia-Salicetti*'**

*Institut National des Télécommunications, Dept. EPH
9 rue Charles Fourier, 91011 Evry, France
**LAFORIA-IBP UA CNRS 1095, Tour 46-00 Boite 169, Université Paris VI
4 Place Jussieu, 75252 Paris Cedex 05, France

Abstract

This work presents a neural post-processor introducing lexical knowledge in a neural predictive system for on-line word recognition [4]. Each word is modeled by the natural concatenation of letter-models corresponding to the letters composing it. Successive parts of a word trajectory are this way modeled by different Neural Networks. A dynamical segmentation allows to adjust letter-models to the great variability of handwriting encountered in the words. Our system combines Multilayer Neural Networks and Dynamic Programming with an underlying Left-Right Hidden Markov Model (HMM). Training was performed on 7000 words from 9 writers, leading to already good results in the letter-labelling process. These results are significantly improved, at the word level, thanks to the use of the post-processor.

1. Introduction

Holistic approaches in handwritten word recognition [2, 3] avoid segmentation of words into letters in order to turn the recognizer more tolerant to local distorsions. Unfortunately, the computational load in these systems is too heavy when using large vocabularies. On the other hand, several analytical approaches segment words into characters *before* word recognition, avoiding to use during recognition the *low-level* context of characters [9]. Those systems are then constrained to compensate for this lack by *high-level* contextual post-processing.

The approach presented in this work is analytical and well suited for large vocabularies. Its segmentation paradigm is the integration of segmentation and recognition, like several other analytical approaches [1, 8]. This paradigm permits to incorporate to the recognition process, the high-order relations existing between characters composing a given word. Also, a main advantage of this approach, is a simple parameter reestimation scheme, known in Speech Recognition as "Segmental K-Means" [6]. It consists in an extension of the Viterbi algorithm [7] to parameter reestimation in HMMs. Besides, the use of neural networks in word-models, permits a simple parameter reestimation, by means of back-propagation.

The following section describes the general functioning of the system and the post-processor introduced in this work. A description of the experimental set-up and results, is given in section 4. The statistical background is explained in section 3.

2. System Overview

2.1. Databases and Preprocessing
Word databases from 9 writers, composed of globally 8781 words, were chosen from a 1000 words' vocabulary. Words were written with a fixed scale on a digitizing tablet that samples the pen trajectory at the frequency of 200 points per second. Data appear as a sequence of (x,y) coordinates, afterwards resampled.

2.2. Feature Extraction
We extract from each word's sequence of points, a sequence of parameter vectors called "frames". Each frame is composed of 15 parameters extracted from a single point. As widely known in on-line handwriting recognition, a robust feature extraction procedure must combine temporal local information with spatial local features. We thus implemented the feature extraction procedure introduced by Manke & al. [5], which consists of a low-resolution image description (9 values) of the spatial context of each point in the trajectory, with also a local temporal description of that point: its local direction and curvature, the relative displacement on the x-axis $(x(t)-x(t-1))$ and its relative position to the baseline on the y-axis. As diacritical marks were far from the middle area of the word, a *rectangular* window centered on each point, was necessary (see *Figure 1*). It shifts along the word's trajectory with a step of one point.

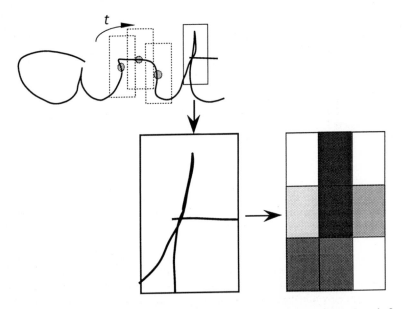

Fig. 1. The number of points in each of the 9 rectangles composing the window is counted, leading to a grey level representation after normalization

2.3. System Specification
A handwritten word is modeled by the concatenation of neural predictive letter-models detailed in a previous work [4], corresponding to the sequence of letters

composing the word. Each letter-model is composed of 8 states, each one consisting of a multilayer neural network of fixed dimensions. A Left-Right topology [8] rules transitions between the set of neural predictors modeling a word. Only transitions from each one to itself or to its right neighbor are permitted.

During learning, *only* the neural predictors in letter-models corresponding to the letters composing a given word, give as output a non-linear prediction on each frame of the word. This prediction is computed frame per frame, each time from a left context of 3 frames [4]. The structure of each neural network is the same: 45 input units code the 3 frames-left context of the current frame (since each frame is described by 15 parameters), 15 units follow in the hidden layer, and 15 units in the output layer code the prediction given by the network on the current frame.

A prediction errors' (the euclidean error between the output of each neural network and the current frame) matrix is computed on each word and processed through Dynamic Programming: every authorized path in the matrix has a global cost, the summation of prediction errors corresponding to that path. Segmentation is performed by the search of the *optimal* path, representing the *least global cost*. Afterwards, frame per frame, the weights of the network belonging to the optimal path, are modified by back-propagating its prediction error on the current frame.

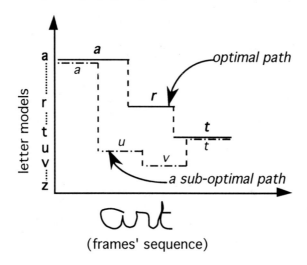

(frames' sequence)

Fig. 2: the optimal path computation and its letter-labelling process (labels "art") during recognition

During recognition, a prediction error matrix is computed on the presented word, but this time, *all* letter-models are incorporated to the matrix. The optimal path represents the least global cost path and allows to attribute a letter-model to each frame composing the word, that is a letter-label (see *Figure 2*). Any other path has a higher global cost and is therefore sub-optimal (see *Figure 2*).

2.4. The Neural Lexical Post-Processor

This work introduces a neural post-processor for robust word recognition. Indeed, the use of a simple edit-distance to a dictionary to compute word recognition rates, only implements a distance measure operating *at the level of letter-labels*. But words judged as "close" in the sense of edit-distance, can be very different at the level of their trajectories. It is why the neural post-processor here presented implements a low-level distance measure. It works on the sequence of labels given by the recognizer by means of 2 modules. The first module is an edit-distance between that sequence and each word in a dictionary. Its output is a list of 50 words, ordered according to their scores. The second module presents the frames of the word to recognize, to the trained neural word-model corresponding to each of those 50 words (see *Figure 3*). For each of them, a prediction error matrix is computed as well as an optimal path. The global cost along each optimal path obtained, is this way associated to the considered neural word-model. In this framework, as *Figure 3* shows, the neural word-model giving the least global cost permits to determine the winning word in the lexicon.

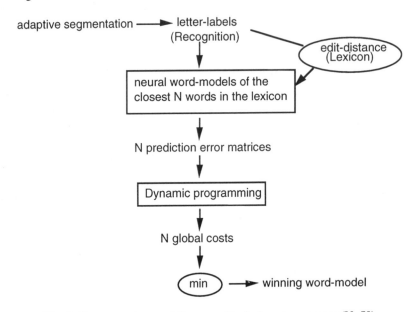

Fig. 3. The recognizer and the neural lexical post-processor (N=50)

3. Statistical Background

We consider an approximate Maximum Likelihood criterion for the model parameters' learning process [4, 6]:

$$\max{}_\lambda\big[\max{}_q[P(O,q\,/\,\lambda)]\big],$$

where λ is the N-states Hidden Markov Model of a given word presented during learning to the system, and q denotes the hidden state sequence. Optimization is thus performed first on segmentation through Dynamic Programming, then by the reestimation of the word-model's parameters via the Back-Propagation algorithm. During recognition, our criterion becomes:

$$\max {}_q[P(O, q \mid \Lambda)]$$

where Λ denotes now the *global* parameter set, including *all* letter-HMMs. A sequence of labels is this way obtained. The latter represents only a first step in word recognition, since a word from a lexicon must be proposed by a recognition system. In this framework, our statistical criterion is:

$$\max {}_W\big[\max {}_q[P(W \mid q, O). P(q \mid O)]\big],$$

where W denotes a word in the considered lexicon. Now the two steps of our recognition process clearly appear: first, the second term's maximization leads to the optimal state sequence q*; then, the first term's maximization permits, through an editing distance, to find the optimal word W* in the considered lexicon.

4. Implementation

4.1. Initializing the Learning Procedure

In the databases, no information is available about the "correct" segmentation of words. One of the strengths of this system is that it finds an initial segmentation *by itself*. To that end, the optimal path computation on each word was constrained *directly* on the cost function. The optimal path was this way forced to remain inside a band around the prediction errors' matrix "diagonal" (only letter-models corresponding to the letters composing the current word are involved during learning). This constraint makes *almost equal parts of the word's trajectory* correspond to *every* network in the word-model. The weight of this forced segmentation is progressively lowered during training. After 3000 words, learning is totally unrestricted, and segmentation evolves freely.

4.2. Results and Conclusions

Our system was trained in the multi-scriptor framework on 7000 words from 9 writers and tested on 1500 words from the same writers. Training is stopped by means of cross-validation, according to the accuracy obtained on letter-labels given by the recognizer. After 21 iterations, the obtained average accuracy in characters of a word (the fraction of correct characters in the sequence of labels obtained, over the total number of characters of the label corresponding to the word presented to the recognizer) is 80.22%, the rates of substitution being 17.56%, of deletion 1.38%, and of insertion 8.17%. These results are very encouraging, since no language model is used up to now, and words are written in an unconstrained style. Word recognition rates given by the neural lexical post-processor (NLPP) are compared in *Table 1* to those obtained by a simple edit-distance between the target word and the sequence of labels given by the recognizer. Three dictionaries are used: the 1000 words dictionary (D1) corresponding to the words in the database, and 2 other dictionaries consisting of the same 1000 words plus a random selection of additional words from the UNIX Spell dictionary. Their sizes are 10600 (D2) and 20200 words (D3). The first 3 proposals in each dictionary are given in *Table 1*.

		D1	D2	D3
without NLPP	1st proposal	84.13%	68.4%	65.93%
	2nd proposal	88.93%	75.67%	73.2%
	3rd proposal	91%	79.53%	76.53%
with NLPP	1st proposal	96.47%	88.6%	85.33%
	2nd proposal	98.07%	91.8%	89.2%
	3rd proposal	98.33%	92.6%	90%

Table 1: word recognition rates with simple edit distance (without NLPP) and with NLPP are compared.

It is thus clear that the use of the NLPP improves significantly the results, since it permits to implement a distance measure between the recognizer's output and the words in a dictionary in a precise way. Moreover, the NLPP could be improved by using an edit-distance *after* obtaining N proposals of letter labels' sequences from the recognizer. These proposals could *guide* the search of the 50 words chosen by the edit-distance in the dictionary, avoiding this way errors still coming from the edit-distance module.

5. References

[1] Bengio Y., Le Cun Y., Nohl C., Burges C.: "LeRec: a NN/HMM hybrid for on-line handwriting recognition", Neural Computation 7, 1289-1303, 1995.
[2] Bercu S., Lorette G.: "On-line handwritten word recognition: an approach based on Hidden Markov Models", IWFHR 93, pp. 385-390.
[3] Farag R., "Word level recognition of cursive script", IEEE Trans. on Comp., Vol. C28, pp. 172-175, 1979.
[4] Garcia-Salicetti S., Gallinari P., Dorizzi B., Wimmer Z., Gentric S.: "From characters to words: dynamical segmentation and predictive neural networks", ICASSP 96, Atlanta, May 1996.
[5] Manke S., Finke M., Waibel A.: "NPen++: a writer-independent, large vocabulary on-line handwriting recognition system", Proceedings of ICDAR 95, pp. 403-408, 1995.
[6] Juang B-H., Rabiner L.R., "The Segmental K-Means algorithm for estimating parameters of Hidden Markov Models", IEEE Transactions on Acoustics, Speech, and Signal Processing, Vol. 38, N°9, pp. 1639-1641, 1990.
[7] Rabiner L.R., Juang B-H.: "Fundamentals of speech recognition", Prentice Hall Signal Processing Series, 1993.
[8] Schenkel M., Guyon I., Henderson D., "On-line cursive script recognition using Time Delay Neural Networks and Hidden Markov Models", ICASSP 94, pp. II637-640, Adelaide, 1994.
[9] Schomaker L.R.B., Teulings H.L.: "Stroke versus character-based recognition of on-line connected cursive script", From Pixels to Features III: Frontiers in Handwriting Recognition, S. Impedovo and J.C. Simon (eds), 1992, Elsevier Science Publishers B.V.

Solving Nonlinear MBPC Through Convex Optimization: A Comparative Study Using Neural Networks

Miguel Ayala Botto[1] and Hubert A.B. te Braake[2] and José Sá da Costa[1]

[1] Technical University of Lisbon, Instituto Superior Técnico
Department of Mechanical Engineering, GCAR/IDMEC
Avenida Rovisco Pais, 1096 Lisboa Codex, Portugal
[2] Control Laboratory, Department Electrical Engineering
Delft University of Technology
P.O. Box 5031, 2600 GA Delft, The Netherlands

Abstract. Typical solutions based on nonlinear constrained optimization-based strategies are hard to find and usually demand for higher level of computation. In this paper two techniques for transforming the initial nonlinear optimization into an approximate convex optimization are presented and tested for a rigid manipulator modeled with a feedforward neural network. The results have shown that the overall performance is enhanced when performing an approximate feedback linearization.

1   Introduction

The basic idea of Model-Based Predictive Control (MBPC) algorithms is to use a model to predict the effect of future control actions on the output of the process. These predictions are then supplied to an optimization routine which minimizes a specific criterion function over a certain horizon. One of the main features of predictive control is that it can provide an easy way to handle process input and output constraints in the design of the controller [8].

Because of their good approximate properties, feedforward neural network models obtained through black-box modeling techniques provide accurate numerical predictions of general nonlinear processes, and so being successfully integrated in predictive control schemes [7]. However, because of their nonlinear nature, the resulting optimization problem is transformed into a non-convex optimization, demanding a high level of computations without the guarantee for finding a feasible solution. It is therefore important to search for different control structures that can yield feasible optimization problems with guaranteed convergence properties and low computational requirements.

In this paper two techniques are presented which transform the original nonlinear optimization into a linear optimization problem totally described by a convex criterion function. The first one is based on the expansion and further linearization over a specified operating trajectory, of the feedforward neural network prediction model. The second technique uses an approximate feedback linearization scheme in order to cancel the highly nonlinear coupled dynamics of the original nonlinear process, resulting in a desired linear model. Because of the nonaffine structure of the neural prediction model, an approximation has to be made resulting in a feasible solution for the complete prediction horizon [2]. A comparison between the loss of optimality and computational burden of both techniques is made considering a nonlinear SIMO continuous-time implement-ation of a single-link rigid manipulator. Notice that the extension of the last technique to the highly nonlinear n-link rigid mainpulators is straighforward, since these are totally feedback linearizable systems [6].

2 Nonlinear MBPC using neural networks

Multilayer feedforward neural networks are parameterized functions with good approximation capabilities. It is possible to find their appropriate parameters by using efficient numerical minimization routines, confirming the neural networks flexibility to represent the behavior of a wide class of nonlinear dynamical sys-tems. A possible choice consists of using the reliable NARX model representing the following nonlinear input–output mapping:

$$\hat{y}_{k+1} = f(\varphi_k) \tag{1}$$

where $\hat{y}_{k+1}$ corresponds to the one-step-ahead output prediction and φ_k re-presents the information vector (or regression vector) known at time instant k, consisting of the collection of past input and output values.

The best strategy for tracking control of nonlinear processes in the presence of input, output or state constraints, assuming no disturbances or uncertainties, consists of solving a nonlinear optimization problem. A schematic configuration of such a control structure is presented in Fig. 1.

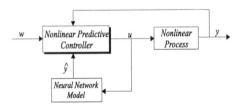

Fig. 1. Nonlinear predictive control

The solution vector from the optimization problem is given by $\tilde{u}^*$, and results from:

$$\tilde{u}^* = \arg\min_{\tilde{u}} J(\tilde{u}) \ , \ J(\tilde{u}) = \sum_{i=1}^{H_p} [w_{k+i} - \hat{y}_{k+i}]^2 + \rho[\Delta\tilde{u}]^2 \tag{2}$$

where k denotes the current time step, H_p is the prediction horizon, vector w corresponds to the desired reference, and $\tilde{u} = [u_k, \ldots, u_{k+H_p-1}]$ represents the controller output over the prediction horizon. It is assumed that the following level and rate inequality constraints are acting on the process input:

$$\begin{cases} \tilde{u}^{min} \leq \tilde{u} \leq \tilde{u}^{max} \\ \Delta\tilde{u}^{min} \leq \Delta\tilde{u} \leq \Delta\tilde{u}^{max} \end{cases} \tag{3}$$

where $\Delta\tilde{u} = u_{k+i-1} - u_{k+i-2}$ for $i = 1, \ldots, H_p$. Usually the receding horizon principle is implemented, which means that despite computing H_p control moves, only the first one is implemented. Due to the nonlinear nature of the neural network prediction model $\hat{y}$, the control vector $\tilde{u}^*$ is the solution of a non-convex optimization problem. This fact is a severe bottleneck to a successful application of this control methodology, since a high computational burden is required without guaranteeing a feasible solution.

In order to avoid such result, two simplifications to this scheme will be adopted, consisting of approximations of the global optimization problem. The basic idea is to transform the general nonlinear optimization problem presented in the previous section, into a linear optimization problem. Linear optimization problems are based on convex criterion functions subjected to linear constraints which can be solved through efficient iterative algorithms. In this case, the optimization problem (2)–(3) is described as the minimization of a quadratic criterion function as:

$$\tilde{u}^* = arg\min_{\tilde{u}} J(\tilde{u}) \, , \, J(\tilde{u}) = \frac{1}{2}\tilde{u}^T H \tilde{u} + c^T \tilde{u} \tag{4}$$

subjected to the following linear constraints:

$$A\tilde{u} \leq b \tag{5}$$

In this case, the optimal solution can always be found, as long as the Hessian matrix H in (4) is positive definite [5].

3 Linear MBPC using neural networks

The most straightforward way in order to obtain the desired quadratic cost function consists of expanding the output predictions of the feedforward neural network model of the process, and then linearize the resulting complete expansion around some chosen operating trajectory. In a compact form, the complete expansion matrix F obtained through successive substitution can be written as [1]:

$$\tilde{\hat{y}} = F(\varphi'_k, \tilde{u}) \tag{6}$$

where $\tilde{\hat{y}} = [\hat{y}_{k+1}, \ldots, \hat{y}_{k+H_p}]^T$ represents the future output predictions and φ' the known state at time instant k. By neglecting the higher order terms in the Taylor's approximation around some operating trajectory $\tilde{u}_0$ (typically taken from the previous optimization), the complete expansion results in the following matrix equation:

$$\tilde{\hat{y}}^l = F(\varphi'_k, \tilde{u}_0) + g(\varphi'_k, \tilde{u}_0)(\tilde{u} - \tilde{u}_0) \tag{7}$$

where the $(H_p{\times}H_p)$ matrix $\boldsymbol{g}(\varphi'_k, \tilde{u}_0)$ represents the transpose of the gradient of $\boldsymbol{F}$ at the operating trajectory. From this result it is easy to construct the desired quadratic criterion function as in (4).

One of the main advantages of this procedure is related to the fact that only a small computational burden is required in order to find a feasible solution. However, if the dynamical behavior is changing considerably during the prediction horizon and if the process is highly nonlinear, then it is most certain that the performed linearization will generate high errors into the optimization routine, resulting in poor closed-loop performance. Additionally the eigenvalues of the resulting Hessian matrix, $H = 2[\boldsymbol{g}(\varphi'_k, \tilde{u}_0)^T \boldsymbol{g}(\varphi'_k, \tilde{u}_0) + \rho]$, must be all positives, otherwise convergence can not be guaranteed.

It is possible, under some conditions [6], to design a feedback controller for a given minimum-phase nonlinear process, such that the resulting composed system behaves like some desired linear system. In this way, the MBPC scheme can also be based on the linear predictions of the resulting linear system, and so enabling the use of quadratic programming methods. The idea beyond feedback linearization consists of applying a static feedback control law to the original nonlinear process, such that the relation between the process output and the new external input is described through some desired linear dynamics. In Fig. 2 this complete scheme is shown.

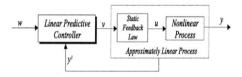

Fig. 2. Linear predictive control using feedback linearization of neural network model

The feedback linearization problem seeks for a control action u_k such that:

$$\hat{y}_{k+1} = f(\varphi'_k, u_k) = \frac{A(q)}{B(q)}v_k = y^l_{k+1} \qquad (8)$$

where v_k corresponds to the new external input, and $A(q)$, $B(q)$ define the linear reference model, with q being the shift operator. Notice that taken u_k from (8) requires solving a numerical problem, with difficult analytical solution. In fact, due to the non-affine structure of the feedforward neural network model, an approximation must be used in order to guarantee an analytical solution to the feedback linearization problem. The proposed idea linearizes through Taylor's expansion the network model at some operating point u_0, resulting in the following linear analytical expression [3]:

$$u_k = \frac{-f(\varphi'_k, u_0) + \frac{\partial f}{\partial u_k}(\varphi'_k, u_0)u_0 + \frac{A(q)}{B(q)}v_k}{\frac{\partial f}{\partial u_k}(\varphi'_k, u_0)} \qquad (9)$$

In this way, provided that the derivative $\frac{\partial f}{\partial u_k}(\varphi'_k, u_0)$ is nonzero, an approximate feedback linearization is obtained. An important consequence of using this linear relation is that the original set of linear constraints on u are directly transformed through equation (9) into a new set of linear constraints on v. This automatically guarantees a feasible solution without constraints violation, for the complete prediction horizon. Otherwise alternative correction methods should be used, or else more restrictive control moves should be adopted [4].

4   Case study

In this section, simulation results of the two predictive control strategies based on the same feedforward neural network model are presented. The nonlinear process considered is a single-link rigid manipulator, with total mass $M = 10$ kg and link size $L = 0.3$ m. Using a black-box modeling approach, a one-hidden layer feedforward neural network model with 3 inputs $(\theta_k, \dot{\theta}_k, \tau_k)$ and 2 outputs $(\hat{\theta}_{k+1}, \hat{\dot{\theta}}_{k+1})$, with sigmoidal activation function, is obtained.

The control problem consists of tracking a reference trajectory in the joint angular position and angular velocity θ and $\dot{\theta}$, respectively, for a cycle motion between $[-1.5, 1.5]$ rad, assuming level input torque constraints $|\tau| \leq 30$ Nm, for a simulation period of 5 sec. with a sampling time of 0.005 sec. The criterion function to be optimized is then stated as:

$$\min_{\tilde{\tau}} \sum_{i=1}^{H_p} \alpha_p [\theta_{k+i}^{ref} - \hat{\theta}_{k+i}]^2 + [\dot{\theta}_{k+i}^{ref} - \hat{\dot{\theta}}_{k+i}]^2 + \rho[\Delta\tau]^2 \tag{10}$$

where θ^{ref} and $\dot{\theta}^{ref}$ designate the desired position and velocity references. The prediction horizon chosen is $H_p = 10$, with $\alpha_p = 1000$ and $\rho = 0.01$. For the linear MBPC with approximate feedback linearization, a first order discrete-time linear system with unitary gain and time constant 0.01 sec. is assumed as the result from the linearization scheme. The results shown in Table 1 compare the angular position and velocity Sum Squared Errors (SSE), the control effort and the number of floating point operations in Matlab (FLOPS) over the entire simulation.

Control strategy	SSE$_\theta$	SSE$_{\dot{\theta}}$	Control effort	# FLOPS
Linear MBPC	3.59	400.68	445	$5.0 \times 10^{+7}$
Linear MBPC with approx. F.L.	1.15	119.61	402	$1.2 \times 10^{+8}$

Table 1. Simulation results for the constrained case

Simulation results have shown that when the input constraints are not active, both linear MBPC schemes behave approximately the same. This result is a consequence of the implementation of a non-exact feedback linearization, due to the non-affine structure of the neural network model. However, when input constraints are active, and if there is computational time left, the feedback linearization scheme is preferable since it enhances the overall closed-loop performance.

5 Conclusions

Two nonlinear MBPC techniques are described in this paper. The first method is nonlinear MBPC based on linearization of the predictions over the complete prediction horizon. The accuracy of this technique is highly dependent on the behavior of the process, since those linearization errors will have a direct influence on the overall closed-loop performance. The other technique integrates a feedback linearization scheme in the predictive control loop. In this way, the resulting approximate linear system is further incorporated into the predictive control scheme, enabling the solution to be found by solving a convex optimization. Because of the non-affine structure of the feedforward neural network prediction model, a linearization of the inner feedback loop is necessary. This approximation originates linearized constraints for the complete prediction horizon, enabling a feasible solution to be found in a fast and reliable way without constraints violation. Simulation results for a single-link rigid manipulator have shown that this technique leads to better closed-loop performance when compared with the previous one. This result encourages the use of such control design in n-link rigid manipulators.

References

1. Botto, Miguel Ayala: Feedback linearization techniques applied to predictive control using neural networks. Control Laboratory, Department of Electrical Engineering, Delft University of Technology (1995) Internal Report R95.042
2. Braake, H.A.B. te, Botto, M. Ayala, Can, H. J. L van, Costa, J. Sá da, Verbruggen, H. B.: Constrained nonlinear model based predictive control. Submitted to 2nd Portuguese Conference on Automatic Control (1996), Oporto, Portugal
3. Braake, H.A.B. te, Can, H.J.L. van, Scherpen, J.M.A., Verbruggen, H. B.: Control of nonlinear chemical processes using dynamic neural models and feedback linearization. Submitted to Computers in Chemical Engineering (1995)
4. Del Re, L.: Hybrid MPC for minimum phase nonlinear plants. Proceedings of 3rd European Control Conference (1995), Rome, Italy, 3561–3566
5. Gill, P. E., Murray, W.: Numerical methods for constrained optimization. Academic Press Inc. (1974), London
6. Isidori, A.: Nonlinear Control Systems: An Introduction. Springer-Verlag (1985) Berlin, Germany
7. Saint-Donat, Jean and Bhat, Naveen and McAvoy, Thomas J.: Neural net based model predictive control. International Journal of Control (1991), vol. 6, No 54, 1453–1468
8. Soeterboek, Ronald: Predictive control: a unified approach. Prentice Hall International (1992), UK, Cambridge

Combining Statistical Models for Protein Secondary Structure Prediction

Yann Guermeur, Patrick Gallinari

LAFORIA IBP-UA CNRS 1095
Université Pierre et Marie Curie
Tour 46-00, Boîte 169
4, Place Jussieu - 75252 Paris cedex 05 France
guermeur,gallinari@laforia.ibp.fr

Abstract: We investigate the problem of combining experts to predict the secondary structure of globular proteins. We first present two different statistical models for this task. We then analyse an efficient linear combination technique, this sheds light on unexplained phenomena frequently encountered in practice for ensemble methods.

1 Introduction

One current challenge of molecular biology is to determine the three-dimensional (3D) structure of proteins. Although it is fully determined by the sequence (primary structure), this problem is still unsolved. The usual way to proceed is to predict first an intermediate one-dimensional description called the secondary structure. This classification task consists in assigning a conformation, α-helix, β-strand or coil, to each residue of a sequence. This problem presents most of the difficulties found in the processing of sequences: long-term dependencies exploitation, noise processing and constraint satisfaction. This paper focuses on the combination of methods for secondary structure prediction. We first present in §2 an original recurrent neural network (NN) architecture. In §3, we describe a statistical model which exploits second order correlations present in the sequence. We then propose in §4 an analysis of linear combination methods which leads to a fast algorithm for combining experts.

2 Neural Architecture

Most of the neural systems applied so far to the prediction of the secondary structure are of the Qian and Sejnowski type [1]. They include two levels: a first network takes as input a segment of adjacent residues and outputs the secondary structure of the central residue in the segment. A second network smoothes these predictions: it takes as input a series of outputs from the first level to make the final decision.

2.1 A Sequence-to-Structure Network

Residues are usually encoded by binary vectors of 20 components, each of which corresponds to an amino acid. The input is then a "window" over the sequence including the central residue plus a left and right context of some neighbours (typically 3 to 9 to each side). The dimension of the input space is thus 20 times the window size. We use here for a more compact encoding a weighted sum of consecutive vectors. Let u_i denote the ith residue of the window encoded in a binary form, and $x \in \mathbb{R}^{20}$:

$$x = \sum_{i=1}^{|W|} w_i u_i \qquad (1)$$

$|W|$ (the window size) residues are thus mapped into one vector x using only $|W|$ adjustable parameters w. The structure of our network is depicted in figure 1. The input is a segment of 17 residues. A first layer computes weighted sums as defined in (1). It has been found empirically that using three distinct sums over the window was more efficient. Recurrent connections with fixed weights from the output to the input allow to implement an iterative scheme in order to consider a context using both past and "future" outputs for predicting y_9 (see notations in figure 1). Correlations between consecutive outputs are then taken into account. This conformational context is stored in a specific data structure and copied into the y' units. An iteration or epoch will consist in the presentation of the set of sequences to the network. For training and testing, the components of the context associated with each residue are initialized to the vector of *a priori* probabilities of the classes. When y_i has been computed, y'_i is replaced by its value, and the context is updated, it is then entirely modified at the end of the epoch.

Back-propagation has been used for training. Learning ends when the network produces a balanced prediction on the validation set, i.e. when the predicted proportions for the three classes on this set are equal to the *a priori* probabilities of the classes.

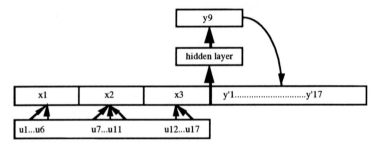

Fig. 1. Network architecture. Thick arrows designate full connections, thin arrows represent one-to-one connections. u denotes the residues, x_1, x_2, x_3 ($x_i \in \mathbb{R}^{20}$) compute respectively a weighted sum for residues 1..6, 7..11, 12..17, y' ($y'_i \in \mathbb{R}^3$) are copies of past outputs of the system corresponding to residues 1 to 17. $y_9 \in \mathbb{R}^3$ is the prediction for u_9.

2.2 A Structure-to-Structure Network
The outputs of the above system are then smoothed via a second classifier, a simple network with extra input cells encoding the charge and hydrophobicity of the residues as defined in [2]. Distinct sets of hidden units are devoted to the two types of inputs.

2.3 Experiments and Results
Database [1] was chosen since numerous methods have been assessed on it. Training and test sets are denoted respectively by setS and setT. A validation set was developed by selecting in another base [3] new sequences with less than 30% pairwise similarity with those of setT. 107 chains were thus gathered (setC). The assignment was done according to DSSP [4]. The following results describe the average behaviour over five runs of the architecture when trained with setS, setC being used for validation, and vice versa. For many biological reasons, the recognition rate cannot account alone for the quality of a prediction. The Matthews' correlation coefficients [5] denoted here C_α, C_β and C_{coil}, are also widely used since they give a measure of the prediction accuracy per structure. The performances compare favourably to those given by Sejnowski (prediction accuracy 64.3%, $C_\alpha=0.41$, $C_\beta=0.31$, $C_{Coil}=0.41$).

Training set	recognition %	C_α	C_β	C_{coil}
setS	64.8	0.43	0.35	0.43
setC	66.0	0.46	0.37	0.43

Table1. Average test performances, for different training sets.

3 A Statistical Module

Simple NNs do not easily exploit higher order features present in the sequence (cf. [1]). The Bahadur expansion allows to do this explicitly [6]. A first attempt to use this technique for secondary structure prediction has been performed in [7], where the authors propose an extension of the original principle that only applies to dichotomies. We have derived a different extension in order to exploit co-occurences of residues in the various positions of the window.

Let x_{ij} be equal to 1 if residue i in the window is amino acid j and 0 otherwise. The content of the window is represented by the matrix $x=[x_{ij}]$, $(1 \le i \le |W|)$, $(1 \le j \le 20)$ ($|W|$ is the window size). The conditional probability of structure S_k given the context is $p(S_k|x) \propto p(x|S_k)p(S_k)$. The second order Bahadur expansion estimates $p(x|S_k)$:

$$p(x|S_k) \approx \hat{p}(x|S_k) = p_1(x|S_k)\left(1 + \sum_{20*i+j<20*l+m} r_{ijlmk} z_{ijk} z_{lmk}\right) \qquad (2)$$

with: $\quad \alpha_{ijk} = E(x_{ij}|S_k), \quad z_{ijk} = (x_{ij} - \alpha_{ijk})/\sqrt{\alpha_{ijk}(1-\alpha_{ijk})}, \quad r_{ijlmk} = E(z_{ijk} z_{lmk}),$

$p_1(x|S_k) = \prod_i \prod_j \alpha_{ijk}^{x_{ij}} (1-\alpha_{ijk})^{1-x_{ij}}$.

z is the normalization of x with 0 mean and unit variance, r is a second order statistic, it measures the co-ocurrence of acids j and m respectively in positions i and l for structure S_k. p_1 denotes the joint distribution of the x_{ij} when they are supposed independant. $\hat{p}$ is thus a second order approximation of p in terms of pairwise correlations. An additional approximation, $\tilde{p}(x|S_k) = \sup(\hat{p}(x|S_k),0)$, guarantees the non-negativity of the estimated values. Set setS + setC was used for training. The performances are significantly superior to those reported in [7].

Training set	recognition %	C_α	C_β	C_{Coil}
setS+setC	66.4	0.45	0.36	0.44

Table 2. Performances of the statistical model on the test set.

4 Ensemble Methods

4.1 Constrained Optimal Linear Combinations

Better classification performances may be obtained by using several classifiers and combining their outputs. The so called "ensemble methods" have been the subject of extensive studies in the field of NNs over these last years. Linear combinations, NNs, majority votes are among the techniques which have been used for this. We have found the former to be superior, when using adequate constraints. We propose below an analysis which sheds light on a frequently observed behaviour of this technique. It allows to implement an efficient algorithm for finding optimal combinations.

Our analysis is asymptotical. We make the assumption that our experts are unbiased. Let f_j $(1 \le j \le P)$ denote the classification function computed by expert j ($f_j(x_i) \in$

$\mathbb{R}^3$) and $Q(v) = \frac{1}{2N}\sum_{i=1}^{N}\left\|y_i - \sum_{j=1}^{P}v_jf_j(x_i)\right\|^2$ the mean squared error (MSE) of the combination.

The linear combination of the experts will be optimal for MSE criterion and unbiased if it is the solution to the following constrained problem:

$$\begin{cases} Minimize\ Q(v) \\ \text{subject to constraint } 1_P^T v = 1 \end{cases} \qquad \text{(Pb. 1)}$$

where 1_P is a column vector of P ones.
In this case, the solution is:

$$v^* = (1_P^T C^{-1}1_P)^{-1}C^{-1}1_P, \qquad Q(v^*) = \tfrac{1}{2}(1_P^T C^{-1}1_P)^{-1},$$

where C is the estimated covariance matrix of the errors of the experts.
It has been frequently observed experimentally, e.g. [8] that using an additional constraint of non-negativity on the $v_i s$ led to significantly lower prediction errors.

Unfortunately, the optimal convex combination can no more be expressed analytically, since it must be computed by constrained iterative methods, such as Uzawa's algorithm. *We suggest here a simple technique for computing a solution, and show its optimality in a simplified case.* Our algorithm is as follows:

> *Iterate*
> > *Compute the optimal solution for (Pb. 1),*
> > *remove the experts with a negative coefficient,*
> *Until the solution is convex.*

Although our method is not optimal in the general case, we observed that it produces in practice the right solution. We propose below an analysis of this phenomenon. For this purpose, an extra hypothesis is made:

The Off- diagonal correlation coefficients are all equal to a constant $\rho > 0$ (Hyp. 1)

This simplifying hypothesis is coherent with the observed behaviour of the experts (see also [7]), since their errors are almost uniformly correlated. The result below can then be shown:

Theorem: under hypothesis (Hyp. 1), the algorithm converges to the solution of the following problem

$$\begin{cases} Minimize\ Q(v) \\ \text{subject to constraints } 1_P^T v = 1 \text{ and } v_i \geq 0 \ (1 \leq i \leq P) \end{cases} \qquad \text{(Pb. 2)}$$

Proof: see appendix.

4.2 Experiments and Discussion

Linear combinations (LC) have been compared to a NN which takes as input the predictions made for 17 consecutive residues. The experts are four NNs (trained respectively on setS and setC, by presenting the sequences from the N-terminus to the C-terminus and conversely), and the statistical module of §3. The training procedure of the combiners is derived from stacked generalization [9]. Learning sets are split into four disjoint parts of roughly equal size. One is removed in each set and experts are trained. Then, the combiners are trained using the outputs of the experts on the unseen training data. This procedure is iterated four times. Finally, the experts are trained again on their entire learning sets. Table 3 gives the performances of the methods.

Combiner	recognition %	C_α	C_β	C_{Coil}
NN	66.8	0.48	0.38	0.45
optimal LC	67.3	0.49	0.37	0.46
convex LC	67.8	0.50	0.37	0.46

Table 3. Relative average performances of ensemble methods.

The variations in the behaviours of the experts between the two training phases explain the poor performances of the NN. Its larger parameterization is useless since the task it learns is significantly biased. Using complex combiners within the framework of stacked generalization is only appropriate when one strictly follows the computer demanding original procedure, or assumes a stable behaviour of the experts.

5 Conclusion

This study highlights some properties of linear and non-linear cooperation techniques applied to a difficult real-world problem, and illustrates their impact on the performances. It offers new insights into the choice of an ensemble method. A natural extension of this work is its transposition for the processing of multiple alignments. The best system developed so far for protein secondary structure prediction [10] takes as input the profile of amino acid substitutions, in place of simple sequences. We are currently working on the problem of weighting the predictions made independently for aligned sequences.

References

[1] Qian, N. and Sejnowski, T.J. (1988). Predicting the Secondary Structure of Globular Proteins Using Neural Network Models. *J. Mol. Biol.*, **202**, 865-884.

[2] Eisenberg, D., Wilcox, W. and Eshita, S. (1987). Hydrophobic moments as tools for analysis of protein sequences and structures. In *Proteins: structure and function*. Edited by James J. L'Italien, Plenum Press, 1987, 425-436.

[3] Colloc'h, N., Etchebest, C., Thoreau, E., Henrissat, B. and Mornon, J.P. (1993). Comparison of three algorithms for the assignment of secondary structure in proteins: the advantages of a consensus assignment. *Protein Engineering*, **vol. 6**, 377-382.

[4] Kabsch, W. and Sander, C. (1983). Dictionary of protein secondary structure: pattern recognition of hydrogen-bonded and geometrical features. *Biopolymers*, **vol. 22**, N°12, 2577-2637.

[5] Matthews, B.W. (1975). Comparison of the predicted and observed secondary structure of T4 phage lysozyme. *Biochim. Biophys. Acta*, **405**, 442-451.

[6] Bahadur, R.R. (1961). A Representation of the Joint Distribution of Responses to *n* Dichotomous Items. In *Studies in Item Analysis and Prediction*, chapt. 9, 158-169, Stanford University Press.

[7] Zhang, X., Mesirov, J.P. and Waltz, D.L. (1992). Hybrid System for Protein Secondary Structure Prediction. *J. Mol. Biol.*, **225**, 1049-1063.

[8] Breiman, L. (1992). Stacked Regressions. Technical Report N° 367, August 1992, Department of statistics, University of California, Berkeley.

[9] Wolpert, D.H. (1992). Stacked Generalization. *Neural Networks*, **vol. 5**, 241-259.

[10] Rost, B. and Sander, C. (1994). Combining Evolutionary Information and Neural Networks to Predict Protein Secondary Structure. *Proteins*, **19**, 55-72.

Appendix: Proof of the theorem in section 4.1

Proving the theorem is equivalent to showing that each negative coefficient in the solution of (Pb. 1) corresponds to a null coefficient in the solution of (Pb. 2).

The computation of $v_{i,P}^*$, component of the solution of problem (Pb. 1), gives:

$$v_{i,P}^* = K_P \prod_{j=1, j\neq i}^{j=P} \sigma_j \left\{ (1+(P-2)\rho) \prod_{k=1, k\neq i}^{k=P} \sigma_k - \rho \left(\sum_{l=1, l\neq i}^{l=P} \prod_{m=1, m\neq l}^{m=P} \sigma_m \right) \right\} \qquad (3)$$

where σ_i is the standard deviation of the error of expert i, and K_P is a positive constant. Let v_P^{**} be the solution of (Pb. 2). We assume that experts are sorted with increasing σ_i. The demonstration goes in three steps:

1) If $v_P^* \notin \mathbb{R}_+^P$, $\exists T \in \{2..P\} / v_{i,P}^* \geq 0$ $(1 \leq i \leq T-1)$, $v_{j,P}^* < 0$ $(T \leq j \leq P)$

This result is established by noticing that the sign of $v_{i,P}^*$ is the sign of the expression between brackets in (3), which decreases for increasing indexes.

2) The strict convexity of the cost function implies that $v_{P,P}^{**} = 0$

- Let $\theta = \min_{i \in \{T..P\}} v_{i,P}^{**} / (v_{i,P}^{**} - v_{i,P}^*)$. If $\forall i \in \{T..P\}, v_{i,P}^{**} > 0$, then $\theta \in]0;1[$, and $w_P = \theta v_P^* + (1-\theta)v_P^{**}$ satisfies the constraints of (Pb. 2). Using the strict convexity of Q, we get $Q(w_P) < \theta Q(v_P^*) + (1-\theta)Q(v_P^{**}) < Q(v_P^{**})$. This is in contradiction with the definition of v_P^{**}, so $\exists i \in \{T..P\} / v_{i,P}^{**} = 0$. If $T=P$, the result is established.

- $2(Q(v_P^{**}) - Q(v_P^*)) = (C(v_P^{**} - v_P^*), v_P^{**}) = (C(v_P^{**} - v_P^*), v_P^{**} - v_P^*)$, i.e. v_P^{**} is the projection of v_P^* on the convex U defined by the constraints of (Pb. 2). According to the projection theorem, $\forall v_P \in U, (C(v_P^{**} - v_P^*), v_P - v_P^{**}) \geq 0$, whence can be derived $\forall v_P \in U, (Cv_P^{**}, v_P - v_P^{**}) \geq 0$.

If we assume $v_{P,P}^{**} > 0$, let $Z \in \{T..P-1\} / v_{Z,P}^{**} = 0$, and let $w_P \in U / w_{i,P} = v_{i,P}^{**}$, $(1 \leq i \leq P-1, i \neq Z)$, $w_{Z,P} = v_{P,P}^{**}$, $w_{P,P} = 0$.

$$(Cv_P^{**}, w_P - v_P^{**}) = \rho(\sigma_Z - \sigma_P)v_{P,P}^{**} \sum_{i \neq Z, i \neq P} v_{i,P}^{**}\sigma_i + (\rho\sigma_Z - \sigma_P)v_{P,P}^{**2}\sigma_P < 0.$$

This proves by contradiction that $v_{P,P}^{**} = 0$.

3) From (3) it can be derived: $\forall i \in \{T..P-1\}, v_{i,P-1}^* < 0$

$$(1+(P-2)\rho)\prod_{k\neq i}\sigma_k - \rho\sum_{l\neq i}\prod_{m\neq l}\sigma_m < 0 \text{ and } \rho(\prod_{m\neq P}\sigma_m - \prod_{k\neq i}\sigma_k) < 0.$$

Thus, $\sigma_P \left\{ (1+(P-3)\rho) \prod_{k\neq i, k\neq P}\sigma_k - \rho \sum_{l\neq i, l\neq P} \prod_{m\neq l, m\neq P}\sigma_m \right\} < 0$ and $v_{i,P-1}^* < 0$.

(This implies that the same demonstration applies again to prove the nullity of all coefficients $v_{i,P}^{**}$ $(T \leq i \leq P-1)$.)

Using RBF-Nets in Rubber Industry Process Control

U. Pietruschka, R. Brause

J.W. Goethe-Universität, Frankfurt a.M., Germany
brause@informatik.uni-frankfurt.de

Abstract

This paper describes the use of a radial basis function (RBF) neural network. It approximates the process parameters for the extrusion of a rubber profile used in tyre production.

After introducing the problem, we describe the RBF net algorithm and the modeling of the industrial problem. The algorithm shows good results even using only a few training samples. It turns out that the „curse of dimensions" plays an important role in the model.

The paper concludes by a discussion of possible systematic error influences and improvements.

1 Introduction

Process control in rubber industry has the smell of a „dirty" industrial branch. This comes not only from the often very dull and dusty rubber and tyre production rooms where the products are „baked" by heat and steam, but also from the fact that the macromolecular proportions of rubber are hard to predict due to their nonlinear character. In the extruder (the melting and form-giving machine) the rubber mixture is heated up to 110°-140°C, compressed with 70-140 bar by a screw conveyor and pressed through a metal mask. On leaving the extruder, the rubber relaxes, that is it expands or shrinks, depending on the mixture, changing therefore its shape in a non-linear manner by 10%-20% up to 50%. Figure 1 shows the basic layout for our example of tyre profile production.

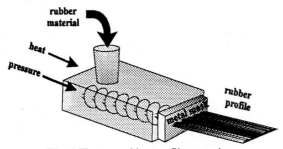

Fig. 1 The tyre rubber profile extrusion

The task of process control consists of estimating the necessary extrusion parameters (i.e. the shape of the extrusion metal mask) for an acceptable rubber product after relaxation. The modeling has to reflect the following facts:

- The rubber expansion pressure and flow within the profile heavily depend whether the neighbor parts of the profile have a high level, or if the neighbor parts are low-leveled. This causes the rubber profile to be also a function of the profile height of the neighbored points.

- Additionally, the extruded rubber profile heights depend nonlinearly on the rubber mixture G, the pressure P by the screw conveyor, on the temperature T, on the extruder type E and on the weight w per meter of the band.
- By the nonlinear form of the screw conveyor the pressure along the profile mask decreases nonlinearly. Therefore, the rubber profile does also depend on the absolute position along the metal mask.

Nevertheless, the whole system is deterministic: the same rubber mixture G with the same mask g(x), temperature T and pressure P result in the same rubber profile r(x), even on a different extruder machine of the same type E.

Up to now, due to the nonlinear nature of the macromolecular mixture this task can not be solved analytically. Instead, specialized people estimate the profile of the original metal mask by their experience with the subject and correct their estimates after experience. This gives a trial-and-error production cycle that causes severe disadvantages for the production business:

- The start for a new product is delayed by the time for 2-3 cycles. Each one takes 4-5 days to make a new mask, install it on the extruder, make an extrusion try, measure the obtained rubber profile and estimate a new metal mask. This delay does not only waste time, money and natural resources, but also increase the production overhead and impedes therefore the production flexibility s everely.
- The experienced employees are tied to this job (which they judge as „boring") without the possibility of a change within the enterprise. Additionally, in the case of illness of an employee or a change to another enterprise, the knowledge is no longer accessible. This causes mayor obstacles for the pr oduction.

Now, in this paper we will show that adaptive process control methods can overcome this kind of problems. They will update the parameters purely based on the final, measured outcome data.

2 An RBF approximation network

In this section we describe the methods for approximating the exact metal mask profile f(x) at location x of the extruder by a two-layer network function F(x). This produces the desired rubber profile r(x), see figure 2. It is well known that such a two-layer neural network can approximate any continuos function to any degree, provided that we have enough neurons in the first layer, e.g. [Xu, Krzyzak, Yuille 1994].

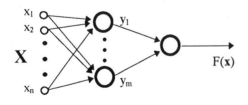

Fig. 2 The activity approximation network

The activity $y = (y_1,...,y_m)$ of the first layer is defined by

$$y_i = S_i(x,c_i) = \exp(-d^2) \qquad i=1..m$$

$$d^2 = |M(x-c)|^2 = (x-c)^T M^T M (x-c)$$

and the second layer with

$$y_0 \equiv 1, \; w_0 \cong \text{bias by}$$

$$F(\mathbf{x}) = F(\mathbf{y}(\mathbf{x})) = \sum_{j=1}^{m} w_j y_j = \mathbf{w}^T \mathbf{y}$$

using m nonlinear basis functions S_i that depend only on the Mahalanobis distance d between the input $\mathbf{x}$ and a neuronal center $\mathbf{c}_i$. For adapting and scaling the elipsoidal input field, we used the scaling equation

$$\mathbf{M}^{NEW} = (\mathbf{I} - \gamma(1-\alpha)(\mathbf{a}\mathbf{a}^T)) \mathbf{M}^{OLD} \qquad \mathbf{a} = (\mathbf{x}-\mathbf{c})/|\mathbf{x}-\mathbf{c}|$$

with the scaling factor α and the learning rate γ, see [Pietruschka, Kinder 1995].

There are principally two approaches to train the network parameters: either we train the two layers separately or as a whole.

The approach of treating the two layers separately, clustering the input space first and then optimizing the weights of the second layer, is fast, but it has some flaws. This gives us a high sample density of output values where we have clusters of input samples, not where the ou tput error is high.

Therefore, we optimize both layers at the same time. To avoid the computational problems of the backpropagation approach we choose a different strategy. We start with the lowest possible complexity of the network and gradually increase the number of neurons in the first layer until the error is sufficiently reduced. This was already proposed for RBF nets, for example by [Schiøler, Hartmann 1992]. We insert the neuron at location $\mathbf{x}_k$, the k-th sample with the maximal error, that has to be compensated by the new m-th neuron. We have to design the width $\mathbf{M}_m$ such that it fits the new basis function in the context of all neighbored neuron basis functions. In contrast to the approach of [Platt, 1995] we do not use gradient descend technique to rearrange all other neurons and adapt all their receptive fields: this is computionally intensive and is the source of new errors. Instead, we stop the adaption process of the new neuron by the criterion of non-significant activation on a data point. Additionally, we reduce the long distance neighborhood influence by a learning rate $\gamma(d)$ which drops with increasing distance from $\mathbf{c}_m$, that is with decreasing activity level, see [Pietruschka, Kinder 1995].

3 Approximating the extrusion process parameters

To apply the approximation algorithm that we developed in the previous sections we have to model the industrial process for the example of tyre production. The main task consists of estimating the profile of a metal mask that extrudes the profile of a rubber band. This band is then cut into a stripe of the perimeter length of a tyre and then glued to the casing. The raw tyre is then „baked" in a metal tyre form for 20 minutes, giving the preliminary profile the ultimate form.

3.1 Modeling the process

Although the extruded rubber profile is a temporary form its desired accuracy is 0.1 mm. This settles the upper limit for our approximation error. In figure 3 a sample profile is shown.

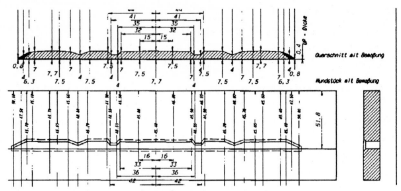

Fig. 3 A rubber profile and the corresponding metal mask

The upper profile is the desired rubber profile; the lower one shows the corresponding rectangular metal mask. On the right hand side a cut through the metal (shaded area) shows the form of the opening (not shaded). Where the rubber flows in, the profile has a wider opening. This corresponds to the dotted line that encircles the profile opening in the metal mask.

The analytical treatment of the nonlinear dependencies is very difficult. Conventional assumptions about energy (i.e. enthalpy) conservation are not valid here. Also the direct measurement of the process parameters like temperature and pressure in the profile are practically limited. In contrast to this, our approach models the system as a whole, avoiding all difference equations and constants which are hard to devise and to measure.

We devided the whole centered profile, depending on the tyre width, into 170-270 points that are placed in the regular distance of d mm. Each point x_i has a desired rubber profile height r(i). The intermediate points are interpolative generated, see figure 4.

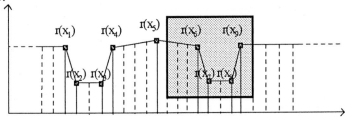

Fig. 4 The intermediate interpolation of the profile and a neighborhood window

Since the influence of the sample points is limited to the neighborhood for a certain rubber profile height r(i) we have only to consider $n=2s+1$ neighbor points

$$g(i) = F(r_{i-s},...,r_i,...,r_{i+s}, i, G, E, P, w,..)$$

By this model we implement a neighborhood window that uses n=2s+1 sampling points around location i. All values r_k for the sampling points outside the profile limits are set to zero.

3.2 Simulation results

For the determination of the two parameters, the number n of neighborhood sampling points, and the distance d between the sampling points, we decided to simulate different configurations in order to get an acceptable choice.

We generated the training set by shifting a window (determined by d and n) by an increment of 1 mm over the profile data of 5 profiles with the same values of G, E and P. This generated 1346 training patterns. The sixth profile was used for the generation of test set of 271 test points. The simulation results generally showed only a very small influence of the position i. So, let us regard other dependencies.

For the expected absolute error for 100 neurons we got different results, depending on the type of network we used. The nets with growing, radially symmetric input regions have in the average 10-90% more error than the growing elipsoidal nets. The best performance of the two types converged by training to the following expected absolute error, depending on the number of sampling points n and the interpoint distance d.

n\d	3 mm	4mm	5mm
7	0,178	0,159	0,187
9	0,167	0,162	0,197
11	0,165	0,206	0,226

It is interesting to see that the error does not automatically decrease or increase when we increase the number of sampling points. There is a configuration of the parameters where we roughly meet the balance and the error becomes quite small. The best results are observed by n=9 and d=4 mm which corresponds to a window size of 32 mm with the expected absolute error of 0.16 mm and the maximal absolute error of 0.56 mm. In figure 5 the test profile, the result of the network and the resulting

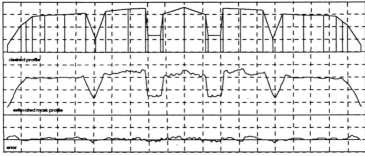

Fig. 5 The wanted profile and the profile produced by the net for n=9, d=4

error is shown for this configuration. The y-axis is scaled up by the factor of three to enhance the visibility of the errors.

4 Discussion and outlook

In the previous sections we have presented an adaptive solution for the problem of unknown process parameters in tyre production. The learning algorithm uses no internal process variables or other intrinsic knowledge but only the measurable external process parameters as the weight per meter and the resulting rubber profile.
This approach avoids many technical and economic disadvantages that were given in the introduction.

Nevertheless, our work also shows that there are still several problems to be solved:

- The current modeling uses the data provided by the production as tuples (wanted rubber profile, successful metal profile). The successful metal profile used for training was obtained after several trials and corrections, that is the training is based on an artificial and not on a real sample of the input/output mapping.

  This problem can be solved by measuring and using only the directly obtained rubber profile data.

An important key for the simulation performance turned out to be the two parameters, the number n of neighborhood sampling points, and the distance d between the sampling points. We can determine the proper choice by balancing the counteracting influences:

- If we choose d too small, we increase the number of necessary sampling points for a certain neighborhood and increase therefore the dimension of the input space. Since we have only a small limited number of training samples, the training becomes very difficult since the input space becomes very sparse. This is known as the „curse of dimensions" [Huber 85]. On the other hand, if we choose d too big, we can lose important information due to undersampling the dependency function.

- If we choose n too big, we meet the same problem of sparseness of the training samples in the input space. Additionally, by increasing to much context information, the generalization ability of the network will be limited. On the other hand, if we limit the window too much, necessary context information which helps to distinguish between different situations is lost.

From the theoretical point of view, this is an interesting situation. By the nature of the problem, we have not hundreds of sample profiles but just a few ones. Nevertheless, we are not aware of an applicable method of determining the optimal d and n to solve the problem of optimal training. Here, the work of theorist is welcome.

5 References

P. Huber: Projection Pursuit; The Annals of Statistics, Vol.13, No.2, pp. 435-475, 1985

U. Pietruschka, M. Kinder: Elipsoidal Basis Functions for Higher-Dimensional Approximation Problems; Proc. ICANN-95, Vol II, Paris 1995, pp.81-85

J. C. Platt; Learning by Combining Memorization and Gradient Descent; NIPS; pp. 714-720;1992.

H. Shiøler, U. Hartmann: Mapping Neural Network Derived from Parzen Window Estimator; Neural Networks, Vol.5, pp.903-909 (1992)

Lei Xu, Adam Krzyzak, Alan Yuille: On Radial Basis Function Nets and Kernel Regression: Statistical Consistency, Convergence Rates, and Receptive Field Size; Neural Networks, Vol. 7, No.4, pp.609-628, (1994)

Towards Autonomous Robot Control via Self-Adapting Recurrent Networks

Tom Ziemke

Neural Computing Group, Dept. of Computer Science, University of Sheffield, UK
&
Connectionist Research Group, Dept. of Computer Science, University of Skövde
Box 408, 54128 Skövde, Sweden
tom@ida.his.se

Abstract

This paper introduces a connectionist architecture for autonomous robot control in which second-order recurrent connections are used to provide a flexible, context-dependent mapping from percepts to actions in order to allow the network to adapt its behaviour continually to its current context and internal state. It is argued that this mechanism, to a higher degree than modular approaches, allows the robot to acquire and adapt complex behaviour autonomously.

1 Introduction & Background

It is commonly agreed that mobile autonomous robots interacting with realistically complex and dynamic environments are facing a complex *control problem*, since they have to be capable of adapting their behaviour to a number of different and varying goals, tasks, situations and requirements whose integration often is far from trivial. Therefore most approaches to the construction of adequate control systems follow the *divide-and-conquer* principle of breaking down the overall control problem into a set of sub-problems each of which might be simpler to solve. Typically this decomposition results in a discrete set of individual and relatively independent functional or behavioural modules organized in some form of hierarchy, e.g. in Brooks *subsumption architecture* (1986).

The concept of true *autonomy* is however not limited to automaticity alone, i.e. the capacity to interact with an environment, it further includes the ability to (continually) learn from this interaction, i.e. *the individual agent's own capacity (at least partly) to form and adapt its principles of behaviour* (Steels, 1995). Hence, a truly *autonomous* robot is facing not only a complex control problem, but also complex *learning problem* (a problem ignored by many approaches), namely how to acquire complex behaviour and continually adapt or extend it.

Due to the fact that the complexity and adaptivity of behaviour than can be achieved by 'simple' artificial neural networks (ANNs) (i.e. feed-forward or simple recurrent networks) is limited, the divide-and-conquer approach to complex learning problems (not restricted to autonomous robot control) has also been adopted in a number of ANN models, resulting in modular and/or hierarchical architectures for specific (e.g. Schyns, 1991) or general learning problems (e.g. Jordan and Jacobs' '(Hierarchical) Mixture of Experts' architecture, 1995).

What however is ignored by virtually all approaches following the divide-and-conquer principle is that

(a) hierarchical/modular decomposition of complex systems into discrete sub-systems does not seem to appear in naturally evolved systems (Beer, 1995), e.g. animals, which nevertheless typically exhibit highly complex, structured and adaptive behaviour,

(b) in a truly autonomous system, according to the above criteria, the 'divide' process, since forming the essential principles/organization of a system's behaviour, would have to be performed (at least partly) by the system itself, whereas in basically in all of the above approaches a rather rigid modularisation of a system is 'engineered' by its designer a priori, and typically is not subject to autonomous adaptation afterwards.

Hence, the work presented here is based on the idea of ANNs which achieve functional/behavioural complexity not through designed complex organization (hierarchical/modular), but through continual self-adaptation of their behaviour to their current context, i.e. networks that *acquire their behaviour as well as its structure/organization* through interaction with their environment.

2 Models & Analysis

2.1 The Architecture

The network architecture used here, mainly inspired by Pollack's *Sequential Cascaded Network* (1991), is similar to Elman's *Simple Recurrent Network* (SRN, 1990), i.e. a three-layer network in which the hidden (or state) layer is fully connected to itself. In addition however there are second-order recurrent connections from the network's state units to the input weights, which should allow the network to learn to adapt its own input weights depending on its own internal state in order to achieve a flexible, context-dependent mapping of percepts to actions.

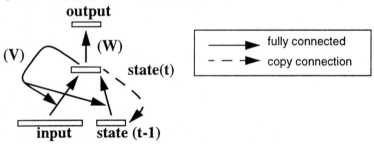

Fig. 1. Self-Adapting Recurrent Network (SARN)

Hence, this architecture can be described by dynamic equations as follows:

$$O(t) = g\left(W \cdot \begin{bmatrix} S(t) \\ 1 \end{bmatrix}\right) \qquad S(t) = g\left(\left(V \cdot \begin{bmatrix} S(t-1) \\ 1 \end{bmatrix}\right) \cdot \begin{bmatrix} I(t) \\ 1 \end{bmatrix}\right)$$

Where g is the logistic function, W is a two-dimensional weight matrix mapping $S(t)$ to $O(t)$, and V is a three-dimensional weight matrix mapping $S(t-1)$ to the next step's input weights.

2.2 The Robot Simulator

The experiments described here have been carried out with the Khepera Simulator (Michel, 1995), a simulated version of the real mobile Khepera robot (Mondada et al., 1993). The robot has two motors, which can independently move forward (in direction of sensors 2 and 3) and backward at different speeds, and eight infrared sensors measuring both ambient light and distance from possible obstacles.

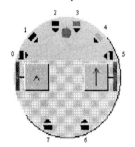

Fig. 2. Simulated Khepera robot

2.3 Learning Algorithm

In the experiments discussed in this paper the control networks have been trained with a modified version of the *complementary reinforcement backpropagation (CRBP) algorithm* (cf. Ackley & Littman, 1990, and Meeden, 1996). CRBP allows to turn the abstract signals of 'reward' and 'punishment' as given during reinforcement learning into precise error measures as they are required by backpropagation learning.

The binary vector B is produced stochastically by interpreting real-valued output R(i) as the probability that B(i) takes on the value 1. If the action resulting from the real-valued output R is punished the network is pushed towards the complement of B by backpropagating the error measure $((1 - B) - R)$, otherwise no error is backpropagated. Hence, in each time step the error at the output layer is backpropagated to update W and the input-to-state weights. Then the new input-to-state weights are used as target output to update the weights (V) between state layer and input-to-state weights.

2.4 Experiments

Initial Experiments

In initial experiments (see (Ziemke, 1996) for details) SARN controllers received input from distance sensors 0 to 5, and were trained on obstacle avoidance in a simple environment. Results showed that significantly better performance (faster learning, lower punishment ratio) was achieved with SARNs than with feed-forward and simpler recurrent controllers (e.g. SRNs), and that SARNs were capable of both

(1) closely approximating finite state automata behaviour, when trained on discrete action selection (i.e. 'turn left', 'turn right' or 'forward') in each time step.

(2) developing continuous, but structured internal state spaces with systematically related input weights, when trained on direct continuous-valued motor control.

Figure 3 shows a) a typical continuous state space (for case (2)) over 5000 time steps, and b) typical trajectories corresponding to left and right turns in the environment.

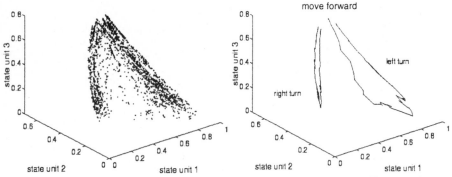

Fig. 3. a) Continuous state space, and b) state space trajectories corresponding to actions

Integrating (More) Complex Behaviours

In addition to the initial experiments, a light source is placed in the environment, and the control network also receives input from the light sensors 0 to 5. Furthermore the network receives one extra input which periodically changes between 1 ('on') and 0 ('off'). In addition to obstacle avoidance the robot should now follow the light gradient towards the light whenever the extra input is 'on', but avoid the light otherwise. In case of conflict between light-related behaviour and obstacle avoidance, light-related behaviour has the higher priority as long as light is perceived. That means that the robot, when the extra input is 'on', should actually hit a light source (or an obstacle close to it) as long as that helps to maximize its light readings.

SARN controllers did in most cases (8 out of 12) learn to control the robot such that (a) the ratio of obstacle-related punishment sank below 10% and (b) the ratio of punishment for incorrectly seeking/avoiding the light sank below 5% overall and below 20% while encountering the light one way or the other. The robot's behaviour could typically be described by the following 'strategy': (1) as long as no light is perceived: avoid obstacles, by turning as soon an obstacle is encountered, (2) if extra input 'off' and light perceived: turn away from the light and move on, and (3) if extra input 'on' and light perceived: turn towards the light, then move forward and hit the light source.

It was found that the control networks in all cases developed structured continuous state spaces and subspaces corresponding to sub-tasks (cf. the above 'strategy') with systematically related input weight sets (s. example in figure 4a). Analysis of the state-to-output weights in this case (not shown here) reveals that there is almost a 1:1 relation between state units and actions. A high activation for state unit 1 alone leads to a right turn, similarly a high activation of state unit 3 will result in a turn to the left. A high activation of state unit 2 alone will lead to a forward move. The following input weight sets (white: positive, black: negative, representative 'snapshots' for continuous groups of input weight sets) show that the control network has successfully learned to handle this problem exploiting the possibility of weight adaptation. Figure 4b shows typical input weights while performing obstacle avoidance only. It can be seen that the network has learned to set the weights for distance sensors (first six units) much higher than those for light sensors (unit 7 to 12). Encountering an obstacle leads to a low activation of state unit 2 (forward move), and depending on strength and location of the sensor readings state units 1 and/or 3 are activated resulting in a corresponding turn.

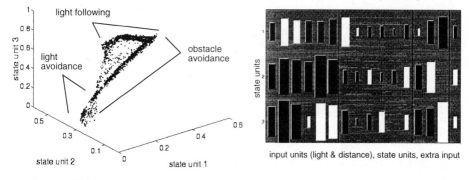

Fig. 4. a) SARN internal state space and subspaces, b) input weights (obstacle avoidance)

An almost completely different set of weights is used for light following (s. figure 5a). Now the weights for the light sensors are set relatively high whereas those for the distance sensors are now lower. Finally, when combining light and obstacle avoidance (s. figure 5b), light and distance sensor weights are of similar magnitude, and they make almost exactly the same use of the state units' effect on the motor outputs.

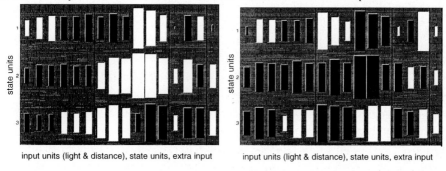

Fig. 5. Input weights, a) light following, and b) light & obstacle avoidance

3 Conclusion

A connectionist architecture for context-dependent self-adaptation of behaviour through second-order recurrent connections has been presented in this paper. The capabilities of this architecture, have been demonstrated and analysed in experiments with rather simple robot control tasks, showing that

(1) the SARNs are capable of learning to develop, depending on the task, both finite-state behaviour as well as structured continuous internal state spaces,

(2) the networks acquired systematic mappings a) between the robot's context and its internal state, i.e. the structure of the internal state spaces reflected the 'structure' of the task, and b) from internal state and sensory input weights, i.e. the robot learned to systematically adapt its mapping from percepts to actions to its current context and internal state,

(3) the robot was capable of exhibiting structured/hierarchical subsumption-

like behaviour (e.g. light following subsuming obstacle avoidance, obstacle avoidance subsuming forward motion), without the underlying mechanism actually being organized that way,

(4) the networks have *autonomously* acquired both their behaviour (cf. above point 2b) and its organization/structure (cf. above points 2a,3) from interaction with their environment.

Future work will include (a) validation on a real robot, (b) evaluation of the architecture with control tasks requiring more complex and diverse behaviour, as it would be necessary for the realization of a complete autonomous agent, and (c) further investigation of the idea of incremental learning in SARNs being realized through development, adaptation and extension of structured internal state spaces.

References

Ackley, D. H. & Littman, M. L. (1990) Generalization and scaling in reinforcement learning, in Touretzky, D. S. (ed.) *Advances in Neural Information Processing Systems*, 550-557, Morgan Kaufmann, San Mateo, CA

Beer, Randall D. (1995) A dynamical systems perspective on agent-environment interaction, *Artificial Intelligence*, **72**, 173-215

Brooks, Rodney (1986) A robust layered control system for a mobile robot, *IEEE Journal of Robotics and Automation*, **2 (1)**, 1986, 14-23

Elman, J. (1990) Finding Structure in Time, *Cognitive Science*, **14**, 179-192

Jordan, Michael I., Jacobs, Robert A. (1995) Modular and hierarchical learning systems, in Arbib, M. (ed.) *The Handbook of Brain Theory and Neural Networks*, MIT Press, Cambridge, MA

Meeden, Lisa (1996) An Incremental Approach to Developing Intelligent Neural Network Controllers for Robots, to appear in *IEEE Transactions on Systems, Man, and Cybernetics*, **26**, special issue on Learning Autonomous Robots

Michel, O. (1995) Khepera Simulator version 1.0 - User Manual, University of Nice-Sophia Antipolis, Valbonne, France, http://wwwi3s.unice.fr/~om/khep-sim.html

Mondada, F., Franzi, E., & Ienne, P. (1993) Mobile robot miniaturisation: A tool for investigation in control algorithms, in *Third International Symposium on Experimental Robotics*, Kyoto, Japan

Pollack, Jordan B. (1991) The induction of dynamical recognizers, *Machine Learning*, **7**, 227-252

Schyns, Phillippe G. (1991) A Modular Neural Network Model of Concept Acquisition, *Cognitive Science*, **15**, 461-508

Steels, Luc (1995) When are robots intelligent autonomous agents?, *Robotics and Autonomous Systems*

Ziemke, Tom (1996) Towards Adaptive Behaviour System Integration using Connectionist Infinite State Automata, to appear in *Proceedings of the Fourth International Conference on the Simulation of Adaptive Behavior*, MIT Press, Cambridge, MA

A Hierarchical Network for Learning Robust Models of Kinematic Chains [*]

Eric Maël

Institut für Neuroinformatik
Ruhr-Universität Bochum, Germany
Email: mael@neuroinformatik.ruhr-uni-bochum.de

Abstract. A hierarchical network for visuo-motor coordination is proposed. The hierarchical approach allows learning geometric models of realistic robots with six or more axes. The network consists of several one-dimensional subnetworks, which learn the coordinate transform and rotation axis for each joint. In our simulation, the network reduces the end-effector error of a 7-axis anthropomorphic robot and 20-axis robot below the visual error.

1 Introduction

A problem still unsolved in biology and robotics is the adaptive visuo-motor coordination of high degree of freedom (DoF) kinematic chains. The two problems are the high degrees of freedom and robustness against sensor errors. The DoF problem results from learning the high dimensional inverse kinematic of a manipulator. A well-known approach of this example, which is restricted to low degrees of freedom, is presented by Ritter *et. al.* [Ri89]. Robustness against sensor errors is a problem if one insists on a control method without feedback (open loop). Both problems can be avoided by confining to learning the forward kinematics and controlling the end-effector with the inverse Jacobian matrix and visual feedback (closed loop). There is neurophysiological and psychological evidence that the motor cortex has a directionally tuned representation [Bu92] similar to the Jacobian matrix and that humans learn in a proximo-distal hierarchical order of their limbs [Pa91] like the forward kinematics. The forward kinematics have the advantage of being naturally separable into simple coordinate transforms for each joint. The high dimensional search space for the whole kinematic chain can be reduced to several low dimensional subspaces, making it possible to learn the kinematics of a realistic manipulator with six or more axes. In contrast to standard robotics, where the whole kinematic chain is described by one transformation matrix which is derived analytically, this approach describes the separated kinematic chain with several numerical coordinate transforms for each joint. This gives the possibility of adapting each joint independently and, in addition, a global geometrical model of the manipulator can be learned, which is essential for avoiding collisions with obstacles. The geometrical model can be

---

[*] Supported by a grant from the DFG (GK KOGNET).

used to calculate the Jacobian matrix to control the end-effector, wrist, or elbow, and it is useful when handling singular joint configurations and redundant manipulators.

2   Network Model

A kinematic manipulator consists of joints and links which must be modeled in the visual space by the network. Any joint with k degrees of freedom can be modeled as k joints of one DoF connected by $k-1$ links of zero length. Therefore, without loss of generality, we choose the number of the subnetworks and joint axes equal to the number of DoF of the manipulator. The advantage of this method is a maximal separation of the forward kinematics.

The inputs of the network are the joint angles and the outputs are the homogeneous transformation matrix and the joint axis for each joint, and a constant tool and base frame. A homogeneous transformation matrix T_B^A describes the position and orientation of a coordinate frame B relative to a frame A [Cr89]. For a joint this matrix depends on the joint angle. To handle this variation, a subnetwork saves several numerical matrices for different joint angles.

Imagine for simplicity a robot arm with two DoF, θ_1 and θ_2, operating in a planar workspace as shown in figure 1. The visual space is symbolized by the cartesian coordinate axis. The shoulder of the robot arm can have a constant offset as represented by the base frame. Two subnetworks representing the shoulder and elbow joint and a constant tool frame represent a gripped object. The visual system observes the robot arm and has to extract the four homogeneous transformation matrices T_{Base}^{Vision}, T_{Elbow}^{Vision}, T_{Wrist}^{Vision} and T_{Tool}^{Vision} relative to the origin of the visual frame. The kinematic chain of the manipulator to be learned by the hierarchical network can than be described as follows:

$$T_T^V(\theta_1, \theta_2) = T_B^V T_E^B(\theta_1) T_W^E(\theta_2) T_T^W \tag{1}$$

To train the subnetworks and the constant tool frame, we have to transform the visual frames into the previous base or joint frame.

$$T_E^B(\theta_1) = T_B^{V^{-1}} T_E^V(\theta_1) \tag{2}$$

$$T_W^E(\theta_2) = T_E^{V^{-1}}(\theta_1) T_W^V(\theta_1, \theta_2) \tag{3}$$

$$T_T^W = T_W^{V^{-1}}(\theta_1, \theta_2) T_T^V(\theta_1, \theta_2) \tag{4}$$

Instead of using the previous visual frame to compute the transform, it is recommendable to use the adapted base frame and the output frames of the previous subnetworks. This method has the advantage of reducing the error accumulation in the kinematic chain and leads to higher accuracy. In the case of an extended arm link, not only the network of the extended link is adapted to the modified kinematic, but also all following subnetworks, which results in faster adaptivity.

The simplest way to learn input-output relations with an adaptive method is a self-organizing map. We choose as subnetwork for each joint an extended model

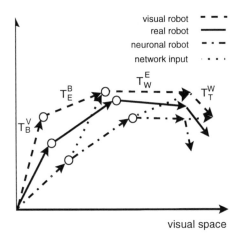

visual robot - - -
real robot ———
neuronal robot - · —
network input · · ·

T_B^V T_E^B T_W^E T_T^W

visual space

Fig. 1. Two DoF real, visual and neural robot in two-dimensional visual space. The network input is a transformation of the visual frame into the previous output frame, which is the hierarchical combined output of the previous subnetworks.

of the growing neural gas (GNG) by B. Fritzke ([Fr95a]). The extension of the GNG is in the associated local homogeneous transformations (LHT) ([Fr95b]). The GNG algorithm with LHT is a supervised network which learns topologies. It incrementally constructs a graph representation of a given data set. In case of one-dimensional joints, the topology is restricted to a closed or open linear graph. The topology is used to adapt the best unit and its topological neighbours. The insertion of units is controlled by a local accumulated approximation error. This error is a combination of position and orientation error between the LHT of the best unit and the desired homogeneous transform. A unit is inserted after a number of adaptation steps between the unit with the maximum accumulated error and its neighbour with the largest error. When a new unit is inserted, its LHT is linearly interpolated between its neighbours. The linear interpolation of homogeneous transformation matrices is also used to adapt the units and to calculate the output transform and the joint axis with the nearest and the second-nearest unit of the subnetwork. Figure 2 shows the processing scheme of the hierarchical network for a two DoF robot.

The network algorithm is divided into three phases: the *initialization phase*, a *training phase* and a *control phase*, with the possibility of extending the training phase after a control phase or to combine the training and control phase. The subnetworks can be trained in parallel or in hierarchical order.

During the initialization phase, each subnetwork has to be initialized with two units. Therefore the visual system has to extract the base, tool and joint frames and transform them with the previous visual frame into the initializing joint frame. These frames have to be saved. The manipulator than makes an incremental move with each joint, and the procedure of visual extraction and transformation must be repeated with the difference of adapting the constant base and tool frame.

During the training phase, the visual system extracts the base, tool, and joint frames. The inputs to the subnetworks are the actual joint angles. The output matrix and joint axis can be calculated through linear interpolation with the

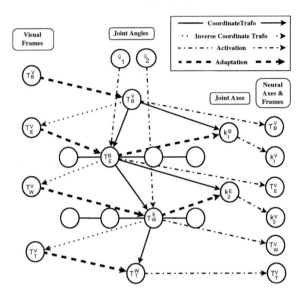

Fig. 2. Processing scheme of the hierarchical network for a two DoF robot.

LHT of the nearest and second-nearest unit for each subnetwork. The visual frames can be transformed into the learning joint frame with the inverse output matrix of the previous subnetworks. The best unit and its neighbours in each subnetwork can be adapted with the corresponding transformed visual frame. Their local error accumulator is then updated. Furthermore, each representation of the joint axis is adapted with the joint axis extracted by interpolation. After some number of iterations, a unit is inserted through interpolation.

During the control phase the visual system extracts the target frame. The inputs to the subnetworks are the actual joint angles. The output matrices for the actual joint angles can be calculated through linear interpolation for each subnetwork. We can calculate the Jacobian matrix geometrically with the neural model of the kinematic chain. This Jacobian matrix has to be inverted in order to calculate an incremental change of joint angles for the given target direction. The direction is given without visual feedback with the neural end-effector frame, or takes it into account if higher accuracy is desired.

3 Simulation Results

We simulate the kinematics of a 7 DoF and 20 DoF robot arm with an arm length of 1000 mm to demonstrate the potential of our network.

The errors in visual frames are simulated by using the exact homogeneous transformation matrix, which is extracted from the simulator, and multiplying it with an error matrix. The absolute visual error is a Gaussian distribution in position and orientation with standard deviations of 100 mm and 10 deg respectivly, which are pessimistic estimates. The direction of the position error

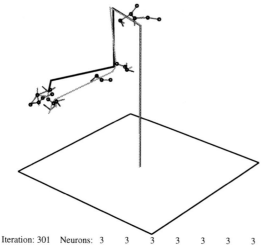

Fig. 3. 7 DoF anthropomorphic robot arm (black line) and 7 subnetworks (open linear graphs with circles) with neural robot model (grey line). The subnetworks are projected into the visual space to show the DoF of each joint. Each unit (circle) represents a local homogeneous transform (LHT). The best unit is marked with the coordinate frame (LHT). The output of the whole network is the neural robot model (grey line).

Iteration: 301   Neurons: 3   3   3   3   3   3   3

and the rotation axis of the orientation error are uniformly distributed. The robot is moving slowly through the whole joint space and all subnetworks are adapted in parallel to the joint angles and visual frames. After 300 iterations a new unit is inserted into each subnetwork.

Figure 3 shows the line drawing of the 7 DoF robot arm and the status of the 7 subnetworks after 301 iterations. Each network consists of an open linear graph with three units. The units are projected into the visual space with the output of the previous networks and a constant offset in direction of the next joint axis. The small grey line drawing illustrates the actual network output. After this small number of iterations, the subnetworks do not sample the whole joint space, but the error of the neural robot model is already in the order of the visual error. Despite of this error, we are able to control the end-effector with the inverted neural Jacobian matrix to a desired target. The accuracy of the final position can be much higher than the absolute visual error and the neural end-effector error. This is possible with an iterative linear trajectory generator with visual feedback. The deviation of the neural Jacobian matrix only leads to a slightly curved trajectory shape, because we recalculate the Jacobian matrix and the direction for the visual end-effector after every incremental movement. The accuracy of the relative position of the visual target and the visual end-effector is much higher than the absolute visual position. Therefore, the error of the neural robot model significantly effects only the accuracy of the open loop trajectory and obstacle avoidance, but this error can be reduced below the visual error.

In Figure 4 the accuracy and adaptivity of the hierarchical network during the training phase is shown. Accuracy and adaptivity are two competitive qualities depending on the adaptation rate. Low adaptation rate leads to high accuracy and high adaptation rate leads to fast adaptivity. The large errors in the beginning of the simulation are caused by extrapolation, when the units of each subnetwork do not yet cover the whole joint space.

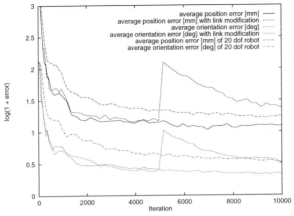

Fig. 4. End-effector error of the 7 DoF and 20 DoF robot, with an arm length of 1000 mm. The standard deviation of the visual input frames are 100 mm and 10 deg. The shoulder link is extended by 100 mm and rotated by 10 deg after 5000 iterations.

4 Conclusion

We propose a hierarchical network for learning geometrical models of high DoF manipulators. This hierarchical approach can solve the problem of high degrees of freedom by learning an adaptive forward kinematic with several one-dimensional subnetworks. In our simulation the simple subnetworks reduce the average end-effector error of the 7 and 20 DoF neural robot model to below the absolute visual extraction error. The positioning error can be of course much lower if we control the end-effector in a closed-loop manner. The network can be easily extended for higher degrees of freedom and scales gracefully.

Acknowledgements

I wish to thank Christoph von der Malsburg and Stefan Zadel for fruitful discussions.

References

[Cr89] J. J. Craig. Introduction to Robotics (2nd ed.), Addison-Wesley Publishing Company, 1989

[Ri89] H. J. Ritter, T. M. Martinetz, K. J. Schulten. Topology-concerving maps for learning visuo-motor-coordination. *Neural Networks*, Vol. 2:, pp. 159-168, 1989

[Pa91] J. Paillard. Brain and Space, Oxford University Press, 1991

[Bu92] Y. Burnod, P. Grandguillaume, I. Otto, S. Ferraine, P. B. Johnson, R. Caminiti. Visuomotor transformations underlying arm movements toward visual targets: A Neural Network model of Cerebral Cortical Operations. *The Journal of Neuroscience*, Vol. 12(4), pp. 1435-1453, April 1992

[Fr95a] B. Fritzke. A growing neural gas network learns topologies. In G. Tesauro, D. S. Touretzky, and T. K. Leen, editors, *Advances in Neural Information Processing Systems*, Vol. 7, pp 625-632. MIT Press, Cambrige MA, 1995

[Fr95b] B. Fritzke. Incremental learning of local linear mappings. In F. Fogelman, editor, ICANN'95: *International Conference on Artificial Neural Networks*, pp. 217-222, Paris, France, 1995. EC2 & Cie.

Context-Based Cognitive Map Learning for an Autonomous Robot Using a Model of Cortico-Hippocampal Interplay

Ken Yasuhara and Hanspeter A. Mallot

Max-Planck-Institut für biologische Kybernetik
Spemannstr.38, 72076 Tübingen, Germany

Abstract

This paper presents an additional module of cortico-hippocampal interplay to our view-based competitive sequence learning scheme of spatial memory. This new model was examined with a mobile robot. In this model both the orthogonalization of input patterns and the integration of object and spatial information are realized by use of a middle-term memory module as a model of hippocampal function. This scheme works well not only for orthogonal input views but also for highly correlated patterns. Even after the memory module is damaged, the robot can take the shortest path to the goal point if enough knowledge has been acquired prior to the damage.

1. Introduction

A scheme for learning a cognitive map of a maze from a sequence of views and movement decisions has been proposed by Schölkopf and Mallot [1]. View graphs can be learned from sequences of views by use of a competitive learning rule which translates temporal sequence (rather than featural similarity) into connectedness (Fig.1). The network takes two inputs, a feature vector representing the view information, and a unique activity in one of a small number of movement units representing the most recent movement decision. The view vectors are mapped to view-cell activity by input weights, which specialize to one view during exploration. View-cells that become winners in subsequent time steps are connected by an intrinsic (auto-associative) weight. This weight is modulated by input from the movement-cell. To examine this theory, experiments with a mobile robot Khepera® were performed in a maze with black and white bar-codes on the ground [2].

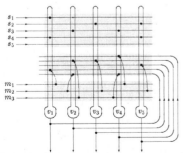

Fig.1 Wiring diagram of the original model
s $1,...,$s J: feature vector corresponding to current view. m $1,...,$m K: motion input. v $1,...,$v N: view cells. Input weights subserve view recognition. Feedback weights represent connections between views. They can be modified by movement facilitation.

We have observed during the experiment that a neuron fires for different views, and several different neurons fire for one view (Fig.2). This phenomenon usually occurs at the beginning of the random exploration and tends to disturb the whole learning process. It may be caused by cross-association between view patterns. We have used both canonical orthogonal input patterns and random patterns. In the random patterns, orthogonality is not guaranteed. With canonical orthogonal patterns we made sure that the learning schema works both in the computer simulation [1] and in the experiment with the robot [2]. With random patterns, however, we had difficulty both in the simulation and robot experiment. Thus we added an orthogonal processing module that can be thought of as a model of hippocampal function.

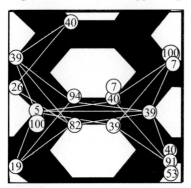

Fig.2. Performance of the old model
The black area shows the open space of the maze. Each corner corresponds to a "view" obtained while the robot is approaching the corner from the opposite corridor. The numbers indicate the winner cell in the view-cell layer that became active when the robot approached this corner. The white solid lines indicate the connections. As one can see, the same neuron (e.g. no.40) fires for different views and several different neurons (e.g. no. 40, 91, 53) fire for one view.

2. Description of the new model

2.1 Interpretation by the brain

To interpret the world, the brain needs at least two kinds of input information, multisensory information, and internal information that depends on the motivational state of the animal. To construct a simple model, we assumed that feature coding or pattern coding of sensory information is processed in the cerebral neocortex and the symbolic coding or concept formation is processed in the hippocapus. By this, Hamming distances among a set of input sensory patterns do not affect the distance among the corresponding symbols. That means that the input sensory patterns can be memorized independently of their similarity. This basic advantage could be used not only for our system but also for other systems that have difficulty memorizing highly correlated patterns.

2.2 Model of hippocampus

2.2.1 Theta rhythm

We assumed that an oscillatory stimulus into the hippocampus changes the state of the hippocampal neural network [3]. One could imagine that the network state moves about from one attractor (equilibrium state) to another staying in one attractor for a period of theta rhythm (Fig.3). The hippocampal neural network has fixed random recurrent connection. The expected number of equilibrium states is $2^{0.4n}$, where n is the number of cells [4]. The sparse activity pattern of the attractor becomes the concept for the sensory input information. The oscillatory wave is generated by use of mutual connections between models of septum and hippocampal theta cells[3] [6](Fig.4).

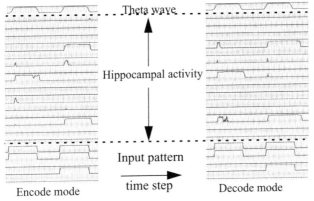

Fig.3 Simulation of EEG of hippocampus
The topmost wave indicates the theta wave. Under it the activity of various hippocampal cells is shown. The lower part indicates the input pattern. For each similar input pattern, an orthogonal code emerged. The horizontal axis is time.

2.2.2 Orthogonalization

One could see this orthogonal code as the middle-term memory (as compared to the short-term of input activity after some object recognition processing). This middle-term memory is transformed into the long-term memory of our original view-cell layer through the competitive sequence learning. In other words, the code becomes the index for the long-term memory. The original view-cell layer takes the code as the input. Because of the orthogonalization ability of the hippocampus, one can expect that the competitive learning does not run into problems. In contrast Kohonen's novelty filter, in which orthogonalization is processed with input information itself by a large matrix computation, this system uses simple Hebbian learning and seems to be biologically plausible. However, because of the conventional associative ability between the input and the orthogonal code, there is a short-coming in this model. It is possible that each pattern is cross-associated to the fixed points. If the input is ambiguous, a wrong attractor is sometimes recalled. However, when the same input is continuously given, the correct attractor can be recalled later. This phenomenon is psychologically plausible, but is a disadvantage in technical applications. To overcome this, we used a fusion of several sensors to increase the attractive force to the correct attractor as described in the next section.

2.2.3 Integration of different sensory information

On the basis of anatomical evidence, we assumed that various sources of sensory information are integrated in the hippocampus. The integration of contextual information about the environment is interpreted by the orthogonal code in the hippocampal module. By acquiring the context, the ambiguity of sensory information is solved. At the same time the attractive force to the correct attractor increases. During the learning phase, the concept in the hippocampus and the various sources of sensory information are associated by a Hebbian rule, while the network state becomes unstable as the oscillatory stimulus increases and the state moves into another stable attractor. During the recalling phase, the encoded concept is decoded and used to look up the long-term memory in the view-cell layer. From psychophysical evidence, it seems that for navigation tasks, metric topology of the environment must be represented in the brain and integrated with the views.

2.2.4 Habit

While the hippocampus works to register new memories and forms long-term memory traces through competitive sequence learning, another path is conditioned. That is the direct synapse connection between the view-cell layer and the sensory input. After sufficient learning, this connection makes "stimulus-response" conditioning possible without the memory module. One could call this path the "habits path" as opposed to the "memory path". During learning this path has no influence on view-cells. But in the background a habit is conditioned. We suggest that behaviour is a combination of automatic responses to the stimulus and actions resulting from memory and expectation.

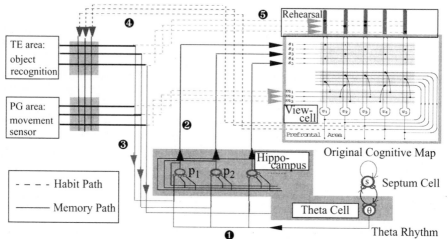

Fig.4 Model of Cortico-Hippocampal Interplay

Concept formation : The mutual connection between the θ-cell and the s-cell (lower right) generates an oscillatory threshold to the p-cells in the memory module. In this way a different attractor is reached in each time period in which a memory is stored (solid line ❶). The emerging orthogonal activity pattern of the memory module (i.e. middle-term memory) is transformed into long-term memory by the competitive sequence learning (solid lines ❷).
Encoding : In a theta wave period, the sensory information (because the movement sensor module is not yet complete, we gave simulated vector as spatial information) is associated with the emerged code by a hebbian rule (solid lines ❸). *Retrieval from memory* : As in the register mode, when the oscillatory stimulus into the p–cell increases, the network state becomes unstable and moves into another attractor. Depending on the input sensory information the state moves to the code attractor, which is registered in the register mode.
Habit : After a view-cell is registered by the memory path, the outputs of the view-cells (see dashed lines ❹) become the teacher signal for the habit path (see dashed lines ❺). As the habit path ❺ activates rehearsal-view-cells in the rehearsal view-cell layer (see above the original module on the right side), both winners of the view-cell layer and rehearsal view-cell layer are compared. The result of the comparison is the supervised signal. Then the strength of the habit path is modified by perceptron learning with the signal in the background until both winners become identical.

3. Experiments with non-orthogonal input

The navigation experiments were performed with a mobile robot, Khepera®, with non-orthogonal bar-codes. The number of view-cells was 100. The visual sensor input was preprocessed to binary code (1 and 0) by a certain threshold without any compression. It was sampled as a vector with dimension 200. The number of views was 12 and the number of places was 7. Because the movement sensor for the path integration was under constructing, we took simulated information as additional spatial information. Learning took approximately 30 minutes to achieve 100 random steps from one place to another. We fixed goal position during the experiments. The result is shown in table 1 below.

3.1 Performance of new model

One result of learning with 100 steps is shown in Fig.5(a). As one can see, there was no overlap among the cells (cf. Fig.2). Fig.5(b) shows the orthogonal sparse codes that the memory module represented during the learning phase. These codes are the index for view-cells. Ten path-planning trials carried out in three different learning sessions (80-110 learning steps each) were performed.

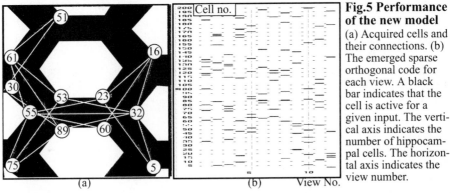

(a)   (b) View No.

Fig.5 Performance of the new model (a) Acquired cells and their connections. (b) The emerged sparse orthogonal code for each view. A black bar indicates that the cell is active for a given input. The vertical axis indicates the number of hippocampal cells. The horizontal axis indicates the view number.

3.2 Lesion experiment

Our new model integrates two forms of information, position (through movements) and object recognition in the memory module. We examined next whether the robot could find the way if the simulated position information was cut. Ten path-planning trials during three different learning sessions (80-110 learning steps) were performed.

Table 1: Comparison of performance

	optimal path found	sub-optimal path	no path found
original model	29%	14%	57%
view information only	40%	30%	30%
view+position information	50%	50%	0%

3.3 Habits formation

At last we examined whether the robot could find its way after learning without the memory module, with only the help of the habit path. At the beginning of learning, different winner cells were activated in both the rehearsal view-cell layer and the view-cell layer. However, they become identical later. Then after 70-125 learning steps in 14 path-planning trials during three different learning sessions, the robot with the deficit found the optimal way in 86% of the trials.

4. Discussion

There is a reason why we used an attractor module. There is a related physiological experiment of delayed matching to sample in area IT in monkey [5]. The most important point is that short-term memory is maintained for each pattern in the form of a stable activity pattern. Moreover, these equilibrium states are not acquired by learning [4]. On the other hand, in most hippocampal units the specific responsiveness is established either instantaneously or very rapidly. This suggests that the coding determining the firing pattern was already present. It could be used at short notice independently of the learning that takes place on first introduction to a new environment [6].

Miller suggests also the existence of the reciprocal connections between the hippocampus and the isocortex which contain a rich sensor repertoire of delay lines including those with total loop-times roughly matching the period of theta rhythms [6]. The mutual entrainment among different areas and hippocampus are further work.

There is also evidence that simple alternation mazes could be learned on the basis of kinesthetic information alone. However, more complex mazes could not be learned without the integrity of visual, olfactory, auditory cues or distance senses [6].

5. Conclusion

1. With an orthogonal memory module, the learning process was improved.
2. The integration of view and simulated position information solves the ambiguity by acquiring contextual information.
3. The habits path accomplished the path-planning task after learning even with damage to the memory module.

6. References

1. Schölkopf, B., Mallot, H. A. View-Based Cognitive Mapping and Path Planning. Adaptive Behavior Vol.3, No.3, 311-348, 1995
2. Mallot, H. A., Bülthoff, H.H., Georg, P., Schölkopf, B., Yasuhara, K. View-based cognitive map learning by an autonomous robot. ICANN'95, 1995
3. Morita, M. Hippocampus Model of Associative Memory. Proc. 27th SICE Ann. Conf., 2, pp.1061-1064, 1988
4. Amari, S., Kurata, K., Akaho, S. Neural Network Models of Long-Term and Short-Term Memory. Technical Report of IEICE, MBE88-143, 1988 (in Japanese)
5. Miyashita,Y., Chang, H.S. Neuronal correlate of the pictorial short-term memory in the primate temporal cortex. Nature, 331, pp.68-70, 1988
6. Miller, R. Cortico-Hippocampal Interplay. Springer-Verlag, Berlin, 1991

An Algorithm for Bootstrapping the Core of a Biologically Inspired Motor Control System[*]

Stefan Zadel

Institut für Neuroinformatik, Ruhr-Universität Bochum, D-44780 Bochum, Germany
zadel@neuroinformatik.ruhr-uni-bochum.de

Abstract. An architecture for a motor control system inspired by biological organisms is outlined. The core of this architecture is a model of the direct kinematics of the articulated chain (AC) under control. The advantage of using the direct kinematics solution to solve the inverse kinematics problem is that the former is separable and can be broken down to low-dimensional problems. A novel algorithm to adaptively learn, in a hierarchical fashion, the direct kinematics solution of an AC with many degrees of freedom (DoF) is presented. The algorithm is designed such that only neurally implementable operations or functions are used. The algorithm is shown to work with an articulated chain with nine DoF. On average, less than 200 iterations per joint are required.

1 A biologically plausible motor control system

An organism has to satisfy two conflicting requirements for controlling its movements: speed and precision. Two basic control strategies exist to achieve this: feedforward and feedback control. Due to delays in a neural system, feedback control is slow in organisms and has a cut-off frequency of a few Hertz. But thanks to its iterative character it is, in principle, good for arbitrary precision. Feedforward control, on the other hand, can be very fast. It is an input-output mapping and can thus, in principle, be implemented by a simple look-up table. To achieve high precision, however, a great number of sample points is necessary, especially if the mapping is nonlinear. The cost of learning time and memory resources grows exponentially with dimension and becomes prohibitive for high-dimensional mappings. The (kinematic) control of an AC asks for a mapping between the six DoF needed to describe the end-effector's position, and n DoF for the n joint angles of the AC[2] [2]. Evidently, feedforward control cannot achieve high precision in motor control. Therefore, the best strategy for an organism would be to combine a feedforward system (fast, but inaccurate) with a feedback system (precise, but slow). This is actually found in organisms [9, 10, 3]: there is feedforward control plus proprioceptive, visual, and tactile feedback.

Neuromorphic designs for feedforward and feedback systems have been proposed (e. g., [8, 4, 1]). But since they often make use of maps sampling the whole control space, they suffer from the problems mentioned above when applied to

---

[*] Supported by grants from the DFG (GK KOGNET) and the German Federal Ministry for Science and Technology (01 IN 504 E9).

[2] $n = 6$ for a common robot arm, and $n \sim 16$ for the AC from a human's head to his finger tips — if only 10 samples per DoF were taken, 10^{16} samples in total would be needed — about 100 times more than there are synapses in our cortex.

high-dimensional tasks. A remedy would be to find a separable formulation, i. e., to break the problem into a product of lower-dimensional problems. For inverse kinematics, such a separation is not in sight. For direct kinematics, a separation is well-known in robotics: the formalism of homogeneous transforms — the transform across a joint can be expressed by a homogeneous matrix dependent only on that one joint angle. The whole solution is then given by the product of these matrices [2]. The chain of matrices forms a natural hierarchy (a matrix affects all those following, but not those preceeding it in the chain). The effect is that the cost of learning time and memory resources increases only *linearly* with the number of DoF, not exponentially as before.

When direct kinematics has been learned, it can be used for feedback control using the Jacobian $\mathbf{J}$ obtained, e. g., by the vector cross product method [2, appendix B]. Alternatively, $\mathbf{J}$ can be approximated by differences instead of differentials (move — virtually — each joint a little and record its effect on the end-effector), which also avoids being trapped by singularities. As inversion of $\mathbf{J}$ might not be implementable in a neurally plausible way, the Jacobian transpose method can be used to produce a movement.[3] Finally, using feedback control, the direct kinematics system can produce successful, though slow directed movements before the feedforward system has been fully configured.

The above approach makes the feedback system, in terms of precision, independent of the feedforward system. It allows having high precision in the feedback system and low precision in the feedforward system, without the latter affecting the former. This permits designing each of the systems as best as possible for its task. For instance, the feedforward system might be implemented with a crude approximation as suggested by psychophysical data [10].

From psychophysics, there is also evidence for a hierarchical organisation of the motor system [7]: Subjects wearing prismatic goggles and allowed to adapt to the visual shift by moving either wrist, elbow, shoulder, or head showed generalization to the joints not moved only in a proximo-distal direction, but not vice versa. This is fully consistent with the view presented above.

2 Algorithm for learning the direct kinematics solution

Principle. Since direct kinematics is separable, learning can be done for each joint independently. A straightforward approach would be to look at all joints and to determine the corresponding link transforms directly, of which an implementation with neural networks is presented in [5]. This does, however, not work with joints that can never be seen, such as the neck. Therefore, one has to come up with a different approach that only needs to see the end-effector. One solution is to learn sequentially joint by joint starting at the most proximal joint [6]. Moving only one joint will move the end-effector on a circle (degenerate cases excepted). From this, the corresponding link transform can easily be established [2, 6]. Below, a different approach is given that also handles degenerate cases. Starting with the first joint and continuing until the last one is reached gives

---

[3] $\Delta\underline{\varphi} \propto \mathbf{J}^T(\underline{\varphi}) \cdot \Delta\mathbf{x}$ [Stefan Schaal, personal communication].

the transforms between all joints. Finally, obtaining the transform from the end-effector to the last joint is straightforward. Sweeping movements covering the whole range of a joint angle give the best learning results. This resembles observations from developmental psychology revealing that humans pass in their first year of life through a phase of rhythmical stereotypies that seem to be transition behavior between uncoordinated activity and complex, coordinated voluntary motor control [11].

Basic idea. Suppose we have an AC with only one joint. Then three transforms are needed to model it: $^J\mathbf{T}_E$ from the end-effector to the joint, $\mathbf{R}_z(\varphi)$ for the joint rotation[4], and $^W\mathbf{T}_J$ from the joint to world coordinates. The model of the AC is $^W\mathbf{T}_E = {}^W\mathbf{T}_J \cdot \mathbf{R}_z(\varphi) \cdot {}^J\mathbf{T}_E$. After adaptation, $^J\mathbf{T}_E$ and $^W\mathbf{T}_J$ will remain constant and changing the joint angle φ will be fully represented by $\mathbf{R}_z(\varphi)$. When not yet adapted, each time φ is changed, the physical end-effector position will be different from the one predicted by the model. This difference can be used to adapt $^J\mathbf{T}_E$ and $^W\mathbf{T}_J$.

Notation. Here, the notation and homogeneous transform matrix formalism of [2] is adopted. The following symbols are used: end-effector transform $^W\mathbf{T}_E$ as the transform from the end-effector frame E to world coordinates W; link transform $^{i-1}\mathbf{A}_i$ of link i; rotation matrix $\mathbf{R}_z(\varphi)$ for a rotation about the z axis by angle φ; link transform $^{i-1}\mathbf{A}_i(\varphi_i) = \mathbf{R}_z(\varphi_i) \cdot {}^{i-1}\mathbf{A}_i$ also dependent on the joint angle φ_i of joint i.

The formulation for the AC when joint i is moved is taken to be

$$^W\mathbf{T}_E = \underbrace{\prod_{j=0}^{i-1} {}^{j-1}\mathbf{A}_j(\varphi_j)}_{^W\mathbf{T}_J} \cdot \mathbf{R}_z(\varphi_i) \cdot \underbrace{{}^{i-1}\mathbf{T}_E}_{^J\mathbf{T}_E} \tag{1}$$

with the offset transform $^{i-1}\mathbf{T}_E$. $^{-1}\mathbf{A}_0(\varphi_0)$ is the transform from base coordinates of the AC to world coordinates. It was introduced to allow an arbitrary position of the AC relative to world coordinates since the z axis of the base frame must be the rotation axis of joint 1. Setting $\varphi_0 \equiv 0$ does not reduce generality.

Procedure. Let n be the number of joints of the AC.

1. *Initialize end-effector transform.* Determine $^W\mathbf{T}_E$.
2. *Joint loop.* For each joint $i = 1, \ldots, n$, perform steps 3 to 8.
3. *Initialize link transform.* $^{i-2}\mathbf{A}_{i-1}$ will now be adapted. Initialize it with zero or some random values.
4. *Adapt offset transform.* Adapt $^{i-1}\mathbf{T}_E$ such that equation (1) is fulfilled.
5. *Move joint i.* Change angle φ_i.
6. *Determine end-effector transform.* (See step 1.)
7. *Adapt link transform.* Adapt $^{i-2}\mathbf{A}_{i-1}$.
8. *Repeat link transform adaptation until converged.* Repeat steps 4 to 7 until $^{i-2}\mathbf{A}_{i-1}$ has converged sufficiently.

---

[4] Here set to be about the z axis by angle φ.

 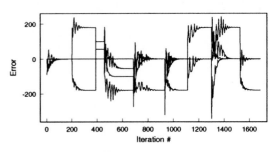

Fig. 1. Left: NEUROS robot shown in a set-up for grasping an object. Right: Progress of the convergence process. The errors of all four D-H parameters are plotted over the accumulated number of iterations. The points where learning switches from one joint to the next can be easily identified. The fundamental frequency of the oscillations stem from the sweep movements.

9. *Determine transform of last link.* Set $^{n-1}\mathbf{A}_n = {}^{n-1}\mathbf{T}_E$.

For the implementation, the link transform representation of Denavit and Harten-berg (D-H) is used [2], i.e., $^{i-1}\mathbf{A}_i = {}^{i-1}\mathbf{A}_i(\vartheta_i, \alpha_i, a_i, d_i)$ and $^{i-1}\mathbf{A}_i(\varphi_i) = {}^{i-1}\mathbf{A}_i(\varphi_i; \vartheta_i, \alpha_i, a_i, d_i) = {}^{i-1}\mathbf{A}_i(\varphi_i + \vartheta_i, \alpha_i, a_i, d_i)$ with the D-H parameters ϑ_i, α_i, a_i, d_i. Some more mathematical details:

To step 4: To adapt $^{i-1}\mathbf{T}_E$, simply rearrange equation (1).

To step 7: After changing φ_i, equation (1) will not be fulfilled unless the adaptation of $^{i-2}\mathbf{A}_{i-1}$ (and $^{i-1}\mathbf{T}_E$) is completed. To adapt $^{i-2}\mathbf{A}_{i-1}$, freeze $^{i-1}\mathbf{T}_E$, and minimize the difference. Rearranging equation (1) yields, say, $^{i-2}\mathbf{A}_{i-1} \stackrel{!}{=} \mathbf{X}$. Find a solution for the parameters ϑ_{i-1}, α_{i-1}, a_{i-1}, d_{i-1} such that $^{i-2}\mathbf{A}_{i-1}$ becomes as similar to $\mathbf{X}$ as possible, e.g., use the delta rule to minimize the total square error [12], which in this case can be done analytically yielding an approximation.

Since homogeneous transform matrices can be easily inverted using scalar products and transposition [2], rearranging equation (1) is neurally implementable.

3 Simulation and results

To test the algorithm, it was simulated on a computer. The kinematics of the AC was given by its D-H parameters, so that it could easily be compared with the model learned. Several small configurations were tested successfully, and also the PUMA robot was learned very well. Here, results using the kinematics of the NEUROS robot (see fig. 1, left), under development at our institute, are presented. It has a stereo camera system on a pan-tilt unit and an arm with seven DoF.

For the simulation, the world coordinate system was taken to be fixed at the pan-tilt unit ("head-centered coordinates"), and the AC from there down to the gripper was simulated. Its D-H parameters are given on the left of table 1: There are nine joints in the chain, i.e., there is an excess of three DoF. Notice that joints 2 ("head" rotation) and 3 ("chest" rotation) are aligned and could therefore be replaced by one joint.

joint	given				learned				number of iterations
i	ϑ_i	α_i	a_i	d_i	ϑ_i	α_i	a_i	d_i	
0	0.	90.	0.	0.	-0.01	89.99	-0.00	0.00	201
1	0.	-90.	0.	0.	-179.99	90.02	0.00	0.00	188
2	0.	0.	0.	-100.	55.54	-0.01	0.00	0.00	67
3	0.	90.	0.	-165.	-55.49	-89.98	0.12	-265.00	229
4	0.	90.	0.	315.	-0.02	-89.89	0.35	315.03	247
5	0.	90.	0.	340.	0.03	-89.91	0.43	340.01	179
6	0.	-90.	0.	0.	0.01	89.99	-0.06	-0.23	188
7	-90.	-90.	0.	340.	-89.95	89.94	0.48	340.10	227
8	0.	90.	0.	0.	-0.18	-89.98	-0.00	-0.13	161
9	0.	0.	0.	100.	–	–	–	–	Σ 1687

Table 1. Results of the NEUROS robot simulation. Angles are given in degrees, lengths in millimeters. The entry on the bottom right is the total number of iterations. There are no entries on the right half of the bottom line since $^{n-1}\mathbf{A}_n$ is simply set as given in step 9 of the algorithm.

The results are presented on the right half of table 1: The algorithm converged on a complementary set of D-H parameters.[5] The end-effector frame, however, is as it was given. Convergence was rather fast and only 1687 iterations were needed to learn the 36 parameters — an average of less than only 200 iterations per joint. The errors are fairly small. Notice that for joints 2 and 3 the results do not represent large errors: As these two joints are aligned, ϑ and d are free, and only the sum of the two rows must be as given — and this condition is met.

The progress of convergence is shown in fig. 1, right. Convergence is quite fast and smooth.

The results presented here are representative for all other configurations tested. The algorithm always converged on a valid solution, and always within a number of iterations comparable to the results presented here. This example has been chosen because of its high dimensionality and redundancy.

4   Discussion

A biologically inspired architecture for kinematic motor control has been proposed and a novel algorithm to bootstrap its core, a model of the direct kinematics of the AC under control, has been presented. The algorithm combines the advantages of two different approaches: In its basic idea it is a type of geometric approach called *screw axis measurement method* or *central point analysis* [6], and in estimating the link transforms it uses a parametric approach[6]. In designing the algorithm, care was taken to use only neurally implementable operations and functions, e.g., complex matrix inversions were avoided. Knowledge about the structure to be learned was built in (e.g., only revolute joints), which results in the numbers of parameters in the system being reduced and convergence made faster. In biology, such knowledge is probably built into organisms by evolution.

---

[5] This depends on initial conditions.

However, a lot of things need to be done to fully implement the architecture. First, the algorithm has to be tested with an acquisition of the end-effector position being disturbed by noise, as is the case in the real world.[6] For a more general approach to the rotation transform, a mapping from the joint actuator control signal (or muscle activation) to the joint angle might be introduced and also be learned. So far the initial bootstrap of the kinematics model has been implemented. But after the first learning only small deviations will occur due to growth or wear and tear. An algorithm must be found to adapt to these changes "on the job". Only kinematics has been considered so far, but I believe that the main idea of the architecture presented can also be applied to tackle the dynamics problem.

Acknowledgements. I wish to thank Christoph v. d. Malsburg, Stefan Schaal and Eric Maël for fruitful discussions, and Eric Maël and Percy Dahm for providing the dimensions of the NEUROS robot.

References

1. D. Bullock, S. Grossberg, and F. H. Guenther. A self-organizing neural model of motor equivalent reaching and tool use by a multijoint arm. *Journal of Cognitive Neuroscience*, 5(4):408–435, 1993.
2. K. S. Fu, R. C. Gonzalez, and C. S. G. Li. *Robotics: Control, Sensing, Vision and Intelligence*. McGraw-Hill, New York, 1987.
3. M. Jeannerod. *The Neural and Behavioural Organization of Goal-Directed Movements*. Oxford University Press, Oxford, 1988.
4. M. Kuperstein. Infant neural controller for adaptive sensory-motor coordination. *Neural Networks*, 4(2):131–145, 1991.
5. E. Maël. A hierarchical network for learning robust models of kinematic chains. In *Proceedings of the International Conference on Artificial Neural Networks (ICANN'96)*, 1996 (to be published).
6. B. W. Mooring, Z. S. Roth, and M. R. Driels. *Fundamentals of Manipulator Calibration*. John Wiley & Sons, New York, 1991.
7. J. Paillard. Motor and representational framing of space. In J. Paillard, ed., *Brain and Space*, ch. 10, pp. 163–182. Oxford University Press, Oxford, 1991.
8. H. Ritter, T. Martinetz, and K. Schulten. *Neuronale Netze: Eine Einführung in die Neuroinformatik selbstorganisierender Netzwerke*. Addison-Wesley, Bonn, 2nd ed., 1991.
9. D. A. Rosenbaum. *Human Motor Control*. Academic Press, San Diego, 1991.
10. J. F. Soechting and M. Flanders. Sensorimotor representations for pointing to targets in three-dimensional space. — Errors in pointing are due to approximations in sensorimotor transformations. *Journal of Neurophysiology*, 62(2):582–594, 595–608, Aug. 1989. (Two consecutive articles).
11. E. Thelen. Rhythmical behavior in infancy: An ethological perspective. *Developmental Psychology*, 17(3):237–257, 1981.
12. B. Widrow and M. E. Hoff. Adaptive switching circuits. *1960 IRE WESCON Convention Record*, pp. 96–104, 1960.

---

[6] Preliminary tests still showed good, though somewhat slower convergence.

Automatic Recalibration of a Space Robot: An Industrial Prototype*

Vicente Ruiz de Angulo and Carme Torras

Institut de Cibernètica (CSIC-UPC). Diagonal 647, 08028-Barcelona. Spain
e-mail: ruiz@ic.upc.es, torras@ic.upc.es

Abstract. We present a neural network method to calibrate automatically a commercial robot after undergoing wear or damage, which works on top of the nominal inverse kinematics embedded in its controller.

1 Introduction

The recalibration of robots installed in unmanned space stations through teleoperation from earth is a very time-consuming task due to communication delays. Within the project CONNY, Daimler-Benz Aerospace proposed an application of maintenance of electronic equipment in a space station that required the automatic recalibration of a 6-dof robot in-situ after wear had occurred. We present here the solution that was implemented in the final demonstrator for the project.

Our starting point was the work of Ritter et al. [2, 1, 5] on learning inverse kinematics from scratch using a hierarchical self-organizing map (SOM). We have modified this model in several ways to suit a more practical setting. Instead of learning the whole mapping, our algorithm learns only the appropriate corrections with respect to the inverse kinematics embedded in the controller, which is thus maintained. The other modifications enhance the cooperation between neurons, speeding up learning by a factor of 70 (near to 2 orders of magnitude).

A detailed description of the methods and results succinctly presented in this paper can be found in [3].

2 Ritter et al.'s approach to learning inverse kinematics

Ritter et al.'s model, as applied to a 5-dof robot, consists of a 3D SOM whose nodes have associated a 2D SOM each. Learning makes the 3D net converge to a discrete representation of the workspace, while the 2D subnet represents the gripper orientation space.

When a given position $\mathbf{u_p}$ and orientation $\mathbf{u_o}$ are supplied as input, the subnet k with input weights $\mathbf{w}_k$ closest to $\mathbf{u_p}$ is selected and, within this subnet,

* This work was partially supported by the ESPRIT III Program of the European Union under the contract No. 6715 (project CONNY). We thank Enric Celaya for hepful discussions, Gabriela Cembrano for support and encouragement, and Conor Doherty for providing an initial version of the extended Kohonen-maps program.

the neuron l with input weights $\mathbf{w}_{kl}$ closest to $\mathbf{u_O}$ is chosen. The joint angles produced for this particular input are then obtained with the expression:

$$\theta' = \theta_{kl} + \mathbf{A}_{kl}((\mathbf{u_p}, \mathbf{u_O}) - (\mathbf{w}_k, \mathbf{w}_{kl})). \tag{1}$$

where θ_{kl} and $\mathbf{A}_{kl}$ are respectively the vector of joint angles and the 5×8 Jacobian matrix associated with the winning neuron kl.

A learning cycle consists of the following four steps:

1. First, the classical Kohonen rule is applied to the weights $\mathbf{w}_k$ and $\mathbf{w}_{kl}$.
2. By applying θ' to the real robot, the end-effector moves to pose $\mathbf{u}' = (\mathbf{u_p}', \mathbf{u_O}')$. The difference between this pose and the desired one $\mathbf{u} = (\mathbf{u_p}, \mathbf{u_O})$ constitutes an error signal that permits applying the LMS rule:

$$\theta^* = \theta_{kl} + \mathbf{A}_{kl}(\mathbf{u} - \mathbf{u}'). \tag{2}$$

3. By applying the correction increment $\mathbf{A}_{kl}(\mathbf{u} - \mathbf{u}')$ to the joints of the real robot, a refined position $\mathbf{u}''$ is obtained. Now, the LMS rule can be applied to the Jacobian matrix by using $\Delta\theta = (\theta'' - \theta')$ as the error signal for $\Delta\mathbf{u} = (\mathbf{u}'' - \mathbf{u}')$:

$$\mathbf{A}^* = \mathbf{A}_{kl} + (\Delta\theta - \mathbf{A}_{kl}\Delta\mathbf{u})\frac{\Delta\mathbf{u}^T}{||\Delta\mathbf{u}||^2}. \tag{3}$$

4. Finally, the Kohonen rule is applied to the joint angles:

$$\theta_{ij}^{new} = \theta_{ij}^{old} + c' \, g_k(i) \, g_{kl}(j) \, (\theta^* - \theta_{ij}), \tag{4}$$

and the Jacobian matrix:

$$\mathbf{A}_{ij}^{new} = \mathbf{A}_{ij}^{old} + c' \, g_k(i) \, g_{kl}(j) \, (\mathbf{A}^* - \mathbf{A}_{ij}), \tag{5}$$

where again c' is the learning rate and $g_k(.)$ and $g_{kl}(.)$ are Gaussian functions centered at $\mathbf{w}_k$ and $\mathbf{w}_{kl}$, respectively, used to modulate the adaptation steps as a function of the distance to the winning neuron. The widths of the Gaussians decrease to zero with time.

3 A new application: Inverse kinematics update

Our application entails learning a mapping from desired robot poses to appropriate pose commands which, when supplied to the controller, lead to the attainment of those desired poses. For the intact robot, this mapping is the identity. After some degradation, this mapping amounts to sending the robot to a fake pose in order for it to reach the desired one. Thus, $\mathbf{u}$ and θ represent for us 6D vectors denoting pose coordinates and pose commands, respectively. This approach avoids the problem of the original application of having a multivalued inverse function, because the controller always chooses the same joint angles for a fixed command. Although we are still learning an inverse function, we now know the workspace shape, so that we can directly place the centers of the cells in a regular grid covering it, in order to minimize the quantization error. These centers do not need to move if the workspace shape does not change and, thus, the first step of applying the Kohonen rule is eliminated from the algorithm.

The direct application of Ritter et al.'s algorithm, with this slight modification, to our particular setting was not as quick as expected, even if the neighborhood widths were tuned in order to maximize learning speed. There are several reasons for this, which will be explained in the following sections together with the modifications triggered by each of them.

4 Separating dependencies

Our mapping is a 6-variate function that depends on 3 position and 3 orientation coordinates of the end-effector. However, the degree of dependency is not the same between all the command components and coordinates. Due to the characteristics of the workspace used (far from singularities and of small size relative to that of the robot) changes in position coordinates influence very slowly the orientation commands, and the same can be said about orientation coordinates with respect to position commands. This advocates for using large gaussians for g_k and g_{kl}. But, on the other hand, position and orientation commands are very sensitive to the real position and orientation coordinates, respectively. This means that, for example, too large neighborhoods in the position coordinates space are counterproductive to learn position commands.

Since there is no good solution to these contradictory interests within Ritter et al.'s framework, we decided to use two different hierarchical networks: one with a narrow g_k and a wide g_{kl} to compute only position commands, and another with a wide g_k and a narrow g_{kl} to compute only orientation commands. Moreover, the position network needs less units in the supernet, while the orientation network saves neurons at the subnet level. This modification leads to a significant increase in learning speed.

As all the subsequent modifications to the algorithm do not distinguish between position and orientation commands, in the remaining of the paper we will follow the notation in Section 2 to refer indistinctly to both networks.

5 What to propagate

Now we show that, in the usual case in which the kinematics change is not drastic, the information propagated to neighboring units can be modified to improve their cooperation.

Coooperation among the θ_{kl}. Ritter et al.'s approach consists in obtaining an estimation θ^* based on $\mathbf{A}_{kl}$ and the first movement attempting to attain $\mathbf{u}$. θ^* is immediately assigned to θ_{kl} and, in general, every θ_{rs} is moved towards the same value θ^* in keeping with the closeness of cells rs and kl. Consider a modification of the learning rule in which the quantity to be propagated is not

θ^*, but the change that θ_{kl} must undergo, that is, $\theta_{rs}^{new} \leftarrow \theta_{rs}^{old} + (\theta^* - \theta_{kl}^{old})$. Thus, (2) and (4) become somewhat simpler:

$$\theta^* = \Delta\theta = \mathbf{A}_{kl}(\mathbf{u} - \mathbf{u}'), \tag{6}$$

$$\theta_{ij}^{new} = \theta_{ij}^{old} + c'\, g_k(i)\, g_{kl}(j)\, \theta^*. \tag{7}$$

It is easy to prove that this kind of cooperation works better than Ritter et al.'s when the new function is more similar to the original one than to a constant function in the proximity of $\overline{\mathbf{w}}_{kl} = (\mathbf{w}_k, \mathbf{w}_{kl})$. To see this, suppose that we are using σ^2-wide neighborhoods, such that $g_k g_{kl}$ is approximately 1 in a spherical ball Ω centered on $\overline{\mathbf{w}}_{kl}$. We depart from a network that has already encoded the function f, so that every cell rs satisfies $\theta_{rs} = f(\overline{\mathbf{w}}_{rs})$. Now we evaluate the changes made to θ_{rs} by steps (2) and (4) (classical version), and (6) and (7) (new version), when trying to learn the new function f'. Let $\theta_{clas}(\mathbf{w})$ and $\theta_{upd}(\mathbf{w})$ be the new values that a hypothetical cell centered on $\mathbf{w}$ would assume as a consequence of the classical and new update versions of the learning rule, respectively. The goodness of the new and the classical versions can be evaluated by the average error they would cause to cells located in the ball Ω:

$$E_{clas} = \int_\Omega (\theta_{clas}(\mathbf{w}) - f'(\mathbf{w}))^2 = \int_\Omega (f'(\mathbf{w}_k) - f'(\mathbf{w}))^2 = \int_\Omega (f'(\mathbf{w}) - k_1)^2$$

$$E_{upd} = \int_\Omega (\theta_{upd}(\mathbf{w}) - f'(\mathbf{w}))^2 = \int_\Omega (f(\mathbf{w}) + (f'(\mathbf{w}_k) - f(\mathbf{w}_k)) - f'(\mathbf{w}))^2 =$$
$$\int_\Omega (f'(\mathbf{w}) - (f(\mathbf{w}) + k_2))^2,$$

where k_1 and k_2 are constants.

Cooperation among the Jacobians. It is not possible to estimate with only two points the ideal Jacobian matrix at $\overline{\mathbf{w}}_{kl}$, but it can be corrected in the direction indicated by the two points. The corrected $\mathbf{A}_{kl}$ matrix is called $\mathbf{A}^*$ and is used as desired matrix by all the $\mathbf{A}_{rs}$. Thus, in all the relevant aspects for us, the problem is the same we encountered with the θ_{rs} update. The corresponding suggested modifications for (3) and (5) are:

$$\mathbf{A}^* = (\Delta\theta - \mathbf{A}_{kl}\Delta\mathbf{u})\frac{\Delta\mathbf{u}^T}{\|\Delta\mathbf{u}\|^2} \tag{8}$$

$$\mathbf{A}_{ij}^{new} = \mathbf{A}_{ij}^{old} + c'\, g_k(i)\, g_{kl}(j)\, \mathbf{A}^*. \tag{9}$$

The discussion is similar to that in the last subsection: The new update version is better than the classical one when the Jacobian function of f' is more similar to $\frac{\partial f}{\partial \mathbf{w}}$ than to a constant matrix.

6   Neighborhood scheduling

When the neighborhood width is large, there is a large number of updated cells per iteration. Few iterations are then required to learn the mapping in all the input space at a coarse level of resolution. Instead, when the neighborhoods are small, the number of cells changing significantly their output in one iteration is very low, and many more iterations are required to make all the cells learn the same number of times as with a larger neighborhood. The neighborhood scheduling proposed by Ritter et al. does not take into account this fact. We will derive now a neighborhood scheduling expressing explicitly all the hypotheses on which it is based. First, we define $L(\mathbf{w}_r, \sigma)$ as the expected learning for cell r in one iteration using neighborhoods of width σ:

$$L(\mathbf{w}_r, \sigma) = \int_{\Omega_I} p(\mathbf{w}) h(\mathbf{w}, \sigma, \mathbf{w}_r) d\mathbf{w} = \frac{\sigma^n (2\pi)^{n/2}}{\text{Volume}(\Omega_I)}, \qquad (10)$$

where Ω_I is the n-dimensional input space of f, $p(\mathbf{w})$ is the probability density in Ω_I, and $h(\mathbf{w}, \sigma, \mathbf{w}_r)$ is the value in $\mathbf{w}_r$ of a neighborhood centered on $\mathbf{w}$ of width σ. The second equality results from assuming Gaussian neighborhoods and a uniform $p(\mathbf{w})$, and it is valid for $\mathbf{w}_r$'s not too close to the border of Ω_I.

Now we must establish how much L should be accumulated along time with each σ. We assume the simplest hypothesis, that is, a cell is visited the same mean number of times with every neighborhood width σ. This means that we must stay with each σ a time inversely proportional to $L(\mathbf{w}_r, \sigma)$.

7 Experimental results

Our recalibration system was tested on a Reiss robot in the space-station mock-up installed in Daimler-Benz Aerospace, Bremen. The robot had to maneuver in a workspace of $50 \times 60 \times 50$ cm with an orientation range of 40 degrees in each dimension. Evaluation with the real robot was necessarily much more restricted than that carried out in simulation. Since the robot integrity was to be preserved, only decalibrations consisting of translations and rotations of the whole robot were tested. The results were in complete agreement with the simulation results for the same situation [3].

Due to lack of space, we will present here only simulation results obtained for a more interesting and representative case: one in which the geometry of the robot undergoes serious distortion. The length of three links were shortened by 1, 1 and 4 cm, respectively, while three joint encoders were shifted by 4, 3, and 4 degrees. This could result, for instance, from link bending and encoder wear. As a consequence, the initial mean average position and orientation error, when executing (1), was 8.3 cm and 4.7 degrees, respectively.

Figure 1 shows the results of performing 200 iterations with our learning system, where supernets and subnets had $3 \times 3 \times 3$ neurons each. Every 4 iterations, learning was interrupted, and the average position and orientation errors over 200 random poses of the workspace were measured using (1). We must remark that these results were obtained with two hierarchical networks which, as a whole, had the same number of neurons as the Ritter et al.'s model we used for comparison. The parameters were selected after a very rough search, using the same initial σ for all neighborhoods and applying a slow linear decrease to α. The original Ritter et al.'s algorithm with optimized parameters required more than 10,000 iterations to get the same combination of orientation and position errors, as reported in [4].

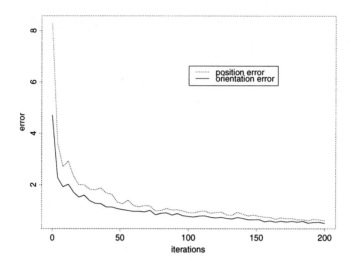

Fig. 1. Evolution of the error along the first 200 iterations. Position and orientation errors are measured in centimeters and degrees, respectively.

8 Conclusions

We have presented a neural network system to recalibrate robots, inspired in Ritter et al.'s work, which can be applied in more practical settings than the original algorithm because of two reasons. First, the system can work without substituting the original controller in commercial robots. And second, the learning is much faster due to the improved cooperation among the learning units. We think that other neural network algorithms based on local units can also benefit from these improvements.

References

1. Martinetz T. and Schulten K.J., 'Hierarchical neural net for learning control of a robot's arm and gripper', *Proc. Intl. Joint Conf. on Neural Networks (IJCNN'90)*, Vol. II, 747–752, San Diego, 1990.
2. Ritter H., Martinetz T. and Schulten K.J., *Neural Computation and Self-Organizing Maps*, New York: Addison Wesley, 1992.
3. Ruiz de Angulo V. and Torras C., 'Automatic recalibration of a space robot: an industrial prototype', *Technical Report IC-DT-03.96*, Institut de Cibernètica (CSIC-UPC).
4. Torras C., Cembrano G., Millán J. del R. and Wells G., 'Neural approaches to robot control: Four representative applications', *Proc. 3rd Intl. Workshop on Artificial Neural Networks (IWANN'95)*, Málaga, June, 1016–1035 (1995).
5. Walter J.A. and Schulten K.J., 'Implementation of self-organizing neural networks for visuo-motor control of an industrial arm', *IEEE Trans. on Neural Networks*, Vol. 4, No. 1, January 1993.

Population Coding in Cat Visual Cortex Reveals Nonlinear Interactions as Predicted by a Neural Field Model

Dirk Jancke, Amir C. Akhavan, Wolfram Erlhagen, Martin Giese, Axel Steinhage
*Gregor Schöner & Hubert R. Dinse

Institut für Neuroinformatik, Theoretische Biologie, Ruhr-Universität Bochum,
*Centre de Recherche en Neurosciences Cognitives, CNRS, Marseille, France

Abstract

We develop population coding ideas toward a general approach to the analysis of cortical function that operationalizes the notion of cooperativity. Neural ensemble activation distributions (population representations) are constructed over a defined stimulus parameter space, in our case the 2-dimensional retinal position of the central visual field. In contrast to classical approaches using receptive field centered stimuli the method presented here requires the stimulation of a whole cell ensemble with an identical common stimulus. The constructed activation distribution allows a quantitative investigation of activation dynamics and cooperative effects, like lateral inhibition and excitatory interaction. We simulated the data with a continuous neural network model as proposed by Wilson & Cowan [14].

1 Introduction

In recent years the functional importance of cooperative effects in visual information processing gains increasing interest in both psychophysical and neurophysiological research [1,5]. Cooperativity leads to essential nonlinear effects in the information processing of sensory stimuli. Nonlinearity is crucial for the occurence of self-organization, which may be the basis for the solution of correspondence problems in visual information processing [9]. Characteristic signatures of such nonlinear phenomena are multistability and hysteresis, both of which have been observed in the visual system [7,13]. The problem is to identify nonlinear components of information processing and to analyze quantitatively nonlinear dynamical phenomena in neurophysiological data.

Classical descriptions of cortical functions in terms of receptive fields appear incomplete in presence of strong cooperative effects [6,10,11]. Moreover, receptive field properties were shown to be highly context dependent [5], to process complex multidimensional tuning properties [12] and complex spatial temporal transfer characteristics [2,3]. All of these extensions of the classical RF-concept are difficult to interpret with respect to their implications for the global behaviour of the neural network.

As one solution of this problem we propose to analyze nonlinear cooperative phenomena at the level of the dynamics of collective neural activation variables. To transform neuronal data into a form that is compatible with neural field models we use a special population coding technique. Population coding was introduced for the analysis in the motor domain [4,8].

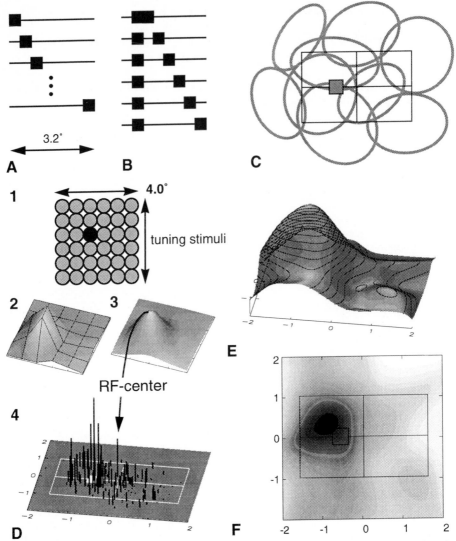

Fig.1.: A: elementary stimuli (squares, 0.4°); **B:** composite stimuli; **C:** The presentation site of a common stimulus (square) is independent from the receptive field location of individual neurons (circles). The frame covers a visual space of 3.2° by 2.0°, the location of the stimulus is indicated by the small square. **D:** Construction of a population representation in a visual stimulus space. The location of each cell's RF-center is determined on the basis of the response planes. **1:** schematical drawing of tuning stimuli; **2.** response plane derived from the single cells responses to the tuning stimuli; **3.** smoothed response plane provides information about RF-center; **4.** individual cell responses are normalized and then placed in the visual field at the location of their RF-centers (arrow from 3 to 4), providing a raw activation distribution. The height of the lines indicate the response strength of individual cells to a stimulus at a location indicated by a white square. **E:** The raw distribution is smoothed and corrected for sample density. **F:** The same data as shown in E, except the population distribution is presented greylevel-coded in 2 dimensions. Axes indicate degrees in visual space.

Population coding in a sensory domain can be regarded as the projection of many single cells' responses to a common stimulus into a space representing the parameter of interest. By means of the construction of a neural activity distribution over a defined functional space our analysis becomes independent of the constraints of anatomical descriptions.

Here we apply the population technique for neurons in cat striate cortex to describe effects of distance dependent interactions of "composite" stimuli. This allows us to compare measured dynamical population representations of these stimuli with representations calculated by superposition of representations of the corresponding single ("elementary") stimuli.

The observed nonlinear effects were simulated using a neural field model. The model is essentially nonlinear and allows to represent cooperative mechanisms like lateral inhibition and self-excitation.

2 Materials and Methods

We recorded responses of single units in the foveal representation in area 17 of 19 cats, with platinum electrodes. Stimuli were presented on a monitor (120 Hz, non-interlaced) at a distance of 114 cm. To all neurons an identical set of collective, flashing stimuli was presented randomly at a fixed foveal position in the visual field. These were of two types: 1) *elementary stimuli*, 8 squares (size 0.4 by 0.4°) randomly flashed along a line of 3.2° length (fig. 1, A), 2) *composite stimuli*, stimuli composed of 2 simultaneously flashed squares separated by various distances (fig. 1, B), presented for 25 ms, at 0.6 Hz., n = 32. The position of the common stimuli was not changed during the entire recording session, irrespective of the receptive field location of individual neurons (non-RF entered approach, fig. 1, C). In addition, the location of the receptive fields of individual cells was quantitatively measured with flashing stimuli (tuning stimuli) using the response plane technique (36 flashing spots of light (0.67° diameter), randomly displayed on a 6 by 6 grid, fig. 1, D1), presented for 25 ms, at 1 Hz, n = 25.

The receptive field (RF) center for each individual cell was calculated offline using its response planes by mapping the response strength of individual cells onto the positions of the corresponding 6 by 6 grid. The response strength was determined by summarizing the responses to a simulus 40 to 65 ms after stimulus onset over 25 repetitions. The RF center was defined as the location of the maximum of a smoothed response plane obtained by convolution with a Gaussian profile (fig. 1, D2 and D3, sigma = 0.67° in visual space).

The firing rate $F_n(s,t)$ of a neuron n to a stimulus s was defined as the average response during 32 stimulus repetitions at time t after stimulus onset. The individual firing rates of the cells were normalized for their maximum fire rates to all tuning stimuli during any single 10 ms time window, 0 -100 ms after stimulus onset.

2.1 Construction of population code

For a given stimulus s, the contribution of each cell n to the population response is its normalized firing rate F_n at time t in parameter space (x,y) centered at its RF-center location $\bar{r}_n$. To achieve an interpolated and smooth activation distribution, spatial lowpass filtering was performed by weighting individual firing rates F_n with a gaussian profile. Thus, the contribution of each cell is given as a gaussian profile g (sigma = $1.0°$ in visual space) in the parameter space (x,y) centered at its RF-center location $\bar{r}_n$ and with a height proportional to its actual firing rate F_n.

This smoothed population representation defines the temporal evolution of the sum activation of all cells for each point $\bar{r} = (x,y)$ in parameter space. To correct for sampling density, the interpolated population activation distribution is divided by a distribution equally constructed from equal individual cell activations of 1.

$$P_{xy}(\bar{r}, t) = \sum_{n=1}^{N} g(\bar{r} - \bar{r}_n) \cdot F_n(s, t) \Bigg/ \sum_{n=1}^{N} g(\bar{r} - \bar{r}_n) \qquad \text{with} \quad \bar{r}_n = (\bar{r}_{n,x}, \bar{r}_{n,y})$$

The result is an interpolated population activation distribution P_{xy} taking into account irregularities in the sampling density (fig. 1B and C).

2.2 Neural Field Model

The neural field model [14] fitted to the data consists of an excitatory and an inhibitory neural field. Each field has a characteristic time constant. Both fields are coupled by cross interaction kernels and have cooperative recurrent intralayer connections. A retinotopic stimulus activation feeds directly into the excitatory field.

Excitatory field:

$$\tau_e \dot{u}(y,t) =$$
$$-u(x, t) + [1 - r_e u(x, t)] \cdot \theta_e \left(\int w_{ee}(x - x') u(x', t) dx' + \int w_{ei}(x - y') v(y', t) dy' + S(x, t) \right)$$

Inhibitory field:

$$\tau_i \dot{v}(y,t) =$$
$$-v(y, t) + [1 - r_i v(y, t)] \cdot \theta_i \left(\int w_{ii}(y - y') v(y', t) dy' + \int w_{ie}(y - x') u(x', t) dx' \right)$$

with $u(x,t), v(x,t)$ activation $\theta_{e,i}$ threshold functions

            $w_{e,i}$ interaction kernels $\tau_{e,i}$ time constants

            $S(x,t)$ stimulus incoupling $r_{e,i}$ refractory time constant

3 Results

3.1 Reconstruction of elementary stimuli

A prerequisite for analyzing interactions of the composite stimuli in terms of parametrical (retinal) space is the accurate reconstruction of each single elementary stimulus. Figure 2 shows the population representation of 6 selected different elementary stimuli. The representations are based on the spike activity of 139 cells in a time interval of 40 to 80 ms after stimulus onset. Each elementary stimulus is indicated as a square. The maximum of the population response displays the actual position of a given stimulus in the visual field with considerable accuracy. Only small deviations from ideal stimulus reconstruction are observable (fig. 3).

3.2 Coherent temporal evolution of the population representation

Using the time-slice technique it is possible to illustrate the temporal evolution of the population representation in discrete time windows (fig. 4). First, there is a gradual and coherent evolution of the population response (composed of a large number of single

Fig. 2. Population representation of 6 different elementary stimuli. Spike activity of 139 cells was averaged over 40 to 80 ms after stimulus onset. The parametric space (central visual field) and the position of each elementary stimulus is shown (0.4 deg. visual field). Population activity was normalized for each single stimulus.

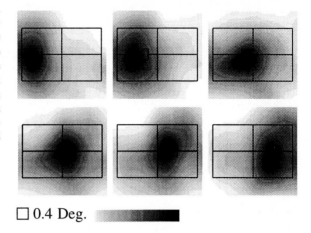

☐ 0.4 Deg.

Fig3. Quantification of reliability of positional accuracy of the population representation. On the abscissa the actual position of each square-center in the visual field is shown. The ordinate indicates the reconstructed position of each stimulus by the maximum of the population activity summed along the y-dimension.

reconstructed position

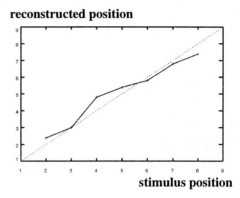

stimulus position

cell responses) to an elementary stimulus over time and space. This is remarkable in view of the complex spatial-temporal structure of the single cells' receptive field. Second, the population representation reaches maximum activity 50 to 60 ms after stimulus onset with a sharp peak at the stimulus location and decays with a much broader activity distribution.

3.3 Nonlinear interactions

3.3a Global inhibition

Global nonlinear interactions were analyzed by comparing the measured population responses to composite stimuli in a time interval of 40 to 80 ms after stimulus onset with the calculated linear superposition of the corresponding elementary stimulus representations. The most striking interaction effect is a distance dependent strong inhibition of the population response. Figure 5 a/b shows two examples of composite stimuli with different spatial separations in the visual space.

3.3b Early excitation and late inhibition

A closer inspection of the above described overall inhibition with respect to a finer temporal resolution reveals a more detailed structure of the interaction effect. First, the population response to composite stimuli reaches its maximum about 10 ms earlier

40 - 50 ms 50 -60 ms 60 -70 ms 70 -80 ms a) * b)

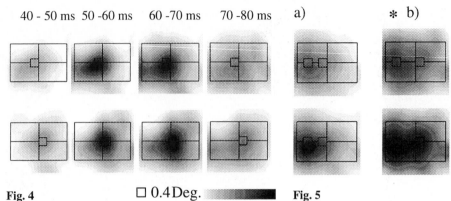

Fig. 4 □ 0.4 Deg. ▬▬▬ **Fig. 5**

Fig. 4. Temporal evolution of the population representation (each normalized) using the time-slice technique for two different locations of the elementary stimuli in the visual field . The population represented within the parametric space (central visual field), shown in timesteps of 10 milliseconds, reveals a coherent build-up and decay of neural activity.

Fig. 5. Upper row shows the integrated (40 - 80 milliseconds after stimulus onset) population representation of composite stimuli. Data were normalized for each example of separation between stimuli. Lower row shows the calculated linear superposition of the single corresponding elementary stimuli.

a. Separation between stimuli: 0.4 deg. visual field; **b**. Separation between stimuli: 0.8 deg. visual field

Asterisk indicates stimulus that is lying closer to the vertical meridian.

Fig. 6. Temporal evolution of the population representations in timesteps of 10 milliseconds evoked by composite stimuli. Data were normalized for each example of spatial separation between stimuli. Upper rows show the measured activity distribution of the population, the lower rows are the calculated superpositions of the single corresponding elementary stimuli. Time steps after stimulus onset are indicated on top. Black arrows indicate early excitation, dotted arrows late inhibition (see text).

a. Separation between stimuli: 0.4 deg. visual field

b. Separation between stimuli: 0.8 deg. visual field

Asterisks indicate stimuli lying closer to the vertical meridian.

0.4 Deg. □

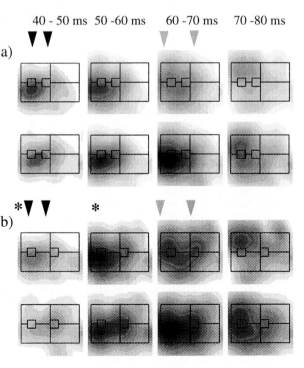

40 - 50 ms 50 -60 ms 60 -70 ms 70 -80 ms

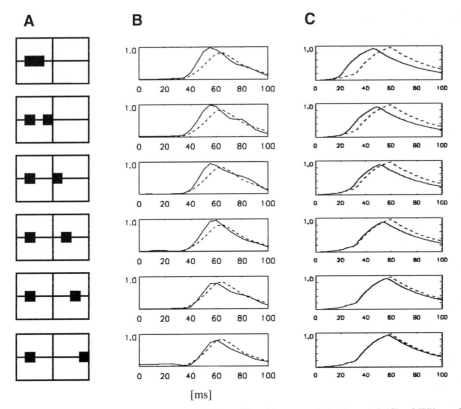

A   **B**   **C**

[ms]

Fig. 7. Comparison of the experimental data (B) with a neural field model (C) of Wilson & Cowan [14]. Solid lines show the time course [milliseconds in B, arbitrary in C] of maximum activity at the position of the left stimulus in the composite stimulus paradigm. As indicated (A), the separation between the squares increases from top to bottom. Dashed lines illustrate the time course of maximum activity at the position of the left square when presented alone.

than the calculated superposition of the corresponding elementary stimuli, which we denote as early excitation (see black arrows, fig. 6 a/b). Second, this early excitation is followed by a late inhibition effect (see dotted arrows, fig. 6 a/b).

Additionally, there seems to be a tendency of the population representation to emphasize in a distance dependent way the stimulus that is lying closer to the vertical meridian of the central visual field. This effect is particularly observable comparing the calculated superposition and the measured population activity in figure 5b and figure 6b (see *).

3.4 Comparing the experimental data with a neural field model

A neural field model based on Wilson & Cowan [14] was fitted to the data in an effort to assess the cooperativity underlying the observed effects. This model predicts the described early excitation and late inhibition interaction effects and their dependence on the spatial separations between the stimuli in the composite stimulus paradigm (fig. 7).

4 Conclusion

We present an analysis of cortical dynamics based on the population coding approach. Our method enables us to construct a collective neural activation distribution (population representation) over a defined parameter space, in our case the retinal spatial coordinates.

This population representation projects information distributed across a large number of neurons, back to abstract functional variables. Since the approach uses the parametric space, it specifies the meaning of the activation of single cells and their complex tuning properties to the global function in the cortex independently of the constraints of cortical maps.

Since the population representation is compatible with neural field models, we were able to fit a Wilson & Cowan model [14] and thus analyze neural activity in terms of lateral inhibition and excitation.

The method used requires a special experimental paradigm. All cells of the sampled ensemble must be stimulated by identical common stimuli independently of the spatial arrangement of the single cells' receptive fields in the visual field (non RF-centered approach). In the case of the observed parameter space (retinal position), this procedure seems to be biological plausible in view of the retinotopic organization and spatial overlap of receptive fields in cat striate cortex: under natural conditions, most of the neurons are unlikely to be stimulated in a receptive field centered mode.

5 References

[1]   Chang SS, Julesz B (1984) Vision Res **24**: 1781 - 1788.

[2]   Dinse HR, Krüger K, Best J (1990). Concepts in Neuroscience (CINS) **1**: 199 - 238.

[3]   Eckhorn R, Krause F, Nelson JI (1993) Biol. Cybern **69**: 37 - 55.

[4]   Georgopoulos AP, Kettner RE, Schwartz AB (1988) J of Neurosc **8**: 2928 - 2937.

[5]   Gilbert CD, Wiesel TN (1990) Vision Res **30**: 1689 - 1701.

[6]   Heggelund P (1981) Exp Brain Res **42**: 99 -107.

[7]   Hock HS, Kelso JAS, Schöner G (1993) J Exp Psych HPP **19**: 63 - 80.

[8]   Lee C, Rohrer WH, Sparks OL (1988) Nature **332**: 357 - 360.

[9]   Lehky SR, Seijnowski TJ (1990) J Neurosc **10** (7): 2281 - 2299.

[10] Movshon JA, Thompson ID, Tolhurst DJ (1978) J Physiol **283**: 79 -99.

[11] Nelson, S.B. (1991) J. of Neurosc. **11** (2): 344 - 356.

[12] Stern E, Aertsen A, Vaadia E, Hochstein S (1993) in: Giles CL, Hanson SJ, Cowan JD (eds) Morgan Kaufmann Publishers.

[13] Williams D, Phillips G, Sekuler R (1986) Nature **324** (20): 253 - 255.

[14] Wilson HR, Cowan HR (1973) Kybernetik **13**: 55 - 80.

Supported by the Deutsche Forschungsgemeinschaft, Nos. Di 334/5 -1, /5 - 3 and Scho 336/4-2

Representing Multidimensional Stimuli on the Cortex

Francesco Frisone and Pietro G. Morasso

DIST - Department of Informatics, Systems and Telecommunication
University of Genova
Via Opera Pia 13, 16145 Genova, Italy
e-mail:friso@dist.dist.unige.it

Abstract. A computational paradigm based on distributed spatial representation provides an unifying framework for dealing with open issues in modelling cortical maps such as the representation of multidimensional stimuli. This paper describes a computational architecture, based on two overlayed Topology Representing Networks, which is shown to reproduce artificially the ocular dominance bands observed from tangential sections of a monkey's right occipital lobe.

1   Introduction

It is observed that brain cortex has a modular organization, in the sense that it is organized into columns communicating by means of a large number of *lateral* connections (i. e. parallel to the cortical surface) [7].

The assumption of the modularity of the cortex implies that, from the point of view of circuitry, the cortex is basically a two dimensional *lattice*. But the observed patterns of lateral connectivity are much more complicated and there is ample evidence that a significant portion of lateral connections is not local. So this suggests that higher dimensional topologies are mapped into a two-dimensional lattice. In this hypothesis the need of non-local lateral connections in the cortex comes from the task of representing higher dimensional stimuli. Such hypothesis is supported by the observation that lattices are ways to map n-dimensional space into a 2-dimensional manifold [2].

Most models [8, 4, 9, 10] of brain cortex assume that lateral connections are local and radially symmetric. However, lateral connections have been observed that are not local at all. The paradigm proposed in this paper, is based from the interaction of two *Topology Representing Networks* (TRN) [11]. By these two networks we map n-dimensional stimuli into a 2-dimensional cortical array.

The power of the mechanism is demonstrated by reproducing ocular dominance bands in the striate cortex as a side-effect of the self-organization process.

2   The Model: two overlayed TRNs

We know that the TRN is a self-organized map that has the property to optimize a distribution of a codebook of prototypes or reference vectors of the feature space over the manifold of the input data.

TRN generate lateral connections (by Hebbian learning rule) that are a good indicator of the dimensionality of the stimuli [3]. But in the TRN lateral connections are locals and this is a hard limitation for representing cortical patterns. The problem is to match the 2-D constraint of locality with the n-D constraint of topology preservation. Our idea is to assume that the formation of the network is driven by two interactive adaptive processes or, dually, that there are two virtual overlayed networks which interact during learning. If we choose the latter explanatory attitude, we can say that the first TRN it is used to represent the cortex frame and the second the topology of the input stimuli. If the dimensionality of the input sensory vector is 2-D, then the two conceptual networks coincide and we obtain patterns similar to those observed in primary sensory cortex (somatosensory and visual).

A cortical map will be modelled hereafter as a lattice of N processing elements (PE) or *filters*, (corresponding to cortical columns) operating in parallel on a common afferent signal or stimulus $\mathbf{x} \in \mathcal{M}_s \subset \mathcal{R}^n$.

$$\mathcal{L} = \{\tilde{p}_i : i = 1, \ldots, N\} \tag{1}$$

Each processing element (or column) is defined by two quantities (or prototype vectors):

$$\tilde{p}_i = \{\pi_i^y; \pi_i^x\} \ . \tag{2}$$

where $\pi_i^y \subset \Re^2$ is the neuron position vector in the cortex, and $\pi_i^x \subset \Re^n$ is its associated weight vector in the input stimulus space. A PE connects itself with another PE by developing a link to this PE. The links (lateral connections) are described by a connection matrix

$$\mathcal{A} = \{a_{ij} : i, j = 1, \ldots, n\} \ , \tag{3}$$

where $a_{ij} = a_{ji} = 1$ if $\tilde{p}_i$ and $\tilde{p}_j$ are connected; $a_{ij} = 0$ otherwise.

Each PE has a selective response or *receptive field*, described by an activation function or *tuning curve* $U_i^x(\mathbf{x})$, that is maximized by a *preferred* input $\mathbf{x} = \pi_i^x$ (usually addressed as the *prototype* vector 'stored' in that *PE*). The *receptive field* of a given PE is defined as the set of input values corresponding to a nonzero activation value.

This kind of architectures is trained[1] into 2 steps with two types of input data:

1. π^y is trained with $\mathbf{y} \in \mathcal{M}_c \subset \Re^2$ (in others words we arrange the prototypes on a two dimensional lattice);
2. π^x are trained with $\mathbf{x} \in \mathcal{M}_s \subset \Re^n$ (as a consequence the network becomes a topographic representation of the stimuli).

In others words we train two TRN (the first corresponding to π^y and the second to π^x) with a shared connection matrix $\mathcal{A}$.

---

[1] For each step the training algorithm used to train the protype vectors and the lateral connections is that showed in [11].

A computational map basically transforms an afferent vector $\mathbf{x}$ into a set of activation values $U^x \in [0,1]^N$; if only one component of U^x differs from zero, the coding of $\mathbf{x}$ is said to be *localized*, whereas if several PEs are active at the same time, the map implements a *population coding* scheme.

Computational maps of this kind have been demonstrated to emerge from competitive learning [12, 1], whose goal is to establish a set of prototype vectors yielding an optimal *quantization* of the input space, so that each PE is tuned to slightly different preferred values.

3   Simulation experiments

The simulation reproduce artificially the patterns of the ocular dominance bands observed from tangential sections of a monkey's right occipital lobe, visualized by the reduced silver method of [6].

In striate cortex, inputs from the two eyes converge onto single cells, accomplishing a sort of reduction in redundancy as two separate views of the world are collapsed into one. It is also know that the cells express a preference for one eye over the other, a preference referred to as ocular dominance. The importance of this property is reflected, for example, in the continuity with which they are mapped across the cortical surface in ways that preserve their neighborhood relationships at the expense of retinotopic position (see figure 1).

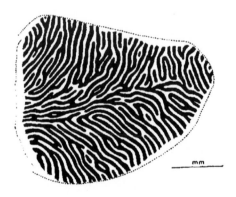

Fig. 1. Ocular dominance bands from tangential sections of a monkey's right occipital lobe, visualized by the reduced silver method [6]. Alternate bands have been inked in (from [5]).

We were interested to test whether the dual maps mechanism explained in the previous section had the power to generate the observed ocular dominance bands. The simulation is structured as follows: The prototype vectors π^y are initialized randomly. The first TRN is trained with a training set consisting of a uniform distribution within a circle or other convex 2D shape (it represent a simplified cortex frame). As a result of learning, the prototypes happen to be arranged on a lattice (typically a 2D triangular grid, with 6 neighbors per prototype, see figure 2a).

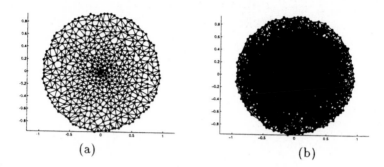

Fig. 2. The first TRN. (a) The lattice of prototype vectors π^y before training the second TRN; (b) the lattice of π^y after have trained the second TRN.

The further step is to train the second TRN with the stimuli. In the numerical simulation the afferent signals to the cortical sheet are characterized by the three dimensional features vector $\mathbf{x}$ where the first two components correspond to retinal coordinates and the third one is a binary variable which identifies the dominant eye (+1 for the left eye and -1 for the right eye). The stimulus $\mathbf{x}$ (embedded in a 3D space) is obtained by randomly sampling an input source with an uniform distribution $p(\mathbf{x})$. The learned prototype vectors π^x are a *topographic representation* of the stimulus space. As a consequence of the second training the first TRN is modified as regards the connections which have grown to the pattern of figure 2b.

The ocular dominance bands are obtained simulating the activity of cells sensitive to one eye. The approach is plausible because sustained cortical activity is accompanied by a host of events (e.g., membrane depolarization, potassium release, glial cell hyperpolarization and swelling, followed by the dilation of capillary beds in response to metabolic demand).

At the end of training the prototype vectors π^x identify which eye the cells are sensitive to. In particular, the ocular dominance of each neuron is determined by the sign of the third component of π_i^x.

Irregular banded patterns are found drawing the contour of a gaussian mixture centered on the prototypes π^y for which their associated prototypes π^x have third component with the same sign. Figure 3 show the result for the right eye sensitive cells.

4   Concluding Remarks

The emergence of biologically consistent ocular dominance bands is just a first step in the direction of a more general theory of the geometrical structure of the

Fig. 3. Irregular banded pattern for the right eye sensitive cells. Pattern is found drawing the contour of a gaussian mixture centered on the prototypes π^y for which their associated prototypes π^x have third component less than 0.

cortex. For example, interesting problems which can be addressed with a similar approach are related with the cortical represetation of the body (*body schema*) [13] and of the external space [14].

References

1. M. Benaim and L. Tomasini. Competitive and Self-Organizing algorithms based on the minimization of an information criterion. In T. Kohonen, K. Makisara, O. Simula, and J. Kangas, editors, *Artificial Neural Networks*, Amsterdam, 1991. North-Holland.
2. V. Braitenberg. *Vehicles - Experiments in Synthetic Psychology*. MIT Press, Cambridge, MA, 1984.
3. F. Frisone, F. Firenze, P. Morasso, and L. Ricciardiello. Application of Topology-Representing Networks to the Estimation of the Intrinsic Dimensionality of Data. In *ICANN95- Int. Conf. on Artificial Neural Networks*, volume 1, pages 323–327, Paris, October 1995.
4. S. Grossberg. *The adaptive brain*. Elsevier, Amsterdam, 1987.
5. D.H. Hubel and D.C. Freeman. Short communications: Projection into the visual field of ocular dominance columns in macaque monkey. *Brain Res.*, 122:336–343, 1977.
6. D.H. Hubel and T.N. Wiesel. Sequence regularity and geometry of orientation columns in the monkey striate cortex. *Journal of Comparative Neurology*, 158:267–293, 1974.
7. E. I. Knudsen, S. du Lac, and S.D. Esterly. Computational maps in the brain. *Annual Review of Neuroscience*, 10:41–65, 1987.
8. T. Kohonen. Self organizing formation of topologically correct feature maps. *Biological Cybernetics*, 43:59–69, 1982.
9. T. Kohonen. The self organizing map. *Proceedings of the IEEE*, 78:1464–1480, 1990.
10. T. Martinetz. Competitive Hebbian Learning Rule forms perfectly topology preserving maps. In S. Gielen and B. Kappen, editors, *ICANN93- Int. Conf. on Artificial Neural Networks*, Amsterdam, 1993.
11. T. Martinetz and K. Schulten. Topology Representing Networks. *Neural Networks*, 3:507–522, 1994.

12. J. Moody and C. Darken. Fast learning in networks of locally-tuned processing units. *Neural Computation*, 1:281–294, 1989.

13. P. Morasso and V. Sanguineti. Self-organizing body-schema for motor planning. *Journal of Motor Behavior*, 26:131–148, 1993.

14. P. Morasso and V. Sanguineti. How the brain can discover the existence of external egocentric space. *Neurocomputing*, 1995. in press.

An Analysis and Interpretation of the Oscillatory Behaviour of a Model of the Granular Layer of the Cerebellum

Lokeshvar N. Kalia

Neural Systems Engineering, Department of Electrical and Electronic Engineering, Imperial College of Science, Technology and Medicine, London, U.K.
email l.n.kalia@ic.ac.uk

Abstract. The granular layer is responsible for producing the representations which convey the state of the sensory and motor environment to the Purkinje cells. In this paper, a model of the granular layer is presented, and methods appropriate to the analysis of commensurate time-delay systems are applied to determine the Hopf bifurcation characteristics of the model.

1 Overview

The granular layer is the receptive surface of the cerebellar cortex. It is responsible for constructing representations of the cerebellum's mossy fibre inputs. This representation is used by the Purkinje cells, whose own outputs are believed to be involved in the exercise of coordinative control. The granular layer recodes mossy fibre signals to facilitate this action. The representations which it produces are the result of the dynamical behaviour that emerges from its network structure. Models of this structure have been developed in order to better understand these processes.

This paper reports on the bifurcation characteristics of one of these models, whose design was based upon a network model used by König and Schillen in their studies of synchronisation phenomena in the visual cortex [2].

The dynamical characteristics of the model suggest ways in which the granular layer may encode information.

2 Model of the Granular Layer

The organisation of the granular layer, at its simplest and most essential, is shown in Figure 1. This depiction takes into account the localising effects of the fractured somatotopic distribution of the layer's mossy fibre afferents [5], and the interaction which occurs between such localised processing regions [1, 5]. This paper describes experiments which are part of an ongoing effort to understand the consequences of such interaction upon the nature of granular layer encoding.

The model examined in this paper is shown in Figure 2. Each excitatory node represents a cluster of granule cells (one localised processing region), each inhibitory node, a (much smaller) cluster of Golgi cells.

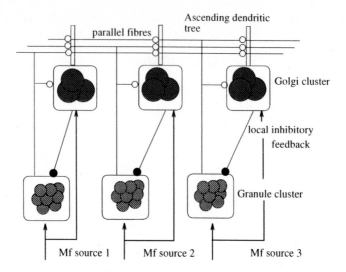

Fig. 1. Simplified organisation of the granular layer.

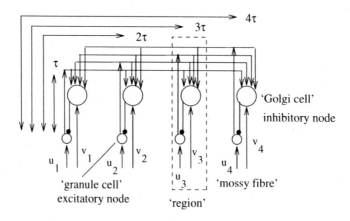

Fig. 2. Circuitry of the granular layer model.

The nodes operated as follows:

$$\dot{x}_i(t) = -\alpha_i x_i(t) + w_i^{gr:mf} u_i + w_{ii}^{gr:go} f(y_i(t - \tau))$$

$$\dot{y}_i(t) = -\beta_i y_i(t) + w_i^{go:mf} v_i + \sum_{j=1}^{n} w_{ij}^{go:gr} f(x_j(t - (|i - j| + 1)\tau))$$

where

x_i is the 'membrane potential' of the ith excitatory node (i.e. the ith region)

y_i is the 'membrane potential' of the ith 'Golgi cell'

u_i is the *constant* 'mossy fibre' input to the ith 'granule cell'

v_i is the *constant* 'mossy fibre' input to the ith 'Golgi cell'

$f(z)$ is a sigmoid function of slope one and threshold zero, 1/1+exp(-z)

τ is the transmission delay between a 'Golgi cell' and the 'granule cell' it inhibits

α_i is the damping constant of the ith 'granule cell' (α_1 is set to unity and all other damping constants are expressed in terms of this)

β_i is the damping constant of the i_{th} 'Golgi cell' (expressed in terms of α_1)

$w^{post:pre}$ is the weight which the post-synaptic cell associates with the pre-synaptic afferent (this is also expressed relative to α_1).

Unlike König and Schillen's model, commensurate delays were used in this model, that is, the delays were distance-dependent and integer multiples of a basic delay-step (τ).

3 Hopf Bifurcations Within the Model

The model shown in Figure 2 is capable of producing stable oscillations provided τ exceeds some critical value (τ_c). When τ equals τ_c, the model produces stable oscillations (frequency ω_c) through a Hopf bifurcation (the eigenvalues of the Jacobian matrix taken about its fixed point equilibrium become purely imaginary, $\pm i\omega_c$). All of the nodes oscillate at the same frequency, but unlike König and Schillen's model, they are not synchronised because different delays are present. Since the point of Hopf bifurcation is an important feature of the network's dynamics, and because König and Schillen did not conduct such an investigation into their own model, this aspect of the model was investigated.

Determination of the bifurcation point for a system of delay-differential equations is not a trivial matter, especially when more than one delay term is involved. Fortunately however, Walton and Marshall [4] have recently described a method for performing this calculation for the case of commensurate-delay systems. Using this method, it was possible to examine the bifurcation point as a function of the input pattern applied to the network.

Figure 3 depicts the results of such an experiment. The network was composed of four regions, was fully-connected, in that each inhibitory node received an input from all of the excitatory nodes and, had uniform weight settings for a given type of afferent. The adjustable parameters were the inputs to the excitatory nodes of regions one and three (u_1 and u_3), and these were systematically advanced from -1 to +1 in 0.05 increments; the inputs to all of the other nodes were kept fixed throughout the successive investigations (the separation of the inputs to the excitatory and inhibitory nodes belonging to the same region, arises from the use of sigmoid functions with unity slope and zero-threshold and is equivalent to a single regional input when (σ, θ) sigmoid functions are employed). For each setting, the fixed-point equilibrium solution was first obtained, and then the method of Walton and Marshall was applied to determine

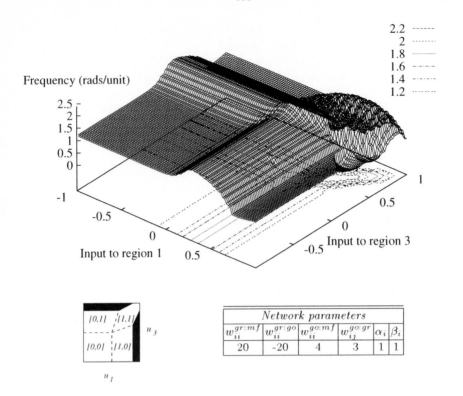

Fig. 3. ω_c-surface for the model as the inputs to regions one and three are varied for each new experiment. The input pattern to the excitatory nodes was therefore *,0.25,*,0.1. All inhibitory nodes received an input of -0.5. Lower-left, shows the regional decomposition of the ω_c surface in terms of the excitatory node activity in regions one and three. Lower-right gives the parameters used in the experiment.

whether this point could be destabilised through a Hopf bifurcation by some τ, and if so, the values ω_c and τ_c at which this occurred.

Of the 1681 input configurations examined, 1365 produced equilibrium solutions in the network which could be destabilised by some τ, to produce sustained oscillations. Those which were immune arose from the areas of the $[u_1, u_3]$ input-space which are shaded black in the lower-left diagram of Figure 3. When the input is in close proximity to these areas, τ_c increases rapidly, reflecting the fact that once they are reached, no delay will suffice to destabilise the fixed-point equilibrium. For this reason, only the ω_c-surface is shown here since, whilst it maintains, in a reciprocal fashion, the trends observed by τ_c over the rest of the input-space, it does not suffer from disproportionate effects at the periphery. The phase-surface (where $\theta_c = \omega_c\tau_c$) is similar in shape to that traced by ω_c and so is not shown on grounds of economy.

The 'parallel fibre' input to the inhibitory node of each region leads to competitive interactions between them, which is reflected in the fixed-point equilibrium solution. The respective equilibria are a measure of the extent to which any node's signal contributes to the network and hence, the extent to which the delays associated with that signal's transmission enter into the network's operation. Hence, the competitive effects which determine the spatial character of network activity, in so doing, proceed to determine the 'pattern of participating delays' (i.e. *effective delays*) which, in turn, characterises the temporal behaviour that is exhibited.

From these arguments, the regional-nature of the ω_c-surface is accounted for in the following way. When the inputs to regions one and three are small $[u_1 \leq 0.0, u_3 \leq 0]$, the activity of the other regions is the dominant feature of the network and they lead to a particular set of effective delays and define its dynamic characteristics accordingly. When one of the regions receives a stronger input (either u_1 or u_2 becomes greater than 0.25, say), then competitive effects elevate the recipient of this stronger input and the network's dynamics are then governed by the delay profile associated with the signals this node transmits to the rest of the network; since these effective delays are different, the ω_c-surface changes. When both regions receive higher inputs, they jointly dominate the spatial pattern of activity and so jointly determine its dynamical properties. The competitive principles which underlie this behaviour are evident in the contours shown in the surface-plot: when only one of the regions (i.e. regions one or three) is in receipt of strong input, the orientation of the contours is fixed by that particular input; the orientation shifts when both regional inputs contribute jointly.

The areas in $[u_1, u_3]$ space which are not prone to delay-destabilisation occur because they lead to equilibria which are sufficiently high that they minimise the dynamic range of the excitatory 'post-synaptic' to them. Although not actually reached in the input-space traversed by this experiment, another 'immune' region looms beyond the [1,1] area of the input domain. This time however, it is an elevation in the inhibitory node's equilibrium position and a consequent diminution in its dynamic range about that position, which is to blame.

4   Application of the Model to the Granular Layer

Naturally, the model is a long way removed from the actual granular layer, however, it does serve to suggest further avenues of inquiry.

Assuming that the Hopf bifurcation characteristics may be taken as an indicator of the broader dynamics of the system, then the results described in this model suggest that input patterns are mapped to certain oscillatory frequencies. Since the output of the nodes is considered analogous to neuronal mean firing rates, then oscillations could be looked upon as being comparable to burst-like behaviour. This suggests the hypothesis that the granular layer uses *temporal structure* to recode mossy fibre signals. Furthermore, that the temporal structure of local representations is modified by activity occurring elsewhere in the network as a result of regional interaction. This may be useful from the point of view of coordination.

Obviously more detailed models are needed in order to investigate these proposals further. With this in mind, work is currently being undertaken to examine the effects of using time-varying inputs; a more complex model is also being explored which uses more realistic neuronal representations.

5 Summary

This paper is addressed at two audiences: those interested in the granular layer of the cerebellum, and those interested in König and Schillen's model of cortical oscillations. The results of the Hopf bifurcation analysis and the application of Walton and Marshall's method are directly relevant to the latter audience. The proposal it makes with respect to investigation of the cerebellum, is that the granular layer uses the *temporal structure* of granule cell signalling to encode regional information.

References

1. F.Chapeau-Blondeau and G. Chauvet. A neural network model of the cerebellar cortex performing dynamic associations. *Biological Cybernetics*, 65:267–279,1991.
2. P.König and T.B.Schillen. Stimulus-dependent assembly formation of oscillatory responses I: Synchronisation. *Neural Computation*, 3:155–166, 1991.
3. W.Singer and C.M.Gray. Visual feature integration and the temporal correlation hypothesis. *Annual Review of Neuroscience*, 18:555–86, 1995.
4. K.Walton and J.E.Marshall. A direct method for TDS stability analysis. Proceedings of the Institute Electrical Engineers, Part D: Control Theory and Applications. 134:101–107, 1987.
5. W.Welker. Spatial organization of somatosensory projections to granule cell cerebellar cortex: functional and connectional implications of fractured somatotopy. In J.S.King, editor, *New Concepts in Cerebellar Physiology* pages 239–280. Alan R. Liss, 1987.

The Cerebellum as a "Coupling Machine"

Franz Mechsner and Günther Palm

University of Ulm, Department of Neural Information Processing,
D-89069 Ulm, Germany

Abstract. Recently, one of the authors [9] proposed a theory that the cerebellum works according to a hitherto unknown principle of motor control. The theory claims, that a simple learning process in the cerebellum, the "coupling operation", is powerful enough to explain the role of the cerebellum in the acquisition and performance of complex motor skills. Already existing theories [1,2,7,8] usually suppose that the task of the cerebellum is to produce fixed, unalterable motor control patterns in response to exactly defined inputs or "contexts". In contrast, the proposed theory explains how training leads to the ability to perform skilled, highly coordinated, finely tuned *situation-specific movements which have never been exercised or done before*. According to the model, the cerebellum influences only complex voluntary movements, not simple ones.

1. The Fundamental Cerebellar Operation

The cerebellum is involved in the control of complex voluntary movements. Its role in the motor system seems to have several aspects [3]: First, it seems to allow training and performance of *faster* motions and movement sequences than it is possible without this brain part. Second, the cerebellum enhances *spatial motor coordination*, i.e., for instance, the number of muscles that can be controlled simultaneously. Movements can be executed more *precisely* and with more *variability*, depending on the intention and situation. So far it has not been possible to develop a coherent theory which accounts for this diversity of effects.

In the following we propose a *simple* fundamental operation that might account for *all* of these aspects in a biologically plausible way.

For the sake of simplicity it is assumed - following Marr [8] - that every motion is a combination of "elementary movements" (EMs). Imagine for instance a thumb bending or tricep activation being such "EMs". Suppose that there is a brain part called "cognitive apparatus" which can - even without cerebellar intervention - combine EMs rather freely, depending on the task. Note: In contrast to Marr an EM is not considered a fixed item here but most variable in time and intensity. These parameters have always to be controlled by the cognitive apparatus.

Due to the bounded control capacity of the cognitive apparatus [10], there are upper limits to certain parameters of movements. Important parameters undergoing such limitations are first, the number of EMs that can be carried out simultaneously. second, motion speed, and third, speed accuracy. Imagine a particular motion being executed by synchronously activating 5 EMs for, let's say, 200 ms with certain intensities. Assume furthermore that - with inactive cerebellum - 5 EMs is the upper limit in the case of that motion. This means: First, if performed faster the coordinated motion becomes unmanageable. Second, it is impossible to synchronously add any further EMs and keep control. Third, it is impossible to enhance accuracy at will.

The present theory claims that it is possible to overcome these limitations if the cerebellum comes into play. The fundamental cerebellar operation is to "couple" EMs by means of a learning process according to the following rule:

- *If a set of EMs has often been performed simultaneously the participating EMs exert a strong excitatory influence on each other when performed together.*

With these "couplings" higher coordinated movements can be performed. To appreciate this, consider the cognitive apparatus driving the 5 "coupled" EMs with roughly the same input pattern as before training. Because the simultaneous EMs activate each other the coordinated movement is performed stronger and faster. By reducing the intensity of the driving pattern the cognitive system can drive the movement at a slower pace as well. While saving handling capacity in this way the cognitive system can simultaneously add new EMs and keep control.

The option of higher motion speed - also in the case of sequences - and the gain of extra handling capacity are the main effects of the cerebellar "coupling" operation. Together these effects allow for a much broader range of flexibility in movement performance, in time as well as in space. On the one hand, higher and higher coordinated movements can be trained step by step by adding new EMs to the coupled set. On the other hand the extra control capacity can be used to vary and combine movements more freely, to better adapt them to changing circumstances and velocities, to tune them more precisely - or simply to perform motions with less attention.

Abandoning the simplistic picture of the motor system given so far one can appreciate the power of the cerebellar "coupling" operation even better. For instance one can realize that there are many mechanisms to train and perform highly coordinated movement sequences, in particular fast ones: Due to temporal distributivity in motor planning and control, to axon conductance time spans and delayed feedback the cerebellum can "couple" neuronal motor activities that influence muscle activities which are not synchronous. These issues are considered in more detail in [9].

2. Outline of the Network Model

The network model is inspired by the theories of cerebellum put forward by Marr [8] and Albus [1]. However, it is different in many respects allowing for the establishing and flexible use of "couplings" as outlined.

2.1. Cerebellar Neurons

Most of the cerebellar microanatomy was established by Ramón y Cajal [11]. A great deal of physiological data has been worked out thoroughly by Eccles et al. [4]. Basic facts first established in these pioneering works will be used here without explicit quotations.

The cerebellum consists of the *cerebellar nuclei* and the *cerebellar cortex*. The cerebellar nuclei provide the output of cerebellum which is exitatory. They can influence planning and execution of movements. Cerebellar nuclei are excited by input axons from outside the cerebellum called *mossy fibers* and *climbing fibers*. The nuclei are inhibited by *Purkinje cells* wich provide the only output of the cerebellar cortex.

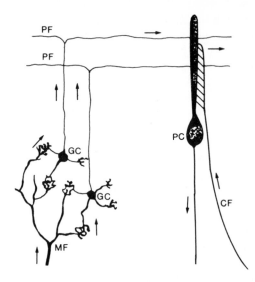

Fig. 1. Main components of the cerebellar network. Purkinje cells (PC) inhibit cerebellar nuclear cells (not depicted here). Climbing fibers (CF) excite PCs. Mossy fibers (MF) excite granule cells (GC). The vertical axon segments of GCs are called parallel fibers (PF) which excite PCs (redrawn and modified from Ramón y Cajal, 1911).

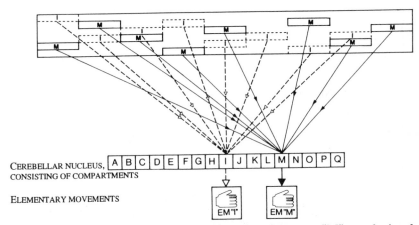

Fig. 2. The model cerebellum. Two cerebellar modules "I" and "M" are depicted, both consisting of 8 cortical microzones projecting to a particular nuclear compartment that influences an elementary movement. For instance, the M-Module consists of 8 M-Microzones, one M-compartment and exerts excitatory influence on the elementary movement "M" (middle finger bending). The I-Module excites or intensifies the elementary movement "I" (index finger bending). The other 15 modules are not depicted explicitly here.

Purkinje cells receive input from outside the cerebellum via the two paths that already have been mentioned (see Figure 1): 1. *Mossy fibers* excite numerous *granule cells* in the lowest layer of the cerebellar cortex under the sheet of Purkinje cell bodies. From every granule cell a tiny axon mounts up, branches in T-form, and runs some millimeters in two opposite directions. All of these axon branches are aligned in parallel, crossing dendritic trees of Purkinje cells perpendicularly, and thus are called *parallel fibers*. When recording Purkinje cell activity one usually finds spike trains with 70-120 Hz frequencies. These so called "simple spikes" are elicited by parallel fiber activity. 2. *Climbing fibers* directly excite Purkinje cells. Every Purkinje cell is influenced exclusively by one single climbing fiber. The climbing fiber powerfully excites the Purkinje cell, eliciting single, isolated pulses which are easily distinguishable from simple spikes. These so-called "complex spikes" occur in very low frequencies (up to not more than 3-4 Hz).

Many inhibitory interneurons are intercalated in the cerebellar network. *Golgi cells*, excited by both mossy and parallel fibers, inhibit granular cells. *Basket cells* and *stellate cells*, both activated by parallel fibers, inhibit Purkinje cells.

2.2. The Model Cerebellum

The simplified model cerebellum consists of a cerebellar cortex and a single nucleus (see Figure 2). Its working units are cerebellar modules, each with possible excitatory influence on a particular elementary movement (EM). Thus the "M"-module can intensify EM "M" which may correspond to, let's say, "middle finger bending".

The output of the modules is provided by "compartments" of the nucleus, neuronal populations of uniform behaviour that can activate a particular EM. (This claim is inspired by Houk [7]). There are as many compartments as there are EMs. A huge number of Purkinje cells exclusively project to each compartment. These Purkinje cells are arranged in smaller populations separated from each other, called microzones. It is supposed that 8 microzones project to each compartment.

2.3. The Cerebellar Initial State

Nuclear neurons in the model are "spontaneously" active with 100 Hz frequency. This activity can only be varied by Purkinje cell inhibition. The activity of a particular compartment is inhibited exactly down to zero if all of the Purkinje cells projecting to it are active with 100 Hz frequency. With all these Purkinje cells silent the compartment is active with 100 Hz thus exerting maximum influence to its EM. One single microzone is claimed to be of little influence.

Purkinje cells are driven by parallel fiber input. It is claimed: All mossy fibers exciting the granular cells in the microzones of a given module indicate the execution of the module's EM (for some experimental evidence see [5]). Parallel fibers conduct this information to the Purkinje cells of neighbouring microzones. Assume for the time being that Purkinje cells of a given microzone are solely driven by parallel fibers originating in neighbouring microzones.

Following Marr [8] it is assumed that parallel fiber activity is - by means of Golgi cells - restricted to 1% of the possible activity. In addition it is claimed that the level of parallel fiber activity remains strictly constant all the time all over the cerebellum. (However, the parallel fiber activity depicts everywhere the percentages of - changing - mossy fiber activities giving rise to it). There is an "initial state" of cerebellum where all synapses from parallel fibers to Purkinje cells have maximum weight. Thus all of the Purkinje cells are constantly active with 100 Hz inhibiting all nuclear neurons to zero: In its initial state the cerebellum does not exert any influence, because the Purkinje cell activity never changes.

2.4. The Learning Process

Only due to learning processes the initial state described in 2.2. can change and the cerebellum can have effects. As first suggested by Albus [1] and experimentally supported by Ito [6] it is assumed that active synapses from parallel fibers to a particular Purkinje cell are slightly weakened if a climbing fiber signal occurs at roughly the same time. Thus the nuclear neurons can become active and stimulate the module's particular EM. However, recall that one single Purkinje cell and even one single microzone are claimed to be of negligible influence on the nuclear compartment.

It is assumed: all Purkinje cells in a module receive the same kind of climbing fiber input. All climbing fibers of a given module indicate the execution of the particular EM influenced by this module (for experimental evidence see [5]).

Consider two EMs, "M" (middle finger bending) and "I" (index finger bending) being often executed together. The effect: Those parallel fiber synapses signalling the execution of "M" to "I"-Purkinje cells are weakened. The same holds for those parallel fiber synapses signalling the execution of "I" to "M"-Purkinje cells. Finally the execution of I diminishes M-Purkinje cell activity and the execution of M diminishes I-Purkinje cell activity in all M- and I-microzones wich are neighbours.

If all M- and I-microzones were neighbours, the described learning process would "couple" EMs M and I as defined in Section 1: Execution of EM M would activate the I-compartment and vice versa. However, a "coupling" of two EMs as described can be a mostly undesired effect - imagine for instance a index finger bending that invariably leads to a middle finger bending.

Yet the cerebellum is only able to effectively couple sets of *many* simultaneous EMs. This happens because the microzones of a given module receive dissimilar parallel fiber input. Imagine 17 possible EMs A, B, ..., Q and thus 17 corresponding cerebellar modules A, B, ..., Q. Each module contains 8 microzones. It is claimed: Microzones are distributed rather at random over the cortex. Every microzone has 4 neighbours rather at random selected out of the possible types A, B, ..., Q.

Thus the 8 sets of microzones neighbouring the 8 M-microzones may be :

(BFIN) (ADKL) (CELQ) (BGKO)
(EJOP) (DFIQ) (AGHJ) (CHNP)

Note: All types of microzones are neighbours of M-microzones, but only twice each.

Neighbours that are connected to M-Purkinje cells via weakened parallel fiber synapses are denoted by bold, underlined letters. Synchronous exercising of EMs M and I then leads to

(BF<u>I</u>N) (ADKL) (CELQ) (BGKO)
(EJOP) (DF<u>I</u>Q) (AGHJ) (CHNP)

Thus only two of the M-microzones become silent if I is executed in isolation. The influence of one or two single microzones on the corresponding nuclear compartment is claimed to be negligible. The consequence: It is not possible for "few" EMs to make a cerebellar module work.

Now suppose that a motion consisting of the 6 simultaneous EMs C, D, H, I, M and O is frequently performed. Concerning the M-microzones, this training procedure leads to

$$(BF\underline{I}N) \quad (A\underline{D}KL) \quad (\underline{C}ELQ) \quad (BGK\underline{O})$$
$$(EJ\underline{O}P) \quad (\underline{D}F\underline{I}Q) \quad (AG\underline{H}J) \quad (\underline{C}\underline{H}NP)$$

This time all of the 8 M-microzones become silent when the set of these 6 EMs is carried out. Thus EM M - middle finger bending - is strongly stimulated by the other EMs and vice versa: All EMs of the trained set exert a strong excitatory influence on each other when performed together. This means:

- *The model cerebellum performs exactly the fundamental "coupling" operation as defined and explained in Section 1.*

If only one of the "coupled" EMs, let's say "I", is executed, there is no effect. Altogether the consequence is: The cerebellum can only influence complex movements not simple ones. Experiments performed by Thach et al. [12] support the view that this is indeed the case.

The model as outlined in this paper accounts mainly for the spinocerebellum. It is possible to expand these considerations also to the pontocerebellum and vestibulocerebellum. In Mechsner [9] the proposed theory as well as experimental evidence is described and discussed in much more detail.

References

1. Albus, J.S. (1971) A theory of cerebellar function. Math Biosc 10, 25-61.
2. Braitenberg, V. (1983) The cerebellum revisited. J Theor Neurobio, 2, 237-241.
3. Dow, R.S. & Moruzzi, G. (1958) The physiology and the pathology of the cerebellum. Minneapolis: University of Minnesota Press.
4. Eccles, J.C., Ito, M. & Szenthagothai, J. (1967) The cerebellum as a neuronal machine. Berlin, Heidelberg, New York: Springer.
5. Ekerot, C.F., Jörntell, H. & Garwicz, M. (1995) Functional relation between corticonuclear input and movements evoked on microstimulation in cerebellar nucleus interpositus anterior in the cat. Exp Brain Res 106, 365-376
6. Ito, M. (1992) Physiology of the cerebellum. In A. Plaitakis (Ed.), Cerebellar degenerations: clinical neurobiology. Boston: Kluwer Academic Publishers.
7. J.C. Houk (1987), Model of the cerebellum as an array of adjustable pattern generators, in: M.Glickstein et al (eds.), Cerebellum and neuronal plasticity, Plenum Press New York, 249-26
8. Marr, D. (1969) A theory of cerebellar cortex. J Phys 202, 437-470.
9. Mechsner, F. (1996) A new theory of cerebellar function. (Connection Science, in press).
10. Miller, G.A. (1956) The magical number seven, plus or minus two: some limits on our capacity for processing information. Psych Rev 63, 91-97.
11. Ramón y Cajal, S. (1911) Histologie du système nerveux de l'homme et des vertébrés, translated by L. Azoulay. Madrid: Instituto Ramón y Cajal, edn. 1972.
12. Thach, W.T., Goodkin H.P. & Keating J.G. (1992) The cerebellum and the adaptive coordination of movement. Ann Rev Neurosc 15, 403-442.

The Possible Function of Dopamine in Associative Learning: A Computational Model

Daniel Durstewitz and Onur Güntürkün
AE Biopsychology, Ruhr-University of Bochum, D - 44780 Bochum, Germany

The neuromodulator dopamine is critically involved in different procedures of instrumental learning and working memory. Based on physiological data, the present study investigates the effects of dopamine on neural network behavior. We demonstrate that dopamine suppresses interference with previously learned patterns and may enable fast learning of new contingencies or associations in biologically significant contexts.

Introduction and Basic Assumptions

Various neuromodulators like dopamine (DA) play an important role in attentional processes and learning. The DA system consists of a relatively small number of dopaminergic midbrain neurons (cell groups A8, A9 and A10) projecting diffusely to various limbic and motor areas. Major targets of dopaminergic afferents are the basal ganglia, the olfactory, entorhinal and anterior cingulate cortex, the prefrontal cortex and, in primates, the motor and premotor cortices [3,12]. After lesions in the DA system or blockade of DA receptors, severe deficits in various operant conditioning paradigms, reversal learning, and delayed response learning occur [2,11,16,23,24]. Some authors suggested response perseveration as a possible cause of the observed deficits (e.g. [23,24]). The DA mibrain neurons are active in new and biologically significant situations for which no 'ready-to-run' behavioral programs exist and new sensory-motor couplings have to be learned, e.g. when food or food predicting stimuli appear in new behavioral contexts [17,18]. This activity vanishes when the organism has learned an appropriate response. The DA signal itself most probably does not contain specific information about the sensory situation. Instead, neural activity in the dopaminergic target areas seems to be modulated in a way that makes fast learning of new associations in biologically significant situations possible or at least easier.

Whenever new representations or associations have to be learned, the participating neurons or cell assemblies have to be held active for some time so that long term modifications like LTP or LTD can develop. We assume that synchronous oscillations of the relevant neural populations are a prerequisite of long term learning (cf. [21]). In the hippocampus and olfactory cortex, oscillations in the theta frequency band (5 to 12 Hz) have been shown to occur during learning and sampling of new information [7,9]. Tetanic stimulation at approximately 140-200 ms intervals seems to be optimal for the induction of LTP in hippocampus [7]. In the neocortex, it has also been shown that in the majority of layer V pyramidal cells intrinsic oscillations with theta frequency can be induced by a short stimulation at depolarized membrane levels [20]. Single spike oscillatory activity proceeds independently for some seconds. We assume that neurons or neural nets can be switched from a regular processing mode to a learning mode, where intrinsic oscillatory activity occurs. This switch may be induced by a neuromodulator like one of the monoamines. Indeed, one author shows that dopamine can induce the theta rhythm in hippocampus (cited in [22]). Herein, we investigate the possible function of dopamine in associative learning within cortex by neural modeling.

Single Neuron Model and Network Architecture

We chose a single compartment model, which is sufficient to demonstrate our functional assumptions. Basically, we include four different types of membrane currents: Leakage currents with constant conductance g_{leak}, excitatory and inhibitory synaptic currents, an after hyperpolarizing (AHP) current, which represents different active potassium currents and is triggered by a spike event, and a depolarizing after potential (DAP), which is turned on only in the learning mode. This latter current is responsible for the intrinsic

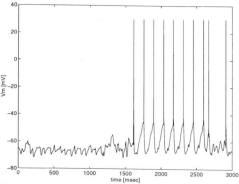

Fig. 1. Spike train of a simulated neuron. At about 1600 ms the oscillatory mode was induced.

oscillations. To generate the oscillatory behavior with theta frequency we assumed a prolonged AHP and a constant conductance g_{DAP}, which become activated upon the first spike event in the learning mode. As shown in fig. 1, the resulting spike train looks very like the ones found in neocortical neurons [20]. Synaptic conductances and the AHP conductance are modeled by alpha functions of the form

$$(*) \quad g_{syn(ijc)}(t) = v_c \, w_{ij}(t_k) \, t \, / \, \tau_{syn(c)} \, \exp(-t \, / \, \tau_{syn(c)}) \, ,$$

where v_c denotes a general synaptic efficiency, $w_{ij}(t_k)$ a synapse specific weight modifiable by learning and $\tau_{syn(c)}$ time to peak. Synaptic conductances are convoluted with the respective presynaptic spike trains $S_j(t) \in \{0;1\}$ and integrated over all homotypical synapses. Currents are given by Ohm's law and sum up to give rise to the membrane potential which is governed by the following differential equation:

$$-C \frac{dV_i}{dt} = \sum_c I_{syn(ic)}(t) + I_{leak(i)}(t) + I_{AHP(i)}(S_i, t) + I_{DAP(i)}(S_i, t) + I_{inj(i)}(t) \, ,$$

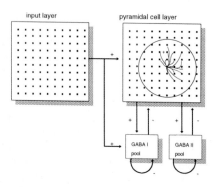

where C denotes membrane capacity. An action potential $S_i(t) = 1$ is delivered whenever the membrane potential $V_i(t)$ exceeds an exponentially decaying threshold $\Theta_i(t)$.

For the present purpose, we devised a neural net according to data from neocortex [6] (fig. 2). The net consists of a two-dimensional layer of 100 'pyramidal cells' which make excitatory synaptic contacts with all other cells

Fig. 2. Network model (see text for explanation).

within a certain radius. Furthermore, there are two pools of GABAergic neurons which are regarded here as homogeneously reacting cell populations. The GABAergic neurons are driven by the pyramidal cells. Vice versa, they form inhibitory synaptic contacts with all pyramidal neurons. We distinguish between fast and strong GABAergic effects (pool I) acting via $GABA_A$ receptors and Cl^- channels, and slow and more weak effects (pool II) acting via $GABA_B$ receptors and K^+ channels. Finally, there is recurrent inhibition within the GABAergic pools. To present patterns to the network in a 'realistic' manner, an input matrix of the same dimensions as the pyramidal cell layer is connected to the latter, innervating both, pyramidal cells and the GABA I pool. Patterns on the input layer are evoked by current injection. It can be shown that the net is capable of stabilizing and achieving synchronous oscillations of patterns not learned previously, despite variance in the synaptic weights and noisy inputs. This effect is due to the weak and long-lasting inhibition provided by pool II and the excitatory couplings. In contrast, the fast inhibitory feedback by pool I prevents that different neural representations collapse in time.

Physiological Effects of Dopamine

In vivo, dopamine mainly seems to inhibit target cells [4,8,19,26]. In our model, we included three different effects of dopamine, which have been repeatedly confirmed and seem to be of major importance:

(1) Dopamine reduces excitatory synaptic transmission [14,15]. In different studies a reduction of glutamate induced excitatory PSPs of up to 30-50 % has been demonstrated after iontophoretic or bath application of dopamine. The AMPA as well as the NMDA component seem to be affected [14,15].

(2) Dopamine reduces the excitability of innervated target neurons, i.e. it increases the threshold for spike generation [4,5,25].

(3) Dopamine seems to favor the emergence of long term depression (LTD) compared to long term potentiation (LTP) [13]. This effect may partly (but not completely, see [13]) be due to the reduced Ca^{++} current through NMDA receptor channels.

Direct hyperpolarizing or depolarizing actions of dopamine have also been described, but these usually seem to be rather small [14,15] and are not considered here. The major cortical targets of dopamine afferents are pyramidal cells [3,10]. Therefore, in our model, only the pyramidal cell layer is affected by DA.

Learning Rule and Model Parameters

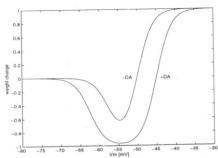

Fig. 3. LTP/LTD-curves. Upper trace: without DA; lower trace: modified by DA.

We now investigated learning in the net described above under conditions with and without simulated dopamine effects. For long term learning we used a rule suggested by Artola and Singer [1]. They propose that the induction of LTD and LTP depends on the calcium level in the postsynaptic spine, which in turn depends on presynaptic activity and the level of postsynaptic membrane depolarization. In particular, when transmitter is released presynaptically, there is a lower postsynaptic potential threshold, above which LTD is induced, and a higher threshold, above which LTP is induced due to sufficient activation of NMDA channels and the respective calcium influx. Fig. 3 shows the learning functions we used in our simulations (cf. [1]). The weight change $\Delta w_{ij}/\alpha$ depends on the highest postsynaptic membrane potential (V_m) in a 8 ms-range (cf. [21]) around the arrival time of the presynaptic spike. The decreased NMDA currents and the increased probability of LTD under dopamine action were modeled by an elevated LTP threshold and a reduced LTD threshold (see fig. 3). The depressed excitatory synaptic transmission was modeled simply by reducing the respective synaptic efficiency v_{exc} (see (*)) by the factor 1/2. Finally, the reduced excitability was modeled by shifting (resting) spike threshold of the pyramidal neurons from -50 mV in the regular mode to -45 mV during dopamine action. The dopaminergic system was activated only during learning of new associations, as empirical findings suggest [17,18]. Also, during this phase, pyramidal neurons were switched from the regular to the oscillatory mode. Some important parameters of the simulations are: resting membrane potential -65 mV, membrane time constant 20.0 ms, time to peak of the AHP conductance 10.0 ms in the regular mode and 20.0 ms in the oscillatory mode, time to peak of the excitatory synaptic conductances and the inhibitory pool I synapses 5.0 ms, time to peak of the pool II synaptic conductances 40.0 ms.

Simulation Results

To demonstrate the functional effects of dopamine, we trained the net on a 'reversal' learning task. This type of experiment is depicted in fig. 5. In two successive learning periods two different sets of target patterns, P1 and P2 and P3 and P4 (fig. 4), respectively, were presented to the net. Two test patterns, T1 and T2 (fig. 4), which are contained in both patterns P1 as well as P3 and P2 as well as P4,

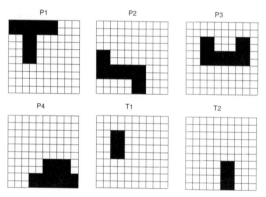

Fig. 4. Target and test patterns (see text).

respectively, were presented before, after and between the two learning periods during the regular mode. The learning (oscillatory) phases lasted for 1000 ms. At the start of the simulation, the mean of the gaussian distributed synaptic weights was adjusted so that the activation of any random pattern might activate some other neurons. The capability of the test stimuli to activate selectively only those neurons which were part of the target patterns, measures the success

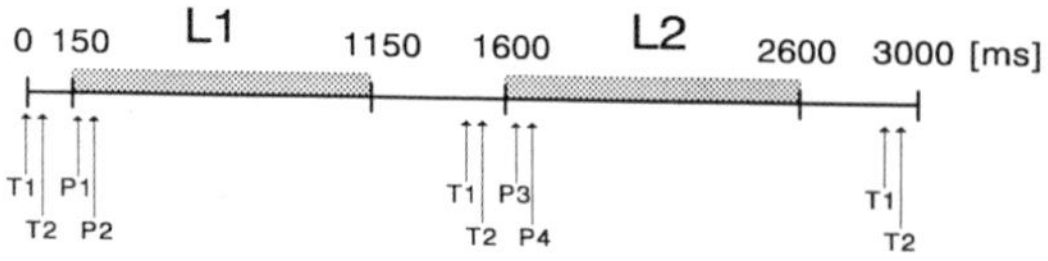

Fig. 5. Illustration of the reversal learning experiment: P1, P2, P3, P4, T1, T2: times, at which target and test patterns were presented. L1, L2: the two learning periods.

of learning. All neurons which became active within 15 ms after a test stimulus appeared on the pyramidal cell layer were considered. In one simulation, dopamine effects were present during both learning periods, in the other, dopamine effects were omitted during the second learning phase. In the first simulation, the net managed the reversal learning, i.e. it recalled patterns P1 and P2 completely upon presentation of T1 and T2 after the first learning phase (L1), and reacted with patterns P3 and P4 to the same test stimuli after the second learning phase (L2) (100 % recall, 0 % noise). In the second simulation without dopamine during L2, interference with the previously learned patterns made the acquisition of new patterns impossible. Instead, the net showed "perseveration", i.e. it reacted with the previously learned patterns P1 and P2 upon the final presentation of T1 and T2 (fig. 6). The reason for this is that the elicitation of P3 and P4 on the pyramidal cell layer during L2 evokes P1 and P2 due to the insufficient suppression of neural activity without dopamine. This in turn leads to a dissociation of T1 from the rest of P3 and of T2 from the rest of P4. Therefore, the coherence of P3 and P4 is lost, i.e. the neurons of these patterns are no longer oscillating synchronously, and the old patterns are preserved. If only the effects of dopamine on synaptic modifications are omitted during L2 whereas the depressive effects on neural activity are included, the previously established connections between T1 and the rest of P1 and between T2 and the rest of P2 are not sufficiently weakened. This results in the activation of a conglomerate of P1 and P3 upon presentation of T1 and of P2 and P4 upon presentation of T2 (values for the third test phase: P1: 100 % recall, 17.95 % noise; P2: 77.27 % recall, 17.95 % noise; P3: 100 % recall, 17.95 % noise; P4: 100 % recall, 11.54 % noise).

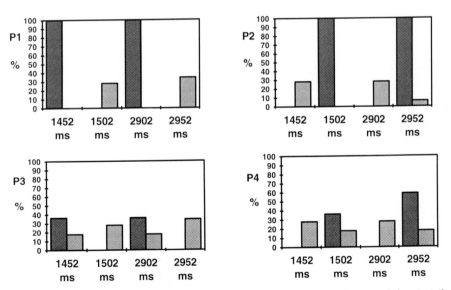

Fig. 6. Percent recall (dark bars) of the target patterns at the four post-training test time points (1452 ms and 2902 ms = activation of T1, 1502 ms and 2952ms = activation of T2). The respective light bars indicate the percentage of non-target units active in the same time intervals. Note that 36.36 % of recall are due to the test patterns alone and therefore indicate no learning.

Conclusion

We conclude that the function of dopamine is 1) to reduce interference with previously established associations and to suppress noise during the learning phase, and 2) to aid relearning or 'forgetting' of 'false' and irrelevant associations. We suggest that dopamine by these means acts as a kind of mnemonic filter or amplifier for new biologically relevant information. It enables fast behavioral adaptations to unexpected environmental demands and changing contingencies. If the dopaminergic system is not active, old patterns will be retrieved and stabilized depending on their similarity with the input pattern.

References

1. Artola A, Singer W (1993). TINS 16: 480-487.
2. Beninger RJ (1993). In JL Waddington (Ed.), D1:D2 dopamine receptor interactions, pp. 115-157. London: Academic Press.
3. Berger B, Gaspar P, Verney C (1991). TINS 14: 21-27.
4. Bernardi G, Cherubini E, Marciani MG, Mercuri N, Stanzione P (1982). Brain Res 245: 267-274.
5. Calabresi P, Mercuri N, Stanzione P, Stefani A, Bernardi G (1987). Neurosci 20: 757-771.
6. Douglas RJ, Martin KAC (1990). In GM Shepherd (Ed.), The synaptic organization of the brain, 3rd ed., pp. 389-438. Oxford: Oxford University Press.
7. Eichenbaum H, Otto T, Cohen NJ (1992). Behavioral Neural Biology 57: 2-36.
8. Ferron A, Thierry AM, Douarin CLE, Glowinski J (1984). Brain Res 302: 257-265.
9. Gluck MA, Granger R (1993). Ann Rev Neurosci 16: 667-706.
10. Goldman-Rakic PS, Leranth C, Williams SM, Mons N, Geffard M (1989). Proc Natl Acad Sci USA 86: 9015-9019.
11. Iversen SD, Mishkin M (1970). Exp Brain Res 11: 376-386.
12. Joyce JN, Goldsmith S, Murray A (1993). In JL Waddington (Ed.), D1:D2 dopamine receptor interactions, pp. 24-49. London: Academic Press.
13. Law-Tho D, Desce JM, Crepel F (1995). Neurosci Lett 188: 125-128.
14. Law-Tho D, Hirsch JC, Crepel F (1994). Neurosci Res 21: 151-160.
15. Pralong E, Jones RSG (1993). Eur J Neurosci 5: 760-767.
16. Sawaguchi T, Goldman-Rakic PS (1994). J Neurophysiology 71: 515-528.
17. Schultz W, Apicella P, Ljungberg T (1993). J Neurosci 13: 900-913.
18. Schultz W, Romo R, Ljungberg T, Mirenowicz J, Hollerman JR, Dickinson A (1995). In JC Houk, JL Davis, DG Beiser (Eds.), Models of information processing in the basal ganglia, pp. 233-248. Cambridge, Mass.: MIT.
19. Sesack SR, Bunney BS (1989). J Pharmacol Exp Ther 248: 1323-1333.
20. Silva LR, Amitai Y, Connors BW (1991). Science 251: 432-435.
21. Singer W (1993). Annu Rev Physiol 55: 349-374.
22. Smialowski A, Bijak M (1987). Neurosci 23: 95-101.
23. Sokolowski JD, McCullough LD, Salamone JD (1994). Brain Res 651: 293-299.
24. Sokolowski JD, Salamone JD (1994). Brain Res 642: 20-28.
25. Stanzione P, Calabresi P, Mercuri N, Bernardi G (1984). Neurosci 13: 1105-1116.
26. Verma A, Moghaddam B (1996). J Neurosci 16: 373-379.

Signatures of Dynamic Cell Assemblies in Monkey Motor Cortex

ALEXA **RIEHLE***, SONJA **GRÜN**[#], AD **AERTSEN**[°], JEAN **REQUIN***

*Center for Research in Cognitive Neuroscience, CNRS, 31, ch. Joseph Aiguier,
13402 Marseille Cx 20, France
[#]Department of Physiology, Hadassah Medical School, Hebrew University,
91120 Jerusalem, Israel
[°]Center for Research of Higher Brain Functions, Weizmann Institute of Science,
76100 Rehovot, Israel

It has been suggested that neuronal representations are embedded in dynamically coupled ensembles of simultaneously active neurons, commonly called cell assemblies. Multiple single-neuron recordings in the primary motor cortex of the monkey support the hypothesis that processing of information is reflected in the activity of cell assemblies. Individual neurons may belong to multiple assemblies. Furthermore, synchronized neuronal activity as an expression of the activation of a cell assembly depends on the behavioral context but not always on changes in the firing rate of neurons.

1. Introduction

How the intention to act results in movement is a fundamental question of brain organization. It has been shown by means of single-cell recording techniques in the behaving monkey that neurons of many different brain structures change their activity in relation to various aspects of motor behavior (Hepp-Reymond, 1988; Requin et al., 1992; Humphrey and Freund, 1992). However, the way by which the brain integrates information on external and/or internal events gave rise to the hypothesis that the neural code for higher brain functions resides in the activity of groups of neurons, or "cell assemblies" (Hebb, 1949). The concept of cell assemblies is based on the assumption that representations consist of dynamically coupled ensembles of simultaneously active neurons that are widely distributed over different cortical areas (Braitenberg, 1978; Requin et al., 1988; Gerstein et al., 1989; Palm, 1993; Singer and Gray, 1995). Synchronization of neuronal activity within a narrow time window seems to be a natural candidate for selecting, as members of such an assembly, the neurons to be temporally involved in the same processing operation. Each neuron could therefore participate in different functional groups at different times. Temporal coherence or synchronous firing would in fact be available directly to the brain as a potential neural code. This mode of information processing in cortical structures was proposed for the sensory areas involved in (visual) perception. However, it may be valid as well for motor areas, in which assemblies of synchronously firing neurons would provide the basic module for neuronal representations of action. Detecting a coherent signal that might represent the control of behavioral output from a system as complex as the central nervous system is a difficult task. The evaluation of the relative timing of action potentials in such an ensemble of neurons recorded simultaneously is thus a straight-forward approach to discover the role played by neuronal cooperativity in processing information involved in sensorimotor integration. The aim of this paper was to look for the relationships between synchronization dynamics in the activity of neurons in motor cortical areas and the animal's behavior. We conducted a series of experiments in which the activity of 2 to 7 neurons was recorded simultaneously in the primary motor cortex of two macaque monkeys trained in a complex sensorimotor task.

2. Methods

Monkeys were trained to touch a target, after a preparatory period (PP) of variable duration, on a video display which was equipped with a capacitive touch screen (Micro Touch) (cf. Riehle at al., 1995). To start a trial, the animal had to hold down a lever. The first, preparatory signal (PS) occurred in the middle of the screen indicated by an open circle. After the PP, during which the animal had to press the lever, the second, response signal (RS) was indicated by filling the circle. The animal had then to release the lever and to touch the circle on the screen. Four durations (600, 900, 1200, or 1500 ms) of the PP were presented at random. It is well-known that in such a situation reaction time (RT) decreases with increasing duration of the PP (cf. Requin et al., 1991). Insofar as the presentation of each PP duration is equiprobable, the conditional probability for the RS to occur increases as the time elapses from the PS. After training, animals were prepared for multiple single-neuron recording. A multi-electrode microdrive was used to insert transdurally into the primary motor cortex 7 independently driven, quartz-insulated platinum-tungsten micro-electrodes (impedance: 2 - 3 MOhm at 1000Hz), spaced 330 μm apart (cf. Mountcastle et al., 1991).

Dynamic changes in synchronicity between neurons were analyzed off-line by calculating, trial by trial, the statistically significant epochs of synchronized firing (Grün et al., 1994; for details of the statistical analysis, see Grün, 1996). The simultaneous observation of the occurrences of action potentials elicited by N neurons can be transformed, using an appropriate binning, to N-dimensional joint-activity vectors. They reflect the various constellations of coincident spiking activity across neurons ("raw coincidences"). We have chosen a bin width of 3 - 10 ms (here 5 ms) since the integrative properties of pyramidal neurons of the cerebral cortex may be as narrow as a few milliseconds (Abeles, 1982; Softky and Koch, 1993). Under the null-hypothesis that the N neurons fire independently from each other, the expected number of occurrences of any constellation and its probability distribution can be calculated on the basis of the single neuron firing rates. The statistical significance of the difference between the actually observed and the expected numbers of synchronous events was tested: those occurrences which exceed the significance level of 5% were called "unitary events" (UEs). The observation of UEs was used to define neurons that are momentarily engaged in the activation of a cell assembly. To deal with non-stationarities in the firing rate of neurons, synchronicity was estimated on the basis of small time segments, by using a sliding window (usually 75 - 150 ms, here 100 ms, depending on the variation of the activity of the neurons) that was shifted in 5 ms steps along the data. This timing segmentation was applied to each trial, and the data of corresponding segments in all trials were then analyzed as one stationary data set.

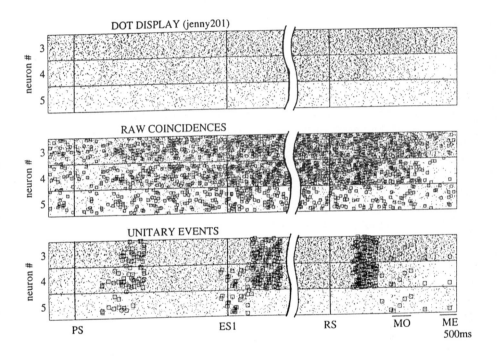

Figure 1: Fast modulation of the interaction between the activity of three neurons recorded simultaneously during 96 trials. Upper panel: Each box represents the activity of a single neuron in which each dot corresponds to the occurrence of an action potential and each row to a trial. Middle panel: the coincidences of spike occurrences of any of the neurons within 5 ms are indicated by squares. Lower panel: The selection of coincident spikes which pass the significance criterion for being "unitary events" (p<0.05, see Methods) are indicated by squares. They are evaluated over a sliding time window of 100 ms. PS: preparatory signal; ES1: first expected response signal (RS); MO: movement onset; ME: movement end. The data of three types of trials are pooled including PP durations of 900, 1200, and 1500 ms. Only the first 800 ms and the last 600 ms of the trials are shown. The temporal interruption is indicated by the white vertical stripe between ES1 and RS, covering thus a duration of 0 to 600 ms, depending on PP duration.

3. Results

Data were obtained during 295 recording sessions in which the activity of 2 - 7 neurons was recorded simultaneously. Of those, 145 sessions, including 359 pairs of neurons, were analyzed using the UE method. In one third (120/359) of the pairs of neurons, corresponding to more than 50% of the analyzed sessions, clusters of UEs were detected, that were loosely time-locked to behaviorally important events. These clusters of UEs were detected in combination or not with phasic changes in the activity of the neurons. During the PP, the occurrence of the RS was expected at several, clearly defined moments, that is at 600, 900, 1200 and 1500 ms, respectively. In 40 pairs of neurons, UEs were time-locked to these moments, and in no case neurons changed their activity at the same time (see Fig. 1, UEs in relation to the expected RS = ES1). However, in 61 pairs of neurons out of 65, in which UEs were detected after the occurrence of RS, neurons did change phasically their activity (see Fig. 1, UEs after the occurrence of RS). Furthermore, in 20 out of 43 sessions, in which the activity of at least three neurons was synchronized, neurons changed rapidly their partnership from one to another assembly. This is demonstrated in Figure 1, in which the activity of three simultaneously recorded neurons is shown. Since cell assemblies are defined by the synchronized activity of their elements, two assemblies were identified, that were alternately active during the task. Three distinct moments of synchronized activity can be observed: After the occurrence of the PS, then after the moment when the first RS was expected to occur (but did not, ES1), and finally after the occurrence of the RS, i.e. during RT. The activity of neurons 3 and 4 was synchronized during all three epochs. The strength of the synchronization increased from epoch to epoch, as seen by the increasing number of UEs. The activity of the same neuron 4 was synchronized with that of neuron 5 about 100 ms before becoming synchronized with the activity of neuron 3, but only at the beginning of the trial, i.e. during the PP. Taking into account when during the trial their synchronized activity occurred, these neurons might be involved in two different aspects of information processing. First, the assembly in which neurons 4 and 5 participate might be involved in the preparatory process responsible for the estimation of the duration of the PP. After the occurrence of the PS, this time estimation process begins before to be interrupted by the possible occurrence of the RS (for instance ES1). Once the RS occurred, this cell assembly is inactivated. Second, the assembly in which neurons 3 and 4 participate might be involved in another aspect of movement preparation, e.g. a process which facilitates the motor command. After the occurrence of the PS, this facilitatory preparatory process is activated and increases with the lengthening of the PP. The largest number of synchronized spikes occurred during RT, i.e. between the presentation of the RS and movement onset.

4. Discussion

There are different ways to provide evidence for the involvement of cell assemblies in information processing, and thus in neuronal representations. First, neurons participate at different times in different cell assemblies by rapidly changing their partnership. This implies the modulation of the effective cooperativity between neurons in relation to the behavioral context (see also Vaadia et al. 1995; Aertsen et al., 1996). Such a modulation was shown in about half of the sessions, in which the activity of at least three neurons was synchronized, as demonstrated in Fig. 1. Second, synchronized activity did not occur at any time during the trial, but at very particular moments, such as the occurrence of a behaviorally relevant signal or even in relation to the expectancy of this signal (e.g. ES1). It has been proposed that increasing the summation of neuronal activity by synchronizing inputs within a range of a few milliseconds - as compared to increasing the discharge rates of several neurons in an uncorrelated way - is a striking mechanism for increasing synaptic efficiency (Abeles, 1991; Softky and Koch, 1993). Third, spike synchronization can be observed in combination or not with a change in neuronal firing rate at the same time. Interestingly, in relation to signal expectancy, thus in relation to an internal, behaviorally relevant event, synchronized neuronal activity occurred preferentially without any change in neuronal discharge frequency. In contrast, after the signal occurred, thus in relation to an external, behaviorally relevant event, synchronized neuronal activity was found to be mostly correlated with a change in discharge frequency. Both the synchronization of neuronal activity and the modulation of the discharge rate could, thus, operate in a complementary way.

5. Acknowledgements

Partial funding was received from CogniSciences, the Centre National de la Recherche Scientifique (CNRS), and the German Bundesministerium für Bildung, Wissenschaft, Forschung und Technologie (BBWFT). Expert technical help was kindly provided by B. Arnaud (mechanical devices), M. Coulmance (computer programs), R. Fayolle (electronic devices), and N. Vitton (animal training and permanent assistance).

6. References

M. Abeles. "Local cortical circuits. An electrophysiological study". Berlin: Springer (1982)

M. Abeles. "Corticonics. Neural circuits of the cerebral cortex". Cambridge: Cambridge University Press (1991)

A. Aertsen, M. Erb, I. Haalman, E. Vaadia. Coherent dynamics in the frontal cortex of the behaving monkey - experimental observations and model interpretation. In: I. Gath, G. Inbar (eds.), "Advances in processing and pattern analysis of biological signals". New York: Plenum Publ. (1996, in press)

V. Braitenberg. Cell assembies in the cerebral cortex. In: R. Heim, G. Palm (eds.), "Theoretical approaches to complex systems". Berlin: Springer, pp. 171-188 (1978)

G.L. Gerstein, P. Bedenbaugh, A.M.H.J. Aertsen. Neuronal asssemblies. IEEE Trans. Biomed. Engin. 36: 4-14 (1989)

S. Grün. Unitary joint-events in multiple-neuron spike activity - detection, significance, and interpretation. PhD Thesis, Fakultät für Physik, Ruhr Universität, Bochum, FRG (1996)

S. Grün, A. Aertsen, M. Abeles, G. Gerstein, G. Palm. Behavior-related neuron group activity in the cortex. Eur. J. Neurosci. Suppl. 7: 11 (1994)

D.O. Hebb. "Organization of behavior. A neurophysiological theory". New York: Wiley & Sons (1949)

M.C. Hepp-Reymond. Functional organization of motor cortex and its participation in voluntary movements. In: H.D. Steklis, J. Erwin (eds.) "Neurosciences: Comparative Primate Biology, Vol. 4". New York: Alan R. Liss, pp. 501-624 (1988)

D.R Humphrey, H.-J. Freund (eds.). "Motor control: concepts and issues". New York: Wiley & Sons (1991)

G. Palm. Cell assemblies, coherence, and corticohippocampal interplay. Hippocampus 3: 219-226 (1993)

V.B. Mountcastle, R.J. Reitböck, G.F. Poggio, M.A. Steinmetz. Adaptation of the Reitböck method of multiple microelectrode recording to the neocortex of the waking monkey. J. Neurosci. Methods 36: 77-84 (1991)

J. Requin, J. Brener, C. Ring. Preparation for action. In R.R. Jennings, M.G.H. Coles (eds.) "Handbook of Cognitive Psychophysiology: Central and autonomous nervous system approaches", New York: Wiley & Sons, pp. 357-448 (1991)

J. Requin, A. Riehle, J. Seal. Neuronal activity and information processing in motor control: from stages to continuous flow. Biol. Psychol. 26: 179-198 (1988)

J. Requin, A. Riehle, J. Seal. Neuronal networks for movement preparation. In: D.E. Meyer, S. Kornblum (eds.) "Attention and Performance XIV", Cambridge, MA: MIT Press, pp. 745-769 (1992)

A. Riehle, J. Seal, J. Requin, S. Grün, A. Aertsen. Multi-electrode recording of neuronal activity in the motor cortex: Evidence for changes in the functional coupling between neurons. In: H.J. Herrmann, D.E. Wolf, E. Pöppel (eds.) "Supercomputing in Brain Research: from Tomography to Neural Networks", Singapore: World Scientific, pp. 281-288 (1995)

W. Singer, C.M. Gray. Visual feature integration and the temporal correlation hypothesis. Annu. Rev. Neurosci. 18: 555-586 (1995)

W. R. Softky, C. Koch. The highly irregular firing of cortical cells is inconsistent with temporal integration of random EPSPs. J. Neurosci. 13: 334-350 (1993)

E. Vaadia, I. Haalman, M. Abeles, H. Bergman, Y. Prut, H. Slovin, A. Aertsen. Dynamics of neuronal interactions in monkey cortex in relation to behavioural events. Nature 373: 515-518 (1995)

Modelling Speech Processing and Recognition in the Auditory System with a Three-Stage Architecture

T. Wesarg, B. Brückner, C. Schauer

Informatics, Federal Institute for Neurobiology
P.O.Box 1860, 39008 Magdeburg, Germany
e-mail:brueckner@ifn-magdeburg.de

Abstract. One approach to the construction of an engineered system for hearing and efficient speech recognition is the modeling of the human auditory system. We applied this approach to our speech recognition tasks using a coupled modeling concept (Fig. 1) which should reproduce this system in a plausible way (Brückner et al. [1]). Starting with a model of signal processing by the cochlea (Kates [4]), our coupled modeling concept contains a lateral inhibitory neural network (LIN) system (Shamma [2]) performing filter operations by spatial processing of the speech evoked activity in the auditory nerve, and a structured formal neural network (Brückner et al. [3]) for learning and recognition of the spectral representations of the speech stimuli provided by the LIN.

1 Introduction

The first stage of our model is a digital time-domain model of the human cochlea which is mainly based on the cochlear model of Kates [4]. Digitalized speech signals obtained by recording the analogue waveform of speech with a condensor microphone and a following 48-kHz ADC conversion, form the input (tympanic membrane pressure) to the middle-ear front-end of the cochlear model. The model output consists of the instantaneous firing rates and spike patterns in the auditory nerve.

The aim of the second stage of the complete model is to find and highlight characteristic properties like peaks and edges in the complex structure of the firing rate patterns in the auditory nerve. For this purpose one of the fundamental neural network topologies, the lateral inhibition based on the LIN model of Shamma [2], is used. A nonrecurrent network is responsible for the detection of spectral peaks in the spatio-temporal pattern of firing rates, and a network with recurrent inhibition is utilized for further sharpening of the averaged firing rate profiles. The extracted features depend on the spectral content and sound intensity of speech stimuli and enable the segmentation of the LIN output.

The third stage is a structured multi-layer formal neural network realized as a modification of the Hypermap Architecture (Brückner et al. [3]). The modified Hypermap network is trained with segmented LIN output of speech signals. Each segment of the LIN output time sequence which represents the properties of a whole syllable or a part of a syllable, is related to one layer in the input vector. The whole time sequence represents the input vector for one complete speech component, i.e. a word or a part of a sentence. After training the network is able to classify untrained speech signals, i.e. can recognize several words spoken by different subjects.

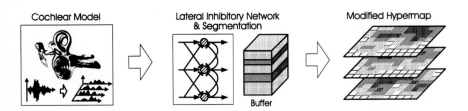

Fig. 1. Overall structure of the three-stage speech processing and recognition model

2 Cochlear Model

A digital time-domain model of the human cochlea similar to the cochlear model of Kates [4] is utilized for modeling the processing of sound in the auditory periphery.

Natural sound processing in the cochlea is realized by the conversion of the sound pressure waveform into neural excitation patterns in the auditory nerve at many different places along the cochlear length dimension. The place-domain continuous cochlea is modeled as 112 discrete sections covering a frequency range of 100 Hz to 16 kHz. Each section consists of the following main structural and functional components:

- the mechanical motion of the cochlea,
- the mechanical-to-neural signal transduction and
- the dynamic range compression concerning cochlear input/output relationships and the adaptive adjustment of the mechanical filter behaviour.

A block diagram for one section of the complete model is shown in Fig.2.

There are three aspects of the **mechanical motion** in the cochlear model (Kates [4]):

- the propagation of sound pressure traveling-waves on the cochlear partition (see Fig. 3) being represented as a cascade of discrete resonant filter sections $T_k(z)$,
- the transformation of the traveling-wave pressure to basilar membrane velocity at a particular cochlear location by the high pass velocity transformation filters $V_k(z)$ and

– the additional mechanical filtering of the velocity output at a particular
location in the cochlea by the second filters $S_k(z)$.

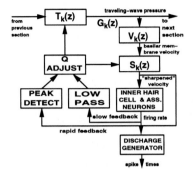

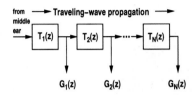

Fig. 2. Block diagram of the k.
section of the complete cochlear
model.

Fig. 3. Traveling-wave propagation in the overall cochlear model
realized as a cascade of digital resonant filter sections $T_k(z)$.

The **mechanical-to-neural signal transduction system** at a particular
cochlear location whose input is the velocity output from the second filter, is
realized by a complex of an inner hair cell and four attached neurons and their
fibres. This complex converts the cochlear partition's motion into a neural firing
rate. A nonhomogeneous Poisson process based discharge generator is used for
the determination of spike times from the neural firing rate.

There is a **dynamic range compression** with a ratio of 2.5:1 concerning
the cochlear input/output relationships under steady-state conditions integrated
into the model. The dynamic range compression mechanism is combined with a
feedback system which consists of a rapid and a slow feedback path.

3 Spatial and Temporal Processing of the Cochlear Model Response

An important function of the auditory system is to discriminate and recognize
complex sounds based on their spectral composition (Shamma [2]). Because of
its complex structure, a speech evoked response pattern of the cochlear model is little suitable as a direct input for learning and classification algorithms.
A preprocessing with neural filter networks is required to detect and highlight
characteristic properties like peaks and edges in the spatio-temporal signal. For
this purpose we implemented the LIN model of Shamma [2] using one of the
fundamental neural network topologies, the lateral inhibition. This topology is
assumed to be involved in auditory sensory reception performed by various nuclei, e.g. the cochlear nucleus. Two types of lateral inhibitory networks are distinguished: a nonrecurrent network with a simple feed-forward architecture (LIN.I),

and a network with recurrent inhibition (LIN.II). Fig.4. illustrates the typical inhibitory and excitatory interconnections of the model.

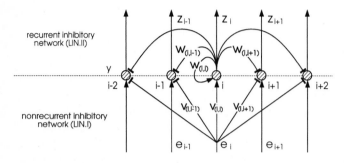

recurrent inhibitory
network (LIN.II)

nonrecurrent inhibitory
network (LIN.I)

Two types of lateral inhibitory interconnections, implemented in the filter stages of the LIN-model with the inhibitory weights V in the nonrecurrent and W in the recurrent network.

Fig. 4. Schematic of a lateral inhibitory network

Both networks are simulated as separate filter stages whose inhibitory weights cause a typical high pass behaviour. The LIN.I filter already detects spectral peaks which represent perceptually significant features of the stimuli such as formants of voiced speech. The recurrent network is used for further sharpening of these features.

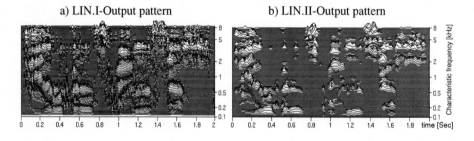

a) LIN.I-Output pattern   b) LIN.II-Output pattern

Fig. 5. Output patterns of LIN.I and LIN.II for the German word /Einundzwanzig/

In our simulations, the sample rate of the auditory nerve response patterns is reduced to ≈ 200 samples/sec at the output of LIN.I. Generally, it is possible to simulate the model with a higher resolution, but this is computationally expensive because of the complicated calculations in the LIN.II filter stage. The sample rate reduction is also based on the physiological observation that phase-locking deteriorates significantly in the response of central auditory neurons, which encode only averaged outputs (Shamma [2]), such as shown in Fig.5. The results of the inhibitory filter processes depend on the spectral content and sound intensity of speech stimuli and allow the segmentation of the LIN.II output.

4 Classification by means of an Artificial Neural Network

The third stage of our model for speech processing and recognition is a structured multi-layer formal neural network realized as a modification of the Hypermap Architecture (Brückner et al. [3]). Unlike the Hypermap Architecture, which was introduced by Kohonen in [5], we suggest a structure with an arbitrary number of levels in the input vector with each level containing the transformed input data. In [3] we described the learning of time-dependent data in form of time sequences. Each part of a time sequence is related to the corresponding level in the input vector which forms a time hierarchy.

In our experiments with speech the speech evoked LIN output is transformed into segments by a segmentation process in order to form the above mentioned time sequence. The input vector consists of a concatenation of these segments. Each segment represents a whole syllable or a part of a syllable like a phoneme. In this sense the whole time sequence represents one complete speech component, i.e. a word or a part of a sentence. The different levels of the input vector are trained and form a hierarchy of encapsulated subsets which define different generalized stages of classification.

The learning algorithm as described in [3] has the following steps:

- Find a first level with a subset S_j of nodes with an error under a given threshold of that level. The sizes of the thresholds should be decreased according to the order of the levels to obtain encapsulated subsets S_j.
- Find the best match $\mathbf{m}_c$ for all nodes in the subset and adapt the weights accordingly.
- Test for a priori knowledge, i.e. in one level is a significant higher error in comparison with the other levels of this input vector, then an adaptive learning algorithm saves these trained states [3].

Classification is achieved by finding the best matching node for each level of the hierarchy and by determining the square mean error of matching. To protect the algorithm from initial disordering the a priori knowledge learning step should start after a given time t_0.

5 Discussion

First of all the digitized speech signals are preprocessed and converted into neural excitation patterns by an implementation of a cochlear model (Kates [4]). Subsequently the model of a lateral inhibitory network with a highpass filter behaviour highlights spectral edges of these patterns and suppresses all other activities. The edges, considered as significant features of the stimuli, are particularly stable with respect to sound level variations, despite of the limited dynamic range of the auditory nerve fibres. Consequently, the discontinuities in the complex overall response texture are largely preserved after the LIN filter process (Shamma [2]). The segmentation process following is a filter algorithm indicating the segments' boundaries. This algorithm is independent of the Hypermap Architecture. The trained Hypermap Network is then able to classify speech components from continuous speech signals spoken by different subjects (Fig.6).

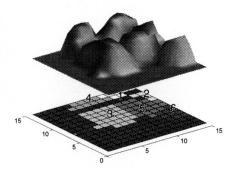

Fig. 6. The global error surface (4. level) after learning of 6 German words in form of sequences of segments is shown. The six words are the numbers 21, 32, 43, 54, 56 and 64 in this order. The network topology forms a single cluster for the initial segments of these words and within this cluster subclusters on the lower levels for the segments followed.

Taking into account the previous studies with conventional self-organizing maps, our presently available data reveal an improved classification [3]. However, there were some problems with "artifacts" in the training set, e.g., words with an unusual accentuation disordered the achieved classification.

With the new a priori knowledge learning algorithm we can solve this problem. First investigations confirm the expected results, thus we can go on optimizing parameters of the algorithm.

Acknowledgement

This work was supported by LSA grant 851A/0023.

References

1. B. Brückner and W. Zander, "Neurobiological modeling and structured neural networks", *Proc. Inter. Conf. Artificial Neural Networks*, Amsterdam, Sept. 13-16, 1993, pp. 43-46.
2. S. Shamma, "Spatial and Temporal Processing in Central Auditory Networks", *C. Koch and I. Segev (eds.): Methods in Neuronal Modeling*, The MIT Press, Cambridge, Massachusetts, pp. 247-289, 1989.
3. B. Brückner, T. Wesarg and C. Blumenstein, "Improvements of the modified Hypermap Architecture for Speech Recognition", *Proc. Inter. Conf. Neural Networks*, Perth, Australia, Nov.22-Dec.1, 1995, vol. 5, pp. 2891-2895.
4. J.M. Kates, "A time-domain digital cochlear model", *IEEE Transactions on Signal Processing*, vol. 39, no. 12, pp. 2573-2592, December 1991.
5. Teuvo Kohonen, "The hypermap architecture", In: T. Kohonen, K. Mäkisara, O. Simula, and J. Kangas, editors, *Artificial Neural Networks*, pp. 1357–1360, Helsinki, 1991. Elsevier Science Publishers.

Binding – A Proposed Experiment and a Model

Jochen Triesch and Christoph von der Malsburg

Institut für Neuroinformatik, Ruhr-Universität Bochum, Germany

Abstract. The binding problem is regarded as one of today's key questions about brain function. Several solutions have been proposed, yet the issue is still controversial. The goal of this article is twofold. Firstly, we propose a new experimental paradigm requiring feature binding, the *"delayed binding response task"*. Secondly, we propose a binding mechanism employing fast reversible synaptic plasticity to express the binding between concepts. We discuss the experimental predictions of our model for the delayed binding response task.

1 Introduction

What is the binding problem? If several objects have to be represented simultaneously in our brains, superposition of the activity patterns may lead to confusion. This effect is called the *"superposition catastrophe"* and the general problem behind it is called the *"binding problem"*. Several solutions have been proposed, only two of which shall be mentioned here. For a recent review of theoretical ideas the reader may refer to [vdM95].

Combination coding cells. The seemingly simplest approach is to introduce *combination coding cells*, which only react to combinations of features like e.g. an object of a particular shape and color at a particular retinal position. The problem with this approach is that it quickly leads to a combinatorial explosion of the number of needed cells. Furthermore, how should cells with these specificities be created? If they were to be learned, many examples of objects of all shapes and colors in all locations would have to be given before the system would work. If they were prewired, most of the connectivity pattern of the network would have to be stored in the genes, an unlikely proposition.

Binding by synchronous activity. This idea has received much attention recently. The signals of neurons representing features of the same object are mutually correlated in time, the signals of neurons representing features of different objects are not correlated or anti-correlated [vdM81]. The binding between cells referring to the same object is expressed in the signal correlations. For a review of experimental data pointing in this direction see [Sin93]. One special version of temporal binding is based on attentional mechanisms [TG80]: The responses to all but one stimulus are suppressed — only the units referring to one object are active. One way of implementing this uses inhibiton between units, although according to this proposal disambiguation is paid for by the restraint that only one stimulus may be represented at any time.

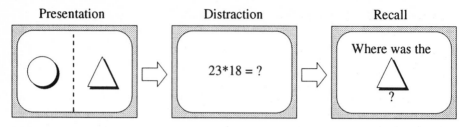

Fig. 1. The delayed binding response task. During the presentation phase the subject is asked to memorize binding relations between concepts, here place on the screen and shape. During a distraction phase the subject has to keep the binding relations (circle–left, triangle–right) in memory. Finally, the subject is asked to report a binding, by complementing a concept given as a cue (in the example the triangular shape is to be complemented by positional information).

2 The delayed binding response task

What would an experiment have to look like in order to test possible binding mechanisms? We propose a *delayed binding response task*, where subjects have to flexibly bind concepts and have to keep the bindings in memory during a distraction task before reporting them.

Behavioural task. The subject (human subject or non-human primate) is located in front of a computer screen, where visual objects are displayed in two different locations (see figure 1). For example the two objects may be a circle in the left half of the display and a triangle in the right half. In this example we have four concepts which have to be bound together: *circle* and *left* as well as *triangle* and *right*. After the subject has memorized the concepts and their binding relations, the stimuli are switched off and the subject performs a distraction task, which prevents him from rehearsing the concepts and their binding relations. After a few seconds one of the (four) concepts is given as a cue, and the subject has to indicate the concept that was bound to it during the presentation phase. If for example, the subject is cued with the triangle he is to indicate that it had been on the right-hand side of the stimulus display (e.g., by moving a joystick to the right).

It is important that the animal is trained (or the human subject instructed) for flexibility in the nature of the concepts and bindings implicit in the stimuli, using pairings such as shape-color, shape-motion, shape-shape, or even auditory or tactile stimuli instead of just shape-position as in our example — the common aspect of all tasks being that a binding is probed by just one of the concepts and the subject is to indicate the other. This is to keep the subject from over-learning the task and forming combination-coding cells specific to it. In the electrophysiological experiments proposed below this flexibility furthermore gives the possibility to select as current stimuli the particular response properties of neurons that the electrodes happen to have found. Our proposed paradigm is similar in many respects to the classical delayed response tasks [FBGR89] [Fus90], although requiring flexibility.

3   A model of binding

An overview of our model is given in figure 2. The model is composed of "units," pools of neurons that respond to one of the stimulus features, or "concepts." Individual neurons in a unit could be concentrated in cortical columns or could be dispersed over several areae, and presumably they are integrated by internal excitatory connections. The units can be activated by sensory stimulation or by queries sent from a center organizing behavior during performance of the experiment (not shown), the center being formed during training or verbal instruction. Units are permanently connected by synaptic links that reflect their logical relationship (mutually exclusive or potentially combined in a single stimulus).

When several stimuli are presented simultaneously, unit activity fluctuates. We interpret this fluctuation as the expression of attention that sequentially focuses on one stimulus at a time. Due to this mechanism, all concepts belonging to a single object or stimulus are activated simultaneously, while concepts belonging to other objects are suppressed. This assumption is supported by many neurophysiological and neuropsychological studies (see, e.g., [MD85] or [Luc94] for a recent review). We do not want to settle on a definite implementation of the attentional mechanism but it seems likely that lateral inhibition in cortical and subcortical structures might be important [Cri84]. At least part of this lateral inhibition may correspond to the inhibitory connections, shown in figure 2, which implement mutual exclusiveness of concepts.

The shifting activity of selective attention realizes temporal binding in our model. In the absence of eye movements we expect for these signals typical switching times in the range of a few tens of milliseconds (for a discussion of time spans typical for covert attention see [TG80]. Other binding paradigms, such as figure-ground separation, may be based on shorter time constants.)

The correlations in activity fluctuations control rapid reversible synaptic plasticity. The existence of fast synaptic plasticity mechanisms has been known for more than thirty years, for a review see [Zuc89]. Theoretical reasons for the importance of fast synaptic mechanisms for information processing in the brain have been emphasized in [vdM81], where a control mechanism is postulated according to which the efficacy of a synapse is rapidly increased if the pre- and postsynaptic neurons fire simultaneously and is rapidly decreased if activity is present but uncorrelated, whereas once signals cease the synapse slowly decays back to a base level, with a time-constant of the order of magnitude of the time span of short-term memory.

In our model for the delayed binding response task the coactivated units belonging to one object quickly strengthen connections between each other during stimulus presentation, which is the mechanism for short-term memory of binding. Later, when (part of) the concepts connected with an object are reactivated the still enhanced connections can reactivate all the concepts belonging to the object, hence giving access to all the information that was bound together.

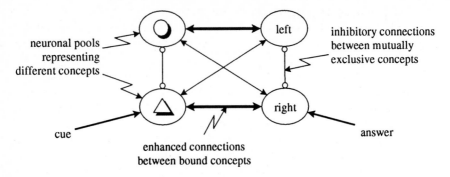

Fig. 2. Binding by enhanced connections between units. The unit pairs circle–left and triangle–right are bound by temporarily enhanced connections between them. Units for mutually exclusive concepts are connected by inhibitory connections, which prevent them from becoming active simultaneously.

Here is how the model would behave in the concrete delayed binding response task illustrated in figure 2. During *presentation* the attentional mechanism would activate either "circle" and "left" or "triangle" and "right" at a time, expressing binding between the concepts. During the times when, e.g., the units representing "circle" and "left" are highly active the synapses between them are strengthened, storing the binding between those concepts. In the *distraction* period the stimuli are switched off and the subject is prevented from rehearsing the concepts and their binding relations. During this time there is no activity in the four neural units (unless the distraction task makes use of their concepts) but the synapses between bound units do remain enhanced.

During *recall*, finally, a triangle is displayed in a central position as a cue and the subject now has to indicate that the triangle had been on the right during the stimulus presentation. With the help of the enhanced connection, the "triangle" unit will now activate the "right" pool, which additionally surpresses the "left" unit. Firing of the "right" unit can easily be imagined to trigger the correct (verbal or other) response to the task.

Note that the presentation of the triangle in a central position will generally also trigger the activiation of "middle" cells, which compete with the desired "right cells". Thus, current stimuli and recalled memories may interfere. We assume that during training of the task or instruction the subject learns to actively surpress such interferences.

4 Experimental Predictions

The model predicts *signal correlations*. Those units that correspond to different concepts present in a given stimulus element fire in a correlated way, whereas units that correspond to concepts belonging to different elements (if present in the stimulus) fire with no correlation or with anticorrelation.

What would be an appropriate recording technique to find and monitor relevant unit activity? As the temporal resolution characterizing the signal correlations to be found is unknown (although we guessed it would be dozens of milliseconds), it would be desirable to record activity with a resolution of a few milliseconds. Candidate techniques could be microelectrode recordings, optical imaging using voltage sensitive dyes, EEG or MEG. We suggest that the spatial resolution should be good enough to pick up signals specific for stimulus features (concepts), perhaps of the order of the diameter of a functional cortical column, between 0.3 and 0.6 mm. Thus the spatial resolution of non-invasive techniques might be inadequate, excluding human subjects from the proposed study.

With microelectrode recordings as an example, a concrete experiment could look like this: While the subject performs the task, simultaneous recordings are made from two neurons selective for different object shapes and from two neurons responsive to different object locations. Suitable shape specific neurons of sufficient complexity may be found in the inferior temporal cortex (IT), and their stimulus specificity may be discovered using a simplified version of the techniques described by K. Tanaka [Tan93]. Neurons selective for different locations in space but relatively unsensitive to stimulus shape could be found in parietal cortex [UH94]. Of course, this is only an example and there are many other ways for choosing stimulus classes and corresponding neurons. However, it is important that the concept pairs (e.g., shape and location) should be chosen such that they are unlikely to be represented by combination coding cells (if they existed then there would be no binding problem). A. Treisman's [TG80] paradigm could be used to test for the existence of combination-coding cells, by showing that the bindings in question are not subject to pop-out but require time proportional to the number of distractors.

A second prediction of the model is *rapid reversible synaptic plasticity*. After memorization during the presentation phase EPSP size of at least some synaptic connections between bound neural units should be increased, while it should remain small or even be diminished between neural units not being bound. To validate this prediction, intracellular recordings from target neurons would be required and individual EPSPs should be discernable (increased compound EPSPs could also be the result of an increased number of active presynaptic neurons). The relevant EPSPs should be identifiable by the correlation of their timing with signals in another unit involved in the task. We fear, however, that a direct test of the prediction rapid reversible synaptic plasticity in the delayed binding response task may be beyond current experimental techniques.

5 Discussion

The temporal binding hypothesis fails to be universally accepted, although the binding problem is widely discussed and recognized as such, and although relevant signal correlations have been found experimentally (as reviewed in [Sin93]). The lingering doubt is fed from two sources, the suspicion that those signal correlations are an insignificant epiphenomenon, and the suspicion that perhaps the brain manages to circumvent the binding problem altogether with the help of combination-coding cells, in spite of all the conceptual difficulties mentioned.

The model we are proposing as a neural basis of the experimental paradigm is minimal and in essence just consists in hypothesizing the presence of appropriate signal fluctuations and rapid reversible synaptic plasticity. Only on the basis of this utmost simplicity and ubiquitous availability of the binding machinery the flexibility can be reached that the experimental paradigm requires. Our model posits rapid reversible synaptic plasticity as basis for short-term memory. This is to be seen in contrast to the idea of reverberating neural activity, as demonstrated in the classical delayed response task [FBGR89] [Fus90]. We feel that neural reverberation is not a viable short-term memory mechanism in situations where flexibility is required.

We recognize that the experiment we propose pushes the state of the art to its limits and that it would constitute an enormous investment from the side of the experimenter, but we feel that an equally enormous scientific reward could make the effort worth while.

References

[Cri84]   F. Crick. Function of the thalamic reticular complex: the searchlight hypothesis. *Proc. of the National Academy of Sciences*, 81:4586–4590, 1984.

[FBGR89] S. Funahashi, C. J. Bruce, and P. S. Goldman-Rakic. Mnemonic coding of visual space in the monkey's dorsolateral prefrontal cortex. *Journal of Neurophysiology*, 61(2):331–349, 1989.

[Fus90]   J. M. Fuster. Inferotemporal units in selective visual attention and short-term memory. *Journal of Neurophysiology*, 64(3):681–697, 1990.

[Luc94]   S. J. Luck. Cognitive and neural mechanisms of visual search. *Current Opinion in Neurobiology*, 4:183–188, 1994.

[MD85]   J. Moran and R. Desimone. Selective attention gates visual processing in the extrastriate cortex. *Science*, 229:782–784, 1985.

[Sin93]   W. Singer. Synchronization of cortical activity and its putative role in information processing and learning. *Annu. Rev. Physiol.*, 55:349–374, 1993.

[Tan93]   K. Tanaka. Neural mechanisms of object recognition. *Science*, 262:685–688, 1993.

[TG80]   A.M. Treisman and G. Gelade. A feature integration theory of attention. *Cognitive Psychology*, 12:97–136, 1980.

[UH94]   L. G. Ungerleider and J. V. Haxby. 'What' and 'where' in the human brain. *Current Opinion in Neurobiology*, 4:157–165, 1994.

[vdM81]   C. von der Malsburg. The correlation theory of brain function. Internal Report 81-2, Dept. of Neurobiology, Max-Planck-Institute for Biophysical Chemistry, Göttingen, Germany, 1981. Reprinted in: Models of Neural networks II, edited by E. Domany, J.L. van Hemmen, and K. Schulten (Springer, Berlin, 1994) Chapter 2, pp. 95–119.

[vdM95]   C. von der Malsburg. Binding in models of perception and brain function. *Current Opinion in Neurobiology*, 5:520–526, 1995.

[Zuc89]   R. S. Zucker. Short-term synaptic plasticity. *Ann. Rev. Neurosci.*, 12:13–31, 1989.

A Reduced Model for Dendritic Trees with Active Membrane

M. Ohme and A. Schierwagen

Institute for Computer Science, University of Leipzig, D-04107 Leipzig

Abstract. A non-uniform equivalent cable model of membrane voltage changes in branching active dendritic trees has been developed. A general branching condition is formulated, extending Rall's 3/2 power rule for passive trees.

1 Introduction

The basic processing units in current models of artificial neural networks (ANN) are mostly poor caricatures of real neurons, although various authors have often stressed the importance of dendritic signal processing for neuron function (see [2] for a short overview). One reason may be the lack of tools to analyze complex branched dendritic trees – only simulations are possible as yet. Especially for network studies reduced models are desirable.

A powerful tool for analyzing passive dendrites is the reduction to an equivalent cable [4] by means of the linear cable theory. Provided certain symmetry conditions are fulfilled, such a tree can be treated as a single cable, and for special cases of the underlying geometry analytical solutions can be derived [5].

However, for several important classes of vertebrate neurons, active ion channels have been positively demonstrated in the dendrites or at least strongly suggested [2; 6]. Also, the reduction to an equivalent cylinder could be helpful for describing signal flow in axonal arbors.

We will show here that the reduction to an equivalent cable is also applicable for trees with active membrane. Starting with the general cable model for dendritic segments we derive branching conditions for those trees which can be collapsed into only one single equivalent cable. These symmetry conditions turn out to be similar to the passive case.

2 Mathematical model of a dendritic tree

General cable model for dendritic segments. The usual mathematical description of the potential distribution in dendritic trees proceeds from the application of one-dimensional cable theory to the tree segments. Representing the cable by a RC ladder network, the cable equation for the transmembrane potential $V(x,t)$ and the axial current $j_a(x,t)$ in a single segment is as follows (x represents distance in axial direction):

$$j_a = -\frac{1}{r_a(x)}\frac{\partial V}{\partial x}, \qquad -\frac{\partial j_a}{\partial x} = j_m = j_c + j_i \tag{1}$$

where $j_m(x,t)$ denotes the membrane current consisting of a capacitive component j_c and a resistive one j_i.

We assume that the current j_i created by the ionic channels in the membrane can be split into a product of a "resting" conductance which depends on the membrane surface at x and a nonlinear potential function $f_0(x, V, u_1, \ldots, u_n)$ reflecting the threshold behavior of the voltage dependent channels as a specific membrane property (i.e. per unit membrane surface). The latter may be time depending so additional variables u_k defined by further first order differential equations have to be included:

$$j_c = c(x) \frac{\partial V}{\partial t} \qquad j_i = g(x) f_0(x, V, u_1, \ldots, u_n) \tag{2a}$$

$$\frac{\partial u_k}{\partial t} = f_k(x, V, u_1, \ldots, u_n) \quad \text{for } 1 \leq k \leq n. \tag{2b}$$

Combining equations (1) and (2a) we obtain for every segment

$$\frac{\partial}{\partial x} \left(\frac{1}{r_a(x)} \frac{\partial V}{\partial x} \right) = j_m = j_c + j_i = c(x) \frac{\partial V}{\partial t} + g(x) f_0(x, V, u_1, \ldots, u_n) \tag{3}$$

Applying the variable transform

$$T = \frac{t}{\tau} \qquad \text{with} \qquad \tau = \frac{c(x)}{g(x)} \qquad \text{and} \tag{4a}$$

$$X = \int_0^x \frac{ds}{\lambda(s)} \qquad \text{with} \qquad \lambda(x) = \frac{1}{\sqrt{r_a(x) g(x)}}, \tag{4b}$$

equation (3) can be rewritten in dimension-less coordinates

$$0 = \frac{\partial^2 V}{\partial X^2} + Q \frac{\partial V}{\partial X} - \frac{\partial V}{\partial T} - f_0(X, V, u_1, \ldots, u_n)$$

$$\frac{\partial u_k}{\partial T} = \tau f_k(X, V, u_1, \ldots, u_n) \quad \text{for } 1 \leq k \leq n \tag{5}$$

with $\qquad Q(X) = \frac{1}{2} \frac{\partial}{\partial X} \ln \left(\frac{g(X)}{r_a(X)} \right)$

The axial current j_a then is given by $\quad j_a(X, T) = -\sqrt{\frac{g(x)}{r_a(X)}} \frac{\partial V}{\partial X}$ $\tag{6}$

Representation of the neuronal tree structure. To represent a reconstructed dendritic tree, we use the graph theoretical concept of a directed, labeled tree. All branching points of the dendrite are then represented by the nodes of the tree with the soma as root point. Some (virtual) branching points have to be included for additional localized input into the dendrite. The edges of the tree correspond to the single dendritic segments.

Any edge π is labeled with the quantities necessary to set up the cable equation for the corresponding cable segment, i.e. length l_π, number n_π of auxiliary variables u_k, functions $r_{a\pi}$, g_π, c_π, f_π and $f_{\pi k}$. Note that the origin of the space variables x and X is the root node.

External current input can enter the tree only by the nodes. So every node K has to be labeled with a current function I_K which depends on the potential (and its time derivatives) at this site, and explicitly from the time. Two simple examples are (1) synaptical input modeled by a change of synaptic conductance and (2) a lumped soma modeled by a RC circuit:

$$I_K(t,V) = \begin{cases} 0 & \text{if } t < \Delta \\ \frac{g_{\max}\tau}{e}(t-\Delta)\exp\left(\frac{\Delta-t}{\tau}\right)(V_{\text{rev}}-V) \end{cases} \quad , \qquad I_K(V,V_t) = \frac{1}{R_K}V + C_K V_t$$

From Kirchhoff's first law (saying that the sum over all currents flowing into a node K vanishes), we yield

$$I_K + \sum_{\pi \in \mathcal{CS}(K)} j_{a_\pi}(X_K,t) - j_{a_{\mathcal{PS}(K)}}(X_K,t) = 0, \tag{7}$$

where $\mathcal{CS}(K)$ denotes the set of all child segments π with K as starting node and $\mathcal{PS}(K)$ stands for the (unique) parent segment ϖ with K as end node. X_K is the electrotonical distance of the the node K from the soma.

To find the potential and current distribution over a given neuronal tree to some external input functions I_K one has to solve equation (5) for all segments under the above boundary conditions (7).

3 Reduced model for an active dendrite

Obviously, the above problem is in general hard to solve. In most cases the only way to get some insight in possible reactions of the system lies in numerical simulations which often remain unsatisfactory because of too many free parameters. Thus any reduction of the original problem to one with less free parameters would be of great help. We will show that the reduction approach for passive dendrites [4] can be generalized to active trees, too. This can be formulated as

Theorem (equivalent cable). *There is a cable which is equivalent to the whole tree in the sense that every solution (potential and current distribution) for this equivalent cable leads to a solution for the tree provided some conditions given below are fulfilled. This means here that the voltage change in time at any point X_π in the tree is equal to the potential change of the equivalent cable at point X with the same electrotonic space coordinate. The sum of axial current flowing through all segments with equal electrotonical distance X from the root is equal to the axial current of the equivalent cable at X. The analog is true for the membrane current between two space coordinates X_1 and X_2.*

The equivalent cable will be described by the same cable equations (5) with parameters l_{eq}, $r_{a_{\text{eq}}}$, g_{eq}, c_{eq}, n_{eq}, f_{eq} and $f_{\text{eq}k}$. Then the conditions for reduction of a neuronal tree are the following:

time constant: $\tau(X)$ has to be constant and independent from the segments and equal to τ_{eq}.

equal length: All terminal nodes T have the same electrotonical distance L from the root S: $X_T = \int_0^{x_T} \frac{dx}{\lambda(x_\pi)} = L = L_{\text{eq}}$.

equal cable model: The functions $f_{\pi i}$ are the same for any segment: $f_{\pi i}(X,V,u_1,\dots,u_n) = f_{\text{eq}i}(X,V,u_1,\dots,u_n)$.

similar cable geometry (generalized $\frac{3}{2}$ branching rule): The same has to be true for the geometrical parameter $Q_\pi(X) = Q_{\text{eq}}(X)$, i.e. there are constants

C_π for every segment π with

$$\sqrt{g_\pi(X)/r_{a\pi}(X)} = C_\pi \sqrt{g_{eq}(X)/r_{a eq}(X)}. \tag{8}$$

At the nodes of the tree an additional condition must be satisfied: the constant C_π belonging to a parent segment π (with $C_{PS(S)} := 1$ for the nonexistent parent segment of the root) has to be equal to the sum of the C_ϖ which belong to all child segments ϖ of π:[1]

$$C_\pi = \sum_{\varpi \in CS(\pi)} C_\varpi \tag{9}$$

similar boundary conditions: External input into the tree has to be symmetrically divided among all places with the same electrotonical distance from the root. That is, for two nodes K and L with the same electrotonical distance $X_K = X_L$ away from the root S the attached current functions have to fulfill:

$$I_K/C_{PS(K)} = I_L/C_{PS(L)}. \tag{10}$$

The input current I_X^{eq} into the equivalent cable at the same electrotonical distance $X = X_K$ is defined by the sum of all inputs at equal electrotonic distance X:

$$I_X^{eq} := \sum_{X_L = X} I_L. \tag{11}$$

Proof of theorem: First we state that any general potential solution of the equivalent cable yields a general solution for all segments $\pi = (K, L)$ – limited to the corresponding space interval $[X_K, X_L]$: $\forall T \geq 0 : V_\pi(X, T) = V(X, T)$. This follows from the assumption that the relevant parameters in the normalized cable equations (5) are equal for all tree segments.

To verify that a potential distribution over the tree which yields from a special solution of the equivalent cable is also a special solution of the tree one has to compute the axial currents in all segments and to prove that the boundary conditions (7) are satisfied.

It follows from the generalized $\frac{3}{2}$ branching rule that for any electrotonical distance X the sum over all C_π is equal to 1 [3].

With this one gets for the input current relation at X with $\pi = PS(K)$:

$$I_K = C_\pi I_X^{eq}, \tag{12}$$

resulting from the definition (11) of I_X^{eq} and the condition (10). The axial current in segment π (see (6)) is computed by:

$$j_{a\pi}(X, T) = -\sqrt{\frac{g_{eq}(x)}{r_{a eq}(X)}} \frac{\partial V}{\partial X} \overset{(8)}{=} -C_\pi \sqrt{\frac{g(X)}{r_a(X)}} \frac{\partial V}{\partial X} = C_\pi j_{a eq}(X, T) \tag{13}$$

Inserting (12) and (13), the boundary condition (7) yields ($\pi = PS(K)$):

---

[1] In the case of a tree with cylindrical segments, i.e. $g_\pi(X)/r_{a\pi}(X) \propto d_\pi^3$, and constant specific cell parameters R_i, R_m and C_m one gets Rall's $\frac{3}{2}$ branching rule[4].

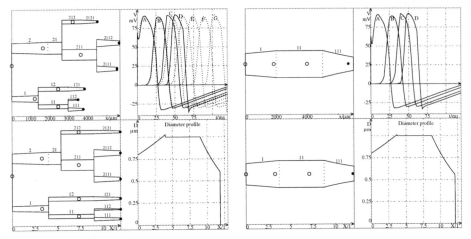

Fig. 1. Comparison of active excitation propagation in a branched neuronal structure (left) and in the corresponding equivalent cable (right). For each case it is shown a) the tree as a dendrogram (left half) in the original (top) and in the normalized space (bottom); b) the profile of the diameter (bottom right) and c) the results of the simulation (top right).

$$I_K + \sum_{\varpi \in CS(K)} j_{a_\varpi}(X_K, T) - j_{a_\pi}(X_K, T) = C_\pi I_X^{eq} + \sum_{\varpi \in CS(K)} C_\varpi j_{a_{eq}}(X+0, T) - C_\pi j_{a_{eq}}(X-0, T)$$

$$\overset{(9)}{=} C_\pi \left(I_X^{eq} + j_{a_{eq}}(X+0, T) - C_\pi j_{a_{eq}}(X-0, T) \right) = 0$$

where ± 0 denotes the left and right limit respectively. So the last term in brackets is the first Kirchhoff's law for the equivalent cable at input site X which vanishes according to our premise. This completes the proof.

4   Example

To illustrate the application of this approach we studied the signal flow in a neuronal structure and its equivalent cable by compartmental simulation (Fig. 1). All segments are described by the FitzHugh-Nagumo system [1], which has one ($n = 1$) auxiliary variable u and a cubic current-voltage relation f in (3). The electrical parameters r_a, g and c result from the standard equations for a nonuniform cable of circular cross-section, here with fixed, slowly increasing (near the soma) or decreasing (at the terminals) exponential diameter function, i.e. $Q(X)$ in (5) is constant in all cases.

Essential for the existence of an equivalent cable are the equal electrotonical distance of the terminal nodes, the synaptical inputs (filled circles in Fig. 1) and the changes of the equation types (here only the segment geometry – represented by $Q(X)$ – changes from segment to segment) from the soma. The realization of this requirements can easily be seen by the representation of the tree in the normalized space (Fig. 1, bottom left).

The cell receives synaptical input (filled circles in the dendrogram) near the dendritic tips (into segments **111, 112, 121, 2111, 2112** and **2121**) with strength proportional to the diameter at the corresponding place to the power $\frac{3}{2}$ (see equations (8) and (10)). The "recording sites" are the soma and some intermediate points (open circles in the dendrogram). In the right top parts of Fig. 1 the resulting signals are shown. The solid lines stand for the potential function at the recording sites (A corresponds to the input sites, D to the soma, B and C combine all the intermediate recording sites). The differences of the potential functions at different recording sites with equal electrotonical distances to the soma are to small to see them in the plot. The same is true for differences between the full tree (left) and the equivalent cable model (right).

If we restrict the input into the cell to only one injection side (synapse on segment 111), we get an excitation propagation to the soma (dashed lines A, B, C and D), similar to the first case, but at every branching point an impulse is send in the hitherto non-excited branches. This leads also to a small time delay for the somatic impulse. In Fig. 1 the potential is drawn only for some recording sites: A corresponds again to the injection site, B to the recording site in segment 11, C corresponding in 1, D at the soma, E in 2, F in 211 and 212 as well as G in 2111, 2112 and 2121. Both records F at 211 and 212 and G at 2111, 2112 and 2121) are not distinguishable in the plot because the subtree of segment 2 can be modeled again by an equivalent cable.

5 Conclusions

The aim of this study was to contribute to the problem of building reduced models. In the literature, Rall's equivalent cable model has been also employed for active trees, however, without strict justification. As an example we mention Mel's clusteron model which consists of a soma and a single dendritic segment. There has been shown that this model with Hebb-type learning rules is able to effectively perform memory tasks [2].

To provide theoretical justification for such reduced models, we developed an "equivalent cable" model for active dendrites, comprising Rall's model for passive trees as a special case. The reduction process has been demonstrated by a simulation of an arbitrarily constructed neuronal tree the segments of which were modeled by the FitzHugh-Nagumo system.

In this way, the analysis and simulation of potential spread in rather realistically modeled neurons can be distinctly simplified. It could therefore be helpful in closing the gap between ANN models and biological neural structures.

Acknowledgements

Supported by Deutsche Forschungsgemeinschaft, Grant No. SCHI 333/4-1.

References

1. R. FitzHugh. Mathematical models of excitation and propagation in nerve. In H. P. Schwan, editor, *Biological Engineering*, pages 1–85. McGraw-Hill: New York, 1969.
2. Bartlett W. Mel. Information processing in dendritic trees. *Neural Computation*, 6, 1994.
3. Michael Ohme. *Modellierung der neuronalen Signalverarbeitung mittels kontinuierlicher Kabelmodelle*. Dissertation, University of Leipzig, Department of Computer Science, 9 1995.
4. W. Rall. Theory of physiological properties of dendrites. *Ann. N. Y. Acad. Sci.*, 96, 1962.
5. A. K. Schierwagen. A non-uniform equivalent cable model of membrane voltage changes in a passive dentritic tree. *J. theor. Biol.*, 141:159–179, 1989.
6. I. Segev. Dendritic processing. In Michael A. Arbib, editor, *The handbook of brain theory and neural networks*. MIT Press, 1995.

Stabilizing Competitive Learning during On-line Training with an Anti-Hebbian Weight Modulation

Stephane Tavitian[1,2], Tibor Fomin[1], and András Lőrincz[1]

[1] Department of Adaptive Systems, Institute of Isotopes,
The Hungarian Academy of Sciences
Budapest, P.O.Box 77, Hungary H-1525
[2] INRIA Sophia-Antipolis
BP 93 - 06902 Sophia-Antipolis Cedex, France

Abstract. Competitive learning algorithms are statistically driven schemes requiring that the training samples are both representative and randomly ordered. Within the frame of self-organization, the latter condition appears as a paradoxical unrealistic assumption about the temporal structure of the environment. In this paper, the resulting vulnerability to continuously changing inputs is illustrated in the case of a simple space discretization task. A biologically motivated local anti-Hebbian modulation of the Hebbian weights is introduced, and successfully used to stabilize this network under real-time-like conditions.

1 Introduction

The full inter-layer connectivity of competitive networks (CN) allows any single correlation in the input to be captured, but it does not ensure that they will account for the emergence of a representation. It is also crucial to let competition modulate the parallel adaptation of the Hebbian weights. This homogeneous development is generally ensured by breaking any temporal correlation within the input space, through the randomization of the order of training samples. This random training order is inconsistent with the usually slow and continuous real-time sensing of the environment.

During on-line training with low rate input changes, one single input may bias competition by giving an advantage to the winner on subsequent similar inputs. Even if the units are properly sharing the activation time, the slow coverage of the space can compromise the representation of the full input set by favoring the most recent ones. On-line CNs have therefore to provide an embedded mechanism to ensure the homogeneity of their weight development. In order to prevent the catastrophic effect of "focused learning", McClelland et al. [8] proposed to introduce a complementary reinstatement structure inspired by a biological functional interpretation of the hippocampus. By intermixing previously learnt patterns with new incoming ones, this module actually reproduces the random training conditions. ART2 [1] networks are also made robust to temporal order by means of their knowledge, according to an intra-categorical criterion and leaving the adjustment of the inter-categorical distances to an adaptive *vigilance* parameter.

The present study addresses the formation of a fixed number of categories through a continuous exploration of a relatively stable environment. In the following, the vulnerability of a biologically plausible competitive network [2] to continuous low rate changes is shown in a minimal task. Section 3 introduces a local anti-Hebbian modulation rule which increases the robustness of this network by limiting indirectly its plasticity in the lower rate region. Relations with similar mechanisms, and biological plausibility are discussed in section 4.

2 Vulnerability of Competitive Learning to Real-Time

2.1 Anti-Hebbian Competition.

The simulations were performed with a single layer of non-linear Hebbian units inhibiting each-other through symmetrical lateral anti-Hebbian connections. This competitive scheme keeps the conceptual advantage of a *local* mechanism, which motivated the original lateral-inhibition based networks, without being pre-wired according to any presumed topology of the space, nor adjusted to the number of competitors.

Let q_{ij} and w_{ik} denote the weights connecting the ith output unit respectively to the jth input unit and to the kth output unit. The dynamics of the response y_i of unit i to the normalized input vector $\underline{x}$, is a non-linear function of the feed-forward activation and the lateral inhibitory feedback from competitors:

$$\tau \dot{y}_i = -y_i + \sigma \left(\sum_j q_{ij} x_j - \sum_{i \neq k} w_{ik} y_k - \theta \right) \tag{1}$$

where σ is a sigmoidal function centered on θ.

The initial values of the feed-forward weights are randomly distributed between zero and one, whereas the lateral connections are initialized to zero. At each learning step, the input and the lateral weights are updated according to the following Hebbian rules:

$$\tau_q \dot{q}_{ij} = \beta x_j y_i - q_{ij} y_i \tag{2}$$

$$\tau_w \dot{w}_{ij} = y_i y_j (\alpha - w_{ij}) \tag{3}$$

being slight modifications of the original learning rules [2]. In order to dissociate effects of different origins, no adaptation of θ was applied.

2.2 A Simple Task in a Minimal Environment.

This network was trained under the minimum conditions allowing to exhibit self-organizing abilities. The environment was a one-dimensional toroidal space covered by an array of thirty equidistant sensors, the inter-sensor distance defining the spatial unit. This space was uniformly populated by normalized Gaussian patterns of identical shapes. The input vector coding the pattern at position p was computed as follows:

$$x_i(p) = exp\left(-\frac{d(p,i)^2}{a^2}\right) / \sum_i x_i(p) \qquad (4)$$

where $d(p,i)$ stands for the Euclidean distance between the real position p and the discrete position of sensor i ($a = 1.25$).

In the case of a uniform density of probability, there is no clue for pattern cluster separation. Therefore, as far as stability is concerned, this apparently simple environment still belongs to the worsts. A classical CN represents such a distribution by partitioning the topological space into equal regions, overlapping each-other as a function of the competition strength. Provided the weights were updated with relaxed values of the y_is, this space discretization was perfectly achieved by this particular network in the random training case ($\theta = 0.4$, $\alpha = 0.5$, $\tau_w = 700$, $\beta = 8$, $\tau_q = 2000$, $t = 300000$). All the receptive fields (RF) were identically shaped and regularly positioned all over the sensory space.

2.3 Slow Motion Induces Instability.

The same input set was then presented through continuous transformations, by moving the same Gaussian pattern across the toroidal space with different low velocities. The velocity range together with the τ value were such that the network followed the input adiabatically. Nevertheless, after training, less regular sets of asymmetric RFs were observed as the velocity decreased below 10 τ_q^{-1} (see Fig 1.a). Within a narrow range of intermediate speeds ($[1.26\ \tau_q^{-1}, 3.0\ \tau_q^{-1}]$), all the RFs started to follow the input, and to spread only over the recently scanned area (see Fig 1.b). Below velocities of 1.0 τ_q^{-1}, only fewer and fewer output units could play a role (see Fig 1.c).

3 Proposed Solution

3.1 Synaptic Modulation.

In order to prevent the sustained synaptic activation to induce a massive growth of a small fraction of the weights, this growth shall be bounded. But simply punishing the winner, or hiding the recently learnt portion of the space, would exclude a subpopulation of output units or input sensors from the competition during an arbitrary period. We propose an intermediate option, whereby a winner is only punished by being hidden the portion of the space it has learnt recently. One local implementation of this scheme is a temporary suppression of the weights following a sustained signal, thus altering the output responses while leaving the knowledge intact. Consequently, we shall distinguish the *learning* weight q_{ij}^0 of a connection, obeying Hebbian rule 2, and its instantaneous factor of transmission q_{ij} that filters the sensory input (equation 1):

$$\frac{dq_{ij}}{dt} = -k_1 y_i x_j q_{ij} + k_2 (q_{ij}^0 - q_{ij}) \tag{5}$$

The first anti-Hebbian term tends to suppress q_{ij} proportionally to the recent correlation of pre- and post-synaptic activities, whereas the second term tends to recover the learnt strength q_{ij}^0. The factors k_1 and k_2 allow to adjust the dynamics and the strength of the suppression.

This rule limits the response of a unit to the same input by time, thus preventing over-learning, without altering the responsiveness to novel patterns. During the suppression process, the same input may be won by other units. Stationary input leads to a silent network.

3.2 Results

The robustness of the network was increased by this mechanism. The random training performances were recovered for velocities as low as $0.8\ \tau_q^{-1}$ (see Fig 1.d), although no other parameters than k_1 and k_2 were optimized. A reasonable space discretization could still be achieved at the velocity of $0.2\ \tau_q^{-1}$. For slower motions, either all the receptive fields were the same and followed the input, or no particular RF structure could emerge at all.

4 Discussion

Related methods. Short-term synaptic plasticity has been used in several works, mainly as an alternative to recurrent connections for the simulation of short-term memory. The proposed cooperation scheme between short and long-term memory is usually based on two separate categories of units. In this paper, an alternative scheme is investigated: short and long-term memory are merely two properties of the same synapses. Short-term negative traces are selectively desensitizing the network to recently seen features. From this point of view, this mechanism could be used as a context-sensitive version of the Kohonen's novelty-filter [3], whose advantage is to leave the long-term knowledge intact.

Generalization. Constant motions have been used in order to quantify the results. Further simulations using more realistic random trajectories that support the efficiency of the approach will be published elsewhere. The stabilizing effect of the anti-Hebbian modulation can easily be generalized to different transformations as well, including scaling or distortion, provided that the related dimension, i.e. size or shape, is coded by the network.

Biological plausibility. Short-term habituation is a well known synaptic phenomenon in neuro-pharmacology [6], and a decade of electro-physiological studies in the infero-temporal cortex revealed that it could influence more than just motor reflexes. Cells were recorded in this high-level visual area, whose responsiveness are selectively but temporarily altered for recently seen stimuli [7]. This effect of repetition was shown to occur between different stimuli as well, proportionally to their similarity [4]. If we assume that this similarity can be assessed at the neural level, by the proportion of recently used synapses among the newly activated ones, then a short-term synaptic depression appears as the simplest explanation of this phenomenon. This local implementation is also consistent with its passive nature, recently demonstrated by Miller [5]. Furthermore, if this temporary suppression has such a stabilizing role as described in this paper, then the need for it would be increased by the highly invariant responses that characterizes high-level visual areas such as the infero-temporal cortex.

5   Conclusion

We have suggested a short-term weight suppression mechanism that extends the use of competitive networks to a wider range of real-time training conditions. It has been speculated that the suggested suppression may be important when learning invariances. We raised the question if the particular short-term response properties of infero-temporal neurons are governed by such a mechanism.

Acknowledgments This work has been partially supported by a European Community HCM fellowship, and OTKA grants # T014330 and T014566. The first author would like to thank Pr. Guy Orban, Pr. Rufin Vogels, and Dr. Marc Van Hulle from the Laboratory of Neuro- and Psychophysiology of the K.U. Leuven (Belgium) for many useful discussions.

References

1. Carpenter, G.A., Grossberg, S.: ART 2: self-organization of stable category recognition codes for analog input patterns. Applied Optics, **26** (1987) 4919–4930
2. Földiák, P.: Forming sparse representations by local anti-Hebbian learning. Biol. Cyber., **64**:2 (1990) 165–170
3. Kohonen, T.: Self Organization and Associative Memory. Springer Verlag, Berlin (1984)

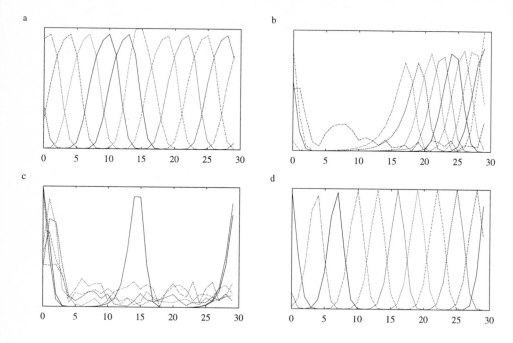

Fig. 1. Receptive fields under different training conditions. **a**: velocity $= 8 \tau_q^{-1}$. **b**: velocity $= 2 \tau_q^{-1}$. **c**: velocity $= 1 \tau_q^{-1}$, "winner-win-all". **d**: velocity $= 0.8 \tau_q^{-1}$, with the suppression mechanism ($k_1 = 0.0005$, $k_2 = 0.00001$).

4. Miller, E.K., Li, L., Desimone, R.: Activity of neurons in anterior inferior temporal cortex during a short-term memory task. J. of Neuroscience **13**:4 (1993) 1460–1478
5. Mille, E.K., Desimone, D.: Parallel neuronal mechanisms for short-term memory Science **263** (1994) 520–522
6. Zucker, R.S.: Short-term synaptic plasticity. Ann. Rev. Neurosci. **12** (1989) 13–31
7. Vogels, R., Orban, G.A.: Activity of inferior temporal neurons during orientation discrimination with successively presented gratings. J. of Neurophysiology **71**:4 (1994) 1428–1451
8. McClelland, J.L., McNaughton, B.L., O'Reilly, R.C.: Why there are complementary learning systems in the hippocampus and neocortex: insights from the successes and failures of connectionist models of learning and memory. Psychological Review **102**:3 (1995) 419–457

Neuro-biological Bases
for Spatio-temporal Data Coding
in Artificial Neural Networks

Gilles Vaucher

SUPÉLEC
Av. de la Boulaie - B.P. 28
35511 Cesson-Sévigné Cedex - France
Gilles.Vaucher@supelec.fr

Abstract. Taking as a starting point Wilfrid Rall's dendritic tree model, well known by neuro-biologists, we propose a spatio-temporal data coding to introduce time in an artificial neuron (AN). This paper describes the biological origin of the coding and its links with the AN. With this type of coding, the algebraic properties of ANs are maintained and applied to sequence processing.

Introduction

There are two ways of minimizing the distance between an artificial and a biological neuron either by simplifying biological models, or by including new biologically plausible properties into the artificial neural cell. Our approach is more particularly based on the second as regards the introduction of time into artificial neural networks (ANNs).

Scientists have considered several ways of introducing time into ANNs [1, 2]. Our solution introduces intrinsic sequence recognition capabilities into each AN by using explicit time coding which is compatible with the algebraic properties of conventional ANs.

1 Algebraic Properties of an Artificial Neuron

In a conventional AN, the *potential* v is given by :

$$v = \sum_j x_j w_j = X^{\mathrm{T}} W \ . \tag{1}$$

X and W are one column arrays filled respectively with the inputs (x_j) and the synaptic weights (w_j). The time constants of the two memories X and W are very different (short term for X, long term for W). An AN will not be able to compute spatio-temporal correlations unless its short term memory includes a *big enough* time constant. Various solutions exist in literature, such as leaky integration [3] or continuous models [4].

Following on from the concept of leaky integration, which has a biological basis, we have placed this approach in the algebraic context of (1). This equation can be read as the scalar product of an input vector **X** and a weight vector **W**. In the context of discrete but asynchronous dynamics, we have looked for a vector space with a scalar product measuring spatio-temporal correlations in order to identify a sequence at neuron level.

2   Wilfrid Rall's Model

Our approach uses Rall's dendritic tree model [5]. Assuming some hypotheses, Rall reduced the electric properties of a biological neuron dendritic tree to a cylindrical cable as a first approximation (*ball and stick* model). Simulations show that the injection of a constant current pulse over a given period produces a post-synaptic potential (PSP) which is propagated to the soma. The further the injection site (*synapse*) from the soma, the lower the PSP upon arrival at the soma level (Fig. 1.a).

When a sequence of such current pulses is injected, the resultant potential at the soma level depends on the correspondence between the temporal order of the pulses and the relative position of the synapses (Fig. 1.b and 1.c). When the synapse positions are defined, the maximum of the resultant potential is large for some sequences and small for others. This means that the model of the dendritic tree works as an elementary sequence detector, and this is the property that we want to introduce into an AN.

3   Spatio-temporal Data Coding

To extract algebraic characteristics from Rall's model, we first reduce the number of degrees of freedom (DOFs) in the PSP representation. In the simulations performed with Rall's model, current pulses which are constant over a given period (rectangular shape : Fig. 2.a) and more realistically designed pulses in biological terms (alpha function : Fig. 2.b) both generate elementary PSPs which are close to the square of an alpha function (Fig. 2.c). All these shapes have two DOFs, depending on signal amplitude and duration. A third DOF must be added to take into account the pulse start date (translation on time axis). In order to simplify the algebraic transposition, we reduce the number of DOFs from three to two, keeping only the amplitude and position (on the time scale) of the maximum. This reduces an elementary PSP to a punctual phenomenon that we call a PPSP (punctual post-synaptic potential) (Fig. 2.d).

In this punctual representation, we define two operations which correspond to two functional characteristics of the dendritic tree : (a) the temporal and spatial *summation* due to the addition of elementary PSPs, see Fig. 4.a, (b) the *propagation* which corresponds to the transformation of a PPS between the injection site and the soma (Fig. 4.b).

Because PPSPs have two DOFs, we formalize them (and the two above operations) using a plane representation. The representation includes two measures

— one for lengths and one for angles. The selected coding associates the amplitude of a PPSP to a length and the delay between two PPSPs (or a PPSP and a time reference) to an angle (Fig. 3). The reasons for these choices are to be found in the summation operation. The sum of two elementary PSPs with in phase maximums gives a PSP in which the maximum has an amplitude equal to the sum of the two maximum amplitudes and a date that is identical to the date of the two summed maximums. With such coding we can link the summation of the elementary PSPs (Fig. 4.a) to an algebraic sum (Fig. 4.c).

The sum of any elementary PSPs does not, however, have the shape of an elementary PSP. The meaning of such a summation when the PPSs are separated by a large delay (Fig. 5) is therefore open to question. In fact, it should be stressed that the defined algebraic sum of PPSPs can be always computed even if the link with the resultant PSP becomes less clear in certain cases. Added to this, there are constraints between the summed PSPs when the time and length constant parameters for the dendritic tree are close to the biological reality modeled by Rall ; signal propagation in the dendritic tree introduces links between delays and fading. Thus, in the Rall's simulation results, the sum of several elementary PSPs produces a shape which *approximatively* resembles an elementary PSP (Fig. 1.b and 1.c).

In the biological world, summation is linked to propagation. In the plane representation, we formalize the latter by a temporal rotation coupled with a decrease in amplitude (Fig. 4.d).

4 Spatio-temporal Data Coding and Artificial Neuron

To link what was just described and the vector formalism of ANNs we consider that each component of vector $\mathbf{X}$ has spatio-temporal characteristics which can be translated in terms of our plane representation. To build a vector space around $\mathbf{X}$ vectors, we need to associate a set of numbers with the plane representation. These numbers must include two DOFs and have a field structure. Because $\mathbf{X}$ components are real numbers from $\mathbb{R}$ in usual ANNs and because the only two-dimensional field including $\mathbb{R}$ is the set of complex numbers $\mathbb{C}$, we naturally identify the numbers searched as complex numbers.

The complex value associated with each component of $\mathbf{X}$ is the result of a temporal summation which corresponds to the afferent activity at a given synapse. It takes no account of synapse plasticity and position (distance from the soma). If a common time reference is defined for all components (e.g. the current instant of time), X codes a sequence which stores the resultant activity of each input and the phase difference between them (Fig. 6). In addition, each component takes account of the propagation operation. Because this operation includes fading, only the most recent past is memorized in the sequence of activities X (short term memory effect).

In ANNs, spatial summation and synaptic weighting are combined with the inputs when the *potential* v is computed (see (1)). Read as a scalar product the potential measures the strength of the link between $\mathbf{X}$ and $\mathbf{W}$.

In Rall's simulations, the dendritic tree *recognizes* certain sequences but not others. We consider that W represents these selected sequences. The scalar product should therefore measure the comparison between the short and the long term memory (X and W). Because the vector space of $\mathbf{X}$ and $\mathbf{W}$ has values in $\mathbb{C}$, we looked at the hermitian product :

$$v = X^{\mathrm{T}} \overline{W} = \sum_j x_j \overline{w_j} = \sum_j \eta_{x_j} \eta_{w_j} \exp \mathrm{i}(\varphi_{x_j} - \varphi_{w_j}) \tag{2}$$

to see how it fits. In (2), v is the result of a spatial summation in which each component x_j is weighted by η_{w_j} (action on the module) and delayed by φ_{w_j} (action on the phase) ; φ_{w_j} takes account of the distance between the input and the soma as far as the time factor is concerned. This scalar product includes the relative layout of the synapses.

5 An Example

To illustrate how W can be a privileged sequence for the hermitian product, we take an example which is close to Rall's simulations. We consider a four-input AN on which the synapsis layout determines : $W = (0.26 \exp(0.64\mathrm{i}); 0.41 \exp(0.46\mathrm{i}); 0.64 \exp(0.25\mathrm{i}); 1)^{\mathrm{T}}$. This sequence corresponds to the reception of pulses reaching the inputs in the following order : $1 \to 2 \to 3 \to 4$. The amplitude of each pulse is one unit and the delay between two successive receptions is one-quarter of a time unit[1].

We calculate v in two cases : the reception of the sequence in the correct order ($X_1 = W$) ; the reception of the sequence in reverse order ($X_2 = (1; 0.64 \exp(0.25\mathrm{i}); 0.41 \exp(0.46\mathrm{i}); 0.26 \exp(0.64\mathrm{i}))^{\mathrm{T}}$). The results are $v_1 = 1.64$ and $v_2 = 0.92 \exp(0\mathrm{i})$ (Fig. 7). As in Rall's simulations illustrated in Figs. 1.b and 1.c, the amplitude of v_1 is larger than the amplitude of v_2. We also note that v_1 is in phase with the latest received pulse[2].

Conclusion

The dendritic tree model proposed here lies between Rall's model which computes three DOFs pulses and McCulloch & Pitts'model in which inputs have

---

[1] The propagation operation P used to determine the components of W is :

$$x = P_{\Delta t}(x_0) = x_0 \underbrace{\exp(-1.8\Delta t)}_{\text{spatial fading}} \underbrace{\exp(\mathrm{i} \arctan \Delta t)}_{\text{temporal rotation}}$$

where x_0 and x correspond to the initial and final (Δt later) PPSPs respectively. If we use the date of the latest received pulse as a common time reference, we can write : $W = (P_{3/4}(1), P_{2/4}(1), P_{1/4}(1), P_0(1))^{\mathrm{T}}$.

[2] The same phase coincidence for v_2 is accidental and depends on the example selected to be as close as possible to Rall's simulations.

only one DOF (amplitude). Adding a second DOF makes it possible to introduce of an intrinsic temporal notion of sequences in an AN, without straying from the discrete model context.

In other studies, the data coding presented here has been incorporated in conventional ANN architectures (multi-layer perceptrons [6], self-organized Kohonen maps [2]). To take advantage of the asynchronous capabilities of the model, we use in [7] the value of the potential v phase to compute the date at which the AN produces output pulses. In [7, 8] we propose a learning algorithm compatible with our spatio-temporal approach.

Such data coding, however, is not very good as a means of taking account of both excitatory and inhibitory effects. This is why we have currently restricted it to excitatory effects but we are now looking at the possible extensions of our model.

Acknowledgments

We wish to thank B. Picinbono and R. Baduel who gave us the opportunity to work on this subject, and A. Baig, B. Jouga, A. Moyon, N. Mozayyani, R. Séguier and B. Vaucher for their assistance in revising this paper.

Figures

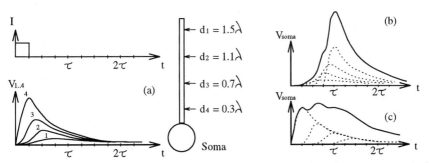

Fig.1. Schematic reproduction of Rall's simulation results. λ and τ are respectively the length and time constants of the cable in the *ball and stick* model. (a) Effect on the somatic potential of the injection of a rectangular current pulse at various distances from the soma. (b) Resultant somatic potential due to the summation of previous effects ; the four pulses are received in the following order $1 \to 2 \to 3 \to 4$ separated by a $\tau/4$ delay. (c) Ditto case b, but the sequence order is : $4 \to 3 \to 2 \to 1$

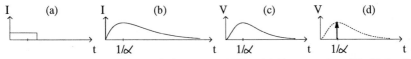

Fig.2. Shapes of current pulses and elementary PSPs. (a) Rectangular. (b) Alpha function ($I = I_0(t/\alpha)\exp(-\alpha t)$). (c) Square of an alpha function. (d) Elementary PSP with the corresponding PPSP.

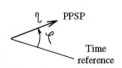

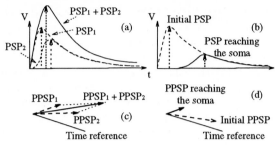

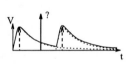

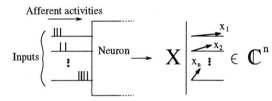

Fig.3. Representation of a PPSP in a plane. The length η corresponds to the amplitude of the PPSP and the angle φ to its date.

Fig.4. (a) PSP summation. (b) PSP propagation. (c) PPSP sum. (The resultant PPSP date lies between the dates of summed PPSPs.) (d) PPSP propagation.

Fig.5. What is the meaning of such a sum ?

Fig.6. The input vector codes a sequence of activities.

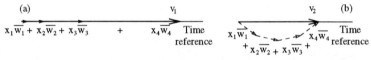

Fig.7. The potential v is the result of a spatio-temporal summation depending on a privileged sequence W. (a) The input sequence is equal to W. (b) The pulses are received in reverse order.

References

1. Chappelier, J.-C., Grumbach, A. : Time in Neural Networks. SIGART Bulletin **5:3** (1994) 3–11
2. Mozayyani, N., Alanou, V., Dreyfus, J.-F., Vaucher, G. : A spatio-temporal data-coding applied to Kohonen maps. ICANN **2** (1995) 75–79
3. Reiss, M., Taylor, J.G. : Storing Temporal Sequences. Neural Networks **4** (1991) 773–787
4. Grossberg, S. : Contour enhancement, short term memory, and constancies in reverberating neural networks. Studies in Applied Mathematics **52** (1973) 217–257
5. Rall, W. : Core conductor theory and cable properties. Handbook of Physiology : The Nervous System **1** (1977) 39–97
6. Roché, E., Vaucher, G. : Perceptron Multi-Couches Étendu aux Corps des Complexes. Valgo **94-2** (1994) 23–37
7. Vaucher, G. : Un modèle de neurone artificiel conçu pour l'apprentissage non-supervisé de séquences d'événements asynchrones. Valgo **93-1** (1993) 66–107
8. Vaucher, G. : Study of a Self-Learning Artificial Neuron Model. ICANN (1993) 204

A Spatio-Temporal Learning Rule Based on the Physiological Data of LTP Induction in the Hippocampal CA1 Network

Minoru Tsukada[1*], Takeshi Aihara[1*], Hide-aki Saito[1*] and Hiroshi Kato[2*]

[1*]*Department of Information & Communication Engineering*
Faculty of Engineering, Tamagawa University
6-1-1, Tamagawagakuen, Machida, Tokyo 194, JAPAN
[2*]*Departmemt of Physiology, Yamagata University, School of Medicine*
e-mail : tsukada@eng.tamagawa.ac.jp
telephone & fax : +81-427-39-8430 & +81-427-39-8858

Abstract

We studied the LTP inducing factors using temporally and spatially modulated stimuli given to the hippocampal neural network. It was found that when the spatial factors were maintained to be constant the positive correlation in the successive inter-stimulus intervals contributes to produce larger LTP. On the other hand, if the temporal factors are kept constant, the spatial coinsidence contributes to poroduce lager LTP. We propose a learning rule by which these experimental results can be consistently interpreted.

1. Introduction

In a biological system it is generally recognized that information presented to a neural network during a learning session is stored in the synapses by adjusting its connecting weight according to a well-defined algorithm. About 50 years ago, Hebb presented an intuitively appealing hypothesis about how a set of biological neurons might learn; a synapse connecting two neurons is strengthened only when both of those neurons fire simultaneously. Recently, it has been shown that a long-lasting enhancement of the efficacy of synaptic transmission can be induced in the hippocampal network following high frequency stimulation of afferent pathways. This long-term potentiation (LTP) of synaptic transmission is presumed to play a crucial role for learning and memory in the brain. However, it is not established as to how the magnitude of LTP is related to the stimulus condition. In other words, it is not known whether the Hebb's rule is enough or not to explain the LTP.

This paper presents the physiological evidence that the LTP is highly sensitive to the spatio-temporal pattern of input stimuli and proposes a theoretical model of spatio-temporal learning rule by which the differences of the LTPs induced by spatio-temporal stimuli of the hippocampal network can be consistently interpreted.

2. Neurophysiological experiments

General methods

Experiments were conducted on hippocampal slices (thickness; 500 μm) prepared from guinea-pigs. The slices were maintained at about 30℃ in an experimental chamber with artificial cerebrospinal fluid. A detailed description of the technique is given by Ito et al. (1989).

Positions of electrodes for the stimulation and the extracellular recording are schematically shown in Fig. 1. A monopolar tungsten electrode was placed at a fixed position (Fig. 1, Stim) in the stratum radiatum of the specific region to stimulate the Schaffer collateral commissural fibers (SC) of CA3. An electrode for extracellular recordings was placed in the pyramidal cell body layer (Fig. 1, Rec).

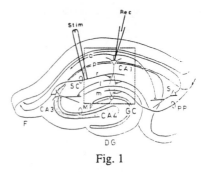

Fig. 1

Optical recording

In the optical recording sessions, slices were stained for 15 minutes with 0.2 mg / ml RH155 in normal medium and then were washed away and recovered for an additional 10 minutes. The position of the stimulating electrode was at a fixed position (Fig. 1, Stim) in the stratum radiatum of the specific region to stimulate the Schaffer collateral commissural (SC) of CA3. The recording area was the square 1.7×1.7 mm of CA1 region. The left side of the square was fiexd at the bounary between CA2 and CA1 area.

The magnitude of induced LTP was estimated by the percentage changes in the amplitude of population spikes generated by constant test stimuli.

A naive slice was used for each stimulus sequence of control-Markov-test. Slices were viewed with $5 \times$ objective. The voltage-sensitive dye signals were recorded with 700 ± 30 nm interference filter. The transmitted light was detected by a 128×128 square array of photodiodes, each has 1.7×1.7 mm receptive area. Each photodiode received light from 14×14 μm area of the microscope objective field, and was coupled to current-to-voltage converter and amplifier (gain; $\times 2000$); we used the HR Delteron 1700 system (Iijima et al., 1991; Fuji Film Co.).

Stimulation

1) Temporal stimuli

An important condition of stimuli is that, different from the "Spatio-Temporal Stimuli" described later, the intensity of the electric pulses to stimulate the Schaffer fibers is fixed at a constant value, expecting that stimulated area might be constant. Exactly, it was adjusted to a half of that producing the maximum amplitude of the population spike. The duration of a stimulus pulse and the total number of pulses were fixed at 0.2 ms and 200, respectively. As a mode of temporal modulation, we selected Markov sequences (MSs) which are characterized by having an identical mean rate (the first order statistics) but different correlation between the successive inter-stimulus intervals (the second order statistics). In our experiments, five types of MS were generated by the following transition matrices

$$SP : \begin{array}{cc} & \begin{array}{cc} s_1 & s_2 \end{array} \\ \begin{array}{c} s_1 \\ s_2 \end{array} & \left(\begin{array}{cc} 0.9 & 0.1 \\ 0.1 & 0.9 \end{array} \right) \end{array}, \quad WP : \begin{array}{cc} & \begin{array}{cc} s_1 & s_2 \end{array} \\ \begin{array}{c} s_1 \\ s_2 \end{array} & \left(\begin{array}{cc} 0.7 & 0.3 \\ 0.3 & 0.7 \end{array} \right) \end{array}, \quad NC : \begin{array}{cc} & \begin{array}{cc} s_1 & s_2 \end{array} \\ \begin{array}{c} s_1 \\ s_2 \end{array} & \left(\begin{array}{cc} 0.5 & 0.5 \\ 0.5 & 0.5 \end{array} \right) \end{array}$$

$$WN : \begin{array}{cc} & \begin{array}{cc} s_1 & s_2 \end{array} \\ \begin{array}{c} s_1 \\ s_2 \end{array} & \left(\begin{array}{cc} 0.3 & 0.7 \\ 0.7 & 0.3 \end{array} \right) \end{array}, \quad SN : \begin{array}{cc} & \begin{array}{cc} s_1 & s_2 \end{array} \\ \begin{array}{c} s_1 \\ s_2 \end{array} & \left(\begin{array}{cc} 0.1 & 0.9 \\ 0.9 & 0.1 \end{array} \right) \end{array}$$

where SP: Strong Positive correlation, WP: Weak Positive correlation, NC: No Correlation, WN: Weak Negative correlation and SN: Strong Negative correlation, and s_1, s_2 stand for a short interval (50 ms) and a long one (950 ms), respectively. These stimuli are diagramatically shown in Fig. 2a.

2) Spatio-Temporal Stimuli

A weak ("w") and a strong ("s") electric pulses were intermixed to stimulate different areas of Schaffer region. Intensity of the stimulus "s" was adjusted to be a half of that producing the maximum amplitude of the population spike to stimulate a relatively large area (schematically shown by the area A+B in Fig. 2c). On the other hand, the stimulus "w" was adjusted to be one tenth of that producing the maximum one to stimulate a small area (area A, Fig. 2c). With stimulus "s", areas A and B are stimulated simultaneously (i.e., with the spatial coincidence), while only the area A is activated by the stimulus "w" (no spatial coincidence between A and B). The four types of stimuli (Type 1-4) with the same mean rate (10 Hz) and the same stimulus number (200 pulses) were generated by combining "w" and "s" with the following transition matrices;

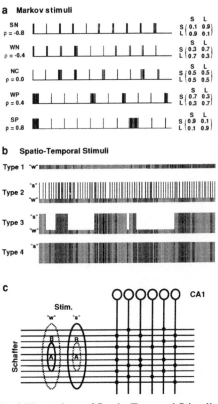

Fig. 2 Illustrations of Spatio-Temporal Stimuli

$$\text{Type 1}: \begin{array}{c} \\ w \\ s \end{array}\begin{pmatrix} w & s \\ 1 & 0 \\ 1 & 0 \end{pmatrix} \quad \text{Type 2}: \begin{array}{c} \\ w \\ s \end{array}\begin{pmatrix} w & s \\ 0.1 & 0.9 \\ 0.9 & 0.1 \end{pmatrix} \quad \text{Type 3}: \begin{array}{c} \\ w \\ s \end{array}\begin{pmatrix} w & s \\ 0.9 & 0.1 \\ 0.1 & 0.9 \end{pmatrix} \quad \text{Type 4}: \begin{array}{c} \\ w \\ s \end{array}\begin{pmatrix} w & s \\ 0 & 1 \\ 0 & 1 \end{pmatrix}$$

Fig. 2b gives a schematic illustration of these stimuli. These stimuli have the following order relationships in the magnitude of spatial coincidence;

Type 4 > Type 3 (= Type 2) > Type 1.

Type 2 and 3 stimuli consisted of the same number of component "s" (100 pulses) and "w" (100 pulses) and are identical in the spatial coincidence, but are different in the time correlations; for Type 3 (strong-positive correlation) long trains of "s" or "w" are prevalent, while for Type 2 (strong-negative correlation) "s" and "w" tend to alternate.

Results

In Fig. 3, we show that the temporal factor is very important to induce LTP.

These results were obtained by extracellular recordings of LTP induced by the five types of Markov sequences. Not only the mean interval but also the correlation between successive inter-stimulus intervals affected the magnitude of LTP. Positively correlated sequences are much more effective in producing LTP, while negatively correlated sequences are relatively ineffective. The magnitude of LTP increased as the serial correlation coefficient of Markov sequences increased.

The "Spatio-Temporal Stimuli" not only confirmed the importance of the temporal pattern for the induction of LTP, but also revealed another interesting aspect of the LTP-inducing factor. The magnitudes of LTP induced by four types of "Spatio-Temporal Stimuli" (Fig. 2b) were estimated in two ways, one using the stimulus "w" and the other using the stimulus "s" as test stimuli. The results are shown in Fig. 4. The order relationships in effectiveness for the LTP induction are Type 1 < Type 2 < Type 3 < Type 4. The reason for that the Type 1 was the least effective and the Type 4 was the most effective is self-evident, i.e., the stimulus intensity is weakest in Type 1 and strongest in Type 4. The difference in the effectiveness of Type 2 and Type 3 stimuli which have the same mean intensity (each consisted of 100 "w" and 100 "s") can be interpreted in the same way as in the case of the LTP induced by the "Temporal Stimuli" (Tsukada, M. 1994). That is, the higher the value of positive correlation, the larger the effectiveness for the LTP induction.

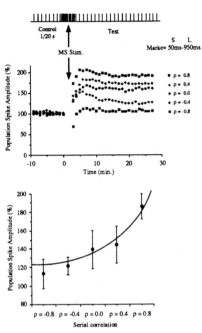

Fig. 3 Temporal-patern dependent LTP

Fig. 5 shows two spatial distribution of the magnitude of the LTP induced by the Type 2 and Type 3 of MS. These results show that the LTP induced by MS of positive correlation distributed over larger area of CA1 region than that induced by MS of negative correlation. On the other hand, stimulus with negative correlation induced LTP in definitely smaller area. This result is consistent with the extracellular observations on the effectiveness of spatio-temporal stimuli showen in Fig. 4: that is, the order of effectiveness is positive correlation > negative correlation. The optical recordings clearly show that the temporal pattern of the Schaffer inputs of CA3 is transformed into different spatial patterns of synaptic activities (which reflects the strength of connecting weights) in the hippocampal CA1 network.

3. A spatio-temporal learning rule

On the basis of the above experimental observations, we propose a mathematical model of learning rule for the induction of long term potentiation (LTP) in the CA1 region of the hippocampus. The model learning rule is basically Hebbian in that the amount of potentiation increases with the frequency of synaptic activation. However,

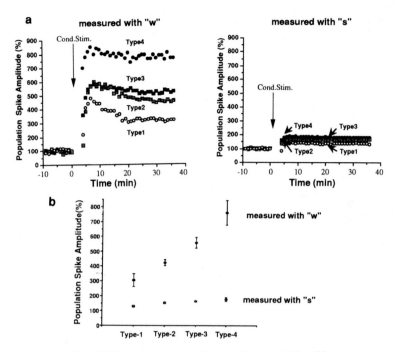

Fig. 4 LTP induced by the "Spatio-Temporal Stimuli"

both the present and the previous (Tsukada, M. 1994) results have strongly suggested that the rule must be sensitive to the correlation between inter-stimulus intervals of the applied stimulus chain (the temporal factor) as well as to the synchronization of input signals from ensembles of neurons impinging upon the target cell (spatial coincidence of inputs) (Teyler and Discenna, 1984). To fulfill the first reguirement, we introduce a two-parameter model. The one parameter I is an elemental quantity of glutamate binding to the post-synaptic NMDA channel. The amount of this binding is thought to be directly related to the LTP (Larson and Lynch, 1988; Bekkers and Stevens, 1989; O'Brien and Fischback, 1986). State of this binding decays exponentially after the time of binding (Lester et al., 1990). We introduced time course of this decay as the second parameter λ. Thus the temporal factor of our model is expressed by

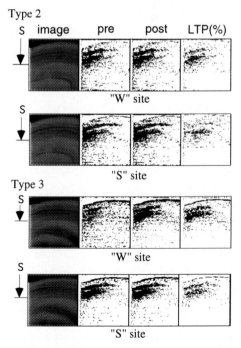

Fig. 5 Optical imaging of LTPs induced by the "Spatio-Temporal Stimuli"

$$T(\{\tau_i\}) = Ie^{-\lambda\tau_1} + Ie^{-\lambda\tau_2} + \ldots\ldots + Ie^{-\lambda\tau_n} \tag{1}$$

where τ_i indicates the time interval between the ith spike arrival and the nth spike arrival.

By performing least squares minimization, we caliculated the best fitting curve to the experimental data (Fig. 3b) of MSs of 50 ms and 950 ms combination using equation [1]. From this curve, I and $1/\lambda$ are estimated as 0.533 and 223, respectively. This time constant of decay is well matched with the value reported in the physiological study (Lester et al., 1990).

To fulfill the second requirement, we introduce "spatial coincidence" to our model. In the model, we assume that the elemental quantity "I" of glutamate binding changes depending on the coincidence between inputs from a certain fiber and from its surroundings. This spatial coincidence is given by eq. [2],

$$I_{ij}(t) = w_{ij}(t)\, x_i(t) \sum_{k \neq i} \{w_{kj}(t)\, x_k(t)\} \tag{2}$$

where $x_i(t)$ is an input from the neuron "i" to the neuron "j" and $x_k(t)$ is an input from neuron "k" to the neuron "j". The parameter $w_{ij}(t)$ is the weight of synaptic connection from the neuron "i" to the target neuron "j", and $w_{kj}(t)$ is connecting weight from the neurons "k" to "j".

$\sum_{k \neq i} \{w_{kj}(t)\, x_k(t)\}$ can be considered as the spatial coincidence factor between the inputs from the neuron "i" from its surrounding inputs.

By combining the above two factors, a spatio-temporal learning rule can be expressed by the following equation. Using discrete parameter m instead of the continuons time interval, and introducing a threshold parameter θ (McNauton et al., 1978),

$$\Delta w_{ij}(t_n) = \eta \left\{ \sum_{m=1}^{n-1} I_{ij}(t_{n-m})\, e^{-\lambda m} - \theta \right\} \tag{3}$$

where $\Delta w_{ij}(t_n)$ is the change in the strength of synaptic connection from neuron "i" to neuron "j", and η is learning-rate.

By this learning rule, different spatio-temporal patterns of the CA3 Shaffer inputs are transformed into different spatial patterns of synaptic connection in the CA1 network. This transformation would work as 'short-term memory' in the hippocampus.

References

Aihara, T., et al. (1996). (to be published).
Bekkers, J. M. and Stevens, C. F. (1989). *Nature* 341, 230-233.
Hebb, D. O. (1949). *John Wiley & Sons.*
Iijima, T., Ichikawa M. and Matsumoto G. (1991). *Neurosci. Res. Suppl* 16, S158.
Ito, K. et al., (1989). *Bio-med Res.* 10, 111-124.
Larson, J. and Lynch, G. (1988). *Brain Res.* 441, 111-118.
Lester, A. J., et al. (1990). *Nature*, 346, 565-567.
McNauton, B. L., et al. (1978). *Brain Res.*, 157, 277-293.
O' Brien and Fischbach, G. D. (1986). *J. Neurosci*, 6, 3275-3283.
Teyler, T. J. and Discenna, P. (1984). *Brain Res. Rev.* 7, 15-28.
Tsukada, M. (1994). *Cybernetics and systems* 25, 189-206.
Tsukada, M., et al. (1994). *Biological Cybernetics* 70, 495-503.

Learning Novel Views to a Single Face Image

Thomas Vetter

Max-Planck-Institut für biologische Kybernetik
Tübingen, Germany, E-mail: vetter@mpik-tueb.mpg.de

Abstract. A new technique is described for synthesizing images of faces from new viewpoints, when only a single 2D image is available. A novel 2D image of a face can be computed without knowledge about the 3D structure of the head. The technique draws on prior knowledge of faces based on example images of other faces seen in different poses and on a single generic 3D model of a human head. The example images are used to learn a pose-invariant shape and texture description of a new face. The 3D model is used to solve the correspondence problem between images showing faces in different poses.

1 Introduction

The 3D structure of an object determines how images of the object change with a change in viewpoint. With viewpoint changes, some previously visible regions of the object become occluded, while other previously invisible regions become visible. Additionally, the configuration of object regions that are visible in both views may change. Accordingly, to synthesize a novel view of an object, two problems must be addressed and resolved. First, the visible regions that the new view shares with the previous view must be redrawn at their new positions. Second, regions not previously visible from the view of the example image must be generated or synthesized. It is obvious that this latter problem is unsolvable without prior assumptions. For human faces such prior knowledge can be learned through extensive experience with other faces.

In recent years, two-dimensional image-based face models have been applied for the synthesis of rigid and nonrigid face transitions [Craw and Cameron, 1991, Poggio and Brunelli, 1992, Beymer et al., 1993, Cootes et al., 1995]. These models exploit prior knowledge from example images of prototypical faces and work by building flexible image-based representations (*active shape models*) of known objects by a linear combination of labeled examples.

In this paper, the *linear object class* approach [Vetter and Poggio, 1996], is improved with a single 3D model of a human head for generating new views of a face. The 3D-model allows a better utilization of the example images and also the transfer of features particular to an individual face, like moles and blemishes, from the given example view into new synthetic views. This is true even when these blemishes, etc., are unrepresented in the "general experience" that the linear class model has acquired from example faces. The limitation of a single 3D head model is the difficulty of representing the variability of head shapes in general, a problem that the linear class model, will allow us to solve.

2 Algorithm

The algorithm can be subdivided into four modules (see also figure 1).

Separation of texture and shape in images of faces: The central part of the approach is a representation of face images that consists of a separate texture vector and 2D-shape vector, each one with components referring to the same feature points – in this case pixels. Assuming correspondence the shape of a face image is represented by a vector $s = (x_1, y_1, x_2, ..., x_n, y_n)^T \in \Re^{2n}$, that is by the x, y displacement of each feature with respect to the corresponding feature in the reference face. The texture is coded as a difference map between the image intensities of the exemplar face and its *corresponding* intensities in the reference face. Such a normalized texture can be written as a vector $\mathbf{T} = (i_1, ...i_n)^T \in \Re^n$, that contains the image intensity differences i of the n pixels of the image. All images of the training set are mapped onto the reference face of the corresponding orientation. This is done separately for each rotated orientation. Automated procedures for this nonlinear normalization are found in the optical flow literature and its application to faces [Bergen et al., 1992, Beymer et al., 1993].

A single 3D head model was used to render the two reference face images for the two different orientations. Additionally the correspondence field between this two reference face images across the view point change was computed using the 3D model.

Module for shape processing: The shape model is based on the linear object class idea [Vetter and Poggio, 1996] and is built on a training set of pairs of images of faces. From each pair of images, the 2D-shape vectors s^r for the "rotated" shape and s^f for the "frontal" shape are computed. Under the assumption, faces form a linear object class and the 2D shape s^r of a given "rotated" view, can be represented by the "rotated" shapes of the example set s_i^r as

$$s^r = \sum_{i=1}^{q} \beta_i s_i^r, \tag{1}$$

then the "frontal" 2D-shape s^f to a given s^r can be computed using β_i of equation (1) and the other s_i^f given through the images in the training set with the following equation:

$$s^f = \sum_{i=1}^{q} \beta_i s_i^f. \tag{2}$$

So far no knowledge of correspondence between equation (1) and equation (2) is necessary (rows in a linear equation system can be exchanged freely). However correspondence (for the reference face given by the 3D model) allows separate linear classes for separate regions of the face and therefore a better utilization of the example images.

Module for texture processing: Two different paths for generating a "frontal" texture given a "rotated" texture are combined. Again the linear object class approach is used to generate a new texture t^f. A given texture t^r is decomposed into the given example textures t_i^r in the same orientation ($t^r = \sum_{i=1}^{q} \alpha_i t_i^r$). The new texture t^f is generated by combining the "frontal" example textures

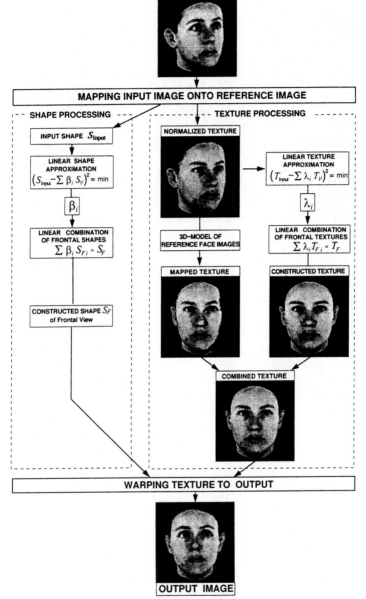

Fig. 1. *Overview of the algorithm for synthesizing a new view from a single input image. After mapping the input image onto a reference face in the same orientation, texture and 2D-shape can be processed separately. The example based linear face model allows the computation of 2D-shape and texture of a new "frontal" view. The single 3D model of the reference face allows additionally texture mapping from the "rotated" to the "frontal" view. Warping the new combined texture along the new deformation field (coding the shape) results in the new "frontal" view as output (lower row).*

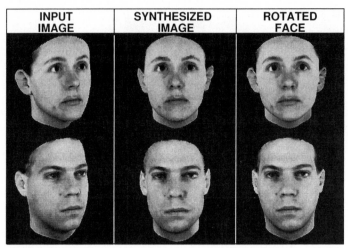

INPUT IMAGE	SYNTHESIZED IMAGE	ROTATED FACE

Fig.2 *Two synthesized frontal views (center column) to a given rotated (24°) image of a face (left column) are shown. The prior knowledge about faces was given through a training set of 99 pairs of images of different faces (not shown) in the two orientations. Additionally a single 3D-head model for the reference face was used to establish correspondence across the view point change. The frontal image of the real face is shown in the right column.*

using the computed weights α_i as follows $\mathbf{t}^f = \sum_{i=1}^{q} \alpha_i \mathbf{t}_i^f$.

This linear texture approach is only satisfactory for regions not visible in the "rotated" texture, since it is hardly able to capture features which are particular to an individual face (e.g. freckles, moles). The single correspondence field available between the reference face images, however, allows a direct mapping from the given normalized "rotated" texture onto the new normalized "frontal" texture.

Final image synthesis: The texture obtained through direct texture mapping across the viewpoint change and the texture obtained through the linear class approach are merged by standard image blending techniques. This new texture is finally warped along the generated new shape vector to the new image representing a new view to the input face image.

3 Results

The algorithm was tested on 100 human faces. For each face, images were given in two orientations (24° and 0°) with a resolution of 256-by-256 pixels and 8 bit.

In a *leave–one–out* procedure, a new "frontal" view of a face was synthesized to a given "rotated" view (24°). In each case the remaining 99 pairs of face images were used to learn the linear 2D-shape and texture model of faces. Figure 2 shows the results for four faces.

The quality of the synthesized "frontal" views was tested in a simple simulated recognition experiment. For each synthetic image, the most similar frontal face image in the data base was computed. For the image comparison, two common similarity measures were used: a) the *direction cosine*; and b) the *Euclidean distance* (L_2). Both measures were applied to the images in pixel representation without further processing. The recognition rate of the synthesized images was 100 % correct, both similarity measures independently evaluated the true "frontal" view to a given "rotated" view of a face as the most similar image.

As the crucial test for the image synthesis, 10 human observers were asked to discriminate the real and synthetic images of a face. Only 6 faces (out of 100) were classified correctly by all subjects . In all other cases the synthetic image was at least by one subject classified as the true image. In average each observer was 74% correct whereas the chance level was at 50%.

4 Discussion

The results of the automated image comparison indicate the importance of the proposed face model for viewpoint independent face recognition systems and demonstrate an improvement over techniques proposed previously [Beymer and Poggio, 1995, Vetter and Poggio, 1996].

The difficulties experienced by human observers in distinguishing between the synthetic images and the real face images indicate, that a linear face model of 99 faces give a good approximation of a new face, it also indicates possible applications of this method in computer graphics. Clearly, the linear model depends on the given example set, so in order to represent faces from a different race or a different age group, the model would clearly need examples of these, an effect well known in human perception (cf. e.g. O'Toole et al., 1994).

References

Bergen, J., Anandan, P., Hanna, K., and Hingorani, R. (1992). Hierarchical model-based motion estimation. In *Proceedings of the European Conference on Computer Vision*, pages 237–252, Santa Margherita Ligure, Italy.

Beymer, D. and Poggio, T. (1995). Face recognition from one model view. In *Proceedings of the 5th International Conference on Computer Vision*.

Beymer, D., Shashua, A., and Poggio, T. (1993). Example-based image analysis and synthesis. A.I. Memo No. 1431, Artificial Intelligence Laboratory, Massachusetts Institute of Technology.

Cootes, T., Taylor, C., Cooper, D., and Graham, J. (1995). Active shape models - their training and application. *Computer Vision and Image Understanding*, 60:38–59.

Craw, I. and Cameron, P. (1991). Parameterizing images for recognition and reconstruction. *Proc. British Machine Vision Conference*, pages 367–370.

O'Toole, A., Deffenbacher, K., Valentin, D., and Abdi, H. (1994). Structural aspects of face recognition and the other-race effect. *Memory and Cognition*, 22:208–224.

Poggio, T. and Brunelli, R. (1992). A novel approach to graphics. Technical report 1354, MIT Media Laboratory Perceptual Computing Section.

Vetter, T. and Poggio, T. (1996). Image synthesis from a single example image. In *Proceedings of the European Conference on Computer Vision*, Cambridge UK.

A Parallel Algorithm for Depth Perception from Radial Optical Flow Fields

Jens Vogelgesang, Alex Cozzi, Florentin Wörgötter

worgott@neurop.ruhr-uni-bochum.de
Inst. für Neurophysiologie, Ruhr-Universität, Bochum

Abstract. While optical flow has been often proposed for guiding a moving robot, its computational complexity has mostly prevented its actual use in real applications. We describe a restricted form of optical flow algorithm, which can be parallelized on chain-like neuronal structures, combining simplicity and speed. In addition, this algorithm makes use of predicted motion trajectories in order to remove noise from the input images.

1 Introduction

Optical Flow has been proposed as a possible cue to guide robot navigation[1, 2, 3]. Unfortunately general optical flow algorithms are unsuitable for hardware implementation or real-time processing as a consequence of their complexity. In this paper we describe a specialized obstacle detector algorithm that is easily implementable in digital hardware and performs at frame-rate speed.

We simplify the problem introducing two constraints, which cover many generic situations:
1. The robot is supposed to move forward, along the optical axis of the camera-system.
2. The scene is supposed to contain only static objects.
Under these hypotheses the optical flow-field has a pure radial structure, with the center of expansion in the middle of the image.

The distance of the objects contained in the scene is computable from the apparent motion of their projection on the image plane and from the camera's characteristics (lens' focal length and CCD size). In our case, exploiting the radial structure of the optical flow, the problem is further simplified from the 2D structure of a generic flow field to the 1D structure along a radius.

2 Simulated Retina and the Explanation of the Technique

We devised an algorithm to implement the approach in the form of a network of idealized neurons. The network can be interpreted as an artificial retina, with each neuron directly connected to a photo-receptor as in Fig. 1.

The computation takes place by neurons exchanging information with each other in the network. Each neuron in our algorithm needs to be connected only to its two directly adjacent neighbors on the same radius. We call the set of neurons on a radius a *neuron chain*.

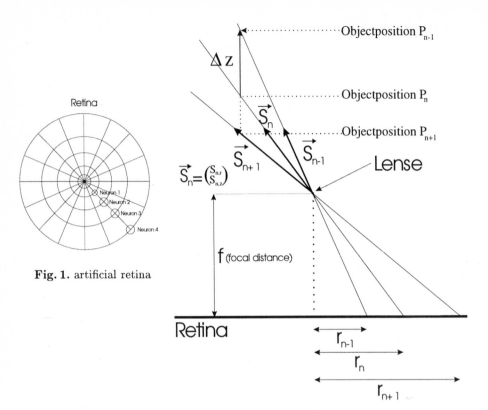

Fig. 1. artificial retina

Fig. 2. geometry of position estimation

Let us assume an object position P_{n-1} at time t_{n-1} projected onto neuron r_{n-1} on the retina. Then the projection of this object will be displaced on the retina to r_n due to the robot motion of which covers a distance of $-\Delta z$ between two camera frames such that the real position of the object will be P_n at time t_n. The equations of a pin-hole camera can be solved in our case to compute the position of an object in world coordinates: (Fig. 2) from the system of geometric equations

$$k \cdot \vec{s_n} + \vec{\Delta z} = l \cdot \vec{s_{n-1}} \tag{1}$$

where k and l are scalar factors for the vectors $\vec{s}$. As solution we obtain the object position (X, Y, Z):

$$\begin{pmatrix} X \\ Y \\ Z \end{pmatrix} = \left(\frac{\Delta z}{s_{n,z}\left(\left(\frac{s_{n,r}}{s_{n-1,r}}\right) - 1\right)} \right) \begin{pmatrix} s_{n,r} \cos \arctan\left(\frac{s_{n,y}}{s_{n,x}}\right) \\ s_{n,r} \sin \arctan\left(\frac{s_{n,y}}{s_{n,x}}\right) \\ s_{n,z} \end{pmatrix} = \Delta z \cdot \vec{P_n} \tag{2}$$

All terms of Eq.2, excepted Δz, which is measured between two camera frames, are constants at each given neuron with:

$$\vec{P_n} = \frac{1}{s_{n,z} \cdot \left(\left(\frac{s_{n,r}}{s_{n-1,r}}\right) - 1\right)} \cdot \begin{pmatrix} s_{n,r} \cdot \cos \arctan\left(\frac{s_{n,y}}{s_{n,x}}\right) \\ s_{n,r} \cdot \sin \arctan\left(\frac{s_{n,y}}{s_{n,x}}\right) \\ s_{n,z} \end{pmatrix} \tag{3}$$

Due to the constant terms, the depth of an object can be determined at the cost of a single multiplication. Solving Eq. 2 is the first step of the algorithm. In order to deal with possible errors introduced by noise or false detections we introduce a second computational step: using the computed position and knowing the actual position of the robot, we can predict where the object will be detected the next time. After having computed the object position (X, Y, Z) at neuron n, we compute the robot position Δz_{pred}, where the neuron $n + 1$ is expected to be activated by this object using Eq. 4. If this excitation does not occur, the old object position was incorrect. If it occurs the old position is confirmed and the object becomes "reliable".

$$\Delta z_{pred} = \sqrt{X^2 + Y^2} \cdot \left(\frac{s_{n,z}}{s_{n,r}} - \frac{s_{n+1,z}}{s_{n+1,r}}\right) \tag{4}$$

Each neuron of our artificial retina is composed of three subcomponents: a photoreceptor, a processing unit and a data storage. The processing unit receives information from the photoreceptor and from the adjacent inner neuron, it sends information to the adjacent outer neuron and can write and read data from the storage and perform simple arithmetic operations as specified by the equations.

3 Defining the Algorithm for a Neuron Chain

Let us suppose that our artificial retina is implemented in a camera, moving forward along its optical axis, in a world containing only one black spot somewhere in the direction of gaze. The computational at process along the neuron chain on which the black spot is projected, can be defined by Fig. 3

1. At position P_0 the neuron 1 detects an object. It communicates the event to neuron 2. (Fig. 3,left)
2. At position P_1 the neuron 2 detects the same object (Fig. 3 middle). Since its memory is not empty it can compute a tentative position (X, Y, Z) for the detected object from the position difference $\Delta Z = 8$ and the known position of the photo-receptors (Eq 2). With this information neuron 2 predicts the distance expected to be covered by the camera motion before the object will be detected by the next neuron ($\Delta Z_{pred} = 6$), and communicates the predicted value and the actual position information to neuron 3.
3. When neuron 3 actually detects the object ($\Delta z = 5$), the prediction can be checked: if the precision is satisfactory (e.g., $\Delta z = \Delta z_{pred} \pm 1$) the computed position of the object is updated and marked as reliable, and all informations

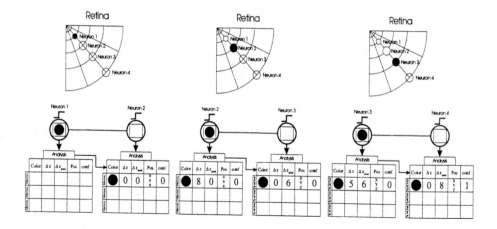

Fig. 3. ideal excitation cycle

are communicated to the next neuron with a reliability value of 1. Otherwise the detected object is considered a new object, and the computation resumes as in 1.

4 Results

Figs. 4-7 show the results obtained from a noisy scene containing three objects at different depths (Fig. 4). Fig. 5 represents the very noisy retina image. Fig. 6 shows the depth information without making use of the predictions while in Fig. 7 only reliable pixels ($conf \geq 3$) are shown and the rough outline of the three objects appears. Depth is coded by gray-scale.

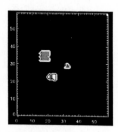
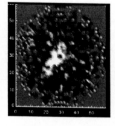
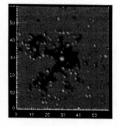
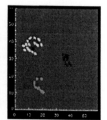

Fig. 4. *first retina image without noise*

Fig. 5. retina image with noise, step 50

Fig. 6. depth map without prediction, step 50

Fig. 7. depth map with prediction, step 50

5 Discussion

The algorithm proposed here can be parallelized with simple computational units on an artificial retina because only a single multiplication is necessary to retrieve the depth information. While this has been suggested before[1] the novel aspect of this study is the use of prediction values to improve the depth maps. Even with such a prediction the arithmetic remains rather simple. The price to pay for this simplicity is the restriction to radial flow-fields. Thus, when changing the direction of motion the algorithm has to be reset. To cover these situations the algorithm should be used in a modular system, where other computationally more expensive depth algorithms[4] are applied to refine the first shot depth maps obtained with our very fast processing scheme. We a currently implementing the algorithm on the robot RHINO of the university of Bonn in cooperation with the computer science department (J. Buhmann and V. Gerdes) and preliminary results indicate that the algorithm will indeed be applicable in real word situations.

6 Acknowledgements

F.W. acknowledges the support of the European community grant ESPRIT BRA 8503 as well as from the Deutsche Forschungsgemeinschaft (Wo 388/5-2).

References

1. Poggio T. & Ancona N., Optical Flow from 1D Correlation: Application to a simple Time-To-Crash Detector /it Massachusetts Institute of Technology (1993).
2. Koenderink J.J. Optic Flow *Vision Res.* **26**, 161-180 (1986).
3. Parado F., Martinuzzi E. Hardware environment for a retinal CCD visual sensor, *Department of Artificial Intelligence, University of Edinburgh, 5 Forest Hill, Edinburg EH1 2QL, Scotland* (1995)
4. Fleet, D., Jepson, A. & Jenkin, M. Phase-based disparity measurement *Comp. Vision, Graphic and Image Proc.*, **53,2**, 198-210 (1991).

Comparing Facial Line Drawings with Gray-Level Images: A Case Study on PHANTOMAS

Wolfgang Konen

ZN GmbH, Bochum,
Germany
WWW: http://www.zn.ruhr-uni-bochum.de

Abstract. We report on an application of neural face recognition algorithms to a task with relevance to forensic investigations: The software tool PHANTOMAS (**phantom** **a**utomatic **s**earch) allows to compare facial line drawings (the German "Phantomzeichnung") with gray-level images of faces. In addition to normal (textual) database search actions, this software tool allows picture-to-picture searches. We present first results on the evaluation of a benchmark on this task. The ranking quality of PHANTOMAS allows to spot the true match belonging to a certain drawing on the average within the upper 2.7% of the database ($N = 103$). It is shown that this is comparable to the human performance on the same data material. Computation time makes it feasible to search online in large databases ($N \approx 10000$). – With the same algorithm it is also possible to classify complex characteristica in faces or facial line drawings, which we demonstrate on the example of gender classification.

1 Introduction

Recognition of faces is a remarkable example for the ability of humans to perform complex visual recognition tasks. Even more striking is the fact that we can reliably identify persons from simple line drawings which have, on the gray-level basis, little in common with the original picture.

To build automated systems with similar capabilities is not only of research interest, but also has an important application in forensic investigations: There one often has the line drawing or a sketch as starting point, but the actual identification task requires that the witness should look through thousands of images – often too much to complete the search successfully. Needed is a tool for automated sorting the database into similar and definitely unsimilar faces, allowing the witness to focus his/her attention on the important cases.

Many algorithms for face recognition have been described in the literature, see e.g. [1, 2, 3]. Most successful algorithms use principles from neural information processing, for example neural filters (e. g. Gabor wavelets) for the early visual processing or neural-net based learning of higher-level invariants [4].

Our algorithm is based on von der Malsburg's Elastic Graph Matching Algorithm [5] and has been successfully applied for the access control system ZN-Face [6] where it has shown to provide robust and secure verification behavior.

1.1 Face recognition by graph matching

The basic face recognition algorithm is described in [6] and is an extension of the Elastic Graph Matching Algorithm [5]; the reader is referred to these articles for a comprehensive description. Here we just mention that faces are stored as flexible graphs or grids (see Fig. 1) with characteristic visual features (Gabor features) attached to the nodes of the graph (*labeled graphs*).

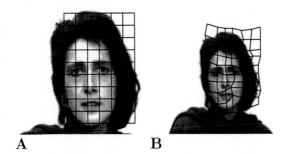

Fig. 1. (A) Labeled graphs are the data representation in *PHANTOMAS*. **(B)** Such graphs can be shifted, scaled and deformed efficiently in the image domain.

A   **B**

The Gabor features are based on the wavelet transform

$$G_{\mathbf{k},\sigma}(\mathbf{x}) = C_{\mathbf{k},\sigma} \exp(-\frac{k^2 x^2}{2\sigma^2}) \exp(i\mathbf{k}\mathbf{x}), \tag{1}$$

and have shown to provide a robust information coding for object recognition [5] (invariance against intensity or contrast changes in the image). Furthermore, Gabor features are less affected by changes in head posture, size and facial expression than raw grey level features.

The matching of a stored graph against an image consists of an optimization of the graph location in the image ("Global Move", GM). We have shown previously [6], that within a range of $\pm 20\%$ variation in scale and aspect ratio it is possible to combine the GM-displacement with simultaneous optimization in scale and aspect ratio. This works without getting stuck in possible local minima and without transforming the Gabor features. We will refer to this joint optimization procedure as 'GM' in the following. [1]

In summary, the face recognition algorithm consists of five basic steps:

(i) Choose a search template and the desired search fundus (which may be the whole database or certain subparts of it),

(ii) Convolve the image with each Gabor Wavelet used in the labeled graph representation,

(iii) With a representative set of graphs from the database we execute the GM. From the best GM, we keep the graph extracted from the search template as "optimally positioned graph" (OPG).

(iv) Match all graphs in the search fundus against the OPG (60 msec/match).

(v) Rank the results according to matching similarity.

---

[1] The whole optimization procedure and its application to verification or database searches has been patented [7].

2 Features of the **PHANTOMAS** system

The PHANTOMAS system hardware consists of a PC and a color scanner as input device for database or search template images. As an option, a color camera with framegrabber may be used for direct digital input.

The software for PHANTOMAS was developped using IDL (Interactive Data Language). This interpreter-type language with many graphic visualization and GUI-capabilities enables platform independence and large flexibility during the development process. The price to pay for this is a somewhat reduced performance in (some) computing intensive task. Therefore, the final product version will have the search kernel routines implemented in C++.

Computational cost: Concerning the number N of persons in the search fundus, there is only one step in the algorithm, namely step (iv), which is linearly dependent on N (complexity $\mathcal{O}(N)$), while the other steps are of complexity $\mathcal{O}(1)$.

The performance times for the unoptimized IDL-code on a Sun Sparc 10/40 are 37 sec for the ($N = 103$)-database and extrapolate to less then 11 min for a ($N = 10000$)-search task. Using a Pentium P130 and optimized search kernel routines we expect a speedup of at least a factor of 2.

The PHANTOMAS software is equipped with a comfortable graphical user interface (GUI) where nearly all actions can be invoked by simple mouse clicks. Among the features of the GUI are the following topics:

- The results of search projects can be presented in many different ways, e.g. (a) with different sortings (rank of similarity, name, randomly), (b) combined with additional matching constraints, (c) with and without descriptive data.
- Interactive recursive search: Given the performance times in the order of 10 min, it is feasible to perform multiple searchs in one session. Example: The result of the first search project may bring up an image which even better matches the witness' memory. In that case a second search can be started with this new image as search template and so on.
- Additional textual search criteria may be defined (e.g. age, male/female, bearded/non-bearded, ...) to constrain the search fundus and speed up the search process.
- Zooming-in on pictures of special interest.

3 Evaluation

3.1 The benchmark

The benchmark undertaken with the system consisted of a rather tough task, namely the comparision of facial line drawings with original photos. The database of original photos is a slight extension of [5] (103 persons, 33 females, 70 males). The pictures were taken in front of an unstructured background and with mostly frontal illumination. However, some pictures had considerable side

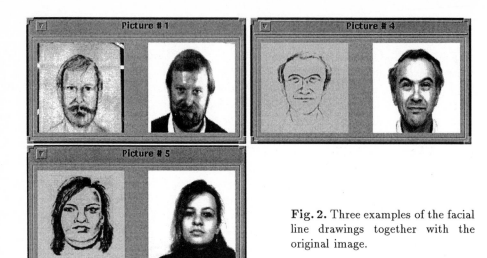

Fig. 2. Three examples of the facial line drawings together with the original image.

illumination (see e.g. picture # 1 in Fig. 2). From 13 of the database pictures, a professional forensic artist produced the "phantom drawings". In these drawings, slight deviations from the original (e. g. moustache, different hair-style) can occur. Some examples are shown in Fig. 2. In most cases the aspect ratios in the drawings and the photos were somewhat different (see inset in Fig. 3).

3.2 Automated similarity ranking

Fig. 3 shows an example result for the top-ranking persons as found by PHAN-TOMAS. Although the gender of the line drawing was not specified as search constraint, PHANTOMAS has picked up dominantly females in the first places. The summary of the ranking results is given in Tab. 1.

As a comparision we show in Tab. 1 also the result of a "naive" algorithm where one tries to rank the line drawings based on the normalized image cross correlation (NC1). Clearly, this approach does not lead to satisfactory results. Even if we use partial information from the graph matching algorithm and warp the input image according to scale and aspect ratio as found by the global move before calculating the normalized correlation coefficient, the results improve only slightly (NC2).

3.3 Comparision with human performance

In order to get a rough feeling for the difficulty of the ranking task, we performed with 5 naive human subjects a similar task. This was not meant to be a comprehensive psychophysical experiment, but only to give an indicator on how well humans can spot the right match. The subjects were asked to rank to each line drawing the 10 most similar photos they found in the database. They were

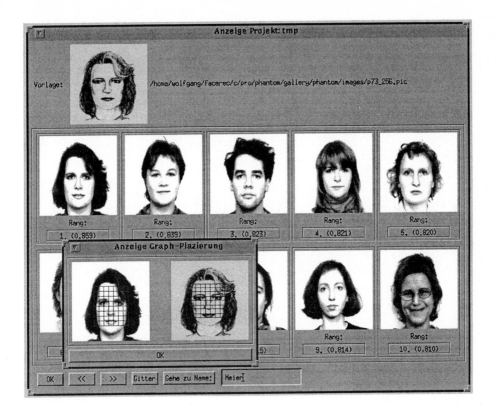

Fig. 3. The result of a PHANTOMAS search on a line drawing (upper left) where the true match is found at place 1 of the ranking list. The small inset shows how the graphs are positioned on the faces in order to account for the different size and aspect ratio.

allowed to pick less than 10 persons only if they were sure, that no one of the non-considered persons was the one actually portrayed by the forensic artist. [2]

As we see from Tab. 1, none of the subjects was able to find all true matches. There is no clear prescription on how to calculate the mean ranking, if some true matches were not found by the subject. We have chosen the following prescription: Since the gender of the line drawings was always correctly classified, even in the 'n/f'-cases, a hypothetical forced choice in the 'n/f'-cases would produce on the average at least the same ranking as random choice ($p = 1/2$) on the same-gender persons:

$$\tilde{R} = \begin{cases} pN_{same-gender} & \text{if true match n/f} \\ R & \text{else} \end{cases} \tag{2}$$

---

[2] It does not make sense to force the subjects to rank *all* 103 database members, since the results would be in most cases non-reproducible.

picture	gender	PH	NC1	NC2	subjects				
					A	B	C	D	E
# 1	m	6	75	50	2	n/f	2	n/f	2
# 2	m	3	100	65	1	1	1	10	n/f
# 3	m	2	3	1	1	1	1	n/f	2
# 4	m	1	7	29	1	1	1	5	1
# 5	f	1	1	14	2	1	1	2	1
# 6	m	1	2	1	1	1	1	3	1
# 7	m	1	6	2	1	1	1	1	1
# 8	m	1	18	4	1	1	1	n/f	1
# 9	m	1	3	1	1	1	1	1	1
# 10	f	13	17	1	n/f	3	n/f	n/f	1
# 11	m	3	25	28	1	1	1	1	1
# 12	m	1	1	2	1	1	1	n/f	1
# 13	f	1	1	1	3	n/f	1	1	1
Median:		1	4.5	2	1	1	1	4	1
Mean:		2.7	19.9	14.6	-	-	-	-	-
Mean w/o # 10:		1.8	20.2	16.4	1.3	-	1.1	-	-
$\langle \tilde{R} \rangle$:		2.7	19.9	14.6	2.1	3.6	1.8	9.9	2.8

Table 1. Performance evaluation: For all 13 facial line drawings a ranking of the 103 database images was performed. Shown is the place R of the true match in the ranking list (1=best match). **PH**: PHANTOMAS, **NC1**: Normalized cross correlation without warping, **NC2**: Normalized correlation with warped images, **A-E**: five different human subjects (n/f: true match was not found).

We have chosen $p = 1/3$ when calculating the expectation value $\langle \tilde{R} \rangle$ in Tab. 1.

3.4 Gender Classification

An interesting feature of the presented face recognition algorithm is its ability to pick up complex facial characteristics (e.g. "is of Basedow-type") which cannot be contributed to a single facial feature alone.

As an example we consider here the male/female distinction. The gender classification has been previously described in the context of Elastic Graph Matching by Wiskott et al. [8] with good results. We show here that with the PHANTO-MAS software and a very simple prescription the composite characteristicum "gender" can be reliably identified even for line drawings:

The search template of a certain search project is classified "female", if the number of females within the first K persons from the ranking list is greater than the random-choice expectation value E_F.

If there are N_F females in a database of N persons in total, then the probability for having r females in a sample of K persons is given by the hypergeometric

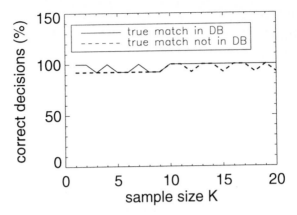

Fig. 4. Gender classification of facial line drawings based on the PHANTOMAS ranking. The "oscillations" between 92% and 100% originate from exactly one of the line drawings, this one being for PHANTOMAS "on the edge" between male and female.

distribution $p(r, K, N)$. The expectation value is

$$E_F = \sum_{r=0}^{K} r\, p(r, K, N) = K\frac{N_F}{N} \tag{3}$$

The results in Fig. 4 show that the correct gender decision is more or less independent from the sample size K and works even if the true match is not in the database. The correct-decision rate of better than 90% is comparable to results reported in other works [8, 4].

The same mechanism should also work for other complex characteristica X as long as the database provides representative material for the both cases "belongs to class X" and "does not belong to class X". This could provide a mechanism for an objective labeling of facial pictures with respect to those characteristica.

4 Discussion

It has been shown that Elastic Graph Matching can successfully recognize faces from facial line drawings. In all cases the true match was within the upper rankings.

For comparision the ranking task was also performed by human subjects. The line drawings difficult for PHANTOMAS (especially # 1 and # 10 in Tab. 1) were also difficult for the human subjects. The overall performance of PHAN-TOMAS and human subjects seems to be comparable although an "average ranking" is difficult to extract from the subjects' results due to the missing va-lues (each subject had at least one "not found"-case). We developped with $\left\langle \tilde{R} \right\rangle$ a simple model for the average ranking in the case of missing values. Tab. 1

shows that PHANTOMAS' $\langle \tilde{R} \rangle$ is close to the median of the human subjects' $\langle \tilde{R} \rangle$.

Why are Gabor wavelets so successful for the recognition of line drawings? We have no clear answer to this question, but one reason might be that dominant features of line drawings are the orientation of bars and step edges, and the Gabor code is also dominated by orientation features.

It has been shown that a similar ranking quality can not be achieved by "naive" approaches (e.g. cross correlation) on the same data material.

The results presented in this case study have of course to be extended to larger databases ($N \approx 10000$ or larger). We are currently designing a pilot project for a law enforcement institution to evaluate a 10 000-person database. A simple extension of the current performance to larger databases leads to the expectation that the average rating performance should be at the 2.7%-level, i.e. instead of 10 000 only 270 images would need to be examined manually.

With representative data material for complex facial characteristics, further research will go into automated labeling of complex face types. The aim is to provide an objective basis for the the assignment of textual labels to faces which in turn will make the textual description of faces more reliable.

References

1. V. Bruce and M. Burton. *Processing Images of Faces*. ABLEX Publishing-Corporation, Norwood, NJ, 1992.
2. M. A. Turk and A. P. Pentland. Face recognition using eigenfaces. In *Proceedings of CVPR'91*. IEEE Press, 1991.
3. M. Bichsel, editor. *Proceedings of the International workshop on automatic face- and gesture-recognition (IWAFGR)*, MultiMedia Laboratory, University of Zurich, Switzerland, 1995.
4. R. Brunelli and T. Poggio. Caricatural effects in automated face perception. *Biological Cybernetics*, 69:235–241, 1993.
5. M. Lades, J. Vorbrüggen, J. Buhmann, J. Lange, C.v.d. Malsburg, R.P. Würtz, and W. Konen. Distortion invariant object recognition in the dynamic link architecture. *IEEE Transaction on Computers*, 42:300–311, 1993.
6. W. Konen and E. Schulze-Krüger. ZN-Face: A system for access control using automated face recognition. In M. Bichsel, editor, *Int. Workshop on Face and Gesture Recognition*, 1995.
7. W. Konen, J.C. Vorbrüggen, and R.P. Würtz. Deutsches Patent Nr. 44 06 020: Verfahren zur automatisierten Erkennung von Objekten. Erteilt am 29.06.1995.
8. L. Wiskott, J.-M. Fellous, N. Krüger, and C. von der Malsburg. Face recognition and gender determination. In M. Bichsel, editor, *Int. Workshop on Face and Gesture Recognition*, 1995.

Geometrically Constrained Optical Flow Estimation by an Hopfield Neural Network

A.Branca, G. Convertino, E. Stella, A. Distante

Istituto Elaborazione Segnali ed Immagini - C.N.R.
BARI - ITALY

Abstract. Sparse optical flow is estimated by an image sequence integrating geometrical constraints induced by camera motion and radiometric similarity, with the aim to obtain useful information about egomotion in planar passive navigation. The OF is estimated matching features (extracted using the Moravec's interest operator) by solving an optimization problem implemented on the Hopfield neural network (HNN).

1 Introduction

Planar passive navigation is the context where the motion of the vehicle is characterized to have mostly a predominant translation component with rotation occurring only around an axis orthogonal to the flat translational plane. So, the 2D motion field, obtained by an image sequence, assumes a typical radial shape, and all vectors radiate from a point, called *focus of expansion (FOE)*, that is a fixed point of the 2D motion field, and is geometrically defined as the intersection of the heading direction with the image plane. A rotation, occurring while the observer is translating, will determine a shifting of the FOE location, by preserving the radial shape. If the OF map is known, two vectors, theoretically, could be enough to locate the FOE in the image, but, in practice, due to the noise, reliable estimations can be obtained by minimizing errors over several vectors. However, this application do not require dense estimation of OF, usually, a sparse, but reliable, OF map can be enough to gather useful information about the observed scene. The estimation of a sparse OF is here formulated as a feature matching problem. The matching methods can be considered as consisting of two steps: *Feature Selection* and *Feature Matching*. In literature several methods to extract feature from images can be found. Those methods can be classified in term of the typology of the selected features: *high-level* and *low-level* features. Usually high-level features (such as lines, curve, closed contours) are considered more reliable for matching problems, but they require a burdensome preprocessing of images that represents the limiting factor for real-time applications [4], [3]. On the contrary, the estimation of low-level features (such as corners, areas at high variance) do not need of a complicate preprocessing, but the matching process is affected by the problem of the ambiguous matches, [8].
In this work we suggest to use low-level features (such as those extracted with the Moravec's interest operator), because their selection make the method more

general and applicable without the necessity to have particular structures in the scene. To manage the problem of ambiguous matches in [1] geometrical constraints (induced by motion) are introduced in a cooperative matching process formulated as an optimization consisting of minimizing an energy function over the features using an Hopfield Neural Network (HNN). The originality of this current work consists in the formulation of a new energy function integrating the approach proposed in [1] with a classical correlation based approach. In order to reduce the computational time and to obtain more reliable matches we propose to consider candidate matches only between features highly correlated.

2   Features Selection and radiometric similarity

The features in the first image are selected by means of the Moravec's interest operator [6]. This operator is particularly interesting for its simplicity. The features are selected as corresponding to areas of high variance. Summarizing, the method works in two stages:

1. The directional variance among neighboring pixels in *four* directions (vertical, horizontal and two diagonal) is computed over a window n×n (in our experiment n = 5) centered on a generic pixel (x, y). The smallest among those four values is called the interest operator value and is considered the variance for point (x, y).
2. Features are chosen where the interest measure has local maxima.

In this way a set of N features: $p_i = (x_i, y_i)$, with $1 < i < N$, in the first image is selected. The best possible candidate matches ($\{p_{ij} = (x_{ij}, y_{ij}) \| j = 1, ..., N_i\}$) for each feature ($p_i = (x_i, y_i)$) are selected in the second image using a correlation based measure (*radiometric similarity*).

3   Motion dependent constraints

Once a first selection of possible matches for each feature is done on radiometric similarity, constraints based on feature trajectories can aid to select for each feature in the first image the best match among those previous considered. These constraints are then used to define the energy function of the HNN used to solve the correspondence problem formulated as an optimization problem. Let us suppose to consider a system (X, Y, Z) fixed with camera and the image plane locate at $Z = f$ and parallel to the plane $X - Y$.

The relationship between the 3D motion vector $\mathbf{V} = (V_X, V_Y, V_Z)$ associated to a point $P = (X, Y, Z)$ and the 2D vector $\mathbf{v} = (u, v)$ obtained as projection of $\mathbf{V}$ on the image plane is given by:

$$
\begin{aligned}
u(x, y) &= \frac{1}{Z}(T_X + xT_Z) = \frac{T_Z}{Z}(x - FOE_x) \\
v(x, y) &= \frac{1}{Z}(T_Y + yT_Z) = \frac{T_Z}{Z}(y - FOE_y)
\end{aligned}
\tag{1}
$$

where, we have assumed that the 3D motion of P is given as $\mathbf{V} = \mathbf{T}$ and $\mathbf{T} = (T_X, T_Y, T_Z)$ is the translational component of 3D velocity. The point $FOE = (FOE_x, FOE_y) = (\frac{T_x}{T_z}, \frac{T_x}{T_z})$ is called *focus of expansion* and in the 2D image plane represents the intersection between the direction of motion and the image plane.

Let us consider two points in the first image indicated as: $p_i = (x_i, y_i)$ and $p_j = (x_j, y_j)$. If we assume that the 2D motion is defined as in (??), the corresponding points in the second image will be given as: $p_{ik} = (x_i + u(x_i, y_i), y_i + v(x_i, y_i))$ and $p_{jl} = (x_j + u(x_j, y_j), y_j + v(x_j, y_j))$.

The two vectors $v_{ij} = p_i - p_j$ and $v_{kl} = p_{ik} - p_{jl}$ will be parallel if the 3D points corresponding to $p_i = (x_i, y_i)$ and $p_j = (x_j, y_j)$ are at the same depth, that is they have the same Z. In real application the condition that points in the 3D world are approximately at the same depth can be considered valid when the distance scene-camera is big and the observed area is small. In any case the (??) is always satisfied by neighborhood vectors.

4 The Point Pattern Matching Problem

The matching problem is formulated as an optimization problem and it is solved in parallel using an HNN [5]. As stated above given N features in the first image we choose $N_1 + N_2 ... + N_N$ points in the second image to represent the matches for the selected features. Each possible match is represented by a neural unit, that is the HNN is constituted of $N_1 + N_2 + ... + N_N$ neurons arranged in N rows of N_i elements. The unit O_{ik}, with $1 < i < N$ and $1 < k < N_i$, represent the match between the features p_i and $p_i k$.

All units give an output in the range $[0..1]$ obtained as follows:

$$O_{ik} = \frac{1}{1 + e^{-\tau I_{ik}}} \tag{2}$$

where I_{ik} represents the internal state of the unit and τ is a positive constant, that defines the slope of the sigmoidal function. The dynamic of a HNN is specified by an updating rule of the internal state defined as:

$$\epsilon \frac{dI_{ik}}{dt} = -I_{ik} + \sum_{jl} W_{ikjl} O_{jl} + B_{ik} \tag{3}$$

where W_{jlik} represents the weight associated to the connection between the units O_{jl} and O_{ik}, and B_{ik} is a bias. These values are chosen on the basis of an energy function associated to the network. The energy function associated to our problem is given as:

$$E = -\frac{K_1}{2} \sum_{ikjl} C_{ikjl} O_{ik} O_{jl} + \frac{K_2}{2} [\sum_i (1 - \sum_k O_{ik})^2] \tag{4}$$

where K_1 and K_2 are two constants used to weight the influence of the corresponding terms in the cost function. The first term of the cost function is used to *measure the compatibility* between the match p_i - p_ik and the match p_j - p_jl. This term is a formalization of the geometrical constraint introduced in section 3. While the second term is used to impose that the sum of the output values on each row is 1. The compatibility between two possible matches is measured as follows:

$$C_{ikjl} = \left(\frac{2}{e^{\lambda(tang(v_{ij}, v_{kl}))^2}} - 1 \right) \frac{1}{Dist(p_i, p_j)} \tag{5}$$

where *tang* indicates the tangent of the angle defined by the vectors v_{ij}, v_{kl} defined in (4). The compatibility function assumes values in the range $[-1..1]$. Negative values are associated to low compatibility and positive values to higher compatibility. The values 1 is assumed when the vectors v_{ij} and v_{kl} are parallel, that is when the geometric constraint defined in section 3 is verified. The constraint term in our method is multiplied for a factor that depends on the distance between the considered matches to introduce a local connection among matches: close matches influence themselves more than far matches. In fact close features in the image have an higher probability to have the same depth and motion and, henceforth, the condition of parallelism of the vectors in (??) has an high probability to be verified. On the basis of (4) the connection values for the network is derived as:

$$W_{ikjl} = K_1 C_{ikjl} - K_2 \delta_{ij} \quad B_{ik} = 2K_2 \tag{6}$$

where $\delta_{ij} = 1$ if $i = j$ and 0 otherwise. So, the dynamic of the network is completely defined. Updating the neuron states according to (3) and (2) an equilibrium point is reached, where $dI_{ik}/dt = 0$, that is a minimum for the *network energy function* (4).

The initial states of the units are initialized to the estimated correlation value. The internal state of the units it is updated synchronously as stated in (3). The output of the units it is computed as in (2). The updating process is stopped when the changes in the neurons states are less than a threshold $\theta = 0.0001$. To each feature $i - th$ of the first image is associated the feature $k - th$ in the second image which neural unit give a higher response with respect the other units in the same row and representing all other candidate matches for the same feature $i - th$.

5 Experimental Results

A first set of experiments has been performed using the synthetic sequence *div-tree* presented in [2] as a test sequence to evaluate the performances of several optical flow estimation techniques. For this sequence the ground-truth 2D motion field is known. Henceforth a quantitative estimation of the error can be derived. In particular, following [2], the angular error between the true 2D velocity vector $\mathbf{v}_c = (x_c, y_c)$ and the estimated vector $\mathbf{v}_e = (x_e, y_e)$ has been computed as:

$$\psi_E = arccos(\mathbf{v_c} \cdot \mathbf{v_e}) \tag{7}$$

where the two vectors, $\mathbf{v}_c = (x_c, y_c)$ and $\mathbf{v}_e = (x_e, y_e)$ are represented as 3D direction vectors $(\mathbf{v}_c = 1/\sqrt{x_c^2 + y_c^+ 1} \cdot (x_c, y_c, 1)^T)$. In figure (1) a typical optical flow map obtained with the proposed method is shown. In figure (2) a graphical

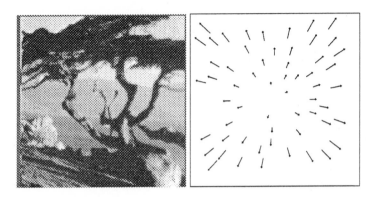

Fig. 1. div-tree. 150×150. The first frame of the sequence and the optical flow computed between the first and the 20th frames. The sequence has been obtained moving the camera along its line of sight (the focus of expansion is in the center of the image) towards a textured tilted plane. The initial focal length is 16 mm and the velocity is constant over the sequence and equal to 0.2 focal lengths per frame.

representation of the average angular error by (7) and the standard deviation for the whole sequence is presented. The camera velocity in this case has been set to the value 0.4 focal lengths per frame, that is its value is double respect to the original one. This camera velocity has been achieved extracting images from the original sequence at a step two. In this way 2 sequence have been obtained, the first one consists of frames (1,3,5,....) and the second consists of frames (2,4,6,...). In this way the motion between two consecutive images is double respect to the original sequence. The results can be considered good compared with [2]. In figures (3) an other experiment performed on the *NASA* sequence, using the first and the 27th frames is shown.

References

1. A.Branca, G.Convertino, A.Distante, Hopfield Neural Network for Correspondence Problems in Dynamic Image Analysis, ICANN95 *International Conference on Artificial Neural Networks*, 1995
2. J.L. Barron, D.J. Fleet, and S.S. Beauchemin, *Performance of Optical Flow Techniques*, Intern. Journal of Computer Vision, Vol. 12, No 1, 43-79, 1994.
3. R. Deriche and O. Faugeras, *Tracking Line Segments*, Image Vision Comput, Vol. 8, 261-271, 1990.

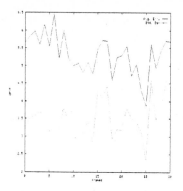

Fig. 2. Average error and standard deviation for the div-tree sequence. Camera velocity equal to 0.4 focal lengths per frame.

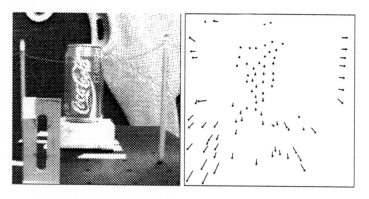

Fig. 3. Nasa-sequence (150 × 150). The sequence is composed by 60 frames (150 × 150), and it has been obtained moving the camera along its line of sight toward the can.

4. H. S. Sawhney and A. R. Hanson, *Trackability as a Cue for Potential Obstacle Identification and 3D Description*, Intern. Journal of Computer Vision, Vol. 11, No. 3, 237-263, 1993.
5. J. J. Hopfield, *Neural computation of decision in optimization problem*, Biol. Cyber., Vol. 52, pp. 141-152, 1985.
6. H.Moravec, *The Stanford Cart and the CMU Rover*, Proc. IEEE, Vol.71, No. 7,pp. 872-878, 1983.
7. N. M. Nasrabadi, C.Y. Choo, *Hopfield Network for Stereo Vision Correspondence*, IEEE Trans. Neural Networks, Vol. 3, No. 1, pp. 5-13, 1992.
8. T. Wu, R. Chellappa, Q. Zheng, *Experiments on Estimating Egomotion and Structure Parameters Using Long Monocular Image Sequence*, Intern. Journal of Computer Vision, Vol. 15, No. 1, 77-103, 1995.

Serial Binary Addition with Polynomially Bounded Weights

Sorin Cotofana and Stamatis Vassiliadis

Delft University of Technology, Dept. Electrical Engineering,
P.O. Box 5031, 2600 GA Delft, The Netherlands

Abstract. This paper presents a new approach to the problem of serial binary addition with feed-forward neural networks. It is shown that the serial binary addition, an important component to a number of applications such signal processing, of two n-bit operands with carry can be done by a feed-forward linear threshold gate based network with polynomially bounded weights, i.e. in the order of $O(n^k)$, associated with small delay and size. In particular, it is shown that the overall delay for the serial addition is $k \log n + \frac{n}{k \log n}$ serial cycles, with the serial cycle comprising a neuron and a latch. The implementation cost of the proposal is in the order of $O(\log n)$, in terms of linear threshold gates, and in the order of $O(\log^2 n)$, in terms of latches. The fan-in is in the order of $O(\log n)$.

1 Introduction

Arbitrary Boolean functions can be computed without learning by feed-forward neural networks [1], with each neuron computing a Boolean function $F(X)$ such that:

$$F(X) = sgn\left(\mathcal{F}(X) = \sum_{i=1}^{n} \omega_i x_i + \psi\right) = \begin{cases} 1 \text{ if } \mathcal{F}(X) \geq 0 \\ 0 \text{ if } \mathcal{F}(X) < 0 \end{cases}. \qquad (1)$$

where the set of input variables and weights associated with the inputs are defined respectively by, $X = (x_1, x_2, \ldots, x_{n-1}, x_n)$ and $\Omega = (\omega_1, \omega_2, \ldots, \omega_{n-1}, \omega_n)$.

Given that Boolean functions are the basis for the computing paradigm, a number of investigations have been reported regarding the possibilities of such an approach in the design of useful Boolean functions. The studies range from symmetric functions[1] [2] to more complex functions, such as multiplication, addition, [3] present in most hardwired computational engines. The investigations mainly concern with the upper/lower bounds for the depth of the networks (worse case delay) and the cost (the size of the network) to be expected in a realization.

While these are extensive studies concerning complex to implement functions, all known results concern parallel circuit implementations. An important

---

[1] A Boolean symmetric function is a Boolean function for which its output value entirely depends on sum of its input values. The Exclusive-Or (parity) of n variables is an example of a symmetric Boolean function.

class of circuit implementations, namely serial implementations of the operations, have not been addressed thus far. Serial implementations constitute an important class of circuit design in that there are numerous applications, such as signal processing [4], that require such design techniques. The reason for such requirements are usually dictated by serial data transmission and implementation cost.

Thus far all the investigations in δ-bit[2] serial architectures assumed logic implementation with technologies that directly implement Boolean gates and no studies have been dedicated to such designs using neural networks.

In this paper we assume LSB first operand reception and investigate δ-bit serial addition in the context of feed-forward neural networks. In our investigation we are mainly concerned in establishing the limits of the circuit designs using threshold based neural networks. That is we are interested in establishing theoretical bounds for delay and size of an implementation. We assume throughout the presentation that small weight sizes are bounded at most by a polynomial size. That is we assume a weight complexity of $O(n^k)$. The main contribution of the paper can be summarized by the followings:

- The addition of two n-bit operands with carry can be performed serially assuming a data transmission rate in order of $O(\log n)$ bits per cycle, in an overall delay of $k \log n + \frac{n}{k \log n}$ serial cycles, with the serial cycle comprising a neuron and a latch.
- The implementation cost of our approach is in the order of $O(\log n)$, in terms of linear threshold gates, and in the order of $O(\log^2 n)$ in terms of latches. The maximum fan-in is in the order of $O(\log n)$.

The paper is organized as follows: in Sect. 2 we introduce the subject and discuss some preliminary results; in Sect. 3 we prove the main results; we conclude with some final remarks.

2   Background and Preliminaries

Generally speaking, the binary addition of two n-bit operands is performed by adding two operands of length n into a single $(n + 1)$-bit number representing the sum. Given that the maximum value of a n-bit number enumerated from $n - 1$ to 0, with the bit enumerated by $n - 1$ being the MSB, is $2^n - 1$ then the binary addition can assume up to $2(2^n - 1)$ different distinct output values. This is equivalent of producing a counter capable of operating on weighted inputs and counting up to the value $2^{n+1} - 2$.

Assuming threshold element feed-forward networks, the least expensive network for Boolean symmetric functions, in terms of linear threshold gates, which

---

[2] Sometime the input data can also come in blocks of δ bits at a time rather than digit by digit or bit by bit. For this type of applications and because it is more general it is more suited to refer to as δ-bit serial computations instead of bit serial or digit serial. We adopt this notation in the remaining of the presentation.

is also capable of producing counting is known to be the Kautz's network [2]. In this paper we will combine binary addition, counters and the Kautz's network for parity function in order to produce δ-bit serial addition.

As an example, the 7 input variables parity function can be computed by a feed-forward network with 3 linear threshold gates. Figure 1 represents this network and also the diagrams for the outputs of the linear threshold gates.

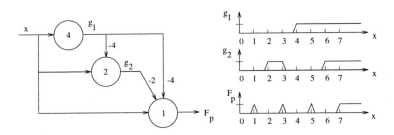

Fig. 1. Feed-Forward Neural Network for 7 Input Variables Parity Function

It can be observed that the output values in Fig. 1 provided by the gates g_1, g_2 and F_p follow the values of the output bits of a $7|3$ counter [5]. This observation leads to an implementation of the $7|3$ counter with only 3 linear threshold gates because we can use the outputs of the gates g_1, g_2 not only to feed-forward the information but also as global network's outputs. This observation holds true also for any $n|r$ counter, where $r = 1 + [\log n]$, because of the following:

- By its definition, any counter's output bit S_i, $i = 0, 1, \ldots, [\log n]$ is equal to 1 inside an interval that includes 2^i consecutive integers, every 2^{i+1} integers, and 0 otherwise. Thus each bit S_i can be described by a periodic symmetric function with the period 2^{i+1}.
- It was shown in [6] that such periodic symmetric functions can be implemented in logarithmic delay and cost with feed-forward neural networks.

In Fig. 2 is depicted the implementation of a $n|r$ counter. Even if the global delay is $1 + [\log n]$ each bit S_i takes the valid value after a delay that is in inverse relation with its significance. Therefore this solution provides first the carry-out of the counter and at the end the LSB of the counter.

If the implementation restriction on the weight sizes are neglected it can be stated that such an $2(2^n - 1)|(n + 1)$ counter is able to perform the addition of two n-bit numbers. This type of counter can be viewed as a serial adder with a delay of $n + 1$. However it is not implementable (2^n exponential weights) and it is rather expensive in terms of delay and in terms of linear threshold gates. In the section to follow we present new schemes for serial addition that substantially improve all of design parameters.

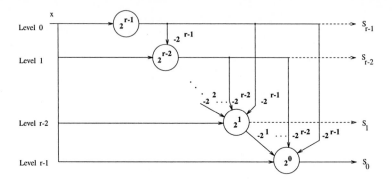

Fig. 2. $n|r$ Counter Feed-Forward Neural Network Implementation

3 Serial Binary Addition

We assume that the operands are applied in a δ-bit serial fashion, δ being an integer greater than or equal to 1. This is equivalent with the partition of the two n-bit operands A and B into $\lceil \frac{n}{\delta} \rceil$ blocks of at most δ bits and assume that the data arrive at the rate of δ-bits per clock period.

If the block pairs are enumerated from 0 to $\frac{n}{\delta} - 1$, with the least significant pair enumerated by 0, the sum bits that correspond to pair i can be computed independently if we know the carry-in into the block i, i.e. the carry-out that results from the summation of the pair $i - 1$. A feed-forward network with $\delta + 1$ linear threshold gates produces the sum bits that correspond to the addition of a pair of blocks. This is because the binary addition of two δ-bit numbers can assume up to $2(2^\delta - 1)$ different distinct values and these values can be produced with a $2(2^\delta - 1)|(\delta + 1)$ counter. Given that the implementation of this counter with linear threshold gates is able to provide the carry-out of a block after the delay of one linear threshold gate we can consider one pair of blocks, i.e. 2δ bits, each clock period. The period of the clock is imposed by the maximum delay of the slowest linear threshold gate in the network. In order to be able to operate in a pipeline environment we have to modify the structure of the counter by adding intermediate pipeline registers between the levels. We have also to add a feed-back line from the output of the first gate to its input register in order to provide the valid carry-in for the next pair of blocks. This modifications are depicted in Fig. 3 for the particular case of the $7|3$ counter, which is equivalent with the addition of blocks of 2 bits with carry (maximum value of 7). In essence the counter in Fig. 3 is able to perform the pipeline addition with carry of pairs of 2-bit blocks. Note that in general for a counter that adds two blocks of δ bits, each input bit enumerated by i, $i = 0, 1, \ldots, \delta - 1$ has assigned a weight value of 2^i.

Given this global scheme we will investigate in the remainder of this section the following issues:

– Assuming a transmission rate of δ-bit per cycle which are the costs and

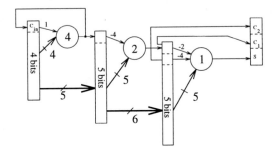

Fig. 3. Pipelined 7|3 Counter

the delay performance of a feed-forward linear threshold based network that implements the serial addition with carry of two n-bit operands[3]?

– What are the consequences of polynomially bounded weight values on the delay, cost and fan-in?

Theorem 1. *Assuming that serial addition is performed with the partitioning of the two operands in $\lceil \frac{n}{\delta} \rceil$ blocks of at most δ bits and that it is implemented with feed-forward network the overall delay is $\delta + \frac{n}{\delta}$ serial cycles. The implementation cost, in terms of linear threshold gates is $\delta + 1$ and $\frac{1}{2}(5\delta^2 + 9\delta) + 2$, in terms of latches. The maximum weight value is 2^δ and the maximum fan-in is $3\delta + 1$.*

Proof. Because we partition the operands in $\lceil \frac{n}{\delta} \rceil$ blocks, each block containing at most δ bits the maximum sum for each block is $2(2^\delta - 1)$. Therefore the sum bits for a pair of blocks can be computed by a $\delta + 1$ stages feed-forward network. The overall delay Δ of the proposed structure is function of the number of stages of the counter, i.e. $\delta + 1$, and of the number of blocks. It is given by (2).

$$\Delta = \delta + 1 + \frac{n - \delta}{\delta} = \delta + \frac{n}{\delta} \ . \tag{2}$$

The structure has $2\delta + 1$ latches at the input in order to be able to memorize the two blocks and the global carry-in, $2\delta + 1 + i$ latches in order to transfer the data between the levels $i - 1$ and i, for $i = 1, 2, \ldots, \delta$ and $\delta + 1$ register bits at the output in order to store the output sum bits. Therefore the overall number of latches is given by:

$$R = 2\delta + 1 + \sum_{i=1}^{\delta}(2\delta + 1 + i) + \delta + 1 = \frac{1}{2}(5\delta^2 + 9\delta) + 2 \ . \tag{3}$$

For this type of feed-forward networks the maximum weight value is 2^δ. The gate with the maximum fan-in is the gate on the level δ of the feed-forward network. For this gate the fan-in is given by the number of bits that participate in one computation step, i.e. $2\delta + 1$, plus the number of gates that are above this gate in the feed-forward network, i.e. δ. This leads to a maximum fan-in of $3\delta + 1$. □

---

[3] We assume the single addition issue and we develop fully pipelined schemes. Consequently the performance we report will improve for back-to-back additions.

Theorem 2. *Assuming polynomially bounded weight values, i.e. in the order of $O(n^k)$, the serial addition with carry of two n-bit operands can be computed in $k \log n + \frac{n}{k \log n}$ serial cycles by a feed-forward neural network with a maximum fan-in in the order of $O(\log n)$. The implementation cost is in the order of $O(\log n)$, in terms of linear threshold gates, and in the order of $O(\log^2 n)$, in terms of latches.*

Proof. Partition the operands into blocks of at most $\delta = k \log n$ bits, with k an integer constant. The maximum weight is given by 2^δ and this partition choice will lead to a maxim weight of $2^{k \log n} = n^k$, i.e. in the order of $O(n^k)$. First implication here is that the data transmission rate is $k \log n$ bits per cycle, dictating the most bits a polynomially bounded weights implementation requires.

The overall delay for the addition in serial cycles is given by $k \log n + \frac{n}{k \log n}$, as a consequence of (2). The cost of the feed-forward network, in terms of neurons, is given by $k \log n + 1$ which is indeed in the order of $O(\log n)$. The cost of the feed-forward network, in terms of latches, is given, as a consequence of (3), by $\frac{1}{2}(5(k \log n)^2 + 9k \log n) + 2$ which is in the order of $O(\log^2 n)$. The maximum fan-in is given by $3k \log n + 1$ and this is in the order of $O(\log n)$. $\square$

4 Conclusions

In this paper we investigated serial binary addition in the context of feed-forward linear threshold gate based networks without learning and with polynomially bounded weights. An overall delay of $k \log n + \frac{n}{k \log n}$ was derived for the addition in the case that the weight values have to be in the order of $O(n^k)$. The implementation cost, in this case, was proved to be in the order of $O(\log n)$, in terms of linear threshold gates, and in the order of $O(\log^2 n)$, in terms of latches. The fan-in is in the order of $O(\log n)$.

Our results are rather important is that they establish the limit of size and delay to be expected in an important class of circuit implementations for the binary addition, an extensively used operation.

References

1. Aleksander, I., Morton, H.: An introduction to neural computing. Chapman and Hall (1990).
2. Kautz, W.: The realization of symmetric switching functions with linear-input logical elements. IRE Transaction on Electronic Computers **10** (1961) 371-378.
3. Siu, K.Y., Bruck, J.: Neural computation of arithmetic functions. Proc. IEEE **78** (1990) 1669-1675.
4. Irwin, M.J., Owens, R.M.: A case for digit serial VLSI signal processing. Journal of VLSI Signal Processing **1** (1990) 321-334.
5. Dadda, L.: Some schemes for parallel multipliers. Alta Frequenza **34** (1965) 349-356.
6. Cotofana, S., Vassiliadis, S.: Periodic symmetric functions with feed-forward neural networks. Proc. NEURAP '95/96 Conference (1996) 215-221.

Evaluation of the Two Different Interconnection Networks of the CNAPS Neurocomputer*

Bertrand GRANADO et Patrick GARDA

Université Pierre et Marie Curie
Laboratoire d'Electronique Analogique et Micro-ondes Boîte 203
4, place Jussieu - 75252 Paris Cedex 05
Tel.: (33 - 1) 44.27.75.07 - Fax.: (33 - 1) 44.27.48.37
email:granado!garda@ufr924.jussieu.fr

1   Introduction

In this paper, we present a general mapping algorithm for MLP onto the CNAPS neurocomputer. We derive an analytical model for the performances of CNAPS for the two major schemes of interconnections between layers. We compare the two interconnection networks of CNAPS for their implementation. Finally we give the performances of two neural systems for character recognition.

2   The CNAPS neurocomputer

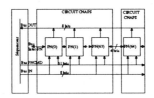

Fig. 1. Computer architecture

CNAPS [1, 5, 8] is a mono-dimensionnal array of processors called PN as shown in figure 1.
There are three bus in this parallel computer, one for transmitting instructions (PNCMD bus), two for broadcasting with a bandwith of 8 bits (OUT and IN), and two for inter-processors communications with a bandwith of 2 bits (Inter-PN). All the simulation were performed with on a CNAPS neurocomputer with 128 PN.

---

* We would like to thank I.E.F and Réseau Doctoral ASM for their support

3 Mapping the MLP on CNAPS

We have evaluated the CNAPS neurocomputer for MLP[2] as they are widely used for pattern recognition. CNAPS was designed to achieve very high performances for their simulation. In MLP, we have two different schemes of connections between layers as shown in figure 2.

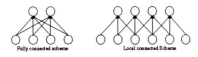

Fully connected scheme   Local connected Scheme

Fig. 2. Full and local connections schemes between layers in MLP

These schemes are the full interconnection and the local interconnection of layers. In the full interconnection, all the neurons of one layer are connected to all the neurons of the other layer. In the local interconnection with a neighborhood of size C, each neuron of the output layer is connected to C neurons of the input layer. Let T_D and T_A be the length of the input layer and output layer respectively. Let r be the size of the overlap between neighborhoods. Then is always possible to add dummy neurons to the layers in order to satisfy the following relationship:

$$T_D = T_A(C - r) + r \tag{1}$$

These different schemes involve different communication volumes. In the general case the global communications volume of the neurons states is $E_i * N_A$, where N_A is the number of input layer neurons. E_i is the number of neurons of the input layer connected to the ith neuron of the output layer. E_i ranges from 1 (local connection with neighborhood of size 1) to N_D (full interconnection).

We introduced a mapping of the layers of a MLP on CNAPS that balances uniformly the workload on the PN. It is well suited for all SIMD computers with a 1-D mesh architecture. The algorithm consist in dividing the MLP into a set of hypercolumn. Each hypercolumn includes at most one neuron of each layer. The set of hypercolumns is divided into groups, each group being mapped to a PN. The hypercolumns are built in such a way that the local interconnections between layers induce local interconnections between hypercolumns. The exchange of the states of neurons going from locally interconnected layers is performed through a local communication between PN. A mapping example is given in figure 3. More details can be found in [3].

---

[2] Multi-Layer Perceptrons

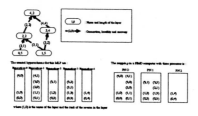

Fig. 3. Hypercolumn Placement : example

4 Implementation

4.1 Software architecture

The simulation of the feeedforward of MLP can be written thanks to three basic procedures [7, 4] :

- Computation of the contribution to the post-synaptic potentials of the neurons of one layer resulting from a local connection to an other layer.
- Computation of the contribution to the post-synaptic potentials of the neurons of one layer resulting from a full connection to an other layer.
- Computation of the neurons states of the neurons of one layer from their post-synaptic potentials.

However, the two procedures computing the post-synaptic potentials require communications between the PN of the CNAPS neurocomputer. Two interconnection networks may be used for these communications: the local inter-PN bus and the global broadcast bus. Thus, we developed two versions of each of these procedures, having a total of 5 basic procedures.

4.2 Optimizations

All these procedures were writtten in CPL, the CNAPS assembly langage, in order to get the highest performances. Moreover, we used CPL to generate microprograms for the PN in order to perform the internal operations of the PN in parallel. In this way we took advantage of several microparallel features of the PN hardware:

- Simultaneous multiply and accumulate operations.
- Pipelining of the communication on the global broadcast bus (there is a 3 stages pipeline).
- Overlap of computations and communications on the broadcast bus.

Table 1. Timing function for the different basic functions

interconnection network	time in number of cycle
	Post-synaptic potential with full connections
broadcast bus	$\alpha + \gamma*NPPD + \varepsilon*NPPD*NPPA + \theta*PD*NPPA*NPPD$
Inter-Processors bus	$\alpha + \varepsilon*NPPD + \theta*NPPD*P + \psi*NPPD*NPPA$
	$+ \phi*NPPD*NPPA*PD$
	Post-synaptic potential with local connections
broadcast bus	$\alpha + \gamma*NPPD + \varepsilon*NPPD*PD + \theta*NPPA + \psi*NPPA*C$
Inter-Processors bus	$\alpha + \gamma*NPPD + \varepsilon*\lceil\frac{C}{2}\rceil + \theta*NPPA + \psi*NPPA*C$
	Computation of the neurons state
	$\alpha + \theta*NPPA$

5 Results

5.1 Analytical models

We introduced an analytical model of the complexity of the 5 basic procedures. Table 2 gives the duration of each procedure as a function of the number NPPD of hypercolumns for the input layer, the number NPPA of hypercolumns for the output layer, the number PD of processors used, and the size C of the neighborhood for local connections. NPPA and NPPD are close to $\lceil\frac{N}{P}\rceil$ where $\lceil\rceil$ is the ceil function, N is the number of neurons of the layer and P is the number of processors in the computer. The values and the expressions α, γ, ε, θ, ψ and ϕ given in table 1 have been estimated both from the number of cycles needed to execute the machine code of each procedure and from experimental measures on executions.

Table 2. Numerical Values for the differents parameters

interconnection network	Time in number of cycle
	Post-synaptic potential with full connections
broadcast bus	$\alpha = 15 + 2*NPPA$; $\gamma = 9$; $\varepsilon = 17$; $\theta = 1$
Inter-Processor bus	$\alpha = 18 + NPPA$; $\varepsilon = 10$; $\theta = 11$; $\psi = 14$; $\phi = 7$
	Post-synaptic potential with local connections
broadcast bus	$\alpha = 38 + 2\lceil\frac{C}{2}\rceil$; $\gamma = 13$; $\varepsilon = 1$; $\theta = 15$; $\psi = 3$
Inter-Processor bus	$\alpha = 39$; $\gamma = 4$; $\varepsilon = 16 + \theta = 15$; $\psi = 3$
	Computation of the neurons state
	$\alpha = 8$; $\theta = 11$

5.2 Asymptotic performance and half-performance length

Then we estimated the r_∞ and $n_{1/2}$ parameters of the CNAPS neurocomputer for the 5 procedures. r_∞ is the maximum or asymptotic performance, which is the maximum performance we can achieve for the procedure. $n_{1/2}$ is the half-performance length which is the length of the layer required to achieve half of the maximum performace [6]. These values are given in Table 3. We use the CPS[3] unit to quantify the speed of CNAPS in the feedforward phase.

$$ r = f\frac{N_c}{N_t} \; CPS $$

where f is the processor frequency, N_c is the number of computed connections, N_t is the number of cycles necessary for computing these connections. Since we can proceed one connection per cycle and per processor, the peak performance r_{peak} of a CNAPS neurocomputer with 128 PN operated at 20 MHz is:

$$ r_{crte} = f\frac{N_c}{N_t} \; CPS = 20.10^6\frac{128}{1} \; CPS = 2,560.10^9 \; CPS $$

Table 3. Values of r_∞ and $n_{1/2}$ for the differents means of communicating and the differents scheme of connections

Scheme of the connection	Interconnection network	r_∞ in MCPS	$n_{1/2}$
Fully connected	Broadcast bus	$\frac{65536}{29} \approx 2259$	$\frac{21982+2\sqrt{222123586}}{290} \approx 179$
	Inter-Processor bus	$\frac{32768}{91} \approx 360$	$\frac{206386+2\sqrt{19302796514}}{910} \approx 532$
Localy connected	Broadcast bus	$\frac{20}{\frac{39}{32}+\frac{3C}{128}}$	$-\frac{-8420-381C-256\lceil\frac{C}{2}\rceil}{156+3C}$
	Inter-Processor bus	$\frac{2560}{19+3C}$	$-\frac{-7405-381C-2048\lceil\frac{C}{2}\rceil}{19+3C}$

6 Benchmarks

We used the simulations of several neural systems as benchmarks to estimate the average performance of CNAPS. These neural systems are derived from applications in the field of handwritten characters recognition. The benchmark programs were written as combinations of the different primitives described in section 4.1. We used two different neural systems designed by two teams actively working on character recognition. The first is a MLP called LeNet [2] developed at AT&T. It includes (4365 Neurons, 1920 full connections and 96522 local connections). The second neural system, called diabolo, is a set of 26 MLP, one for each character. Each MLP includes 1 local connected and 3 fully connected

---

[3] Connections Per Second

layers with 324, 196, 40 and 256 neurons: this results in 816 neurons and 18080 full connections and 5392 local connections per MLP. We did not implement the pre and post processing which are needed by the two neural systems, as we wanted to focus on the performance for neural networks simulations. The two benchmarks are typical of different neural architectures for the same application. LeNet is a very large MLP with numerous layers and receptives fields whereas diabolo is based on a rather large number of medium size fully connected MLP. Table 4 gives the results of the execution of those two benchmarks.

Table 4. Results of LeNet and Diabolos simulations

Name of the MLP	Time of Simulation (ms)	Speed (MCPS)
LeNet	4.8	20.4
Diabolos	24	25.2

7 Discussion

The table 3 shows the impact of the interconnection network on the performances of the 5 basic procedures. The broadcast bus is naturally perfect for fully connected layers. This was a key issue in the design of the CNAPS architecture. For this bus and this connection scheme, the r_∞ performance (2,259 GCPS) is nearly identical to the peak performance (2,560 GCPS), and the $n_{1/2}$ has a small value of 175 neurons per layer. These two parameters provide high performance for fully connected MLPs.

On the other hand the broadcast bus provides low performances for locally connected layers. Let the neighborhood size C be 3. In this case the asymptotic performace, r_∞, is only 46 MCPS, and the half performance vector length is 53. This results from the serialization of the local communications between PN trough the broadcast bus. These parameters provide a drop of performances for most MLP with local connections, when compared to the peak performance r_{peak} of 2,560 GCPS.

Now let us consider the performances provided by the inter-PN bus. When the inter-PN bus is used for communications of fully connected layers, the performances are naturally worse than with the global broadcast bus. From table 3 we see that the asymptotic performance r_∞ is 360 MCPS and the half performance layer length is 532. This leads to performances which an order of magnitude lower than with the broadcast bus for typical MLP. This result from two features. First, the number of cycles required by broadcast operation performed on a ring of PN is proportionnal to the number of PN. Secondly, the inter-PN bus has a low bandwith of 2 bits. On the other hand the inter-PN bus is well suited to the local communication required by local interconnections of layers. Table 3 shows that the asymptotic performance for large values of the size C of the neighborhood is close to 1/3 of the peak performance r_{peak} of the neurocomputer. This results from the architecture of the PN: it is impossible to fetch in a single cycle two operands from memory or two operands from registers. The consequence is that for local connections with inter-PN communications 3 cycles are needed to process each connection.

Finally let us consider the performances achieved on the two neural systems used for character recognition. We observe in table 3 that the average performance is only 1% of the peak performance r_{peak} of the neurocomputer, and about 10% of the performance of the 5 basic procedures. This results from the use of a generic mapping to perform the benchmarks. Better results could achieved by a mapping specific to each MLP, but this would loose the generality of the mapping.

8 Conclusion

In this paper, we presented a general mapping algorithm for MLP onto the CNAPS neurocomputer. We derived an analytical model of the performances of CNAPS for the two schemes, local and full, of interconnections between layers. We compared the two interconnection networks of CNAPS for their implementation. Finally we gave the performance of two neural systems for character recognition. It should be observed that the performances achieved for the simulation of LeNet are only 2 to 4 times slower than those achieved by dedicated hardware such as ANNA.

References

1. Jim Baley and Dan Hammerstrom. Why vlsi implementation of associative vlcns require connection multiplexing. In *Proceedings of the IEEE 2nd Annual International Conference on Neural Network*, pages 112–119, 1988.
2. Y. Le Cun, B. Boser, J.S. Denker, D.henderson, R.E. Howard, W. hubbard, and L.J. Jackel. Handwritten digit recognition with a back-propagation network. In *Neural Information Process and System*, pages 396–404, 1990.
3. Antoine Dupuy and Patrick Garda. Répartition optimale de réseaux de neurones en temps linéaire sur une machine parallèle communicant par bus sécable. In *Actes de renpar'5 - 5ieme rencontres Francophones du Parallélisme*, Mai 1993.
4. Bertrand Granado and Patrick Garda. Une méthode de prédiction des temps de simulation des perceptrons multi-couches sur des machines à parallélisme de données. In *Actes de Renpar'7*, Juin 1995.
5. Matthew Griffin, Gary Tahara, Kurt Knorpp, Ray Pinkham, Bob Riley, Dan Hamerstrom, and Eric Means. An 11 million transistor digital neural network execution engine. In *Proceedings of IEEE International Solid-State Circuits Conference*, 1991.
6. R.W Hockney and C.R Jesshope. *Parallel Computer 2*. Adam Hilger, Bristol and Philadelphia, 1988.
7. Jacques-Olivier Klein, Hubert Pujol, and Patrick Garda. Simulation de la machine de boltzmann en temps réel. *Traitement du Signal*, Juillet 1994.
8. Dean Mueller and Dan Hammerstrom. A neural network systems component. In *Proceedings od International Conference on Neural Network*, pages 1258–1264, 1993.

Intrinsic and Parallel Performances
of the OWE Neural Network Architecture

Nicolas Pican

CRIN-CNRS / INRIA
F-54506 Vandoeuvre-lès-Nancy Cedex, France

Abstract. The OWE (Orthogonal Weight Estimator) architecture is constituted of a main MLP in which the values of each weight is computed by another MLP (an OWE). The number of OWEs is equal to the number of weights of the main MLP. But the computation of each OWE is done independently. Therefore the training and relaxation phases can straightforward parallelized. We report the implementation of this architecture on an Intel Paragon parallel computer and the comparison with its implementation on a sequential computer.

1   Introduction

One of the heaviest problems of ANNs is the execution time, that is often unsuitably high. Parallel implementations of the training and relaxation phases have been studied for many years [3, 2]. One of the most studied parallelization of ANNs is that of the MLP. The main strategy used for its parallelization is to distribute the multiplications of the connection weight matrix W per the neuron output vectors on several processors [1, 4, 10]. Two goals are pursued: the first one is to allow the use of ANNs in real time systems in which the duration of the relaxation phase must be as small as possible. The second one is the trimming of the training phase that can last for several weeks for large training corpora or/and large neural networks sizes.

In this paper we present a straightforward parallel implementation of the OWE architecture, based on the distribution of the modular MLPs composing the OWE neural network. We compare the execution times of this parallel implementation and of the sequential one.

2   The OWE architecture

Our studies [5, 6] have shown that if we remove some inputs, called *external* or *contextual* inputs φ, of a MLP and train it to model a continuous function $f(\mathbf{x}, \varphi)$ all φ be joined, the output of the MLP is a function $g(W, \mathbf{x})$ where each value w_{ij} of the connection matrix W is the expectation of a continuous function $\Omega_{ij}(\varphi)$ over the external input space Φ:

$$w_{ij} = E_\Phi[\Omega_{ij}(\varphi)] \tag{1}$$

In fact, we have shown that the value of the weight of a connection in the MLP is a continuous function only depending of the value of the removed external vector φ. This constation implies that each weight can be independently estimated by the computation of $\Omega_{ij}(\varphi)$. And, because $\Omega_{ij}(\varphi)$ are continuous functions, each one can be modeled by a MLP too.

We now consider our ANN architecture, called OWE architecture, including a main MLP fed by a vector $\mathbf{x}$ in which each connection value $\Omega_{ij}(\varphi)$ is computed by a set of other MLPs (the OWEs) fed by a vector φ.

Each OWE computes:

$$\Omega_{ij} : \Phi \to \Re$$
$$\varphi \to \Omega_{ij}(\varphi)$$

The main MLP computes:

$$g : X \times \Phi \to \Re$$
$$\mathbf{x}, \varphi \to g(\Omega(\varphi), \mathbf{x})$$

2.1 Parallel Relaxation and Training phases

According to this architecture, the relaxation phase, for each presented pattern $(\mathbf{x}, \varphi)$, is done in two sequential phases. The first one is the propagation of the external parameters φ through each OWE to compute each connection value of the main MLP given by the approximation of the $\Omega_{ij}(\varphi)$ function. The propagations of φ through each OWE can be done in the parallel way. After the end of this first phase all the connection values of the main MLP are known. Then the second phase is a classical propagation of the input $\mathbf{x}$ through main MLP.

The training phase begins with a relaxation phase and is followed by a classical backpropagation of the error gradient through the main MLP. When all the gradients are known for all the main MLP connections we use each gradient value as the realization of the OWE output error that is backpropagated through the corresponding OWE. These OWE trainings, called *on-line learning algorithm* [6] can be done in a parallel way too.

Consequently, the first and straightforward parallel implementation of the OWE architecture consists in the distribution of the computation of the $\Omega_{ij}(\varphi)$ functions by the distribution of the OWE neural networks on several nodes.

Let M the number of nodes used in a parallel computer, N the number of connections in the main MLP (the number of OWEs), and T_{Main} and T_{OWE} (respectively T'_{Main} and T'_{OWE}) the execution times of the main MLP and of an OWE for one relaxation phase (respectively for one training phase).

One node is used as the master node for the computation of the main MLP and the management of the OWE message transfers. The other nodes (called *slaves*) are used to manage the OWE computations. Then the global execution time of the OWE architectures for one relaxation phase (T) and one training phase (T') can be approximated by:

$$T = T_R = T_{Main} + \beta T_{OWE} + \mu$$
$$T' = T_R + T_T = T_R + T'_{Main} + \beta T'_{OWE} + \mu'$$

where T_R is the execution time of one relaxation (R), T_T is the execution time of one backpropagation (T). $\beta = \frac{N}{M-1}$ is the number of OWE per node, μ and μ' are the time consumed by message transfers. Test and training corpora must be loaded in memory for a quick relaxation phase succession thus μ, but not μ', depends on two possible data loading strategies. The first one (A), is to minimize the necessary memory used by each slave node for one epoch computation. In this case, $X \times \Phi$ corpus is loaded in the memory of the master node which sends φ vector (q floats, $q = dim(\varphi)$) to each OWE at each relaxation. The second strategy (B), is to minimize μ by loading X corpus in master node memory and Φ corpus in each slave node memory. Thus, at the beginning of a training epoch the master node only sends a random integer value to each slave. This integer is used by each node to initalizate its random generator that will give at each relaxation the current pattern indice used. It is evident that the second strategy is the best in time consumming. But when only one node of a parallel computer has a large memory, this strategy cannot be applied for large corpus. For each strategy, μ can be approximated by:

$$\mu_A = S_{m \to s}(M - 1, q) + S_{s \to m}(M - 1, \beta);$$
$$\mu_B = S_{s \to m}(M - 1, \beta);$$
$$\mu' = S_{m \to s}(M - 1, \beta)$$

where $S_{m \to s}(n, k)$ (resp. $S_{s \to m}(n, k)$) is a continuous increasing function of n and k expressing that the time taken by the transmission of one message of size k octets from the master to n slave nodes (resp. from n slave to the master node). In fact, the importance of the term μ decreases with regard to the size of the OWE neural networks. The bigger there are, the smaller is the part of the communication time in the total execution time.

2.2 Intrinsect performances

These performances are given in terms of number of epochs needed for training and quality of the function modelization of the OWE architecture with respect to a classical MLP. The chosen function is the mathematical function in the boundary space $[-5, 5] \times [-5, 5]$ (Fig. 1) defined by:

$$f(x, y) = xye^{\cos^2 x + \sin^2 y} + \frac{(x + y)^2}{10} - \log(|x + y| + 1) + \sqrt{x^2 + y^2} \quad (2)$$

The training corpus (respectively the test corpus) is composed by $21 \times 21 = 441$ (respectively $101 \times 101 = 10,201$) patterns $(x, y, f(x, y))$ taken from x and $y = \{-5, -4.5, \ldots, 4.5, 5\}$ (respectively taken from x and $y = \{-5, -4.9, \ldots, 4.9, 5\}$). y has been arbitrarily chosen as *external* inputs ($\varphi = y$).

The internal structure of OWE architecture is a 1x12x1 local feedforward MLP with bias neuron (LFF+B) for the main MLP and a 1x6x1 LFF+B (19 connections) for each OWE. There are 37 connections in the main MLP so there are also 37 OWEs. Thus, this OWE architecture includes 703 (37x19) free parameters. y feeds the OWEs and x feeds the main MLP. The internal structure

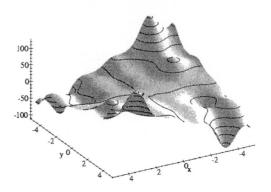

Fig. 1. The desired function $f(x, y)$

of the classical MLP used for comparison, equivalent in terms of number of free parameters (weights), is a 2x30x20x1 LFF+B (**731** connections).

After 5,000 training epochs the mean error left for the classical MLP is 1.13 against 0.83 for the OWE architecture. The speed of convergence during the first epoch trainings is higher for the OWE architecture than for the classical MLP. The OWE architecture reaches 1.13 after only 2400 training epochs when classical MLP reaches 5.4. Other experiments have confirmed these intrinsect performances [7, 8].

2.3 Parallel performances

In this section, we measure the OWE architecture performances in terms of execution time on different numbers of nodes of the parallel implementation. The following table presents nine experiments with nine different OWE architecture sizes.

#	Main MLP	OWEs	N	N_W
#1	1x12x1	1x6x1		703
#2	1x12x1	1x6x3x1	37	1,369
#3	1x12x1	1x12x6x1		4,033
#4	1x12x6x1	1x6x1		2,071
#5	1x12x6x1	1x6x3x1	109	4,033
#6	1x12x6x1	1x12x6x1		11,881
#7	1x24x12x1	1x6x1		6,859
#8	1x24x12x1	1x6x3x1	361	13,357
#9	1x24x12x1	1x12x6x1		39,349

The used parallel computers are two Intel Paragons. The first one (IRISA, Rennes, France) has 59 nodes with: two nodes with 16 and 32 Mo RAM for I/O, one node for boot with 32 Mo RAM, and 56 useable nodes with 16 Mo RAM. The second one (LIP, Lyon, France) has 32 nodes with: one node for I/O with 32 Mo RAM, one node for boot with 32 Mo RAM, and 30 useable nodes including 10 with 32 Mo RAM and 20 with 16 Mo RAM.

For each experiment we measure the execution times for one relaxation (for the strategies A and B) and one training for various numbers of nodes $M - 1 = \{1, ..., 55\}$ (fig. 2 and 3).

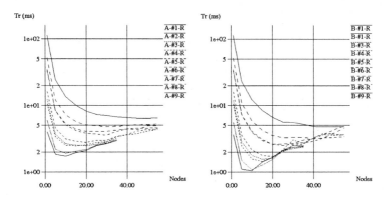

Fig. 2. T_R vs number of nodes, for the strategies A (left) and B (right)

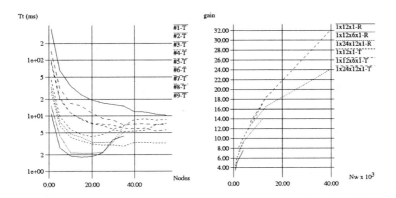

Fig. 3. T_T vs number of nodes (left) and Gain factor vs N_W for strategy B (right)

Figure 3 shows the gain factor in execution time for the relaxation and training phases, between the sequential implementation and the best number of nodes used in the parallel implementation of OWE, with respect to the number of free parameters for each main MLP for the B strategy. The time needed to train the OWE architecture (#1) in the previous section with the B strategy on 15 nodes is 4:30 hours against 14:00 hours on a single node.

3 Discussion

We have presented the results of the parallel implementation of the OWE architecture, that is a main MLP in which each connection weight is computed in parallel by another MLP (an OWE). The OWE architecture is interesting from two points of view. Firstly, its intrinsect performances are at least equal to those of a classical MLP with an equal number of weights. Secondly, its parallel implementation is very straightforward which is very valuable in case of large neural networks. The parallel OWE algorithm complexity is in $O(\beta(N_w/N)^2 + N^2)$ that is always smaller than its sequential implementation ($\beta = N$) $O(\frac{N_w^2}{N} + N^2)$ (for $N > 1$) that is always smaller than for a MLP $O(N_w^2)$ (for $1 \ll N < N_w$). The condition on N can always be reach for large neural networks ($N_w \gg 1$).

We are currently working on a continuous speech recognizer application involving an OWE architecture with a 22x12x6x6 LFF+B for the main MLP and a 44x4x1 LFF+B for each OWE. This architecture has 73,260 connections and the training and testing corpus sizes are respectively 17,000 and 6,000 vectors of dimension 66. 100 training epochs on a sequential computer would take 5 days against 4h30 on the Paragon computer with 30 nodes! Actually, we need more than 100 training epochs; thus it is the parallel implementation that makes the realization of this continuous speech recognizer possible [9].

Acknowledgements

I am grateful to the LIP (Parallel Computer Lab., Lyon, France) and the IRISA (Computer research lab., Rennes, France) for the access to their Intel Paragon computers.

References

1. Anguita, D., Parodi G. and Zunino R.: An Efficient inplementation of BP on RISC-based workstations. *Neurocomputing* **6** (1994) 57–65, Elsevier.
2. Arbib, M.A., Bekey, G., Lyons, D., Sun, R.: Neural architectures and distributed AI. *Workshp proceedings*, Center for Neural Engineering, University of Southern, California, October 19-20, 1993.
3. Barbosa, Valmir C.: Massively parallel models of computation : distributed parallel processing in artificial intelligence and optimisation. Ellis horwood series in Artificial intelligence (col.), Ellis Horwood (ed.), New York, London, Toronto. XII-253 (1993)
4. Kerchoffs, E.J.H, Wedman, F.W. and Frietman, E.E.E.: Speeding up backpropagation training on hypercube computer. *Neurocomputing* **4** (1992) 43–63
5. Pican, N., Fort, J.C, and Alexandre, F.: Lateral Contribution Learning Algorithm for Multi-MLP Architecture. *ESANN'94 Proceedings*, D Facto Brussels Belgium. April 20-22, 1994.
6. Pican, N., Fort, J.C, and Alexandre, F.: An On-line Learning Algorithm for the Orthogonal Weight Estimation of MLP. *Neural Processing Letters*, D Facto Brussels Belgium. **1**(1) (1994), 21–24.
7. Pican, N.: Synaptic efficiency modulations for context integration: The Meta ODWE architecture. In *ESANN'96 Proceedings*, D Facto Brussels Belgium. Bruges April 24-26, 1996.
8. Pican, N.: An Orthogonal Delta Weight Estimator for MLP Architectures. In *ICNN'96 Proceedings*, Washington DC. June 2-6, 1996.
9. Pican, N., Fohr, D., and Mari, J.F HMMs and OWE Neural Network for Continuous Speech Recognition. In Proceedings of *International Conference on Spoken Language Processing, ICSLP*. Philadelphia October 3-6, 1996.
10. Singer, A.: Implementations of artificial neural networks on the connection machine. *Parallel Computing* **14** (1990) 305–315

An Analog CMOS Neural Network with On-Chip Learning and Multilevel Weight Storage

M.Conti, G.Guaitini and C.Turchetti

Dept. of Electronics, Univ. of Ancona, v. B. Bianche, 60131 ANCONA, ITALY

e-mail: maxmc@eealab.unian.it

Abstract

An analog neural network with four neurons and 16 synapses, fabricated in a 1.2 μm n-well single-polysilicon, double-metal process, is presented. The circuit solutions adopted, for on-chip learning and weight storage, particularly simple and silicon area-efficient, are capable of solving the main problems to the implementation of analog neural networks.

1. Introduction

Analog CMOS VLSI technology seems to be very promising for implementing Artificial Neural Networks (ANNs) [1-4].

The main obstacles to the implementation of ANNs with analog VLSI circuits are: i) on chip learning; ii) weight storage; iii) technological tolerances. *On chip learning* is desirable to gain in accuracy. However, learning algorithms such as Backpropagation [3-4], where the weight changes depend on gradient of the error, fail when the error objective function has a great number of local minima as in general occur in practical applications. Circuits for *weight storage* must satisfy essentially two requirements: they must be as simple as possible and they must be able to maintain the information indefinitely. *Technological tolerances* play a key role in analog implementation, because the accuracy attainable in this case is lower than in digital equivalents. Because fabrication tolerances are unavoidable, the only way to overcome this problem is to adopt on chip learning, so that the network is able to learn weights which compensate for technological variations.

In this paper we suggest: i) a stochastic learning approach, ii) a multilevel weight storage solution and iii) an architecture for the neural network capable of solving the problems mentioned above. All these solutions have been adopted and implemented in an analog neural network fabricated with a 1.2 μm CMOS technology.

2. Analog Learning and Multilevel Weight Storage

Let us refer to a given Artificial Neural Network (no dynamical) described by

$$y = \Gamma(w) \, x \tag{1}$$

where $x \in R^p$ is the input vector, $y \in R^q$ the output vector and $\Gamma(w)$ a non-linear operator (in this context it is unnecessary to specify it more precisely) depending on a vector of weights $w \in R^r$.

Supervised learning is equivalent to the following problem:

for a given desired output $\tilde{y}$, change the weights w to find a value w* such that the error $E = f[d(y(w^*), \tilde{y})]$ is minimum, being $d(\,.\,,\,.\,)$ a suitable distance between functions.

Thus learning operation, as involves weight changes with time $w(t, E(t))$, is governed by dynamical equations depending on the instantaneous error E.

In analog neural networks both the time and the variables x, y, w vary continuously, so that learning in this case is represented by a continuous-time differential equation (CTDE). Searching methods based on the gradient of objective function, such as Newton algorithm for instance, are not particularly suited because trial solutions can be trapped into a local minimum. Instead random approaches, such as Simulated Annealing , do not suffer for this drawback since they are able to inspect all the objective function domain. This is equivalent to consider learning procedure as a stochastic process which, in the analog world, is described by a random differential equation. The main properties the procedure must satisfy in this case are: i) the most probable value of weights w must be close to the minimum of E(w) as t→∞; ii) the variance of w must vanish as t→∞.

With these observations in hand, and assuming a low order n for the CTDE to reduce the complexity of implementation, we propose the following random differential equation for weight movement (see also [5] for preliminary results)

$$\frac{dw(t)}{dt} = hv(t) \tag{2a}$$

$$\frac{dv(t)}{dt} = \begin{cases} h[n'(t) - v(t)] & c(t) > 0 \\ 0 & c(t) = 0 \end{cases} \tag{2b}$$

where

$$c(t) = u\left(\frac{dE(w(t))}{dt}\right), \quad E = d(y, \tilde{y}) = |y - \tilde{y}|_{L1} = |e|_{L1} = \sup_X |e(x)|$$

$$n'(t) = \frac{n(t)}{h^2}$$

u(x) is the step function and n(t) a random process vector, whose components are statistically uncorrelated with zero mean and variance vanishing as t→∞. Note that the sup norm has been chosen because it can be easily implemented.

This equation has a very simple meaning. Consider a time instant t, when dE/dt ≥ 0, that is the trajectory of w(t) proceeds in a region where the energy increases, then it is easy too see that the equation becomes:

$$\frac{d^2 w(t)}{dt^2} = -h\frac{dw(t)}{dt} + n(t) \tag{3}$$

This is the well known Langevin's equation describing Brownian motion [6], where n(t) represents a forced term. Thus, in this case, the learning scheme determines the subsequent direction of w randomly, according to eq. (3). It is easy to show that eqs. (2) map into the schematic of fig.1, where the minimum error detector stops the learning phase after an acceptable error has been obtained.

In the learning scheme above the memory elements for weight storage have to be able to maintain a non denumerable (continuous-values) set of stable equilibrium states, because the weights vary continuously in a given interval.

To overcome this problem we suggest a modified learning scheme in which eq. (2a) is replaced by

$$\frac{dw(t)}{dt} = hv(t) + g(w) .$$ (4)

By choosing the function g(w) in such a way that the equation $g(w) = 0$ has n distinct solutions $w_j, j = 1,..,n$, or equilibrium points, with m<n stable points $\{w_j, j = 1,..,m\}$, and assuming that at the end of the learning phase the term v(t) is forced to vanish, the dynamical system $\dot{w} = g(w)$ behaves as a multistable system.

3.   Chip Design

The learning scheme and the weight storage approach suggested above are quite general and can be applied to any neural network. In this paper we will refer to a class of networks, named Approximate Identity Neural Networks (AINNs) and recently proposed in [7], which are particularly suitable to be implemented with analog CMOS circuits. The networks have been developed on the basis of the theory of approximation and are described by

$$y = f(x,w) = \sum_{j=1}^{N} c_j \prod_{i=1}^{m} w_{ij} =$$

$$= \sum_{j=1}^{N} c_j \prod_{i=1}^{m} \left\{ \tanh\left(\frac{n_{ij}\left(x_i - t_{ij}\right) + \sigma_{ij}}{2}\right) - \tanh\left(\frac{n_{Lij}\left(x_i - t_{ij}\right) - \sigma_{ij}}{2}\right) \right\}$$ (5)

A schematic of the learning architecture for a single neuron is depicted in fig.2, and some details for the LEARN block are given in fig.3.

The neural architecture suggested has been implemented using a standard CMOS VLSI technology. Our design methodologies and circuit techniques are based on the assumption that all the MOS transistors operate in the weak inversion (subthreshold) mode, and we assume for the drain current the following model (see, for reference, [8])

$$I_D = I_{ON}e^{\frac{V_{GS}-V_{ON}}{V_T}}\left(1 - e^{\frac{-V_{DS}}{V_T}}\right)\left(1 + \lambda V_{DS}\right)$$ (6)

where λ takes into account the channel modulation effect, while the body effect has been neglected for the sake of simplicity. The choice of working in subthreshold is motivated by the following considerations:
i) the high level of parallelism in neural networks imposes a serious limit on the amount of power that each subcircuit can be allowed to dissipate. Subthreshold operation mode ensures low power dissipation, so that massively parallel architecture can be implemented;
ii) the exponential dependence of I_D on V_{GS} is particularly suitable for the implementation of the required nonlinearities in each neuron, especially for the class of Approximate Identity Neural Networks.
The multistable circuit is the fundamental element for the multilevel weight storage

approach. A schematic of the circuit is given in fig.4a. It essentially consists of four main blocks : i) a current conveyor (transistors M1-M6); ii) the nonlinear two-terminal element f_1 (transistors M7-M15); iii) the non linear two-terminal element f_2 (transistors M16-M25); iv) an output current block (transistors M_{r1} and M_{r2}). The same voltage V_p, generated by M_g and I_M, biases both f_1 and f_2. The output current I_W is a replica of I_1. As it is well known, the behaviour of the current conveyor is such that it results

$$V_1 = V_W \quad , \quad I'_2 = -I_1 \tag{7}$$

for any value of V_W and I_1. If $I_{LEARN}=0$, that is when the LRN signal is low, at equilibrium the current flowing across the capacitor C_W vanishes and $I_{MULT}=0$, so we have

$$I_2 = -I'_2 \quad or \quad I_2 = I_1 \tag{8}$$

Thus the equilibrium points are the solutions of the nonlinear equation

$$f_2(V_W) = f_1(V_W) \tag{9}$$

equivalent to the intersections of the two curves $I_1=f_1(V_W)$, $I_2=f_2(V_W)$. By properly dimensioning the transistors in the layout, the characteristics of f_1 and f_2 turn out to be the ones shown in fig.4b. In this way the shape of $I_{MULT} = I'_2 + I_2$ has 9 intersections with the horizontal axis, corresponding to 5 stable and 4 unstable equilibrium points. This scheme only requires a good matching for the transistors within each multistable circuit, a condition easily satisfied if the devices are placed close enough to each other. Test chips have been fabricated in 1.2 μm n-well single-polysilicon, double-metal CMOS process using CMP facilities. The 5 stable states have been experimentally observed on the chips, and the measured output characteristic of a multistable circuit is shown in fig.4c. A photograph of the chip, which occupies a silicon area of 1300 μm x 1100 μm, is shown in fig.5.

Preliminary measurements performed on these test chips showed a good agreement with the behaviour expected.

References

[1] -   Y.Arima et al., "A Refreshable Analog VLSI Neural Network Chip with 400 Neurons and 40k Synapses", IEEE J. Solid-State Circuits, p. 1854, Vol. 27, Dec. 1992, No. 17.

[2] -   B.L.Barranco et al., "Modular Analog Continuous-Time VLSI Neural Networks with On-Chip Hebbian Learning and Analog Storage", IEEE ISCAS '92, S. Diego, 10-13 May 1992.

[3] -   T.Morie, Y.Amemiya, "An All-Analog Expandable Neural Network LSI with On-Chip Back-Propagation Learning", IEEE J. Solid State Circuits, Vol. 29, No. 9, September 1994, pp.1086-1093.B17.

[4] -   Y.K.Choi, S.Y.Lee, "Subthreshold MOS Implementation of Neural Networks with On-Chip Error Back-Propagation Learning", Proc. of I.J.C.N.N., 1993, pp. 849-852.

[5] -   M.R.Belli, M.Conti, C.Turchetti,"Analog Brownian Weight Movement for Learning of Artificial Neural Networks", Proc. of the European Symposium on Artificial Neural Networks ESANN'95, Brussels, p.75-80, April 19-21 1995.

[6] -   A.Papoulis, "Probability, random variables, and stochastic processes", McGraw-Hill, New York, 1965.

[7] -   M.Conti, S.Orcioni, C.Turchetti, "A Class of Neural Networks Based on Approximate Identity for Analog IC's Implementation", IEICE Trans. on Fundamentals, Japan, Vol. E77-A, No. 6, June 1994, pp. 1069-1079.

[8] -   M.H.Cohen, A.G.Andreou, "Current-Mode Subthreshold MOS Implementation of the Herault-Jutten Autoadaptive Network", IEEE J. Solid State Circuits, Vol. 27, No. 5, May 1992, pp. 714-727.

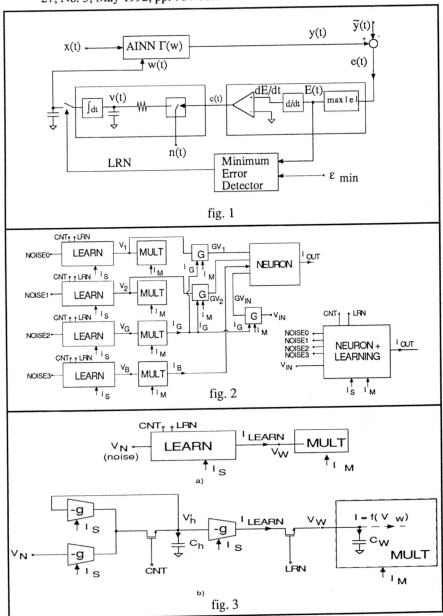

fig. 1

fig. 2

fig. 3

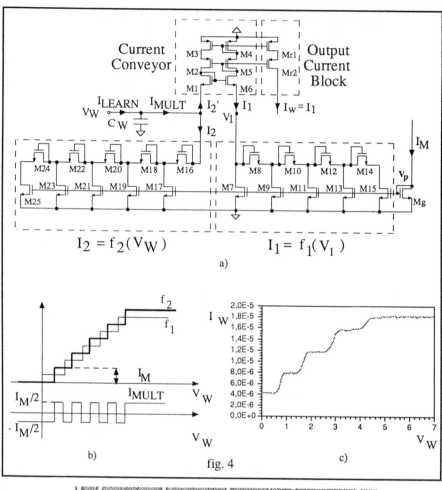

$$I_2 = f_2(V_W) \qquad\qquad I_1 = f_1(V_1)$$

a)

b)

fig. 4

c)

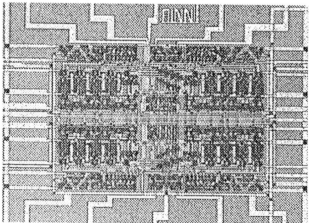

fig. 5

Exponential Hebbian On-Line Learning
Implemented in FPGAs

M. Rossmann[*], T. Jost[*] and K. Goser[*], A. Bühlmeier[+] and G. Manteuffel[++]

[*] University of Dortmund – Department of Microelectronics, D-44221 Dortmund, Germany
[+] University of Bremen - FB 3, D-28334 Bremen, Germany
[++] FBN, D-18196 Dummerstorf, Germany

E-Mail : rossmann@luzi.e-technik.uni-dortmund.de

Abstract. Hebbian learning is a local learning algorithm and allows an on-line adaptation of the weights. Therefore an artificial neural network with built-in hebbian learning is capable of learning on operation. This paper presents the implementation of this algorithm in a digital Field Programmable Gate Array (FPGA). Nonlinearity is introduced by applying nonlinear low-pass-filtering to all input signals as well as using exponential shaped adaptation of weights with different time constants for rising and falling. Employing a complete serial design for the data flow, the implementation of overall 8 synapses in a single FPGA device gets possible. The neuron comprises four conventional synapses with fixed weights and four hebbian synapses for exponential on-line learning. Experiments show the improved performance of this system compared with a linear solution.

1   Introduction

Hebbian learning expresses the ability of an on-line weight variation. The common biological formulation describes the process of learning as strengthening the connection between two neurons when both the pre- and the postsynaptic neuron are excited simultaneously [1]. Composing a unit that consists of a postsynaptic neuron and synapses the learning algorithm only requires local signals. Thus it is especially suitable for VLSI integration regardless whether it is analog or digital. A first approach has been done in a Field Programmable Gate Array (FPGA) with linear updating of the weights as described in [2]. The FPGA represents an ideal platform for prototyping and feasibility tests. Experiments show interesting results when applying this implementation to a real sensory-mechanical system equipped with ultrasonic and tactile sensors [3]. Here, the time constants for the learning process have to be larger than decay terms of the weight to obtain the demanded behaviour of correlating two sensory signals and avoiding to hit obstacles when moving forward. Using a nonlinear variation of the weight update, the general performance of the network can be increased. Therefore, an updating rule is implemented using an exponential shaped function with different time-constants for rising (learning) and falling (forgetting) that is psychologically motivated [4]. In order to obtain comparable results we use the same external testboards as for the linear model and the same FPGA device type XILINX-3090. Concluding experiments show much better performance applying this approach to the system than the linear model as it is shown in section 4.

2 Theory

The central algorithm for on-line Hebbian learning with exponential shaped updating of the weight is given by the following discrete time rule

$$h_j(k+1) = \begin{cases} h_j(k) + a_j\big(h_{max} - h_j(k)\big) & : & Sig_{oth} \geq \Theta_{oth} \text{ and } Out_{lpf} \geq \Theta_{lpf} \\ h_j(k) - b_j h_j(k) & : & Sig_{oth} \geq \Theta_{oth} \text{ and } Out_{lpf} < \Theta_{lpf} \\ h_j(k) - c_j h_j(k) & : & Sig_{oth} < \Theta_{oth} \end{cases} \quad (1)$$

Here, h_j, h_{max}, k, a_j, b_j and c_j denote the weight of the j-th synapse, the maximum weight, the discrete time-index and the time constants for learning, active forgetting and long-term forgetting, respectively. The terms Sig_{oth}, Θ_{oth}, Out_{lpf} and Θ_{lpf} indicate the sum of all synapse outputs except the updated one, the threshold value for this sum, the output of the synapse after nonlinear low-pass-filtering and the threshold value for this signal. The output of the neuron is computed by clipping the weighted sum of all synaptic outputs. A detailed description of the whole mathematical foundation can be found in [2].

3 Serial Design

An implementation of the neural model in an FPGA is possible using parallel integration and results in high speed processing. However, the realization of mathematical computation in a parallel way is quite resource consuming and not applicable to FPGA in our case [2]. Therefore a serial data flow is applied to the whole structure leading to a radical decrease of resource consumption. Regarding the possible bandwidth of the serial implementation the speed of computation is sufficient for experiments with the reference mechanical system.

Principle of feedback loop

To achieve a design that is suitable for FPGA implementation a feedback loop with switched feedback factors for rising and falling is used as depicted in fig. (1). Thus the time constants for the rising and the falling edge of the exponential response are different. A comparator is needed to detect whether the output is larger (falling case) or smaller (rising case) than the input. This comparator switches the feedback factor β to β_r or β_f for rising and falling, respectively. This approach shows the following advantages: Applying the serial data flow the local shift register is used to store the signal for the feedback loop and acts as a delay at the same time. Besides, the comparator, the adder and the feedback loops are also performed as one-bit units. With the constraints for

$$\beta_{r,f} = 2^{k_{r,f}} \quad (2)$$

the different feedback factors can be realized by hard-wired pickoffs at the shift register with no additional effort [2].

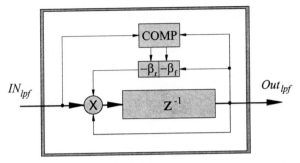

Fig. 1. Low pass filter with nonlinear feedback loop. The delay unit in the centre is realized as a shift register and serves as a local memory. The comparator switches the two different feedback factors β_r and β_f for the rising or falling edge, respectively.

The time constants of this structure strongly depend on the feedback factors as well as on the clock frequency. A mathematical expression for linear feedback filters with an infinite impulse response (IIR) is given in [5]. The time constants for the rising and falling edges are determined by

$$\tau_{r,f} = \frac{k'}{\beta_{r,f}} \cdot \frac{1}{f_g} \tag{3}$$

with τ, β, k' and f_g denoting the time-constant, the feedback factor, the number of clock cycles for one complete computation and the global clock frequency, respectively. The indices r and f mark the quantities for rising and falling. With a fixed k' the term $\beta_{r,f}$ and f_g determines the time-constant $\tau_{r,f}$.

Application to Short Term Memory and Weight Update

The application of the nonlinear feedback is applicable to the short term memory realized as a low-pass-filter as well as to the update units for the weights. In both cases an exponential shaped function has to be implemented. The schematic in fig. (1) shows the abstract concept for the low-pass-filter.

For storing and updating the weight, the structure can be used in a similar way. When learning is activated, the input to the unit is switched to the maximum weight h_{max} whereas for active forgetting the minimum weight $h = 0$ is applied to the input. An explicit storage of the signal is again not necessary because the delay unit acts as a memory at the same time. Long-term forgetting is not implemented caused by the enormous amount of weight discretization when using high clock frequencies.

The advantage of this design is obvious: A single FPGA device allows the implementation of overall 8 synapses with four conventional and four adaptive synapses. The conventional inputs have fixed weights whereas the adaptive synapses

770

Multiplication, Weight-Storage and Update   Output-Confinement

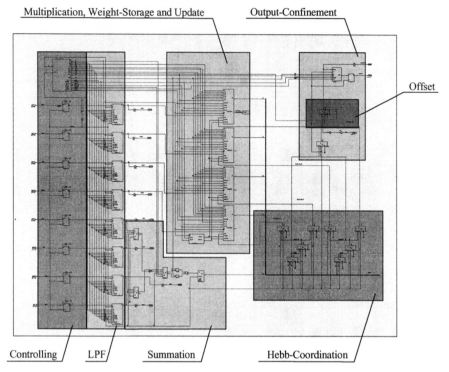

Offset

Controlling   LPF   Summation   Hebb-Coordination

Fig. 2. Top schematic for 8 synapses (left) and one output (top right) with marked areas for controlling, filtering, summation, learning and storage

implement the exponential Hebbian on-line learning. Because the basic structure of the neuron has already been described in [2] with more detailed block diagrams we here present the top schematic for the whole implementation as depicted in fig. (2). The shift register in the low-pass-filter amounts to 19 bit while the weight memory is discretized in 30 bit. Testing has been performed up to 1 Mhz clock frequency successfully and was confined by external hardware serving the necessary bit-serial data flow for input and converting the bit-serial output back to analog signals.

4 Experimental Analysis

A reference sensory mechanical system as described in [3] serves as a platform for testing purposes. Both the linear and the exponential shaped learning functions are applied. The system is equipped with sonar sensors as well as with bumpers. The output of the sonar signal is connected to two Hebbian synapses and increases linearly when an object moves towards it. When the object hits the bumper it excites the neuron by applying a constant signal to two excitatory synapses. This excitation forces the system to move backwards. As the excitation *and* the output of the Hebbian synapse are coincident and exceed the threshold level shown in eqn. (1) the weight increases as depicted in fig. (3).

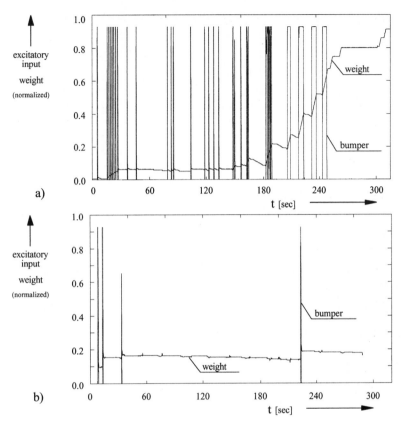

Fig. 3: Plots of measured excitatory input and weight vs. time with linear (a) and exponential adaptation (b)

Although the time constants are the same for both the linear and the exponential learning, the linear adaptation leads to a weight that is not sufficient to make the system move back on occurrence of only the sonar signal. This fact results in an occurrence of the bumper signal as shown in fig. (3a). In contrast the exponential adaptation leads to a response to the sonar signal after hitting the bumper only three times as depicted in fig. (3b). After this the system moves back when the sonar detects an object without touching the bumper. This behaviour is due to the different weight adaptation slope (see eqn. (1)). Caused by active forgetting, the weight slightly decreases and leads to a bumper signal after approximately 220 sec.

An interesting effect shows fig. (3a) in the left part: After recognizing that the weight did not increase any longer, the learning was forced by pushing the bumper manually. As the weight was exceeding a certain level the sonar signal was sufficient for further increase of the weight due to self-stimulation. This effect takes place because the sonar signal is connected to 2 Hebbian synapses and so *coincidence is inherent*. So the learning condition in eqn. (1) is always met by only the sonar signal and the weight increases even on absence of an input at the excitatory synapses.

Resource Consumption FPGA (XC3090)							
	total	2 *linear* Hebbian synapses Controller		2 *exponential* Hebbian synapses Controller		4 *exponential* Hebbian synapses	
CLBs	320	294	(92%)	309	(97%)	320	(100%)
Unbonded IOBs	74	0	(0%)	0	(0%)	74	(100%)
Bonded IOBs	70	47	(67%)	57	(81%)	64	(91%)

Tab. 1. Resource consumption for several design level

5   Resource Consumption

Three different design are realized implementing the serial data flow. All show four conventional synapses with fixed weights and the characteristics shown in table 1. Using all Configurable Logic Blocks (CLB) and all unbonded Input/Output Blocks (IOB) four Hebbian synapses can be implemented. A unit for external controlling purposes is not implemented caused by the exploited resources.

6   Conclusions and Future Aspects

This paper presents principles of the implementation of the exponential Hebbian learning algorithm in a digital FPGA. The application to a mechanical system shows the fundamental behaviour of the realization. A system consisting of several FPGA-neurons is realized at the moment. A free configurable system including a number of neurons is in preparation as well and will be applied to a more dynamic system in the near future. Furthermore an analog implementation in VLSI is in preparation.

7   Acknowledgement

The authors would like to thank the German Research Society (Deutsche Forschungsgemeinschaft) for their support.

References

[1]   Hebb, D. O., *The organization of behaviour*, J. Wiley, New York, 1949

[2]   Rossmann *et. al.*, *Implementation of a Biologically Inspired Neuron Model in FPGA*, Proc. of MicroNeuro'96, Lausanne, pp 322-330, 1996

[3]   Bühlmeier, A., *Analog Neural Networks in Autonomous Systems*, PhD thesis, Universität Bremen, Germany, forthcoming

[4]   Kimble, G., Garmecy, N., *Principles of General Psychology*, Ronald Press, New York, 3. ed., 1968

[5]   Tietze, U., Schenk, Ch., *Halbleiter-Schaltungstechnik*, Springer Verlag, Berlin, Heidelberg, New York, 9. ed., 1991

An Information-Theoretic Measure for the Classification of Time Series

Christian Schittenkopf[1,2] and Gustavo Deco[2]

[1] Technische Universität München, Arcisstraße 21, 80290 Munich, Germany
[2] Siemens AG, Corporate Research and Development, ZFE T SN 4,
Otto-Hahn-Ring 6, 81730 Munich, Germany

Abstract. We present a practicable procedure which allows us to decide if a given time series is pure noise, chaotic but distorted by noise, purely chaotic, or a Markov process. This classification is important since the task of modelling and predicting a time series with neural networks is highly related to the knowledge of the memory and the prediction horizon of the process. Our method is based on a measure of the sensitive dependence on the initial conditions which generalizes the information-theoretical concept of Kolmogorov-Sinai entropy.

1 Introduction

The first step in modelling or even predicting a time series should always be an analysis of the underlying dynamical system in the sense of a coarse classification. On the one hand, it should become clear how complex or random the time series is and on the other hand, we would like to know how far into the future predictions are meaningful at all. For instance, if the signal is white noise, there are not any correlations to detect and to use for forecasting. If the dynamical system is governed by chaos, the whole past influences the future behaviour, and only short-term predictions will be possible. In the presence of noise an estimate of the noise level might be very helpful in determining the maximal possible accuracy of our model of the time series. If the system is Markovian, we know that the memory is restricted to a finite number of steps.

To this aim we use the information-theoretical concept of conditional entropy. A generalization of the famous Kolmogorov-Sinai entropy [1] is the tool which allows for the classification of a given time series as "pure noise", "purely chaotic", "chaotic but distorted by noise" or "Markovian". The method is based on a measure of the sensitive dependence on the initial conditions which is approximated by the calculation of various correlation functions.

In Sect. 2 we review the fundamental concepts of information theory. A short summary of the topic of correlation functions is given in Sect. 3. In Sect. 4 we explain our method to distinguish different classes of processes. Section 5 includes our experimental results on computer-generated and real-world data sets.

2 Information Theory

The source entropy of a symbol sequence is defined as

$$h^\beta = \lim_{n \to \infty} H^\beta_{(n+1|n,\cdots,1)} \tag{1}$$

where $H^\beta_{(n+1|n,\cdots,1)}$ denotes the conditional entropy of the $n+1$-th symbol given the previous n symbols. The Kolmogorov-Sinai (KS) entropy or metric entropy of a dynamical system is defined as the supremum of the source entropy over all partitions β, i. e.

$$h(1) = \sup_\beta h^\beta. \tag{2}$$

The importance of the metric entropy lies in the fact that a finite and non-zero value of $h(1)$ is characteristic for chaotic time series. The generalization to arbitrary lookaheads p is given by

$$h^\beta(p) = \lim_{n \to \infty} H^\beta_{(n+p|n,\cdots,1)} \text{ and } h(p) = \sup_\beta h^\beta(p) \tag{3}$$

where $H^\beta_{(n+p|n,\cdots,1)}$ is the conditional entropy of a prediction p steps ahead given the previous n symbols and $h(p)$ denotes the generalized KS entropy. $h(p)$ is the maximal possible uncertainty about a future symbol given the whole past. This definition and the profile of $h(p)$ as a function of p are independent of the partition, and therefore they characterize the inherent information flow of the dynamical system. In [3, 4] we proved that for all single-humped maps and for all hyperbolic chaotic systems $h(p) = ph(1)$ holds which means that the prediction uncertainty of a point p steps ahead given the whole past increases linearly with p. An important aspect of (3) is the fact that in general the partition maximizing the uncertainty about a future symbol is unknown. Therefore one usually uses a grid of boxes and drives the box sizes ϵ to zero, i. e.

$$h(p) = \lim_{\epsilon \to 0} h(p, \epsilon). \tag{4}$$

3 Generalized Correlation Functions

The numerical estimation of the generalized KS entropy via the above definitions is rather problematic since one cannot do meaningful statistics for very fine grids of boxes. Therefore we used a variation of the original correlation functions described in [2]. Given a scalar time series $\{x_t\}, t = 1, 2, \cdots, N$, we constructed a higher-dimensional embedding using time delay coordinates

$$\mathbf{x}_t^{n,\tau,p} = (x_t, x_{t+\tau}, \cdots, x_{t+(n-1)\tau}, x_{t+(n-1+p)\tau}), \tag{5}$$

where τ is the time delay, n denotes the sufficiently high embedding dimension, p is the lookahead, and $t = 1, 2, \cdots, \hat{N} = N - (n - 1 + p)\tau$. We then define

$$c_t^{n,\tau,p,\hat{N},\epsilon} = \frac{1}{\hat{N}} \sum_{\hat{t}=1}^{\hat{N}} \Theta(\epsilon/2 - \|\mathbf{x}_t^{n,\tau,p} - \mathbf{x}_{\hat{t}}^{n,\tau,p}\|), \tag{6}$$

where $\Theta(.)$ denotes the Heaviside function. The generalized KS entropy of the time series is approximated via [2]

$$h(p) = 1/\tau \lim_{\epsilon \to 0} \lim_{n \to \infty} \lim_{\hat{N} \to \infty} \frac{1}{\hat{N}} \sum_{t=1}^{\hat{N}} \log_2 \frac{c^{n,\tau,p,\hat{N},\epsilon}}{c^{n-1,\tau,1,\hat{N},\epsilon}}. \tag{7}$$

More precisely, we did not perform $\lim_{\epsilon \to 0}$ but studied the behaviour of $h(p) = h(p,\epsilon)$ in dependence of ϵ, which reveals much more "information" about the underlying system.

4 Distinguishing Different Dynamics

We now use the results of Sect. 2 and 3 to explain our method. In Fig. 1 schematic illustrations of the dependence of the generalized KS entropy on the lookahead p and on the box size ϵ are displayed for different classes of processes. In the following we give some explanations to each field and try to summarize the qualitative behaviour of the processes in a formula, respectively.

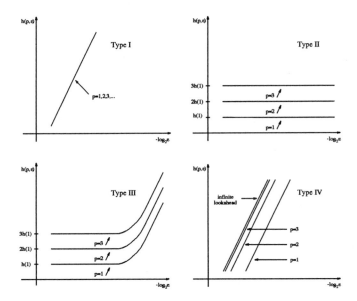

Fig. 1. A schematic illustration of the information flow depending on the lookahead p and on the box size ϵ for white noise (top left), chaos (top right), noisy chaos (bottom left) and a class of Markov processes (bottom right).

In Fig. 1 on the top left we see that the information flow of a stochastic time series (white noise) is characterized by two facts: First, the smaller the box size, the larger the uncertainty since the metric entropy is infinite for a stochastic process. Second, the uncertainty does *not* depend on the lookahead p, because all the information about the system is lost in the first step $(p = 1)$.

$$\text{Type I: } \forall p : h(p, \epsilon) \simeq -\log_2(\epsilon). \tag{8}$$

The situation for a chaotic process is illustrated in the top right-hand corner. If the box size ϵ is small enough, the numerical estimate of the (finite) KS entropy $h(1)$ will be almost constant which results in a horizontal line $(p = 1)$. This typical property also holds for larger lookaheads since we have $h(p) = ph(1)$. Additionally, this equation implies that the distance between successive, horizontal lines equals $h(1)$.

$$\text{Type II: } h(p, \epsilon) \simeq ph(1). \tag{9}$$

The most interesting case for real world applications is a chaotic time series distorted by noise. Looking at the illustration (bottom left) we see that we have a "mixture" of the pictures above. If the precision of our measurements concerning the underlying system is low, we will suspect that the process is purely chaotic. However, lowering ϵ reveals the presence of noise. The horizontal lines suddenly turn upwards, and the entropies diverge. The "knee" in the curves appears at a resolution which is approximately equal to the noise level. Not only do we therefore gain the knowledge that the time series is chaotic and noisy but we also get an estimate of the smallest resolution below which measurements are meaningless.

$$\text{Type III: } h(p, \epsilon) \simeq \begin{cases} ph(1) & : \quad \epsilon \gg 0 \\ -\log_2(\epsilon) + ph(1) & : \quad \epsilon \downarrow 0 \end{cases} \tag{10}$$

Finally we want to look at a class of Markov processes (bottom right), namely the so-called autoregressive (AR) models of order q. The information flow is characterized by the fact that the generalized KS entropies diverge for $\epsilon \to 0$ and that the distance between the lines decreases to zero for $p \to \infty$. More details will be given in the following section (logistic map). We call this behaviour type IV.

5 Experimental Results

We did experiments with the logistic map and with the monthly sunspot time series. To check the validity of our computations we did control experiments with three kinds of surrogate data sets. Method 1 solely consists in mixing up the original time series, i. e. changing the chronological order of the data points. The second procedure is a rescaling of the values in the original time series in a way that they are gaussian. The third way to generate surrogate data consists in approximating the properties of the original time series by an AR model of first order.

5.1 The Logistic Map

We have done several numerical experiments for a logistic map with unknown probability distribution: $x_{t+1} = 3.9x_t(1 - x_t)$. We used 250000 data points to be able to do calculations for very small resolutions ϵ. Figure 2 (left) shows approximations of $h(p, \epsilon)$ for $p = 1$ (lower curve) and $p = 2$ (upper curve). We certainly have behaviour of type II. A calculation of the Lyapunov exponent ($\lambda = h(1) \approx 0.70$) shows that the approximations of $h(1)$ and $h(2)$ are excellent.

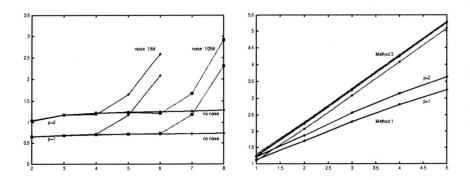

Fig. 2. (Abscissa: $-\log_2 \epsilon$, ordinate: $h(p, \epsilon)$) (Left) A comparison of the generalized KS entropies for the purely chaotic and for the noisy, chaotic case (Right) Diverging entropies for two surrogate data sets generated by method 1 and 3.

To deal with a more realistic situation we distorted the logistic map with additive, uniform noise of the form $[-c; c]$ with $c = 2^{-6}$ and $c = 2^{-8}$, respectively, and we did again calculations for $p = 1$ and $p = 2$. As expected behaviour of type III can be seen (Fig. 2 (left)). For large box sizes we cannot distinguish between the purely chaotic and the noisy, chaotic time series. However, if we reach the noise level, the entropies start to diverge. The lower curves in Fig. 2 (right) are the mean value of our calculations' results for three surrogate data sets of method 1. As expected the generalized KS entropy diverges which is in total contrast to the deterministic case. The upper curves in Fig. 2 (right) show the information flow for a first order AR model of the logistic map. The divergence of the generalized KS entropy is even more obvious. We see that the distance between the lines gets smaller as the lookahead increases (type IV). It can be shown that the curves tend toward the function $-\log_2 \epsilon + 0.292$.

5.2 The Monthly Sunspot Data

The yearly sunspot time series has been extensively studied in the literature, but we chose the monthly values[3] since the number of data points (2819) is larger.

---

[3] The authors want to thank the Sunspot Index Data Centre, Brussels, Belgium, for making the monthly sunspot time series available to them.

Figure 3 (left) shows the dependence of the uncertainty on the box size ϵ for lookaheads $p = 1, 2, 3$ and embedding dimensions $n = 1, 2, 4$ ($\tau = 1$), respectively. We see that the generalized KS entropy diverges and that the uncertainty increases slightly with the lookahead which is typical behaviour of type IV. Additionally, one realizes that for large embedding dimensions meaningful statistics can not be done since the correlation functions turn towards zero very quickly.

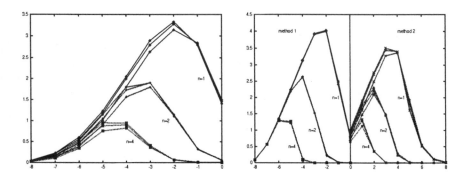

Fig. 3. (Abscissa: $-\log_2 \epsilon$, ordinate: $h(p, \epsilon)$) (Left) The information flow for the monthly sunspot time series (Right) The information flow for two surrogate data sets generated by method 1 and 2.

We also did several control experiments with surrogate data sets generated by method 1 and method 2 (Fig. 3 (right)) with the same values of p and n. As expected the entropies do not increase with the lookahead for the shuffled time series (type I). The behaviour of the information flow of the surrogate data set of method 2 indicates that there is no nonlinearity owing to the measurement procedure. Summarizing we notice that the monthly sunspot time series has a Markovian information flow.

References

1. Eckmann, J. P., Ruelle, D.: Ergodic Theory of Chaos and Strange Attractors. Rev. Mod. Phys. **57** (1985) 617–656
2. Leven, R. W., Koch, B. P., Pompe, B.: Chaos in dissipativen Systemen (1994) Akademie-Verlag
3. Schittenkopf, C., Deco, G.: Exploring the Intrinsic Information Loss in Single-humped Maps by Refining Multi-Symbol Partitions. Physica D (in press)
4. Deco, G., Schürmann, B., Schittenkopf, C.: Evolution of the Uncertainty in Chaotic Systems: An Information Dynamics Approach. Proceedings of the 1995 International Symposium on Nonlinear Theory and its Applications, Las Vegas, NV (1995) 481–484

Transformation of Neural Oscillators

Rainer Malaka and Jörg Berdux

Universität Karlsruhe, Institut für Logik, Komplexität und Deduktionssysteme
76128 Karlsuhe, Germany

Abstract. In this paper we outline a method of model transformation for neural oscillators defined by a set of ordinary differential equations and a non-linearity. The transformation sets the parameters such that the transformed oscillators have the same phase and frequency in the state of harmonic balance. In simulations we show that these transformed neural oscillators not only behave equivalently in the state of harmonic balance of one single oscillator, but also mainly equivalent in stationary, oscillatory or chaotic activation states of networks of such elements.

1 Introduction

A variety of models exist that simulate the temporal activity of neurons by oscillators, which generate periodical activation maxima followed by "refractory" activation pauses. It has been shown that, similar to networks with spiking neurons [7], networks with coupled oscillators are able to synchronize, de-synchronize, and to show feature binding [8]. Their collective dynamics are able to mimic those measured in local field potentials [10, 2, 5]. Moreover, such oscillators can also simulate chaotic or quasi-periodic attractors similar to those observed in the olfactory bulb of rats [3, 9, 10].

As with all kinds of mathematical modeling, there is a tradeoff between simplification and adequateness in neuronal modeling. On the one hand, simple models are necessary for feasibility of mathematical analysis and for effective computer simulations. On the other hand, more complex models incorporate more details and are more accurate. It is thus a hard problem to find the right level of abstraction and it is often not easy to decide which neuronal model to use in order to simulate a particular information processing.

In this study, we show that information processing with oscillatory neuronal elements does not crucially depend on the particular neuronal model. We propose transformations that set the parameters of an oscillatory neuronal model in a way that its emergent network behavior is equivalent to that of another model described by a different set of differential equations. Thus many features of information processing in those neural networks are basically not due to the particular neuronal model, but rather properties of the network connectivity.

2 Neural Oscillators and their Transformation

In this paper, we want to investigate oscillators that are modeled by ordinary differential equations (ODEs) and non-linear output functions. One prominent model is the KI oscillator proposed by Freeman and colleagues [3, 6] which incorporates two elements G and M with activation state variables g and m defined by

$$\frac{d^2g}{dt^2}(t) + A\frac{dg}{dt}(t) + B\,g(t) = B\,(w_{mg}f(m(t)) + I_g) \tag{1}$$

$$\frac{d^2m}{dt^2}(t) + A\frac{dm}{dt}(t) + B\,m(t) = B\,(w_{gm}f(g(t)) + I_m) \tag{2}$$

where A and B are constants determining the dynamics of each element, f is a non-linear output function and $w_{mg} > 0$ is a positive weight from the M element to the G element and $w_{gm} < 0$ is an inhibitory weight in reverse direction. External input to the elements, I_g and I_m, can come from sensory input or from other oscillators in the network.

However the elements G and M in the KI model stand for granule cells and mitral cells of the olfactory bulb, they do not simulate single neuronal activity, but are supposed to reflect the activity of *cell assemblies* or as said in Freeman's work *neural masses*.

The model specification given in Eqs.(1,2) use second order ODEs and a non-linear output function f. There is a variety of ODEs and non-linearities f that may define oscillators. Some of these models can be transformed into each other and thus they represent classes of functionally equivalent elements. For this purpose, it is more convenient to express the ODEs as their Laplace transform and to distinguish between activation function and output function.

Using these conventions, one neuron y with summed input x which is modeled according to Freeman's equations (1,2) is now represented as

$$y(t) = f(a(t)) \tag{3}$$
$$A(s) = FD2(s)X(s) \tag{4}$$
$$FD2(s) = \frac{1}{1 + 2dTs + T^2s^2} \tag{5}$$

where $y(t)$ corresponding to $m(t)$ or $g(t)$ is the neural output, which results from application of the non-linearity f to the activation $a(t)$. $A(s)$ is the Laplace transform of $a(t)$, a complex function, that can be expressed as the product of a transfer function $FD2(s)$ and the input $X(s)$ which are defined in the complex frequency domain (s domain). FD2 is a second order fractional delay function which is determined by parameters T and d. If $1 \leq d$, i.e. the muting parameter is high, which is the case for Freeman's model, FD2 can be split into two first order fractional delay functions:

$$FD2(s) = \frac{1}{1 + T_1s}\frac{1}{1 + T_2s} = FD1_{T_1}(s)FD1_{T_2}(s) \tag{6}$$

with positive real-valued time constants T_1 and T_2. Using this notation arbitrary complex neuronal activation functions can be build as the product of $X(s)$ and a transfer function $U(s)$. The transfer function $U(s)$ can be composed as the product of any number of fractional time delays (FD1, FD2), pure time delays

$PD(s) = e^{-Ts}$, proportional transfer functions $P = k$ and other linear time-invariant transfer functions. In the following we classify the resulting neurons by their transfer function, thus Freeman's neuron is a FD2 neuron or a FD1(2) neuron, saying that its transfer function is a FD2 function, or, if we split it, the sequence of two FD1 functions.

Under the assumption of harmonic balance, i.e., the oscillator is approximated by a sine function, we can substitute the nonlinear output functions by so called describing functions in the frequency domain. An oscillator consisting of two such neuronal elements may oscillate in harmonic balance[1], if each element can perform a phase shift $(\angle U(i\omega))$ of $-\pi/2$, i.e., the ODEs defining the transfer functions have to be at least second order ODEs[2]. Besides the phase shift $\angle U(i\omega) = -\pi/2$, the transfer function is characterized by its amplification factor $|U(i\omega)|$.

If we now have a given oscillator consisting of two elements with the same transfer function U, and want to transform it to an oscillator defined by another set of ODEs, we first determine the target transfer function U' by Laplace transformation. In order to have equivalent transfer characteristics of both models in the state of harmonic balance, we have to solve the two equations

$$\angle U(i\omega) = \angle U'(i\omega) = -\pi/2 \tag{7}$$

$$|U(i\omega)| = |U'(i\omega)|. \tag{8}$$

The solution of (7) and (8) yields a parameter set for the target transfer function U' (such as time constants etc.) such that both oscillators have the same characteristics in the state of harmonic balance.

3   Simulation Results

The method described above, allows us for given phase and frequency of one oscillator in the state of harmonic balance, to set the parameters of the other oscillator in a way that phase and frequency stay the same. Thus at least in the state of harmonic balance amplitude, frequency, and phase are the same for both oscillators. An example of such a transformation is given in Fig.1. The figure shows data from three types of activation functions: (a) an activation function which is a sequence of four first-order fractional time delays, thus a FD1(4) model, (b) the Freeman model FD1(2) with two fractional time delays, and (c) a model incorporating a fractional time delay and a pure time delay transfer function, a FD1/PD model. The models have been transformed in order to have the same frequency, phase and amplitude in the state of harmonic balance. The activation traces for input values 4 and 5 show that each model has, in fact, the same overall dynamics, even in states that converge to a fixpoint attractor

---

[1] Whereas we assume that the output function does not cause a phase shift.

[2] Thus two simple FD1 neurons (Hopfield-Tank neurons) are not able to oscillate in harmonic balance, but FD2 neurons, FD1(2) neurons or FD1/PD neurons may oscillate in harmonic balance.

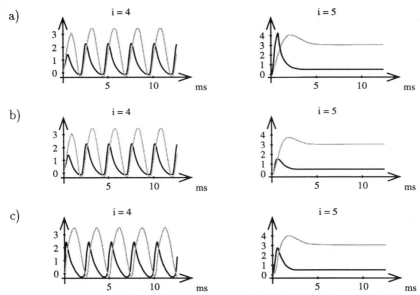

Fig. 1. Activation trajectories of the M cells (black lines) and the G cells (grey lines) for different input values i are displayed for a (a) FD1(4) model (b) FD1(2) model, (c) FD1/PD model, where the model parameters of each model are set to values that result in mutually equivalent phase and frequency in the state of harmonic balance using the technique described in the text.

($i = 5$). An additional computation of the Lyapunov exponents [4] has been performed to investigate the dynamic attractor states of the models in more detail. All three models show similar major Lyapunov exponents in the full range of the input, i.e., the largest exponents which mainly determine the system dynamics are almost identical (Fig.2b). The smaller components, however, differ and thus indicate differences in the fine-scale dynamics which can be also seen in the different onset-peaks in the activation diagrams for the fixpoint attractor (Fig.1).

To further investigate the differences of the various models, an analysis of larger networks is necessary. In a further study, we investigated larger networks with 12 oscillators in an auto-associative pattern retrieval task [1]. We used FD1(4), FD2(2), and FD1/PD elements. The parameters of each type of elements have been set to have equivalent behaviors in the state of harmonic balance. The results showed that the bifurcation points are the same for each model. In each phase, the network converges into a certain kind of attractor [1].

Thus the transformations together with the simulation results demonstrate that the dynamic behavior of these networks does not crucially depend on the transfer function. For a given parameter set for one transfer function, the parameters of another transfer function can be set to values that lead to analogous oscillators with either transfer function. The resulting transformed oscillators show very similar dynamic properties not only as single oscillators, but also as

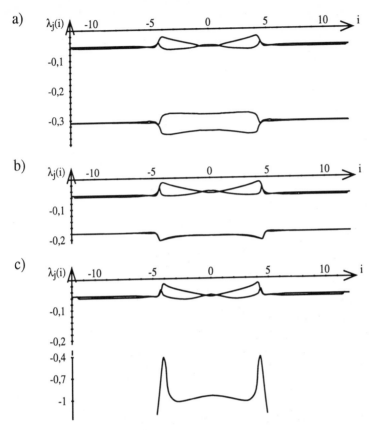

Fig. 2. Input-dependent Lyapunov diagrams are shown for a (a) FD1(4) model (b) FD1(2) model, (c) FD1/PD model with the same parameters as in Fig.1.

networks. Although the transformation was done under the assumption of harmonic balance, the resulting dynamics are still comparable even in stationary, oscillatory or chaotic activation states.

4 Conclusions

In this paper we investigated neural oscillators consisting of two neuronal elements that are defined by a set of ordinary differential equations (ODEs) and a non-linearity. We proposed a new transformation method which, for a given oscillator model defined by a particular set of ODEs, determines a parameter set for another oscillator model which is defined by another set of ODEs. The resulting target oscillator then has the same transfer characteristics in the state of harmonic balance as the original oscillators.

The actual activity traces of both original and target oscillator differ only in minor aspects, e.g, the smaller Lyapunov exponents, but their collective dynamics are qualitatively the same. However the transformation keeps properties only in the state of harmonic invariant, the transformed oscillators show basically equivalent dynamics even in stationary states and show the same bifurcation points in larger networks of coupled oscillators.

It is now possible to adjust the model parameters of a KI oscillator with a given type of neuronal elements in a way that it elicits the same phase, amplitude and frequency as another KI oscillator with another neuronal type. This provides a framework of uniform network transformation and model classification on the basis of the oscillatory behavior of one oscillator. Moreover, the simulations show that as long as small differences in the activity traces are not investigated, the most simple and convenient model can be used and many different neural oscillator models are mutually equivalent.

Acknowledgments

This work was supported by DFG grant Me672/5-3.

References

1. J. Berdux. Synthese und Analyse rückgekoppelter künstlicher neuronaler Netze. Diplomarbeit, Universität Karlsruhe, 1995.
2. R. Eckhorn, R. Bauer, W. Jordan, M. Brosch, W. Kruse, M. Munk, and H.J. Reitboeck. Coherent oscillations: A mechanism of feature linking in the visual cortex? multiple electrode and correlation analysis in the cat. *Biol. Cybern.*, 60:121–130, 1988.
3. W.J. Freeman. Simulation of chaotic EEG patterns with a dynamic model of the olfactory system. *Biol. Cybern.*, 56:139–150, 1987.
4. K. Geist, U. Parlitz, and W. Lauterborn. Comparison of different methods for computing lyapunov exponents. *Progress of Theoretical Physics*, 83(5):875–893, 1990.
5. C.M. Gray, P. König, A.K. Engel, and W. Singer. Oscillatory responses in cat visual cortex exhibit inter-columnar synchronization which reflects global stimulus properties. *Nature*, 338:334–337, 1989.
6. S. Jakubith and W. J. Freeman. Bifurcation analysis of continuous time dynamics of oscillatory neural networks. In A. Aertsen, editor, *Brain Theory*. Elsevier, New York, Amsterdam, London, Tokyo, 1993.
7. R. Malaka and U. Kölsch. Pattern segmentation in recurrent networks of biologically plausible neural elements. In C.H. Dagli, B. Fernández, J. Ghosh, and S.R.T. Kumara, editors, *Intelligent Engineering Systems Through Artificial Neural Networks*, volume 4, pages 639–644, New York, 1994. ASME Press.
8. C. von der Malsburg and W. Schneider. A neural cocktail-party processor. *Biol. Cybern.*, 54:29–40, 1986.
9. C.A. Skarda and W.J. Freeman. How brains make chaos in order to make sense of the world. *Behavioral Brain Science*, 10:161–195, 1987.
10. Y. Yao and W.J. Freeman. Model of biological pattern recognition with spatially chaotic dynamics. *Neural Networks*, 3:153–170, 1990.

Analysis of Drifting Dynamics with Competing Predictors

J. Kohlmorgen[1], K.-R. Müller[1], K. Pawelzik[2]

[1] GMD FIRST, Rudower Chaussee 5, 12489 Berlin, Germany
[2] Inst. f. theo. Physik, Universität Frankfurt, 60054 Frankfurt/M., Germany

Abstract. A method for the analysis of nonstationary time series with multiple modes of behaviour is presented. In particular, it is not only possible to detect a switching of dynamics but also a less abrupt, time consuming drift from one mode to another. This is achieved by an unsupervised algorithm for segmenting the data according to the modes and a subsequent search through the space of possible drifts. Applications to speech and physiological data demonstrate that analysis and modeling of real world time series can be improved when the drift paradigm is taken into account.

1 Introduction

Modeling dynamical systems through a measured time series is commonly done by the reconstruction of the state space using time-delay coordinates [11, 15]. The prediction of the time series can then be accomplished on this basis by e.g. training neural networks [6, 17]. If, however, a system operates in multiple modes and the dynamics is *drifting* or *switching*, standard approaches like simple multi-layer perceptrons are likely to fail to represent the underlying input-output relations. Such time series can originate from many kinds of systems in physics, biology and engineering. Phenomena of this kind are observed in e.g. speech [14], brain data [12, 9], or dynamical systems which switch attractors [2].

For time series from switching dynamics, we already proposed a tailored framework in which an ensemble of neural network predictors specializes on the respective operating modes [3, 8, 13]. Related approaches can be found in [1, 16]. We now extended the ability to describe a mode change not only as a switching but – *if appropriate* – also as a drift from one predictor to another during a certain time interval. Our results indicate that natural signals include drift, which underlines their potential relevance to further applications.

2 Detecting Drifts in the Dynamics

The detection and analysis of drifts is performed in two steps. First, we apply an unsupervised segmentation method, which we already introduced earlier [3, 8, 13]. As a prerequisite of this method, mode changes should occur infrequent, i.e. between two mode changes the dynamics should operate stationary in one mode for at least a certain number of time steps. Thereby, we obtain a (hard) segmentation of the given time series into different modes together with prediction experts for each mode. In case of a drift between two modes, the interval, in which the drift occurs, tends to be subdivided into several parts, because a single predictor is not able to handle the nonstationary drift (see Fig.1). The second step consists in applying a new segmentation algorithm that also allows to model drifts between two stationary modes by combining the two predictors, f_i and f_j, of the respective modes. We model such a drift by a weighted superposition

$$f(x_t) = (1 - a(t))\, f_i(x_t) + a(t)\, f_j(x_t),\tag{1}$$

where $a(t)$ runs from zero to one in the interval $[t_a, t_b]$. For simplicity, the characteristics of the drift that we consider in the experiments in section 3 is a linear transition,

$$a(t) = \frac{t - t_a}{t_b - t_a}.\tag{2}$$

Also nonlinear transitions, e.g. of sigmoidal shape, might be used. The appropriate shape depends on the specific problem considered and prior information can be used.

The proposed algorithm makes a complete search for the optimal segmentation that also takes the described drift into account. The search is performed in the following way: For a given time series, which is not necessarily the training set used in the first step, each of the previously trained experts makes a prediction at every time step, which yields a matrix of expert outputs vs. time steps. This matrix can then be used to compute the total prediction errors for all possible segmentations of the time series, including drifts of arbitrary length but with a fixed characteristics, e.g. eq.(2). The best segmentation with the lowest mean squared prediction error (MSE) is taken as the result (see Fig.1). This problem can be efficiently solved using dynamic programming, implemented similar to the Viterbi algorithm for Hidden Markov Models (HMM) [14]. In fact, we also use a penalty term that is proportional to the number of mode changes in a given segmentation. This is in analogy to a high probability in HMM's to remain in each state and a low probability to switch to another state. In this context, the resulting segmentation can be interpreted as the most likely predictor (state) sequence – allowing drift – that could have generated the given time series under the assumption that mode changes occur infrequent.

3 Application to Speech and Physiological Data

To show the practical relevance of modeling drifts in the dynamics, we now present two examples of applications to real world data: (a) speech data and (b) physiological data of the sleep onset of a human.

Speech The speech data to be segmented are 16-dimensional mel-scale FFT coefficients obtained from continuously spoken vowels (AEIOU, single speaker) at a sampling rate of 16kHz (Fig.2(a)). Thus, in this case, prediction is done

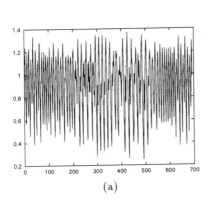

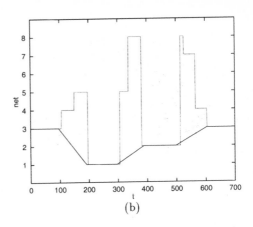

(a)

(b)

Fig. 1. (a) A time series generated by the well-known Mackey-Glass delay differential equation [7]. Three operating modes, A, B, and C, are established by using different delays, $\tau = 17, 23, 30$. After 100 time steps, the dynamics is drifting from mode A to mode B. The drift takes another 100 time steps. Then, the dynamics runs stationary in mode B for the following 100 time steps, whereupon it is drifting to mode C, etc. (b) The segmentation that is achieved after the first (*dotted line*) and the second step (*solid line*) of the algorithm. Shown is the sequence of nets as a function of time. Note that the solid line between 100 and 200 denotes a drift from net 3 directly to net 1. The first segmentation is obtained on the basis of the low-pass filtered prediction errors, which cannot represent the drift. The second segmentation results from the new algorithm, which perfectly reproduces the behaviour of the dynamics in (a).

in the 16-dimensional Fourier space and not directly on the speech signal itself. Eight predictors are used and our hard competition algorithm (step one) is performed on a 16×4 dimensional input vector since a time-delay embedding of $m = 4$ is chosen (cf. [8]). After training, five networks represent A,E,I,O,U respectively (see dotted line in Fig.2(b)), while two nets (1 and 8) specialize on silence and one network (5) does not contribute to the prediction at all. We then applied the new segmentation algorithm using all previously trained predictors (step two). Four of the six transients (solid line in Fig.2(b)) are found to be modeled as linear drifts from one predictor to the other taking 30 ms from A to E, 110 ms from E to I, 80 ms from I to O, and 30 ms from U to silence. The two others – the speech onset of the A, and the transition from O to U – are still modeled as switches from one time step to the next (10ms). This new segmentation has a significantly higher prediction accuracy at the transients than the first one (only there they differ). In particular, the MSE for predicting the 110 ms segment of the E/I-transient is reduced by the factor 0.38, the MSE of the 80 ms I/O-transient yields the factor 0.64. Note that our results imply that the linear drift is a more appropriate model for the transients of the vowel data than a rapid switching of modes.

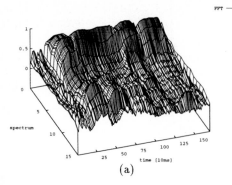

FFT —

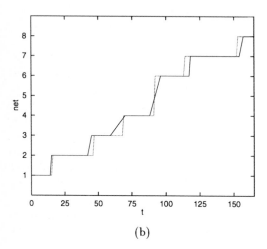

(a)

Fig. 2. (a) 16-dimensional mel-scale FFT coefficients (spectrum) of the continuously spoken vowels AEIOU, plotted over time [10ms]. (b) The segmentation of the time series after the first (*dotted line*) and the second step (*solid line*) of the algorithm.

(b)

Wake/Sleep Data In a previous investigation we analyzed physiological data recorded from the wake/sleep transition of a human (cf. [9]). The objective was to provide an unsupervised method to detect the sleep onset and to give a detailed approximation of the signal dynamics with a better time resolution, ultimately to be used in diagnosis and treatment of sleep disorders. From the neurophysiological point of view, the sleep onset corresponds to a reorganization of the neuronal network in the reticular formation of the brain stem, which regulates and integrates cardiovascular, respiratory and somatomotor systems and vigilance [5]. This reorganization can be thought of as "switching to a different mode" of its dynamics.

We applied our new segmentation algorithm to data from e.g. horizontal eye movements (electrooculogram, EOG) and obtained a better modeling of the dynamical system. The new algorithm yields a drift of 5s at the sleep onset and a more precise segmentation in the further parts of the time series. At the points, where our old and new segmentation do not coincide, we observe that the prediction error can be reduced by a factor of 0.66, when the drift paradigm is incorporated.

Because we find a better predictability, if we consider a drift model for the data, drift must necessarily be a more appropriate model for the wake/sleep data than simple switches. Further details go beyond the scope of this contribution and we refer the reader to our forthcoming publication [10].

4 Discussion and Outlook

It was demonstrated that the modeling of natural data can be improved, when drift is taken into account. Vice versa, our analysis shows that drift is actually found in natural systems and it is therefore relevant to concentrate on this aspect of data description.

By construction, if neither drifting nor switching are inherent in the data set, our algorithm will simply model the stationary signal without any segmentation. This holds equivalently for the case where the data is generated by mode switches, i.e. also in this case no drifts are detected. Clearly, we are at this stage limited to special assumptions about the drift, e.g. linearity or other fixed forms of the transition eq.(2). However, we are free to incorporate all our prior knowledge about the systems investigated into the transition assumptions. If we have no a priori information, as, for example, in the case of physiological systems, where the dynamical state transitions are far from being understood, the goal must be to go one step further and to extract the type of transition from the data in an unsupervised manner and in this way to gain more information for the biological modeling of the underlying physiology. Our future work is therefore dedicated to the extraction of arbitrary drifts.[3]

Acknowledgment: K.P. acknowledges support of the DFG (grant Pa 569/1-1). We thank J. Rittweger for the physiological data and for fruitful discussions.

References

1. Cacciatore, T.W., Nowlan, S.J. (1994). Mixtures of Controllers for Jump Linear and Non–linear Plants. NIPS'93, Morgan Kaufmann.
2. Kaneko, K. (1989). Chaotic but Regular Posi-Nega Switch among Coded Attractors by Cluster-Size Variation, Phys. Rev. Lett. **63**, 219.
3. Kohlmorgen, J., Müller, K.-R., Pawelzik, K. (1994). Competing Predictors Segment and Identify Switching Dynamics. ICANN'94, Springer London, 1045–1048.
4. Kohlmorgen, J., Müller, K.-R., Pawelzik, K. (1995). Improving short-term prediction with competing experts. ICANN'95, EC2 & Cie, Paris, 2:215-220.

---

[3] Further information on related research can be found at:
http://www.first.gmd.de/persons/Mueller.Klaus-Robert.html.

5. Langhorst, P., Schulz, B., Schulz, G., Lambertz, M. (1983). Reticular formation of the lower brainstem. A common system for cardiorespiratory and somatomotor functions: discharge patterns of neighbouring neurons influenced by cardiovascular and respiratory afferents. J.A.N.S. 9: 411-432.

6. Lapedes, A., R. Farber (1987). Nonlinear Signal Processing Using Neural Networks: Prediction and System Modelling. Technical Report LA-UR-87-2662, Los Alamos National Laboratory, Los Alamos, New Mexico.

7. Mackey, M., Glass, L. (1977). Oscillation and Chaos in a Physiological Control System, Science **197**, 287.

8. Müller, K.-R., Kohlmorgen, J., Pawelzik, K. (1995). Analysis of Switching Dynamics with Competing Neural Networks, *IEICE Transactions on Fundamentals of Electronics, Communications and Computer Sciences*, E78-A, No.10, 1306-1315.

9. Müller, K.-R., Kohlmorgen, J., Rittweger, J., Pawelzik, K. (1995). Analysing Physiological Data from the Wake-Sleep State Transition with Competing Predictors, in NOLTA 95: Las Vegas Symposium on Nonlinear Theory and its Applications.

10. Müller, K.-R., Kohlmorgen, J., Rittweger, J., Pawelzik, K., in preparation

11. Packard, N.H., Crutchfield J.P., Farmer, J.D., Shaw, R.S. (1980). Geometry from a Time Series. *Physical Review Letters*, 45:712-716, 1980.

12. Pawelzik, K. (1994). Detecting coherence in neuronal data. In: Domany,E., Van Hemmen, L., Schulten, K., (Eds.), *Physics of neural networks*, Springer.

13. Pawelzik, K., Kohlmorgen, J., Müller, K.-R. (1996). Annealed Competition of Experts for a Segmentation and Classification of Switching Dynamics, *Neural Computation*, 8:2, 342-358.

14. Rabiner, L.R. (1988). A Tutorial on Hidden Markov Models and Selected Applications in Speech Recognition, Proc. IEEE, Vol **77**, pp. 257–286.

15. Takens, F. (1981). Detecting Strange Attractors in Turbulence. In: Rand, D., Young, L.-S., (Eds.), *Dynamical Systems and Turbulence*, Springer Lecture Notes in Mathematics, **898**, 366.

16. Weigend, A.S., Mangeas, M. (1995). Nonlinear gated experts for time series: discovering regimes and avoiding overfitting, University of Colorado, Computer Science Technical Report, CU-CS-764-95.

17. Weigend, A.S., Gershenfeld, N.A. (Eds.) (1994). *Time Series Prediction: Forecasting the Future and Understanding the Past*, Addison-Wesley.

Inverse Dynamics Controllers for Robust Control: Consequences for Neurocontrollers

Csaba Szepesvári[†‡] and András Lőrincz[†]

[†] Department of Adaptive Systems, Institute of Isotopes of the
Hungarian Academy of Sciences, Budapest, P.O. Box 77, Hungary, H-1525
[‡] Bolyai Institute of Mathematics, University of Szeged,
Szeged, Aradi vrt. tere 1, Hungary, H-6720

Abstract. It is proposed that controllers that approximate the inverse dynamics of the controlled plant can be used for on-line compensation of changes in the plant's dynamics. The idea is to use the very same controller in two modes at the same time: both for static and dynamic feedback. Implications for the learning of neurocontrollers are discussed. The proposed control mode relaxes the demand of precision and as a consequence, controllers that utilise direct associative learning by means of local function approximators may become more tractable in higher dimensional spaces.

1 Introduction

Neurocontrollers typically realise static state feedback control where the neural network is used to approximate the inverse dynamics of the controlled plant [1]. In practice it is often unknown *a priori* how precise such an approximation can be. On the other hand, it is well known that in this control mode even small approximation errors can lead to instability [2]. The same happens if one is given a precise model of the inverse dynamics, but the plant's dynamics changes. The simplest example of this kind is when a robot arm grasps an object that is heavy compared to the arm. This problem can be solved by increasing the stiffness of the robot, i.e., if one assumes a "strong" controller. Industrial controllers often meet this assumption, but recent interest has grown towards "light" controllers, such as robot arms with air muscles that can be considerably faster [3]. There are well-known ways of neutralising the effects of unmodeled dynamics, such as the σ-modification, signal normalisation, (relative) dead zone, and projection methods, being widely used and discussed in the control literature (see for example, [2]). Here we propose a new method where the feedforward controller is used in a parallel feedback operation mode thus resulting in robust control. The unattributed statements will be presented elsewhere.

2 Preliminaries

Let $\mathbf{R}^{m \times n}$ denote real $m \times n$ matrices. We say that a real matrix $\mathbf{A}$ admits a generalised inverse[1] if there is a matrix $\mathbf{X}$ for which $\mathbf{A}\mathbf{X}\mathbf{A} = \mathbf{A}$ holds [4]. For

---

[1] Sometimes it is called the pseudo- inverse, or simply the inverse of matrix $\mathbf{A}$.

convenience, the generalised inverse of a non-singular matrix $\mathbf{A}$ will be denoted by $\mathbf{A}^{-1}$. Assume that the plant's equation is given in the following form [5]:

$$\dot{\mathbf{q}} = \mathbf{b}(\mathbf{q}) + \mathbf{A}(\mathbf{q})\,\mathbf{u} \tag{1}$$

where $\mathbf{q} \in \mathbf{R}^n$ is the state vector of the plant, $\dot{\mathbf{q}}$ is the time derivative of $\mathbf{q}$, $\mathbf{u} \in \mathbf{R}^m$ is the control signal, $\mathbf{b}(\mathbf{q}) \in \mathbf{R}^n$, and $\mathbf{A}(\mathbf{q}) \in \mathbf{R}^{n \times m}$. We assume that the domain (denoted by D) of the state variable $\mathbf{q}$ is compact and is simply connected; that $n \leq m$, and for each $\mathbf{q} \in D$ the rank of matrix $\mathbf{A}(\mathbf{q})$ is equal to n; that is, the matrix is non-singular. As a consequence the plant is strongly controllable. In this case the inequality $n < m$ means that there are more independent actuators than state vector components, i.e., the control problem is redundant. Another kind of redundancy, or ill- posedness occurs when $n > m$ in which case even $\mathbf{A}^{-1}$ is non-unique. Further, we assume that both of the matrix fields, $\mathbf{A}(\mathbf{q})$ and $\mathbf{A}^{-1}(\mathbf{q})$ are differentiable w.r.t. $\mathbf{q}$.

Let $\mathbf{v} = \mathbf{v}(\mathbf{q})$ be a fixed n dimensional vector field over D. The *speed field tracking task* is to find the static state feedback control $\mathbf{u} = \mathbf{u}(\mathbf{q})$ that solves the equation

$$\mathbf{v}(\mathbf{q}) = \mathbf{b}(\mathbf{q}) + \mathbf{A}(\mathbf{q})\mathbf{u}(\mathbf{q}). \tag{2}$$

Conventional tasks, such as the *point to point control* and the *trajectory tracking* tasks cannot be exactly rewritten in the form of speed field tracking and speed field tracking is more robust against noise then these conventional tasks. Speed fields must be carefully designed if $\mathbf{A}(\mathbf{q})$ is singular (as in the case of a robot arm).

Given the plant's dynamics by Equation (1) the inverse dynamics of the plant may be written as follows:

$$\mathbf{p}(\mathbf{q}, \dot{\mathbf{q}}) = \mathbf{A}^{-1}(\mathbf{q})\Big(\dot{\mathbf{q}} - \mathbf{b}(\mathbf{q})\Big) + \Big(\mathbf{E} - \mathbf{A}^{-1}(\mathbf{q})\mathbf{A}(\mathbf{q})\Big)\mathbf{y}(\mathbf{q}, t), \tag{3}$$

where $\mathbf{E}$ is the unit matrix and $\mathbf{y} = \mathbf{y}(\mathbf{q}, t)$ is an arbitrary function. Of course, the control signal $\mathbf{u}(\mathbf{q}) = \mathbf{p}(\mathbf{q}, \mathbf{v}(\mathbf{q}))$ solves the speed field tracking control task. *In the following we will look at the main value of the inverse dynamics, i.e., we assume that* $\mathbf{y}(\mathbf{q}, t) = 0$ *and thus we let* $\mathbf{p}(\mathbf{q}, \mathbf{v}) = \mathbf{A}^{-1}(\mathbf{q})\Big(\dot{\mathbf{q}} - \mathbf{b}(\mathbf{q})\Big)$. This assumption simplifies the calculations and can be supported by appropriate learning algorithms.

3 Dynamic State Feedback control

The approximation errors of the inverse dynamics can be viewed as permanent perturbation to the plant's dynamics. Thus we assume that instead of Equation (1) the plant follows

$$\dot{\mathbf{q}} = \hat{\mathbf{b}}(\mathbf{q}) + \hat{\mathbf{A}}(\mathbf{q})\mathbf{u}, \tag{4}$$

where $\hat{\mathbf{A}}(\mathbf{q})$ is another nonsingular matrix field. Let us first assume that we seek a static state feedback compensatory control signal, $\mathbf{w} = \mathbf{w}(\mathbf{q})$, such that the control signal $\mathbf{u}(\mathbf{q}) + \mathbf{w}(\mathbf{q})$ solves the original speed field tracking problem for the perturbed plant.

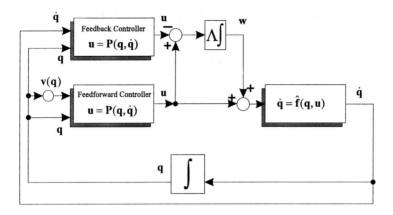

Fig. 1. SDS Control by doubling the role of the inverse dynamics controller

This can be solved by letting $\mathbf{w}(\mathbf{q})$ satisfy the equation $\mathbf{u}(\mathbf{q}) = \mathbf{p}\left(\mathbf{q}, \mathbf{v}\left(\mathbf{q}, \mathbf{w}(\mathbf{q})\right)\right)$, where $\mathbf{v}(\mathbf{q}, \mathbf{w}) = \dot{\mathbf{q}} = \hat{\mathbf{b}}(\mathbf{q}) + \hat{\mathbf{A}}(\mathbf{q})\left(\mathbf{u}(\mathbf{q}) + \mathbf{w}\right)$. Although $\mathbf{w}(\mathbf{q})$ can be explicitly expressed from this it contains terms like $\hat{\mathbf{A}}^{-1}(\mathbf{q})$ and $\hat{\mathbf{b}}(\mathbf{q})$ and thus to estimate $\mathbf{w}(\mathbf{q})$ on- line is approximately the same as retaining the adaptivity of the feedforward controller. This problem can be alleviated by introducing dynamic state feedback for estimating the compensatory control signal. The simplest corresponding error-feedback law is to let $\mathbf{w}$ change until $\mathbf{w}(\mathbf{q})$ satisfies the optimality condition:

$$\dot{\mathbf{w}} = \Lambda\left(\mathbf{u}(\mathbf{q}) - \mathbf{p}(\mathbf{q}, \mathbf{v}(\mathbf{q}, \mathbf{w}))\right)$$
$$\dot{\mathbf{q}} = \hat{\mathbf{b}}(\mathbf{q}) + \hat{\mathbf{A}}(\mathbf{q})\left(\mathbf{u}(\mathbf{q}) + \mathbf{w}\right), \tag{5}$$

where Λ is a fixed positive number, the gain coefficient of dynamic feedback. If the speed of the plant is measurable then Equation System (5) can be realised by a compound control algorithm. The block diagram of the compound controller is given in Fig. 1. The compound controller will be called as the **S**tatic and **D**ynamic **S**tate Feedback Controller (SDS Controller).

If $\mathbf{S}$ is a symmetric real matrix then let $\lambda_{\min}(\mathbf{S})$ denote by the minimal eigenvalue of $\mathbf{S}$. Let $\|\cdot\|$ denote the Euclidean norm and let $\mathbf{z} = \hat{\mathbf{A}}(\mathbf{q})\mathbf{w} - \left(\mathbf{v}(\mathbf{q}) - \hat{\mathbf{v}}(\mathbf{q})\right)$. Simple calculation yields that $\dot{\mathbf{q}} = \mathbf{v}(\mathbf{q}) + \mathbf{z}$ and thus $\mathbf{z}$ is the error of tracking $\mathbf{v}(\mathbf{q})$. In order to guarantee the robustness of tracking it is sufficient to guarantee that the error of tracking is small. Conditions are given in the following theorem:

THEOREM 3.1 *Assume that the perturbation of* $\mathbf{A}(\mathbf{q})$ *can be decomposed as* $\hat{\mathbf{A}}(\mathbf{q}) = \mathbf{D}(\mathbf{q})\mathbf{A}(\mathbf{q})$. *Suppose that* $\mathbf{A}(\mathbf{q})$, $\mathbf{b}(\mathbf{q})$, $\mathbf{v}(\mathbf{q})$ *and* $\mathbf{D}(\mathbf{q})$, $\hat{\mathbf{b}}(\mathbf{q})$ *have continuous first derivatives and the constants* $a = \inf\{\,\|\mathbf{A}(\mathbf{q})\| \mid \mathbf{z} \in D\,\}$, $d = \inf\{\,\|\mathbf{D}(\mathbf{q})\| \mid \mathbf{z} \in D\,\}$, *and* $\lambda = \inf\{\,\lambda_{\min}(\mathbf{D}(\mathbf{q}) + \mathbf{D}^T(\mathbf{q})) \mid \mathbf{q} \in D\,\}$ *are positive. Then for all* $\epsilon > 0$ *there exists a gain* Λ *and an absorption time* $T > 0$ *such that for all* $\mathbf{z}(0)$ *that satisfy* $\|\mathbf{z}(0)\| < K\Lambda$ *it holds that* $\|\mathbf{z}(t)\| < \epsilon$ *provided that* $t > T$ *and the solution can be continued up to time* t. *Here* K *is a fixed positive constant and* $\mathbf{z}(0)$ *denotes the initial value of* $\mathbf{z}$. *Further,* $\Lambda \sim O(1/\epsilon)$.

For convenience $\mathbf{D}(\mathbf{q}) + \mathbf{D}^T(\mathbf{q})$ will be called the *Symmetrised Perturbation matrix* and will be abbreviated to *SP-matrix*. We say that a perturbation of Equation (1) is *non-invertive* or *uniformly positive definite* if λ is positive. The proof is based on a modification of Liapunov's Second Method with the semi-Liapunov function $V(\mathbf{x}) = \mathbf{z}^T\mathbf{z}$. If $\mathbf{z}$ satisfies the conclusions of Theorem 3.1 then it is said to admit the property of uniform ultimate boundedness (UUB).

Discussion

It is an important question whether the theory can be applied to real world situations. Along these lines it was shown that the SDS Control is stable when an idealised robot arm grasps or releases an idealised object (i.e., the mass of the end point changes). More recently, we have successfully applied this scheme to control a chaotic bioreactor with a severely mismatched inverse dynamics model. These results will be published elsewhere.

In neurocontrol the inverse dynamics is modelled by a neural network that may be inherently incapable of realising the inverse dynamics with zero error (i.e., the structural approximation error is non-zero). It is known that mere approximate direct inverse control does not guarantee the (ultimate) boundedness of the tracking error. Can then the boundedness of tracking error be guaranteed if one applies SDS Control? Assume that we approximate $\mathbf{A}^{-1}(\mathbf{q})$ and $\mathbf{b}(\mathbf{q})$ by $\mathbf{P}(\mathbf{q})$ and $\mathbf{s}(\mathbf{q})$, respectively. Then, of course, $\mathbf{A}(\mathbf{q})$ "approximates" $\mathbf{P}^{-1}(\mathbf{q})$. Now turning everything upside down let us imagine that the inverse dynamics of our controller is "exact", i.e., the plant's equation is given by $\dot{\mathbf{q}} = \mathbf{P}^{-1}(\mathbf{q}) + \mathbf{s}(\mathbf{q})$ and Equation (1) is thought of as the perturbed system. We can apply Theorem 3.1 and obtain stability provided that beside some smoothness conditions $\inf_{\mathbf{q}} \lambda_{\min}(\mathbf{D}^T(\mathbf{q}) + \mathbf{D}(\mathbf{q})) > 0$ holds, where $\mathbf{D}(\mathbf{q}) = \mathbf{A}(\mathbf{q})\mathbf{P}(\mathbf{q})$. If $\lambda = \inf_{\mathbf{q}} \lambda_{\min}(\mathbf{D}^T(\mathbf{q}) + \mathbf{D}(\mathbf{q})) > 0$, we say that *the controller represents the inverse dynamics of the plant signproperly.* Under this condition we have that for large enough gains the tracking error is UUB and the ultimate bound on error is proportional to $1/\Lambda$. This also means that *no matter how imprecise the controller is the error of tracking may be arbitrarily decreased by choosing large enough Λs.* The positivity of the symmetrised perturbation matrix follows if $\mathbf{P}$ approximates $\mathbf{A}^{-1}$ sufficiently closely. Consequently, the number of parameters of the estimator can be reduced, which may revitalise the use of fast learning local approximator based neural networks as direct inverse controllers. Besides fast learning the advantage of local approximators is that *a priori* knowledge of the control manifold can be used in training, e.g. when using geometry representing networks one may introduce neighbour training without the loss of generality provided that the inverse dynamics function is smooth.

If the SP-matrix is uniformly positive definite then the use of SDS Control seems to be advantageous during the learning phase. However, if the initial controller does not represent the plant (semi-)sign properly then care is needed. There are two ways to achieve signproper initial estimates. First, one may initialise the controller so that it realises the everywhere zero function, or one may prelearn a 0th stage model until it becomes signproper. Then still one has to ensure that the learning algorithm preserves the signproper nature of approximation. For example, monotone learning admits this property.

When learning is *decoupled* from the control of plant (there is no reference signal during learning) then neither tracking, nor the tracking error and consequently nor SDS Control can be constructed. Thus in the following assume that learning is drived by a reference signal.

When *variational learning* is used (see, e.g. [6]), i.e., when the error of tracking is used for training the controller then SDS Control may delay or even may upset the identification of the inverse dynamics, since the inverse dynamics model may seem more precise than it is in reality. Another problem is that overcompensation may render the learning process unstable: large errors in trajectory tracking may be caused by both the overcompensation and the imprecise feedforward control signal. Consequently, this learning method should be cautiously used together with SDS Control. One solution may be to use temporal backpropagation that takes into account the double role played by the inverse dynamics of the controller by forcing weight sharing. Further research is needed to clarify this point.

In the case of *associative learning*, however, when the training data are given in the form of action-response pairs, it is possible to apply SDS Control during learning, but then the effective control signal should be used in to achieve proper learning. Note, that during learning the tracking of the desired trajectory is more precise with *SDS* Control than without it. This means that neither stability, nor the identification process is affected by SDS Control.

In the proofs we assumed that the perturbation is stationary, i.e., it does not change with time, which is unrealistic. It was shown however, that almost the same method can be applied to nonstationary perturbations. This means that if the changes are slow, i.e., these terms are bounded, then for large enough Λ one may retain the ultimate boundedness of the error signal. However, it is mentioned that in real world applications these changes are usually fast (e.g., when a robot arm grasps a heavy object). In such a case the error signal may become very large and the system may leave the stable region. This can be handled for example by the projection method.

Noise can be handled just as non-stationary perturbations provided that it has bounded amplitude and bandwidth. Note, that to the contrary to linear controllers the present method is capable of compensating such noise that enters the system at the inputs of the controller. However, if the noise enters the system just before the point where the compensatory vector is integrated, i.e., the noise affects $\dot{\mathbf{w}}$, then the system may easily become unstable. This problem, however, is not peculiar to our system, but is shared by every *dynamic* state feedback controller.

4 Conclusions

The so called SDS Control Mode was proposed to compensate inhomogeneous, non-linear, non-additive perturbations of non-linear plants that admit inverse dynamics. Such perturbations arise, for example, when a robot arm grasps or releases a heavy object. The SDS Controller is composed of two identical copies of an inverse dynamics controller. One copy acts as the original closed loop controller while the other identical copy is used to develop the compensatory signal. The advantage of this compound controller is that it can develop a control signal for compensating unseen perturbations and structural approximation errors and thus can control the plant more precisely than the closed loop feedforward controller alone. This relaxes the number of parameters required to achieve a given precision in control and thus may enable the use of fast learning local approximator based neural networks, that are otherwise known to suffer from combinatorial explosion in the dimension of the state space. SDS Control was implemented on a fully self- organising neural network controller that exploits only local, associative direct inverse estimation methods. Preliminary computer simulations show that compensation is fast for non-linear plants, which follows from the sketched theory for linear plants.

5 Acknowledgements

We are grateful to Prof. András Krámli for his invaluable comments and suggestions. This work was partially founded by OTKA grants T017110, T014330, T014566, and US-Hungarian Joint Fund Grant 168/91-A 519/95-A.

References

1. W.T. III Miller, R.S. Sutton, and P.J. Werbos, editors. *Neural Networks for Control.* MIT Press, Cambridge, Massachusetts, 1990.
2. R. Ortega and T. Yu. Theoretical results on robustness of direct adaptive controllers: A survey. In *Proc. 10th IFAC World Congress*, pages 26–31, Munich, 1987.
3. P. van der Smagt and K. Schulten. Control of pneumatic robot arm dynamics by a neural network. In *Proc. of the 1993 World Congress on Neural Networks*, volume 3, pages 180–183, Hillsdale, NJ., 1993. Lawrence Erlbaum Associates, Inc.
4. A. Ben-Israel and T.N.E. Greville. *Generalized Inverses: Theory and Applications.* Pure and Applied Mathematics, Wiley-Interscience. J. Wiley & Sons, New York, 1974.
5. A. Isidori. *Nonlinear Control Systems.* Springer-Verlag, Berlin, 1989.
6. K. Narendra and K. Parthasarathy. Gradient methods for the optimization of dynamical systems containing neural networks. *IEEE Trans. Neural Networks*, 2(2):252–262, 1991.

A Local Connected Neural Oscillator Network for Pattern Segmentation

Hiroaki Kurokawa and Shinsaku Mori

Dept. of E.E. Keio University 3-14-1 Kouhoku-ku Hiyoshi Yokohama 223 JAPAN
kuro@mori.elec.keio.ac.jp

Abstract. This paper proposes the local connected neural oscillator network which is useful for image processing. In this proposed network, each neural oscillator employs a learning method to control its phase and frequency. Since the learning method has an ability to control the phase adjustment of each neural oscillator, the information expression in the phase space is achievable. Furthermore, it is considered that the network has the ability to achieve real time image processing and information processing in a time series. The learning method is possible under the assumption that synapses have plasticity. However, since it is supposed that only the feedback synapse has plasticity, we can construct a network with high simplicity.

1 Introduction

The oscillatory phenomena in the artificial/biological neural networks is one of the most remarkable issue on the study of neural networks. The detailed analysis of its dynamics have reported in recent work [1][2][6]. On the other hand, the study of coupled oscillatory neural network are interested in the view of engineering and biology [3]-[5]. Since it is considered that the neural oscillator has an ability to achieve information processing such as an associative memory or image processing, it is also significant issue to show the method to control the oscillation in the artificial neural networks. Taking into account of these studies, we had proposed a learning method for the synchronization of the oscillatory neural network as shown in following section [7].

In this paper, we propose the local connected neural oscillator network which is useful for image processing. Actually the successful system of local connected network for image processing has already proposed by DeLiang Wang [3]. However since the oscillators in his local connected network synchronizes and inhibits *mutually*, it would be hard to control the phases of these oscillators to desired phases in the phase space. On the contrary, in the local connected network which proposed in this paper, since the phase of each neural oscillator is *controlled* accurately with our learning method [7], it is realized that the information expression in the phase space. Furthermore, it is considered that the network has the ability to achieve real time image processing and information processing in a time series. In the simulation, we show that the proposed network is useful for the pattern segmentation.

2 Oscillatory neural networks and learning rule

Figure 1 shows the oscillatory neural network model which is composed of two excitatory neurons. Each neuron has the positive or negative synaptic weight, and only the neuron N_1 has the positive feedback weight. Dynamics of the internal states of these neurons N_1 and N_2 in this oscillatory neural network are given by,

$$\begin{cases} \tau_1 \frac{dU_1}{dt} = -U_1 + w_{12}f(U_2) + T_1 f(U_1) + S_1 \\ \\ \tau_2 \frac{dU_2}{dt} = -U_2 + w_{21}f(U_1) + S_2 \end{cases} \tag{1}$$

Where U_1 and U_2 are internal states of the neurons, τ_1 and τ_2 are time constants, w_{12}, and w_{21} indicate synaptic weight, T_1 indicates a feedback weight of N_1, and S_1 and S_2 are the external input of N_1 and N_2 respectively. The function $f(U_k)$ is activation function of the neuron which is given by

$$f(U_k) = 1/(1 + \exp(-\lambda U_k)). \tag{2}$$

Where λ is the coefficient to manage the gradient of the function.

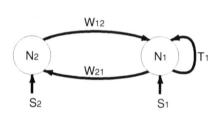

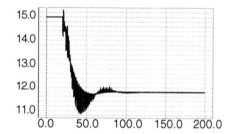

Fig.1. A neural oscillator   **Fig.2.** An example of changing of T_1

We can find the equilibrium of this system by means of substituting 0 for dU_k/dt in the left hand of the equation(1). While the synaptic weight and feedback weight are adjusted to appropriately, we can find that the unique equilibrium is unstable and a limit cycle exist around the equilibrium. Therefore, the state of this system (U_1, U_2) must converge to the limit cycle and the convergence ability never depend on any initial states of (U_{10}, U_{20}). Hence, the internal state of the neurons in this neural network, U_1 and U_2 are oscillatory. Characteristics of the oscillation depends on all parameters in the neural oscillator.

Since we assume that only the feedback synapse, T_1, has the plasticity, it is worth to discuss how change the frequency of the neuron's state with variation of the value of T_1. From the computer simulation, it had been known that the frequency is increased as T_1 increases [7]. However, it had been also known that there exist the bounded region to satisfy the condition of the oscillation. Hence, the feedback parameter T_1 must be adjusted to the value in this region to keep oscillatory.

In our previous study [7], we defined the parameter which indicate the state of the neuron in the oscillation as follows.

$$\varphi_1 = \begin{cases} Cos^{-1}\left(U_1/\sqrt{(U_1^2 + U_2^2)}\right) (U_2 \geq o) \\ \\ -Cos^{-1}\left(U_1/\sqrt{(U_1^2 + U_2^2)}\right) (U_2 < o) \end{cases} \tag{3}$$

Where we also defined that the region of $Cos^{-1}(x)$ is $(0.0, \pi)$ for $-1 \leq x \leq 1$. In short, the φ_1 shows the phase of the oscillation of neural oscillator. By means of the definition of φ_1 by eq. (3), we can compare the state of the neuron with the state of the objective periodic function in a same system of coordinates.

The learning method which we proposed [7] is realized by means of the update rule of T_1 which is given by,

$$\Delta T = \varepsilon_1(\varphi_1 - \varphi_I + 2\pi\theta) + \varepsilon_2(\frac{d\varphi_1}{dt} - \frac{d\varphi_I}{dt}). \tag{4}$$

Where ε_1 and ε_2 are positive coefficient and φ_I is the phase of the objective function. The first term of eq. (4) shows the difference of the phase of oscillations to the phase condition which is desired. In eq. (4), $2\pi\theta$ is the desired phase difference. On the other hand, the second term of eq. (4) shows the difference of velocity of the phase transition of the oscillations.

We can easily obtain the solution of differential form of φ_1

$$\frac{d\varphi_1}{dt} = -\frac{1}{U_1^2 + U_2^2}(U_2\frac{dU_1}{dt} - U_1\frac{dU_2}{dt}) \tag{5}$$

According to equation(1), we can obtain the update rule of feedback parameter T_1.

$$\begin{aligned}
T_1(t + \Delta t) &= T_1(t) + \Delta T \\
&= T_1(t) + \varepsilon_1(\varphi_1 - \varphi_I + 2\pi\theta) \\
&\quad -\varepsilon_2\frac{1}{U_1^2 + U_2^2}\{U_2(-U_1 + w_{21}f(U_2) + T(t)f(U_1) + S_1) \\
&\quad -U_1(-U_2 + w_{12}f(U_1) + S_2) - \frac{d\varphi_I}{dt}\}
\end{aligned} \tag{6}$$

The learning rule has the ability to synchronize the oscillation of neural oscillator with the external objective periodic function. An example of the learning is shown in fig. 2 and fig. 3.

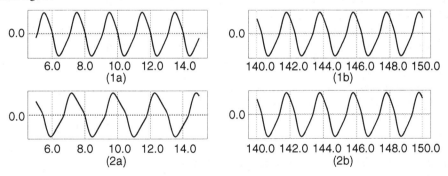

Fig.3. A synchronization of neural oscillator
(1a,b) A time series of internal state of objective function. (2a) A time series of internal state of learning neural oscillator, learning is not applied. (2b)Learning is applied.

3 Local Connected Neural Oscillator Network

The paper proposes local connected neural oscillator network. Figure 4 shows the $N \times M$ proposed local connected neural oscillator network. In fig. 4, open and filled circle indicate neural oscillator. The filled circle oscillator does not have plastic synapse and never change its way of oscillation. It only gives an objective periodic function to other oscillators. On the other hand, each oscillators which is indicated by open circle has plastic feedback synapse and the learning is possible. In short, the learning is applied to open circle oscillator to synchronize to the filled circle oscillator. In the simulation, we apply the proposed network to the pattern segmentation. The patterns which will be segmented are presented to a 8×8 matrix of oscillators as shown in fig. 5. The external inputs of each oscillator which indicated by filled square are $S_1 = 5.0, S_2 = -5.0$(oscillatory) and also the external inputs of each oscillator which indicated by open square are $S_1 = 0.0, S_2 = 0.0$(non-oscillatory). The number in the square indicates the oscillator number. A specific phase difference to the objective function(i.e. $2\pi\theta$ in eq. 4) is initially assigned to each open circle oscillator. The parameter θ_{ij} which is assigned to the Oscillator O_{ij} is given by,

$$\theta_{ij} = \frac{iN + j}{NM}. \tag{7}$$

In the network, there are two kinds of local connection. One is the connection between a open circle and the filled circle and the other is between open circles. The connections between open and filled circle oscillators are to compare the objective function for the learning. Also the connections between the open circle oscillator are to propagate the value of θ_{ij}. These connections, T_{ij}, are defined as follows.

$$T_{ij,kl} = J(U_{1ij})J(U_{1kl}) \tag{8}$$

Where $J(U_{1ij})$ is 1 when the oscillator O_{ij} oscillates and $J(U_{1ij})$ is 0 when the oscillator O_{ij} does not oscillate. The goal of the θ_{ij} propagation is that the oscillators which connected each other have same value of θ_{ij} by θ_{ij} propagation. Where "O_{ij} and O_{kl} are connected" means that the value of $T_{ij,kl}$ is 1. We show the one of the rules of θ_{ij} propagation in the following.

– Search the oscillating oscillator O_{ij} in order
– Propagate θ_{ij} to connected oscillator $O_{kl}, O_{mn}, \ldots$ and change $\theta_{kl}, \theta_{mn}, \ldots$ to θ_{ij}
– The θ which has changed never change its value again

Note that these connection does not mean the synapse which propagate the output to others. By means of such construction of local connected network, patterns on the $N \times M$ matrix are segmented and identified by phase differences.

4 Simulation results

In this section, we show the simulation results of pattern segmentation. In these simulation, parameters are defined as follows: $\tau_1 = \tau_2 = 1.0$, $w_{12} = -25.0$, $w_{21} = 10.0$, and $\lambda = 2.0$. We also define the coefficients ε_1 and ε_2 are 0.001 and 0.01, respectively. The initial value of the feedback parameter T_1 is 15.0. In all simulations, differential equations are calculated using fourth-order runge-kutta method. Figure 6 shows the waveform of the O_I and the oscillators which consists segment2 in the pattern2. We can observe the synchronization among the oscillators $O_{50}, O_{51}, O_{52}, O_{61}, O_{62}$ and O_{72}. Furthermore the synchronized oscillation keeps a constant phase difference from the oscillation of O_I. By means of computer simulations, we also confirm that such a synchronization conform to the other segment of pattern1 and pattern2. Figure 7 and 8 show the phase differences among the oscillation of the O_I and synchronized oscillation in each segment on same matrix. We can observe that the synchronized oscillations keep a constant phase difference from the oscillation of O_I and never be in-phase oscillation each other. According to these simulation,we can conclude that the proposed local connected neural oscillator network is the useful system for the pattern segmentation and it can identify each pattern by means of phase differences.

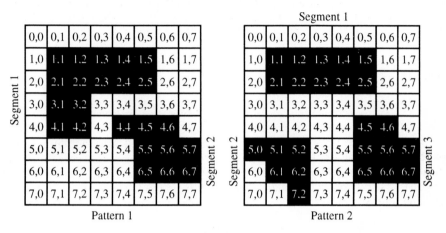

Fig.5. Segmented patterns on 8×8 matrix

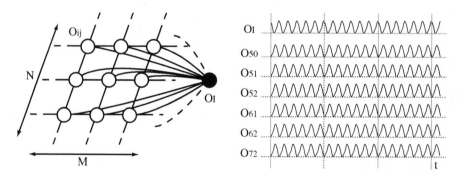

Fig.4. Neural oscillator network

Fig.6. The waveforms of the O_I and the oscillators which consists segment2 in the pattern2

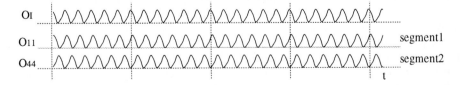

Fig.7. The waveforms of the O_I and synchronized oscillation in each segment on pattern1

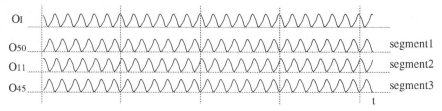

Fig.8. The waveforms of the O_I and synchronized oscillation in each segment on pattern2

5 Conclusion

In this paper, we proposed the local connected neural oscillator network and also show the application to pattern segmentation. Since the learning method which is employed in the proposed network has an ability to control the phase adjustment of each neural oscillator, it is realized that the information expression in the phase space. We believe that the proposed network will be one of the basics to achieve real time information processing and/or information processing in a time series. Simulation results show the efficiency of the proposed network.

References

1. Shun-ichi Amari, "Random Nets Consisting of Excitatory and Inhibitory Neuron-like Elements" IEICE Transactions vol.55-D pp.179-185 (1972), (in Japanese)
2. Shun-ichi Amari, "Characteristics of Random Nets of Analog Neuron-Like Elements", IEEE Transactions on Systems, Man, and Cybernetics, vol. SMC-2, no.5 pp. 643-657 (1972)
3. DeLiang Wang, "Emergent Synchrony in Locally Coupled Neural Oscillators", IEEE Transactions on Neural Networks, vol.6, no.4 pp. 941-948 (1995)
4. Ch. von der Malsburg and W. Schneider, "A Neural Cocktail-Party Processor", Biol. Cybern., vol.54, pp. 29-40 (1986)
5. Ch. von der Malsburg and Joachim Buhmann, "Sensory segmentation with coupled neural oscillators", Biol. Cybern., vol.67, pp. 233-242 (1992)
6. Kenji Doya and Shuji Yoshizawa, "Motor Pattern Memory in Neural Network" IEICE technical report MBE87-141(1987), (in Japanese)
7. Hiroaki Kurokawa and Shinsaku Mori,"A Learning method for the Synchronization of the Oscillatory Neural Network" (submitted to 1996 World Congress on Neural Networks)
8. R.Hecht Nielsen,"Neurocomputing", Addison Wesley (1990)
9. Hirofumi Nagashino and Yohsuke Kinouchi, "An Oscillatory Input Makes a Neural Network Nonoscillatory", Proc. of International Symposium on Nonlinear Theory and Its Application (NOLTA'95), pp.1137-1140

Approximation Errors of State and Output Trajectories Using Recurrent Neural Networks

Binfan Liu and Jennie Si

Arizona State University, Tempe, AZ 85287

Abstract. This paper addresses the problem of estimating training error bounds of state and output trajectories for a class of recurrent neural networks as models of nonlinear dynamic systems. We present training error bounds of trajectories between the recurrent neural network models and the target systems. The bounds are obtained provided that the models have been trained on N trajectories with N independent random initial values which are uniformly distributed over $[a, b]^m \in R^m$.

1 Introduction

Recurrent neural networks have received considerable attention in recent years in the area of modeling nonlinear dynamic systems. Many training schemes have been proposed for training recurrent neural networks. Among them are recurrent back-propagation [1] [4], back-propagation through time [3] [6], and real-time recurrent learning [7]. Extensions of the back-propagation method to nonlinear dynamic systems composed of linear time-invariant dynamic systems and multilayer neural networks are discussed in [2]. However, a quantitative evaluation of the performance of nonlinear dynamic neural network models is still lacking.

We consider in the present paper to use a class of recurrent neural networks as models of nonlinear dynamic systems. The networks consist of two static feedforward multilayer networks. Training error bounds of state and output trajectories between the recurrent network model and the target system are presented. Toward this end, we assume the model has been trained on N finite duration trajectories initiated at N random values which are uniformly distributed over $[a, b]^m \in R^m$.

2 Problem Formulation and Notations

In this paper, we consider a class of nonlinear dynamic systems given by

$$\dot{x}(t) = g(x(t), u(t)), \tag{1}$$
$$y(t) = h(x(t)), \quad t_0 \leq t \leq t_0 + T, \tag{2}$$

where state x belongs to a convex compact subset X of R^m, control u belongs to a class of admissible controls U, $g(x, u)$ is a Lipschitz continuous function in x over $X \times U$, and $h(x)$ is a Lipschitz continuous function over X. When g and h in (1) and (2) are unknown, a recurrent network is constructed by employing two static feedforward multilayer networks to approximate the mappings g and h, denoted by N1 and N2, respectively. Therefore, the recurrent neural network can be represented by

$$\dot{\hat{x}}(t) = \hat{g}(\hat{x}(t), u(t); w_f), \tag{3}$$

$$\hat{y}(t) = \hat{h}(\hat{x}(t); w_h), \quad t_0 \leq t \leq t_0 + T, \tag{4}$$

where $\hat{g}$ and $\hat{h}$ are the outputs of feedforward networks N1 and N2, respectively. Symbols w_f and w_h denote the adjustable weights of these networks. A block diagram of this recurrent model is given in Fig. 1. The activation functions of N1 and N2 are either sigmoid functions or Gaussian functions.

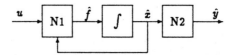

Fig. 1. Recurrent Neural Network Model

Let $s(t; x_0)$ and $r(t; x_0)$ denote the state and output trajectories with control u of (1) and (2) corresponding to an initial value $x(t_0) = x_0$, evaluated at t. Let $\hat{s}(t; x_0)$ and $\hat{r}(t; x_0)$ denote the state and output trajectories of (3) and (4) corresponding to an initial value $\hat{x}(t_0) = x_0$.

The problem addressed in this paper is to estimate bounds for state trajectory errors $\|s(t; x_0) - \hat{s}(t; x_0)\|$ and output trajectory errors $\|r(t; x_0) - \hat{r}(t; x_0)\|$ between (1) (2) and (3) (4) after the recurrent neural network model has been trained for N trajectories with N random initial values.

Throughout the paper, let $\|x\| = max_i|x_i|$ where $x \in R^m$, and x_i be the i-th component of x, $i = 1, 2, \cdots, m$. The probability of an event A is denoted by $P(A)$. Symbol $E\{\cdot\}$ denotes the expectation of its argument.

3 Training Error Bounds of Trajectories

Before addressing our results, we need the following lemma.

Lemma 1. *Let $x^{(1)}, \cdots, x^{(N)}$ and $x^{(*)}$ be independent and uniformly distributed random variables over $[a, b]^m \in R^m$. Then,*

$$E\{\min_k \|x^{(*)} - x^{(k)}\|\} = (b - a) \int_0^1 (1 - (1 - z^2)^m)^N dz.$$

Proof. Let α and $\beta \in R$ be independent random variables each distributed uniformly over $[a, b]$. Let $\psi = |\alpha - \beta|$. We then have the probability distribution function F_ψ of ψ given by

$$F_\psi(z) = \begin{cases} 0, & z < 0; \\ 1 - (1 - \frac{z}{b-a})^2, & 0 \leq z \leq b - a; \\ 1, & z > b - a. \end{cases} \tag{5}$$

Let $\zeta_k = \|x^{(*)} - x^{(k)}\|$ and $\xi = \min_k \zeta_k$. The distribution function F_ξ of ξ is

$$
\begin{aligned}
F_\xi(z) &= P(\min_k \zeta_k \le z) \\
&= 1 - P(\min_k \zeta_k > z) \\
&= 1 - P(\zeta_1 > z)P(\zeta_2 > z) \cdots P(\zeta_N > z) \\
&= 1 - (1 - F_{\zeta_1}(z))(1 - F_{\zeta_2}(z)) \cdots (1 - F_{\zeta_N}(z)),
\end{aligned}
$$

where F_{ζ_k} is the distribution function of ζ_k, $k = 1, 2, \cdots, N$. Since $\zeta_1, \zeta_2, \cdots, \zeta_N$ are identically distributed random variables, we obtain that

$$
F_\xi(z) = 1 - (1 - F_\phi(z))^N, \tag{6}
$$

where $F_\phi(z)$ is the distribution function of $\phi = \|\mu - \nu\|$, μ and $\nu \in R^m$ are independent and uniformly distributed random variables over $[a, b]^m$. Let μ_i and ν_i, $i = 1, 2, \cdots, m$, be the i-th component of μ and ν, respectively. Random variables μ_i and ν_i are independently and uniformly distributed over $[a, b]$. Hence, the distribution function F_ϕ can be computed by

$$
\begin{aligned}
F_\phi(z) &= P(\max_i |\mu_i - \nu_i| \le z) \\
&= P(|\mu_1 - \nu_1| \le z)P(|\mu_2 - \nu_2| \le z) \cdots P(|\mu_m - \nu_m| \le z) \\
&= F_{\phi_1}(z)F_{\phi_2}(z) \cdots F_{\phi_m}(z), \tag{7}
\end{aligned}
$$

where F_{ϕ_i} is the distribution function of random variable $\phi_i = |\nu_i - \mu_i|$, $i = 1, 2, \cdots, m$. These ϕ_i's are identically distributed random variables and their distribution functions are given in (5). Thus, $F_\phi(z) = (F_\psi(z))^m$. Substituting it into (6), we obtain that

$$
F_\xi(z) = 1 - (1 - (F_\psi(z))^m)^N. \tag{8}
$$

Therefore, the expectation of ξ is given by $E\{\xi\} = \int_0^{b-a} f_\xi(z)dz$, where f_ξ is the probability density function of ξ. Using integration by parts, we have that

$$
E\{\xi\} = zF_\xi(z)|_0^{b-a} - \int_0^{b-a} F_\xi(z)dz = (b - a)\int_0^1 (1 - (1 - z^2)^m)^N dz.
$$

This completes the proof of the lemma. $\qquad\qquad\qquad\qquad\qquad\qquad\qquad\square$

Our first result gives a training error bound of state trajectories of the recurrent neural network model and the target system with an initial value in $[a, b]^m$. We obtain the bound assuming that the recurrent neural network has been trained for N trajectories with N independent random initial values in $[a, b]^m$. The trained state trajectory error $\|s(t; x^{(k)}) - \hat{s}(t; x^{(k)})\|$ is less than or equal to a constant δ for $k = 1, 2, \cdots, N$.

Theorem 2. *Let the recurrent neural network (3) and (4) be trained on N (≥ 1) trajectories for initial values $x^{(k)}$ such that $\|s(t; x^{(k)}) - \hat{s}(t; x^{(k)})\| \leq \delta$, $k = 1, 2, \cdots, N$. Let $x^{(1)}, x^{(2)}, \cdots, x^{(N)}$ and $x^{(*)}$ be independent and uniformly distributed random variables over $[a, b]^m \in R^m$. Let $e_s(t) = s(t; x^{(*)}) - \hat{s}(t; x^{(*)})$ be the error of state trajectories between (1) and (3) with initial value $x^{(*)}$. Then, there exists a constant C such that*

$$E\{\|e_s(t)\|\} \leq C \int_0^1 [1 - (1 - z^2)^m]^N dz + \delta.$$

Proof. Let Ω be the compact set defined as $\Omega = \{(x, u) : x \in X, u \in U\}$. Because $g(x, u)$ satisfies a Lipschitz condition in x over Ω, there exists a constant k_g such that for all $(x, u), (\bar{x}, u) \in \Omega$, $\|g(x, u) - g(\bar{x}, u)\| \leq k_g \|x - \bar{x}\|$. Because sigmoid functions and Gaussian functions are Lipschitz continuous, $\hat{g}$ is Lipschitz continuous in the compact set Ω. Therefore, there exists a constant $k_{\hat{g}}$ such that for all $(x, u), (\bar{x}, u) \in \Omega$, $\|\hat{g}(x, u) - \hat{g}(\bar{x}, u)\| \leq k_{\hat{g}} \|x - \bar{x}\|$. Let $e^{(k,j)}(t) = s(t; x^{(k)}) - s(t; x^{(j)})$ and $\hat{e}^{(k,j)} = \hat{s}(t; x^{(k)}) - \hat{s}(t; x^{(j)})$. We have that

$$\frac{d}{dt} e^{(k,j)}(t) = g(s(t, x^{(k)}), u) - g(s(t, x^{(j)}), u),$$

with initial value $e^{(k,j)}(t_0) = s(t_0; x^{(k)}) - s(t_0; x^{(j)}) = x^{(k)} - x^{(j)}$. Therefore, $e^{(k,j)}$ is governed by

$$e^{(k,j)}(t) = x^{(k)} - x^{(j)} + \int_{t_0}^t (g(s(\tau; x^{(k)}), u) - g(s(\tau; x^{(j)}), u)) d\tau.$$

It follows that

$$\|e^{(k,j)}(t)\| \leq \|x^{(k)} - x^{(j)}\| + \int_{t_0}^t k_g \|e^{(k,j)}(\tau)\| d\tau.$$

Using the Bellman-Gronwall lemma [5], we obtain that

$$\|e^{(k,j)}(t)\| \leq \|x^{(k)} - x^{(j)}\| \exp(k_g T), \tag{9}$$

and, similarly,

$$\|\hat{e}^{(k,j)}(t)\| \leq \|x^{(k)} - x^{(j)}\| \exp(k_{\hat{g}} T). \tag{10}$$

Now, consider the trajectory error $e_s(t) = s(t; x^{(*)}) - \hat{s}(t; x^{(*)})$ between (1) (2) and (3) (4) for a random initial value $x(t_0) = \hat{x}(t_0) = x^{(*)}$ in $[a, b]^m$. For any k, $1 \leq k \leq N$,

$$\begin{aligned}
\|e_s(t)\| &\leq \|s(t; x^{(*)}) - s(t; x^{(k)})\| + \|s(t; x^{(k)}) - \hat{s}(t; x^{(k)})\| \\
&\quad + \|\hat{s}(t; x^{(k)}) - \hat{s}(t; x^{(*)})\| \\
&= C_1 \|x^{(*)} - x^{(k)}\| + \delta,
\end{aligned} \tag{11}$$

where $C_1 = \exp(k_g T) + \exp(k_{\hat{g}} T)$. Inequality (11) holds for all $k = 1, 2, \cdots, N$. Hence, the inequality $\|e_s(t)\| \leq C_1 \min_k \|x^{(*)} - x^{(k)}\| + \delta$ must hold. It follows that $E\{\|e_s(t)\|\} \leq C_1 E\{\min_k \|x^{(*)} - x^{(k)}\|\} + \delta$. Therefore, the theorem is concluded by applying Lemma 1 with $C = (b - a)C_1$. $\qquad \square$

Remark. The bound obtained in Theorem 2 is less than $(C+\delta)$ and decreases to δ as $N \to \infty$ (see Fig. 2), where δ is the training error for the N state trajectories which have been trained using N random initial values.

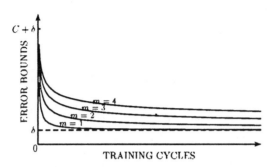

Fig. 2. Error Bounds for State Trajectories

The following theorem concerns a training error bound for output trajectories between (2) and (4).

Theorem 3. *Let the recurrent neural network (3) and (4) be trained on N (≥ 1) trajectories for initial values $x^{(k)}$ such that $\|r(t; x^{(k)}) - \hat{r}(t; x^{(k)})\| \leq \delta$, $k = 1, 2, \cdots, N$. Let $x^{(1)}$, $x^{(2)}$, $\cdots$, $x^{(N)}$ and $x^{(*)}$ be independent and uniformly distributed random variables in $[a, b]^m \in R^m$. Let $e_r(t) = r(t; x^{(*)}) - \hat{r}(t; x^{(*)})$ be the error of output trajectories between (2) and (4) with initial value $x^{(*)}$. Then, there exists a constant D such that*

$$E\{\|e_r(t)\|\} \leq D \int_0^1 [1 - (1 - z^2)^m]^N dz + \delta.$$

Proof. Because h and $\hat{h}$ satisfy Lipschitz conditions, there exist constants k_h and $k_{\hat{h}}$ such that $\|h(x) - h(\bar{x})\| \leq k_h\|x - \bar{x}\|$ and $\|\hat{h}(x) - \hat{h}(\bar{x})\| \leq k_{\hat{h}}\|x - \bar{x}\|$, for all $x, \bar{x} \in X$. Hence, for any k ($k = 1, 2, \cdots, N$), we have that

$$\|e_r(t)\| = \|h(s(t; x^{(*)})) - \hat{h}(\hat{s}(t; x^{(*)}))\|$$
$$\leq \|h(s(t; x^{(*)})) - \hat{h}(\hat{s}(t; x^{(k)}))\| + \|h(s(t; x^{(k)})) - \hat{h}(\hat{s}(t; x^{(k)}))\|$$
$$+ \|\hat{h}(\hat{s}(t; x^{(k)})) - \hat{h}(\hat{s}(t; x^{(*)}))\|$$
$$\leq k_h\|s(t; x^{(*)}) - s(t; x^{(k)})\| + \delta + k_{\hat{h}}\|\hat{s}(t; x^{(k)}) - \hat{s}(t; x^{(*)})\|. \qquad (12)$$

From (9) and (10), inequality (12) becomes

$$\|e_r(t)\| \leq D_1\|x^{(*)} - x^{(k)}\| + \delta, \qquad (13)$$

where $D_1 = k_h \exp(k_g T) + k_{\hat{h}} \exp(k_{\hat{g}} T)$. Inequality (13) holds for all $k = 1, 2, \cdots, N$. It follows that $\|e_r(t)\| \leq D_1 \min_k \|x^{(*)} - x^{(k)}\| + \delta.$

By Lemma 1, we obtain that

$$E\{\|e_r(t)\|\} \le D \int_0^1 [1 - (1 - z^2)^m]^N dz + \delta,$$

where $D = (b - a)D_1$. This completes the proof of the theorem. $\qquad\square$

4   Conclusion

In the present paper, we have estimated training errors for a class of recurrent neural networks as models of nonlinear dynamic systems. We have presented training error bounds for state trajectories (Theorem 2) and output trajectories (Theorem 3) between the recurrent neural network model and the target system. The bounds are given in terms of m and N, where N is the number of trajectories for which the model has been trained using N independent random initial values which are uniformly distributed over $[a, b]^m \in R^m$.

Acknowledgment

The research is supported in part by EPRI under contract RP8015-03, by NSF under grant ECS9553202 and by Motorola.

References

1. Almeida, L.: A learning rule for asynchronous perceptrons with feedback in a combinatorial environment. Proceedings of the First International Conference on Neural Networks. (1987) 609-618.
2. Narendra, K., Parthasarathy, K.: Gradient methods for the optimization of dynamical systems containing neural networks. IEEE Transactions on Neural Networks. 2 (1991) 252-262.
3. Pearlmutter, B.: Learning state space trajectories in recurrent neural networks. Neural Computation. 1 (1989) 263-269.
4. Pineda, F.: Generalization of back-propagation to recurrent neural networks. Physical Review Letters. 59 (1987) 2229-2232.
5. Sontag, E.: Mathematical Control Theory. New York: Springer-Verlag, (1990).
6. Werbos, P.: Backpropagation through time: What it does and how to do it. Proceedings of IEEE. 78 (1990) 1550-1560.
7. Williams, R., Zipser, D.: A learning algorithm for continually running fully recurrent neural networks. Neural Computation. 1 (1989) 270-280.

Comparing Self-Organizing Maps

Samuel Kaski and Krista Lagus

Helsinki University of Technology
Neural Networks Research Centre
Rakentajanaukio 2C, FIN-02150 Espoo, Finland

Abstract. In exploratory analysis of high-dimensional data the self-organizing map can be used to illustrate relations between the data items. We have developed two measures for comparing how different maps represent these relations. The other combines an index of discontinuities in the mapping from the input data set to the map grid with an index of the accuracy with which the map represents the data set. This measure can be used for determining the goodness of single maps. The other measure has been used to directly compare how similarly two maps represent relations between data items. Such a measure of the dissimilarity of maps is useful, e.g., for analyzing the sensitivity of maps to variations in their inputs or in the learning process. Also the similarity of two data sets can be compared indirectly by comparing the maps that represent them.

1 Introduction

The self-organizing map (SOM) [4, 5] algorithm forms a kind of a nonlinear regression of an ordered set of reference vectors m_i, $i = 1, \ldots, N$, into the data space $\Re^n$. Each reference vector belongs to a map unit on a regular map lattice. In exploratory data analysis (data mining) with the SOM the aim is to extract and *illustrate* the essential structures within a statistical data set by a map that, as a result of an unsupervised learning process, follows the distribution of the data in the input space. Each data sample is mapped to the unit containing the most similar reference vector, whereby the relations of the data samples become reflected in geometrical relations (order) of the samples on the map. The density of the data points in different regions of the input space (reflected in the distances between the reference vectors of neighbor units) can be visualized with gray levels on the map display [6, 9].

2 Measures of Goodness of Maps

Measures are needed for choosing good maps from a sample set of maps resulting from a stochastic learning process, or for determining good learning parameters for the maps.

2.1 Previously Proposed Measures

The accuracy of a map in representing its input can be measured with the average *quantization error*, i.e., the distance from each data item to the closest reference vector. If also the distance from the reference vectors of the *neighbors* (units that lie within a specified radius on the map grid) of the winner is incorporated [5], the measure becomes sensitive to the local orderliness of the map. Although these two measures are necessary in guaranteeing that the map represents the data set well, they cannot be used to compare maps with different stiffnesses since they favor maps with specific neighborhood radii.

Several orderliness measures have been proposed that compare the relative positions of the reference vectors in the input space with the positions of the corresponding units on the map lattice (e.g., [1]). As has been pointed out by Villmann et al. [10], however, these measures cannot distinguish between folding of the map along nonlinearities in the data manifold and folding within a data manifold. The former is a highly desirable property whereas the latter causes discontinuities in the mapping from the input space to the map grid, which may be undesirable in some applications.

A more sensitive measure [10] computes the adjacency of the "receptive fields", or cells in the Voronoi tessellation, of the different map units *within the data manifold*. In a perfectly ordered map only units that are neighbors on the map lattice may have adjacent receptive fields. A possible problem with this measure is that noise or nonrepresentative inputs may easily cause some receptive fields to be erroneously judged as adjacent within the manifold.

Kiviluoto [3] has used a more gradual measure of the adjacency of the receptive fields: The proportion of samples for which the nearest and the second nearest units reside in non-neighboring locations on the map. Even this measure does not, however, consider the *extent* of the discontinuities in the mapping from the input space to the map grid.

Kraaijveld et al. [6] have compared different mapping methods by computing the accuracies with which a given data set can be classified in the mapped spaces. Although their goodness measure is not sufficiently general for our purposes since it requires classified input samples, the way they computed distances between data points has been found useful also in our studies.

2.2 A Novel Measure

We formed a measure that combines an index of the continuity of the mapping from the data set to the map grid with a measure of the accuracy of the map in representing the set (the quantization error). For each data item x we compute the distance $d(x)$ *from x to the second nearest reference vector $m_{c'(x)}$* passing first from x to the best matching reference vector $m_{c(x)}$, and thereafter *along the shortest path* to $m_{c'(x)}$ through a series of reference vectors. In the series each reference vector must belong to a unit that is an immediate neighbor of the previous unit. If there is a discontinuity in the mapping near x, such a distance along the map from unit $c(x)$ to $c'(x)$ is in general large, whereas if the units are neighbors the distance is smaller.

The distance $d(x)$ can be expressed more formally as follows: Denote by $I_i(k)$ the index of the kth unit on a path along the map grid from unit $I_i(0) = c(x)$ to $I_i(K_{c'(x),i}) = c'(x)$. In order for the function I_i to represent a path along the map grid the units $I_i(k)$ and $I_i(k+1)$ must be neighbors for $k = 0, \ldots, K_{c'(x),i} - 1$. Using these notations the distance $d(x)$ is

$$d(x) = ||x - m_{c(x)}|| + \min_i \sum_{k=0}^{K_{c'(x),i}-1} ||m_{I_i(k)} - m_{I_i(k+1)}|| \, . \tag{1}$$

The goodness C of the map is defined as the average (denoted by E) of the distance over all input samples (low values denote good maps),

$$C = E[d(x)] \, . \tag{2}$$

In simulations with a simple data set (Fig. 1) C measured a satisfactory combination of the continuity of the mapping and the quantization error, a result not obtainable with the previously proposed methods.

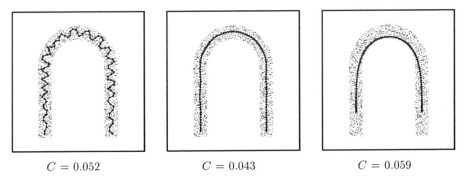

$$C = 0.052 \qquad\qquad C = 0.043 \qquad\qquad C = 0.059$$

Fig. 1. The goodness measure C of SOMs with varying stiffnesses produced by varying the final neighborhood width in the learning process. The input (small dots) came from a two-dimensional, horseshoe-shaped distribution. The reference vectors of the 100-unit, one-dimensional SOMs are shown in the input space as large black dots, with lines connecting reference vectors belonging to neighbor units. The best (lowest) value of C is yielded by the SOM in the middle that covers all of the horseshoe without folding unnecessarily.

3　A Novel Measure of Dissimilarity of Maps

For a given data set there may exist several different representations that are all useful for different purposes. Therefore it may not always be sensible to compare the goodnesses of the maps as was done in Sec. 2.2. It might in any case be useful to know *how* different the maps are from each other. A measure of the dissimilarity of maps could be used, e.g., for detecting outlier maps or for analyzing the sensitivity of the maps for variations in the inputs or in the learning process.

We define the dissimilarity of two maps, L and M, as the average (normalized) difference in how they represent the distance between two data items. The *representational distance* $d_L(x, y)$ between the pair (x, y) of data samples, represented by map L, is defined as follows. The distance is computed along the shortest path which passes through the best matching reference vectors $m_{c(x)}$ and $m_{c(y)}$, and through a series of reference vectors. In the series the units corresponding to each successive pair of reference vectors must be immediate neighbors. Using the notation introduced in Sec. 2.2, denote by $I_i(k)$ the index of the kth unit on a path from $I_i(0) = c(x)$ to $I_i(K_{c(y),i}) = c(y)$. The distance between samples x and y on map L is then

$$d_L(x, y) = \|x - m_{c(x)}\| + \min_i \sum_{k=0}^{K_{c(y),i}-1} \|m_{I_i(k)} - m_{I_i(k+1)}\| + \|y - m_{c(y)}\|, \quad (3)$$

and the dissimilarity of maps L and M is defined to be

$$D(L, M) = E\left[\frac{|d_L(x, y) - d_M(x, y)|}{d_L(x, y) + d_M(x, y)}\right]. \quad (4)$$

Here the expectation E is estimated over all pairs of data samples (x, y) in a representative set. To reduce the computational complexity of the measure the reference vectors of one or all of the maps can be used as the representative set.

It can be shown that D is a dissimilarity measure in the mathematical sense. To demonstrate that D does indeed measure the dissimilarity of maps we have applied it in a case study to compare maps that had progressively more different input data sets (Fig. 2).

4 A Demonstration of the Use of the Dissimilarity Measure

Assume a scenario where SOMs are used by several parties to explore their data sets and to present summaries of the data. The parties could be individual people, institutions, or software agents, and the data sets might consist of information about any specific topic area, e.g., encoded documents or economical statistics (cf. [2, 5]). The parties might make the SOMs accessible through, for example, the Internet as advertisements or reports of their work, although they might not want to open their data sets for public use, e.g., due to confidentiality or the size of the data.

The SOMs are representations of the knowledge, or "expertise", inherent in the data sets of the parties. It might therefore be of interest for the parties to assess the similarity of their SOMs. We have demonstrated the use of the measure D (4) in comparing maps describing different phonemes (Fig. 3). Maps taught with similar data sets (e.g., /m/ and /n/) were found to be more similar than maps taught with dissimilar sets (e.g., /m/ and /s/).

The significance of the measured dissimilarity between two maps could be assessed by computing the probability that the maps represent the same data set, for example using a nonparameteric statistical test. The baseline distribution of the dissimilarities, under the hypothesis that the maps have been taught with the

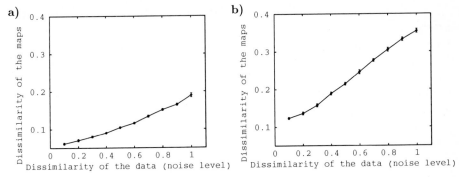

Fig. 2. Demonstration of a sensitivity analysis using the dissimilarity measure. Varying amounts of noise were added to a data set that consisted of 39 indicators for each country in a set of 78 countries, describing different aspects of their welfare [2]. The dissimilarity D between the SOMs taught with noisy data and a SOM taught with the original data set was computed when (a) the maps were of equal size (13 by 9 units) and had equal learning parameters (the final width of the neighborhood was two), and (b) when the map taught with the noisy data was different in size (16 by 7 units) and had different learning parameters (final neighborhood width was one instead of two). In both cases the dissimilarity D of the maps increased when the dissimilarity of their inputs increased. The bars in the figure denote the standard errors of the means of ten distances computed between maps that had different random input sequences while learning. The noise level is the standard deviation of the i.i.d. Gaussian noise. The variance of each data dimension was normalized to unity.

same data set, can be formed by teaching a set of maps with different (stochastic) input sequences. Also different stochastically chosen learning parameters and initial states can be used if the learning procedures of the maps are unknown.

5 Discussion

We have proposed for the comparison of SOMs two measures that are suitable especially for data mining applications. In data mining the map lattice must for illustratory purposes be regular and of a low dimension, whereby neither a perfectly topography preserving mapping [7] nor matching of the dimensions of the map and the input space [8] would be useful in general.

The proposed measure of the goodness of a map can be used to choose maps that do not fold unnecessarily in the input space while representing the input data distribution.

The measure of the dissimilarity of two maps can be used to compare directly how the maps illustrate relations between data items.

In the measures, the representational distances between data points are computed in the input space along paths following the "elastic surface" formed by the SOM. Such distances reflect the perceptual distance of data items on a map display, on which distances between neighboring reference vectors have for data mining purposes been illustrated with gray levels.

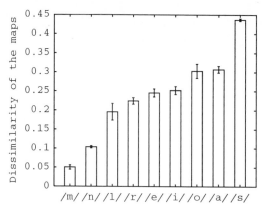

Fig. 3. Demonstration of the use of the dissimilarity measure D for comparing SOMs representing different data sets. The sets consisted of 20-dimensional short-time cepstra collected around the middle parts of phonemes of one male speaker (over 900 samples in each class). For each data set, 10 maps of the size of 6 by 4 units were taught using different random input sequences. The average of the distances between those maps and a common reference map are shown in the figure, together with the standard deviations. The reference map was chosen (based on the goodness measure C) from a batch of maps representing the set /m/.

References

1. Bauer, H.-U., Pawelzik, K. R.: Quantifying the neighborhood preservation of self-organizing feature maps. IEEE Tr. Neural Networks **3** (1992) 570–579
2. Kaski, S., Kohonen, T.: Exploratory data analysis by the self-organizing map: Structures of welfare and poverty in the world. *In* Neural Networks in the Capital Markets, World Scientific (to appear)
3. Kiviluoto, K.: Topology preservation in self-organizing maps. *In* Proc. ICNN96, IEEE Int. Conf. on Neural Networks (to appear)
4. Kohonen, T.: Self-organized formation of topologically correct feature maps. Biol. Cybern. **43** (1982) 59–69
5. Kohonen, T.: Self-Organizing Maps. Springer, Berlin (1995)
6. Kraaijveld, M. A., Mao, J., Jain, A. K.: A non-linear projection method based on Kohonen's topology preserving maps. *In* Proc. 11ICPR, 11th Int. Conf. on Pattern Recognition. IEEE Comput. Soc. Press., Los Alamitos, CA (1992) 41–45
7. Martinetz, T., Schulten, K.: Topology representing networks. Neural Networks **7** (1994) 507–522
8. Speckmann, H., Raddatz, G., Rosenstiel., W.: Considerations of geometrical and fractal dimension of SOM to get better learning results. *In* M. Marinaro and P. G. Morasso, eds, Proc. ICANN94, Int. Conf. on Artificial Neural Networks. Springer, London (1994) 342–345
9. Ultsch, A., Siemon, H. P.: Kohonen's self organizing feature maps for exploratory data analysis. *In* Proc. INNC90, Int. Neural Network Conf. Kluwer, Dordrecht (1990) 305–308
10. Villmann, T., Der, R., Martinetz, T.: A new quantitative measure of topology preservation in Kohonen's feature maps. *In* Proc. ICNN'94, IEEE Int. Conf. on Neural Networks. IEEE Service Center, Piscataway, NJ (1994) 645–648

Nonlinear Independent Component Analysis by Self-Organizing Maps

Petteri Pajunen

Helsinki University of Technology
Laboratory of Computer and Information Science
Rakentajanaukio 2 C, FIN-02150, Espoo, Finland
e-mail: Petteri.Pajunen@hut.fi

Abstract.

Linear Independent Component Analysis considers the problem of find-ing a linear transformation that makes the components of the output vector statistically independent. This can be applied to blind source separation, where the input data consist of unknown linear mixtures of unknown in-dependent source signals. The original source signals can be recovered from their mixtures using the assumption that they are statistically inde-pendent. More generally we can consider nonlinear mappings that make the components of the output vectors independent. We show that such a mapping can be approximately realized using self-organizing maps with rectangular map topology. We apply these mappings to the separation of nonlinear mixtures of sub-Gaussian sources.

1 Introduction

Statistical independence is a considerably stronger requirement than uncorrelat-edness. Its usefulness arises from the fact that it can often be justified by physical considerations. In [1] the concept of Independent Component Analysis (ICA) is formalized. A direct application of ICA is blind source separation (BSS), where source signals are recovered from their linear mixtures. Such blind techniques have applications for example in array processing, communications, speech pro-cessing, and medical signal processing. Recently, BSS has become an active research area of unsupervised neural learning. Several neural algorithms that use nonlinearities instead of explicit higher-order statistics have been proposed; see for example [2, 3, 4, 5]. These algorithms try to find a linear transformation that makes the components of the output vectors statistically independent.

One natural generalization of the linear ICA and BSS is to consider the prob-lem of finding more general, nonlinear mappings that would yield statistically independent outputs. This extension has already been considered in a few papers [6, 7, 8, 9, 10]. However, the proposed algorithms are not completely satisfying from a neural network point of view, because they are either complicated and/or require explicit computation of higher-order moments.

In this paper we justify that rectangular self-organizing maps can be used to construct mappings that make output vectors approximately statistically inde-pendent. This is applied to the blind separation of nonlinearly distorted mixtures of sub-Gaussian sources as originally suggested in [11].

2   ICA and Blind Source Separation

Consider a random vector $s \in \mathbb{R}^m$ and the corresponding joint density p. The density p is said to be *factorizable* if the following holds:

$$p(\mathbf{s}) = p_1(s_1) \cdot p_2(s_2) \cdots p_m(s_m) \tag{1}$$

where p_i is the marginal density of the ith component s_i of $\mathbf{s}$.

The independence of the joint density is the only assumed a priori information in ICA. Consider a random vector $s \in \mathbb{R}^m$ which has a factorizable density. In linear BSS we observe linearly transformed random vectors $x = \mathbf{A}s$ where the mixing matrix $\mathbf{A}$ is constant, non-singular and usually square matrix. The aim is to find a separating matrix $\mathbf{B}$ which gives good estimates of vectors $\mathbf{s}$:

$$\hat{s} = y = \mathbf{B}x. \tag{2}$$

The source separation problem is generally difficult, because both $\mathbf{A}$ and $\mathbf{s}$ are unknown. However, using the strong but realistic assumption that the components of $\mathbf{s}$ are mutually independent, it is possible to estimate the matrix $\mathbf{B}$ [1]. There remains some indeterminacy in $\mathbf{B}$ due to invariance properties of factorizable densities: a random vector with a factorizable density remains factorizable if it is multiplied by a permutation matrix and/or a diagonal matrix. This means that the recovered signals could have arbitrary variances and ordering, but their waveforms are recovered.

Now consider the case of a nonlinear mixing function F. In general, the separating function G is nonlinear (if it exists) and we have a more serious indeterminacy: the factorizable vector $\mathbf{s}$ remains factorizable if any nonlinear componentwise mapping is applied, i.e.

$$x = [F_1(s_1), F_2(s_2), \ldots, F_m(s_m)]^T. \tag{3}$$

This means that the waveforms of the original signals cannot necessarily be recovered.

3   Self-Organizing Maps

A self-organizing map (SOM) [12] performs a mapping from input space to an array of nodes in the output space. The positions of the nodes are fixed and each node is represented by a reference vector in the input space. Input vectors are mapped by finding the closest reference vector with respect to some distance function. The image of the input vector is the corresponding node on the map.

By considering the coordinates of the nodes in the output space we can think of SOM as a mapping from input space to the map (the output space). Using some suitable interpolation method, we can regard SOM as a homeomorphism from input space to the map.

3.1 Statistical Properties of Self-Organizing Maps

In SOM the distribution of the reference vectors in the input space can be roughly described using the density

$$c \cdot p(\mathbf{x})^{\alpha} \qquad (4)$$

where p is the input space density and α is the magnification factor. If $\alpha = 1$ then the weight vectors are distributed approximately according to the input space density. Then we have the important property that each weight vector on the map is equally likely the winner, i.e. lies closest to a random input vector. This implies that the joint density on the map is uniformly distributed. Since we consider rectangular maps, the final implication is that the map density is also statistically independent. This follows from the fact that for data distributed uniformly inside a rectangle, the marginal distributions in the directions of the sides of the rectangle are mutually independent. Learning rules that can be used to make the magnification factor equal to one have been recently proposed in [13, 14].

3.2 Application to Blind Source Separation

The self-organizing map can be used to estimate the inverse of the mixing function F by considering the set of observed mixture vectors $\mathbf{x}(t)$ as the input vectors to the SOM. Then the coordinates of the winner neuron on the map would define the output vector $\mathbf{y}(t)$ for each $\mathbf{x}(t)$. Using interpolation on the map this function can be made continuous.

Even though the self-organizing map can produce roughly statistically independent output vectors, this property does not guarantee that the components of the output vectors are good estimates for the source signals. With some heuristic constraints, however, it seems that the sources can be separated at least roughly. The mixture density should be of such a shape that a rectangular map can naturally adapt to it, and this natural adaptation should provide the correct separation. One reasonable class of problems to which this description applies consists of sub-Gaussian sources that are mixed linearly and then mildly distorted with nonlinearities. A sub-Gaussian signal has a density which has a flatter shape than Gaussian density. Formally stated, a sub-Gaussian signal x has the property $E[x^4] - 3(E[x^2])^2 < 0$. The expression on the left is called the *kurtosis* of x and it equals zero for Gaussian signals.

Due to the flat shape of sub-Gaussian densities the linear mixture of sub-Gaussian sources gives rise to a roughly rectangular density. If this density is distorted nonlinearly, it seems that the inverse G of the resulting nonlinear mixing function F can be roughly learned by the self-organizing map. Then the estimates of the original source signals are obtained by mapping mixture vectors $\mathbf{x}(t)$ to the converged map. The coordinates of the winning neuron provide the estimated sources $y_i(t), i = 1, \ldots, M$.

4 Experiments

We have made preliminary experiments using two different sub-Gaussian source signals. In the following example, the source signals consist of a sinusoid and uniformly distributed white noise.

First, the sources $\mathbf{s}(t)$ were linearly mixed using a mixing matrix

$$\mathbf{A} = \begin{pmatrix} 0.6 & 0.4 \\ 0.4 & 0.6 \end{pmatrix} \tag{5}$$

The resulting vectors $\mathbf{v}(t) = \mathbf{A}\mathbf{s}(t)$ were then distorted by a nonlinearity f applied to each component, which yields the mixture vectors

$$\mathbf{x}(t) = [f(v_1(t)), f(v_2(t)), \dots, f(v_m(t))]^T. \tag{6}$$

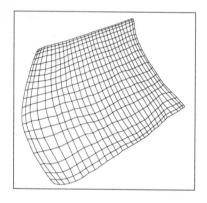
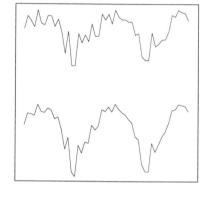

Fig. 1: Left: the converged map.   Right: mixed sources.

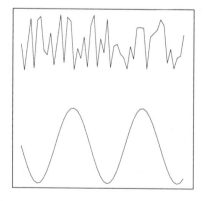
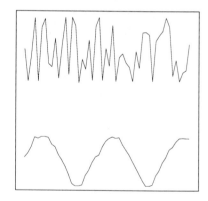

Fig. 2: Left: original source signals.   Right: separated source signals.

The vectors $\mathbf{v}(t)$ were first whitened so that their covariance matrix becomes the identity matrix. Effects of whitening are discussed in [15].

Fig. 1 shows the converged map and the mixed sources, where a linear mixture has been distorted by $f(x) = \log(x + 1.5)$. The results are shown in Fig. 2. Note that the signs of separated signals have been changed in some cases to make comparisons with original source signals easier.

5 Discussion

We have demonstrated that a self-organizing map can be used to construct nonlinear mappings that make output vector components approximately statistically independent. This is applied to nonlinear blind source separation. The experiments made on two sub-Gaussian sources show that at least in simple cases the sources can be recovered. There are, however, limitations to the applicability of SOM to blind source separation. Perhaps the most important restriction is the distribution of the source signals. These must be sub-Gaussian to ensure roughly rectangular joint density. Extending the use of SOM to super-Gaussian sources is currently being studied. Further discussion on applying SOM to blind source separation can be found in [11, 15].

References

[1] P. Comon, "Independent component analysis – a new concept?," *Signal Processing*, vol. 36, no. 3, pp. 287–314, 1994.

[2] S. Amari, A. Cichocki, and H. Yang, "A new learning algorithm for blind signal separation," in *Proc. of Neural Information Processing Systems (NIPS 95)*, (Denver, USA), MIT Press, November 1995.

[3] J. Karhunen, E. Oja, L. Wang, R. Vigario, and J. Joutsensalo, "A class of neural networks for independent component analysis," Report A28, Helsinki Univ. of Technology, Lab. of Computer and Information Science, October 1995. Submitted to a journal.

[4] A. Bell and T. Sejnowski, "An information-maximisation approach to blind separation and blind deconvolution," *Neural Computation*, vol. 7, no. 6, pp. 1004–1034, 1995.

[5] J. Karhunen, "Neural approaches to independent component analysis and source separation," in *Proc. of ESANN'96 (4th European Symposium on Artificial Neural Networks)*, (Bruges, Belgium), pp. 249–266, April 1996.

[6] G. Burel, "Blind separation of sources: A nonlinear neural algorithm," *Neural Networks*, vol. 5, no. 6, pp. 937–947, 1992.

[7] G. Deco and D. Obradovic, *An Information-Theoretic Approach to Neural Computing*. New York: Springer-Verlag, 1996.

[8] G. Deco and W. Brauer, "Nonlinear higher-order statistical decorrelation by volume-conserving neural architectures," *Neural Networks*, vol. 8, no. 4, pp. 525–535, 1995.

[9] L. Parra, G. Deco, and S. Miesbach, "Statistical independence and novelty detection with information preserving nonlinear maps," *Neural Computation*, vol. 8, pp. 260–269, 1996.

[10] L. Parra, "Symplectic nonlinear component analysis," in *NIPS-95, Advances in Neural Information Processing Systems 8*, (Cambridge, MA), MIT Press, 1996. (in press).

[11] P. Pajunen, A. Hyvärinen, and J. Karhunen, "Nonlinear blind source separation by self-organizing maps," in *ICONIP'96*, (Hong Kong), September 1996. Submitted.

[12] T. Kohonen, *Self–Organizing Maps*, vol. 30 of *Springer Series in Information Sciences*. Springer–Verlag, 1995.

[13] H. Bauer, R. Der, and M. Herrmann, "Controlling the magnification factor of self-organizing feature maps," *Neural Computation*, 1996. to appear.

[14] M. Van Hulle, "Globally-ordered topology-preserving maps with a learning rule performing local weight updates only," in *Neural Networks for Signal Processing V*, pp. 95–104, IEEE Press, 1995.

[15] M. Herrmann and H. Yang, "Perspectives and limitations of self-organizing maps in blind separation of sources," in *ICONIP'96*, (Hong Kong), September 1996. Submitted.

Building Nonlinear Data Models with Self–Organizing Maps

Ralf Der[1], Gerd Balzuweit[1], Michael Herrmann[2]

[1] University of Leipzig, Institute of Informatics
P. O. Box 920, 04109 Leipzig, Germany
[2] RIKEN, Lab. of Information Representation
2-1 Hirosawa, Wako-shi, 351-01 Saitama, Japan

Abstract. We study the extraction of nonlinear data models in high dimensional spaces with modified self-organizing maps. Our algorithm maps lower dimensional lattice into a high dimensional space without topology violations by tuning the neighborhood widths locally. The approach is based on a new principle exploiting the specific dynamical properties of the first order phase transition induced by the noise of the data. The performance of the algorithm is demonstrated for one- and two-dimensional principal manifolds and for sparse data sets.

1 Introduction

Artificial neural networks provide convenient tools for reconstructing nonparametric data models from noisy data. Consider data representing a functional relation $y = f(x)$ corrupted by noise, i.e. the observations are given by $y = f(x) + \eta$, where η is the noise which may or may not depend on x. Building a data model means extracting the systematic term $f(x)$ from the noisy data. If the noise also affects the x values the problem transforms into the construction of principal manifolds (PMs) that model data distributions by providing a curvilinear coordinate system for a dimension reduced representation of the data. PMs, principal curves e.g., may be defined self-consistently by the requirement that each point on the PM is the average of the data points projecting to it, cf. [1]. Thus, the mean square deviations (MSD) of the data from the PM are minimized locally at each base point. Averaged over the input distribution, however, the PM is only a stationary point with respect to the MSD [1], i.e. for some directions in the function space the PM belongs to, it is stable, for others unstable. So far there exist no explicit criteria for the existence, uniqueness and stability of PMs. It is known from the simulation of many data sets [1] that stability is guaranteed by local averaging the span of average being guided *globally* by cross validation [1].

Kohonen's algorithm provides a nonlinear mapping of some lattice (representing the physical positions of neurons) into the data manifold in input space. Thus, a piecewise linear approximation of the desired curvilinear coordinate system is established (cf. [3, 4]). Whether or not these coordinates reflect the main features of the data or, more precisely, whether they are principal curves of the data becomes in the context of self-organizing maps (SOMs) a stability problem. For this problem – and in accordance with the empirical observations for principal manifolds – it could be shown [3, 5] that stability is ensured by a sufficient range of local collaboration, i.e. by local averaging over a sufficiently wide span. This indicates the principal manifold are saddle points with respect to the MSD which are stable with respect to long ranged variations and unstable for short ranged variations. Averaging or neighborhood interaction stabilizes PMs because the short ranged deviations are suppressed. In [3, 4] stability problems have not been addressed, i.e. the neighborhood parameter has been assumed to vary only in the stable range. Since on the other hand the averaging disrupts the representation of specific data features, and thus tends to increase the MSD, the neighborhood coherence should not be to long-ranged.

The present paper focuses on a choice of this parameter that compromises between the requirements of smoothness and small MSD error. We start from Kohonen's unsupervised learning rule cf. [2] to develop a simple and robust algorithm for learning a good approximation of the principal manifold. As compared to the HS algorithm, it has the advantages of being on-line learning, not dimension specific and uses a *locally* defined width for the smoothing operation. This is important for data distributions with locally varying variance of the noise (multiplicative noise). Most importantly these widths are self-regulating. Our approach rests on a new principle which exploits the specific dynamical properties of the first-order phase transition [3, 5] induced by the noise. The approach is shown to work also for sparse data sets and should therefore be favorable also in the case of high-dimensional inputs.

2 Maps with locally adaptive smoothness

Let us assume that the data embedded in some d-dimensional space $\mathcal{X}$ actually are scattering about some manifold of the lower dimension $D < d$. Then Kohonen's algorithm may be used to map the data topographically onto a D-dimensional lattice $\mathcal{A}$, where the lattice points $\mathbf{r} \in [1, N]^D$ may be considered as the physical positions of N^D neurons. Upon presentation of a data vector $\mathbf{x}$, the learning step for the synaptic vectors $\mathbf{w_r}$ is

$$\Delta \mathbf{w_r} = \varepsilon \, h_{\mathbf{r},\mathbf{r}'} \left(\mathbf{x} - \mathbf{w_r} \right), \tag{1}$$

where the position $\mathbf{r}'$ of the winner (best matching) neuron is determined by $\mathbf{r}' = \arg\max_{\mathbf{r}} \|\mathbf{x} - \mathbf{w_r}\|$. The neighborhood function $h_{\mathbf{r},\mathbf{r}'} = \left(2\pi\sigma^2\right)^{-\frac{D}{2}} \exp\left(-(\mathbf{r} - \mathbf{r}')^2/(2\sigma^2)\right)$ defines the range of cooperativity between neurons which controls the smoothness of the map: Its radius of curvature ρ obeys $\rho > 2d$ where d is about the Euclidean distance between neurons in input space which are a distance σ apart on the lattice. (The normalization factor $\left(2\pi\sigma^2\right)^{-\frac{D}{2}}$ was introduced in order that the average force exerted on $w_{\mathbf{r}}$ is independent on σ). For varying width of the data scattering we need the smoothness of the map to be defined locally. This may be achieved by an individual neighborhood $\sigma_{\mathbf{r}}$ for each neuron, so that

$$h_{\mathbf{r},\mathbf{r}'} = \left(\frac{1}{\sqrt{2\pi}\tilde{\sigma}_{\mathbf{r}}} \right)^D \exp \left(-\frac{(\mathbf{r} - \mathbf{r}')^2}{2\tilde{\sigma}_{\mathbf{r}}^2} \right) \tag{2}$$

where $\tilde{\sigma}_{\mathbf{r}} = \min \{\sigma_{\mathbf{r}}, \sigma_{\mathbf{r}'}\}$ with the additional constraints $1 \leq \tilde{\sigma}_{\mathbf{r}} \leq \sigma_{\max}$.

The crucial point now is the determination of the local values $\sigma_{\mathbf{r}}$. The principal manifold can be mapped topographically onto the lattice because of the dimensions match by definition. However, with the data points scattering about the PM we have to compromise between two options. On the one hand the lattice should be mapped tightly to the PM which requires a small σ (curvature) if the PM is nonlinear. On the other hand the stiffness and hence σ should be sufficiently large in order to avoid the folding due to the dimensional mismatch.

For the definition of the optimal $\sigma_{\mathbf{r}}$ we exploit the dynamics of the phase transitions induced by the scattered data points. Namely, if σ is lowered below a value σ^{crit} the representation of the main data feature by the map gets distorted by other 'noisy' features. The transition has has been shown earlier [5] to proceed through a either of two phases, one preserving and one violating topology. Since the latter is more pronounced, emerging folding into secondary features will be signaled immediately by topology violations. Concentrating on the first and second winner the criterion for the occurrence of a fold is $\alpha > 1$ where $\alpha = \|\mathbf{r}' - \mathbf{r}''\|$ is the distance between the first and second winner in the lattice.

In contrast to [6], where the neighborhood width has been updated following an energy function resulting in the occurrence of local minima and a slow convergence, our approach here consists in *keeping $\sigma_{\mathbf{r}}$ fluctuating* around its critical value $\sigma_{\mathbf{r}}^{\mathrm{crit}}$. The result of the algorithm is given in terms of sliding averages over the fluctuations rather than a convergent network state. For this purpose we decrement $\sigma_{\mathbf{r}}$ at each step as

$$\Delta \sigma_{\mathbf{r}} = -\frac{1}{NT_\sigma} \sigma_{\mathbf{r}} \quad \forall \mathbf{r} \tag{3}$$

and reset whenever $\alpha > 1$ the σ's in the vicinity of the topology distortion as

$$\sigma_{\mathbf{r}} := \max \left(\sigma_{\mathbf{r}}, \alpha \, \exp \left(-\frac{2(\mathbf{r} - \mathbf{R})^2}{\alpha^2} \right) \right), \quad \text{where } \mathbf{R} = \frac{1}{2}(\mathbf{r}' + \mathbf{r}''). \tag{4}$$

As a result, the map fluctuates around the PM due to the phase transition taking place each time the critical value σ^{crit} is crossed. In order to average over the fluctuations each neuron keeps a second pointer $\overline{\mathbf{w}}_{\mathbf{r}}$ obtained by the moving average

$$\Delta \overline{\mathbf{w}}_{\mathbf{r}} = \frac{1}{KNT_\sigma} (\mathbf{w}_{\mathbf{r}} - \overline{\mathbf{w}}_{\mathbf{r}}) \tag{5}$$

over the fluctuations, where K is of the order of 10. The $\overline{\mathbf{w}}_{\mathbf{r}}$ provide in most cases a very good first order data model. Further improvements depend on the task. In the case of modelling a functional relationship (see introduction) one

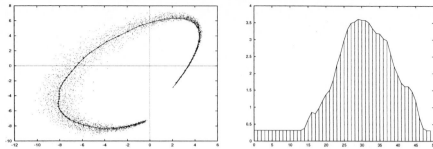

Fig. 1. The map of a two-dimensional data distribution of varying scattering width onto a one-dimensional chain of 50 neurons (left). Final values of σ_r along the neural chain produced by enforcing topology preservation (right).

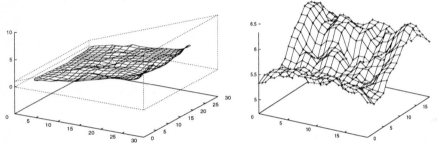

Fig. 2. Embedding a two-dimensional neural lattice in a three-dimensional data set (left). Values of the neighborhood widths $\sigma_{\mathbf{r}}$ as a function of the net position $\mathbf{r}$ (right).

may use the $\overline{\mathbf{w}}_{\mathbf{r}}$ to investigate the properties of the noise η in order to improve the model. For the PM case, an essential improvement consists in using the $\overline{\mathbf{w}}_{\mathbf{r}}$ as starting positions for a final step in the sense of the iterative HS algorithm. This can be implemented more easily by monitoring directly the averages over the data in each domain. Hence, instead of $\overline{w}_{\mathbf{r}}$ each neuron gets a second pointer $\overline{v}_{\mathbf{r}}$ measuring the local average of the input data which may deviate from $\overline{\mathbf{w}}_{\mathbf{r}}$ if the principal manifold is curved. $\overline{v}_{\mathbf{r}}$ is updated if neuron $\mathbf{r}$ is the winner as $\Delta \overline{v}_{\mathbf{S}} = \frac{1}{KT_\sigma} (\mathbf{v} - \overline{v}_{\mathbf{S}})$ The set $\{\overline{v}_{\mathbf{r}} \mid \mathbf{r} = 1, ..., N\}$ are the final result of the algorithm, i. e. they represent the PM in input space.

3 Sparse data sets

The above algorithm hinges on the abundance of data points which signal the folding via the topology violations. This may fail if the number of data points is small. For this case, a very sensitive criterion for the emergence of the critical fluctuations was found to be a wavelet transform [7] of the map. For a one-dimensional SOM we use the Gabor transform

$$g_{r'} = \frac{1}{\sqrt{2\pi}u_{r'}} \left\| \sum_{k=1}^{N} w_{r'} \, \exp\left(\frac{-(k-r')^2}{2u_{r'}^2}\right) \exp\left(-i \, k\omega_{r'}\right) \right\| \qquad (6)$$

where both the frequency $\omega_{r'}$ of the kernel and the width are functions of the current values of $\sigma_{r'}$ so that the kernel is always in resonance with potential

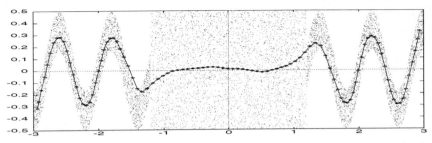

Fig. 3. Mapping a noisy *sin*-wave onto a chain of neurons. The strong noise in the center of the data distribution is clipped by the map.

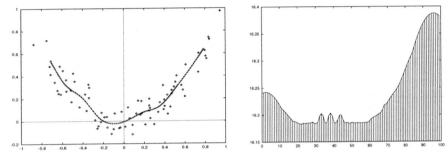

Fig. 4. Using wavelet transform to control the neighborhood widths in a sparse data set. Display of the emerged map (left) and σ_r-values (right).

foldings. At the critical point $\sigma = \sigma^{\mathrm{crit}}$ the wavelength of the emerging folds is $\lambda = 4.04\sigma l$, where l is the average distance between the neurons in that region, cf. [3]. Choosing $w_{r'} = u_{r'} = 4\sigma_{r'}$ causes $g_{r'}$ to jump by an order of magnitude when $\sigma_{r'}$ drops below $\sigma_{r'}^{\mathrm{crit}}$. Hence, $g_{r'}$ is the desired sensitive criterion for detecting the onset of the phase transition. In the algorithm we use (3) as before. If for the winner $g_{r'}$ exceeds a small threshold we use $\alpha = \kappa\sigma_{r'}$ in (4), observing $1 \leq \alpha \leq \sigma_{\max}$, where $\kappa = 1.2$ is an empirical factor. In the simulations a control of κ is obtained from monitoring the fluctuations of σ which optimally should stay in the region of a few percent.

4   Numerical simulations

We have applied the above algorithms to map both one- and two-dimensional lattices into higher-dimensional input spaces with inhomogeneous data distributions of effective dimension $D = 1$ or $D = 2$, respectively. The local scattering of the data points around the central manifold varied by up to an order of magnitude. In all simulations the algorithms were stable and produced the desired results. In particular the cooling time T_σ could be varied by more than an order of magnitude without instability problem. Parameters used in the simulations were $T_\sigma = 1500$, $\sigma_{\max} = N/3$, and $\varepsilon = 0.15$ (kept constant during the simulations. It should be noted that such relatively high values of ε allow for fast convergence and improve the efficiency of the average (5)), which in turn compensates the fluctuation in the final map $\overline{\mathbf{w}}_\mathbf{r}$. The first and the second example (Figs. 1 and 2, respectively) are carried out by the basic algorithm explained in

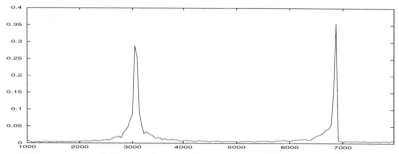

Fig. 5. Time course of the wavelet transform of one of the neurons during 7000 steps.

the second section. In the third example shown in fig. 4 we use wavelet transform to adjust the σ_r. A data set of 80 points and a chain of 100 neurons are chosen. In practical computation it is easy to use FFT for computing the Gabor transform function.

5 Conclusion

The present contribution is dedicated to the problem of building nonlinear data models with a modified Kohonen learning rule which learns its own parameters. In particular, the neighborhood range of the output units, which determines the smoothness of the produced maps, is an essential prerequisite for the formation of principal manifolds in the input data set. Our algorithm solves the task of the determination of the neighborhood parameter stably and in a local manner. This is of importance for processing sparse data sets with spatial heterogeneities. Therefore, the algorithm appears to be well suited for more complex tasks. In particular, the algorithm was found to be also stable with higher-dimensional (up to $d = 8$) inputs. The application to real world data is underway.

ACKNOWLEDGEMENT: One of the authors (RD) gratefully acknowledges the hospitality of RIKEN (Tokyo) he received during a visit in February 1996.

References

1. T. Hastie, W. Stuetzle, Principal curves. Journal of the American Statistical Association 84, 502–516 (1989)
2. T. Kohonen, *The Self-Organizing Map*, Springer-Verlag, 1995.
3. H. Ritter, T. Martinetz, K. Schulten, *Neural Computation and Self-Organizing Maps*, Addison-Wesley, 1992.
4. F. Mulier, V. Cherkassky, Self-Organization as an Iterative Kernel Smoothing Process. *Neural Computation* **7**, 1165-1177, 1995.
5. R. Der, M. Herrmann, Critical Phenomena in Self-Organized Feature Maps: A Ginzburg-Landau Approach, *Phys. Rev.* **E 49**:6, 5840–5848, 1994.
6. M. Herrmann, "Self-Organizing Feature Maps with Self-Organizing Neighborhood Widths", *Proc. 1995 IEEE Intern. Conf. on Neur. Networks*, 2998–3003, 1995.
7. C. K. Chui, *An Introduction to Wavelets*, Academic Press, 1992.

A Parameter-Free Non-Growing Self-Organizing Map Based upon Gravitational Principles: Algorithm and Applications

J. S. Lange, H. Freiesleben

Institut für Kern- und Teilchenphysik, Technische Universität Dresden,
D-01062 Dresden, Germany

Abstract. The Kohonen Algorithm was extended by *(a)* a coupling between node and input space based upon gravitational principles and *(b)* a controling mechanism based upon two-point correlation functions. The extended algorithm avoids optimization of the learning rate and neighbourhood parameters. Applications are given for both a 3-dimensional example data set with mixed topologies and a 13-dimensional data set of a particle physics experiment.

1 Introduction

The self-organizing map (SOM), according to Kohonen [1], is one of the most successful vector quantization (VQ) algorithms and is used in many applications due to its ability of unsupervized learning. Recently, VQ algorithms were developed which dynamically create and destroy nodes during training, such as the Growing Cell Structures [2] and the Growing Neural Gas [3]. Nevertheless in many applications it is necessary to use a non-growing self-organizing map. On the one side many applications require a pre-defined, fixed granularity of quantization in the input space corresponding to the fixed number of nodes. On the other side hardware platforms like the Adaptive Solutions CNAPS and the Siemens SYNAPSE-I require at least a fixed maximum number of nodes due to a fixed number of processors.

One of the disadvantages of each VQ algorithm is the necessity of optimization of the learning rate, the neighbourhood parameters as well as the time dependance of both [4]. Thus, one may try to develop a *parameter-free* algorithm with a self-controling SOM to avoid parameter optimization.

2 The Algorithm

2.1 The Kohonen Algorithm

The Kohonen map is an arrangement of nodes $k = 1, ..., K$ in the node space (in this case chosen as one-dimensional), each of them representing a N-dimensional vector $\underline{v}(k)$ in the input space. During training phase for each N-dimensional input event vector $\underline{e}$ out of a sample of E input vectors the winner node is given by the specific node k_w related to the vector $\underline{v}(k_w)$ with the smallest Euklidian distance $r(k) = |\underline{e} - \underline{v}(k)|$ in the input space, $r(k_w) = min[r(k)]$. The vector of the winner node is modified according to

$$\underline{v}'(k_w) = \underline{v}(k_w) + \varepsilon \cdot (\underline{e} - \underline{v}(k_w)) \tag{1}$$

with the learning rate ε. The vector of each neighbour node k_n of the winner node is modified according to

$$\underline{v}'(k_n) = \underline{v}(k_n) + \varepsilon \cdot F(d) \cdot (\underline{e} - \underline{v}(k_n)) \tag{2}$$

with the Gaussian function

$$F(d) = exp(-d^2/2\sigma^2) \tag{3}$$

including the distance d between k_w and k_n in the node space and a spreading width σ. The parameters ε and σ are time-dependent in order to freeze out the map. In general, according to [4] a linear time dependence for both ε and σ can be chosen which leads to four free parameters ε_i, ε_f, σ_i and σ_f (i denoting *initial* and f *final*).

In a former analysis of a particle physics experiment [5], [6] the final network status turned out to depend strongly upon these parameters, even if the number of training steps was massively increased up to 10^7.

2.2 The Coupling between node space and input space

In the Kohonen algorithm the node space is a fixed grid with time-independent distances between the nodes, i.e. there is no dynamics in the node space. Cluster formation in nature reveals a different principle: the contraction of a cluster of mass modifies the space-time grid around the cluster according to the principles of general relativity. Applying this principle to the SOM, firstly the grid of nodes in a SOM can formally be treated like the space-time grid. Secondly, a coupling between the node space and the input space can be introduced. If a cluster contracts in the input space, the grid in the node space is deformed. This deformation leads to an attraction of an event in the cluster's neighbourhood. For a mathematical description a modified formula of the general relativity [7] was used which describes the modification of a radial unit length $dx=1$ in a gravitational field. In a distance r to a pointlike mass, the length dx is contracted according to

$$dx(r) = 1 - \frac{\alpha}{2r} \tag{4}$$

with a constant parameter α fixed by the theory of general relativity. To avoid singularities, this formula was modified to

$$dx(r) = 1 - \frac{\alpha'}{\alpha' + r} \tag{5}$$

with a new parameter α'. The parameter r corresponds to the Euklidean distance between the winner node and a neighbour node in the input space

$$r = |\underline{v}(k_w) - \underline{v}(k_n)| \quad . \tag{6}$$

When applying Eq. 5 to the algorithm, the Gaussian function Eq. 3 is replaced by

$$F(d) = exp(-(d \cdot dx)^2/2\sigma^2) \quad .$$ (7)

2.3 The Two-Point Correlation Function

In order to construct a parameter-free algorithm, the network itself must have the ability both to test its actual status and to adapt the actual modification strength $\underline{v}'(k_n)/\underline{v}(k_n)$ according Eq. 2. This test can be performed via the two-point correlation function $\mathcal{P}(r)$ which can be defined for both node and input vector distances

$$r_{node} = |\underline{v}(k) - \underline{v}(k')| \quad \text{for} \quad k, k' = 1, ..., K, k \neq k'$$ (8)

$$r_{input} = |\underline{e}_i - \underline{e}_{i'}| \quad \text{for} \quad i, i' = 1, ..., E, i \neq i' \quad .$$ (9)

$\mathcal{P}(r)$ denotes the probability for finding a certain value r among all node vector pairs or all input vector pairs, respectively. Technically, for all pairs a histogram with r-bins is filled and normalized afterwards. The evaluation of the network status is performed via the validation function

$$\mathcal{P}_{val}(r) = \quad [\quad \mathcal{P}_{node}(r) - \mathcal{P}_{input}(r) \quad]_{norm} \quad .$$ (10)

$\mathcal{P}_{val}(r)$, which is normalized to a range $-1 \leq \mathcal{P}(r) \leq 1$, is used to recognize differences in the r_{node} and r_{input} distributions. $\mathcal{P}_{input}(r)$ is only computed once at the beginning of the training, $\mathcal{P}_{node}(r)$ and $\mathcal{P}_{val}(r)$ are evaluated once per epoche. In the algorithm, $\mathcal{P}_{val}(r)$ is added as a correction factor in Eq. 5

$$dx(r) = 1 - \frac{\alpha'}{\alpha' + r(1 - \mathcal{P}_{val}(r))} \quad .$$ (11)

r again is given according to Eq. 6. If there are too many nodes related to a specific r-bin (relatively compared to the r_{input} distribution), then $\mathcal{P}_{val}(r)$ is positive, dx is small and $F(d)$ is large. In effect, there is a strong adaption of the node vector of the neighbour node k_n in order to push the node into another r-bin. If there are (relatively) too few nodes within a certain r-bin, $\mathcal{P}_{val}(r)$ is negative, dx is large and $F(d)$ is small. The node vector adaption is small and the node remains in the r-regime. It is obvious that the two-point correlation extension cannot be used without the cluster-gravitation extension. It should be noted that two-point correlation functions are used very successfully in analyses of cluster formation in galaxy distributions.

2.4 Parameters

Obviously, the four Kohonen network parameters and the new parameter α' are still implemeted in the algorithm. But there is no parameter optimization necessary anymore to achieve a "good" final network state. For *each* of the presented applications (see below) the *same* parameter set was used:

$$\varepsilon_i = 1.0, \quad \varepsilon_f = 0.0, \quad \sigma_i = 1.0, \quad \sigma_f = 1.0, \quad \alpha' = 1.0 \ .$$

Due to the fact that the quality of the final network setup does not depend upon these parameters, it is allowed to use the term "*parameter-free*". It should be noted that these parameters correspond to a "bad" final network status in case of using the Kohonen algorithm alone (c. f. Fig.1, top left).

3 Applications

3.1 Learning of Topologies

The complete algorithm was tested with a data set corresponding to a popular example: 3-dimensional input vectors were randomly generated within a 3-dimensional, 2-dimensional and 1-dimensional substructure. Originally, this example was designed to show the advantages of growing node structures [2], [3], because the final node structures consist of 1-, 2- and 3-dimensional substructures themselves. In our case the example is used to show the mapping of the 3-dimensional input structure onto a 1-dimensional node structure (node chain) and thus reducing the dimension, i.e. in contrast to growing node structures the dimension of the node structure is fixed. The result for the different methods is shown in Fig. 1. The quality of the mapping increases obviously when using the presented extensions to the Kohonen algorithm.

3.2 Application in Particle Physics

One of the experimental goals being pursued by means of the TOF-spectrometer [8] at the COSY accelerator at the Forschungszentrum Jülich is the measurement of pp-bremsstrahlung (pp→ppγ) to get direct information on the nucleon-nucleon interaction. A first experiment was performed in September 1995 with T_{beam}=348 MeV [9]. The TOF-spectrometer detects the two protons; the photon has to be reconstructed via four-momentum conservation. The standard *1-dimensional* offline method for reaction identification is the calculation of the mass of the undetected, third particle for each event [9]. Due to finite detector resolution events of the rare pp-bremsstrahlung process should be visible as a broadened peak at m_γ=0 in the one-dimensional mass spectrum. This peak is burried by background reactions, such as pion production pp→ppπ⁰ and pp→pnπ⁺, which can only be extracted in a *high-dimensional* analysis exploiting all kinematic information. With a SOM including the complete algorithm (13-dimensional input space, 30.000 input events, 100 nodes) it was possible to exclude the background so that a peak at m_γ=0 gets visible (c. f. Fig.2). Again, no parameter optimization was necessary. This peak remains unvisible using analysis methods exploiting *2-dimensional* cuts on several combinations of 2 out of the 13 input parameters. The time which formerly was necessary for the parameter optimization in a similar analysis [5], [6] (several weeks for 14.000 test runs, 5 min CPU time each) was reduced to typically 18 h CPU time for *one* training phase including the full algorithm.

3.3 Summary

We have shown that it is possible to introduce an extension to the Kohonen algorithm which replaces parameter optimization by a self-controling method. The success of this method was demonstrated in both a mixed topology example and an analysis of a particle physics experiment. The most time consuming step in the extended algorithm (evaluation of $\mathcal{P}_{input}(r)$) scales with $\mathcal{O}(E^2)$ (E being the number of input events, typically $E \simeq 10^4$) has to be performed only once per training. Therefore the implementation of the algorithm is no problem on a standard workstation (≥ 25 MFlops).

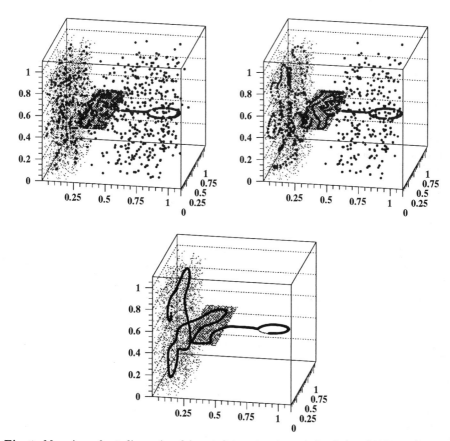

Fig. 1. Mapping of a 3-dimensional input data set onto a node chain of 1000 nodes using different methods (axes: normalized input parameters). *Top left*: Original Kohonen algorithm. *Top right*: Kohonen algorithm plus cluster-gravitation. *Bottom*: Kohonen algorithm plus cluster-gravitation plus two-point correlation function extension. Training was performed with parameters according 2.4 in each case ("bad" parameter set)

832

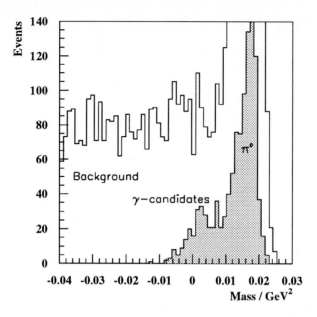

Fig. 2. Mass spectrum of the training events of the particle physics application (c. 3.2). *White:* All events including background. *Gray:* Events extracted by the SOM Cut. A peak at $m_\gamma = 0$ is visible.

References

1. Kohonen, T.: Self-Organization and Associative Memory. Springer Series in Information Sciences **8**, 1984, pp. 125
2. Fritzke, B.: Growing Cell Structures – a self-organizing network for unsupervised and supervised learning. Neural Networks **7** (1994) 1441-1460
3. Martinetz, T., Schulten, K.: A ,,neural-gas" network learns topologies. Artificial Neural Networks, Editors: Kohonen, T., Mäkisara, K., Simula, O., Kangas, J., North-Holland, Amsterdam, 1991, pp. 397
4. Behrmann, K.-O.: Leistungsuntersuchung des "Dynamischen Link-Matchings" und Vergleich mit dem Kohonen-Algorithmus. Diploma Thesis, Institut für Neuroinformatik, Ruhr-Universität Bochum, 1993
5. Lange, J. S.: Extraktion von Bremsstrahlungsereignissen in Proton-Proton-Reaktionen durch Anwendung künstlicher neuronaler Netze. Ph.D. Thesis, Institut für Experimentalphysik I, Ruhr-Universität Bochum, 1994
6. Lange, J. S. et al.: Cluster-Gravitation - An Extension to the Kohonen Map for the Identification of pp-Bremsstrahlung at COSY. 4th Int. Workshop on Software Engineering, Artificial Intelligence and Expert Systems for High Energy and Nuclear Physics AIHENP 95, Pisa, Apr 4-8, 1995, (in press)
7. Einstein, A.: Die Grundlage der allgemeinen Relativitätstheorie. Ann. d. Phys. **49** (1916) 769
8. The COSY-TOF-Collaboration: Physica Scripta **48** (1993) 226
9. Hermanowski, P.: Ph. D. Thesis, Institut für Experimentalphysik I, Ruhr-Universität Bochum, 1996, (in preparation)

Topological Maps for Mixture Densities

F. Anouar[1], F. Badran[1] et S.Thiria[1,2]

1 CEDRIC, Conservatoire National des Arts et Métiers, 292 rue Saint Martin - 75 003 PARIS
2 Laboratoire d'Océanographie et de Climatologie (LODYC), Université de PARIS 6. 4 Place Jussieu, 75005 PARIS
E-mail : anouar@asimov.cnam.fr

Abstract
We propose in this paper a new learning algorithm which uses a probabilistic formalism of topological maps. This algorithm approximates the density distribution of the input set with a mixture of normal distributions. We show that, under certain conditions, the classical Kohonen algorithm is a specific case of this algorithm, therefore allowing a probabilistic interpretation of Kohonen maps.

1. Introduction

This paper deals with a probabilistic formalism and proposes a new learning algorithm PSOM (Probabilistic Self Organizing Map) which maximizes the likelihood function. Due to this probabilistic formalism, each neuron represents a gaussian function, and the learning algorithm estimates both mean and variance of each gaussian. Under some particular hypothesis, this algorithm is closely related to the classical SOM (Self Organizing Map) algorithm [Kohonen 95].

The remainder of this paper proceeds as follows. We first introduce the batch algorithm used to develop the probabilistic algorithm PSOM in the second section. Then we draw up the analogy with the SOM algorithm. The last section is devoted to simulation results.

2. Batch algorithm for topological maps

Let us first introduce the notations we used. Let D be the data space ($D \subset \Re^n$) and $App = \{z_i \; ; \; i=1,...,N\}$ the training set ($App \subset D$). We denoted by (C) a map of M neurons, this map is assumed to have a neighborhood system controlled by a kernel function K(.). The distance $\delta(c, r)$ between two neurons (c) and (r) of the map is the length of the shorter path between c and r. To each neuron c, we associate a weight vector $W_c = (w_c^1, w_c^2, \cdots, w_c^n)$ called reference vector, thus $W = \{W_c; c = 1 \cdots M\}$ is the set of all reference vectors.

A Batch algorithm

Let ϕ be an affectation function $\phi : D \to C$. ϕ defines a partition $F_\phi = \{F_c; c = 1 \cdots K\}$ (where $F_c = \{z; \phi(z) = c\}$) of the data space.

We present a batch version of SOM algorithm [Luttrel 94]. This algorithm proceeds iteratively to optimize an error function. Each iteration is composed of two steps : an optimization step followed by an affectation step.

• In our case the optimization step involves the minimization of an error function given by: $E(\phi, W) = \sum_{z_i \in App} d(z_i, \phi(z_i))$ where $d(z, c) = \sum_{r \in C} K(\delta(c, r))\|z - W_r\|^2$

The minimum of $E(\phi, W)$ with respect to W can be obtained for :

$$W_c^* = \frac{\sum_{r \in C} K(\delta(c, r))Z_r}{\sum_{r \in C} K(\delta(c, r))n_r} \quad \text{where } Z_r = \sum_{z_i \in App \cap F_r} z_i \text{ and } n_r = \text{card}(App \cap F_r).$$

• the affectation step uses the affectation function ϕ_W: $\phi_W(z) = \arg\min_{c \in C} d(z, c)$.

For a given neighborhood function K(.), it is proved that this iterative proceeding (by affectation and minimization) implies convergence to a stationary point after a limited number of iterations. The stationarity is reached at a local minimum of $E(\phi, W)$. An on-line training procedure of this algorithm can be designed and underlines the relationship between this algorithm and the SOM. The main difference between the two algorithms relies on the way the winner is chosen.

3. Approximation of the density function for topological maps

We now present a new algorithm for topological maps. This algorithm deals with the problem of mixture density learning. It is based on a probabilistic formalism.

3.1. Probabilistic formalism
We use now a three layers architecture to modelize the map [Luttrel 94]. The input layer has n neurons receiving input vector z. The initial map C is duplicated in two similar maps C_1 and C_2 provided with the same topology as C. C_1 and C_2 will be respectively the second and the third layer .

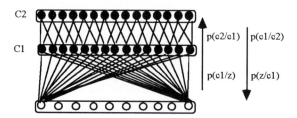

Fig. 1. Three layers architecture

This type of structure is called a folded Markov chain [Luttrel 94]. The Markov property is given by : $p(c_2 / z, c_1) = p(c_2/c_1)$ and $p(z / c_1, c_2) = p(z / c_1)$. Using the Bayes rule, we obtain the joint probability :

$$p(z) = \sum_{c_2} p(c_2)p_{c_2}(z) \text{ where } p_{c_2}(z) = \sum_{c_1} p(c_1 / c_2)p(z / c_1)$$

We assume that the probability density has the following form :

$$p(c_1 / c_2) = \frac{1}{T_{c_2}} K(\delta(c_1, c_2)) \text{ where } T_{c_2} = \sum_r K(\delta(c_2, r)),$$

$p(z/c_1) = f_{c_1}(z)$, where f_{c_1} is a gaussian density with a mean vector W_{c_1} and a

covariance matrix $\Sigma_{c_1} = \sigma_{c_1}^2 I$, thus $p_{c_2}(z) = \frac{1}{T_{c_2}} \sum_{r \in C_1} K(\delta(c_2, r)) f_r(z)$.

Given these assumptions, the global density $p(z)$ is a mixture of the density functions $p_{c_2}(z)$. The learning procedure we propose maximizes the likelihood function, and estimates the mixture parameters.

3.2. Parameters estimation

To deal with the maximization of the likelihood function, the EM algorithm [Dempster and al.77] is an efficient method. But we chose a partitioning method [Diday 80] [Schroeder 76] which allows to develop an algorithm operating like the batch algorithm discussed below. We can afterwards show that the SOM algorithm can be derived from the on-line version of our algorithm.

Let us assume that each observation z_i is generated in a deterministic way from one of the M mixtures. We define the affectation function χ by : $\chi(z) = \arg\max_{c_2} p_{c_2}(z)$.

χ depends on the current parameters and will be noted χ^k at the kth iteration. Assuming that the observations are independents the likelihood is :

$$p(z_1, z_2, \cdots, z_N, \chi, W, \sigma) = \prod_{i=1}^{N} p_{\chi(z_i)}(z_i).$$

We minimize the log-likelihood function $E(\chi, W, \sigma)$ according to the affectation functions χ and to the parameters W and σ :

$$E(\chi, W, \sigma) = \sum_{i=1}^{N} -Ln\left(\sum_{r \in C} K(\delta(\chi(z_i), r)) f_r(z_i, W_r, \sigma_r) \right) = \sum_{i=1}^{N} E_i(\chi, W, \sigma)$$

As for the batch algorithm, the learning procedure is iterative and composed of two steps. At the iteration k :

The affectation step uses the current function χ to partition the input space,

The optimization step updates the parameters by setting the derivatives to zero and yields the following new parameters :

$$W_r^k = \frac{\displaystyle\sum_{i=1}^{N} z_i K(\delta(r, \chi^{k-1}(z_i)) \frac{f_r(z_i, W_r^{k-1})}{p_{\chi^{k-1}(z_i)}(z_i)}}{\displaystyle\sum_{i=1}^{N} K(\delta(r, \chi^{k-1}(z_i)) \frac{f_r(z_i, W_r^{k-1})}{p_{\chi^{k-1}(z_i)}(z_i)}} \tag{1}$$

$$\sigma_r^k = \frac{\displaystyle\sum_{i=1}^{N} \left\| W_r^{k-1} - z_i \right\|^2 K(\delta(r, \chi^{k-1}(z_i)) \frac{f_r(z_i, W_r^{k-1})}{p_{\chi^{k-1}(z_i)}(z_i)}}{n \displaystyle\sum_{i=1}^{N} K(\delta(r, \chi^{k-1}(z_i)) \frac{f_r(z_i, W_r^{k-1})}{p_{\chi^{k-1}(z_i)}(z_i)}} \tag{2}$$

This algorithm is summarized as follows :

PSOM algorithm

Initialization : Choose randomly the initial parameters W^0 et σ^0 and the associated affectation function χ^0.
Iterative step : at the iteration k , W^{k-1} et σ^{k-1} are known,
Minimization step :
Compute the new parameters W^k and σ^k according to (1) and (2).
Affectation step :
Compute the new affectation function χ^k associated to W^k et σ^k
Repeat the iteration step until stabilization.

3.3. Convergence proof:

The minimization step at the kth iteration minimize $E(\chi^{k-1}, W, \sigma)$ according to the parameters W and σ for a fixed affectation function χ^{k-1} , we obtain the following inequality : $E(\chi^{k-1}, W^k, \sigma^k) \le E(\chi^{k-1}, W^{k-1}, \sigma^{k-1})$.

The affectation step at the kth iteration minimize $E(\chi, W^k, \sigma^k)$ with respect to χ for a given parameters W^k and σ^k and gives: $E(\chi^k, W^k, \sigma^k) \le E(\chi^{k-1}, W^k, \sigma^k)$

Combining the two inequalities we obtain: $E(\chi^k, W^k, \sigma^k) \le E(\chi^{k-1}, W^{k-1}, \sigma^{k-1})$.

The set of partitions is finite and each partition obtained by χ^k has a unique associated parameters W^k and σ^k, thus the number of expected values of $E(\chi^k, W^k, \sigma^k)$ is finite. Since the log-likelihood function $E(\chi, W, \sigma)$ decreases at each iteration, the algorithm converges in a limited number of iterations. The stationary point is a local minimum of $E(\chi, W, \sigma)$.

3.4. Analogy with the SOM algorithm

If we assume that all covariance matrix are equal to $\sigma^2 I$ where σ is a constant, an on-line version of the PSOM algorithm is given by substituting the minimization step by

the following update rule: $W_r^k = W_r^{k-1} - \varepsilon \dfrac{\partial E_i}{\partial W_r}$ (3)

Using an uniform function for $K(.)$ yields a clear analogy with the SOM algorithm.

We assume that the kernel function is equal to $\dfrac{1}{N_c}$ if $\delta(r, c) \le d$ and zero otherwise,

where Nc is the number of neurons r such that $\delta(r, c) \le d$. The rule (3) becomes :

(4)
$$\begin{cases} W_r^k = W_r^{k-1} - \varepsilon \left(\dfrac{f_r(z_i)}{P_{\chi^{k-1}(z_i)}(z_i)} \right)(W_r^{k-1} - z_i) & \text{if} \quad \delta(r, \chi^{k-1}(z_i)) \le d \\ W_r^k = W_r^{k-1} & \text{otherwise} \end{cases}$$

Where $p_{c_2}(z_i) = \dfrac{1}{N_{c_2}} \sum_{r, \delta(r,c) \le d} f_r(z_i)$.

The updating parameters in (4) operates like a SOM update rule. The main difference is that the computation of the kernel function in (4) depends on the observations and not on the map as in the SOM algorithm. Thus the updating rule is more sensitive to the local concentration of the observations. These two algorithms become similar when the constant σ is sufficiently large. This gives a probabilistic interpretation of the SOM algorithm: SOM is a PSOM in which all covariance matrix are equal to $\sigma^2 I$ where σ is sufficiently large.

4. Simulation results

We tested the batch version of the PSOM algorithm on two sets of simulated data. The first data base (data1) has 800 examples with uniform density in [-12,12]x[-12,12]. The second data base (data2) has 900 examples and is generated according to a mixture of 9 gaussians. All simulations use squared topological maps. For data1 the map has 10x10 neurons, and for data2 it has 6x6 neurons.

All simulations use a temperature parameter which decreases as training proceeds in the following form : $(0.6)^{\frac{it}{it_f}}$ where 'it' is the current iteration and 'it$_f$' is the total number of iterations. The kernel function is defined by: $K_T(u) = \frac{1}{T} e^{\frac{-u^2}{T^2}}$.

For each experiment, we show the results after learning and we display the reference vectors and the corresponding topological map.

Figures 2(a) and 3(a) show the topological maps for the two data bases after training. On these maps, we can notice the topology preserving mapping. Figures 2(b) and 3(b) show the estimated variances. Each variance is represented by a circle centered on the weight vector and with a radius equal to the standard deviation .

(a)   (b)

Fig. 2. (a) Represents the map after learning with the PSOM algorithm for data1. (b) Displays variances with circles centered on the reference vectors.

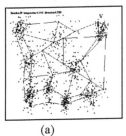
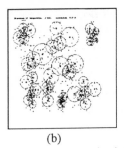

(a)   (b)

Fig. 3. a) Represents the map after learning with the PSOM algorithm for data2.
(b) Displays variances with circles centered on the reference vectors.

Figures 2(b) and 3(b) show the influence of the variance parameters. These parameters fit the local distribution, thus they are small in region with high density and great in region with weak density.

We tested the PSOM algorithm for classification. The data set has been labeled with three labels so that the different classes overlap. Then we assign to each weight vector one of the three label. The assignment uses the majority vote method. For the PSOM algorithm, the assignment and the classification uses the distance $\|z - Wi\|^2 / 2\sigma_i$ between the observation z and the neuron i. We generated a test set of 300 examples. The PSOM algorithm recognizes all examples, on the other hand the SOM algorithm misclassifies 21 points which are situated in the overlapping region. The PSOM algorithm classify easily this simulated data so that we are studying the classification of real data.

5. Conclusion

We have developed a new probabilistic learning algorithm for topological map which estimates parameters of a density mixture using likelihood estimation. We show that this algorithm allows a probabilistic interpretation of topological maps. Then we investigated the relationship between an on-line version of this algorithm and the classic SOM algorithm.

References

[1] Dempster A. P., Laird N., M., Rubin D., B, 1977 "Maximum likelihood from incomplete data via the EM algorithm" Journal of the Royal Statistical Society, B, 39, 1-38.
[2] Diday E. et collaborateurs, 1980. *Optimisation en Classification Automatique.* Editeur INRIA.
[3] Luttrel S.P , 1990. "Derivation of a class of training algorithms". *IEEE Trans. Neural Networks* 1, 229-232.
[4] Luttrel S.P , 1994. "A Bayesian Analysis of Self-Organizing Maps". *Neural Computing* vol 6, n° 5,
[5] Kohonen T. 1994. *Self-Organizing Maps.* Springer.
[6] Schroeder A. , 1976. "Analyse d'un mélange de distribution de probabilité de même type". *RSA.* vol 24, n° 1,

A Novel Algorithm for Image Segmentation Using Time Dependent Interaction Probabilities

R. Opara and F. Wörgötter

Dept. of Neurophysiology, Ruhr-Universität, 44780 Bochum, Germany

Abstract. For a consistent analysis of a visual scene the different features of an individual object have to be recognized as belonging together and separated from other objects and the background. Classical algorithms to segment a visual scene have an implicit representation of the image in the connection structure. We propose a new model that uses an image representation in the time domain, operating on stimulus dependent latencies. Such stimulus dependent temporal differences are observed in biological sensory systems. In our system they will be used to define the interaction probability between the different image parts. The gradually changing pattern of active image parts will thereby lead to the assignment of the different labels to different regions which leads to the segmentation of the scene.

1 Introduction

The segmentation of a visual scene is a fundamental process of early vision, where elementary features are grouped together into discrete objects and objects are segregated from each other and the background. In the brain of the higher vertebrates it has been suggested that this could be achieved by synchronization between cells [3], [5], [7]. It has also been supposed that temporal differences of neuronal signals could play an important role in the perception of higher vertebrates [2], [6], [8].

In this paper we present an labeling algorithm that utilizes stimulus dependent temporal differences (latencies) to segment visual scenes. This temporal structure is the only representation of the image in our system.

We will first give an overview of the used labeling model. Then we will show an example how the system segments a visual scene containing a shaded surface, a square and a disk. Finally we will discuss the results.

2 The Model

For the task of image segmentation a labeling algorithm is used which is based on the interaction of labels in order to minimize the energy of the system as given in Eq. 1. The energy function is not related to observables of a physical system, but only a quantity to describe the interaction of labels. The dynamics of the system will tend to assign the same label to units representing one object and a different label to units representing another object.

The units are arranged on a two dimensional lattice of size $N = N_x x N_y$. Each unit $i = (i_x, i_y)$ can take k different labels $\sigma_i \in \{1..k\}$.
The energy of the system is depending on the label configuration and the connection strength between units.

$$E_{tot}(t) = \sum_{i=1}^{N} \sum_{\substack{j \\ \|j - i\| < d_{max}}} -w_{i,j} \delta_{\sigma_i \sigma_j} P_{i,j}^{B}(t) \tag{1}$$

N	: number of units at the two dimensional lattice ($N = N_x x N_y$),
σ_i, σ_j	: labels of unit i and j,
$w_{i,j}$	: connection strength between unit at location i and j,
$P_{i,j}^{B}(t)$	: time dependent binary coupling term $P_{i,j}^{B} \in \{0,1\}$ which indicates, if two units are interacting at anyone point in time or not,
$\delta_{\sigma_i \sigma_j}$	: kronecker function,
$\|j - i\|$	: distance of unit i and j,
d_{max}	: maximal allowed range of interaction.

In classical label algorithms for image segmentation [4] the connection strength $w_{ij} = w_{ji}$ between unit i and j defines the similarity between two locations i and j. In our approach we will show that it is sufficient to have a constant connection strength for all units, while the probability for an interaction P_{ij}^{B} is given in the time domain.

Thus, the coupling term $P_{ij}^{B}(t)$ adds an additional time dependence to the dynamics of the system. At first each unit is assigned a time dependent probability described by

$$P_i(t) = \alpha \exp(-\frac{(t - t_i)^2}{s_i^2}). \tag{2}$$

t_i is the characteristic latency of unit i given as a function of the contrast C of pixel i ($t_i \to \infty$, if $C \to 0$). The probability for an interaction P_{ij}^{I} of unit i and j is given by the product of the corresponding probabilities P_i (Eq. 2) of unit i and P_j of unit j. For a given point in time we have $P_{ij}^{I}(t) = P_i(t) P_j(t)$. Finally we restrict the approach to binary interactions $P_{ij}^{B} \in \{0, 1\}$. The probability that a binary interaction at time t is actually taking place ($P_{ij}^{B} = 1$) is given by $P_{i,j}^{I}(t)$. In Fig. 1 the interaction probability is shown as a function of time for three units having different characteristic latencies.

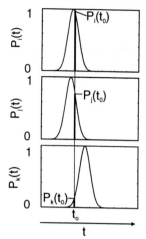

Two units (i, j) have rather similar latencies while the third unit (k) has a much longer latency. The probability of interaction $P^I(t)$ at time t_0 is much larger for unit i and j as compared to unit i and k (Fig. 1). The effect of the time dependent interaction term is that units representing bright objects are responding earlier than units which represent darker objects. This yields to a temporal separation of the different parts of the input image. In Fig. 2 it is shown that the temporal structure by itself is sufficient to segment a visual scene (the connection strengths w_{ij} are constant for all units).

Fig. 1. The probability P for the interaction of three units i,j and k, as a function of time.

3   Results

As an example Fig. 2b shows the label distribution of the units if a stimulus (Fig. 2a) is given to the system. The stimulus consists of two objects with uniform distributed gray values g a square ($\bar{g}_{square}$=240) and a disk $\bar{g}_{disk} = 40$), a shaded surface ($g_{min} = 60$, $g_{max} = 220$) and a background ($\bar{g}_{background} = 128$).
The simulation starts with a random label configuration thus each label is represented by nearly the same number of units (Fig. 2c).
During the first iterations only the most salient objects (low latency) are processed (square and bright parts of the shaded surface) (Fig. 2b iteration 10 and 350 and Fig. 2d). According to the dynamics of the system the units of the square will receive the same label. This is indicated in Fig. 2c. The number of assigned units to the label *square* is increased after a few iterations. The plateau for the square (Fig. 2c) in the beginning, however, also contains units from other regions which by chance have the *square* label. These will eventually be removed and the plateau reaches its final level. Due to the latency the parts with a lower luminance are not processed at this time (Fig. 2d).
At later times the case is quite different. Now the bright objects are processed with a small probability, while the darker image parts are processed with a high probability (Fig. 2b, 2d).
One remarkable effect occures after $\sim$ 350 iterations (Fig. 2), where the probability distributions of the shaded surface and the background are overlapping. In spite of the overlapping probability distributions, the background and the shaded surface are segmented with different labels, except for a small part at the border of background and shaded surface. The assignment of different labels to the background and the shaded surface occures due to the history of the

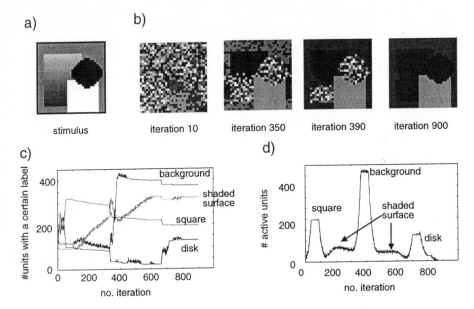

a)

b)

stimulus   iteration 10 iteration 350 iteration 390 iteration 900

Fig. 2. a: The stimulus consists of two objects with a uniform distributed gray value, a shaded surface and the background b: Label distribution of the system at different times. The labels are coded as gray values. Units with the same label (same gray value) belong to the one object, while different gray levels indicates the assignment to different objects. c: The number of units assigned to a certain label. d: The number of units which interact at a given point in time.

processing of the objects and the small overlapping area.

In standard algorithms for image segmentation [4] such smooth transitions between two objects are also hard to segment, because at this area the similarity between the objects is nearly the same as within an object.

4 Discussion

In our conceptual framework we show that image segmentation can effectively be realised in the time domain. Therefore no implicit representation of the image in the connection structure is needed. Features with a high contrast are thereby favored and will be processed earlier than objects that do not "jump to the eye". This mechanism thereby mimics human perception [1] but more importantly it efficiently limits the information flow which needs to be evaluated at any point in time.

To this end we were only concerned with the separation of image parts by contrast. With only a few restrictions other features (color, spatial frequency etc.) could also be included in the system. The integration of other features would lead to a more robust segmentation.

Acknowledgements

The authors acknowledges the support of the Deutsche Forschungsgemeinschaft (grant WO388 4-2, 5-2).

References

1. Burgi, P. Y. and Pun, T. Figure-ground separation: evidence for asynchronous processing in visual perception? *Perception* **20**, 69, (1991).
2. McClurkin, J. W., Optican, L. M., Richmond, B. J. and Gawne, T. Concurrent processing and complexity of temporally encoded neural messages in visual perception. *Science* **253**, 675-677, (1991).
3. Eckhorn, R., Frien, A., Bauer, R., Woelbern, T. & Kehr, H. High frequency (60-90 Hz) oscillations in primary visual cortex of awake monkey. *NeuroReport*, **4**, 243-246 (1993).
4. Geman, D., Geman, S., Graffigne, C. & Dong, P. Boundary detection by constrained optimization. *IEEE Trans. Pattern Analysis Machine Intelligence*, **12**(7), 609-628, (1990).
5. Gray, C.M., König, P., Engel, A.K. & Singer, W. Oscillatory responses in cat visual cortex exhibit inter-columnar synchronization which reflects global stimulus properties. *Nature*, **338**, 334-337 (1989).
6. Hopfield, J. J. Pattern recognition computation using action potential timing for stimulus representation. *Nature*, **376**, 33-36, (1995).
7. v.d. Malsburg, C. The correlation theory of brain function. Int. report 81-2, Dept. of Neurobiol. Max-Planck-Institute for Biophysical Chemistry, Göttingen, (1981).
8. Wörgötter, F., Opara, R., Funke, K. & Eysel, U. Utilizing latency for object recognition in real and artificial neural networks. *NeuroReport*, **7**, 741-744 (1996).

Reconstruction from graphs labeled with responses of Gabor filters*

M. Pötzsch[1], T. Maurer[1], L. Wiskott[1]** and C. v.d. Malsburg[1,2]

[1] Ruhr-Universität Bochum, Institut für Neuroinformatik, 44780 Bochum, Germany
[2] University of Southern California, Dept. of Computer Science and Section for
Neurobiology, Los Angeles, CA

Abstract. The work presented is part of a larger effort to build a general object recognition system. Objects as well as human faces are represented by graphs labeled with Gabor filter responses. We describe an optimal method to reconstruct images from such graphs. Two examples of how this can be used to analyze the object representation or to compensate for its deficiencies are presented. Since the reconstruction method is formulated generally for an arbitray set of linear filters, it can also be applied to data produced by other systems, artificial or biological.

1 Introduction

Our point of departure is an object recognition system [1, 2, 3], which uses Gabor filters as basic features. The region surrounding a given pixel in the image is represented by the responses of a set of Gabor filters of different frequencies and orientations, all centered at the same pixel position. This set of responses is called a *jet*. Objects are represented by graphs whose nodes are labeled by jets, and whose edges describe topographical relations (see Fig. 3a). An object is identified and located by Elastic Graph Matching (EGM), which is a simple algorithmic caricature of Dynamic Link Matching, a neural model based on synchrony coding of feature binding and rapid reversible synaptic plasticity [4]. The system has been successfully applied to face recognition and face segmentation [1, 3], as well as object recognition in complex scenes [5].

Here, we describe how an image can be reconstructed from a graph labeled with jets. In contrast to the method presented in [6], our reconstruction is optimal in the sense that all information is preserved, and it is general, because it is not restricted to rectangular graphs. In Section 2, we describe a reconstruction scheme for an arbitrary set of linear filters and explain certain aspects of it by the example of Gabor filters. In section 3, the reconstruction is applied to labeled graphs, and the results are discussed in the last section.

---

* Supported by the German Federal Ministry of Science and Technology
** currently at the Computational Neurobiology Laboratory, The Salk Institute for
Biological Studies, San Diego, CA

2 Reconstruction

A simple model for the cell responses in the visual cortex is the linear receptive field. Thus, the response J_ν of neuron ν can be written as

$$J_\nu = (\mathcal{T}I)_\nu = \int I(\mathbf{x})\, p_\nu(\mathbf{x})d\mathbf{x}, \tag{1}$$

with the shape of its receptive field given by $p_\nu(\mathbf{x})$ and the image $I(\mathbf{x})$ as a stimulus. This operation can also be interpreted as transformation $\mathcal{T}$ of an image. Since the intergral computes a projection of the image onto filter p_ν, p_ν is also called a *projection function*.

Furthermore, it has been shown that receptive field profiles of *simple cells* in the visual cortex V1 can be modeled by Gabor filters $\psi_{\mathbf{k},\mathbf{x}_0}$ as projection functions [7]:

$$\psi_{\mathbf{k},\mathbf{x}_0}(\mathbf{x}) = \frac{\mathbf{k}^2}{\sigma^2}\, \exp\left(-\frac{\mathbf{k}^2(\mathbf{x}_0 - \mathbf{x})^2}{2\sigma^2}\right)\, \exp\left(i\mathbf{k}(\mathbf{x}_0 - \mathbf{x})\right), \tag{2}$$

where $\mathbf{k}$ is the main frequency of the filter and $\mathbf{x}_0$ specifies the location of the filter. The filters have the shape of complex waves (third factor) restricted by a Gaussian envelope function (second factor). Thus, the complex–valued ψ_ν (with $\nu := (\mathbf{k}_\nu, \mathbf{x}_{0,\nu})$) is composed of an even (cosine-type) and an odd (sine-type) part. The first factor compensates for the frequency-dependent decrease of the power spectrum in natural images [8]. (There is an additional correction for the DC-value of the filters which is not shown here.)

Since the transformation $\mathcal{T}$ is linear, the optimal reconstruction of the image from the values J_ν is linear as well and given by a well-known concept of linear algebra:

$$I^{\mathcal{R}}(\mathbf{x}) = \mathcal{R}\mathbf{J} = \sum_\nu J_\nu b_\nu(\mathbf{x}), \tag{3}$$

where $\mathcal{R}$ symbolizes the reconstruction operation and $b_\nu(\mathbf{x})$ are appropriate basis functions.

If the projection functions were orthogonal and normalized ($\int \bar{p}_\nu(\mathbf{x})\, p_\rho(\mathbf{x})d\mathbf{x} = \delta_{\nu\rho}$) the basis functions would simply be $b_\nu(\mathbf{x}) = \bar{p}_\nu(\mathbf{x})$. Since they are not orthogonal in case of Gabor functions, their affinity must be taken into consideration by using a particular linear combination of the projection functions:

$$b_\nu(\mathbf{x}) = \sum_\rho (P^{-1})_{\nu\rho}\, \bar{p}_\rho(\mathbf{x}), \quad \text{with} \quad P_{\nu\rho} := \int \bar{p}_\nu(\mathbf{x})\, p_\rho(\mathbf{x})d\mathbf{x}. \tag{4}$$

The matrix coefficients $P_{\nu\rho}$ are dot products of the projection functions. In case of Gabor filters, these products can be obtained analytically.

Gabor filters can also be used as basis functions ($b_\nu = \psi_\nu$) [9], in which case the values $\tilde{J}_\nu$ have to be computed by minimizing $\|I(\mathbf{x}) - \sum \tilde{J}_\nu \psi_\nu\|^2$. Since the $\tilde{J}_\nu$ are defined by the reconstruction formula, they differ from those computed by

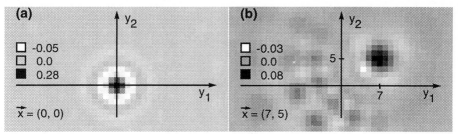

Fig. 1. The projector $\mathcal{P} = \mathcal{RT}$ shown in this figure was derived from a set of 32 Gabor filters as projection functions all centered at the same position $\mathbf{y}_0 = (0,0)$. The figure shows how the original image contributes to the reconstructed image $I^{\mathcal{R}}$ at a certain point $\mathbf{x}$ ((0,0) in (a) and (7,5) in (b)). E.g., in (b), the contributions come mainly from the gray values around the corresponding point $\mathbf{y} = (7,5)$ but also from the gray values at all other points within the extent of the Gabor filters. Formally speaking, the kernel $c(\mathbf{x}, \mathbf{y})$ of the projection process $I^{\mathcal{R}}(\mathbf{x}) = \mathcal{P}I(\mathbf{y}) = \int c(\mathbf{x}, \mathbf{y})I(\mathbf{y})d\mathbf{y}$ is shown: (a) $c((0,0), \mathbf{y})$, (b) $c((7,5), \mathbf{y})$. (The wave vector $\mathbf{k} = (k, \varphi)$ of the applied Gabor filters takes on four different frequency values k and eight different orientation values φ.)

Eq. (1) and the image can be reconstructed directly by $I^{\mathcal{R}} = \sum \tilde{J}_\nu \psi_\nu$. However, in the sense of the findings in [7], this approach cannot serve as a model for the cell responses in sensory cortex.

The reconstruction defined by Eqs. (3) and (4) is perfect in the sense that the amount of information in the reconstructed image $I^{\mathcal{R}}(\mathbf{x})$ is equal to that of the transformed data $\mathbf{J}$. That also implies that the values $\mathbf{J}$ can be identically recalculated by transforming $I^{\mathcal{R}}$, i.e., the compound operation $\mathcal{TR}$ is the identity operator[1].

The compound operator $\mathcal{P} = \mathcal{RT}$ is called a *projector* and it projects images into the "space" of images which can be represented by the chosen set of projection functions p_ν. $\mathcal{P}$ satisfies the condition for being a projector ($\mathcal{P}^2 = \mathcal{P}$), for $\mathcal{TR}$ equals the unity operator. Figure 1 shows such a projector for the set of Gabor filters all centered at the same positions.

The reconstruction formulas presented can only be applied to a linear independent set of projection functions, because otherwise the determinant of P vanishes and P cannot be inverted. In other words, the transformed data of a linear dependent set of p_ν is redundant. This means at least one p_μ can be

---

[1] The compound operation $\mathcal{TR}$ is the identity operator, for

$$(\mathcal{TRJ})_\mu = \int I^{\mathcal{R}}(\mathbf{x})\, p_\mu(\mathbf{x})d\mathbf{x} = \sum_\nu \left(\int b_\nu(\mathbf{x})\, p_\mu(\mathbf{x})d\mathbf{x} \right) \cdot J_\nu$$

$$= \sum_\nu \left(\sum_\rho (P^{-1})_{\nu\rho} \underbrace{\int \bar{p}_\rho(\mathbf{x})\, p_\mu(\mathbf{x})d\mathbf{x}}_{=P_{\rho\mu}} \right) \cdot J_\nu = \sum_\nu \delta_{\nu\mu} J_\nu = J_\mu.$$

represented by a linear combination of (a subset of) the other projection functions and thus, the corresponding value(s) J_μ can be omitted for the purpose of reconstruction without loss of information.

3 Application to labeled graphs

The object recognition system described in [1, 2, 3] represents objects as well as human faces by graphs whose nodes are labeled by jets. The amount of information contained in such a representation can be visualized by reconstruction. Eq. (3) and (4) provide an optimal reconstruction for any arbitrary set of linear filter responses. However, since the coefficients most important for reconstructing a given image point are contained in the jet nearest to that point, and the nodes are sufficiently far apart, the reconstruction can be approximated by using the components of just that jet independent of others. This has the advantage that only a small set of basis functions has to be computed, which can be applied to all nodes (see Fig. 3b as an example). Each so called *local* reconstruction is restricted to a Voronoi area around its location, because no interaction between the basis functions of adjacent jets is taken into account. This approximated reconstruction differs little from the exact one, and can be computed much faster (in about 4 sec. –3.5 sec. to compute the basis functions and 0.5 sec. to apply them– compared to about 1 hour needed for the exact version on a Sparc 20).

Visualization of the information stored in a labeled graph may help expose deficiencies in the object representation. It may even help to compensate for these deficiencies. As an example, consider the nodes located near the outline of an object. Their jets are not only influenced by the object itself but also by the background, because of the spatial extent of the filters. A simple linear tranformation on the image, namely the multiplication with a 2-dimensional Heavyside-function properly located, could suppress the background locally. Let us denote this transformation by Θ. In order to perform the corresponding transformation directly on a jet, one simply has to concatenate reconstruction, background suppression, and retransformation to form a single linear transformation $\mathcal{L} = \mathcal{T}\Theta\mathcal{R}$, which can then be applied to a jet (see Fig. 2c) [10, 11].

For the task of finding a face in an image without attempting to identify the person, a graph containing the knowledge about several other faces is used [3]. For this purpose, each node is labeled with a *bunch* of jets, each jet extracted from the image of a different person but at the same facial landmark (such as the tip of the nose or the left eye). Each bunch may, for example, contain jets from 100 different persons. Thus, it covers a wide variety of shapes for a single landmark. Such a graph is called a *bunch graph*.

During the process, in which the bunch graph is matched onto an image with an unknown face, the jet fitting best to the presented face is automatically selected in each bunch. A reconstruction of an image from these jets fitting best leads to a *phantom face*; a face created from transformed data taken from several different persons (see Fig. 3c). Phantom faces have already been introduced in [3], where they have been used for the determination of facial attributes. How-

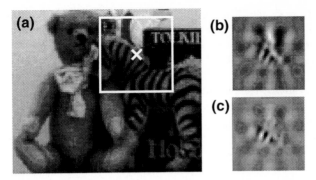

Fig. 2. Background suppression: **(a)** Scene of toys (a zebra in front of a bear and a book). **(b)** Reconstruction from the jet marked in (a). **(c)** Reconstruction of a modified version of the jet marked in (a). Knowing the outline of the zebra the considered jet can be linearly transformed to suppress the influence of the background.

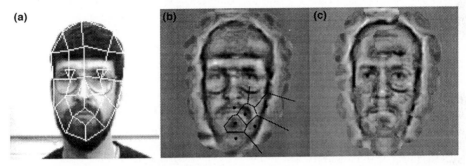

Fig. 3. Reconstruction from a labeled graph: **(a)** A face with a graph. **(b)** Reconstruction from the labeled graph taken from the image in (a). The reconstruction is achieved by a local reconstruction of each jet restricted to a Voronoi area around its location. Some of these areas are indicated by lines. **(c)** The phantom face: a reconstruction of the jets in a bunch graph which fitted best to the image (a). The bunch graph contained jets from about 100 different persons (not including the one shown in (a)). **(b,c)** As the DC-part of the chosen filters vanishes, the absolute grey levels are missing in the reconstructed images.

ever, they have not been created from jets themselves (by reconstructing them). Instead, patches of the corresponding original images have been used.

4 Discussion

Sensory data, e.g., visual or auditory, are preprocessed in the cortex as in many artificial systems. Usually the chosen filters applied for preprocessing are designed to simplify the perceptual task by transforming the original data appropriately. Visualization of the remaining amount of information after preprocessing is a great help in understanding and analyzing an artificial or even biological system. We have presented a reconstruction method (optimal in the sense that all information is preserved) for data which have been computed by a set of projection functions (or linear filters). The method has been applied to an arbitrary set of Gabor filters, and an approximating variant of it to labeled graphs. Two aspects of an object recognition system have been analyzed by visulisation: the background problem concerning the representation of objects, and the detection of faces. In the latter case, phantom faces have been created to show what information is used to find a face without regard to identity.

References

1. M. Lades, J.C. Vorbrüggen, J. Buhmann, J. Lange, C. v.d. Malsburg, R.P. Würtz, and W. Konen. Distortion invariant object recognition in the dynamic link architecture. *IEEE Trans. Comput.*, 42(3):300–311, 1993.
2. J. Buhmann, M. Lades, and C. v.d. Malsburg. Size and distortion invariant object recognition by hierarchical graph matching. In *Proc. of the IJCNN International Joint Conference on Neural Networks*, pages II 411–416, San Diego, June 1990. IEEE.
3. L. Wiskott, J.M. Fellous, N. Krüger, and C. v.d. Malsburg. Face recognition and gender determination. In *Proc. of the International Workshop on Automatic Face-and Gesture-Recognition, IWAFGR 95*, Zurich, June 1995.
4. C. v.d. Malsburg. The correlation theory of brain function. Internal report, 81-2, Max-Planck-Institut für Biophysikalische Chemie, Göttingen, FRG, 1981. Reprinted in E. Domany, J.L. van Hemmen, and K.Schulten, editors, *Models of Neural Networks II*, pages 95–119. Springer, Berlin, 1994.
5. L. Wiskott and C. v.d. Malsburg. A neural system for the recognition of partially occluded objects in cluttered scenes. *Int. J. of Pattern Recognition and Artificial Intelligence*, 7(4):935–948, 1993.
6. R.P. Würtz. *Multilayer Dynamic Link Networks for Establishing Image Point Correspondences and Visual Object Recognition*, volume 41 of *Reihe Physik*. Verlag Harri Deutsch, Frankfurt a. Main, 1995. PhD thesis.
7. J.P. Jones and L.A. Palmer. An evaluation of the two-dimensional gabor filter model of simple receptive fields in cat striate cortex. *Journal of Neurophysiology*, 58(6):1233–1258, 1987.
8. D.J. Field. Relations between the statistics of natural images and the response properties of cortical cells. *Optical Society of America*, 4(12):2379–2394, 1987.
9. J.D. Daugman. Complete discrete 2-d gabor transforms by neural networks for image analysis and compression. *IEEE Transactions on Acoustics, Speech, and Signal Processing*, 36:1169–1179, 1988.
10. M. Pötzsch, N. Krüger, and C. v.d. Malsburg. Improving object recognition by transforming Gabor filter responses. *Network: Computation in Neural Systems*, 7(2), 1996.
11. M. Pötzsch. Die Behandlung der Wavelet-Transformation von Bildern in der Nähe von Objektkanten. Internal Report (IR-INI) 94-04, Institut für Neuroinformatik, Ruhr-Universität Bochum, 1994.

Edge Information: A Confidence Based Algorithm Emphazising Continuous Curves[*]

O. Rehse[1], M. Pötzsch[1] and C. von der Malsburg[1,2]

[1] Ruhr-Universität Bochum, Institut für Neuroinformatik, 44780 Bochum, Germany
[2] University of Southern California, Dept. of Computer Science and Section for Neurobiology, Los Angeles, USA

Abstract. We introduce an algorithm which emphasizes edges that form continuous curves. The algorithm avoids binary decisions, simply complementing edge information with confidence values. Our algorithm is based on Mallat's method of edge detection. Continuity is detected by the combination of responses of Mallat filters of different scales. Due to its structure the algorithm can profitably be used as part of a modular computer vision system.

1 Introduction

Typical problems in computer vision, such as segmentation or object recognition, are often approached by combining information from different visual submodalities [1, 2, 3]. Early binary decisions in single modules must be avoided to allow for a meaningful interaction of information from different modalities [4]. Single modules are unable to interpret a situation on their own and cannot reliably decide which information to pass on. On the other hand, appropriate reduction of the flood of information created by a single modality is a crucial step towards an integrated interpretation in the different modules.

In this paper we discuss this problem for a typical vision module, edge extraction and processing. Our point of departure is a system which is based on confidence values instead of binary decisions. Here we describe an extension of a system which emphasizes edges that are part of continuous curves.

As the final goal of image analysis is to gain information about the physical world implicit in the image, it is obvious that the extraction of meaningful edges is very important: Different causes acting in the environment — shadows, textures, boundaries of objects — cause the same effect, namely abrupt changes of grey levels. Only a certain subset of these edges yield information useful to the solution of a given perceptual task. For instance, for the purpose of segmentation, edges corresponding to contours of objects are to be extracted while others disturb.

One common procedure for edge extraction is to first apply a filter operation and then a binary decision process to select significant edges. For example, Marr and Hildredth [5] used as filter the second derivative of a Gaussian, the width

---

[*] Supported by a grant from the German Federal Ministry for Science and Technology.

of which sets the scale on which the structures in the image are analysed. In the second step, local edge elements were assigned by a binary decision to those positions of the preprocessed image where zero-crossings were detected [6].

In distinction to this method, the edge extraction process introduced by Mallat [7] can be used to assign continuous confidence levels to edge elements. Filters based on the first derivatives of a Gaussian in the horizontal and vertical directions are applied in the first processing step. They create two-dimensional results, one for each filter. The Euclidian norm of the responses of the two filters at each point (also called *absolute value*) can be interpreted as measuring the strength of an edge, for it is proportional to the gradient of the gray level. Thus, such a value exists everywhere in the image and can be assigned as a confidence value for the appearence of a local edge element at a pixel. In analogy to the method by Marr and Hildredth the width of the Gaussian determines the size of the structure analysed. Usually filters for several sizes are applied to get a *multi-scale* analysis. A second processing step extracts the local maxima of the edge strength, called modulus maxima, from the preprocessed data [8].

A further step of edge processing aims to reduce the amount of the resulting information. One possible way of doing this is to group the local edge elements into curves as structures on a bigger scale and extract those structures which fulfill certain conditions such as smoothness or parallelity. Much successful work has been devoted to this complex problem [9, 10, 11]. A common characteristic of many approaches is the use of local edge information based on zero-crossings and strict selection of continuous curves.

We here present an algorithm which works on the confidence-based data created by Mallat's preprocessing. Instead of immediately eliminating edges, their confidence values are modified to emphasize certain configurations. Since we want to perform object segmentation, the algorithm emphasizes the confidence of local line elements to be part of a continuous curve. We achieve this by combining responses of Mallat filters of different scales. Although multi-scale image analysis is commonly used in the literature [12], this concept for the detection of continuity is novel as far as we know.

In the next section we describe our approach and its prerequisites. In section 3 we present examples of processed images. Finally, we discuss the algorithm and applications.

2 The algorithm

As described in the introduction [7], the extraction of edges from intensity images $I(\mathbf{x})$ according to Mallat is formulated as a convolution with filters $\psi_s^{(\nu)}(\mathbf{x})$

$$\mathbf{T}_s(\mathbf{x}) = f(\mathbf{x}) * \begin{pmatrix} \psi_{s_i}^{(\mathrm{h})}(\mathbf{x}) \\ \psi_{s_i}^{(\mathrm{v})}(\mathbf{x}) \end{pmatrix}$$

where $*$ represents a convolution, h and v stand for horizontal and vertical, and s_i ($s_i = s_0\, 2^i, i \in N$) represents the width of a Gaussian, the derivatives of which are used as filters. The absolute value $a_i(\mathbf{x}) = |\mathbf{T}_{s_i}(\mathbf{x})|$ and the phase $\varphi_i(\mathbf{x})$ measure strength and direction of an intensity change at scale s_i and position $\mathbf{x}$. Thus, the a_i are interpreted as confidence values for the appearence of a local line element. Additionally, we introduce a flag $f_i(\mathbf{x})$ which indicates the appearence of a modulus maximum at a certain scale and position. For the calculation of modulus maxima see [8].

Edge information can be very helpful to improve the performance of other vision modules. Visual features such as the optical flow often loose their usefulness near boundaries of objects so that, for instance, segmentation based on these features has poor local resolution. Keeping this in mind, our goal should be to supply the other vision modules with a maximum of localised edge information. As the scale increases with the index i, the filter responses assigned to $i = 0$ fulfill this condition best. They contain, however, the largest amout of confusing information, representing the finest detail in the image. By increasing the scale one is able to reduce this flood of information but the filter responses loose their locality. This trade-off can be solved by combining the information of filter responses at different scales. The chosen kind of combination emphasizes those local edge elements which are part of a continuous curve.

For this purpose, we search a counterpart at scale s_n for each edge at scale s_0 . Since there is no one-to-one mapping between edges at different scales, localization of edge information on coarser scales being imprecise, one has to search a local area for an appropriate counterpart. This is done with the help of a function $\mathcal{F}$ which measures the degree of the similarity between an edge at scale s_0 and an edge considered as a possible counterpart at scale s_n. The one with maximal similarity to that at scale s_0 is chosen.

Formally speaking, the similarity function between an edge at scale s_0 and position $\mathbf{x}_0$ and an edge at scale s_n and position $\mathbf{x}_n$ reads

$$\mathcal{F} = \frac{a_n(\mathbf{x}_n)}{a_n^{\max}} \, \exp(-\frac{|\mathbf{x}_n - \mathbf{x}_0|^2}{\sigma^2}) \, |\cos(\varphi_n(\mathbf{x}_n) - \varphi_0(\mathbf{x}_0))| \, ,$$

where $a_n^{\max}$ marks the maximal confidence at the scale s_n. The confidence values assigned to the scale s_0 are modified by

$$a_0^{\mathrm{mod}}(\mathbf{x}_0) = a_0(\mathbf{x}_0) \left(\max_{\mathbf{x}_n \in \mathcal{A}(\mathbf{x}_0)} \mathcal{F}(\mathbf{x}_0, \mathbf{x}_n, a_n(\mathbf{x}_n), \varphi_0(\mathbf{x}_0), \varphi_n(\mathbf{x}_n)) \right)^{\alpha}$$

with a circular area

$$\mathcal{A}(\mathbf{x}_0) = \{\mathbf{x}_n | (|\mathbf{x}_n - \mathbf{x}_0| < d)\} \quad \text{and} \quad d, \sigma, \alpha \in \mathbb{R}^+ \, .$$

The modified confidence of the scale s_0 at a location $\mathbf{x}_0$ is hence calculated by multiplication of the unmodified confidence with the similarity with the best fitting counterpart, confidence a_0 being most weakened by lowest similarity. The search for the counterpart in the area $\mathcal{A}$ can be limited to modulus maxima positions to reduce the computation time. The similarity function $\mathcal{F}$ consists of three factors: The first is proportional to the confidence a_n of the edge considered at scale s_n (relative to the maximal confidence at this scale). The second factor measures the dependence of the distance between the location of the edge considered at scale s_0 and the one at scale s_n. It leads to a low similarity value if the distance is large, the parameter σ controlling this dependence. The third factor reflects the difference of the direction between of the edges at scale s_0 and scale s_n. If the edges are perpendicular to each other this term produces its lowest value.

The parameter α controls the sensitivity of the confidence values $a_0(\mathbf{x}_0)$. If it comes to an interaction of different modules, this parameter tunes edge information given by this module relative to other vision modules.

Since the similarity values lie in the interval $[0; 1]$, the confidence values are decreased (or remain unchanged). The important information is in the ratios between confidence values and not in their absolute values, and the weakening of some confidence values is equivalent to the strengthening of others.

3 Results

Fig. 1 shows results for three sample images. The second row shows the confidence values extracted by Mallat filters from the original images given in the first row. As these confidence values are related to the scale s_0 they represent many line elements belonging to small detail (e.g., caused by texture or noise), only some of which are part of smooth extended curves. In the third row, one can find the modified confidence values as a result of our algorithm. It is evident that smooth curves are emphasized. The three examples differ mainly in the scale s_n which determines the size of the structures being emphasized.

The scale s_4 was selected in the first two examples (elephant and head) for emphasizing the dominating steady curves. As the resulting edge information mainly contains the outline of the objects it can assist in solving the segmentation problem. No binary decision is made concerning the exact placement of the outline. Rather, confidence values of the appearance of continuous curves are given which represent information about the outline as well as other important lines. As the interesting structures in the third example (house) are of smaller size (e.g., windows), a smaller scale (here, s_2) was selected. The parameters d and σ are chosen in a way to adapt the size of the Mallat filters related to the scale s_n. The influence parameter α is set to 1 for all examples. As the Mallat responses can represent the confidence for the appearence of only one local line element at a certain position, problems may occur at corners or intersections. This can be seen, e.g., at the top left corner of the second example (head).

4 Discussion

We have presented an algorithm which is based on a method of edge detection introduced by Mallat. It emphasizes those edges in an image which are part of a smooth curve. This is achieved by combining filter responses of different scales, thus combining information on edge locality provided by smaller scale filters with information on curve continuity represented on coarser scales.

n=4; sigma=2; d=4   n=4; sigma=2; d=4   n=2; sigma=1; d=2

Fig. 1. Results. first row: intensity images; second row: confidence extracted by Mallat filters; third row: confidence modified by our algorithm. Parameter values are given below each column.

Using confidence values and avoiding early decisions on relevance of edges for a given task is important because a single visual module is not able to interpret an image on its own. Rather, a successful interpretation needs the integration of information from different vision modules. For this purpose the concept of confidence is crucial.

As area-based vision modules which, for instance, search for common textures, grey level or motion, loose their exactness near the boundaries of areas, our edge-based module can improve image interpretation decisively. We plan to apply our algorithm to enhance both a segmentation system [1, 2, 3] and an edge-based object recognition system [13] currently being developed in our laboratory.

References

1. Jan C. Vorbrüggen. *Zwei Modelle zur datengetriebenen Segmentierung visueller Daten*, volume 47 of *Reihe Physik*. Verlag Harri Deutsch, Thun, Frankfurt am Main, 1995.
2. C. Eckes. *Integration daten- und modellgetriebener Segmentierung in natürlichen Bildern*. Diplomarbeit, Universität Dortmund, 1995.
3. Christian Eckes and Jan C. Vorbrüggen. Combining data-driven and model-based cues for segmentation of video sequences. *World Congress on Neural Networks, San Diego, USA*, 1996.
4. D. Marr. Early processing of visual information. *Phil. Trans. R. Soc. Lond. B*, 275:483–524, 1976.
5. D. Marr and E. Hildreth. Theory of edge detection. *Proc. R. Soc. Lond. B*, 207:187–217, 1980.
6. D. Marr, T.Poggio, and S.Ullman. Bandpass channels, zero-crossings, and early visual information processing. *J. Opt. Soc. Am.*, 69:914–916, 1979.
7. S. Mallat. A theory for multiresolution signal decomposition: The wavelet representation. *IEEE Trans. PAMI*, 11:674–693, 1987.
8. S.Mallat and S.Zhong. Characterization of signals from multiscale edges. Technical report, Courant Institute, NYU, New York, USA, 1991.
9. Rakesh Mohan and Ramakant Nevatia. Using perceptual organization to extract 3-d structures. *IEEE Trans. PAMI*, 11(11):1121–1138, 1989.
10. P. Parent and S. W. Zucker. Trace inference, curvature consistency and curve detection. Technical Report CIM-86-3, Computer Vision and Robotic Lab., McGill University, 1986.
11. S. W. Zucker, R. Hummel, and A. Rosenfeld. An application of relaxation labeling to line and curve enhancement. *IEEE Trans. on Computers*, 26:394–403, 1977.
12. D. Marr. *Vision*. W.H. Freeman and Co, San Francisco, first edition, 1982.
13. E. Kefalea and C.v.d. Malsburg. Object classification based on contours with Elastic Graph Matching. *(work in progress)*.

Neural Model of Cortical Dynamics in Resonant Boundary Detection and Grouping

Heiko Neumann, Petra Mössner

Abt. Neuroinformatik, Fakultät f. Informatik, Universität Ulm,
D–89069 Ulm, Germany; hneumann@neuro.informatik.uni-ulm.de

Abstract. A new model for visual boundary detection and contour grouping is presented that is based on functional elements of resonant matching of activation between neural layers in cortical architecture. The model architecture relates to visual cortical areas V1 and V2 which are bidirectionally interconnected via feedforward as well as feedback projections. It is suggested that their functionality is primarily determined by the measurement and integration of signal features that are continuously matched against neural codes of expectancies generated by the long-range integration of coherent activity. The net effect produces grouping and illusory contour completion at model V2 as well as context-sensitive shaping of orientation tuning and selectivity of receptive fields at model V1 layer. A pilot implementation of the model architecture has been successfully tested on various test stimuli.

1 Introduction

Grouping of visual elements of fragmented layout is a key functional mechanism for the segmentation of surface outlines. The detection of boundary contours contributes to the generation of an abstract representation of the visual scene. Several approaches to grouping have been developed to model neural mechanisms in biological and computer vision (e.g. [6, 4, 7, 12] and [11, 14]).

We propose a novel functional architecture that has been inspired by the principles of adaptive resonance theory ([5]). The architecture consists of two major stages of feedforward and feedback processing each of which having different sublayers. The input to the recurrent network of resonant processing is generated by orientation selective units for local contrast detection. Subsequent processing is organized as a bidirectionally connected structure of F_1 and F_2 layers, which functionally correspond to cortical levels V1 and V2, respectively. Afferent activity generated at V1 feeds forward to V2 where it is integrated from different spatial branches in the visual field. Conjunctive arrangement of oriented input activity in turn generates an activation that is fed back to gate activities of orientation selective units in V1. This mechanism implies a new computational theory in which V2 acts as a code layer such that the input signal activation in V1 is tested against "expected" activity in a much broader visual context ([5, 10]). The gating mechanism is closely related to the hypothesized linking fibres suggested to model mechanisms of activity synchronization for feature linking (e.g. [1]). The net effect of gating and positional and orientational competition shapes the orientation selectivity of cells in a context-sensitive manner ([3]). Furthermore, inconsistent activity that fails to match the V2 feedback "expectancy" will be suppressed. Contributions to grouping and long-range completion are thus predicted to be generated at the level of V2 ([13]) primarily due to increased receptive field size and non-linear input integration.

2 Outline of the Computational Architecture

The input activity distribution to the considered functional architecture is generated by oriented contrast detection at the complex cell level (denoted by $x_{i\varepsilon}$, for location i and orientation ε). The specific mechanisms for their generation will not be discussed here (but see e.g. [9]). The architecture consists of two main network layers, namely F_1 and F_2 (see Fig. 1(a)), each of which is organized in two sublayers. Processing in F_1-layer currently consists of two processing stages: *(1)* Local competition (on-center/off-surround feedback processing) generates activity based on resonant matching between feedforward and feedback streams, and *(2)* local competition of "match activity" from the previous stage to enhance spatial and orientation contrast. In the first sublayer $w_{i\varepsilon}^{(1)}$-activity denotes the activation generated in response to the combination of the bottom-up feedforward (denoted by x-activities) and the top-down feedback stream (denoted by y-activities):

$$\frac{d}{dt}w_{i\varepsilon}^{(1)} = -Dw_{i\varepsilon}^{(1)} + (U - w_{i\varepsilon}^{(1)})(x_{i\varepsilon} + Cx_{i\varepsilon}f(y_{i\varepsilon}^{(2)})) - w_{i\varepsilon}^{(1)}\sum_{j\in\mathcal{N}_i^y}\left(\sum_{\gamma}f(y_{j\gamma}^{(2)})\cdot\psi_{\varepsilon\gamma}^-\right)\cdot\lambda_{ij}^{(y-)},$$

(1)

where $f(.)$ denotes a signal function for the transformation of feedback activity. In the next sublayer $w_{i\varepsilon}^{(2)}$-activity is defined by

$$\frac{d}{dt}w_{i\varepsilon}^{(2)} = -Ew_{i\varepsilon}^{(2)} + (F - Gw_{i\varepsilon}^{(2)})\sum_{\gamma}w_{i\gamma}^{(1)}\cdot\psi_{\varepsilon\gamma}^+ - w_{i\varepsilon}^{(2)}\sum_{j\in\mathcal{N}_i^w}\left(\sum_{\gamma}w_{j\gamma}^{(1)}\cdot\psi_{\varepsilon\gamma}^-\right)\cdot\lambda_{ij}^{(w-)}. \quad (2)$$

The constants D and U in eqn. 1 and E, F and G in eqn. 2 denote the parameters of the shunting equations to control passive decay and saturation levels. Equation 2 is solved at equilibrium for $G = 0$. One-dimensional weighting functions $\psi_{\varepsilon\gamma}^{\pm}$, both centered at target orientations (with $\psi_{\varepsilon\gamma}^+$ more narrowly tuned than $\psi_{\varepsilon\gamma}^-$) realize an on-center/off-surround interaction scheme in orientation space between orientations ε and γ. The inhibitory weighting function $\lambda_{ij}^{(w-)}$ realizes a spatial coupling (low-pass filter) for locations j in the neighborhood of i. The constant C (eqn. 1) denotes a scaling factor for the transformed excitatory feedback activation.

In the subsequent F_2-layer curvilinear arrangements of oriented contrast activation will be integrated based on the evaluation of a much broader spatial context. Processing in F_2-layer again consists of two hierarchically organized sublayers: *(1)* Long-range cooperative interaction integrates contrast activation in a curvilinear arrangement, and *(2)* local intra-orientational on-center/off-surround competition (between spatial locations) and activity normalization over orientations generate a locally contrast-enhanced activation distribution. The integration mechanism at the first F_2-sublayer realizes a spatial *relatability* measurement

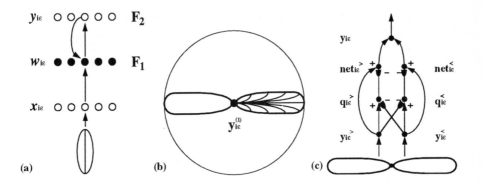

Fig. 1. *Components of the neural architecture; (a) layout of coarsely defined model structure with basic network layers F_1 and F_2, (b) definition of spatial layout of bipole cells (the target cell receives input from two spatial subfields), (c) definition of the micro-circuit which implements the spatial interaction scheme for activity accumulation from different branches.*

(see [8]). The mechanism is based on a bipole cell (Fig. 1(b), see e.g. [6, 7]) in which two non-linearly interacting subfields integrate activity from two different half-planes in the visual field. The integration utilizes an 'association field' ([2]) which is subdivided into two spatially coincident antagonistic ON- and OFF-subfields. The support of a given activation in an orientation field is determined on the basis of an evaluation of a compatibility function such as in probabilistic relaxation schemes ([15]). The spatial outlines of the bipole integration fields are displayed in Fig. 2. The activities $y_{i\varepsilon}^{\rightarrow}$ and $y_{i\varepsilon}^{\leftarrow}$ represent the spatially integrated orientation responses in the left and right subfield of the bipole, respectively. The cooperation mechanism depicted in Fig. 1(c) is realized by a micro-circuit of non-linear feedforward disinhibitory interaction of conjunctive input activation. Pooling of equilibrated responses of individual bipole branches at the final stage is denoted by an additive equation, namely $\frac{d}{dt}y_{i\varepsilon}^{(1)} = -y_{i\varepsilon}^{(1)} + \text{net}_{i\varepsilon}^{\rightarrow} + \text{net}_{i\varepsilon}^{\leftarrow}$. At steady-state, the net response for the non-linear integration results in

$$y_{i\varepsilon}^{(1)} = y_{i\varepsilon}^{\rightarrow} y_{i\varepsilon}^{\leftarrow} \frac{\frac{2}{B} + y_{i\varepsilon}^{\rightarrow} + y_{i\varepsilon}^{\leftarrow}}{\frac{2}{B^2} + \frac{1}{B}(y_{i\varepsilon}^{\rightarrow} + y_{i\varepsilon}^{\leftarrow}) + y_{i\varepsilon}^{\rightarrow} y_{i\varepsilon}^{\leftarrow}}. \tag{3}$$

The constant B is used to control the effectivity of interactions between both input branches. This demonstrates that the bipole generates a net response only if both subfields receive positive net input activation.

The processing stage in the second sublayer consists of a spatial competition between activations in a local spatial neighborhood over all orientations. Thus the individual activity generated at one position for one orientation is normalized w.r.t. the spatially blurred activation of orientation columns. The resulting activity is defined by

$$\frac{d}{dt}y_{i\varepsilon}^{(2)} = -Hy_{i\varepsilon}^{(2)} + (K - My_{i\varepsilon}^{(2)})y_{i\varepsilon}^{(1)} - y_{i\varepsilon}^{(2)} \sum_{j \in \mathcal{N}_i^y} \sum_{\gamma} y_{j\gamma}^{(1)} \cdot \lambda_{ij}^{(y-)}. \tag{4}$$

The constants H, K and M denote the parameters of the shunting equation. Equation 4 is solved at equilibrium with $M = 0$. This activity generated in the F_2-layer is fed back to layer F_1 to match oriented contrast measures $x_{i\varepsilon}$ and generate $w_{i\varepsilon}^{(1)}$-activations.

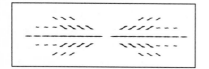 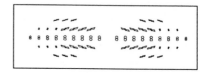

Fig. 2. *Outline of spatial integration in bipole cells based on a spatial weighting function ('association field') and a compatibility function in the feature domain; ON-field (left) denotes spatial orientations that support the target orientation (bipole center) in an excitatory fashion, OFF-field (right) denotes orientations which activations provide inhibitory contributions to the support function.*

3   Simulation Results

The new proposed network architecture has been implemented and tested on a set of stimuli. All linear equations have been solved at equilibrium; Eqn.1 has been solved numerically using a 4th order Runge Kutta scheme with adaptive step size control.

In this initial phase of the model evaluation, input data has been synthesized at the complex cell stage (x-activities). In particular, the input energy at each spatial location is equally distributed over all orientations of the orientation field. Hence, such an initialization generates maximal orientation uncertainty in a model hypercolumn at a given spatial location. We tested the competence of the network for different spatial arrangements of input activation. The goal of this initial testing was to evaluate the self-organizing properties of the network response based on the *spatial arrangement* of salient configurations of input activation alone. Figure 3 shows the results of processing for a rectangular pattern of input activity (left). The equilibrated net activations in the F_1 and F_2 layers are shown. The final representation shows that properties such as activity suppression in homogeneous regions, border enhancement and context sensitive shaping of receptive field sensitivity occur at the model V1 stage (F_1). At the model V2 stage (F_2) cells are selective to orientations of the shape outline. A line grating pattern is used as input in Fig.4. The results show the influence of context sensitivity on RF profiles and the ability to bridge boundary gaps. Thus, the network is capable to generate activity patterns that correspond to illusory contours.

4   Summary

A new model architecture has been developed for visual boundary detection and grouping. The functional organization is based on the principles of adaptive resonance. We postulate that the role of V2 → V1 feedback processing can be understood as a process of dynamic code testing and prediction. The net effect produces grouping and illusory contour completion. In particular, the results of model simulations demonstrate that *(1)* the orientation selectivity of V1 cells is sharpened depending on the spatial arrangement and visual context, *(2)* orientation selective contrast measurements are enhanced at boundaries of spatially homogeneous stimulus arrangements and suppressed in the interiors, *(3)* oriented cells at model V2 stage only respond to curvilinear input arrangements of activity at both branches of bipole cells, *(4)* model V2 cells show fine tuning for orientation selectivity and generate subjective contours to bridge gaps in arrangements of oriented contrast. Based on these initial test cases we intend to systematically investigate preprocessed input luminance patterns of the type used as stimuli in psychophysical experiments.

References

1. R. Eckhorn, H.J. Reitboeck, M. Arndt, and P. Dicke. Feature linking via synchronization among distributed assemblies: Simulations of results from cat visual cortex. *Neural Computation*, 2:293 – 307, 1990.
2. D.J. Field, A. Hayes, and R.F. Hess. Contour integration by the human visual system: Evidence for local 'association field'. *Vision Research*, 33(2):173 – 193, 1993.
3. C.D. Gilbert and T.N. Wiesel. The influence of contextual stimuli on the orientation selectivity of cells in primary visual cortex of the cat. *Vision Research*, 30(11):1689 – 1701, 1990.
4. A. Gove, S. Grossberg, and E. Mingolla. Brightness perception, illusory contours and corticogeniculate feedback. *Visual Neuroscience*, 1995. (in press).
5. S. Grossberg. How does a brain build a cognitive code? *Psychological Review*, 87:1 – 51, 1980.
6. S. Grossberg and E. Mingolla. Neural dynamics of perceptual grouping: Textures, boundaries, and emergent segmentation. *Perception and Psychophysics*, 38(2):141 – 171, 1985.
7. F. Heitger and R. von der Heydt. A computational model of neural contour processing: Figure-ground segregation and illusory contours. In *Proc. 4th Int. Conf. on Computer Vision, ICCV-93*, Berlin, May 11-14 1993.
8. P.J. Kellman and T.F. Shipley. A theory of visual interpolation in object perception. *Cognitive Psychology*, 23(2):141 – 221, 1991.
9. J.J. Koenderink and A.J. van Doorn. Generic neighborhood operators. *IEEE Transactions on Pattern Analysis and Machine Intelligence*, PAMI-14(6):597 – 605, 1992.
10. D. Mumford. On the computational architecture of the neocortex II: The role of cortico-cortical loops. *Biological Cybernetics*, 65:241 – 251, 1991.
11. P. Parent and S.W. Zucker. Trace inference, curvature consistency, and curve detection. *IEEE Transactions on Pattern Analysis and Machine Intelligence*, PAMI-11(8):823 – 839, 1989.
12. W.D. Ross, S. Grossberg, and E. Mingolla. A neural model of illusory contour formation in V1 and V2. In *Proc. ARVO'95 (Investigative Ophtalmology and Visual Science, PGM# 2187)*, 1995.
13. R. von der Heydt and E. Peterhans. Mechanisms of contour perception in monkey visual cortex. I. Lines of pattern discontinuity. *The Journal of Neuroscience*, 9(5):1731 – 1748, 1989.

14. S.W. Zucker, A. Dobbins, and L. Iverson. On the computational neurobiology of curve detection. In *Proc. British Machine Vision Conference (BMVC90)*, pages xvii – xxiii, Oxford (GB), Sept. 24-27 1990.

15. S.W. Zucker, R.A. Hummel, and A. Rosenfeld. An application of relaxation labeling to line and curve enhancement. *IEEE Transactions on Computers*, C-26(4):394 – 403, 1977.

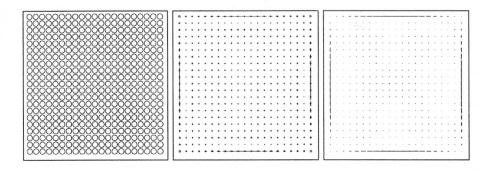

Fig. 3. *Processing results for a rectangular shape: Spatial input activation of equal magnitude for all orientations of a hypercolumn (left); equilibrated $w^{(2)}$-activity in F_1 layer displays context dependent tuning of cells along borders of the square arrangement, orientation uncertainties at corners and reduced activity in the interior (center); equilibrated $y^{(2)}$-activity in F_2 layer displays computed grouping information (after bipole integration) fed back to F_1 layer (right). Note, that the reduction of activity near corners depends solely on the shape of the spatial coupling in the bipole for activity integration.*

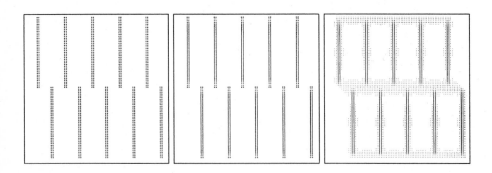

Fig. 4. *Processing results for a bar grating: Spatial arrangement of a synthetically generated distribution of input activity (left); $w^{(2)}$-activity distribution in F_1 layer (center) shows minimized orientation uncertainty for individual bars with uncertainty at line ends; $y^{(2)}$-activity distribution in F_2 layer (right) displays strong grouping activity along bar outlines and more shallow integration along illusory contour completion orthogonal to line ends.*

A Neural Clustering Algorithm for Estimating Visible Articulatory Trajectory

Fabio Vignoli, Sergio Curinga, Fabio Lavagetto

DIST University of Genova
Via Opera Pia 13 Genova 16145 Italy
Tel: ++39 10 3532802 , Fax: ++39 10 3532948
E-mail: bafio@dist.unige.it

Abstract. The bimodal acoustic-visual nature of speech establishes sound correlations between its audio component and the corresponding articulatory information associated to the time-varying geometry of the vocal tract. In this paper we propose an estimation structure consisting of a simplified Time-Delay Neural Network (TDNN) working on 4-5 dimensional cepstrum trajectories provided by a preceding clusterization layer based on a Self Organizing Map (SOM). The use of this pre-processing layer has allowed an effective non-linear clusterization of cepstrum vectors thus simplifying of one order the complexity of the resulting system while maintaining unchanged the global estimation performances. The achieved results are shown in terms estimation precision and robustness with reference to previously published results.

1 A direct approach to articulatory estimation

The objective of any direct approach to articulatory estimation is the design of a suitable mechanism for mapping a predefined acoustic representation of speech into a corresponding motor space representation. No intermediate explicit recognition or classification is required since it is assumed that all the necessary processing is embedded in the conversion mechanism itself.

Extensive experimentations on normal hearing and hearing impaired subjects [2] [3] have clearly demonstrated that if, on one hand, phonemes can be associated rather easily to well defined mouth configurations (called "visemes"), the inverse association is usually troublesome since the same posture of the mouth can correspond to different phonemes. As an example the "bilabial" viseme is associated to different phonemes like /m,p,b/ and the "velar" viseme is associated to different phonemes like /k,g/. Moreover, deep investigation on the articulatory dynamics [4] [5] stress the role played by the coarticulatory phenomena which describe the effects on articulation due to past acoustic outputs (backward coarticulation) and to future going-to-be-produced acoustic information (forward coarticulation).

Since no exhaustive study on coarticulatory dynamics is still available, an approach to speech articulatory description passing through the phonemic level results extremely complex and, at least, formally incomplete. On the contrary, a more viable solution seems that of performing the articulatory estimation directly on suitable spectral, or spectral-equivalent, domains obtainable after short-term speech analysis.

According to this latter approach, estimates of the articulatory parameters are typically obtained by means of vector quantizers or neural networks [7] [8] based on preliminary learning procedures for training the system to associate acoustic speech representations to coherent visual informations. In all these methodologies, however, the association phoneme-viseme is performed without any explicit coarticulation modeling.

In [1] a method for overcoming this limitation has been proposed, based on the use of a Time-Delay Neural Network (TDNN) whose neurons have finite memory so that their output represents the response to the weighted sum of a variable number past inputs. In continuity with this work, further research has been carried out, which is here reported, on the simplification of the neural estimation structure in order to decimate the complexity at learning time.

2   Reducing the computational complexity

The basic idea which has inspired this work was to reduce the complexity of the task performed by the TDNNs in [1] where they had to process 12-dimensional cepstrum trajectories defined over time intervals of 260 milliseconds. In that approach, in fact, it was shown experimentally that a good coarticulatory modeling for the italian language was possible only if the memory of the conversion neural system employed guaranteed the inclusion of this minimum time window. Since each 20 ms speech segment was represented by means of the first 12 cepstrum coefficients, the network had to process 13 12-dimensional vectors for providing an articulatory estimate.

To achieve the aimed simplification, the estimation processing has been subdivided into two subsequent steps each of them relying on neural structures, though of different kind, encharged of performing smaller tasks. The phylosophy was to maintain the use of a TDNN stage of processing at the second step, while inserting beforehand a stage of preprocessing for reducing the dimensionality of the vectors input to the TDNN. The first stage of processing, based on a SOM, has been therefore conceived for clustering 12-dimensional cesprum vectors into a topological space of lower dimensionality in which the original cepstrum waveform was replaced by the coordinates of the most excited neurons.

3   Experimental Setup

3.1   Audio processing

Speech has been segmented into 10 ms intervals and 12 LPC-derived cepstrum coefficients, computed over voiced segments and have been presented in input to a 12x12 Self-Organizing Map.

The SOM performs a sort of Vector Quantization (VQ) and, in this case, has been used to make a reduction of dimensionality in the space of the cepstrum coefficients. This means that the TDNN layer has to make decisions on surface of lower complexity than in the case of previous papers [1]: this results in better performances achieved on the training set.

The 2D coordinates of the most responding neuron to 4 consecutive inputs have been ordered into a 8-dimensional vector representative of a 40 ms speech segment. This has been done in consideration of the fact that the articulatory parameters used as target sequences for training the TDNNs in the second stage had been extracted from 25 Hz video. Then, the 8-dimensional vectors provided by the SOM map have been normalized to the range [-1, 1] and presented in input to a bank of 5 TDNNs each of them encharged of estimating one specific mouth articulatory parameter.

These networks have 8 neurons in the input layer, 4 in the first hidden layer, 3 in the second hidden layer and 1 in the output layer. Perceptrons have 2-order memory in the first hidden layer, 4-order memory in the second hidden layer and 6-order memory in the output layer.

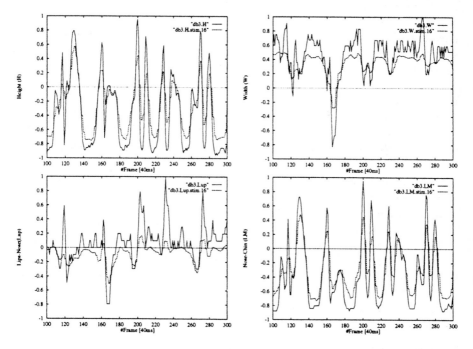

Fig. 1. Trajectories of the APs H,W,LM,Lup estimated through the proposed method.

3.2 Visual processing

Articulatory parameters (AP), as described in table 1 have been extracted from a sequence of images (25 frame/s) with a simple threshold-based algorithm applied to different color spaces like RGB and YUV ,and ,have been then normalized to the range $[-1, 1]$ for their use with TDNNs.

AP	Meanings
W	horizontal distance between the corners of the mouth
w	horizontal distance between the internal corners of the mouth
H	vertical distance between the external contours of the lips
LM	vertical distance between the chin and the nose
Lup	vertical distance between the external contour of the upper lip and the nose

Table 1. Articulatory parameters extracted from the sequence of images

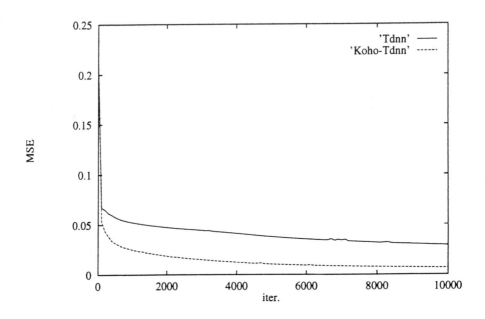

Fig. 2. Comparison between the target-pattern Mean Square Error during the learning phase in case of the 1-stage method (Tdnn) and the 2-stage proposed method (Koho-Tdnn).

4   Experimental results

Comparisons have been done with the estimation structure employed in [1], based on a single TDNN stage with 12-dimensional cepstrum inputs, and consisting of 12 neurons in the input layer, 8 neurons with 2-order memory in the first hidden layer, 3 neurons with 3-order memory in the second hidden layer and 1 neuron with 4-order memory in the output layer.

Sequence Time [s]	Frames	Learning Time [min]	MSE
64	3200	135	0.032
64	1600	20	0.006

Table 2. Data comparison between the old method (TDNN only) and the new one (KOHO-TDNN) [Results obtained on a SILICON GRAPHICS Indigo2 Impact]

In the proposed method we use LPC-derived cepstrum coefficients that have been computed within a 10 ms time window instead of a 20 ms one, in order to detect also short time phonemes.

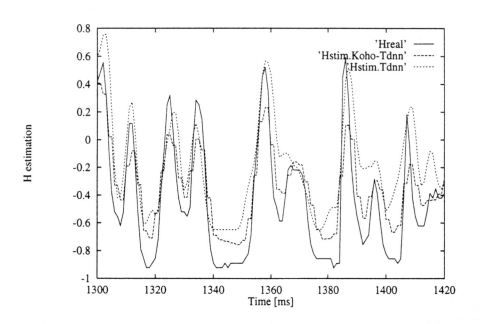

Fig. 3. Comparison between the actual trajectory of the articulatory parameter H (Hreal) and the trajectories estimated by means of the 1-stage (Hstim.Tdnn) and 2-stage proposed method (Hstim.Koho-Tdnn).

Another problem ,which have been solved through the new method, is the difference in time frequency between pattern data (LPC-derived cepstrum coeff.) and target data (articulatory parameters) in the training phase . The former was computed every 20 ms and the latter (that depend on the 25 Hz video frame rate) were computed every 40 ms. This situation resulted in a loss of correlation in the phase of training because we presented in input two pattern data associated to the same duplicated target reference, and in a higher computational complexity. In contrast now we have trained half pattern/target frames (see table. 2).

Besides providing a one order reduction of the computational complexity with respect to the single-stage approach, the 2-stage architecture has also achieved a lower pattern-target MSE has shown in Figure 2 for the articulatory parameter H. As far as the capability to track the modes and dynamics of the articulatory trajectories is concerned, as shown in Figure 1, the 2-stage architecture provides higher mode resolution and an improved detection of the mouth closures, counterbalanced by an increased amplitude distortion in positive peaks reproduction.

References

1. F.Lavagetto," Converting Speech into Lip Movements: A Multimedia Telephone for Hard of Hearing People" IEEE Trans. on RE, Vol.3, n.1, 1995, pp. 90-102.
2. A.Q. Summerfield, "Use of Visual Information for Phonetic Perception", Phonetica, Vol.36, pp.314-331, 1979.
3. E. Owens, B. Blazek, "Visems Observed by Hearing-Impaired and Normal-Hearing Adult Viewers", Journal of Speech and Hearing Research, vol.28, pp.381-393, 1985.
4. C.A. Fowler "Coarticulation and Theories of Extrinsic Timing", Journal of Phonetics, 1980.
5. O. Fujimura "Elementary gestures and temporal organization. What does an articulatory constraint means?" in The cognitive representation of speech, North Holland Amsterdam, pp. 101-110, 1981.
6. A.P. Benguerel, M.K. Pichora-Fuller, "Coarticulation Effects in Lipreading", Journal of Speech and Hearing Research, Vol.25, pp.600-607, 1982.
7. S. Morishima, H. Harashima, "A Media Conversion from Speech to Facial Image for Intelligent Man-Machine Interface", IEEE Journal on Sel. Areas in Comm.,vol.9, N.4, pp. 594-600,1991.
8. B.P. Yuhas, M.H. Goldstein Jr. and T.J. Sejnowski, "Integration of Acoustic and Visual Speech Signal Using Neural Networks", IEEE Communications Magazine, pp. 65-71, 1989.

Binary Spatter-Coding of Ordered K-tuples

Pentti Kanerva

RWCP[1] Neuro SICS[2] Laboratory

SICS, Box 1263, S–164 28 Kista, Sweden

e-mail: kanerva@sics.se

Abstract. Information with structure is traditionally organized into records with fields. For example, a medical record consisting of name, sex, age, and weight might look like (Joe, male, 66, 77). What 77 stands for is determined by its location in the record, so that this is an example of local representation. The brain's wiring, and robustness under local damage, speak for the importance of distributed representations. The Holographic Reduced Representation (HRR) of Plate is a prime example based on real or complex vectors. This paper describes how spatter coding leads to binary HRRs, and how the fields of a record are encoded into a long binary word without fields and how they are extracted from such a word.

1 Introduction

Nested compositional structure is fundamental to high-level mental functions, such as language and analogy. Accordingly, modeling these functions with neural nets requires that the structures be represented in a form suitable for neural nets. What that means, is discussed in depth by Hinton (1990).

Information with structure—for example, a data base—is traditionally organized into *records* with *fields*. The meanings of the fields are implied by the record type. For example, a medical record for a person might consist of name, sex, age in years, and weight in kilograms. A record for JSmith might look like (Joe, male, 66, 77). What 77 stands for is determined by its location in the record, so that this is an example of local representation. Local representation is common in artificial neural nets.

Various distributed representations have been used in cognitive modeling (e.g., Murdock, 1982; Rumelhart & McClelland, 1986; Pollack, 1990; Smolensky, 1990; Touretzky, 1990). Many of them are summarized in Plate (1994). Since the present proposal has much in common with the Holographic Reduced Representation of Plate, I review it first, and then explain the present proposal in the same terms.

2 Holographic Reduced Representation of Plate

Plate's (1994) Holographic Reduced Representation (HRR) is distributed. The entity corresponding to a record is called a *frame* and is realized by an N-dimensional vector of real numbers, where N can be in the thousands. The vector has no fields although it is composed of the same information as is the record: name = Joe, sex = male, age = 66, and weight = 77. The composition is done as follows. Each of the eight items 'name', 'Joe', 'sex', ..., '77' is encoded by a random N-vector with independent, normally distributed components (0 mean, $1/N$ variance). The N-vectors are

---

[1] Real World Computing Partnership

[2] Swedish Institute of Computer Science

written here in boldface. The relation 'name = Joe' is encoded by the circular convolution of the two vectors **name** and **Joe** and is written as **name∗Joe** (Plate's symbol includes a circle around the ∗). The circular convolution of two N-vectors **x** and **a** is an N-vector **u** defined by $\mathbf{u} = \mathbf{x} \ast \mathbf{a}$, where $u_n = \sum_k x_k a_{n-k \bmod N}$. We say that **name∗Joe** *binds* the *filler* 'Joe' to the *role* 'name'. Other words for a role and a filler are 'attribute' and 'value', and 'variable' and 'value'. The *instantiated* frame is then the vector sum of the four role–filler bindings, which is normalized:

$$\mathbf{JSmith} = \langle \mathbf{name{\ast}Joe} + \mathbf{sex{\ast}male} + \mathbf{age{\ast}66} + \mathbf{weight{\ast}77} \rangle \qquad (1)$$

where $\langle \ldots \rangle$ stands for normalization.

How are bindings decoded; how do we find the name, for example? When the information is stored in a conventional record, we simply extract the value from the appropriate field. However, when the binding is encoded with convolution, the decoding is also done with convolution. In fact, the involution of **x**, written as $\mathbf{x}^*$, is an approximate inverse of **x**, so that

$$\mathbf{x}^* \ast \mathbf{u} \approx \mathbf{a} \text{ if } \mathbf{u} = \mathbf{x} \ast \mathbf{a} \qquad (2)$$

(the involution of **x** is defined by $x_n^* = x_{N-n \bmod N}$, $n = 0, 1, \ldots, N-1$). Thus,

$$\mathbf{name}^* \ast (\mathbf{name} \ast \mathbf{Joe}) \approx \mathbf{Joe} \qquad (2')$$

Furthermore, **name** can be extracted by convolving the entire frame for JSmith:

$$\mathbf{name}^* \ast \mathbf{JSmith} \approx \mathbf{Joe} \qquad (3)$$

because convolution distributes over addition, and the sum $\mathbf{name}^* \ast (\mathbf{sex} \ast \mathbf{male}) + \mathbf{name}^* \ast (\mathbf{age} \ast \mathbf{66}) + \mathbf{name}^* \ast (\mathbf{weight} \ast \mathbf{77})$ acts as Gaussian noise.

Because decoding is approximate, an additional mechanism is needed for exact results. This is accomplished with a *clean-up memory,* which is an autoassociative memory that stores all currently defined code-vectors as fixed-point attractors. Thus, $\mathbf{name}^* \ast \mathbf{JSmith}$ is used as an input to the clean-up memory, to retrieve **Joe**.

Plate describes another distributed representation, related to the above, in which frames are realized by N-dimensional vectors of complex numbers. He calls them 'HRRs in the frequency domain'. I will use the terms 'real HRRs' and 'complex HRRs' for the two, and will show how spatter-coding gives us binary HRRs.

3   The Spatter-Coding of Sets

Previously, I have discussed the binary spatter-coding of sets (Kanerva, 1994, 1995). The codewords are random N-bit vectors, where N is large—usually in the thousands. A codeword is like a pointer in a computer program, in that it can represent all sorts of things, for example, an object, a concept, a feature, a combination of features, an element, a set, an attribute, a value, an attribute–value pair, a data structure. New codewords are created by combining existing codewords.

As above, codewords (N-vectors) will be written in boldface (e.g., **a**, **b**, …, **z**). They will also be used as names for the things they represent, as in "**a** is an element of the set **t**."

Sets are encoded as follows. If the set T consists of the elements **a**, **b**, …, **c**—i.e., $T = \{\mathbf{a}, \mathbf{b}, \ldots, \mathbf{c}\}$—its N-bit codeword, **t**, is gotten by adding the vectors **a**, **b**, …, **c** and by thresholding the resulting sum vector. This will be denoted by

$$\mathbf{t} = [\mathbf{a} + \mathbf{b} + \ldots + \mathbf{c}]$$

where the brackets [...] represent thresholding.

If the probability of 1s in the codewords and the threshold are chosen properly, the composition is *recursive*, meaning that the probability of 1s in **t** is the same as in the elements **a**, **b**, ..., **c**, and that such composition can be iterated indefinitely.

4 The Spatter-Coding of K-tuples: HRRs in the Binary Domain

Spatter-coding leads to particularly simple binary HRRs when the following conditions are satisfied:

1. The spatter code is recursive.
2. The recursive probability equals 0.5 (i.e., 0s and 1s occur equally likely in the codewords).

Consequences of these conditions are that

3. The number of elements, K, in the spatter-coded sets is odd.
4. The threshold is symmetric, meaning that it bisects the range $\{0, 1, 2, \ldots, K\}$ of the bitwise sum (the threshold is $K/2$).

When condition 4 is satisfied, Boolean Exclusive-OR (XOR, $\otimes$) distributes over thresholded sum, so that

$$\mathbf{h} \otimes \mathbf{t} = \mathbf{h} \otimes [\mathbf{a} + \mathbf{b} + \ldots + \mathbf{c}] = [\mathbf{h} \otimes \mathbf{a} + \mathbf{h} \otimes \mathbf{b} + \ldots + \mathbf{h} \otimes \mathbf{c}] \tag{4}$$

This allows XOR to be used for composing and decomposing binary HRRs.

Why is (4) true? Let $a, \ldots, c$ be the nth bits of the vectors **a**, ..., **c**, and let s be their sum ($s = a + \ldots + c$) and t the thresholded sum ($t = 0$ if $s \leq K/2$, and $t = 1$ otherwise); and let h be the nth bit of **h**. If $h = 0$, XORing with h leaves everything unchanged, and so (4) holds. If $h = 1$, XORing with h complements the K bits $a, \ldots, c$ so that $h \otimes a + \ldots + h \otimes c = K - s$. Since s and $K - s$ fall on different sides of the threshold $K/2$ (they cannot fall on the threshold because K is odd), the effect is the same as complementing t, and again (4) holds.

Binary HRRs are constructed similarly to real and complex HRRs. A set is encoded by superposition—that is, by adding together the elements and by normalizing the sum. The sum of binary vectors is normalized by thresholding, so that the vectors produce a recursive spatter code. In all three cases—real, complex, and binary—the result is a mean vector whose expected length is the same as that of the set's elements.

In real HRRs, fillers are bound to roles by circular convolution (e.g., **name∗Joe**). In binary HRRs, bitwise XOR is used. Fillers **a**, **b**, ..., **c** are thus bound to roles **x**, **y**, ..., **z** by $\mathbf{x} \otimes \mathbf{a}$, $\mathbf{y} \otimes \mathbf{b}$, ..., $\mathbf{z} \otimes \mathbf{c}$, and the binary HRR is the normalized mean of these bound pairs. Its bits are determined by the majority rule, with ties broken at random. This can be expressed as the thresholded sum $[\mathbf{x} \otimes \mathbf{a} + \mathbf{y} \otimes \mathbf{b} + \ldots + \mathbf{z} \otimes \mathbf{c} (+ \mathbf{R})]$, and it is the spatter-coding of the *(ordered) K-tuple* or *list* (**a**, **b**, ..., **c**) if the roles **x**, **y**, ..., **z** encode positions of elements. A random vector **R**—different **R** at different times—has to be added when the number of bound pairs is even, and its effect is to break possible ties. The binary HRR for JSmith now becomes

$$\textbf{JSmith} = [\textbf{name} \otimes \textbf{Joe} + \textbf{sex} \otimes \textbf{male} + \textbf{age} \otimes \textbf{66} + \textbf{weight} \otimes \textbf{77} + \textbf{R}] \qquad (5)$$

which is essentially identical to equation (1) for real HRRs.

The key to HRRs is their decoding. Recall that an element bound to $\textbf{x}$ in a real HRR is extracted approximately by convolving it with $\textbf{x}^*$ (see eqns. 2–3). Because binary $\textbf{x}$ is its own inverse under XOR, an element bound to it with XOR is extracted by XORing with $\textbf{x}$:

$$\textbf{x} \otimes \textbf{u} = \textbf{a} \ \text{if} \ \textbf{u} = \textbf{x} \otimes \textbf{a} \qquad (6)$$

(the corresponding eqn. 2 for real HRRs is approximate). Because XOR distributes, it also extracts individual elements from binary HRRs:

$$\textbf{x} \otimes [\textbf{x} \otimes \textbf{a} + \textbf{y} \otimes \textbf{b} + \ldots + \textbf{z} \otimes \textbf{c}] = [\textbf{x} \otimes \textbf{x} \otimes \textbf{a} + \textbf{x} \otimes \textbf{y} \otimes \textbf{b} + \ldots + \textbf{x} \otimes \textbf{z} \otimes \textbf{c}] \qquad (7)$$

$$= [\textbf{a} + \textbf{x} \otimes \textbf{y} \otimes \textbf{b} + \ldots + \textbf{x} \otimes \textbf{z} \otimes \textbf{c}]$$

$$\approx \textbf{a}$$

This extraction is approximate because $\textbf{x} \otimes \textbf{y} \otimes \textbf{b} + \ldots + \textbf{x} \otimes \textbf{z} \otimes \textbf{c}$ acts as random noise (cf. eqn. 3).

Typical to HRRs is that elements extracted from them are approximate or noisy. Table 4 in Kanerva (1995) allows us to estimate the amount of noise that must be cleaned up when binary HRRs are decoded. The table gives the relative distance between the codewords for a set and one of its elements—i.e., the probability that their corresponding bits differ—as a function of set size. The relevant values in the table are $d_1 = 0.25, 0.313, 0.344, 0.363, 0.377, 0.387$, and they correspond to set sizes $K = 3, 5, 7, 9, 11, 13$, respectively. That these values equal the noise to be cleaned up, can be seen from equation (7), since $\textbf{a}$ is an element of the K-element set $\{\textbf{a}, \textbf{x} \otimes \textbf{y} \otimes \textbf{b}, \ldots, \textbf{x} \otimes \textbf{z} \otimes \textbf{c}\}$, which is spatter-coded by $\textbf{x} \otimes [\textbf{x} \otimes \textbf{a} + \textbf{y} \otimes \textbf{b} + \ldots + \textbf{z} \otimes \textbf{c}]$. For example, elements extracted from binary HRRs that encode five role–filler pairs, will have about 31.3% noise. An autoassociative sparse distributed memory (Kanerva, 1988) can be used as a clean-up memory for binary HRRs, but the amount of noise that it can tolerate might limit the set size to five.

5 Discussion

We have seen that the idea of a Holographic Reduced Representation of Plate (1994) can be extended into the binary domain. Typical of these representations is that they encode ordered K-tuples with random, high-dimensional vectors without fields. The dimensionality remains constant when higher level structure is built from lower level ones, so that HRRs pose no limit on the number of levels that are practical to represent.

The spatter-coding of K-tuples leads to binary HRRs. The logic of binary and real HRRs is the same, so that the extensive discussion and results of Plate apply, by and large, also to binary HRRs.

Plate arrives at complex HRRs (in the frequency domain) through the Fourier transform of real HRRs. Binary spatter-coding is another route to the same end. Its bitwise XOR suggests a coordinatewise multiplication operation that distributes over addition, is bounded, and is easily inverted. Multiplication of numbers on the unit circle of the complex plane is a natural candidate. In fact, binary HRRs are equivalent to complex HRRs whose vector components are restricted to the values $\pm 1 + 0i$.

Decoding of HRRs is noisy and requires a clean-up memory. Autoassociative outer-product (matrix) memories are natural for the purpose because their stable states attract, so that convolution-based memories and matrix memories are complementary. The amount of noise that matrix memories can clean up, sets a relatively low limit on the number of vectors that can be combined at once into a decomposable HRR. Such combining provides a mathematical model for mental organization by chunking, and the limit is mathematical evidence for the limit on the number of mental items that people chunk together at once (Miller, 1956).

Binary HRRs are desirable on several accounts:
- Binary vectors and XOR are simple to build or simulate;
- Their simplicity helps make the underlying principles easily understood; and
- Nature is likely to use simple representations and operations if they work.

The binary domain is blemished by the requirement that the number of elements in the spatter-coded sets needs to be odd. This can be remedied by adding a random noise vector into sets with even number of elements before spatter-coding them (see eqn. 5). This can be thought of as a part of the normalizing of binary HRRs. Similar normalizing can be used if the vectors become too sparse or too dense as the result of many iterations of spatter-coding.

References

Hinton, G.E. (1990) Mapping part–whole hierarchies into connectionist networks. *Artificial Intelligence* **46**(1–2):47–75.

Kanerva, P. (1988) *Sparse Distributed Memory.* Cambridge, Mass.: MIT Press.

Kanerva, P. (1994) The Spatter Code for encoding concepts at many levels. In M. Marinaro and P.G. Morasso (eds.), *ICANN '94, Proceedings International Conference on Artificial Neural Networks* (Sorrento, Italy), vol. 1; 226–229. London: Springer–Verlag.

Kanerva, P. (1995) A family of binary spatter codes. In F. Fogelman–Soulie and P. Gallineri (eds.), *ICANN '95, Proceedings International Conference on Artificial Neural Networks* (Paris, France), vol. 1; 517–522. Paris: EC2 & Cie.

Miller, G.A. (1956) The magical number seven, plus or minus two: Some limits on our capacity for processing information. *Psychological Review* **63**(2):81–97.

Murdock, B.B. (1982) A theory for the storage and retrieval of item and associative information. *Psychological Review* **89**(6):316–338.

Plate, T.A. (1994) Distributed representations and nested compositional structure. PhD thesis. Graduate Department of Computer Science, University of Toronto. (Available on the Internet at Ftp-host: ftp.cs.utoronto.ca as Ftp-file: /pub/tap/plate.thesis.ps.Z)

Pollack, J.P. (1990) Recursive distributed representations. *Artificial Intelligence* **46**(1–2):77–105.

Rumelhart, D.E, and McClelland, J.L. (1986) On learning the past tenses of English verbs. In J.L. McClelland and D.E. Rumelhart (eds.), *Parallel Distributed Processing 2: Applications*; 216–271. Cambridge, Mass.: MIT Press.

Smolensky, P. (1990) Tensor product variable binding and the representation of symbolic structures in connectionist systems. *Artificial Intelligence* **46**(1–2):159–216.

Touretzky, D.S. (1990) BoltzCONS: Dynamic symbolic structures in connectionist networks. *Artificial Intelligence* **46**(1–2):5–46.

A Hybrid Approach to Natural Language Parsing

Christel Kemke

Computer Science Department, University College Dublin, Belfield, Dublin 4, Ireland
e-mail: kemke@ollamh.ucd.ie

Abstract. In this paper we suggest an approach to combine principles and techniques from traditional and connectionist paradigms in order to overcome the specific problems of traditional and connectionist parsers and to simultaneously preserve their advantages. The described approach - which is deeply hybrid - is based on the idea of a dynamic generation of a neural network according to the rules of a context-free grammar and driven by the input. The neural network represents the derived parse-tree or chart of the input structure, and is involved actively in the parsing process by activation flow. The parsing is truly incremental, hypothesizes expected structures, and gives estimates for the 'probailities' of the expected structures. The hybrid parsing method has been used successfully in three different natural language processing systems (PAPADEUS, INKAS, and INKOPA).

1   Introduction

Symbol-based parsers well known in Artificial Intelligence (AI) and Computational Linguistics (CL) provide highly developed tools and techniques, but they suffer from certain difficulties, for example to process ambiguous sentences or ungrammatical structures. Connectionist parsers, on the other hand, have problems with representing recursive structures, processing sequences, and handling variables but they can have certain advantages like fault-tolerance and distributed representation of syntactic and semantic knowledge.

Based on existing work in connectionist parsing and traditional parsers we developed three different systems (PAPADEUS, INKAS, and INKOPA)[1] which incorporate principles from both paradigms, i.e. symbolic parsers and neural network processing principles, and address some of the typical problems of symbolic and connectionist approaches, respectively.

The main common characteristic of all three systems is the dynamic generation of the parse tree, according to the rules of a context-free grammar, and realized in form of a neural network. Using this hybrid approach to parsing first of all ensures that the problems of variable binding or recursion typical for neural

---

[1] For more details on PAPADEUS, INKAS, and INKOPA see [16, 13, 14, 12]. The three systems have been developed in cooperation with the author by H. Kone (INKAS, INKOPA) and C. Schommer (PAPADEUS) for their Masters Studies at the University of the Saarland, Germany.

network approaches do not arise; second, the systems are more appropriate to deal with uncertainties and exceptions, like ambiguity of words and bindings (in PAPADEUS), and ungrammatical structures (in INKAS and INKOPA), than traditional parsers.

This paper focusses on a description of the basic principles and processing in the hybrid parsing approach used in these systems. [2]

2 Design Decisions for the Hybrid Parsing Approach

We attempted to bring together the advantages of traditional parsers and neural networks, in order to solve the usual problems of neural network parsers, and to provide a basis for solving some typical problems of traditional parsers. Work of both areas, i.e. traditional and connectionist parsing, has been taken into account in the development of the hybrid parsing approach, although the primary motivation was to investigate neural networks for natural language processing.

2.1 Concepts taken from the Traditional Paradigm

Context-free grammars (CFGs) were chosen to be the central structure for parsing. CFGs are well investigated in CL and they are suitable and often used in parsing tasks which don't incorporate very complicated inter- and intrasentential relations.

We used Chart-Parsing as the basic parsing theory because it seemed well-suited for the kind of processing performed by neural networks and Chart-Parsing has also been used by Waltz & Pollack in their work on word sense disambiguation.

2.2 Concepts used in the Connectionist Paradigm

In the work on connectionist parsing, some approaches are heavily based on traditional parsing techniques such as [15, 4]. They use the same structures and processes as the equivalent traditional parsers and thus merely reimplement a symbol-oriented parser in neural networks without gaining real advantages.

On the other hand, there are approaches that are more closely related to the neural network paradigm, for example work of Jain & Waibel [5, 8, 7, 6], Waltz & Pollack [17], and also Cottrell [3], Bookman [1, 2], and St. John & McClelland [9]. The work of Jain & Waibel is based on recurrent neural networks which are divided into several levels of representation, i.e. a word level, where words are described by features, the syntax level which represents syntactical units, and the structure level representing semantical relations between the phrases. The networks developed by Waltz & Pollack are also divided into several layers of representation, i.e. the syntax layer, the lexical layer, and the context layer. The main disadvantage of these parsers is their dependence on fixed structures and their elative inability to process unlimited, recursive structures.

---

[2] For details on the use of this approach in disambiguation see [16, 13], and for ungrammatical structures see [14, 12].

3   Basis for the Developed Hybrid Parsing Systems

The approach suggested in the following takes the above mentioned traditional and neural network based approaches into account[3] but allows unlimited recursion on the one hand and distributed processing on the other hand.

The formal basis for all three parsing systems are networks of finite automata which are suitable and well-defined structures for representing neural networks. The networks of finite automata used for modelling neural networks are special, parameterized structures based on a prototypical neuron model (see [10, 11]). The parameters determining the network are 1) the structure of the underlying graph defining the weighted connections between neurons, and 2) parameters of the single neurons like the threshold value, a value for the resting-potential, values for minimal and maximal activation states, and learning and forgetting functions in case of adaptive neurons. Common to all neurons in the network are the processing functions, i.e. state transition and output function which describe the computation of the activation state for the next time step and the output of the neuron, depending on the actual values of the above parameters of the formal neuron.

In connectionist systems for natural language processing, formal neurons can represent different entities or concepts but they are mainly used to model syntactic linguistic items, e.g. grammatical categories, and sometimes semantical items, like concepts or microfeatures for the description of the meaning of a word or concept. In the general approach to hybrid parsing described below, neurons are only used to represent syntactic items, like words, categories, and terminals and non-terminals in the grammar, and neural networks mainly represent grammar rules and charts.[4]

4   The Hybrid Parsing Approach

4.1   Rough Overview of the Hybrid Parsing Approach

The parsing essentially follows the grammar rules in a manner similar to a traditional parser, but in the systems we developed the grammar rules correspond to (sub-)networks, i.e. small (partial) neural networks. Partial neural networks or sub-networks corresponding to grammar rules are activated by the inputs and integrated or 'merged' during the parsing-process in order to build a more complex network which finally constitutes the parsing-tree or chart for the whole input structure.

The parsing in the developed systems proceeds incrementally; the network is built up according to the construction of the parse tree. Thus, the usual problems with variable binding or with representing recursive structures do not arise.

---

[3] The development of PAPADEUS, which performs lexical and syntactical disambiguation, was mostly influenced by the work of Waltz & Pollack, whereas INKAS and INKOPA are closely related to the approach of Jain & Waibel.

[4] In PAPADEUS, neurons represented in addition concepts, microfeatures, and contexts.

For the parsing mainly two kinds of syntactical knowledge is necessary: lexical rules and grammatical rules; the grammatical rules correspond to (partial) neural networks. The parsing process is described in more detail in the following section.

4.2 Description of the Hybrid Parsing Approach

For each grammar rule, e.g. S → NP VP, there is a corresponding sub-network consisting of a number of nodes representing the non-terminals, e.g. u1 for S, u2 for NP, and u3 for VP, and weighted links between these nodes constituting the directed graph (a tree-structure) underlying the neural network and representing the relation between the non-terminals. The father node in this tree-structure represents the left hand side of the rule, e.g. S, and the child nodes represent the right hand side of the rule, e.g. NP and VP. For the example above, the father node is u1 representing the non-terminal S; the child node u2 represents NP and u3 VP; u2 and u3 are linked to u1 by weighted connections directed to u1.

The construction of the network proceeds as follows: The input sentence is read word by word. When a new word is read, the lexical rules are tested in order to see to which lexical item the word belongs (this is for example a certain verb stem like *come* instead of the form *comes*). In this way we cope with different morphological appearances of one word.

Next, the deflected input word is matched against the grammatical rule base in order to find the grammatical category of the word, e.g. a name *Peter* is recognized as a Noun N, or *come* is categorized as a verb according to the rule V → *come*.

If the category is determined, the network proceeds with the parsing process. Iteratively, the grammatical rules are matched to the yet determined non-terminals. If the actual input word is, for example *Peter's*, the lexical item is *Peter*, the category is Noun N, and the rules with the non-terminal N on the right hand side are matching, e.g. NP → N. If a rule matches, a copy of the corresponding (partial) neural network is generated.[5]

Based on the neural paradigm, activation is then passed on from the right hand side of the rule to the left hand side, e.g. from N to NP. Since the right hand side of the rule is complete, i.e. only N, the NP-node will be fully activated, i.e. 1.0. If the activation of a node on the left hand side, e.g. NP, is strong enough, it can activate again other matching rules/sub-networks, e.g. NP activates S → NP VP. The sub-networks, i.e. NP → N and S → NP VP, are in the next phase merged, i.e. the corresponding non-terminal nodes are collapsed, in this example the NP-nodes, and the result of the merging process is one complex network corresponding to the actual chart or interim parse-tree. Due to activation flowing from the NP node to the S node when the network is extended by the sub-network representing the rule S → NP VP, the father node S will become slightly activated, i.e. 0.5. That means that in this state of processing, the network 'hypothesizes' a sentence structure (S) and 'expects' a verb phrase (VP).

---

[5] In the following we also say 'activate a network' to denote the generation of a copy of a partial neural network.

This activating of rules/sub-networks, matching, and merging is repeated for every new input. If the next input word is, for example *comes*, the rules with the non-terminal V on the right hand side are matching, e.g. VP $\rightarrow$ V, and again the corresponding sub-network is activated. Since a parse-tree containing a non-terminal VP was just constructed and the VP has not yet been activated, the new sub-tree VP $\rightarrow$ V will be merged with the old complex parse-tree / network. Then, activation is passed from VP to S, the sentence-node. The S-node thus becomes fully activated, i.e. 1.0, and the input sequence is recognized as a complete, well-formed sentence.

If there is no further input, the parsing is finished; in case there is further input, the parsing may proceed, using other interim structures which have been constructed during the process.

5 Short Discussion of the Hybrid Parsing Approach

The parsing is truly incremental in the sense that it constructs partial parsing trees as far as the input is processed, and it hypothesizes expected structures. The use of neural networks and activation flow between nodes enables the system to display partially parsed structures with estimates for the real appearance of hypthesized sub-structures. In the example above the value of 0.5 for activation of the S-node could be interpreted as an estimate for the 'probability' that the sentence will be completed and the sentence symbol will be derived.

It has to be noted that in the current parsing process based on neural networks, the actual sequencing of the input words is not taken into account. This is due to the fact that the neural networks in this approach do not distinguish the sources of their inputs, i.e. inputs coming from different childnodes are processed in parallel, in the same manner. Thus, the order implicitly contained in the grammar rules is not preserved in the neural network. This can be advantageous in order to gain a certain robustness for processing, for example, spontaneous dialogues in which ungrammatical structures might be used. But if the sequentiality has to be taken into account, an additional mechanism has to be integrated. The basic idea is to prevent that a child-node becomes active before its 'predecessor-node', i.e. the child-node which is prior in the sequence, is active. This is done by inserting a so-called 'precondition link' which prevents that the father-node becomes active if the child nodes are not activated in the correct order.

6 Conclusions

In this paper we suggested a hybrid approach to natural language parsing which combines principles of traditional parsing techniques and neural networks.

The hybrid parsing process involves an incremental, dynamic construction of the neural network where the neural network corresponds to the (partial) parse tree and at the same time is contributing to the parsing process by activation flow between nodes. The hybrid parsing approach displays an expectation of further structures and gives a kind of estimate for the 'probablity' of certain structures, based on activation passing in the neural network.

The hybrid approach to parsing has been implemented and successfully used in the systems PAPADEUS, INKAS and INKOPA.

References

1. L. A. Bookman. A microfeature based scheme for modelling semantics. In *Proc. of the 10 th IJCAI*, pages 611–614, Milan, Italy, 1987.
2. L. A. Bookman. A connectionist scheme for modelling context. In D. Touretzky, G. Hinton, and T. Sejnowski, editors, *Connectionist Models Summer School*. Morgan Kaufmann, Palo Alto, 1988.
3. G. W. Cottrell. *A Connectionist Approach to Word Sense Disambiguation*. Morgan Kaufmann, 1989.
4. M. Fanty. Context-free parsing in connectionist networks. Technical Report TR 174, Computer Science Dept., University of Rochester, Rochester, NY, 1985.
5. A. N. Jain. Generalization performance in PARSEC - a structured connectionist parsing architecture. In J. E. Moody, S. J. Hanson, and R. P. Lippman, editors, *Proc. NIPS 4*. Morgan Kaufmann, 1992.
6. A. N. Jain and A. H. Waibel. A connectionist parser aimed at spoken language. Technical report, School of Computer Science, Carnegie Mellon University, 1989.
7. A. N. Jain and A. H. Waibel. Incremental parsing by modular recurrent connectionist networks. In D. S. Touretzky, editor, *Proc. NIPS 2*. Morgan Kaufmann, 1990.
8. A. N. Jain and A. H. Waibel. Robust connectionist parsing of spoken language. In *ICASSP-90*, 1990.
9. M. F. St. John and J. L. McClelland. Applying contextual constraints in sentence comprehension. In D. Touretzky, G. Hinton, and T. Sejnowski, editors, *Connectionist Models Summer School*. Morgan Kaufmann, Palo Alto, 1988.
10. C. Kemke. Modelling neural networks by means of networks of finite automata. In *Proc. 1st ICNN*, 1987.
11. C. Kemke. Neural network modelling by means of networks of finite automata. In A. Prieto, editor, *Artificial Neural Networks*, pages 48–53. 1991.
12. C. Kemke and H. Kone. Incopa - an incremental connectionist parser. In *Proceedings World Congress on Neural Networks*, volume 3, pages 41–44, 1993.
13. C. Kemke and C. Schommer. Papadeus - parallel parsing of ambiguous sentences. In *Proceedings WCNN*, volume 3, pages 79–82, 1993.
14. H. Koné. INKOPA - Ein inkramenteller konnektionistischer Parser für natürliche Sprache. Diplomarbeit, FB Informatik, Universität des Saarlandes, 1993.
15. H. Schnelle. The challenge of concrete linguistic description: Connectionism, massively parallel distributed processing, net-linguistics. Technical Report GENET-24, Sprachwissenschaftliches Institut, Ruhr-Universität, Bochum, F.R.G., 1987.
16. C. Schommer. PAPADEUS - Ein inkramenteller konnektionistischer Parser mit einer parallelen Disambiguierungskomponente. Diplom-Arbeit, Fachbereich Informatik, Universität des Saarlandes, 1993.
17. D. L. Waltz and J.B. Pollack. Massively parallel parsing: A strongly interactive model of natural language interpretation. *Cognitive Science*, 9(1):51–74, 1985.

A Self-Organizing Neural Network Approach for the Acquisition of Phonetic Categories

Kay Behnke
Kay.Behnke@mpi.nl

Peter Wittenburg
Peter.Wittenburg@mpi.nl

Max Planck Institute for Psycholinguistics, Postbus 310
6500 AH Nijmegen (The Netherlands)

Abstract. We present a neural network approach to the process of acquisition of phonetic categories in infants. In our approach we investigate the question to what extend the development of phonetic categories can be described by a self–organizing process. Simulation results show that with digitized speech as input, the network is able to learn representations of the vowel categories in the input set.

1   Introduction

Results from speech perception experiments show that newborns possess basic speech perception capacities which function without a prior period of prolonged exposure to language input (e.g. Bertoncini, Bijeljac-Babic, Blumstein & Mehler, 1987). During further development, the native language has an impact on the speech perception capacities and infants learn about the regularities that characterize their native language. These regularities are important when learning that different speech sounds should be treated as belonging to the same category, in distinguishing among different words, or in specifying boundaries between different words. E.g. there are reasons to believe that knowledge of regularities in the sound structure of the native language is crucial for segmenting speech and recognizing words (e.g. Church, 1987).

In our theory, the impact of the native language is based on the development of phonetic categories which constitute memory representations of the characteristics of the ambient language. One can regard the map of phonetic categories as a filter which abstracts from details in the incoming speech stream and tunes the recognition system to the sound structure of the native language.

In this paper we present a new neural network approach. We ask to what extent the development of phonetic categories can be described by a self–organizing process.

2   Assumptions about the speech input

The input data consisted of isolated spoken CVCV–words in which the consonant and vowel remain constant in a word. As consonants we used /b/, /d/, /l/, /m/, and /p/ and as vowels we used /a/, /e/, and /u/. The data was sampled with a sample frequency of 16 kHz and smoothed by a Hamming window of length 256. A Fourier Transformation of order 8 converted the sampled speech data to the frequency domain. We transformed the power spectrum to an auditory–like 16–dimensional spectral representation according to an Acoustical Band Spectrum (ABS) algorithm (Hermansky, 1990) and normalized each vector to a length of one.

One crucial point of our theoretical model is the assumption that the nature of input to the phonetic map (in which the phonetic categories eventually develop) is not static but increases gradually in resolution during development. We assume that infants are not able to use all details contained in the speech signal from the beginning on, but instead that perception is constrained by an "energy filter". According to our pre–processing algorithm, each element in a vector corresponds to the energy in a particular frequency band. Therefore, we computed the length of each vector before normalization and set it equivalent to the energy value of an input vector.

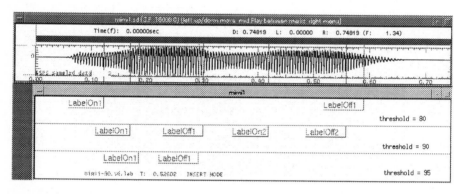

Fig. 1.: Speech signal of the utterance /mimi/

Figure 1 shows the speech signal of the utterance /mimi/ in connection with labels for three different energy thresholds: 80, 90, and 95. The speech signal which lies in the area marked by a pair of labels contains speech vectors which energy values are above the corresponding threshold value. The figure shows that centers of vowels correspond to regions of high energy in the speech signal. A statistical analysis of our input data shows that a high energy threshold restricts the space which serves as input for the phonetic map (see table 1). In our view, this decrease in complexity enables the infant to build basic phonetic categories which act as perceptual "anchors" when resolution, and therefore complexity of the input increases.

There comes only little support from psycholinguistic experiments for our assumption. This is because there is no technique to test auditory thresholds in infants younger than three months. Nevertheless, the data with infants older than five months shows that the absolute auditory thresholds are higher than

threshold	Euclidean distance to mean vector of category /a/				
	< 0.025	0.025 − 0.05	0.05 − 0.075	0.075 − 0.1	> 0.1
80	1.28%	63.17%	17.34%	6.0%	12.21%
90	6.24%	74.93%	9.56%	3.03%	6.24%
95	12.74%	74.42%	8.74%	1.89%	2.21%

Table 1.: The input vectors of each input file were labeled according to the vowel of the utterance, e.g. if the input file represents the utterance /baba/, the input vectors of that file were labeled by /a/. After filtering the data, we computed the mean vector for each input category and the distance of each input vector to its mean vector. The data shows that the higher the energy threshold during filtering the smaller the variance of the input vectors to the its mean vector.

for adults (Aslin, 1987). Additional support comes from experiments which investigated the *visual* system of infants (Atkinson & Braddick, 1981). Acuity and contrast sensitivity are around one-thirtieth of adult levels in neonates and improves by a factor of three or four by the fifth month. These findings let us conclude that the assumption we have made is reasonable.

3   Why a new network approach?

During the last few years, the practical effectiveness of the Self–Organizing Map (SOM) model as an unsupervised clustering tool has been demonstrated across several applications (e.g. Ritter & Kohonen, 1989). In addition, the SOM has served as a starting point for further developments of topology–preserving feature maps, like the "neural–gas" algorithm of Martinetz (Martinetz & Schulten, 1994) in which the specification of a network topology is not necessary anymore, and the Laterally Interconnected Synergetically SOM (LISSOM) of Sirosh and Miikkulainen (1994) which introduced the development of lateral connections into the network structure and investigated how they affect the self–organization of afferent connections.

We investigated an SOM with regard to its suitability for our purposes and to a "dynamic" input space in particular. We tested this by defining three squared regions of equal size and equal probability in the two–dimensional input space $[-1, 1] \times [-1, 1]$. After $20,000$ simulation steps, the initially–random weight vectors in the SOM described a mapping of the three input categories to the map of units. At this point, we added a new input category of same size and same probability as the previous ones to the input space. Our results show that the addition of a new category during simulation requires a reset of the functions which determine the neighborhood of a winning unit and the learning rate. However, a reset partly or totally destroys existing category representations in the map which is in contradiction to our theoretical model and results of psycholinguistic experiments.

Our conclusion is that the same is true for the "neural–gas" algorithm and for the LISSOM model. Similar to the SOM, both approaches (inherently) make use of a neighborhood function which determines the number of units which significantly change their afferent weights with each simulation step. The scope of this function has to decrease during simulation so that the input categories can be represented by the network units. However, the function has to be reset if a new category is added to the input space, again resulting in the destruction of existing representations.

In summary, the self–organizing neural network approaches described so far are not suitable for our purposes. We are looking for an algorithm which is able to represent different categories in the input space in a self–organizing manner but which is flexible enough to react to changes in the input space without influencing already developed representations of the categories. In the following we describe a new self–organizing neural network approach which has these characteristics.

4 Network architecture and learning paradigm

The architecture of the neural network consists of a two-dimensional map of processing units in which each unit receives the same input vector at a simulation step. The input to a unit (i, j) is weighted by an afferent weight vector $\mathbf{aff}_{ij}$ which is initially located at a random point in input space. Each unit in the network has excitatory and inhibitory lateral connections to some other units in the map. Excitatory lateral connections exist between the current unit and its direct neighbors. In contrast, inhibitory lateral connections connect a random number of units which have a distance in the map greater than the parameter *repelling radius* with the current unit (see figure 2(a)).

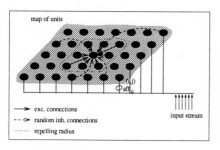

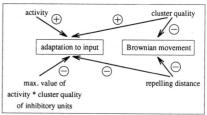

(a) Sketch of the neural network architecture.

(b) Factors which influence the two learning processes.

Fig. 2.: Network architecture and learning rule factors

The learning rule consists of two independent processes: (1) an adaptation process in which the afferent weight vectors are adapted in the direction of the current input vector according to Hebbian learning, and (2) a randomization process ("Brownian movement") in which each afferent weight vector is slightly adapted in a random direction independently of the current input vector. Figure 2(b) shows the factors which have positive and negative influence on each of the two processes, respectively. Positive influence means that a high value of a factor strengthens the corresponding process. In contrast, a negative influence means that a high value weakens the process.

In our model, the learning process is mainly dependent on three factors: (1) the *activity* of a unit, which describes the correlation between the afferent weight vector and the current input vector, (2) the *cluster quality* of a unit, which describes the correlation between neighboring afferent weight vectors, and (3) the highest value of *activity* * *cluster quality* of all inhibitory units to a unit. The combination of these factors means that only units which are sensitive for the current input vector, which build a potential cluster, and which are not inhibited by an existing cluster are adapted in the direction of the input vector (for further details, see Behnke, 1996).

5   Simulation results

We generated three sets of speech files. The first set contained only utterances of the form $C/e/C/e/$ (in which C represents one of the five consonants), the second set contained only utterances of the form $C/e/C/e/$ and $C/u/C/u/$, and the third set contained utterances of all possible CVCV–words. We started the simulation with the first set and changed at simulation step 10,000 to input set two and at simulation step 20,000 to input set three. Figure 5 shows the simulations results after 50,000 simulation steps.

Six clusters have been developed after training. Three for the input category /a/, two for the category /u/, and one for the category /e/. The distribution of the cluster quality at simulation steps 10,000, 20,000, and 50,000 show that during training new clusters develop dependent on the input set. However, the development of new clusters or the change in input space has no influence on already developed clusters in the map.

6   Conclusions

The results show that the network approach is able to build representations for each of the three vowel categories and is flexible enough to react on changes in input space without influencing already developed representations of the categories. According to our theoretical model that means that infants are able to learn these representations by a self–organizing process. However, our statistics about the input data show a strong overlap between different phoneme categories and therefore indicate that not all phonemes can be represented by such a self–organizing process. We are currently investigating our network approach with the seven long vowels of the Dutch vowel system. If the development of phonetic categories is an initially self–organizing process, no prior discriminating of the

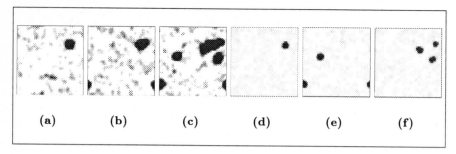

Fig. 3.: Simulation results: distribution of the average cluster quality in the map after **(a)** 10,000, **(b)** 20,000, and **(c)** 50,000 simulation steps; distribution of the average activity in the map for a speech vector representing **(d)** the /e/ in the utterance /meme/, **(e)** the /u/ in the utterance /bubu/, and **(f)** the /a/ in the utterance /dada/.

input is required. Given the intrinsic complexity of speech it is interesting to ask whether learning can be achieved in this way, or whether additional discrimination processes will be required. Our analysis of the learning of seven rather than just three vowels is addressing this issue.

References

Aslin, R. N. (1987). Visual and auditory development, *in* J. D. Osofsky (ed.), *Handbook of Infant Development*, J. Wiley & Sons, New York, pp. 5–97.

Atkinson, J. & Braddick, O. J. (1981). Acuity, contrast sensitivity and accommodation in infancy, *in* R. N. Aslin, J. R. Alberts & M. R. Petersen (eds), *The Development of Perception: Psychobiological Perspectives, vol. 2. The visual system*, Academic Press, New York, pp. 245–277.

Behnke, K. (1996). A self–organizing neural network approach for the acquisition of phonetic categories during the first year of life, *Technical report*, Max Planck Institute for Psycholinguistics.

Bertoncini, J., Bijeljac-Babic, R., Blumstein, S. E. & Mehler, J. (1987). Discrimination in neonates of very short CVs, *The Journal of the Acoustical Society of America* **82**(1): 31–37.

Church, K. W. (1987). Phonological parsing and lexical retrieval, *Cognition* **25**: 53–69.

Hermansky, H. (1990). Perceptual linear predictive (PLP) analysis of speech, *The Journal of the Acoustical Society of America* **87**(4): 1738–1752.

Martinetz, T. & Schulten, K. (1994). Topology representing networks, *Neural Networks* **7**: 507–522.

Ritter, H. & Kohonen, T. (1989). Self–organizing semantic maps, *Biological Cybernetics* **61**: 241–254.

Sirosh, J. & Miikkulainen, R. (1994). Cooperative self–organization of afferent and lateral connections in cortical maps, *Biological Cybernetics* **71**: 66–78.

Inward Relearning: A Step Towards Long-Term Memory

Sylvie Wacquant, Frank Joublin

Institut für Neuroinformatik
Lehrstuhl für Theoretische Biologie
Ruhr-Univ. Bochum, D-44780 Bochum, Germany
e-mail:wacquant@neuroinformatik.ruhr-uni-bochum.de

Abstract. Artificial neural networks are often used as models of biological memory because they share with the latter properties like generalisation, distributed representation, robustness, fault tolerance. However, they operate on a short-term scale and can therefore only be appropriate models of short-term memory. This limitation is known as the so-called catastrophic interference: when a new set of data is learned, the network totally forgets the previously trained sets. To palliate these restrictions, we have developed an algorithm which enables some types of neural network to behave better in the longer term. It requires local networks where the representation takes the form of prototypes (as example, we utilize a RBF network). These prototypes model the previously learned input subspaces. During the presentation of the new input subspace, they can be inwardly manipulated such as to enable a "relearning" of a part of the internal model. In order to show the long-term capabilities of our heuristic, we compare the results of simulations with those obtained by a multi-layer network in the case of a typical psychophysical experiment.

1. Introduction

Connectionist models of memory are all confronted to the so-called *plasticity-stability dilemma* [5]. This situation leads to very few accurate long-term memory (LTM) models [7], [5],[8]. The problem is the following: the memory system must be plastic to acquire new significant input while remaining stable in responses to previously learned items. Indeed, when confronted with new inputs, most models suffer from the so-called *catastrophic interference* described in [6], [18]. This effect can be characterized by the abrupt and radical forgetting of previously learned informations due to the interaction with newly learned informations.

This *catastrophic* interference is particularly shown in multilayered perceptrons (MLP) trained by some backpropagation-like learning algorithms. In this case, the forgetting effect is mainly due to the distributed nature of the representation [8]. Each new input influences most of the weights of the network through the backpropagation of the output error and previously learned patterns become then corrupted. A few works have addressed this issue, in [6] and [18] the authors analyse this effect and make systematic studies, varying parameters in backpropagation algorithms. They conclude that *catastrophic interference* cannot be satisfactorily avoided in MLP networks.

Learning systems based on local technique seem to be a better approach to build long-term memories. In this kind of network, information is processed at the level of a small group of neurons in the neighbourhood of a single neuron also called *prototype*. These networks quantize the input space to be learned by distributing *prototypes* using algorithms like self-organizing maps (SOM) [13], or vector quantization technique. They also suffer from *catastrophic interference* but in a more restricted way. This is due to the limited subset of neurons affected by the learning rule. This limited effect is observed in any network whose memory relies partially on *prototypes* such as radial basic function networks (RBF) [17].

Some authors ([8], [18], [11], [14]) have reduced the effects of *catastrophic interference* in MLP by reducing the receptive fields of neurons in the hidden layers in order to supply them with a more local processing. The price to pay for these improvements in the long-term behaviour is longer training periods and higher number of hidden nodes to allow good generalisation performances. They showed [1] the existence of a trade-off between resistance to *catastrophic interference* (local representations) and generalisation (distributed representations).

Pedagogical strategies belong to the class of solutions which can reduce the forgetting of old items but they are based on the *rehearsal* of part of the whole data set. This prerequisite is not always available to incremental learning problems.

In the following parts of this paper, we present a heuristic based on a growing RBF network which allows to overcome the problem of *catastrophic interference* by periodically consolidating the internal model of the LTM. The performance of the resulting LTM is then compared to MLP models in a psychological experiment showing human memory performances in *retroactive interference*.

2. Inward Relearning Mechanisms

2.1 Some Psychological Aspects of Inward Relearning

Very often, one can have clues on the properties of processes by observing their dysfunctions. It is often the case when studying the distortions of human long-term memory. According to psychological observations, it seems that events are coded by already existing structures reflecting our past experience [12]. In 1955, Kay conducted an experiment which illustrates the extent to which a person's memory is influenced by his own interpretation of material. He presented a text to a group of subjects. Once a week during one month, they were asked to write what they remember and then to listen again the genuine text. They recalled their own reproductions much more accurately than what was read to them (despite the possibility of corrections during the following reading). More generally, many studies on LTM describe the problems of interferences in witness reports [2]: the degree of forgetting is a function of the similarity between the material to be remembered and the interfering material. The accuracy of the internal model seems to influence the accuracy of the long-term memorisation. This idea underlies the principle of inward relearning: the LTM builds itself by fusing new data with its own model of old ones such as to form a new internal model of both.

2.2 Learning Ambiguities

Our model of LTM is based on a RBF network. This means that the internal representation performs a quantization of the input space where each quantization vector is called a *prototype*. Each *prototype* is tuned to a limited sub-space of the input space. When a new input comes, only a few *prototypes* (at least one) respond and learning occurs by modifying the position of the *prototypes* and the range of their associated sub-space. Non-responding *prototypes* remain unchanged. We call this restricted number of modified neurons *ambiguous prototypes* to highlight their high contribution to the current learning and thus their high risk to forget their previous selectivity. In order to retain as much as possible of the previous model of the input space, the parameters of the *ambiguous prototypes* are temporary memorised before beginning to learn the new input space. These prototype parameters, being of same nature as input vectors, can be seen as partial samples of the previous input spaces: they will be used as new inputs to the LTM network to minimize the forgetting effects. *Inwnrd relearning* consists then in presenting both the new inputs and the *ambiguous prototypes* to the LTM.

In fact, the training strategy takes into account that in incremental learning problems, old input data are no more available and that the closest evidences of their values are modelled and stored into the network *prototypes*. The part of the LTM which interferes with the new input space and the new inputs are stored temporarily in the equivalent of a short-term memory, called STM (cf Fig. 1a). This temporary memory is also justified by the fact that "items encoded in groups are significantly more resistant to forgetting than items learned individually" [18]. Elements stored in this STM are then presented to the network alternatively (cf Fig. 1b): this enables the memorisation of new inputs by adapting the model of old ones.

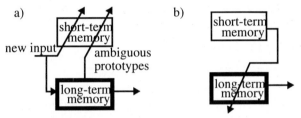

Fig. 1. Tandem of memories: the STM stores both new data and the model of old data which is in conflict with new ones.

2.3 Inward Relearning Heuristic

Our network is composed of three layers:

- one input layer, L1, representing the different input samples,,
- one layer of RBF units, L2, mapping the input spaces with a growing structure,
- one layer of simple perceptron units, L3, classifying the prototypes of the previous layer.

The main layer L2 is an incremental version of RBF, called GNGN (Growing Neural Gas Network) which has the advantage to adapt its size to the problem size and to work with a local learning paradigm (for details see [9]). Training consists in finding an optimal placement of units to obtain a good vector quantization of the input space. It contributes thus to provide a good classification on the last layer. The so called "best matching" unit j for an input vector I corresponds to the unit of L2 whose weight vector W_j (associated to input links of j) is the nearest from the input vector I.

Inward relearning is based mainly on informations provided by the layer L2. We present the algorithm in the case of a supervised classification of a set of input subspaces. Suppose $\bigcup_{i \in \{1, 2..., k\}} S_i$ is a set of input subspaces which have already been trained. The next input subspace S_{k+1} is then learned by the following procedure:

1)We present all samples of S_{k+1} to the network, without learning. It enables to determine the best matching neuron for each sample. The attributes of all the best matching neurons, weight vectors and corresponding network outputs generated on the last layer L3, are stored in P_{k+1} (cf. (Eq. 1)).

$$P_{k+1} = \{ (w_j, o_j) \,|\, j \in BM(S_{k+1}) \} \qquad \text{(Eq. 1)}$$

where $BM(S_{k+1})$ is the best matching neurons set, when all elements of S_{k+1} are

presented directly after being trained on $\displaystyle\bigcup_{i \in \{1, 2.., k\}} S_i$,

where o_j is the answer of the network to the presentation of w_j as input vector.

The ambiguous prototypes, called $BM(S_{k+1})$ correspond to the part of the old data model which interferes with the new data S_{k+1}.

2)P_{k+1} as well as S_{k+1} are memorised in a temporary memory, called STM.
3)Learning is performed on all elements of the STM. This training may be modulated by varying the proportion between training time of P_{k+1} and of S_{k+1}. In our simulation, they are learned with the same frequency. The learning of P_{k+1} is what we call *inward relearning*.

3. Example of Behaviour

We test the *inward relearning* algorithm on a typical example where neural networks usually fail and show effects of catastrophic interference [6], [18], namely the psychological experiment of Barnes and Underwood [3]. Subjects of the experiment were trained to associate a list A of nonsense syllables to a list B of adjectives. As soon as all the eight pairs of A-B are learned, subjects are then trained to associate the same list A to another list of adjective C (still for eight pairs of associations). In our simulation, following [6], a syllable or an adjective is represented by a random pattern of 10 bits. The succession of the two associations is characterized by a change of context: this context is constant during one association training and is also coded by a random pattern of ten bits. The neural network learns first to associate A-b to B and then A-c to C, b and c being respectively the context of A-B and A-C training. We did the simulation with the same decision parameters as in [6]: a recognition threshold of 0.1 and a best matching criterion. Fig. 2. presents a comparison of the results of this experiment during the learning of the association A-C. In psychology, the forgetting effect observed on the recall of the first list is called *retroactive interference*: learning of the C responses interfered with the recall of the B responses. But even after an amount of A-C training sufficient to yield over 90% of correct responses on the A-C list, subjects still performed at a level better than 50% of correct responses on the A-B list. In contrast, the backpropagation network ability is already reduced to 0% of recognition on the A-B list after three learning trials (20% of correct responses on the A-C list) as reported in [6]. For a local network (GNGN), the performance is better: for 20% of correct responses on the A-C list, the network still shows 41% of correct responses on the A-B list and for 90% of correct responses on the A-C list, the performance decreases to 11% of recognition on the A-B list. With an *inward relearning strategy*, the network, called LTM-GNGN, behaves similarly to the results obtained in the psychophysical experiment. For 20% of correct responses on the A-C list, the network responds to A-B list with 62% of correct responses and for 90% of correct responses on A-C, 57% of the A-B list is still well memorized (average on 100

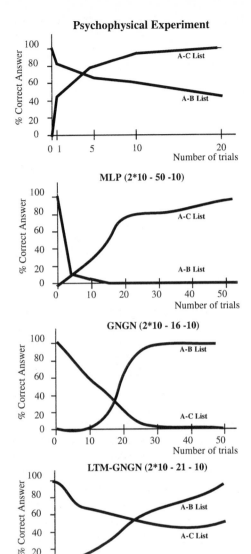

Psychophysical Experiment

MLP (2*10 - 50 -10)

GNGN (2*10 - 16 -10)

LTM-GNGN (2*10 - 21 - 10)

Fig. 2. Comparisons of results on the psychophysical experiments of Barnes and Underwood [3]: a) genuin results of experiments on people. b) results from McCloskey and Cohen [6] with a multi-layered perceptron (MLP). c) results from a GNGN without inward learning. d) result from the LTM-GNGN network.

tests).

In the LTM-GNGN network, we usually observe a first phase where the learning rate of A-B decreases. Then in a second phase, comparatively shorter, it increases. It reaches finally a stabilised level as the A-C learning achieves 100% of correct responses. This development may be explained by a kind of *restructuring process*, the internal model of the old data is adapted to cope with the new data. It creates during its transformation worse results before it reaches an optimised state. This may roughly correspond to the process of *accommodation* as defined by J. Piaget [16], i.e. the transformation of the internal frame to fit current data. Inward relearning enables thus the design of a dynamical internal model.

4. Conclusion

From a biological point of view, inward relearning was already suggested by Marr [15]. He made the hypothesis of the reactivation of networks when receiving no inputs from environments. He points out that this type of consolidation could be directed by the hippocampal system which would store experiences as they happen during the day, and then replays its memories to the neocortex overnight. He interprets that as a way to provide data for the category formation and integration process [7].

More generally, three main types of properties may characterise our heuristic. It first allows that the system develops continuously a representation adapted to current and past environment while being always available. Secondly, in opposition to complete rehearsing of the past data, it needs only to retain, in the STM, a constant limited amount of information (the number of *ambiguous prototypes* which varies in a restricted range). This provides a gain of space (size of the short-term memory) and of time. However, despite of the performance of our model, it is not yet possible to recover all the data previously learned. This limitation, which can also be observed in biology does not seem to be an obstacle to the incremental learning of behaviours. Moreover, by controlling the amount of internal relearning and the selection of ambiguous prototypes, it is possible to focus the attention of the system on particularly important informations or to forget selectively uninteresting ones. This would be the third interesting feature of inward relearning. At last, this procedure yields a stabilisation of the previous representation while concurrently permitting a further learning of new input spaces. This point will be formally explored in future works. From a more general point of view, these simulations give some more hints of the importance of working on the internal representation.

5. Acknowledgement

We would like to thank C. Lorin for the helpful corrections of early versions of this manuscript as well as C. Goerick and the anonymous reviewers for their advises on the last version.

Reference

[1] V. R. de Angulo & C. Torras, "On-Line Learning with Minimal Degradation in Feedforward Networks", *IEEE Trans. on Neural Networks*, 6(3), May 1995.

[2] A.D. Baddeley, *theory and practice*, Allyn & Bacon, Boston, 1990.

[3] J.M. Barnes , B.J. Underwood, *Journ. of Exp. Psychol.*, 58: 97-105, 1959.

[4] C. Cachin, "Pedagogical pattern selection strategies", *Neural Networks*, 7: 175-181, 1994.

[5] G.A. Carpenter, S. Grossberg, "Massively parallel architecture for self organizing neural pattern recognition machine", *Comp. Vis., Graph., and Image Proc.*, 37: 54-115, 1987.

[6] M. McCloskey, N.J. Cohen, "Catastrophic Interference in Connectionist Networks: the Sequential Learning Problem", In G.H. Bower (Ed.), *The psychology of learning and motivation*, New York: Acad. Press, 109-165, 1989.

[7] J.L. McClelland, B.L. McNaughton, R. C. O'Reilly, "Why there are complementary Learning Systems in the Hippocampus and Neocortex: insights from the successes and failures of connectionist models of learning and memory", *Technical Report PDP.CNS.94.1*, Carnegie Mellon University, March 1994.

[8] R.M. French, "Using semi-distributed Representations to overcome catastrophic forgetting in connectionist Network", *Proc. of the 13th Annual Cogn. Sci. Soc. Conf.*, Hillside, NJ: Lawrence Erlbaum, 173-178, 1991.

[9] B. Fritzke, "Growing cell structures - A self-organizing network for unsupervised and supervised learning", *Neural Networks*, 7(9): 14441-1460, 1994.

[10] M.J.A. Howe, *Introduction to human memory*, Harper & Row, 81-83, 1970.

[11] P. Kanerva, *Sparse Distributed Memory*, Cambridge, MA: MIT Press, 1988.

[12] H. Kay , "Learning and retaining verbal material", *Brit. Journ. of Psych.*, 44: 81-100, 1955.

[13] T. Kohonen, "Self-organized formation of topologically correct feature maps", *Biological Cybernetics*, 43: 59-69, 1982.

[14] J. K. Kruschke, "ALCOVE: an exemplar-based connectionist model of category learning", *Psychological Review*, 99(1): 22-44, 1992.

[15] D. Marr, "Simple memory: a theory for archicortex", *The Philos. Trans. of the Royal Soc. of London*, 262: Series B, 23-81, 1971

[16] J. Piaget, *La psychologie de l'enfant*, Armand Colin, 1947.

[17] T. Poggio, F. Girosi, "A theory of networks for approximation and learning", *MIT Technical report* No 1140, 1989.

[18] R. Ratcliff, "Connectionist Models of Recognition Memory: Constraints Imposed by Learning and Forgetting Functions, *Psychological Review*, 97(2): 285-308, 1990.

Neural Modelling of Cognitive Disinhibition and Neurotransmitter Dysfunction in OCD

Jacques Ludik[1] and Dan J. Stein[2]

[1] Department of Computer Science,
[2] Department of Psychiatry,
University of Stellenbosch, Stellenbosch 7600, South Africa

Abstract. In this paper an Elman recurrent neural network model of *obsessive-compulsive disorder* (OCD) is developed to provide a simulation of the relationship between the cognitive disinhibition and serotonin/dopamine dysfunction that characterize this disorder. Cognitive disinhibition in OCD is apparent when OCD patients are compared with other anxiety disorder patients on a Temporal Stroop test, with OCD patients showing reduced negative priming. Alterations of the color gain parameter, the context gain parameter, and maximum cycle number were made in order to simulate changes in monoamine neutransmitter function. The recurrent network model was able to simulate reduced cognitive inhibition as well as serotonergic and dopaminergic dysfunction in OCD.

1 Introduction

This paper aims to develop a neural network model of obsessive-compulsive disorder (OCD) in order to provide an integrated account of the cognitive and neurotransmitter dysfunction that characterize this disorder. The rationale for this project is found in recent advances in two disparate areas of investigation - the psychobiology of OCD and neural network modelling.

From a neurochemical perspective, there is strong evidence that OCD is mediated by the serotonergic system, as well as by additional neurotransmitters such as the dopamine system (see for example, [10]). From a neuroanatomical perspective, there is strong evidence that OCD is mediated by cortical-basal ganglia-thalamic-cortical neuronal circuits (see [9]). From a neuropsychological perspective, a number of dysfunctions have been documented in OCD (see [7]). Enright and Beech in [2] have characterized neuropsychological dysfunction in OCD in terms of decreased cognitive inhibition. Despite these advances, the question of how to integrate biological and psychological data in OCD has not been adequately answered. A comprehensive psychobiological integration is unlikely to emerge from any one particular empirical study - rather this would seem to require a theoretical synthesis of existing work. In this paper, we suggest that neural net modelling provides a means of pursuing such a synthesis in a methodological way.

2   Experimental Data

Several studies have documented neuropsychological dysfunction in OCD, with impaired set-shifting being a consistent finding (e.g., [6]). However, the majority of these investigations have not differentiated between neuropsychological dysfunction in OCD and other psychiatric disorders. In an interesting sub-set of this work, however, Enright and Beech in [2] found that OCD differs markedly from other anxiety disorders on tests of negative priming, particularly when the complexity of these tests is increased. This experimental paradigm was therefore chosen to provide simulation data for the neural network developed here.

Enright and Beech also presented a more complex negative priming task (which might be dubbed the Temporal Stroop). In this task, the stimuli were 10 words, two drawn from each of 5 semantic categories. Subjects were instructed to ignore the green word in each pair and to categorize the red word into one of the five semantic categories (animal, furniture, etc.). Stimuli were presented under one of five conditions:

- attended repetition (AR) - the attended priming stimulus was identical to the subsequent probe (e.g. red DOG, red DOG)
- attended semantic (AS) - the attended priming stimulus was semantically related to the subsequent probe (e.g. red DOG, red CAT)
- control (CO) - the attended and ignored priming stimuli were unrelated to the subsequent probe (e.g. red DOG/green CHAIR, red FOOT)
- ignored semantic (IS) - the ignored distractor prime was semantically related to the subsequent probe (e.g. red CHAIR/green DOG, red CAT).
- ignored repetition (IR) - the ignored distractor prime was identical to the subsequent probe (e.g. red CHAIR/green DOG, red DOG)

There were 40 randomized trials in each of the five conditions. The amount of negative priming was calculated by substracting the mean reaction time of the CO condition from the mean reaction time of the IR condition. The amount of semantic negative priming was calculated by subtracting the mean reaction time of the CO condition from the mean RT of the IS condition.

In the Temporal Stroop, Enright and Beech in [2] found that anxiety disorder patients demonstrated negative priming in both the repetition priming and the semantic negative priming condition (longer reaction times to previously ignored stimuli). In contrast, OCD patients failed to show any priming effects in the repetition priming condition and demonstrated reduced negative priming in the semantic priming condition (i.e., facilitation in the semantic priming condition or shorter reaction times to previously ignored stimuli). The heightened difference between OCD patients and anxiety disorder controls on the semantic negative priming task would seem to reflect the increased complexity of this task.

3   A Recurrent Neural Network Model

A recurrent neural network was developed in order to simulate normal performance on the semantic negative priming task. A recurrent model was chosen as these are particularly useful for the storage and recognition of temporal processes. The Elman network (see [1]) was chosen as this incorporates feedback from hidden units, effectively allowing access to an internal state – see Figure 1.

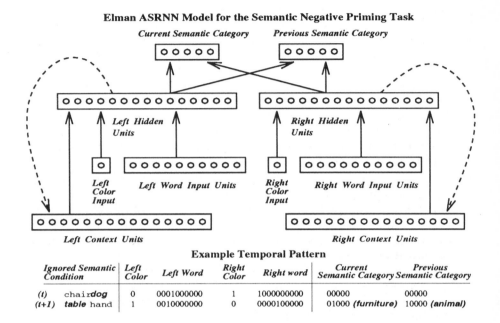

Elman ASRNN Model for the Semantic Negative Priming Task

Example Temporal Pattern

Ignored Semantic Condition		Left Color	Left Word	Right Color	Right word	Current Semantic Category	Previous Semantic Category
(t)	chair**dog**	0	0001000000	1	1000000000	00000	00000
(t+1)	**table** hand	1	0010000000	0	0000100000	01000 *(furniture)*	10000 *(animal)*

Fig. 1. Elman Network Model for the Temporal Stroop Task

3.1 Training Method

A temporal training pattern consists of two input-target patterns, where the first input-target pattern represents the first pair of red/green words, and the second input-target pattern represents the second pair of red/green words. The training data consisted of 400 input patterns and thus 200 temporal patterns, with 40 randomized trials in each of the five conditions.

The Elman network was trained with a back-propagation learning algorithm with a learning rate of 0.1. No noise was added to the net input during training. The network was trained until the RMS error was less than 0.005 and the percentage of correct responses on the training set was 100%. A response was classified as correct if the current and previous semantic category output units with maximum activation values matched their respective desired targets. This was achieved after 6000 epochs.

3.2 Simulating the Temporal Stroop Task

A single simulation consisted of a single run through an independent test set of 200 temporal patterns. The following were the processing steps for a particular temporal pattern in a single simulation [8] :

(a) After presenting the first pair of red/green words, the hidden and output units were activated in a normal single processing step. (b) For the second pair of red/green words, the context layers contained a copy of the previous hidden unit activation values and a gradual buildup of activation over time was allowed for the hidden and output units.

During a simulation, a noise term (which was sampled independently for each unit and on each cycle from a normal Gaussian distribution) was added to the computation of the net input of each hidden and output unit of the network in order to simulate the variability of human subjects in the task.

For very high response thresholds, maximum values in the asymptotic output pattern are sometimes not able to become greater or equal to the response threshold. To deal with such a situation, a maximum cycle number was introduced to determine the maximum number of cycles after which the gradual buildup of activation over time should be finished in order to get a response.

Reaction time was implemented in the Elman network as a function of the number of cycles it takes to accumulate the specified amount of activation according to the reponse threshold.

3.3 Lesioning the Network

Three kinds of lesions were made to the network, each representing changes to a single monoamine neurotransmitter system. First, gain of the context module was decreased in order to simulate dopaminergic dysfunction (i.e., the effective activation values of the context units was reduced by a multiplier). There is good evidence that dopamine has a modulatory effect on the responsivity of cells in prefrontal cortex, where it mediates the representation of goals (see [3]). For example, dopaminergic lesions may result in increased stereotypy in pre-clinical paradigms, or in impaired execution of goals with consequent loss of executive functioning in patients. In the neural network described here, the context module played a specific role in representing information about differences between previous and current goals.

Second, gain of the color module was decreased in order to simulate serotonergic dysfunction. There is good evidence that the serotonin modulatory effects in frontal cortex plays a role in harm assessment. Thus in pre-clinical and clinical paradigms serotonergic dysfunction characteristically results in impaired impulse control, leading to increased other-directed and self-directed aggression (see [5]). In the neural network described here, the color module played a specific role in representing information about the color of the current stimulus, effectively ensuring the representation of information about the irrelevance of the current stimulus (green being a color that must be avoided (ignored), and red being a color that must be focused on).

Third, the maximum time cycle during network training was increased in order to simulate noradrenergic dysfunction. Noradrenergic neurons have a significant role in the modulation of alertness and anxiety. In the neural network developed here, the maximum time cycle played a specific role in determining the extent of alertness under which simulation takes place. That is, under conditions of increased alertness the simulation is effectively undertaken with greater concentration and focus, with a longer time allowed for processing of each set of input.

4 Results

Results reported are the mean of 50 single simulations for each combination of parameters. Results were tabulated for normal simulation, reduction of context gain, reduction of color gain, and combined reduction in context and color gain (see Table 1). Reduction of either context or color gain results in slower reaction times and more error responses than in the normal case.

Lesion Type	Normal	Color	Context	Combined	OCD
Color Gain	1.0	0.6	1.0	0.6	
Context Gain	1.0	1.0	0.8	0.8	
Reaction Times (msec)					
Attended Repetition (AR)	424.7 (6.4)	436.4 (8.0)	578.8 (7.9)	561.1 (7.1)	447.4
Attended Semantic (AS)	455.6 (7.4)	487.8 (8.7)	579.6 (6.3)	597.6 (4.1)	558.1
Control (CO)	463.6 (9.1)	475.0 (7.9)	605.2 (5.5)	590.3 (8.4)	602.0
Ignored Semantic (IS)	469.6 (10.0)	472.0 (8.0)	597.5 (7.4)	569.4 (7.3)	565.9
Ignored Repetition (IR)	456.3 (9.9)	480.4 (7.4)	607.4 (4.5)	590.6 (6.3)	602.1
Derived Priming Data					
RNP = IR - CO	-7.3	5.4	2.2	0.4	0.2
SNP = IS - CO	6.0	-3.1	-7.7	-20.9	-36.1
% Correct Performance	100.0	99.8	98.9	99.1	97.1

Table 1. Mean Results of 50 Simulations for Normal Performance, Only Context Gain Reduction, Only Color Gain Reduction, and Combined Context and Color Gain Reduction compared to OCD experimental results obtained by using a 2-way mixed ANOVA to analyze the data of 36 OCD subjects (see [3]). (SNP - Semantic Negative Priming - SNP; RNP - Repetition Negative Priming; Standard deviation in brackets; time average rate = 0.6; response threshold = 0.8; and random variance = 0.25)

Increasing the maximum cycle number under ordinary conditions results in further increase in semantic negative priming, while increasing the maximum cycle number in OCD during context or color gain reduction results in further decrease in semantic negative priming. See [8] for a more complete justification of the equivalence of parts of the model to cognitive disinhibition and monoamine neurotransmitter dysfunction.

5 Discussion

A neural network simulation of a semantic negative priming task, a test on which patients with OCD have reduced negative priming, but in which anxiety disorder patients do not, allowed an investigation of the effects of lesioning of each of the monoamine neurotransmitter systems. The incorporation of a temporal component in this task demanded a particularly complex neural network, which arguably led to a more comprehensive connectionist simulation of neurotransmitter dysfunction than has previously been offered. Lesions of the Elman recurrent neural network that corresponded to dopaminergic and serotonergic dysfunction resulted in reduced semantic negative priming, while lesions of the network that corresponded to noradrenergic dysfunction resulted in magnification of effects that had been present in the absence of the lesion.

References

1. Elman JL: Finding structure in time. Cognitive Science (1990) **14**:179–211
2. Enright SJ, Beech AR : Reduced cognitive inhibition in obsessive-compulsive disorder. Br J Clin Psychology (1993) **32**:67–74.
3. Goldman-Rakic PS: Circuitry of primate prefrontal cortex and regulation of behavior by representational memory. Handbook of Physiology - The Nervous System (1989) **5**:373-417.
4. Pauls DL, Leckan J: The inheritance of Gilles de la Tourette's syndrome and associated behaviors: evidence for autosomal dominant transmission. N Engl J Med (1986) **315**:993–997.
5. Soubrie P: Reconciling the role of central serotonin neurones in human and animal behavior. Behav Brain Sci (1986) **9**:319-364.
6. Stein DJ, Hollander E: A neural network model of obsessive- compulsive disorder. J Mind Behavior (1994) **15**:25–40.
7. Stein DJ, Hollander E, Cohen L: Neuropsychiatry of Obsessive-Compulsive Disorder, in Hollander E, Zohar J, Marazzati D, Olivier B (eds.), Current Insights in Obsessive-Compulsive Disorder. Wiley: Chichester, (1994).
8. Stein DJ, Ludik J: A Neural Network of OCD: Modelling Cognitive Disinhibition and Neurotransmitter Dysfunction, Internal Report, Dept of Psychiatry/Computer Science, University of Stellenbosch, 1996.
9. Wise S, Rapoport JL: Obsesssive-compulsive disorder: is it basal ganglia dysfunction, in Rapoport JL (ed.), Obsessive-Compulsive Disorder in Children and Adolescents. Washington, DC: American Psychiatric Press, (1989).
10. Zohar J, Insel TR: Obsessive-compulsive disorder: psychobiological approaches to diagnosis, treatment, and pathophysiology. Biol Psychiatry (1987) **22**:667–687.

Confluent Preorder Parser As Finite State Automata

HO, Kei Shiu Edward   CHAN, Lai Wan

Department of Computer Science and Engineering,
The Chinese University of Hong Kong, Shatin, N.T., Hong Kong
email: ho052@cs.cuhk.edu.hk, lwchan@cs.cuhk.edu.hk

Abstract. We present the Confluent Preorder Parser (CPP) in which syntactic parsing is achieved via a holistic transformation from the sentence representation to the desired parse tree representation. Simulation results show that CPP has achieved excellent generalization performance and is capable of handling erroneous sentences and resolving syntactic ambiguities. An analysis is presented which elucidates the operations of CPP as governed by a finite state automaton. The parsing is interpreted as a series of decision makings during the process. Based on this formalism, syntactic parsing and generalization by CPP can be explained in terms of state transitions.

1 Introduction

In language processing, syntactic parsing addresses the determination of the hierarchical relationship between the terminals (or more specially, words) in a sequential sentence. For example, by classifying the words in the sentence "the boy takes the apple on the table", we form a sequence ⟨ D N V D N P D N ⟩ where the terminal D stands for determiner, N for noun, V for verb and P for preposition. Upon successful parsing of this sequence according to some grammatical rules, the parse tree in Figure 1 is produced, where np stands for noun phrase, vp for verb phrase, pp for prepositional phrase and s for sentence.

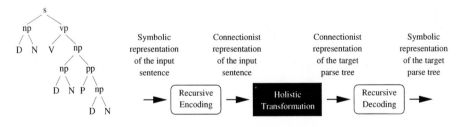

Fig. 1. Parse Tree of the Sentence ⟨ D N V D N P D N ⟩

Fig. 2. Holistic parsing paradigm

Conventional AI approaches represent sentences and parse trees as symbolical hierarchical data structure and the parsing problem is treated as a rational inference process which involves multiple intermediate steps and decision making (including backtracking) as governed by a well-defined algorithm. Compared with the symbolic AI methods, connectionist approaches offer a robust alternative. They are good at inductively learning from examples and can generalize naturally to unseen sentences. To exploit these advantages, efforts have been paid to integrate the connectionist techniques into the symbolic processing framework, giving rise to "hybrid" parsers such as the "Massively Parallel Parsing" system [11], the PARSEC model [8], the CDP approach [9], the SPEC architecture [10], the SCAN architecture [14] and the recurrent network models [4, 5, 13].

2 Confluent Preorder Parser (CPP)

In this paper, syntactic parsing is accomplished via a **holistic transformation** [1, 2] in which the connectionist representation encoding the input sentence is directly mapped to the connectionist representation encoding the target parse tree (Figure 2). Compared with other connectionist parsers, the **Confluent Preorder Parser (CPP)**, using this paradigm, succeeds to produce good generalization [6]. It uses two SRAAM networks for the encoding and decoding. The holistic transformation is made to be an identity mapping, so that the two SRAAMs form one dual ported SRAAM [3], with one single hidden layer shared between two SRAAM networks [12] – the SRAAM Sentence Encoder and the SRAAM Preorder Traversal Encoder (Figure 3). To ensure good generalization, two techniques were used; linearization and confluent inference. Instead of encoding the hierarchical representation of the parse tree, we used a linearized representation – its preorder traversal. For example, the tree in Figure 1 gives ⟨ s np D N vp V np np D N pp P np D N ⟩ upon preorder traversal. Confluent Inference [3] is used so that error from the output layer of either network back propagates to the input layer of both encoder. In this way, identical representation is developed for the sentence and the preorder traversal.

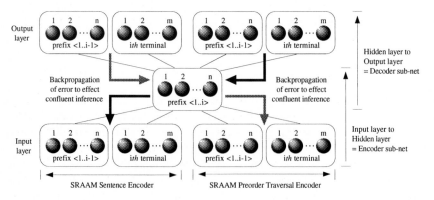

Fig. 3. Confluent Preorder Parser (CPP)

2.1 Performance of CPP

In this paper, the following context-free grammar from [12] was used :

Sentence	Noun Phrase	Verb Phrase	Prepositional Phrase	Adjectival Phrase
s → np vp s → np V	np → D ap np → D N np → np pp	vp → V np vp → V pp	pp → P np	ap → A ap ap → A N

Each syntactically well-formed sentence (and its parse tree) is generated using s as the root. We apply the production rules to 5 levels, a total of 112 sentences are obtained. Among them, 80 sentences are randomly selected for training while the rest are reserved for testing. The length of the longest training sentence and that of the longest testing sentence are both equal to 13. We implemented CPP on the DECmpp computer. It required less than 30 seconds to finish one epoch and totally, it took about 2,000 cycles for convergence. Two versions of CPP have been examined, CPP1 and CPP2. CPP1 is trained with complete sentences only, whereas CPP2 is trained to parse phrases to produce the corresponding sub-parse-trees, in addition to complete sentences.

Details of the mechanism and the performances of CPP1 and CPP2 were reported in [6]. We showed that they have good generalization ability, the ability to recover from erroneous sentences and detecting syntactic ambiguities. Table 1 summarized the result.

Table 1. The performance of CPP1 and CPP2.

	CPP1	CPP2
Generalization	Training : 100 %	Training : 100 %
	Testing : 81.25 %	Testing : 93.75 %
Error recovery ability	wrong terminal : 91.21 %	wrong terminal : 94.72 %
	missing terminal : 69.64 %	missing terminal : 70.93 %
	extra terminal : 46.62 %	extra terminal : 50.80 %
	swapped terminal : 63.65 %	swapped terminal : 66.88 %
Syntactic disambiguation	1 ambiguity : 98.24 %	1 ambiguity : 99.65 %
	3 ambiguities : 91.89 %	3 ambiguities : 97.31 %

3 A Finite State Automata Model of Syntactic Parsing

In this section, we perceive CPP as a finite state automaton and analyze its characteristics. To examine the internal mechanism of the parser, we tested CPPs on parsing each of the 80 training sentences and decoded the intermediate hidden layer activation after reading each terminal (which represents a sentence prefix). These activation patterns are perceived as the states of the parser. The identities of the states are defined by the decoding's results (which are sequences of terminals that may or may not represent meaningful preorder traversals of parse trees). When two states give the same result upon decoding, they are treated as the same state. A state is said to be a final state if it is decoded to give a syntactically well-formed parse tree or sub-parse-tree. The number of states found in CPP1 and CPP2 are listed in the middle column in Table 2.

Table 2. The number of states that were identified when 80 training sentences and 32 testing sentences were parsed by each of CPP1 and CPP2

	States identified when the training sentences are parsed		Additional states identified when testing sentences are parsed	
	CPP1	CPP2	CPP1	CPP2
Starting State (#)	1	1	+0	+0
Intermediate States	89	77	+22	+20
Final States	92	106	+16	+19
Total No. of States	182	184	+38	+39

These states, together with the possible transitions between them, effectively define a finite state automaton [7] which encodes the grammatical knowledge/regularity necessary for the parsing function. Thus intuitively, the holistic parsing process of a sentence can be analyzed as a rational inference procedure governed by a finite state automaton, with the intermediate decisions reflected by the outputs of the states "visited" by the parser during the processing of the sentence. For example, the state transition sequence resulted in parsing the training sentence ⟨ D A N P D N V D A N ⟩ by CPP1 has been depicted in Figure 4, with the identities (or interpretations) of the states involved summarized in Table 3. Darkened nodes are used to represent the final states. For convenience of illustration, each state is labeled by a unique number[1].

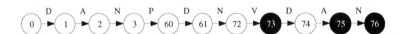

Fig. 4. Syntactic parsing perceived as state transitions. The final states 76 outputs the parse tree (((D (A N)) (P (D N))) (V (D (A N)))).

Table 3. The identities of the states in Figure 4, where "..." represents a sequence or sub-sequence that cannot be interpreted as a well-formed tree or sub-tree

State	Decoded Sequence (Parse Tree)	State	Decoded Sequence (Parse Tree)
0	the starting state		
1	...	72	((D (A N)) (P (D N))) ...
2	...	73	(((D (A N)) (P (D N))) V)
3	(D (A N)) ...	74	...
60	((D (A N)) V) ...	75	(((D (A N)) (P (D N))) (V (D N)))
61	...	76	(((D (A N)) (P (D N))) (V (D (A N))))

---

[1] These numbers indicate the relative order in which different states are extracted (for example, the first state found for CPP1 will be assigned the label "1").

3.1 Characteristics of the Finite State Automata Extracted

Several characteristics of this finite state automaton were observed.

- If the parser simply remembered the training cases as independent sentences and parse trees by brute force, the total number of states should have been 234. In our results, we obtained 182 and 184 for CPP1 and CPP2 respectively, as reported in Table 2. The state transition sequences corresponding to the parsing of two similar sentences have overlapped in certain degree, thus some intermediate states have been shared. This implies that a certain degree of abstraction of grammatical knowledge has been achieved.

- Given a 100% success rate in training, each of the 80 training sentences should correspond to a unique final state, giving a total of 80 final states. However, 12 extra final states have been identified in CPP1, each corresponds to a syntactically well-formed parse tree or constituent phrase of some kind. And for some intermediate states (e.g. states 3, 60 and 72 in Table 3), their interpretations show traces of partial sub-parse-trees. This can be attributed to the use of confluent inference. Although CPP1 has only been exposed to complete sentences, the parser has, to some extent, induced their internal structure. Intuitively, grammatical knowledge has been acquired.

- The construction of the finite state automaton is deterministic. The definitions of the states are uniquely defined by the set of weights of the trained CPP which are affected by how the network is trained. Thus for CPP, the number of states which determines the performance of CPP will be affected by the training environment. We repeated our experiments with different initial parameters and the number of states vary only within a range of 10%.

3.2 Explaining Generalization

Interestingly, the generalization capability of CPP can also be explained by the finite state automata model. In addition to the final states corresponding to the training sentences, the training process also organizes the representational space (which is the hidden layer of CPP) in such a way that another set of states can be decoded to output the correct parse trees of the testing sentences. In our case, when 32 testing sentences (26 of which can be correctly generalized) were parsed by CPP1, 16 additional final states were identified. The other 10 final states were among the 12 extra final states that were obtained during the parsing of the 80 training sentences. For example, when the testing sentence ⟨ D A N P D N V D N ⟩ was parsed by CPP1 (see Figure 5), the final state 75 had already been identified during the parsing of the training sentence ⟨ D A N P D N V D A N ⟩ (see Figure 4). For CPP2, 19 additional final states were identified by parsing the 32 testing sentences (see Table 2).

4   Conclusion and Discussion

In this paper, we have demonstrated that CPP can successfully tackle syntactic parsing. As revealed by the simulation, CPP has achieved excellent generalization performance and robust processing. We analyzed CPP (in general, this technique is applicable to any recurrent models in symbolic computation) as a finite state automaton. We examined the number of states and their transitions,

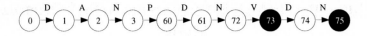

Fig. 5. Parsing the testing sentence ⟨ D A N P D N V D N ⟩ by CPP1. The final state 75 had already been identified when the training sentence ⟨ D A N P D N V D A N ⟩ was parsed.

and revealed how knowledge is captured and processed in CPP. In addition, holistic parsing as implemented by CPP can also be perceived as if a rational inference process is involved.

References

1. D. S. Blank, L.A. Meeden, and J. B. Marshall. Exploring the symbolic/subsymbolic continuum: A case study of RAAM. In J. Dinsmore, editor, *The Symbolic and Connectionist Paradigms: Closing the Gap*, pages 113–148. Lawrence Erlbaum Associates, Hillsdale, NJ, 1992.

2. D. J. Chalmers. Syntactic transformations of distributed representations. In N. Sharkey, editor, *Connectionist Natural Language Processing*, pages 46–55. Intellect Books, 1992.

3. L. Chrisman. Learning recursive distributed representations for holistic computation. *Connection Science*, 3:345–366, 1991.

4. J. L. Elman. Finding structure in time. *Cognitive Science*, 14:179–211, 1990.

5. C. Giles, C. Miller, D. Chen, H. Chen, G. Sun, and Y. Lee. Learning and extracting finite state automata with second-order recurrent neural networks. *Neural Computation*, 4(3):393–405, 1992.

6. K. S. Ho and L. W. Chan. Syntactic parsing using RAAM. In *Proc. World Congress on Neural Networks*, volume 1, pages 485–488, 1995.

7. J. E. Hopcroft and J. D. Ullman. *Introduction to Automata Theory, Languages, and Computation*. Addison-Wesley, Reading:Mass, 1979.

8. A. N. Jain and A. H. Waibel. Incremental parsing by modular recurrent connectionist networks. In D. S. Touretzky, editor, *Advances in Neural Information Processing Systems 2*, pages 364–371, San Mateo, CA, 1990. Morgan Kaufmann.

9. S. C. Kwasny and K. A. Faisal. Symbolic parsing via subsymbolic rules. In J. Dinsmore, editor, *The Symbolic and Connectionist Paradigms: Closing the Gap*, pages 209–236. Lawrence Erlbaum, Hillsdale, NJ, 1992.

10. R. Miikkulainen. Subsymbolic case-role analysis of sentences with embedded clauses. Technical Report AI93-202, Department of Computer Sciences, the University of Texas at Austin, 1993.

11. J. Pollack and D. Waltz. Massively parallel parsing: A strongly interactive model of natural language interpretation. *Cognitive Science*, 9:51–74, 1985.

12. J. B. Pollack. Recursive distributed representations. *Artificial Intelligence*, 46:77–105, 1990.

13. R. Watrous and G. Kuhn. Induction of finite state languages using second-order recurrent networks. In J. Moody, S. Hanson, and R. Lippmann, editors, *Advances in Neural Information Processing Systems 4*, pages 309–316, San Mateo, CA, 1992. Morgan Kaufmann.

14. S. Wermter. *Hybrid Connectionist Natural Language Processing*. Chapman & Hall, London, 1995.

Modeling Psychological Stereotypes in Self-Organizing Maps

Thomas Crämer, Josef Göppert, Wolfgang Rosenstiel

University of Tübingen, Sand 13, D-72076 Tübingen, Germany

Abstract. Cognitive psychology defines stereotypes as categories associating information according to perceived similarity rather than according to logical criteria [1]. Self-organizing maps (SOMs) [2] are used as models of human perception in an experiment on stereotype formation. Two groups of SOMs were trained with coded newspaper information, one with information from a conservative newspaper, the other with information from a progressive one. The SOMs performed well in mapping the major differences between the newspapers (r=0.99). Further, each group generated distinct associations of dissimilar information which could be interpreted as 'artificial stereotypes'. The SOM is suggested as a research tool for cognitive research on stereotype change.

1 Introduction

The motivation for the present study is an incident which was reported in the Los Angeles Times on February 2, 1987 concerning a friend of mine. A scientist was arrested by the police on the way to his lab because the officers doubted the information in his passport that identified him as a doctor of neurobiology. This seems to be an absurd process until the information is added that the scientist is black and wears long dreadlocks. With this additional piece of information the situation becomes intuitively understandable: The officer was deceived by his own prejudice. 'Prejudice' and 'Intuition' are properties of the human mind that are difficult to capture formally. 'Intuition' presents the greatest strength of the human mind, 'prejudice' presents its greatest weakness. The two concepts are related since both allow quick responses without requiring elaborate logical information processing. Cognitive psychology introduces the term 'stereotype' to refer to this processing strategy. According to Fiske and Taylor (1991) it is defined as "a cognitive structure that represents knowledge about a concept or type of stimulus, including its attributes and the relations among those attributes" [1 p. 98]. An important characteristic of stereotypical information processing is the fact that associative links between stimuli are established rather arbitrarily based on perceived similarity. No clear cut objective criteria exist nor is any logical processing involved in associating information. Stereotypes act as 'cognitive shortcuts' when quick decisions are required. However, they present prejudices when false associations lead to serious errors, such as the scientist's arrest.

In the present study the problem of stereotyping will be addressed utilizing self-organizing maps (SOM) [2] with standard training by applying a special evaluation procedure. Similar results could be obtained with modified SOM training as presented in [3, 4]. The SOM is chosen for its unsupervised learning which is biologically more plausible than supervised learning in backpropagation type nets, LVQ nets or hybrid forms. The SOM is chosen over biologically more complex ART-architectures for its simplicity. In future research ART-architectures could be used as well.

2 Experiment

The SOM is applied as a rough and extremely simplified model of human perception in the process of reading the newspaper. The SOM organizes input vectors according to similarity rather than logical criteria. This corresponds to the principle of stereotypical information processing as defined in cognitive psychology. In order to investigate SOM behavior under 'realistic' conditions, actual newspaper information was coded. Two different newspapers were used in order to isolate the effects of different information formats. One newspaper, BILD (Axel Springer Verlag AG, Berlin) is known to be politically conservative and is best described as a yellow press publication. The other newspaper, TAZ ('taz, die tageszeitung'. Verlagsgenossenschaft e.G. Berlin) caters to a more progressive intellectual audience. The newspapers were coded between June 26, and July 2, 1995. Each individual or group of individuals mentioned by the newspapers was coded according to gender, nationality, occupation, etc.. Seven variables were coded for each of them (see appendix). The final data set comprised a total of 888 individuals or groups. Preliminary analysis revealed the clearest structuring of the SOM's output space for the variables 'gender', 'nationality', and 'occupation' (paid or unpaid position). The differences between the two newspapers were significant with respect to all three variables.

Due to the initialization with random numbers, identical SOMs yield individual differences even if trained on the same data set. To control for individual differences, two experimental groups of SOMs were formed each comprising 20 members. All SOMs were composed of seven input neurons (one for each variable) and an output layer of 10 x 10 neurons. One group of SOMs was trained exclusively with BILD-data, the other exclusively with TAZ-data.

3 Results

3.1 SOM Data Mapping

The output layer of each trained SOM exhibits clusters of neurons with similar activation. The locations and shapes of clusters vary individually from one SOM to another. The size of a cluster is measured by counting output neurons with the same activation. Aggregated cluster sizes are listed in table 1.

Cluster	BILD Mean	St.dev	TAZ Mean.	St.dev	t (38 df)
Unpaid Non-Ger. Male[1]	12.15	2.286	15.1	1.609	-4.599**
Paid Non-German Male	21.4	2.458	24.8	2.135	-4.552**
Unpaid German Male	13.95	2.479	13.8	2.713	0.176
Paid German Male[2]	34.6	2.728	29.25	2.736	6.036**
Unpaid Non-German Female	3.65	1.108	4.3	1.054	-1.853
Paid Non-German Female	3.05	0.921	1.65	0.792	5.025**
Unpaid German Female	6.95	1.396	5.65	1.769	2.515*
Paid German Female[3]	4.25	1.299	5.45	1.802	-2.355*
	100		100		

1) $\chi^2 = 40.367**$; 2) $z = -1.96*$; 3) $\chi^2 = 38.489*$; χ^2-tests with 19 df
df= degrees of freedom; *) p<0.05; **) p<0.01

Table 1. Cluster sizes in each group of trained SOMs

Due to differences in newspaper information formats significant differences in cluster sizes occur between the two groups of nets. Cluster size corresponds to the percentage of the respective group in the newscoverage. The cluster representing 'German men in paid positions' is significantly larger in SOMs of the BILD-treatment group. This corresponds to the fact that this group plays a predominant role in the conservative newspaper's news coverage. Non-German men make up a larger portion of the progressive news coverage. Hence the respective clusters are larger in SOMs of the TAZ-treatment group. All these findings are quite intuitive. The performance of the SOM in mapping the newspaper information can be assessed from the correlation coefficients for the neural and the statistical data analysis (see fig. 1). It yields the linear relationship between the mean cluster sizes in table 1 and the means of the respective category in the raw data.

$$r_{Bild}=0.9862$$

$$r_{TAZ}=0.9899$$

Fig. 1. Correlation coefficients for neural and statistical data analysis

For both newspapers the correlation coefficient amounts to approximately r=0.99 representing a 99 percent correspondence between neural and statistical data analysis. This result shows that SOMs are accurate in mapping the raw data. All major differences between the two newspapers were represented in the activation patterns of the SOMs.

3.2 Artificial Stereotypes

The cognitive notion of the 'stereotype' denotes the association of information based on perceived similarity rather than applying logical criteria. Stereotypical information processing is highly sensitive to information context. In our experiment we present the three variables of interest in a larger context of seven values. The SOM is forced to project the higher dimensional information onto the two dimensional structure of the map.

How can 'perceived similarity' be measured in SOMs? In SOMs 'similarity' is represented by spacial proximity of clusters on the output layer. Neighboring clusters are 'perceived' to be more similar than are remote clusters. Thus, adjacent clusters are recorded as 'perceptionally similar'. The length of a border between two clusters indicates *how* similar they are perceived to be and hence measures the strength of their associative connection. Table 2 presents the length of boundaries between clusters which share only one common property and differ in two respects. The figures represent the number of bordering neurons between two given clusters.

If stereotype formation was not a function of neural organization we would expect significant differences only in terms of the sizes of clusters, not in their spatial organization. No matter how many unpaid non-German males were covered in each newspaper, their relative position towards unpaid German women should be the same. However significant differences in spatial organization do occur. SOMs trained on BILD-data, for example, locate the two above mentioned clusters next to each other, while SOMs trained on TAZ-information locate the clusters at remote ends of the output space. All differences between the experimental groups of SOMs were tested for significance. Significant χ^2 values are listed. The first column in table 2 (see following page) indicates the shared property of neighboring clusters. The last column states which group of SOMs generated the given association significantly more frequently.

At first glance the results seem rather counter intuitive. SOMs trained with progressive newspaper information perceive similarity between dissimilar groups primarily based on gender or nationality. SOMs trained on conservative BILD-data exclusively use economic criteria (paid or unpaid position) to establish similarity. This is surprising since the tabloid BILD is known for its conservative ideology in which nationality plays an important role while women play an old fashioned one. The progressive newspaper TAZ on the other hand tries to keep a balance by emphasizing women and foreigners in its news coverage. Why do 'progressive' SOMs organize their world along the lines of gender and nationality while 'conservative' SOMs organize their world according to economic criteria? A possible explanation for this surprising finding could be the fact, that existing disparities can only be highlighted by explicitly distinguishing the antagonistic groups. The desire to rectify differences in the real world can therefore lead to increased perception of these differences.

Common Property	Dissimilar Properties	BILD Mean (St.dev.)	TAZ Mean (St.dev.)	χ^2	SOM
Gender					
male	unpaid n-Ger.				
	paid German	0.5 (0.81)	0.35 (0.79)		
	paid non-Ger.				
	unpaid Ger.	0.2 (0.68)	0.3 (0.64)		
female	unpaid n-Ger.				
	paid German	0.25 (0.54)	0.15 (0.48)		
	paid non-Ger.				
	unpaid Ger.	0.15 (0.36)	0.4 (0.58)	53.33**	TAZ
Nationality					
non-German	unpaid male				
	paid female	0.45 (0.74)	0.4 (0.58)		
	paid male				
	unpaid female	0.15 (0.36)	0.55 (0.5)	38.82**	TAZ
German	unpaid male				
	paid female	0.2 (0.4)	0.8 (1.29)	207.5**	TAZ
	paid male				
	unpaid female	0.9 (0.99)	0.65 (0.91)		
Occupation					
unpaid	non-Ger. male				
	German fem.	0.75 (1.04)	0.25 (0.54)	75.65**	BILD
	German male				
	non-Ger. fem.	0.55 (0.67)	0.2 (0.4)	55.94**	BILD
paid	non-Ger. male				
	German fem.	0.2 (0.4)	1.05 (1.32)	218.44**	TAZ
	German male				
	non-Ger. fem.	0.55 (0.59)	0.2 (0.4)	43.44**	BILD
χ^2-tests with 19 degrees of freedom; *) $p<0.05$; **) $p<0.01$					

Table 2. Clusters perceived as similar based on only one common property

4 Conclusion

Two conclusions can be drawn from the results of the present study:

1) Self-organizing maps perform well in mapping differences among newspapers.

2) The information formats of different newspapers lead to distinct associations.

Following the cognitive definition these can be understood as 'artificial stereotypes'. If the results of the present experiment are verified in a psychological experiment with human newspaper readers, the SOM can be introduced into cognitive psychology as a powerful new tool for research into stereotyping. With its assistance, effects of different information formats can be tested 'in vitro'. Eventually strategies may be developed to alter stereotypes and prevent prejudices from forming.

5 Self-Organizing Properties of Human Cognition

In the so called "nature versus nurture controversy" [5 p. 372] two approaches compete to account for human behavior. The biological approach views the brain as an organ shaped by evolution that determines behavior and preferences genetically. Stereotypes are viewed as an inborn defense mechanism against fictuous enemies. The competing environmental approach imagines the brain to be a biological von Neumann computer, the internal states of which are determined exclusively by its environmental input. Stereotypes are viewed as faulty programs or modules. Both approaches are deterministic and they both search for the causes of mental processes outside the mind. Neurocomputing offers a new perspective. Neurons are an example of "the interaction of nature and nurture at the molecular-genetic level" [5 p. 372]. Based on a few simple physical properties, neurons once placed next to each other start stimulating each other and categories evolve in self-organizing fashion. Moralizing appeals against thinking in categories turn out to be appeals against the very essence of thinking itself. Neurally based thinking is not perceivable other than in categories. Artificial neural nets used as "virtual guinea pigs" in psychological laboratory experiments may help us gain an understanding of the self-organizing processes underlying stereotype formation and stereotype change.

Appendix

Seven variables were used as training input, coded as follows:
Gender: 1 = female; 0 = neutral or unknown; -1 = male
Nationality: 1= German; 0 = unknown; -1 = non- German
Number: 1 = singular; 2 = plural
Position: 0 = unknown; 1 = rank and file; 2 = functionary; 3 = leading position
Paid Position: 1 = yes; 0 = unknown; -1 = no
Political Leaning:1 = conservative; 0 = neutral or unknown; -1 = progressive
Evaluation: 1 = positive; 0 = neutral; -1 = negative

References

1. Fiske, S. T. and Taylor, S. E. Taylor: Social Cognition. Second Edition. McGraw-Hill, New York 1991.
2. Kohonen, T.: Self-Organized Formation of Topology Correct Feature Maps. Biological Cybernetics, 34: 59-69, 1982.
3. Göppert, J. and Rosenstiel, W.: Dynamic Extensions of Self-Organizing Maps. In Proc. of ICANN'94, Sorrento, Italy,. Springer Verlag 1994.
4. Göppert, J. and Rosenstiel, W.: Selective Attention and Self-Organizing Maps. In Proc. of NEURAP'94. IUSPIM, Marseille, France 1994.
5. Lloyd, D. and Rossi, E. L. (eds.): Ultradian Rhythms in Life Processes. An Inquiry into the Fundamental Principles of Chronobiology and Psychobiology. Springer Verlag, New York 1992.